NATURAL
HAZARDS

The new revised fifth edition of *Natural Hazards* remains the go-to introductory-level survey intended for university and college courses that are concerned with earth processes that have direct, and often sudden and violent, impacts on human society. The text integrates principles of geology, hydrology, meteorology, climatology, oceanography, soil science, ecology, and solar system astronomy.

The textbook explains the earth processes that drive hazardous events in an understandable way, illustrates how these processes interact with our civilization, and describes how we can better adjust to their effects. Written by leading scholars in the area, the new edition of this book takes advantage of the greatly expanding amount of information regarding natural hazards, disasters, and catastrophes. The text is designed for learning, with chapters broken into small consumable chunks of content for students. Each chapter opens with a list of learning objectives and ends with revision as well as high-level critical thinking questions. A Concepts in Review feature provides an innovative end-of-chapter section that breaks down the chapter content by parts: reviewing the learning objectives, summary points, important visuals, and key terms. New case studies of hazardous events have been integrated into the text, and students are invited to actively apply their understanding of the five fundamental concepts that serve as a conceptual framework for the text. Figures, illustrations, and photos have been updated throughout.

The book is designed for a course in natural hazards for nonscience majors, and a primary goal of the text is to assist instructors in guiding students who may have little background in science to understand physical earth processes as natural hazards and their consequences to society.

Edward A. Keller is a professor in the Environmental Studies and Earth Sciences Departments at University of California at Santa Barbara, USA.

Duane E. DeVecchio is a research professor in the School of Earth and Space Exploration (SESE) at Arizona State University, USA.

FIFTH EDITION

NATURAL HAZARDS

EARTH'S PROCESSES AS HAZARDS, DISASTERS, AND CATASTROPHES

**EDWARD A. KELLER AND
DUANE E. DeVECCHIO
WITH ASSISTANCE FROM
ROBERT H. BLODGETT**

Routledge
Taylor & Francis Group

LONDON AND NEW YORK

Fifth edition published 2019
by Routledge
2 Park Square, Milton Park, Abingdon, Oxon, OX14 4RN

and by Routledge
711 Third Avenue, New York, NY 10017

Routledge is an imprint of the Taylor & Francis Group, an informa business

First edition published by Pearson Education, Inc. 2006
Fourth edition published by Pearson Education, Inc. 2015

British Library Cataloguing-in-Publication Data
A catalogue record for this book is available from the British Library

Library of Congress Cataloging-in-Publication Data
Names: Keller, Edward A., 1942– author. |
DeVecchio, Duane E. (Duane Edward), 1970– author.
Title: Natural hazards: earth's processes as hazards, disasters, and catastrophes / Edward A. Keller and Duane E. DeVecchio.
Description: Fifth edition. | New York: Routledge, 2019. |
"Fourth edition published by Pearson Education, Inc. 2015"—
T.p. verso. | Includes bibliographical references and index.
Identifiers: LCCN 2018028016| ISBN 9781138058415
(hardback: alk. paper) | ISBN 9781138057227 (paperback: alk. paper) |
ISBN 9781315164298 (eBook)
Subjects: LCSH: Natural disasters.
Classification: LCC GB5014 .K45 2019 | DDC 551—dc23
LC record available at https://lccn.loc.gov/2018028016

ISBN: 978-1-138-05841-5 (hbk)
ISBN: 978-1-138-05722-7 (pbk)
ISBN: 978-1-315-16429-8 (ebk)

Typeset in Janson Text
by codeMantra

Visit the companion website: www.routledgetextbooks.com/textbooks/9781138057227/

Printed in Canada

Brief Contents

Contents

6 Flooding 208

⑦ Mass Wasting 256

8   Subsidence and Soils 298

(11) Coastal Hazards 438

About the Authors

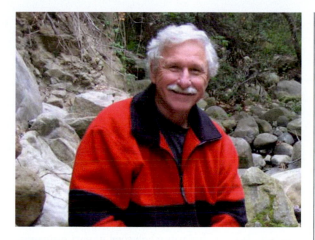

EDWARD A. KELLER

Ed Keller is a professor, researcher, writer, and most importantly, mentor and teacher to undergraduate and graduate students. Ed's students are currently working on earthquake hazards, how waves of sediment move through a river system following a disturbance, and geologic controls on habitat to endangered southern steel-head trout.

Ed was born and raised in California. He received Bachelor's degrees in geology and mathematics from California State University at Fresno and a Master's degree in geology from the University of California at Davis. It was while pursuing his Ph.D. in geology from Purdue University in 1973 that Ed wrote the first edition of *Environmental Geology*, a text that became the foundation of an environmental geology curriculum in many colleges and universities. He joined the faculty of the University of California at Santa Barbara in 1976 and has been there ever since, serving multiple times as chair of both the Environmental Studies and Hydrologic Science programs. In that time he has been an author on more than 100 articles, including seminal works on fluvial processes and tectonic geomorphology. Ed's academic honors include the Don J. Easterbrook Distinguished Scientist Award, Geological Society of America (2004); the Quatercentenary Fellowship from Cambridge University, England (2000); two Outstanding Alumnus Awards from Purdue University (1994, 1996); a Distinguished Alumnus Award from California State University at Fresno (1998); and the Outstanding Outreach Award from the Southern California Earthquake Center (1999).

Ed and his wife Valery, who brings clarity to his writing, love walks on the beach at sunset and when the night herons guard moonlight sand at Arroyo Burro Beach in Santa Barbara.

DUANE E. DEVECCHIO

Duane DeVecchio is a research professor in the School of Earth and Space Exploration (SESE) at Arizona State University, where he is deeply engaged in undergraduate education and teaches a broad range of courses to students completing degrees in both the geological sciences and environmental studies. Outside of the classroom, Dr. DeVecchio serves on both the SESE undergraduate education committee and the curriculum redesign committee, where he collaborates with coworkers to develop student learning strategies and curriculum designed to develop student critical thinking skills and scientific literacy, which are particularly important today when cable television and the Internet offer accessibility to vast amounts of information, yet the validity of this information is often questionable or misleading.

Duane grew up in California where he completed his Bachelor's degree at San Francisco State University before moving on to Idaho State University where he completed a Master's degree. In 2009 he graduated with his Ph.D. from University of California Santa Barbara where he honed his teaching skills while conducting a wide range of research in the geological sciences. Dr. DeVecchio has a broad field-based background in the geological sciences, and likes to share stories about his many months living alone in mobile trailers in the mountains and deserts of the western United States. For his Master's degree and post-Master's research he conducted structural and stratigraphic analyses as well as numerical dating of volcanic and volcaniclastic rocks in southeastern Idaho and the central Mojave Desert of California, which record the Miocene depositional and extensional histories of these regions. His dissertation and current research focus on the timing and rates of change of Earth's surface due to depositional and erosional processes that result from climate change and tectonics. When he is not teaching or conducting research, he's hang gliding, but he also enjoys rock climbing, whitewater rafting, snow boarding, and camping.

Preface

Natural Hazards: Earth Processes as Hazards, Disasters, and Catastrophes, Fifth Edition, is an introductory-level survey intended for university and college courses that are concerned with earth processes that have direct, and often sudden and violent, impacts on human society. The text integrates principles of geology, hydrology, meteorology, climatology, oceanography, soil science, ecology, and solar system astronomy. The book is designed for a course in natural hazards for nonscience majors, and a primary goal of the text is to assist instructors in guiding students who may have little background in science to understand physical earth processes as natural hazards and their consequences to society.

In revising the fifth edition of this book we take advantage of the greatly expanding amount of information regarding natural hazards, disasters, and catastrophes. Since the fourth edition was published, many natural disasters and catastrophes have occurred. Becoming the wettest tropical cyclone in U.S. recorded history, Hurricane Harvey in 2017 produced peak accumulations of rainfall of more than 1.5 m (~60 in.) over a three-day period causing catastrophic flooding in the Houston metropolitan area; earthquakes killed more than 8,000 people in Nepal in 2015, and were responsible for widespread destruction in Mexico in 2017; wildfires in northern and southern California in 2017 destroyed more than 7,000 buildings and killed 46 people; and the deadliest U.S. landslide on record killed 43 people in Washington state in 2014. Most of these disasters have a common denominator—they were not unexpected and the effects of these catastrophes could have been minimized had the population been better prepared and the warnings from the scientific community been heeded. In other words, they were largely, in terms of lives lost and property damaged, disasters caused by humans.

On a global scale, climate change is causing glaciers, ice caps, and permafrost to melt; the atmosphere and oceans to warm; and sea levels to rise more rapidly than originally forecast. These changes are caused in part by human activities, primarily the burning of fossil fuels, which releases vast quantities of carbon dioxide and other gases into the atmosphere each day. The interaction between humans and earth processes has never been clearer, nor has the need for understanding these processes as hazards for our economy and society ever been greater. This edition of *Natural Hazards* seeks to explain the earth processes that drive hazardous events in an understandable way, illustrate how these processes interact with our civilization, and describe how we can better adjust to their effects.

A central thesis to this text is that we must first understand that earth processes are not, in and of themselves, hazards. Earthquakes, floods, volcanic eruptions, wildfires, and other processes have occurred for millennia, indifferent to the presence of people. Natural processes become hazards when they impact humanity. Ironically, it is human behavior that often causes the interactions with these processes to become disasters or, worse, catastrophes. Most important is the unprecedented increase in human population in the past 50 years linked to poor land-use decisions.

In addition to satisfying a natural curiosity about hazardous events, there are additional benefits to studying natural hazards. An informed citizenry is one of our best guarantees of a prosperous future. Armed with insights into linkages between people and the geologic environment, we will ask better questions and make

the 5 fundamental concepts

1
Science helps us predict hazards.

Natural hazards, such as earthquakes, volcanic eruptions, landslides, and floods, are natural processes that can be identified and studied using the scientific method. Most hazardous events and processes can be monitored and mapped, and their future activity predicted, on the basis of frequency of past events, patterns in their occurrence, and types of precursor events.

2
Knowing hazard risks can help people make decisions.

Hazardous processes are amenable to *risk analysis*, which estimates the probability that an event will occur and the consequences resulting from that event. For example, if we were to estimate that in any given year, Los Angeles has a 5 percent chance of a moderate earthquake, and if we know the consequence of that earthquake in terms of loss of life and damage, then we can calculate the potential risk to society.

better choices. On a local level we will be better prepared to make decisions concerning where we live and how best to invest our time and resources. On a national and global level we will be better able to advise our leaders on important issues related to natural hazards that impact our lives.

Major New Material in the Fifth Edition

The fifth edition benefited greatly from feedback from instructors using the previous edition, and many of the changes reflect their thoughtful reviews. New material for the fifth edition includes the following:

- **Active, Modular Learning Path.** The fifth edition of *Natural Hazards* is designed for learning with each chapter broken into smaller, more consumable, chunks of content for students. Each chapter opens with a list of learning objectives and students can check their mastery of these objectives with new **Check Your Understanding** questions at the end of each section. Each chapter concludes with a new **Concepts in Review** section: an innovative end-of-chapter section that breaks down the chapter content by section—reviewing the learning objectives, summary points, important visuals, and key terms. All chapters end with high-level critical thinking questions.

- **Applying the Fundamental Concepts.** Every chapter begins with a captivating hazard story that applies the five fundamental concepts that serve as a conceptual framework for the text. We then close each chapter by recapping this hazard story and, through assessment, invite the students to actively apply their understanding of the five fundamental concepts.

- Current hazard coverage on the Tohoku, Japan earthquake and tsunami in 2011, Hurricane Sandy in New Jersey and New York City in 2012, the tornado in Moore, Oklahoma in 2013, historic flooding in Houston in 2017, and the 2017 fire season in California.

- New in-chapter **case studies** and **chapter openers** that highlight hazardous events have been integrated into the text, providing coverage of the most recent natural disasters on earth.

- Updated art program showcases new figures, illustrations, and photos throughout. Each image has been reviewed for accuracy and relevance, focusing on its educational impact.

Distinguishing Features of the Fifth Edition

We have incorporated a number of features designed to support the student and instructor.

A BALANCED APPROACH

Although the interest of many readers will naturally focus on natural hazards that threaten their community, state, or province, the globalization of our economy, information access, and human effects on our planet require a broader, more balanced approach to the study

---

3

Linkages exist between natural hazards.

Hazardous processes are linked in many ways. For example, earthquakes can produce landslides and giant sea waves called tsunamis, and hurricanes often cause flooding and coastal erosion.

4

Humans can turn disastrous events into catastrophes.

The magnitude, or size, of a hazardous event as well as its frequency, or how often it occurs, may be influenced by human activity. As a result of increasing human population and poor land-use practices, events that used to cause disasters are now often causing catastrophes.

5

Consequences of hazards can be minimized.

Minimizing the potential adverse consequences and effects of natural hazards requires an integrated approach that includes scientific understanding, land-use planning and regulation, engineering, and proactive disaster preparedness.

of natural hazards. A major earthquake in Taiwan affects trade in the ports of Seattle and Vancouver; the economy of Silicon Valley in California affects the price of computer memory in Valdosta, Georgia, and Halifax, Nova Scotia. Because of these relationships, we provide examples of hazards from throughout the United States as well as throughout the world.

This book discusses each hazardous process as both a natural occurrence and a human hazard. For example, the discussion of earthquakes balances the description of their characteristics, causes, global distribution, estimated frequency, and effects with a description of engineering and nonstructural approaches to reduce their effects on humans, including actions that communities and individuals can take.

The five concepts introduced in Chapter 1 and revisited in every chapter are designed to provide a memorable, transportable, and conceptual framework of understanding that students can carry with them throughout their lives to make informed choices about their interaction with and effect upon earth processes.

CASE STUDIES: SURVIVOR STORIES, PROFESSIONAL PROFILES, AND CLOSER LOOKS

Many of the chapters contain personal stories of someone who has experienced the effects of a hazardous event, such as a flood, earthquake, or wildfire, a profile of a scientist or other professional who has worked with a particular hazard, and Closer Look boxes that use life events and data to enhance the understanding and comprehension of not only the hazard but also the mitigation that coincides with the hazard. Most of us in our lifetimes will experience (directly or indirectly) a flood, wildfire, volcanic eruption, tsunami, or major hurricane, and we are naturally curious as to what we will see, hear, and feel. For example, a scientific description of wildfire does not convey the fear and anxiety when a fire is rushing toward your home. Likewise, the stream gauge records for the Rio Grande River do not give us the sense of excitement and fear felt by Jason Lange and his friends when their spring-break canoe trip in West Texas nearly turned to tragedy (see Case Study 6.1: Survivor Story). To fully appreciate natural hazards we need both scientific knowledge and human experience. As you read the survivor stories, ask yourself what you would do in a similar situation, especially once you more fully understand the hazard. Knowledge from reading this book could save your life someday, as it did for Tilly Smith and her family on the beach in Phuket, Thailand, during the Indian Ocean tsunami (see Case Study 4.2: Indonesian Tsunami).

People study and work with natural hazards for many reasons—scientific curiosity, monetary reward, excitement, or the desire to help others deal with events that threaten our lives and property. As you read the professional profiles, think about these people's motivation, the type of work that they do, and how that work contributes to increasing human knowledge or to saving people's lives and property. For example, for many years Bob Rasely worked as a geologist with the U.S. Natural Resources Conservation Service studying what happens to hillslopes after wildfires in Utah (see Case Study 7.5: Professional Profile). For Bob, geology had long been both a vocation and an avocation. He maintains an intellectual curiosity as to how Earth works and a practical goal of predicting the likelihood and location of debris flows following a fire. Working with other state and federal scientists, Bob develops plans for hillslope recovery and helps communities downslope from burned land establish warning systems to protect lives and property. Nearly all the survivor stories and professional profiles are based on interviews conducted exclusively for *Natural Hazards* by Kathleen Wong, a science writer from Oakland, California, and Chris Wilson, a former journalism student at the University of Virginia.

For the Instructor

The Instructor Resource Center provides a collection of resources to help teachers make efficient and effective use of their time. All digital resources can be found in one well-organized, easy to access place. The IRC (download only) includes:

- Pre-authored Lecture Outline PowerPoint presentations, which outline the concepts of each chapter with embedded art and can be customized to fit instructors' lecture requirements

- The test bank includes over 1,000 multiple-choice, true/false, and short-answer test questions on the science of natural hazards and their effects on the world and humankind. Questions are correlated against chapter-specific learning outcomes and Bloom's Taxonomy to help teachers better map the assessments against both broad and specific teaching and learning objectives. TestGen software for both PC and Mac.

The Instructor Resource content is available online at www.routledgetextbooks.com/textbooks/ 978113805 7227/#instructors.

For the Student

This title has recently been acquired by Taylor & Francis. Due to rights reasons, any multimedia resources will not be available from Taylor & Francis. Mastering Geology incorporating Hazard City will still be available for purchase from Pearson.

Acknowledgments

Many individuals, companies, and agencies have helped make this book a success. In particular, we are indebted to the U.S. Geological Survey and the National Oceanic and Atmospheric Administration for their excellent natural hazard programs and publications.

To authors of papers cited in this book, we offer our thanks and appreciation for their contributions. Without their work, this book could not have been written. We must also thank the thoughtful scholars who dedicated valuable time reviewing chapters of this book. The following reviewers' comments helped us significantly improve the fifth edition:

Stephen P. Altaner, *University of Illinois*
Allison Beauregard, *Northwest Florida State College*
Hassan Boroon, *California State University at Los Angeles*
Patrick Burkhart, *Slippery Rock University*
Andrew Curtis, *University of Southern California*
Tim Dolney, *Penn State University-Altoona College*
Sean Fitzgerald, *University of Massachusetts*
Jennifer Haase, *Purdue University*
Michael Hamburger, *Indiana University*
Brian A. Hampton, *Michigan State University*
Brenda Hanke, *Northern Kentucky University*
James Hibbard, *North Carolina State University*
Jean Hoff, *St Cloud State University*
Richard W. Hurst, *California State University at Los Angeles*
Bridget James, *San Francisco State University*
Blair Larsen, *Utah State University*
Todd Leif, *Cloud County Community College*
Susan McGeary, *University of Delaware*
Debra Metz, *Delaware County Community College*
Stephen Nelson, *Tulane University*
William Newman, *University of California at Los Angeles*
Scott Nowicki, *University of Nevada at Las Vegas*
Beth Rinard, *Tarleton State University*
Sarah de la Rue, *Purdue University at Calumet*
Isabelle Sacramento Grilo, *San Diego State University*
Lisa Skinner, *Northern Arizona University*
Jennifer Snyder, *Delaware County Community College*
Christine Stidham, *Stony Brook University*
Michael Wacker, *Florida International University*
Robin Whatley, *Columbia College Chicago*

Thanks also to the following reviewers who helped us put together the first, second, third and fourth editions:

Bob Abrams, *San Francisco State University*
David Best, *Northern Arizona University*

Beth Nichols Boyd, *Yavapai College*
Katherine V. Cashman, *University of Oregon*
Jennifer Cole, *Northeastern University*
Bob Davies, *San Francisco State University*
John Hermance, *Brown University*
Jean M. Johnson, *Shorter College*
Jamie Kellogg, *University of California at Santa Barbara*
Stephen Kesler, *University of Michigan at Ann Arbor*
Guy King, *California State University at Chico*
Timothy Kusky, *Saint Louis University*
Gabi Lakse, *University of California at San Diego*
Thorne Lay, *University of California at Santa Cruz*
William P. Leeman, *Rice University*
Alan Lester, *University of Colorado at Boulder*
Lawrence McGlinn, *Rensselaer University*
Rob Mellors, *San Diego State University*
Richard Minnich, *University of California at Riverside*
Stephen A. Nelson, *Tulane University*
Mark Ouimette, *Hardin Simmons University*
Gigi Richard, *Mesa State College*
Jennifer Rivers, *Northeastern University*
Tim Sickbert, *Illinois State University*
Don Steeples, *University of Kansas*
Paul Todhunter, *University of North Dakota*
Robert Varga, *College of Wooster*
Tatiana Vislova, *University of St. Thomas*
John Wyckoff, *University of Colorado at Denver*

Special thanks to Kathleen Wong and Chris Wilson for locating and interviewing many of the natural hazard professionals and survivors. All of us are especially grateful to these people who told us their stories. Thanks also to Jim Kennett for reviewing the case study on the Younger Dryas Boundary event; to Joel Michaelsen for reviewing the chapters on meteorology, hurricanes, and extratropical cyclones; and to Rob Thieler for his review of the case study on Folly Island. Ed would like to thank Tanya Atwater, William Wise, and Frank Spera for their assistance in preparing the content on plate tectonics, rocks and minerals, and impacts, respectively. Thanks to Kenji Satake, David Cramer, Gordon Wells, Theresa Carpenter, and David Rogers for providing photographs and graphics for use in the fourth edition.

Special thanks are extended to Professor Bob Blodgett who as coauthor worked very hard on the first two editions. Much of the fifth edition reflects his excellent contributions.

We are particularly indebted to our publisher and editors at Taylor & Francis. First and foremost we are grateful to Andrew Mould and Egle Zigaite for their outstanding coordination work, Veronica Jurgena for her great developmental editing and many suggestions, and Donna Mulder and Bill Norrington for editing. Our appreciation is extended to Gina Cheselka and Janice Stangel in production. We would also like to thank photo researcher Marta Johnson, Director of Development Jennifer Hart, Andrew Troutt and his team at Precision Graphics, Rebecca Dunn at CodeMantra, and Editorial Assistant Sarah Shefveland for all of their hard work.

Last, but certainly not least, Ed would like to thank his wife Valery for her encouragement and support.

Duane would like to offer special thanks to his family and friends, and to all of the students who continue to ask good questions, reinforcing and reminding him of the importance of science today.

Edward A. Keller

Santa Barbara, California

Duane E. DeVecchio

Tempe, Arizona

Nearly 190,000 homes were
destroyed or damaged,
killing nearly a
quarter of a million
people

1

Introduction to Natural Hazards

Earthquake in Haiti, 2010: A Human-Caused Catastrophe?

One of the primary principles emphasized in this book is that, as a result of increasing human population and poor land-use choices, what were once disasters are now catastrophes in many cases. Haiti has been recognized for many years as an environmental catastrophe waiting to happen. Haiti's population has increased dramatically in recent decades, and about 90 percent of this mountainous country has been deforested (Figure 1.1). The country is extremely vulnerable to hurricanes and other high-intensity storms, causing frequent runoff flooding and landslides on bare slopes. Haiti is also vulnerable to large earthquakes, with three major earthquakes occurring there since 1750.

On January 12, 2010, an earthquake killed more than 200,000 Haitians and injured over 300,000 more. This was an appalling loss of life for an earthquake of magnitude 7.0 (see chapter-opener photograph). For example, the magnitude 7.1 Loma Prieta earthquake that struck the San Francisco area in 1989 caused 63 deaths and injured about 4,000. Why did two earthquakes of about the same magnitude cause such vastly different casualty counts?[1,2]

The poorest country in the Western Hemisphere, Haiti's annual income per person is only a small fraction of that in the United States. Two years before the 2010 earthquake, within a span of about a month, four tropical storms and hurricanes struck in 2008. Land denuded of trees responded quickly to torrential downpours, and hillslopes went crashing into homes. By September 2008 when a hurricane hit, Haiti was devastated. Nearly 800 people had died, and much of the country's harvest of food crops had been destroyed. The situation was grim.

< **Collapsed buildings in Port-au-Prince, capital of Haiti, resulting from the January 12, 2010, earthquake that** killed more than 200,000 people. The extensive damage and loss of life were in part a result of the high population density in the capital and poorly constructed buildings unable to withstand the earthquake shaking. *(Jorge Silva/Reuters)*

LEARNING Objectives

Natural processes such as volcanic eruptions, earthquakes, floods, and hurricanes become hazards when they threaten human life and property. As population continues to grow, hazards, disasters, and catastrophes become more common. An understanding of natural processes as hazards requires some basic knowledge of earth science.

After reading this chapter, you should be able to:

LO:1 Explain the difference between a disaster and a catastrophe.

LO:2 Discuss the role of history in the understanding of natural hazards.

LO:3 Discuss the components and processes of the geologic cycle.

LO:4 Apply the scientific method to a natural hazard of your choice.

LO:5 Synthesize the basics of risk assessment.

LO:6 Explain how much of the damage caused by natural hazards is often related to decisions people make before, during, and after a hazardous event.

LO:7 Explain why the magnitude of a hazardous event is inversely related to its frequency.

LO:8 Summarize how natural hazards are linked to one another and to the physical environment.

LO:9 Give reasons why increasing population and poor land-use practices compound the effects of natural hazards and can turn disasters into catastrophes.

LO:10 Explain how events we view as hazards provide natural service.

LO:11 Summarize links between climate change and natural hazards.

(a)   (b)

⋀ **FIGURE 1.1 Deforestation in Haiti**
(a) Steep mountain slopes in southern Haiti where trees have been removed to burn and make charcoal as a cooking fuel. About 30 million trees were planted in the 1980s at a cost exceeding $20 million, paid for with U.S. aid, but by 2008, nearly all of them had been cut down. Without the trees, soil erosion accelerated, damaging farmland and making slopes more vulnerable to landslides (especially when linked to earthquake shaking) and floods (water runs off more quickly and in greater quantities than from forested land). *(©Getty Images/Robin Moore)* (b) A small shack at the base of a deforested slope in the same area as (a). There are a few banana trees, but agricultural opportunity is severely limited by soil erosion. *(Julio Etchart/Alamy)*

Prior to the earthquake, about 85 percent of the people in the capital of Port-au-Prince lived in slum conditions in concrete buildings that could not withstand intense earthquake shaking. Half the people did not have toilets, and only about 30 percent had access to clean water. When the earthquake struck, nearly 190,000 homes were destroyed or damaged, killing nearly a quarter of a million people and leaving over 2 million more homeless with poor sanitation and water quality. In the immediate aftermath of the quake, about 1.5 million people lived in camps exposed to storms and flooding.[3] The city was covered in 19 million cubic meters (25 million cubic yards) of rubble and other debris (enough to fill about 4 million dump trucks) where homes, government buildings, and schools once stood. The result was a humanitarian disaster with too many untreated injuries, too little food, and a predictable outbreak of infectious diseases, including cholera (causing nearly 6,000 deaths) and diarrhea.

Once the damage from the 2010 earthquake was surveyed, it was found that some buildings that had not collapsed were evidently designed and built to withstand earthquakes and had functioned adequately. If more of the buildings in Haiti had been constructed properly, incorporating earthquake-resistant design and construction, there would have been much less damage. Problems observed ranged from construction of heavy, unsupported block walls to a lack of rebars (thin steel rods that strengthen concrete). Concrete that was poorly

compacted and utilized substandard materials, such as marine sands that corroded more easily than river sands, was another problem. Because of these poor construction practices, buildings in Port-au-Prince and other areas could not withstand the shaking from the magnitude 7.0 event.[4]

Thus, the answer to our question of why there were so many deaths from the earthquake is easy to find. A very heavy human footprint can be seen in the catastrophic loss of life. Had the buildings been constructed properly, many more lives would have been spared. If the population of Haiti had not grown so fast, with so many young people (thousands of schools collapsed during the earthquake), there would have been fewer deaths.

During the seven years since the earthquake, Haiti has not recovered, and in October of 2016, Hurricane Matthew hit the island, killing about 1,000 people and causing extensive destruction in many communities in more rural southern Haiti. People killed by the storm had to be buried in mass graves. A cholera outbreak occurred as people consumed contaminated water. The earthquake set the stage for extensive damage and loss of life as many people were living in poorly constructed shelters at risk from high winds and flooding. The land was vulnerable to erosion as a result of deforestation, and the hurricane destroyed crops and damaged the ability of the land to produce food. Thus, the catastrophic earthquake and lack of recovery rendered the country more liable to damage from the hurricane in a cascade of declining conditions.

As Haiti rebuilds with international support, ways must be found to cost-effectively build the next generation of structures to better withstand seismic shaking and the high winds of storms. The consequences of another damaging earthquake or hurricane is painful to contemplate. Haiti must be assisted to take appropriate steps and be proactive in hazard preparation.[4]

1.1 Why Studying Natural Hazards Is Important

Since 1995, the world has experienced the deadliest tsunami in recorded history, caused by a massive Indian Ocean earthquake; another devastating tsunami in Japan caused by one of the largest and costliest earthquakes in recorded history; catastrophic flooding in Pakistan, Venezuela, Bangladesh, Thailand, and central Europe; a volcanic eruption that shut down international airports for more than a week; and deadly earthquakes around the world. At the same time, North America has experienced catastrophic hurricanes on the Gulf Coast, along the Atlantic Coast, and in Guatemala and Honduras; record-setting wildfires in western Canada, Arizona, Colorado, Utah, California, and the high plains of Kansas and Texas; the worst outbreak of tornadoes in U.S. history; a record-matching series of four hurricanes within six weeks in Florida and the Carolinas; a paralyzing ice storm in New England and Quebec; record-setting hail in Nebraska; and a rapid warming of the climate, especially (but not limited to) Alaska, northern Canada, and Arizona. These events are the result of enormous forces that are at work both inside and on the surface of our planet. In this book, we will explain these forces, how they interact with our civilization, and how we can better adjust to their effects. Although we will describe most of these forces as *natural hazards*, we can at the same time be in awe of and fascinated by their effects.

PROCESSES: INTERNAL AND EXTERNAL

In our discussion of these natural hazards, we will use the term *process* to mean the physical, chemical, and biological ways by which events, such as volcanic eruptions, earthquakes, landslides, and floods affect Earth's surface. Some of these processes, such as volcanic eruptions and earthquakes, are the result of internal forces deep within Earth. Most of these internal processes are explained by the theory of plate tectonics, one of the basic unifying theories in science. In fact, *tectonic plates*, large surface blocks of the solid Earth, are mapped by identifying zones of earthquakes and active volcanism (see Chapter 2).

Other processes associated with natural hazards result from external forces that are at or very near Earth's surface. For example, energy from the sun warms Earth's atmosphere and surface, producing winds and evaporating water (see Chapter 9). Wind circulation and water evaporation are responsible for forming Earth's climatic zones and for driving the movement of water in the *hydrologic cycle*. These forces are in turn directly related to hazardous processes, such as violent storms and flooding, as well as coastal erosion. Still other external processes, such as landsliding, are the result of gravity acting on hillslopes and mountains. Gravity is the force of physical attraction to a mass—in this case, the attraction of materials on Earth's surface to the entire mass of Earth. Because of gravitational attraction, objects on hillslopes or along the bed of a river naturally tend to move downslope. Gravity results in water that falls as precipitation on mountain slopes, moving downhill on its way back to the ocean.

Thus, the processes we consider to be hazards result from natural forces such as the internal heating of Earth or external energy from the sun. The energy released by natural processes varies greatly. For example, the average tornado expends about 1,000 times as much energy as a lightning bolt, whereas the volcanic eruption of Mount St. Helens in May 1980 expended approximately a million times as much energy as a lightning bolt. The amount of solar energy Earth receives each day is about a trillion times that of a lightning bolt. However, it is important to keep in mind that a lightning bolt may strike a tree, igniting a tremendous release of energy in a forest fire, whereas solar energy is spread around the entire globe.

Events such as earthquakes, volcanoes, floods, and fires are natural processes that have been occurring on Earth's surface since long before it was populated by humans. These natural processes become hazardous when human beings live or work in their path. We often use the terms *hazard*, *disaster*, and *catastrophe* to describe our interaction with these natural processes.

HAZARD, DISASTER, OR CATASTROPHE

A **natural hazard** is a natural process and event that is a potential threat to human life and property. The process and the events themselves are not a hazard but become so because of human use of the land. A **disaster** is a hazardous event that occurs over a limited time span within a defined area. Criteria for a natural disaster are (1) 10 or more people are killed, (2) 100 or more people are affected, (3) a state of emergency is declared, and (4) international assistance is requested. If any one of these

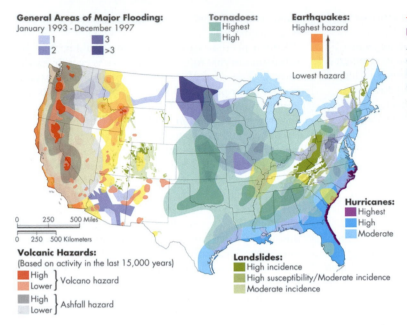

General Areas of Major Flooding:
January 1993 - December 1997
■ 1   ■ 3
■ 2   ■ >3

Tornadoes:
■ Highest
■ High

Earthquakes:
Highest hazard

Lowest hazard

Hurricanes:
■ Highest
■ High
■ Moderate

0   250   500 Miles
0   250   500 Kilometers

Volcanic Hazards:
(Based on activity in the last 15,000 years)
■ High }
■ Lower } Volcano hazard
■ High }
■ Lower } Ashfall hazard

Landslides:
■ High incidence
■ High susceptibility/Moderate incidence
■ Moderate incidence

< **FIGURE 1.2 Major Hazards in the United States** Areas of the United States at risk for earthquakes, volcanoes, landslides, flooding, hurricanes, and tornadoes. Almost every part of the country is at risk for one of the six hazards considered here. A similar map or set of maps is available for Canada. *(From the U.S. Geological Survey)*

applies, an event is considered a *natural disaster*.[5,6] A **catastrophe** is a massive disaster that requires significant expenditure of money and a long time (often years) for recovery to take place. Hurricane Katrina, which flooded the city of New Orleans and damaged much of the coastline of Mississippi in 2005, was the most damaging, most costly catastrophe in the history of the United States. Recovery from this enormous catastrophe has taken years and still continues in parts of New Orleans.

Natural hazards affect the lives of millions of people, and all areas of the United States are at risk from more than one hazardous Earth process (Figure 1.2). Not shown in Figure 1.2 are the areas prone to blizzards and ice storms; the coastlines that have experienced tsunamis during the past century; the areas regularly affected by wildfires; the regions that have experienced drought, subsidence, or coastal erosion; or the craters made by the impacts of asteroids and comets. No area is considered hazard-free.

From 1996 to 2015, natural disasters such as earthquakes, floods, and hurricanes have killed about 1.3 million people worldwide, an average of about 65,000 people per year. Earthquakes and tsunamis accounted for 57 percent of the deaths, storms 18 percent, extreme temperature (hot or cold) 12 percent, and floods 11 percent.[7] Financial loss from natural disasters now exceeds $50 billion per year, with individual years capable of being much higher, and that figure does not include social losses such as loss of employment, mental anguish, and reduced productivity. Two individual disasters, a hurricane accompanied by flooding in Bangladesh in 1970 and an earthquake in China in 1976, each claimed well over 300,000 lives. The Indian Ocean tsunami in 2004 resulted in at least 230,000

deaths, and another hurricane that struck Bangladesh in 1991 claimed 145,000 lives (Figure 1.3). Hurricane Katrina in 2005 destroyed much of the coastal development in Mississippi and Louisiana and caused the flooding of New Orleans. Katrina killed more than 1,600 people and inflicted about $250 billion in damages, making the storm the largest financial catastrophe from a natural hazard in U.S. history. An earthquake in Pakistan in 2005 claimed more than 80,000 lives, destroyed many thousands of buildings, and caused extreme property damage (Figure 1.4). These catastrophes, along with the Haiti earthquake introduced in the chapter opener, were

∧ **FIGURE 1.3 Killer Hurricane**
Flooded fields and villages adjacent to a river in Bangladesh the day after the 1991 hurricane struck the country, killing approximately 145,000 people. *(Defence Visual Information Center)*

∧ FIGURE 1.4 Devastating Earthquake
People searching for victims in a 10-story building that collapsed in a major earthquake in Islamabad, Pakistan, in 2005. By the onset of winter, more than 87,000 people had died and more than 3 million were homeless. (©Getty Images/FAROOQ NAEEM/Stringer)

from 1980 to 2015; also shown are the events, by name and location, with the largest economic damage). Although there are several hundred disasters from natural hazardous events each year, only a few are classified as a great catastrophe—one that results in deaths or losses so great that outside assistance is required.[8] Of disasters since the 1990s, flooding and storms caused about 61 percent of disasters and about 56 percent of the total number of people affected by disasters, while earthquakes caused about 57 percent of the deaths, and, over the same period, countries with medium to low income suffered most from floods and storms. High-income countries suffered some of the greatest economic losses but the lowest number of deaths. The losses could have been even greater were it not for improvements in warning systems, disaster preparedness, and sanitation following disasters.[8–12] Nevertheless, economic losses have increased at a faster rate than have the number of deaths. Figure 1.6 shows the disasters in 2015. Notice that about half of the total disasters (44 percent) occurred in Asia where about two-thirds of the people (4.44 billion) on Earth live. By contrast, the United States, with about 0.33 billion people, experienced about 6 percent of the disasters.[8–12]

caused by natural processes and forces that have always existed. Hurricanes are caused by atmospheric disturbance, and Earth's internal heat drives the movement of tectonic plates, which causes earthquakes and, sometimes, tsunamis. The impact of these events is directly connected to human population density and land-use patterns.

During the last few decades, there has been a significant increase in the number of catastrophes and disasters worldwide (Figure 1.5: Disasters and catastrophes

DEATH AND DAMAGE CAUSED BY NATURAL HAZARDS

When we compare the effects of various natural hazards, we find that those that cause the greatest loss of

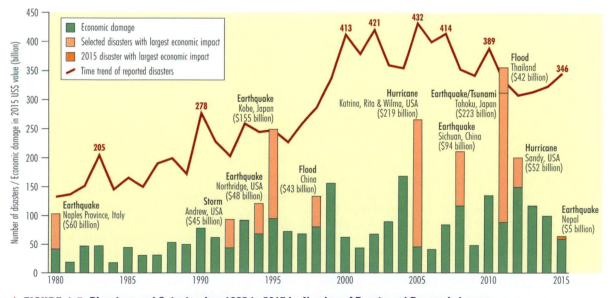

∧ FIGURE 1.5 Disasters and Catastrophes 1980 to 2015 by Number of Events and Economic Losses
Several examples of events with very large economic costs are listed by date, type of hazard, and location.
(Center for Research on the Epidemiology, 2016. Disaster data: A balanced perspective. Cred Crunch Issue No. 41)

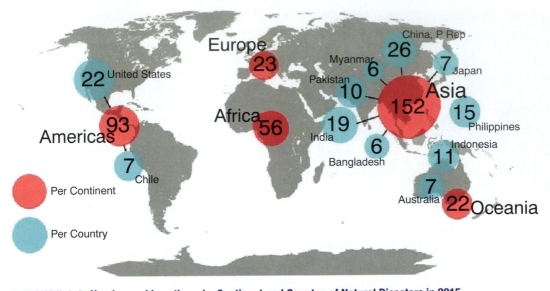

∧ **FIGURE 1.6 Number and Locations, by Continent and Country, of Natural Disasters in 2015**
Notice that about half are in Asia. *(Center for Research on the Epidemiology, 2016. Disaster data: A balanced perspective. Cred Crunch Issue No. 41)*

human life are not necessarily the same as those that cause the most extensive property damage. The largest number of deaths each year is associated with tornadoes and windstorms, although heat waves, lightning, floods, and hurricanes also take a heavy toll. Loss of life from earthquakes can vary considerably from one year to the next because a single great quake can cause tremendous human loss. For example, in 1994, the large, but not great, Northridge earthquake in the Los Angeles area killed some 60 people and inflicted at least $20 billion in property damage. The next great earthquake in a densely populated part of California could inflict $100 billion in damages while killing several thousand people.[11]

Natural disasters cost the United States multibillion dollars annually; the average cost of a single major disaster may exceed $500 million. The cost in 2104 was about $25 billion. About 200 weather-related disasters and catastrophes have occurred in the United States since 1980. The total economic losses were about $1.1 trillion (an average of about $33 million per year). Hurricane Katrina cost about $250 billion (damages and other economic losses). Table 1.1 lists deaths (476) and economic loss (about $4 billion) from U.S. weather-related disasters in 2015. Floods claimed the most lives (176), but other hazards, including several types of storms, extreme temperature, and coastal beach rip currents, all were consistent killers. Because the population is steadily increasing in high-risk areas, such as along coastlines, we can expect this number to continue to increase significantly.[12]

∨ **TABLE 1.1 Deaths and Damages from Weather-related Natural Disasters in the United States for 2015**

Event	Deaths	Total Damages (millions of dollars)
Lightning	27	16
Tornado	36	320
Thunderstorm (wind)	41	268
Extreme cold	53	3
Extreme heat	45	0
Flood	176	2,748
Rip current (coastal)	56	0
Hurricane	14	52
Winter storm	20	530
Ice	0	59
Avalanche	8	0

Source: www.nws.noaa.gov.

An important aspect of all natural hazards is their potential to produce a catastrophe, which has been defined as any situation in which the damages to people, property, or society in general are sufficient that recovery and/or rehabilitation is a long, involved process. Natural hazards vary greatly in their potential to cause a catastrophe (see Table 1.2). Floods, hurricanes, tornadoes,

▼ **TABLE 1.2 Potential for Humans to Influence Selected Natural Hazards in the United States**

Hazard	Occurrence Influenced by Human Use	Potential to Produce a Catastrophe
Flood	Yes	High
Earthquake	Yes	High
Tsunami	No	High
Landslide	Yes	Moderate
Volcano	No	High
Coastal erosion	Yes	Low
Expansive soils	No	Low
Hurricane	Perhaps[a]	High
Tornado and windstorm	Perhaps[a]	High
Lightning	Perhaps[a]	Low
Drought	Perhaps[a]	Moderate
Frost and freeze	Perhaps[a]	Low
Heat wave	Perhaps[a]	High
Wildfire	Yes	High
Extraterrestrial[b] impact	No	Low

[a] Weather-related hazards are listed as perhaps being influenced by human processes because the role of climate change is not well understood.

[b] Extraterrestrial impact is in a separate class because small impacts are frequent, and a rare large impact may produce the largest catastrophes possible on Earth.

Source: Based on White, G. F., and Haas, J. E., *Assessment of Research on Natural Hazards*. Cambridge, MA: MIT Press, 1975.

earthquakes, volcanic eruptions, large wildfires, and heat waves are the hazards most likely to have a high potential to create catastrophes. Landslides, because they generally affect a smaller area, have only a moderate potential to produce a catastrophe. Drought also has a moderate potential to produce a catastrophe because, although drought may cover a wide area, there is usually plenty of warning time before its worst effects are felt. Hazards with a low potential to produce a catastrophe include coastal erosion, frost, lightning, and expansive soils.[13]

The effects of natural hazards change with time because of changes in patterns of human land use, as well as climate change. Urban growth can influence people to develop on marginal lands, such as steep hillsides and floodplains. This trend is a particular problem in areas surrounding major cities in developing nations where urbanization is proceeding rapidly. In addition to increasing population density, urbanization can also change the physical properties of earth materials by influencing drainage, altering the steepness of hillslopes, and removing vegetation. Changes in farming, forestry, and mining practices can affect rates of erosion and sedimentation, the steepness of hillslopes, and the nature of vegetative cover. Climate change in the United States and the world is affecting natural hazards. As the world warms, warming oceans feed more energy into storms, causing an increase in storm intensity. Rising sea levels and larger waves from more intense storms associated with global warming are flooding land and increasing coastal erosion. Climate change is also causing drought to become more common and more extreme in the arid and semiarid regions of Earth (see Chapter 12). Overall, damage from most natural hazards in the United States is increasing, but the number of deaths from many hazards is decreasing because of better prediction, forecasting, and warning.

1.1 CHECK your understanding

1. Differentiate between natural hazards, disasters, and catastrophes.

2. Which natural hazards in the United States take the most lives each year, and which are most costly from an economic perspective?

3. Which hazards have taken the most of the lives worldwide and in the United States in the past two decades?

4. How and why are land-use change and global warming influencing natural hazards?

1.2 The Role of History in Understanding Hazards

A fundamental principle for understanding natural hazards is that they are repetitive events, and, therefore, the study of their history provides much-needed information for any hazard reduction plan. You read in the chapter opener that poor building practices in Haiti led to the great loss of life when the earthquake struck. Whether we are studying earthquakes, floods, landslides, or volcanic eruptions, knowledge of historic events and the recent geologic history of an area is vital to our understanding and assessment of the hazard. For example, if we wish to evaluate the flooding history of a particular river, one of the first tasks is to identify floods that have occurred in the historic and recent prehistoric past. Useful information can be obtained by studying aerial photographs and maps as far back as the record allows. In our reconstruction of previous events, we can look for evidence of past floods in stream deposits. Commonly, these deposits contain organic material such as wood or shells that may be radiocarbon dated to provide a chronology of ancient flood events. This chronology can then be linked with the historic record of high flows to provide an overall perspective of how often the river floods and how extensive the floods may be. Similarly, if we are studying landslides in a particular area, an investigation of both historic and prehistoric landslides is necessary to better predict future landslides. Geologists have the tools and training to "read the landscape" and evaluate prehistoric evidence for natural hazards. Linking the prehistoric and historic records extends our perspective of time when we study repetitive natural events.

In summary, before we can truly appreciate the nature and extent of a natural hazard, we must study in detail its historic occurrence, as well as any geologic features that it may produce or affect. These geologic features may be landforms, such as channels, hills, or beaches; structures, such as geologic faults, cracks, or folded rock; or earth materials, such as lava flows, meteorites, or soil. Any prediction of the future occurrence and effects of a hazard will be more accurate if we can combine information about historic and prehistoric behavior with a knowledge of present conditions and recent past events, including land-use changes.

To fully understand the natural processes we call hazards, some background knowledge of the geologic cycle processes that produce and modify earth materials, such as rocks, minerals, and water, is necessary. In the next few sections, we will discuss the geologic cycle and then introduce five concepts that are fundamental to understanding natural processes as hazards.

1.2 CHECK your understanding

1. Why is recent geologic history as well as human history important in the study of natural hazards?

1.3 The Geologic Cycle

Geologic conditions and materials largely govern the type, location, and intensity of natural processes. For example, earthquakes and volcanoes do not occur at random across Earth's surface; rather, most of them mark the boundaries of tectonic plates. The location of landslides, too, is governed by geologic conditions. Slopes composed of a weak rock, such as shale, are much more likely to slip than those made of a strong rock, such as granite. Hurricanes, although not themselves governed specifically by geology, will have differing effects, depending on the geology of the area

they strike. An understanding of the components and dynamics of the geologic cycle will explain these relationships.

Throughout much of the 4.6 billion years of Earth's history, the materials on or near Earth's surface have been created, maintained, and destroyed by numerous physical, chemical, and biological processes. Continuously operating processes produce the earth materials, land, water, and atmosphere necessary for our survival. Collectively, these processes are referred to as the **geologic cycle**, which is really a group of subcycles that includes:

> the tectonic cycle
> the rock cycle
> the hydrologic cycle
> biogeochemical cycles.

THE TECTONIC CYCLE

The term *tectonic* refers to the large-scale geologic processes that deform Earth's crust and produce landforms such as ocean basins, continents, and mountains. Tectonic processes are driven by forces deep within Earth. To describe these processes, we must use information about the composition and layering of Earth's interior and about the large blocks of the solid Earth that we call *tectonic plates*. The *tectonic cycle* involves the creation, movement, and destruction of tectonic plates. It is responsible for the production and distribution of rock and mineral resources invaluable to modern civilization, as well as hazards such as volcanoes and earthquakes. The tectonic cycle and its linkages to hazards is the subject of Chapter 2.

THE ROCK CYCLE

Rocks are aggregates of one or more *minerals*. A mineral is a naturally occurring, crystalline substance with defined properties (see Appendix A for a discussion of minerals). The *rock cycle* is the largest of the geologic subcycles, and it is linked to all the other subcycles. It depends on the tectonic cycle for heat and energy, the biogeochemical cycle for materials, and the hydrologic cycle for water. Water is then used in the processes of weathering, erosion, transportation, deposition, and lithification of sediment.

Although rocks vary greatly in their composition and properties, they can be classified into three general types, or families, according to how they were formed in the rock cycle (Figure 1.7); rocks and their properties are discussed in Appendix B. We may think of the rock cycle as a worldwide rock recycling process driven by Earth's internal heat, which melts the rocks

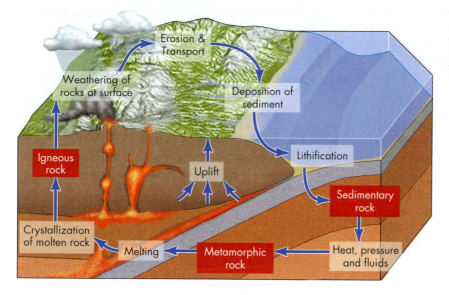

subducted in the tectonic cycle (see Chapter 2). *Crystallization* of molten rock produces *igneous rocks* beneath and on Earth's surface. Rocks at or near the surface break down chemically and physically by *weathering* to form particles known as *sediment*. These particles vary in size from fine clay to very large pieces of boulder-sized gravel. Sediment formed by weathering is then transported by wind, water, ice, and gravity to depositional basins, such as the ocean. When wind or water currents slow down, ice melts, or when material moving downward due to gravity reaches a flat surface, the sediment settles and accumulates by a process known as *deposition*. The accumulated layers of sediment eventually undergo *lithification* (conversion to solid rock), forming *sedimentary rocks*. Lithification takes place by cementation and compaction as sediment is buried beneath other sediment.

With deep burial, sedimentary rocks may be metamorphosed (altered in form) by heat, pressure, or chemically active fluids to produce *metamorphic rocks*. Metamorphic rocks may be buried to depths where pressure and temperature conditions cause them to melt, beginning the entire rock cycle again. As with any other Earth cycle, there are many exceptions or variations from the idealized sequence. For example, an igneous or metamorphic rock may be altered into a new metamorphic rock without undergoing weathering or erosion, or sedimentary and metamorphic rocks may be uplifted and weathered before they can continue on to the next stage in the cycle. Finally, there are other sources of sediment that have a biological or chemical origin and types of metamorphism that do not involve

deep burial. Overall, the type of rock formed in the rock cycle depends on the rock's environment.

The rock cycle has many links to natural hazards. Different rock types have different composition, and the different composition is often linked to specific processes. For example, limestone is composed of calcite ($CaCO_3$) which dissolves in weak soil and carbonic acid to form collapse pits known as sink holes (see Chapter 8). Shales are very common sedimentary rocks composed of fine particles. Shales are a red flag to those working with landslide hazards (see Chapter 7).

THE HYDROLOGIC CYCLE

The movement of water from the oceans to the atmosphere and back again is called the *hydrologic cycle* (Figure 1.8). Driven by solar energy, the cycle operates by way of evaporation, precipitation, surface runoff, and subsurface flow, and water is stored in different compartments along the way. These compartments include the oceans, atmosphere, rivers and streams, groundwater, lakes, and ice caps and glaciers (Table 1.3). The *residence time*, or estimated average amount of time that a drop of water spends in any one compartment, ranges from tens of thousands of years or more in glaciers to nine days in the atmosphere.

As you can see from studying Table 1.3, only a very small amount of the total water in the cycle is active near Earth's surface at any one time. Although the combined percentage of water in the atmosphere, rivers, and shallow subsurface environments is only about

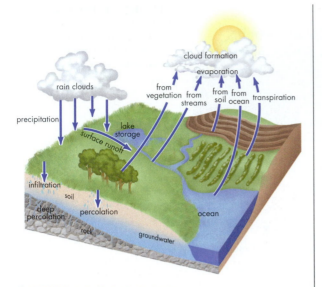

⋀ FIGURE 1.8 Hydrologic Cycle
Idealized diagram showing the hydrologic cycle's important processes and transfer of water. *(Based on the Council on Environmental Quality and Department of State.* The Global 2000 Report to the President, *Vol. 2., 1980)*

0.3 percent of the total, this water is tremendously important for life on Earth and for the rock and biogeochemical cycles. This surface or near-surface water helps move and sort chemical elements in solution, sculpt the landscape, weather rocks, transport and deposit sediments, and provide our water resources.

The hydrologic cycle is closely related to most natural hazards:

> Surface runoff is the primary factor in river flooding (see Chapter 6)

> Groundwater is one of the most important factors in landslides and subsidence (see Chapters 7 and 8)

> Water is important in volcanic processes (see Chapter 5)

> Water injected with other materials is an important factor in human-induced earthquakes (see Chapter 3)

> Water in motion in the oceans and atmosphere is an important factor in coastal erosion and violent storms (see Chapters 9–11)

> Water vapor is an important factor in atmospheric processes that produce storms. As a result, it is a significant factor in climate change and changes in hazardous events (see Chapter 12).

BIOGEOCHEMICAL CYCLES

A *biogeochemical cycle* is the transfer or cycling of a chemical element or elements through the atmosphere (the layer of gases surrounding Earth), lithosphere (Earth's rocky outer layer), hydrosphere (oceans, lakes, rivers, and groundwater), and biosphere (the part of the Earth where life exists). It follows from this definition that biogeochemical cycles are intimately related to the tectonic, rock, and hydrologic cycles. The tectonic cycle provides water from volcanic processes, as well as the heat and energy required to form and change the earth materials transferred in biogeochemical cycles. The rock and hydrologic cycles are involved in many processes that transfer and store chemical elements in water, soil, and rock.

Biogeochemical cycles can most easily be described as the transfer of chemical elements through a series of storage compartments or reservoirs (e.g., air, soil, groundwater, vegetation). For example, carbon (in the

⋁ TABLE 1.3 The World's Water Supply (Selected Examples)

Location	Surface Area (km²)	Water Volume (km²)	Percentage of Total Water	Estimated Average Residence Time
Oceans	361,000,000 (139,000,000 mi.²)	1,230,000,000 (295,000,000 mi.³)	97.2	Thousands of years
Atmosphere	510,000,000 (197,000,000 mi.²)	12,700 (3047 mi.³)	0.001	9 days
Rivers and streams	–	1,200 (288 mi.³)	0.0001	2 weeks
Groundwater; shallow	130,000,000 (50,000,000 mi.²)	4,000,000 (960,000 mi.³)	0.31	Hundreds to many thousands of years to a depth of 0.8 km
Lakes (freshwater)	855,000 (330,000 mi.²)	123,000 (29,500 mi.³)	0.009	Tens of years
Ice caps and glaciers	28,200,000 (10,900,000 mi.²)	28,600,000 (6,860 mi.³)	2.15	Up to tens of thousands of years and longer

Source: Data from U.S. Geological Survey.

form of CO_2) is exhaled by animals, enters the atmosphere, and is then taken up by plants. When a biogeochemical cycle is well understood, the rate of transfer, or *flux*, among all the compartments is known. However, determining these rates on a global basis is a very difficult task. The amounts of such important elements as carbon, nitrogen, and phosphorus in each compartment, and their rates of transfer among compartments, can be only approximately known. The carbon cycle (CO_2) is of particular importance to hazards as it is associated with global warming, which is changing climate and the nature and intensity of weather-driven processes, such as intense storms.

1.3 CHECK your understanding

1. Define the geologic cycle and describe its subcycles.

2. Why is the geologic cycle pertinent to understanding natural hazards?

1.4 Fundamental Concepts for Understanding Natural Processes as Hazards

The five concepts described next are basic to an understanding of natural hazards. These fundamental concepts serve as a conceptual framework for our discussion of each natural hazard in subsequent chapters of this book. Most chapters begin and end with a case study that evaluates a natural hazard with respect to each of the five fundamental concepts described below.

1 ━━━━━━━━━━━━━━━━━

Science helps us predict hazards.

SCIENCE AND NATURAL HAZARDS

Science is a body of knowledge that has accumulated from investigations and experiments, the results of which are subject to verification. The method of science, often referred to as the scientific method, has a series of steps. The first step is the formulation of a question. With respect to a hazardous event, geologists may ask: Why did a landslide occur that destroyed three homes? In order to explore and answer this question, geologists will spend time examining the slope that failed. They may notice that a great deal of water is

emerging from the base of the slope and landslide. If the geologists also know that a water line is buried in the slope, they may refine the question to ask specifically: Did the water in the slope cause the landslide? This question is the basis for a hypothesis that may be stated as follows: The landslide occurred because a buried water main within the slope broke, causing a large amount of water to enter the slope, reducing the strength of the slope materials and causing the landslide. Often, a series of questions or multiple hypotheses are tested.

The hypothesis is, thus, a possible answer to our question and is an idea that can be tested. In our example, we may test the hypothesis that a broken water main caused a landslide by excavating the slope to determine the source of the water. In science, we test hypotheses in an attempt to disprove them. That is, if we found that there were no leaking water pipes in the slope on which the landslide occurred, we would reject the hypothesis and develop and test another hypothesis. Use of the scientific method has improved our understanding of many natural earth processes, including flooding, volcanic eruptions, earthquakes, hurricanes, and coastal erosion. In our scientific study of natural processes, we have identified where most of these processes take place, their range of magnitude, and how frequently they occur. We have also mapped the nature and extent of hazards. Coupled with knowledge of the frequency of past events, we can use such maps to predict when and where processes such as floods, landslides, and earthquakes will occur in the future. We have also evaluated patterns and types of precursor events. For example, prior to large earthquakes, there may be foreshocks and, prior to volcanic eruptions, patterns of gas emissions may signal an imminent eruption.

HAZARDS ARE NATURAL PROCESSES

Since the dawn of human existence, we have been obligated to adjust to processes that make our lives more difficult. We humans apparently are a product of the Pleistocene ice ages, which started about 2 million years ago (Table 1.4). The Pleistocene epoch and Holocene epoch (last 11,000 years) experienced rapid climatic changes—from relatively cold, harsh glacial conditions as recently as a few thousand years ago to the relatively warm interglacial conditions we enjoy today. Learning to adjust to harsh and changing climatic conditions has been necessary for our survival from the very beginning.

It has recently been suggested that we may be in a new epoch known as the Anthropocene epoch. The idea for an Anthropocene epoch is that humans have so profoundly changed Earth that a new unit of geologic time is necessary to define it. To be a new unit of geologic time there needs to be a strong signal in earth materials (sometimes called a "golden spike") that is present in the Earth's rock record, such as the dinosaur extinction 65 million years ago (Table 1.4). The signal for the Anthropocene must be a geologic marker in the distant future. The concept is controversial, and there is no agreement on when (or if) a Holocene-Anthropocene boundary should be defined. Candidates for the boundary, include the Industrial Revolution (mid-eighteenth century), the extensive use of plastics (plastic pollution) starting in the early twentieth century, or about 1950 when radioactive elements were dispersed across the globe by nuclear bomb experiments. Geologic time is so long that it is difficult, perhaps impossible, to define a new unit of time that just started.

Events we call natural hazards are natural Earth processes. They become hazardous when people live or work near these processes and when land-use changes such as urbanization or deforestation amplify their effect. To reduce damage and loss of life, it is imperative to identify potentially hazardous processes and make this information available to planners and decision makers. However, because the hazards that we face are natural and not the result of human activities, we encounter a philosophical barrier whenever we try to minimize their adverse effects. For example, when we realize that flooding is a natural part of river dynamics, we must ask ourselves if it is wiser to attempt to control floods or to simply make sure that people and property are out of harm's way when they occur.

Although it is possible to control some natural hazards to a certain degree, many are completely beyond our control. For example, although we may have some success in temporarily preventing damage from forest fires by using controlled burns and advanced firefighting techniques, we will never be able to prevent earthquakes. In fact, we may actually worsen the effects of natural processes simply by labeling them as hazardous. Rivers will always flood. Because we choose to live and work on floodplains, we have labeled floods as hazardous processes. This label has led to attempts to control floods. Unfortunately, as we will discuss later, some flood control measures actually intensify the effects of flooding, thereby increasing the very hazard we are trying to prevent (see Chapter 6). The best approach to hazard reduction is to identify hazardous processes and delineate the geographic areas where they occur. Every effort should be made to avoid putting people and property in harm's way, especially for those hazards, such as earthquakes, that we cannot control.

FORECAST, PREDICTION, AND WARNING OF HAZARDOUS EVENTS

Predicting Changes in the Earth System The idea that "the present is the key to the past," called **uniformitarianism**, was popularized in 1785 by James Hutton (referred to by some scholars as the father of geology) and is heralded today as a fundamental concept of Earth sciences. As the name suggests, uniformitarianism holds that processes we observe today also operated in the past (flow of water in rivers, formation and movement of glaciers, landslides, waves on beaches, uplift of the land from earthquakes, and so on). Uniformitarianism does not demand or even suggest that the magnitude (amount of energy expended) and frequency (how often a particular process occurs) of natural processes remain constant with time. We can infer that for as long as Earth has had an atmosphere, oceans, and continents similar to those of today, the present processes were operating.

In making inferences about geologic events, we must consider the effects of human activity on the Earth system and what effect these changes to the system as a whole may have on natural Earth processes. For example, small streams with drainage areas of a few to several tens of square kilometers will flood, regardless of human activities, but human activities, such as

the 5 fundamental concepts

1

Science helps us predict hazards.

Natural hazards, such as earthquakes, volcanic eruptions, landslides, and floods, are natural processes that can be identified and studied using the scientific method. Most hazardous events and processes can be monitored and mapped, and their future activity predicted, on the basis of frequency of past events, patterns in their occurrence, and types of precursor events.

2

Knowing hazard risks can help people make decisions.

Hazardous processes are amenable to *risk analysis*, which estimates the probability that an event will occur and the consequences resulting from that event. Estimating the risk for hazard scenarios for an earthquake, flood, hurricane, wildfire, or other hazardous event is a proactive step in minimizing impacts of hazards.

paving the ground in cities, increase runoff and therefore increase the magnitude and frequency of flooding. That is, floods of a particular size are more frequent following the paving, and a particular rainstorm can produce a larger flood than before the paving. Therefore, to predict the long-range effects of flooding, we must be able to determine how future human activities will change the size and frequency of floods. In this case, *the present is the key to the future*. For example, when environmental geologists examine recent landslide deposits (Figure 1.9) in an area designated to become a housing development, they must use uniformitarianism to infer where there will be future landslides, as well as to predict what effects urbanization will have on the magnitude and frequency of future landslides. We will now consider linkages between processes in what is called environmental unity or "you can't do just one thing."

The principle of **environmental unity** states that one action causes others in a chain of actions and events. For example, if we remove the native vegetation from a steep slope or build a large structure on top of the slope, a landslide may occur. The slide mass may move into a canyon and dam a stream, which will cause the stream flow to back up, forming a lake. The dam of landslide material could be overtopped and quickly eroded, causing a flood. Alternatively, the dam may become saturated and collapse, producing a muddy flow of water and debris downstream as a destructive debris flow. Either event could destroy homes and perhaps kill people down the valley. Thus, modifying and destabilizing the slope set off a chain or series of events that changes the environment where people live. Therefore, when we consider the potential effects of a particular event, we need to think about what other events could happen and take appropriate precautions.

A **prediction** of a hazardous event such as an earthquake involves specifying the date, time, and size of the event. This is different from predicting where or how often a particular event such as a flood will occur. A **forecast**, on the other hand, has ranges of certainty. The weather forecast for tomorrow may state there is a 40 percent chance of showers. Learning how to predict hazardous events so we can minimize human loss and property damage is an important endeavor. For some natural hazards, we have enough information to predict or forecast events accurately. When there is insufficient information available, the best we can do is locate areas where hazardous events have occurred and infer where and when similar future events might take place. If we know both the probability and the possible consequences of an event occurring at a particular location, we can assess the risk that the event poses to people and property, even if we cannot accurately predict when it will next occur.

The effects of a hazardous event can be reduced if we can forecast or predict it and if we can issue a warning. Attempting to do this involves most or all of the following elements:

> Identifying the location where a hazardous event is likely to occur
> Determining the probability that an event of a given magnitude will occur
> Observing any precursor events
> Forecasting or predicting the event
> Warning the public.

3

Linkages exist between natural hazards.

Hazardous processes are linked in many ways. For example, earthquakes can produce landslides, and giant sea waves (called tsunamis) and hurricanes often cause flooding and coastal erosion.

4

Humans can turn disastrous events into catastrophes.

The magnitude, or size, of a hazardous event, as well as its frequency or how often it occurs, may be influenced by human activity. As a result of increasing human population and poor land-use practices, events that used to cause disasters are now often causing catastrophes.

5

Consequences of hazards can be minimized.

Minimizing the potential adverse consequences and effects of natural hazards requires an integrated approach that includes scientific understanding, land-use planning, and regulation, engineering, and proactive disaster preparedness.

▼ TABLE 1.4 Geologic Time with Important Events

Era	Period	Epoch	Million Years before Present	Life	Events / Earth	Million Years before Present	True Scale (Million Years before Present)
Cenozoic (Neogene / Paleogene)	Quaternary	Holocene	0.0114	• Extinction event • Modern humans	Formation of Transverse Ranges, CA Ice age	1.8	Cenozoic
		Pleistocene	1.8	• Early humans	Formation of Andes Mountains		
	Tertiary	Pliocene	5.3	• Grasses	Collision of India with Asia forming Himalayan Mountains and Tibetan Plateau		
		Miocene	23	• Whales			
		Oligocene	34	• Extinction event			
		Eocene	56	• Mammals expand	Rocky Mountains form		Mesozoic
		Paleocene	65	• Dinosaur extinction, extinction event		65	
Mesozoic	Cretaceous		146	• Flowering plants • Birds	Emplacement of Sierra Nevada granites (Yosemite National Park) • Supercontinent Pangaea begins to break up		
	Jurassic		200	• Extinction event • Mammals • Dinosaurs			
	Triassic		251	• Extinction event • Reptiles	• Ice age	251	Paleozoic
Paleozoic	Permian		299	• Coal swamps • Extinction event	Appalachian Mountains form		
	Carboniferous		359	• Trees			
	Devonian		416	• Land plants • Extinction event			
	Silurian		444	• Fish			
	Ordovician		488	• Explosion of organisms with shells			
	Cambrian		542		• Ice age	542	
	Precambrian time		2500	• Multicelled organisms	• Ice age		Precambrian
			3500	• Free oxygen in atmosphere and ozone layer in stratosphere			
			4000	• Primitive life (first fossils)	• Oldest rocks		
			4600		• Age of Earth	4600	4600

[1] Many scientists believe that not all dinosaurs became extinct but that some dinosaurs evolved into birds.

Landslide

⋀ FIGURE 1.9 Urban Development
The presence of a landslide on this slope suggests that the slope is not stable, and that further movement may occur in the future. This is a "red flag" for future development in the area. *(Edward A. Keller)*

Location For the most part, we know where a particular kind of event is likely to occur, and we can accurately map where such events have occurred (see Appendix C for a discussion of maps and images useful in hazard analysis). On a global scale, the major zones for earthquakes and volcanic eruptions have been delineated by mapping (1) where earthquakes have occurred, (2) the extent of recently formed volcanic rocks, and (3) the locations of recently active volcanoes. On a regional scale, we can use past eruptions to identify areas that are likely to be threatened in future eruptions. This risk has been delineated for several Cascade Range volcanoes such as Mt. Rainier in the Pacific Northwest, as well as for several volcanoes in Japan, Italy, Colombia, and elsewhere. On a local scale, detailed mapping of soils, rocks, groundwater conditions, and surface drainage may identify slopes that are likely to fail or where expansive soils exist. In most cases, we can predict where flooding is likely to occur from the location of the floodplain and from mapping the extent of recent floods.

Probability of Occurrence Determining the probability of a particular event occurring at a particular location within a particular time span is an essential part of a hazard prediction (see Appendix D for a discussion of how geologists determine the all-important chronology necessary to evaluate the time element of

hazard evaluation and probability of occurrence). For many rivers, we have sufficiently long records of flow to develop probability models that can reasonably predict the average number of floods of a given magnitude that will occur in a decade. Likewise, droughts may be assigned a probability on the basis of past rainfall in the region. However, these probabilities are similar to the chances of throwing a particular number on a die or drawing an inside straight in poker—the element of chance is always present. Although the 10-year flood may occur on average only once every 10 years, it is possible to have several floods of this magnitude in any one year, just as it is possible to throw two straight sixes with a die.

Precursor Events Many hazardous events are preceded by *precursor events*. For example, the surface of the ground may creep (i.e., move slowly for a long period of time) prior to an actual landslide. Often, the rate of creep increases up to the final failure and landslide. Volcanoes sometimes swell or bulge before an eruption and, often, earthquake activity increases significantly in the area. Foreshocks or unusual uplift of the land may precede earthquakes.

Identification of precursor events helps scientists predict when and where a major event is likely to happen. Thus, measurements of landslide creep or swelling of a volcano may lead authorities to issue a warning and evacuate people from a hazardous area.

Forecasting and Prediction

When a forecast of an event is issued, the certainty of the event is given, usually as the percent chance of something happening. When we hear a forecast of a hazardous event, it means we should be prepared for the event.

It is sometimes possible to accurately predict when certain natural events will occur. Flooding of the Mississippi River, which occurs in the spring in response to snowmelt or large regional storm systems, is fairly common. Hydrologists with the National Weather Service usually can predict when the river will reach a particular flood stage or water level. When a hurricane is spotted far out to sea and tracked toward the shore, we can predict when and where it will likely strike land. Large ocean waves, called tsunamis, that are generated by undersea earthquakes, landslides, and other disturbances, may also be predicted once they form. In the past several decades, the prediction of the arrival time for tsunamis in the Pacific Ocean has been fairly successful. Even a short-notice prediction of a hazardous event, such as a tornado, motivates us to act to reduce potential consequences before the event happens.

Warning

After a hazardous event has been predicted or a forecast has been made, the public must be warned. The flow of information leading to the **warning** of a possible disaster—such as a large earthquake or flood—should move along a predefined path (Figure 1.10). People do not always welcome such warnings, especially when the predicted event does not come to pass. In 1982 geologists advised that a volcanic eruption near Mammoth Lakes, California, was quite likely.

The eruption did not occur, and the advisory was eventually lifted, but this advisory caused loss of tourist business and apprehension for the residents. In July 1986, a series of earthquakes occurred over a four-day period near Bishop, California. The quakes began with minor magnitude 3.0 tremors and culminated in a damaging magnitude 6.1 shock. Investigators concluded that there was a high probability of a larger quake in the vicinity in the near future and issued a warning. Local business owners, who feared the loss of summer tourism, thought that the warning was irresponsible; in fact, the predicted quake never materialized.

Incidents of this kind have led some people to conclude that scientific predictions are worthless, and that advisory warnings should not be issued. Part of the problem is poor communication between the investigating scientists and the news media. Newspaper, television, Internet, and radio reports may fail to explain the evidence or the probabilistic nature of natural hazard prediction, leading the public to expect completely certain statements about what will happen. Although scientists are not yet able to predict volcanic eruptions and earthquakes accurately, it would seem that they have a responsibility to publicize their informed judgments. An informed public is better able to act responsibly than an uninformed public, even if the subject makes people uncomfortable. Ship captains, who depend on weather advisories and warnings of changing conditions, do not suggest that they would be better off not knowing about an impending storm, even though the storm might veer and miss the ship. Just as weather warnings have proved useful for planning, official warnings of hazards such as earthquakes, landslides, and floods are also useful to people deciding where to live, work, and travel.

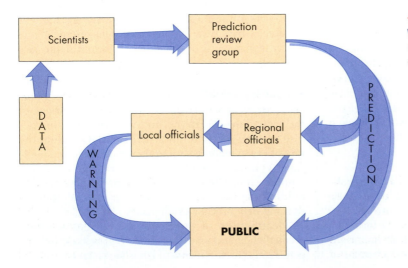

◀ **FIGURE 1.10 Hazard Prediction or Warning**
Possible flow path for issuance of a prediction or warning for a natural disaster.

Consider once more the prediction of a volcanic eruption in the Mammoth Lakes area of California. The location and depth of earthquakes suggested to scientists that molten rock was moving toward the surface. In light of the high probability that the volcano would erupt and of the possible loss of life if it did, it would have been irresponsible for scientists not to issue an advisory. Although the predicted eruption did not occur, the warning led the community to develop evacuation routes and consider disaster preparedness. This planning may prove useful, for it is likely that a volcanic eruption will occur in the Mammoth Lakes area in the future. The most recent event occurred only 600 years ago! As a result of the prediction, the community is better informed than it was before and, thus, better able to deal with an eruption when it does occur.

2

Knowing hazard risks can help people make decisions.

Before rational people can discuss and consider adjustments to hazards, they must have a good idea of the risk that they face in various circumstances. The field of risk assessment is rapidly growing, and its application in the analysis of natural hazards probably should be expanded.

RISK AS A CONCEPT

The **risk** of a particular event may be simply defined as the product of the probability of that event occurring multiplied by the consequences should it occur.[14] Consequences (damages to people, property, economic activity, public service, and so on) can be expressed on a variety of scales. If, for example, we are considering the risk of earthquake damage to a nuclear reactor, we may evaluate the consequences in terms of radiation released, which then can be related to damages to people and other living things. In any such assessment, it is important to calculate the risks for various possible events (scenarios), for example, for earthquakes of various magnitudes. A large event has a lower probability than does a small one, but its consequences are likely to be greater.

Thinking deeper about risk (Figure 1.11), we have come to realize that the hazard and the probability of occurrence of hazards includes both the natural process (earthquake, flood, etc.) and the human processes that can change the probability. For example, urbanization can change the magnitude and frequency of flooding, and human-induced climate change is increasing the intensity of storms. Likewise, consequences are

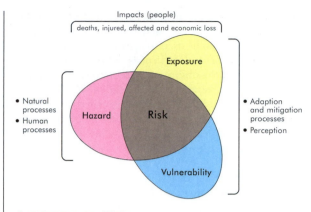

∧ FIGURE 1.11 Risk
Defined as the intersection of hazard, exposure, and vulnerability. *(Modified after IPCC, 2014. Climate change 2014: Impacts, summary, and vulnerability. Technical Summary)*

complex, including the concepts of resiliency, exposure, vulnerability, adaption, mitigation, hazard perception, and social media. Thus, risk can be thought of as the product of hazard, exposure, and vulnerability.[15] Resiliency is the ability to absorb damage and recover from a hazardous event. Exposure refers to elements (people and property) at risk from a natural hazard. Vulnerability is the degree of loss due to exposure to a hazardous event such as a flood or storm (the inverse of resiliency). Vulnerability may vary. For example, a person living in an landslide-prone home who seldom leaves home is more vulnerable to the landslide hazard than other family members who attend school or leave the home each day to work in a nearby town or city. Adaption refers to options that may be taken to reduce exposure and vulnerability, such as insurance, evacuation, engineering to strengthen infrastructure and protect people, and land-use planning to avoid hazardous areas. Mitigation means to reduce the effects of something and is often used by scientists, planners, and policy makers in describing disaster preparedness efforts. For example, after earthquakes and floods, water supplies may be contaminated by bacteria, increasing the spread of diseases. To mitigate the effects of this contamination, a relief agency or government may deploy portable water treatment plants, disinfect water wells, and distribute bottled water.

Vulnerability and exposure can also be reduced by using online resources and social media. For example, during changing conditions of a hazardous event such as a flood, people may use their smartphones to alert friends and family of what is going on (when floodwater will arrive and how deep the flooding is likely to be) and how to respond, such as when to evacuate or how to protect property. Many people have smartphones, and warnings can be issued by emergency agencies and others that a tsunami may arrive soon from a distant earthquake at beaches and coastal communities. Warnings may also be used to alert people to landslides or storms that have closed roads or where emergency help is needed. Possibilities are numerous. During the disastrous 2011 Bangkok flood, social media is thought to have helped reduce damage to property by as much as 37 percent by providing real-time flood conditions of flood location and depth that allowed people to move some possessions out of harm's way.[16]

Determining *acceptable risk* is also complicated, for the risk that society or individuals are willing to take depends on the situation. Driving an automobile is fairly risky, but most of us accept that risk as part of living in a modern world. On the other hand, for many people the acceptable risk from a nuclear power plant is very low because they consider almost any risk of radiation poisoning unacceptable. Nuclear power plants are controversial because many people perceive them as high-risk facilities. Even though the probability of an accident owing to a geologic hazard such as an earthquake may be quite low for most power plants, the consequences could be devastating, resulting in a relatively high risk.

Institutions, such as the government and banks, approach the topic of acceptable risk from an economic point of view rather than from a personal perception of the risk. For example, a bank will consider how much risk it can tolerate with respect to flooding. The federal government may require that any property receiving a government loan cannot have a flood hazard risk that exceeds 1 percent per year; that is, the property must be protected up to and including the 100-year flood.

On an individual level, it is important to recognize that you have some degree of choice regarding the level of risk you are willing to live with. For the most part, you can choose where you are willing to live. If you move to the North Carolina coast, you should realize you are putting yourself in the path of potentially deadly hurricanes. If you choose to live in Los Angeles, it is highly likely that in your lifetime you will experience an earthquake. So why do people live in hazardous areas? Perhaps you were offered an excellent job in North Carolina, or the allure of warm weather, mountains, and the ocean drew you to Los Angeles. Whatever the case, we as individuals must learn to weigh the pros and cons of living in a given area and decide whether or not it is worth the risk. This assessment should consider factors such as the level of potential devastation caused by an event, the frequency of the event, and the extent of the geographic area at risk, and it should compare these factors to the potential benefits of living in the high-risk area. In this way, we determine our own acceptable risk, which may vary from person to person.

A frequent problem of risk analysis is a lack of reliable data for analyzing either the probability or the consequences of an event. It can be difficult to assign probabilities to geologic events such as earthquakes and volcanic eruptions because the known record of past events is often inadequate.[13] Similarly, it may be difficult to determine the consequences of an event or series of events. For example, if we are concerned with the consequences of releasing radiation into the environment, we need a lot of information about the local biology, geology, hydrology, and meteorology, all of which may be complex and difficult to analyze. Despite these limitations, risk analysis is a step in the right direction. As we learn more about determining the probability and consequences of a hazardous event, we should be able to provide the more reliable analyses necessary for decision making.

RISK MANAGEMENT

Management of risk from natural hazards is part of the broader field of land-use planning and hazard management programs for cities, counties, states, and the United States. The national agency is the Federal Emergency Management Agency (FEMA). Natural hazard plans for cities and counties have names such as Emergency Management Plan, and they include discussions of hazard assessment for specific events, such as earthquake, flood, and wildfire; hazard mitigation, which includes specific actions to reduce exposure and vulnerability to natural hazards; and action taken for recovery following a natural hazard event.

Hazard management is becoming more complex because increasing human population has resulted in more people in cities and rural localities living in areas where natural hazards occur (although no place is completely free from exposure to natural hazards), and global change is altering the intensity and occurrence of violent storms and heat waves linked to other hazards, such as wildfire and landslides. Complex natural hazard problems are often resistant to clear problem definitions and easily identified solutions. Some natural hazard management issues involve social, political, and cultural issues that make problem solutions difficult if not impossible to resolve. Such problems are sometimes labeled as a **wicked problem**.[17] Here, the term "wicked" is not used in the familiar and historic sense of something immoral or evil or in the more modern usage of the term to indicate something pleasurable or desirable. Rather, it is used to indicate resistance to a solution. Wicked problems do not have simple solutions. Not all complex natural hazard management problems are wicked, but those with strong social or political implications may be. Science alone is not able to mitigate wicked problems. Progress (positive change) on wicked problems requires the use of science, politics, economics, and social justice. Approaches to wicked problems include:[17] decision making at regional to local levels, including government and non-government organizations and private ownership; decision making across several administrative levels in regard to water, land, and air; balancing differences in economic, social, political, and scientific norms and values of the various stakeholders; and, perhaps most important, use of adaptive management.

Adaptive management is a science-based active process that recognizes the uncertainty of future outcomes; the limited usefulness of predictions; and the importance of learning through field experience, monitoring, reevaluating, and being flexible.[17]

Examples of wicked problems associated with natural hazards include natural hazards such as coastal erosion. Coastal erosion is a complex problem with a difficult solution because those living near or visiting beaches often have competing values. Property owners wish to protect their private property, often at the expense of the beach. People visiting the beach for recreation such as swimming or walking wish to protect the beach, even at the expense of property owners. If there is a beach park with grass, picnic tables and areas designated for social gatherings merging with the beach, people using these areas may well value the grass more than the beach itself. The agencies responsible for managing a beach may belong to the city, county, or state, and the issues involved may include water and air quality as well. There are no easy answers to this wicked problem. Therefore, best management is often incremental management (including adaptive management) that needs to be flexible, monitored, reevaluated, and, possibly, changed over time.

3 ━━━━━━━━━━━━━━━━━━━

Linkages exist between natural hazards.

Many of the hazards are linked. For example, hurricanes are often associated with flooding, and intense precipitation associated with hurricanes can cause erosion along a coast and landslides on inland slopes. Volcanic eruptions on land are linked to mudflows and floods, and eruptions in the ocean are linked to tsunamis. Wildfire that denudes vegetation is often linked to soil erosion, landslides, and flooding as runoff from burned slopes increases (see Chapter 13). Climate change with warming oceans and rising sea levels are linked to more intense storms such as hurricanes, as well as to accelerated coastal erosion (see Chapter 12).

One disaster can influence others as a cascade of conditions. One disaster can impact another disaster that soon follows after the first. For example, an earthquake or flood that causes significant loss of life and property damage can lead to pollution of water supplies, resulting in an outbreak of water-borne disease. Consider a severe storm, such as a tropical cyclone with high winds and torrential rain that damages or destroys homes in an area with little resiliency. People are forced into temporary shelter (perhaps tents). If another severe storm occurs before repairs and new housing is made available, then the loss of life may well be worse than in the first storm.

4 ━━━━━━━━━━━━━━━━━━━

Humans can turn disastrous events into catastrophes.

Early in the history of our species, our struggle with earth processes was probably a day-to-day experience. However, our numbers were neither great nor concentrated, so losses from hazardous earth processes were not very significant. As people learned to produce and maintain a larger and, in most years, more abundant food supply, the population increased and became concentrated near sources of food. The concentration of population and resources also increased the effect of periodic earthquakes, floods, and other potentially hazardous natural processes. This trend has continued, so that many people today live in areas that are likely to be damaged by hazardous earth processes or are susceptible to the effect of such processes in adjacent areas. An increase in population puts a greater number of people at risk and also forces more people to settle in hazardous areas, therefore increasing the need for planning in order to minimize losses from natural disasters.

EXAMPLES OF DISASTERS IN DENSELY POPULATED AREAS

Mexico City is the center of North America's largest and the world's third most populous metropolitan area. Approximately 22 million people are concentrated in an area of about 2300 km^2 (~890 mi.^{2}). Families in this area average five members, and an estimated one-third of the families live in a single room. The city is built on ancient lakebeds that amplify the shaking in an earthquake, and parts of the city have been sinking several centimeters per year due to pumping of groundwater. The subsidence has not been uniform, so the buildings tilt and are thus even more vulnerable to earthquake shaking.[18] In September 1985, Mexico endured a magnitude 8.0 earthquake that killed about 10,000 people in Mexico City.

Another example comes from the Izmit, Turkey, earthquakes in 1999. These quakes killed more than 17,000 people because they took place near a heavily populated area where many buildings were poorly constructed.

One reason that the Mexico City and Izmit earthquakes caused such great loss of life was that so many people were living in the affected areas. If either of these quakes had occurred in less densely populated areas, fewer deaths would have resulted.

POPULATION GROWTH AS A FACTOR IN HAZARDS

The world's population has more than tripled in the past 70 years. Between 1830 and 1930, the world's population doubled from 1 to 2 billion people. By 1970, it had nearly doubled again, and as of 2016, there were about 7.4 billion people on Earth. How did we get to be 7+ billion strong?

The increase in the number of people on our planet can be related to various stages of human development. When we were hunter-gatherers, our numbers were small, population growth rates were low, and population density was far less than 1 person per km^2 (~0.39 mi^2). With the development of agriculture, population growth rates increased by several hundred times and population density increased about 30 times as the result of a stable food supply but still was considerably less than one person per km^2. During the early industrial period (A.D. 1600 to 1800), growth rates increased again by about 10 times, and density of persons increased to about seven persons per km^2. Since the Industrial Revolution, with modern sanitation and medicine, the growth rates increased another 10 times. Human population reached 6 billion in 2000, and by 2016 it exceeded 7.4 billion. That is 1.5 billion new people in only 16 years. By comparison, total human population reached 1 billion only about A.D. 1800, after more than 40,000 years of human history! Today, the density of the human population is much better known and is tremendously variable (it probably always was very variable, but we do not have the data to know). For example, according to the World Bank (2012), in Australia human population density or persons per km^2 (~0.39 mi^2) is about 3, United States 34, Spain 93, China 140, United Kingdom 260, Haiti 360, and India 410. For the entire world, the density is about 50 persons per km^2. The main conclusion is that very recently (in the past 100 years), human population and density has increased dramatically.

Future Population Growth Because Earth's population is increasing exponentially, many scientists are concerned that in the twenty-first century it will be impossible to supply resources and a high-quality environment for the billions of people added to the world population. Increasing population at local, regional, and global levels compounds nearly all hazards, including floods, landslides, volcanic eruptions, and earthquakes.

In the future, we may be able to mass-produce enough food from nearly landless agriculture or to use artificial growing situations. However, enough food does not solve the problems of the space available to people and maintaining or improving their quality of life. Some studies suggest that the present population is already above a comfortable carrying capacity for the planet. *Carrying capacity* is the maximum number of people Earth can hold without causing environmental degradation that reduces the ability of the planet to support the population.[19]

The news regarding population growth is not all bad—for the first time in the past 50 years, the rate of growth is decreasing. Growth may have peaked at 85 million people per year in the late 1980s, but by 2012, the increase was still about 85 million new people per year.[20,21] Every second, about 4.5 people are born somewhere on Earth, and 1.8 people die. Thus, twice as many people are born than die. We are experiencing what is known as *population momentum* as more people are living longer as a result of improved sanitation, nourishment, water quality, and medicine. Eventually, death rates will increase as the population ages and fall more in line with birth rates. From an optimistic point of view, it is possible that our global population of 6 billion persons in 2000 may not double again. Although population growth is difficult to project because of variables such as agriculture, sanitation, medicine, culture, and education, by the year 2050, human population is forecast to be about 10 billion. Most of the population growth in the twenty-first century will be in developing nations. India will likely have the greatest population of all countries by 2050, which will be about 18 percent of the total world population; China will have about 15 percent. By 2050, these two countries will have more than a third of the total world population![21]

The explosive increase in population during the twentieth century was due to *exponential growth* of the number of people on Earth, which means that the population does not grow each year by the addition of a constant number of people; rather, it grows by the addition of a constant percentage of the current population. Exponential population growth results in higher population densities (people per square kilometer), more exposure of people to hazardous natural processes, increased pollution, reduced availability of food and clean drinking water, and a greater need for waste disposal and energy resources. A significant question is: How many people can Earth support?

There is no easy answer to the population problem, but the role of education is paramount. As people (particularly women) worldwide become more educated, the population growth rate tends to decrease. As the rate of literacy increases, population growth is reduced. Given the variety of cultures, values, and norms in the world today, it appears that our greatest hope for population control is, in fact, through education.[19] However, until the growth rate is zero, population will continue to grow. If the rate of growth is reduced to 0.7 percent per year, which is about two-thirds the current rate of 1.1 percent, human population will still double in 100 years. Having discussed the link between human population and hazards, we will now turn to an important link between how often hazardous processes occur and how big these hazardous events are.

MAGNITUDE AND FREQUENCY OF HAZARDOUS EVENTS

The *impact* of a hazardous event is in part a function of the amount of energy released, that is, its *magnitude*, and the interval between occurrences, that is, its *frequency*. Its impact is also influenced by many other factors, including climate, geology, vegetation, population, and land use. In general, the frequency of an event is inversely related to the magnitude. This is the **magnitude–frequency concept**. Small earthquakes, for example, are more common than large ones. A large event, such as a massive forest fire, will do far more damage than a small, contained burn. However, such events are much less frequent than smaller burns. Therefore, although planners need to be prepared for large, devastating events, the majority of fires suppressed will be smaller ones.

As an analogy to the magnitude–frequency concept, consider the work of logging a forest done by resident termites, human loggers, and elephants. The termites are numerous and work quite steadily, but they are so small that they can never do enough work to destroy all the trees. The people are fewer and work less often, but being stronger than termites, they can accomplish more work in a given time. Unlike the termites, the people can eventually fell most of the trees (Figure 1.12). The elephants are stronger still and can knock down many trees in a short time, but there are only a few of them and they rarely visit the forest. In the long run, the elephants do less work than the people and bring about less change. In our analogy, it is humans who, with a moderate expenditure of energy and time, do the

▲ FIGURE 1.12 Human Scale of Change
Human beings, with our high technology, are able to down even the largest trees in our old-growth forests. The lumberjack shown here is working in a national forest in the Pacific Northwest. *(Inga Spence/Alamy)*

most work and change the forest most drastically. Similarly, natural events with a moderate energy expenditure and moderate frequency are often the most important shapers of the landscape. For example, most of the sediment carried by rivers in regions with a subhumid climate, like most of the eastern United States, is transported by flows of moderate magnitude and frequency. However, there are many exceptions. In arid regions, for example, much of the sediment in normally dry channels may be transported by rare high-magnitude flows produced by intense but infrequent rainstorms. Along the barrier island coasts of the eastern United States, high-magnitude storms cause the most change to the beaches.

Land use may directly affect the magnitude and frequency of events. Four deadly catastrophes resulting from natural hazards were Hurricane Mitch and the flooding of the Yangtze River in China, both in 1998; the Indonesian tsunami in 2004 that killed about 230,000 people; and Hurricane Katrina in 2005 that killed at least 1,600 Americans. Hurricane Mitch, which devastated Central America, caused approximately 11,000 deaths, whereas the floods in the Yangtze River resulted in nearly 4,000 deaths. Land-use changes made the damage from these events particularly severe. For example, Honduras has lost nearly one-half of its forests, and an 11,000 km[2] (~4200 mi.[2]) fire occurred in the region prior to the hurricane, destroying vegetation cover. As a result of deforestation and the fire, hillsides washed away, and with them went farms, homes, roads, and bridges. In central China, the story is much the same; in recent years, the Yangtze River basin has lost about 85 percent of its forest as a result of timber harvesting and conversion of land to agriculture. These land-use changes have probably made flooding on the Yangtze River much more common than it was previously.[22]

5 ──────────────────────────

Consequences of hazards can be minimized.

The ways in which we deal with hazards are too often primarily *reactive*—following a disaster, we engage in search and rescue, firefighting, and providing emergency food, water, and shelter. There is no denying that these activities reduce loss of life and property and need to be continued. However, a move to a higher level of hazard reduction will require increased efforts to *anticipate* disasters and their effects. Land-use planning that limits construction in hazardous locations, hazard-resistant construction, and hazard modification or control (such as flood control channels) are some of the adjustments that anticipate future disastrous events and may reduce our vulnerability to them.[23]

Consequences of hazards obviously include damage directly from a hazard event such as a flood, landslide, or earthquake, and these costs in property and lives may be very significant—but they may not result in the greatest losses. Often it is what happens before, during, and after a hazardous event occurs that determines most of the losses. Consider total losses A + B = C, where C is the total loss; A represents the direct losses from the event; and B consists of losses related to human actions including, for example, inadequate engineering of dikes or not building homes, schools, and other buildings to withstand earthquake shaking; not wanting to or unable to spend funds for disaster planning that result in supplies, water, food, and medicine being quickly delivered after an earthquake or other disastrous event has occurred; not having public safety plans ready to be implemented; or incompetency, greed, or just being lazy and not doing the right thing or even knowing what to do (see Case Study 1.1: Professional Profile).[24]

REACTIVE RESPONSE: IMPACT OF AND RECOVERY FROM DISASTERS

The effect of a disaster on a population may be either direct or indirect. *Direct effects* include people killed, injured, dislocated, or otherwise damaged by a particular event. *Indirect effects* are generally responses to the disaster. They include emotional distress, donation of money or goods, and payment of taxes levied to finance the recovery. Direct effects are felt by fewer individuals, whereas indirect effects affect many more people.[25,26]

The stages of recovery following a disaster are emergency work, restoration of services and communication lines, and reconstruction (Figure 1.13). We can see these stages in recovery activities following the 1994 Northridge earthquake in the Los Angeles area. Restoration began almost immediately with the repair of roads and utilities, using funds from federal programs, insurance companies, and other sources that arrived in the first few weeks and months after the earthquake. The damaged areas then moved quickly from the restoration phase to what is known as the reconstruction stage, which lasted until 2000. The recovery time from a disaster can vary from a few to many years. Northridge has recovered in less than a decade, while in Haiti the recovery following the 2010 was delayed by the 2016 hurricane, and the country remains still early in the recovery process which will take decades and require massive international aid.

As we move through the reconstruction stage, it is important to remember lessons from two disasters: the 1964 Anchorage, Alaska, earthquake and the flash flood that devastated Rapid City, South Dakota, in 1972. Restoration following the Anchorage earthquake began almost immediately in response to a tremendous influx of dollars from federal programs, insurance companies, and other sources approximately one month after the earthquake. Reconstruction moved rapidly as everyone tried to obtain as much of the available funds as possible. In Rapid City, the restoration did not peak until approximately 10 weeks after the flood, and the community took time to carefully think through the best alternatives. As a result, Rapid City today uses land on the floodplain in an entirely different way—it is now a greenbelt with golf courses and other such facilities—and the flood hazard is much reduced.[13,25,26] Conversely, in Anchorage, the rapid restoration and reconstruction were accompanied by little land-use planning. Apartments and other buildings were hurriedly constructed across areas that had suffered ground rupture; the ground was prepared for rebuilding by simply filling in the cracks and regrading the land surface. By ignoring the potential benefits of careful land-use planning, Anchorage has made itself vulnerable to the same type of earthquake that struck in 1964.

In the Northridge case, the effects of the earthquake on highway overpasses and bridges, buildings, and other structures have been carefully evaluated. The goal has been to determine how more vigorous engineering standards for construction of new structures or strengthening of older structures can be implemented

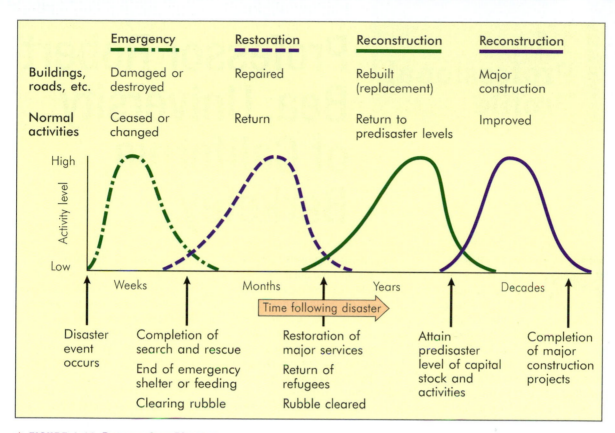

	Emergency	Restoration	Reconstruction	Reconstruction
Buildings, roads, etc.	Damaged or destroyed	Repaired	Rebuilt (replacement)	Major construction
Normal activities	Ceased or changed	Return	Return to predisaster levels	Improved

Activity level — High / Low

Weeks — Months — Years — Decades

Time following disaster

Disaster event occurs	Completion of search and rescue	Restoration of major services	Attain predisaster level of capital stock and activities	Completion of major construction projects
	End of emergency shelter or feeding	Return of refugees		
	Clearing rubble	Rubble cleared		

∧ FIGURE 1.13 Recovery from Disaster

Generalized model of four stages of recovery following a disaster. The first stage after a disaster is a state of emergency, in which normal activities cease or are changed. The next stage in the restoration is a return to normal activities, although probably not at predisaster levels. This is followed by the stages of reconstruction and improvement. Finally, major construction and development are underway, and normal activities have returned. Lengths of the various stages are highly variable, depending on the nature of the disaster and available funding. The entire process may take a decade or longer. *(Modified from Kates, R. W., and Pijawka, D. 1977. From rubble to monument: The pace of reconstruction. In Disaster and Reconstruction, ed. J. E. Haas, R. W. Kates, and M. J. Bowden, pp. 1–23. Cambridge, MA: MIT Press)*

during the reconstruction stage (Figure 1.13). Future moderate to large earthquakes are certain to occur again in the Los Angeles area. Therefore, we need to continue efforts to reduce earthquake hazards.

ANTICIPATORY RESPONSE: AVOIDING AND ADJUSTING TO HAZARDS

The options we choose, individually and as a society, for avoiding or minimizing the effects of disasters depend in part on our perception of hazards. A good deal of work has been done in recent years to try to understand how people perceive various natural hazards. This understanding is important because the success of hazard reduction programs depends on the

attitudes of the people likely to be affected by the hazard. Although there may be adequate perception of a hazard at the institutional level, this perception may not filter down to the general population. This lack of awareness is particularly true for events that occur infrequently; people are more aware of situations such as brush or forest fires that may occur every few years. Standard procedures, as well as local ordinances, may already be in place to control damage from these events. For example, homes in some areas of southern California are roofed with shingles that do not burn readily, they may have sprinkler systems, and their lots are frequently cleared of brush. Such safety measures are commonly noticeable during the rebuilding phase following a fire.

1.1 Professional Profile

Professor Robert Bea, University of California, Berkeley

Professor Robert Bea (Figure 1.1.A) is widely recognized as perhaps the most influential, and sometimes controversial, contributor to our understanding of disasters. Professor Bea has an interesting background in civil engineering and has spent time in both industry and academia. He has published literally hundreds of papers and is a recognized expert in risk management. Much of his research has been focused on industrial catastrophes, such as oil blowouts and exploding oil platforms. However, he has also commented extensively on natural hazards, giving particular attention to what happens before, during, and after events such as hurricanes and floods.

Professor Bea has stated that, following the Occidental oil drilling platform explosion in 1988 that killed over 160 people, he realized that catastrophes and disasters have complex parts related to the poor decisions people make. Sometimes the event is natural and sometimes an engineering failure, but why it occurs, how it occurs, and the role of humans in the event are points he constantly hammers home. For example, a volcanic eruption in a remote part of the world is a big event, and, if there are not many people around, it will more or less be considered

a large damaging event perhaps to ecosystems but not in terms of financial losses to people. On the other hand, should a large volcanic eruption occur near a large city, it would be considered a disaster or catastrophe, and then taking steps to better understand volcanoes and having emergency plans in place become paramount to minimizing total effects of the eruption. Bea has

pointed out that often 20 percent of the financial losses in a disaster may be directly related to the event itself, but the majority of the cost of catastrophes could be greatly reduced by improving human systems that include warnings and better understanding of the hazard.[24]

Bea understands that we have some really big cities in harm's

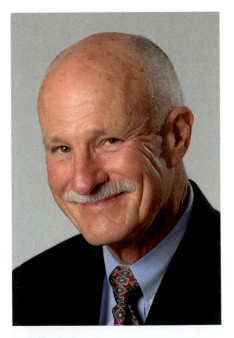

⋀ FIGURE 1.1.A
Professor Robert Bea, University of California, Berkeley is an expert on disasters and their causes. *(Courtesy of Robert Bea)*

way of natural processes, such as flooding and hurricanes, but the decisions we make prior to events and other decisions made afterward can reduce the ultimate consequences of the events by many times. He points out that we need to pick carefully the areas we wish to defend, but be prepared in other instances to move out of harm's way or spend the necessary funds for proper engineering structures that really might protect a city such as New York or New Orleans. With respect to New Orleans, Bea has a personal history there. He states that he and his family lost everything in Hurricane Betsy in 1965, including all of his memorabilia such as wedding photographs, marriage license, and so forth. He points out that 40 years later, following Katrina, he went back to the same place and, while there was a new home built on the site, people were again digging out from the hurricane. That affected him emotionally, as he realized that we could do much more to prevent disasters and catastrophes through better planning, management, and just plain fortitude in doing what is necessary to save money down the road. He points out that you can fix an environment of a big city now, or you can do it later when the cost may be 100 times as much.

Bea also raised early the possibility of an unusual storm that would hit New York, and eventually Superstorm Sandy devastated the coastline. He had indicated earlier that the city of New York could build offshore barriers to protect the city from storm surge, and that storm doors could cover openings to underground systems and bridges. In other words, he proposed floodproofing the city. If that had been done, it probably would have cost $15–20 billion, but the damages from Superstorm Sandy total several times that amount, and the city remains vulnerable.

On the West Coast of the United States, there is the large delta system of the Sacramento River. Levees are old and in poor repair, but these levees protect the ninth-largest economy in the world. Should these fail, then millions of people will be without power or fresh water, perhaps for months. Bea and other scientists have pointed out that this could happen at any time from a megaflood, the likes of which last occurred about 1860, or an earthquake, or a combination of events. California planners at the highest levels are aware of the research that points to a potential catastrophe, and still nothing has been done. Is this because the state simply doesn't have the money? Perhaps it's because people just don't believe that such events will occur because they have never seen them in their lifetime. It seems as if California is saying, "Yes, this could happen, and we're simply going to bear the loss." According to Bea and others, this is not an acceptable solution. Solving the delta problem would cost billions of dollars but nowhere near the estimated cost should the system fail. And rest assured that, without some sort of maintenance and a program to reestablish the integrity of the system, it will eventually fail, and the losses could be hundreds of times the cost of doing a fix now.[24]

Professor Bea has pointed out that some of the major problems are related to people, institutions, companies, and government. Often people and institutions take the easy road, putting maintenance or solving a problem of a critical area on the back burner until an event happens—and then everyone acts really surprised. It may be a surprise to some, but not to people like Bea, who study natural disasters and their occurrence. Bea points out that it is not natural events or engineering failures that are the chief culprits producing the catastrophes; rather, it is the failure of human and organizational systems along with, in some cases, inadequate science, complicated by conflicting egos and people being too lazy to do the things they know they should. The key to reducing catastrophic events is not only to better understand their occurrence and the science behind the event but also to greatly improve our institutions, safety protocols, and disaster preparedness that should take place before, during, and after events. For large cities and critical facilities such as power plants, more attention must go toward providing fixes for potential catastrophes that, while costly now, will save that cost many times over when the unexpected event occurs. Of course, these events are not really surprising or unusual, only perhaps rare, but they should be expected and planned for.

One of the most environmentally sound adjustments to hazards involves **land-use planning**. That is, people can avoid building on floodplains, in areas where there are active landslides, and in places where coastal erosion is likely to occur. In many cities, floodplains have been delineated and zoned for a particular land use. With respect to landslides, legal requirements for soil engineering and engineering geology studies at building sites may greatly reduce potential damages. Damages from coastal erosion can be minimized by requiring adequate setback of buildings from the shoreline or sea cliff. Although it may be possible to control physical processes in specific instances, land-use planning to accommodate natural processes is often preferable to a technological fix that may or may not work.

Insurance is another option that people may exercise in dealing with natural hazards. Flood insurance is common in many areas, and earthquake insurance is also available. Just because insurance is available, however, doesn't mean it is practical (or ethical) to build in an earthquake-prone area without careful engineering, knowing that you will most likely have to use your insurance at some point. In fact, because of large losses following the Northridge earthquake, several insurance companies announced they would no longer offer the insurance.

Evacuation is an important option or adjustment to the hurricane hazard in the states along the Gulf of Mexico and along the eastern coast of the United States. Often, there is sufficient time for people to evacuate, provided they heed the predictions and warnings. However, some people refuse to leave their homes, and if people do not react quickly and the affected area is a large urban region, then evacuation routes may be blocked by residents leaving in a last-minute panic.

Disaster preparedness is an option that individuals, families, cities, states, or even entire nations can implement. Of particular importance is the training of individuals and institutions to handle large numbers of injured people or people attempting to evacuate an area after a warning is issued.

Attempts at *artificial control of natural processes* such as landslides, floods, and lava flows have had mixed success. Seawalls constructed to control coastal erosion may protect property to some extent, but over a period of decades, they tend to narrow or even eliminate the beach (see Chapter 11). Even the best-designed artificial structures cannot be expected to adequately defend against an extreme event, although retaining walls and other structures to defend slopes from landslides have generally been successful when well designed. Even the casual observer has probably noticed the variety of such structures along highways and urban land in hilly areas. Structures to defend slopes have a limited effect on the environment and are necessary where artificial cuts must be excavated or where unstable slopes impinge on human structures. Common methods of flood control are channelization and construction of dams and levees. Unfortunately, flood control projects tend to provide floodplain residents with a false sense of security; no method can completely protect people and their property from high-magnitude floods (see Chapter 6).

An option chosen all too often is simply bearing the loss caused by a natural disaster. Many people are optimistic about their chances of making it through any sort of disaster and, therefore, will take little action in their own defense. This response is particularly true for those hazards—such as volcanic eruptions and earthquakes—that are rare in a given area. Regardless of the strategy we use to minimize or avoid hazards, it is imperative that we understand and anticipate them and their physical, biological, economic, and social effects.

1.4 CHECK your understanding

1. Describe the five fundamental concepts.
2. How is risk from natural hazards evaluated?
3. What is uniformitarianism, and how does it help us understand natural hazards?
4. Differentiate among precursor events, forecasting, predictions, and warnings.
5. Explain what is meant by the magnitude–frequency concept.
6. Why are population growth and land use important in understanding consequences of natural hazards?
7. Differentiate between reactionary responses and anticipatory responses.
8. When we say consequences of hazards are as simple as A + B = C, what do we mean?
9. Explain Robert Bea's message in his Professional Profile.
10. Define what is meant by a wicked problem.

1.5 Many Hazards Provide a Natural Service Function

It is ironic that the same natural events that take human life and destroy property also provide us with important benefits, sometimes referred to as *natural service functions*. For example, periodic flooding of the Mississippi River supplies nutrients to the floodplain and creates the fertile soils needed for agriculture. Flooding, which causes erosion on mountain slopes, also delivers sediment to beaches and flushes pollutants from estuaries in the coastal environment. Landslides may bring benefits to people when the debris forms dams, creating lakes in mountainous areas. Although some landslide dams will collapse and cause hazardous downstream flooding, if a dam remains stable, it can provide valuable water storage and an important aesthetic resource.

Volcanic eruptions have the potential to produce real catastrophes but also provide us with numerous benefits. They often create new land, as in the case of the Hawaiian Islands, which are completely volcanic in origin (Figure 1.14). Nutrient-rich volcanic ash may settle on existing soils and quickly become incorporated, creating fertile soil suitable for both crops and native plants. Earthquakes can also provide us with valuable services. When rocks are pulverized during an earthquake, they may form an impervious clay zone known as *fault gouge* along a geological fault. In many places, fault gouge has formed natural subsurface barriers to groundwater flow; these barriers then pool water or create springs that are important sources of water. In addition, most of the hydrocarbon traps that contain the world's petroleum resources are produced by deformation along faults that create large zones where oil and gas accumulate, without which the Industrial Revolution would never have happened (Figure 1.15). Finally, earthquakes are important in mountain building and, thus, are directly responsible for many of the scenic landscapes of the western United States.

1.5 CHECK your understanding

1. What is meant by the phrase *natural service functions*?

2. List some examples of natural service functions of hazards such as floods, earthquakes, landslides, and volcanic eruptions.

(a)

(b)

⋀ FIGURE 1.14 New Land from Volcanic Eruption
New land being added to the island of Hawaii. (a) The cloud of steam and acidic gases in the central part of the photograph is where hot lava is entering the sea. *(Edward A. Keller)* (b) Close-up of an advancing lava front near the gas cloud. *(Edward A. Keller)*

◄ FIGURE 1.15 Hydrocarbon Trap
Faulting in the Santa Barbara Channel has produced a linear fault trap, along which oil rigs at the surface are situated above. *(Nik Wheeler/ Corbis)*

1.6 Global Climate Change and Hazards

Global change in climate is one of the defining science and social issues of the twenty-first century. The science issues are largely settled; human processes (especially burning fossil fuel) is adversely impacting the global climate system (see Chapter 12). In recent years, we have seen the mass movement of people from agriculture areas of Syria suffering from serious drought that lowers food production for cities. Apart from social problems incurred by the mass immigration of people to other regions, such as Western Europe and the Americas, this has led to civil war.[27]

Global warming, a symptom of climate change, is likely affecting the incidence of hazardous natural events such as storms, landslides, drought, and fires (see Chapter 12). How might climatic change affect the magnitude and frequency of these events? With global warming, sea level is rising as the heating of seawater expands the volume of the ocean, and glacial ice is melting. As a result, coastal erosion will increase. Climate change may shift food production areas, as some places receive more precipitation and others receive less. Deserts and semiarid areas will likely expand, and warmer northern latitudes could become more productive. Such changes could lead to global population shifts, which might bring about wars or major social and political upheavals.

Global warming is feeding more energy from warmer ocean water into the atmosphere; this energy is likely to increase the severity and frequency of hazardous weather, such as thunderstorms and tornadoes, and increase the intensity of hurricanes. In fact, this trend may already be underway—an analysis of global extreme-weather events by a United Nations panel indicates that since the 1950s there has been an increase in heavy precipitation events in midlatitude regions and, since the 1970s, a likely increase in the intensity of hurricanes.[28]

1.6 CHECK your understanding

1. In what ways might global climate change influence the occurrence of natural hazards such as coastal erosion, extreme weather events, and droughts?

CONCEPTS in review

1.1 Why Studying Natural Hazards Is Important

LO:1 Explain the difference between a disaster and a catastrophe.

- Natural hazards are responsible for causing significant death and damage worldwide each year. Processes that cause hazardous events include those internal to the Earth, such as volcanic eruptions and earthquakes that result from Earth's internal heat, and those external to the Earth, such as hurricanes and global warming, which are driven by energy from the sun.
- Natural processes may become hazards, disasters, or catastrophes when they interact with human processes.
- Central to an understanding of natural hazards is awareness that hazardous events result from natural processes that have been in operation for millions and possibly billions of years before humans experienced them. These processes become hazards when they threaten human life or property and should be recognized and avoided.

KEY WORDS
catastrophe (p. 5), disaster (p. 5), natural hazard (p. 5)

1.2   The Role of History in Understanding Hazards

LO:2   Discuss the role of history in the understanding of natural hazards.

- Hazards involve repetitive events. Thus, a study of the history of these events provides much-needed information for hazard reduction. A better understanding and more accurate predictions of natural hazards comes from integrating historic and prehistoric information, present conditions, and recent past events, including land-use changes.

1.3   The Geologic Cycle

LO:3   Discuss the components and processes of the geologic cycle.

- Geologic conditions and materials largely govern the type, location, and intensity of natural processes. The geologic cycle creates, maintains, and destroys earth materials by physical, chemical, and biological processes.
- Subcycles of the geologic cycle are the tectonic cycle, rock cycle, hydrologic cycle, and various biogeochemical cycles. The tectonic cycle describes large-scale geologic processes that deform Earth's crust, producing landforms such as ocean basins, continents, and mountains. The rock cycle may be considered a worldwide Earth material recycling process driven by Earth's internal heat, which melts the rocks subducted in the tectonic cycle. Driven by solar energy, the hydrologic cycle operates by way of evaporation, precipitation, surface runoff, and subsurface flow of water. Bio-geochemical cycles can most easily be described as the transfer of chemical elements through a series of storage compartments or reservoirs, such as air or vegetation.

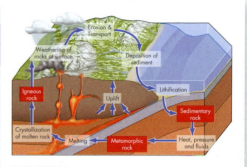

KEY WORD

geologic cycle (p. 10)

1.4   Fundamental Concepts for Understanding Natural Processes as Hazards

LO:4   Apply the scientific method to a natural hazard of your choice.

LO:5   Synthesize the basics of risk assessment.

LO:6   Explain how much of the damage caused by natural hazards is often related to decisions people make before, during, and after a hazardous event.

LO:7   Explain why the magnitude of a hazardous event is inversely related to its frequency.

LO:8   Summarize how natural hazards are linked to one another and to the physical environment.

LO:9   Give reasons why increasing population and poor land-use practices compound the effects of natural hazards and can turn disasters into catastrophes.

KEY WORDS

environmental unity (p. 15), **forecast** (p. 15), **land-use planning** (p. 27), **magnitude–frequency concept** (p. 23), **prediction** (p. 15), **risk** (p. 19), **uniformitarianism** (p. 14), **warning** (p. 18)

- Five fundamental concepts establish a philosophical framework for studying natural hazards:
    1. Hazards are predictable from scientific evaluation.
    2. Risk analysis is an important component in our understanding of the effects of hazardous processes.
    3. Linkages exist between different natural hazards, as well as between hazards and the physical environment.
    4. Hazardous events that previously produced disasters are now producing catastrophes.
    5. Consequences of hazards can be minimized.
- Environmental unity states that one action often leads to others in a sequence (you can't do just one thing).
- Human population growth and poor land-use decisions are turning former disasters into catastrophes.
- The magnitude–frequency concept states that the larger the magnitude of a natural process, the less frequently it occurs.
- Consequences of a natural hazard event such as a flood, hurricane, or earthquake can produce loss of life and property. Human decisions and actions before, during, and after a hazardous event can greatly increase or decrease the consequences.

1.5   Many Hazards Provide a Natural Service Function

LO:10   Explain how events we view as hazards provide natural service.

- Natural service functions refer to benefits provided by a particular process; for example, flooding provides fertile floodplain soil, and volcanic eruption produces new land.

1.6   Global Climate Change and Hazards

LO:11   Summarize links between climate change and natural hazards.

- Global climate change with global warming will likely affect the frequency and intensity of thunderstorms and tornadoes and the intensity of hurricanes. Rising sea level will increase coastal erosion.
- With global warming, deserts and semiarid regions will expand, and droughts will be more common.
- With warming, the frequency and intensity of wildfire will both likely increase.

CRITICAL thinking questions

1. How would you use the scientific method to test the hypothesis that sand on the beach comes from the nearby mountains?

2. It has been argued that we must control human population because otherwise we won't be able to feed everyone. Even if we could feed 10 to 15 billion people, would we still want a smaller population? Why or why not?

3. Considering that events we call natural hazards are natural processes that have been occurring on Earth for millions of years, how do you think we should go about trying to prevent loss of life from these events? Think about the choices society has in terms of attempting to control and prevent hazards or attempting to keep people out of harm's way.

4. Provide an example or two that support the hypothesis that human processes and the appropriate and inappropriate decisions that people make are potentially more important to total losses from a hazardous event than the natural event itself.

5. Global warming is a major concern today. Discuss how global warming might influence the magnitude or frequency of hazardous events, disasters, or catastrophes caused by natural hazards.

6. What was the role of science in understanding the 2010 Haiti earthquake? Could this event have been expected?

7. What other hazards are linked to the 2010 earthquake? Will the effect of these hazards increase or decrease in the future? What do you need to know to determine the likelihood of such an increase or decrease?

8. Human processes are part of the equation of determining the consequences of future large earthquakes in Haiti. What human processes could be implemented in order to reduce the impact of future earthquakes?

9. What do we mean when we say the risk from a natural hazard event is the product of the probability, exposure, and, vulnerability? Provide a hypothetical example to illustrate your answer.

References

1. **USGS**. 2010. Magnitude 7.0 Haiti region. http://earthquake.usgs.gov.

2. **USGS**. 2009. October 17, 1989 Loma Prieta earthquake. http://earthquake.usgs.gov.

3. **Disaster Emergency Committee**. 2013. Haiti earthquake facts and figures. www.dec.org.uk/haiti-earthquake-facts-andfigures. Accessed 7/14/13.

4. **Eberhard, M. O.**, and **four others**. 2010. The Mw 7.0 of January 12, 2010: USGS/EERI Advance reconnaissance team report. U.S. Geological Survey Open-File Report 2010–1048. Executive Summary. Washington DC.

5. **Hoyvis, P., Below, R., Scheuren, J.-M.**, and **Guha-Sapir, D**. 2007. *Annual disasters statistical review: Numbers and trends 2006*. Brussels, Belgium: University of Louvain, Center for Research on the Epidemiology of Disasters (CRED).

6. **Renner, M.**, and **Chafe, Z**. 2007. *Beyond disasters*. Washington, DC: World Watch Institute.

7. **Center for Research on the Epidemiology**. 2016a. Poverty and death: Poverty Mortality 1996–2015. Cred Crunch Issue No. 44.

8. **Center for Research on the Epidemiology**. 2016b. Disaster data: A balanced perspective. Cred Crunch Issue No. 41.

9. **Guha-Sapir, D., Hargitt, D.**, and **Hoyois, P**. 2004. *Thirty years of natural disasters 1974–2003: The numbers*. Brussels, Belgium: University of Louvain, Center for Research on the Epidemiology of Disasters (CRED).

10. **Center for Research on the Epidemiology.** 2013. Disaster data: A balanced perspective. Cred Crunch Issue No. 32.

11. **Guha-Sapir, D.**, and **Hoyola, P**. 2012. *Measuring the human and economic impact of disasters*. Government Office for Science, United Kingdom.

12. **Crossett, K. M., Culliton, T. J., Wiley, P. C.**, and **Goodspeed, T. R**. 2004. *Population trends along the coastal United States: 1980–2008*. National Ocean Service, National Oceanic and Atmospheric Administration.

13. **White, G. F.**, and **Haas, J. E**. 1975. *Assessment of research on natural hazards*. Cambridge, MA: MIT Press.

14. **Crowe, B. W**. 1986. Volcanic hazard assessment for disposal of high-level radioactive waste. In *Active tectonics*, ed. Geophysics Study Committee, pp. 247–60. National Research Council. Washington, DC: National Academy Press.

15. **Alexander, D**. 2002. *Principles of emergency planning and management*. New York: Oxford University Press.

16. **Allaire, M. C.** 2016. Disaster loss and social media: Can online information increase flood resiliency? *Water Resources Research* 52: 7408–23.

17. **DeFries, R.,** and **Nagendra, H.** 2017. Ecosystem management as a wicked problem. *Science* 356: 265–69.

18. **Jones, R. A.** 1986. New lessons from quake in Mexico. *Los Angeles Times*, September 26.

19. **Brown, L. R., Flavin, C.,** and **Postel, S.** 1991. *Saving the planet*. New York: W. W. Norton & Co.

20. **Population Reference Bureau.** 2000. *World population data sheet*. Washington, DC: Population Reference Bureau.

21. **Smil, V.** 1999. How many billions to go? *Nature* 401: 429.

22. **Abramovitz, J. N.,** and **Dunn, S.** 1998. Record year for weather-related disasters. *Vital Signs Brief* 98–5. Washington, DC: World Watch Institute.

23. **Advisory Committee on the International Decade for Natural Hazard Reduction.** 1989. *Reducing disaster's toll*. National Research Council. Washington, DC: National Academy Press.

24. **Marsa, L.** 2013. Robert Bea, Master of disaster. *Discovery*, June.

25. **Kates, R. W.,** and **Pijawka, D.** 1977. From rubble to monument: The pace of reconstruction. In *Disaster and reconstruction*, ed. J. E. Haas, R. W. Kates, and M. J. Bowden, pp. 1–23. Cambridge, MA: MIT Press.

26. **Costa, J. E.,** and **Baker, V. R.** 1981. *Surficial geology: Building with the Earth*. New York: John Wiley.

27. **Kelley, C. P. and four others** 2015. Climate change in the Fertile Crescent and implications of the recent Syrian drought. *Proceedings of the National Academy of Sciences* 112: 3241–46.

28. **Trenberth, K. E. and 11 others** 2007. Observations: Surface and atmospheric climate change. In *Climate change 2007: The physical science basis*, contribution of Working Group I to the fourth assessment report of the Intergovernmental Panel on Climate Change, ed. S. Solomon, D. Qin, M. Manning, Z. Chen, M. Marquis, K. B. Averyt, M. Tignor, and H. L. Miller. New York: Cambridge University Press.

In about 20 million years the cities will be side by side

2

Written with the assistance of Tanya Atwater

APPLYING the 5 fundamental concepts
Two Cities on a Plate Boundary

1 Science helps us predict hazards.

Like the way scientific investigation of the San Francisco earthquake led to an understanding of the earthquake hazard of the San Andreas fault, Plate Tectonic Theory, developed in the 1960s, provided a testable model from which scientists could understand the distribution and predict the occurrences of several destructive natural geologic hazards linked to plate tectonics.

2 Knowing hazard risks can help people make decisions.

Although the internal structure of Earth and plate tectonics are not natural hazards in a strict sense, many of the most destructive natural hazards are directly and indirectly related, and therefore understanding these processes can help guide your decisions in how to minimize your risk from related natural hazards.

Internal Structure of Earth and Plate Tectonics

Two Cities on a Plate Boundary

California straddles the boundary between two tectonic plates. That boundary between the North American and Pacific plates is the notorious San Andreas fault (Figure 2.1). A *fault* is a fracture along which one plate has moved relative to the other, and the San Andreas fault is a huge zone of fracturing hundreds of kilometers long. Two major cities, Los Angeles to the south and San Francisco to the north, are located on opposite sides of this fault. In 1906, movement on the San Andreas fault caused a major earthquake that devastated San Francisco, destroying about 80 percent of the city and killing 3,000. At that time, earthquakes were not understood and most believed they resulted from underground explosions. However, scientific investigations of

◀ **SAN FRANCISCO FOLLOWING THE 1906 EARTHQUAKE**
Fires, set off by earthquake ruptured gas mains, raged across the city for days following the 1906 earthquake. More than 25,000 buildings were destroyed by the fires including those in the financial district, shown here. *(NARA National Archives and Records Administration, Photographer: Chadwick, H. D.)*

The surface of Earth would be much different—relatively smooth, with monotonous topography—if not for the active tectonic processes within Earth that produce earthquakes, volcanoes, mountain chains, continents, and ocean basins. In this chapter, we focus directly on the interior of Earth.

After reading this chapter, you should be able to:

LO:1   Describe the basic internal structure and processes of Earth.

LO:2   Summarize the various lines of evidence that support the theory of plate tectonics.

LO:3   Compare and contrast the different types of plate boundaries.

LO:4   Explain the mechanisms of plate tectonics.

LO:5   Outline how plate tectonics has changed the appearance of Earth's surface over time.

LO:6   Compare and contrast the two fundamental processes that drive plate tectonics.

LO:7   Link plate tectonic processes to natural hazards.

LEARNING Objectives

3

Linkages exist between natural hazards.

The movement of tectonic plates is directly linked to earthquakes and most volcanic activity. These hazards are in turn linked to tsunamis and mass wasting and, because plate tectonics is responsible for creating high topography, plate movement is indirectly linked to weather patterns and severe weather, flooding, erosion, and climate change.

4

Humans can turn disastrous events into catastrophes.

In 1906, the combined population of San Francisco and Los Angeles was less than 1 million. Today, more than 17 million people live in these two cities at the boundary between the Pacific and North American plates. When the "Big One" does strike, population growth alone may be responsible for a catastrophe.

5

Consequences of hazards can be minimized.

As with California's strict building codes, which enable man-made structures to withstand ground shaking from earthquakes, in the following chapters you will learn that the consequences of other hazards associated with plate tectonics can be minimized.

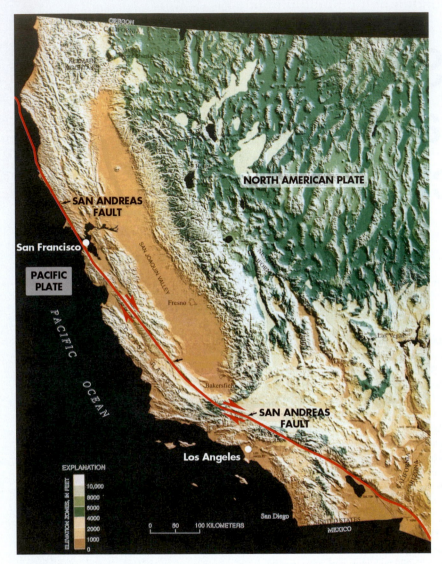

< FIGURE 2.1 San Andreas Fault
False color topographic map of the western United States showing the location of the San Andreas fault in California. Red arrows show relative plate motions on either side of the fault. *(Based on R. E. Wallace/ National Earthquake Information Center. U.S.G.S.)*

caused by earthquakes. Older structures must be *retrofitted*— changes made to their structure—to withstand the shaking, and many people purchase earthquake insurance in an attempt to protect themselves from "the Big One."

Los Angeles is on the Pacific plate and is slowly moving toward San Francisco, which is on the North American plate. In about 20 million years, the cities will be side by side. If people are present, they might be arguing over which is a suburb of the other. Of course, there will still be a plate boundary between the Pacific and North American plates 20 million years from now, because large plates have long geologic lives, on the order of 100 million years. However, the boundary may not be the San Andreas fault. The plate boundary will probably have moved eastward, and the topography of what is now California may be somewhat different. In fact, some earthquake activity, such as the large 1992 Landers earthquake east of the San Andreas fault, may mark the beginning of a shift in the plate boundary.

The Five Fundamental Concepts introduced in Chapter 1 are applied here to our discussion of damaging earthquakes on the San Andreas fault, which result from the movement of tectonic plates (see Applying the Fundamental Concepts, above). Each of the following chapters will similarly begin with an example of applying the fundamental concepts to an

the 1906 earthquake led to the identification of the San Andreas fault and a new understanding of the cause of earthquakes and consequent ground shaking.

Many of the moderate to large earthquakes in the Los Angeles area are on faults related to the San Andreas fault system. Most of the beautiful mountain topography in coastal California near both cities is a direct result of processes related to movement on the fault. However, this beautiful landscape comes at a high cost to society. Since 1906, earthquakes on the San Andreas fault system or on nearby faults, undoubtedly influenced by the plate boundary, have cost hundreds of lives and many billions of dollars in property damage. Constructing buildings, bridges, and other structures in California is more expensive than elsewhere because these must be designed to withstand ground shaking

introductory case study, and at the end of each chapter the case study will be revisited in detail and you will have an opportunity to apply the five fundamental concepts yourself.

2.1 Internal Structure of Earth

Earth is a complex dynamic planet with a layered internal structure that in some ways resembles a chocolate-covered cherry. That is, Earth has a rigid outer shell, a solid center, and a thick layer of gooey material between the two that moves around as a result of dynamic internal processes. These internal processes have incredibly important impacts on the surface of Earth. They are responsible for the largest landforms: continents and ocean basins. The configuration of the continents and ocean basins in part controls the oceans' currents and the distribution of heat carried by seawater in a global system

that affects climate, weather, and the distribution of plant and animal life on Earth. Finally, Earth's internal processes are also responsible for regional landforms including mountain chains, chains of active volcanoes, and large areas of elevated topography, such as the Tibetan Plateau and the Rocky Mountains. The high topography that includes mountains and plateaus significantly affects both global circulation patterns of air in the lower atmosphere (weather) and longer-term climate patterns, thereby directly influencing all life on Earth. Thus, our understanding of the internal processes of Earth is much more than simply academic interest. These processes are at the heart of producing the multitude of environments shared by all living things on Earth.

Earth Layers Earth has a radius of about 6370 km (~3960 mi.). The internal layers of Earth are shown and described in Figure 2.2. We can consider the internal structure of Earth in two fundamental ways:

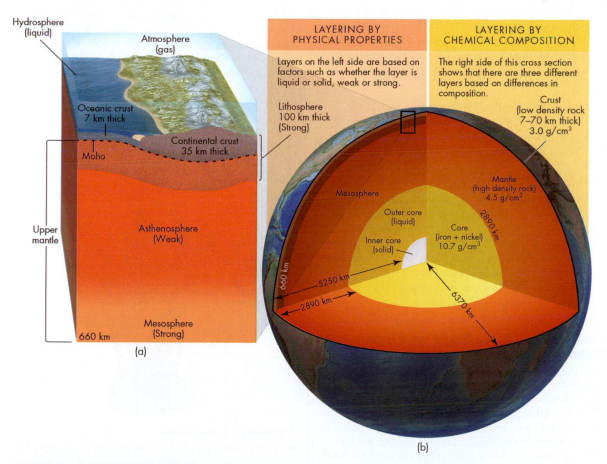

∧ FIGURE 2.2 Earth and Its Interior
(a) Idealized diagram showing the internal structure of Earth and its layers extending from the center to the surface. (b) Notice that the lithosphere is a strong layer that includes the entire crust and part of the upper mantle. The asthenosphere, located entirely within the mantle, is a weak layer.

1. By composition and density (heavy or light).
2. By physical properties (e.g., solid or liquid, weak or strong).

Our discussion will explore the two ways of looking at the interior of our planet. Some of the basic components of Earth's structure[1] are:

> A solid **inner core** more than 1200 km (~750 mi.) thick, roughly the size of the moon but with a temperature about as high as the temperature of the surface of the sun.[2] The inner core is believed to be primarily metallic, composed mostly of iron (about 90 percent by weight), with minor amounts of elements such as sulfur, oxygen, and nickel.

> A liquid **outer core** with a thickness of just over 2200 km (~1370 mi.) and a composition similar to that of the inner core. The average density of the inner and outer core is approximately 10.7 grams per cubic centimeter (~0.39 pound per cubic inch). The maximum near the center of Earth is about 13 g/cm^3 (~0.47 lb/in.^{3}). By comparison, the density of water is 1 g/cm^3 (~0.04 lb/in.^{3}) and the average density of Earth is approximately 5.5 g/cm^3 (~0.2 lb/in.^{3}).

> The **mantle**, nearly 2900 km (~1800 mi.) thick, surrounds the outer core and is composed mostly of solid iron- and magnesium-rich silicate rocks. The average density of the mantle is approximately 4.5 g/cm^3 (~0.16 lb/in.^{3}), which is less than half that of the underlying core.

> The **crust**, with variable thickness, is the outer rock layer of Earth. The boundary between the mantle and crust is known as the Mohorovičić discontinuity (also called the **Moho**). The Moho separates the lighter rocks of the crust, which have an average density of approximately 3.0 g/cm^3 (~0.10 lb/in.^{3}), from the more dense rocks of the mantle below.

Continents and Ocean Basins Have Significantly Different Properties and History

Within the uppermost portion of the mantle, near the surface of Earth, our terminology becomes more complicated. For example, the cool, strong outermost layer or sphere of Earth is also called the **lithosphere** (*lithos* means "rock"). It is much stronger and more rigid than the material underlying it, the **asthenosphere** (*asthenos* means "without strength"), which is a hot and slowly flowing layer of relatively weak rock (Figure 2.2). The lithosphere averages about 100 km (~62 mi.) in thickness, ranging from a few kilometers (~1 to 2 mi.) thick beneath the crests of mid-ocean ridges to

about 120 km (~75 mi.) beneath ocean basins and 20 to 400 km (~13 to 250 mi.) beneath the continents. The crust is embedded in the top of the lithosphere. Crustal rocks are less dense than the mantle rocks below, and oceanic crust is slightly denser than continental crust. Oceanic crust is also thinner: The ocean floor has a uniform crustal thickness of about 6 to 7 km (~3.7 to 4.4 mi.), whereas the crustal thickness of continents averages about 35 km (~22 mi.) and may be up to 70 km (~44 mi.) thick beneath mountainous regions. Thus, the average crustal thickness is less than 1 percent of the total radius of Earth and can be compared to the thin skin of a tangerine. Yet it is this layer that is of particular interest to us because we live at the surface of the continental crust.

In addition to differences in density and thickness, continental and oceanic crusts have very different geologic histories. Oceanic crust of the present ocean basins is less than approximately 200 million years old, whereas continental crust may be several billion years old. Three thousand kilometers (~1865 mi.) below us, at the core–mantle boundary, processes may be occurring that significantly affect our planet at the surface. Scientists speculate that gigantic cycles of convection occur within Earth's mantle, rising from as deep as the core–mantle boundary up to the surface and then falling

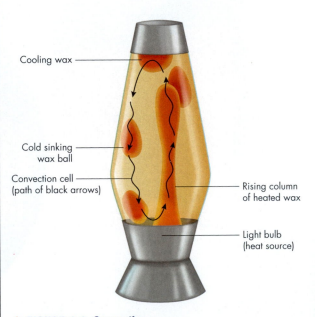

∧ FIGURE 2.3 Convection
Annotated photograph of a lava lamp illustrating the concept of convection. Heating at the base of the lamp causes the wax to rise buoyantly due to a decrease in density. Away from the heat source, the wax cools and becomes more dense, sinking back to the bottom and completing the convection cell.

back again. The concept of **convection** is illustrated by using a lava lamp (Figure 2.3). In the base of the lamp is a light bulb that provides heat to the so-called lava in the lamp, which is actually a type of wax. Heating causes the wax at the bottom of the lamp to expand, which lowers its density. Once the density of the wax is less than that of the fluid in the lamp, it will rise toward the top. As it rises and moves away from the heat source, the wax cools, begins to contract, and its density increases. Once its density is greater than that of the fluid it will sink towards the bottom, where it will be heated again and the process is repeated. This temperature-driven circulation is known as a *convection cell*.

Although it is a bit more difficult to comprehend convection cells operating in solid rock, it has been suggested that convection cells operate within Earth's mantle. Mantle convection is fueled by Earth's internal heat, which results from the original heat of formation of the planet, heat generated by crystallization of the core, and heat supplied by radioactive decay of elements (such as uranium) scattered throughout the mantle. A complete convection cycle in the mantle may take as long as 500 million years.[1] The simplest way to understand mantle convection is to understand plate tectonics, which can be thought of as the surface manifestation of this process.

2.1 CHECK your understanding

1. What are the major differences between the inner and outer cores of Earth?
2. In what ways are the major properties of the lithosphere different from those of the asthenosphere?

2.2 Plate Tectonics

The term *tectonics* refers to the large-scale geologic processes that deform Earth's lithosphere, producing landforms such as ocean basins, continents, and mountains. Tectonic processes are driven by forces within Earth. These processes are part of the tectonic system, an important subsystem of the Earth system.

MOVEMENT OF THE TECTONIC PLATES

The lithosphere is broken into large pieces called *lithospheric plates* or *tectonic plates*[3] that are in motion due to the weak character of the underlying asthenosphere, which is able to flow. The world map in Figure 2.4a shows the locations and names of the major tectonic plates; the red arrows indicate the relative movement of each plate with respect to any adjacent plate. Processes associated with the creation, movement, and destruction of these plates are collectively known as **plate tectonics**. Plate tectonics is responsible for several of the most devastating natural hazards discussed in this book, including earthquakes and volcanoes.

It is important to notice that all of the major plates include both a part of a continent and a part of an ocean basin (Figure 2.4a). Even the Pacific plate, which is mostly composed of oceanic lithosphere, includes a small piece of the North American continent, as discussed earlier (see Figure 2.1). Some plates are very large and some are relatively small, although they are all significant on a regional scale. For example, the Juan de Fuca plate off the Pacific Northwest coast of the United States (Figure 2.4a), which is relatively small, is responsible for many of the earthquakes in northern California and all of the volcanoes in Oregon and Washington. Because there is relative movement between the edges of adjacent lithospheric plates, plate boundaries are geologically active areas. Specifically, most earthquakes are associated with plate boundaries, and the edges of plates can easily be identified by plotting the distribution of global seismicity on a world map (Figure 2.4b). Over geologic time, plates are formed and destroyed, cycling materials from the interior of Earth to the surface and back again at these boundaries. The continuous recycling of tectonic plates is collectively called the *tectonic cycle*.

Plate Tectonic Theory As the rigid lithospheric plates move over the asthenosphere, they carry the continents embedded within them.[4] The idea that continents move is not new; it was first suggested by German scientist Alfred Wegener in 1915. His primary evidence for **continental drift** was based on the congruity of the shape of continents, and the similarity in the fossils found in South America and Africa, which are now separated by the Atlantic Ocean (see Section 2.4). Although Wegener's hypothesis was correct, it was not taken seriously because he could not explain the mechanism by which the continents drifted around Earth's surface. The explanation came to light in the late 1960s, when **seafloor spreading** was discovered. In seafloor regions called **mid-ocean ridges** or **spreading centers**, tectonic plates move away, or diverge from one another. Divergence causes the plates to grow because hot rock from Earth's interior rises to fill the void space and then cools, adding new oceanic lithosphere to the plate edge (Figure 2.5). While oceanic lithosphere is created along some plate boundaries (spreading centers), it is destroyed at others, where cool, dense lithosphere sinks into the asthenosphere at **subduction zones** (Figure 2.5). Thus, continents do not move *through* oceanic crust; rather, they are *carried along with it* as the crust moves either away from spreading centers or toward subduction zones. Because the rate of production of new lithosphere at spreading centers is balanced by consumption at subduction zones, the size of Earth remains constant, neither growing nor shrinking.

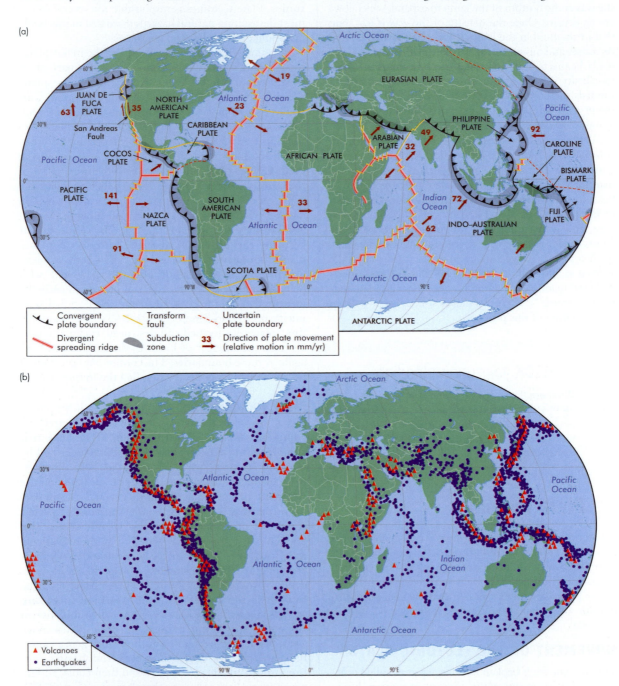

∧ FIGURE 2.4 Earth's Plates
(a) Map showing the major tectonic plates, plate boundaries, and direction of relative plate movement.
(b) Map showing location of volcanoes and earthquakes. Notice the strong correlation between the locations of earthquakes and volcanoes on this map and the locations of plate boundaries in (a). *(Based on Christopherson, R., Geosystems, 2nd ed. Englewood Cliffs, NJ, 1994; Macmillan and Hamblin, W. K.,* Earth's Dynamic Systems, *6th ed. New York: Macmillan, 1992)*

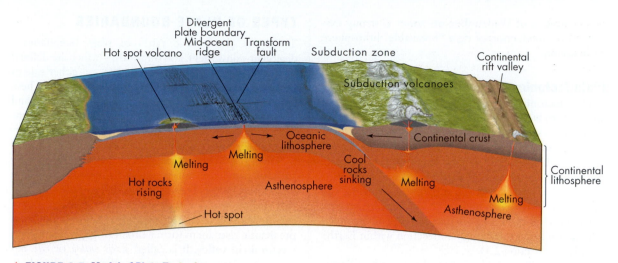

∧ FIGURE 2.5 Model of Plate Tectonics
Block diagram illustrating important plate tectonic features. New oceanic lithosphere is being produced at the mid-ocean ridge (divergent plate boundary). Elsewhere, oceanic lithosphere returns to the interior of Earth at a subduction zone (convergent plate boundary).

Sinking Plates Generate Volcanoes and Earthquakes

The concept of a lithospheric plate sinking into the upper mantle at a subduction zone is shown in detail in Figure 2.6. Interaction between the cold down-going, or subducting, oceanic crust and the hot asthenosphere causes rocks at a depth of about 150 km (~90 mi.) to begin melting, generating large volumes of molten rock called *magma* (see Chapter 5). The less dense magma rises to the surface through the overriding plate, producing a chain of volcanoes along the subduction zone, These volcanic chains, like the one that rings the Pacific Basin, are known as a *volcanic arc*. The path of the descending plate (or *slab*, as it sometimes is called) into the upper mantle is clearly marked by earthquakes. As the oceanic plate subducts, earthquakes are produced both between it and the overriding plate and within the interior of the subducting plate. Earthquakes occur because the sinking lithospheric plate is relatively cooler and stronger than the surrounding asthenosphere; this difference causes rocks to break and seismic energy to be released (see Chapter 3).[5]

Descending plates may enter a subduction zone at very shallow or nearly vertical angles. Scientists can determine this angle by plotting the locations of shallow, intermediate, and deep earthquakes that occur in the down-going slab (Figure 2.6). These inclined planes of earthquakes are called **Wadati–Benioff zones**. The

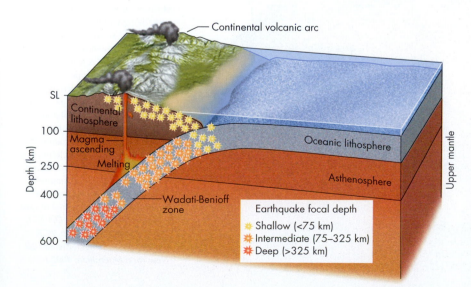

< FIGURE 2.6 Subduction Zone
Idealized diagram illustrating the linkage between subduction zone processes and the occurrence of volcanoes and earthquakes. Interaction between the subducting oceanic lithosphere causes melting and volcanism on Earth's surface. Destructive earthquakes are common at subduction zones and occur over a broad range of depths. The inclined array of earthquake foci from shallow to deep that delineates the descending lithospheric plate is known as the Wadati–Benioff zone (named after the two scientists who discovered it).

very existence of Wadati–Benioff zones is strong evidence that subduction of rigid "breakable" lithosphere is occurring.[5]

Plate Tectonics Is a Unifying Theory
The theory of plate tectonics is to geology what Darwin's origin of species is to biology: a unifying concept that explains an enormous variety of phenomena. Biologists now have an understanding of evolutionary change; geologists understand how Earth works: the direction of plate movement, distribution of earthquakes and volcanoes, similarities among fossils on different continents, and changes in Earth's magnetism. In geology, we are still seeking the exact mechanism that drives plate tectonics, but we think it is most likely convection within Earth's mantle (see Section 2.1).

Figure 2.7 illustrates hypothetical convection cells that may drive plate tectonics, where hot molten rocks from deep in Earth's mantle buoyantly rise toward the base of the lithosphere. Divergence of the plates along mid-ocean ridges causes some magma to leak out onto the seafloor and hot rocks below the surface to cool, adding new oceanic lithosphere to the edge of the plates. Over the next 10–200 million years this newly formed oceanic lithosphere will be eventually conveyed away from its source of heat at the divergent plate boundary and will cool. Once the ocean lithosphere has cooled enough so that its density is similar to that of the underlying asthenosphere, it will be able to sink back down into the mantle at a subduction zone, completing the convection cell (Figure 2.7).

TYPES OF PLATE BOUNDARIES

There are three basic types of plate boundaries—divergent, convergent, and transform—which are defined by the relative movement of the plates on either side of the boundary (see Figure 2.4a and Table 2.1). These boundaries are not narrow cracks, as shown on maps and diagrams,[5] but are zones that range from a few to hundreds of kilometers across. Plate boundary zones are narrower in oceanic crust and broader in continental crust.

Divergent boundaries occur where neighboring parts of plates are moving away from each other and new lithosphere is being produced. Typically this process, called seafloor spreading, occurs at mid-ocean ridges. Mid-ocean ridges form when hot material from the mantle rises up to form a broad ridge, typically with a central rift valley. It is called a *rift valley*, or *rift*, because the plates are diverging, pulling the lithosphere apart and splitting, or rifting, it. The system of mid-ocean ridges along divergent plate boundaries forms linear submarine mountain chains that are found in virtually every ocean basin on Earth (Figure 2.8). Although the peaks of the mountain chain are as high as 2500 meters (~8200 ft.) above the surrounding seafloor, this range is completely concealed below sea level, except on the island of Iceland (see Chapter 5). The mid-ocean ridge mountain chain may not contain the tallest mountains, but with a length of 65,000 km (~40,400 mi.) this submarine range is 10 times longer than any mountain range on land. Although much less common, divergent plate boundaries and rift valleys can develop above sea

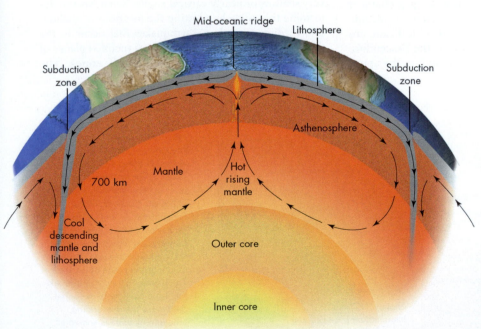

< FIGURE 2.7 Plate Movement Model of plate movement and mantle convection. The outer layer (or lithosphere) is approximately 100 km (~62 mi.) thick and is stronger and more rigid than the underlying asthenosphere, which is a hot and slowly flowing layer of relatively low-strength rock. The mid-ocean ridge is a spreading center where plates pull apart, drawing up hot, buoyant material into the gap. Conveyed away from the ridge, the plates cool and become dense and are able to descend back into the mantle at subduction zones, completing the convection system. *(From Grand, S. P., "Mantle Shear Structure Beneath the Americas and Surrounding Oceans," Journal of Geophysical Research 99: 11591–621, 1994. Based on Hamblin, W. K., Earth's Dynamic Systems, 6th ed. New York: Macmillan, 1992.)*

▼ TABLE 2.1 Types of Plate Boundaries: Dynamics, Results, and Examples

Plate Boundary	Plates Involved	Dynamics	Results	Examples	Natural Hazards
Divergent	Usually ocean	Spreading. The two plates move away from each other and molten rock rises up to fill the gap.	Mid-ocean ridge forms and new oceanic lithosphere is created and added to each plate.	Mid-Atlantic Ridge, which defines the African and North American plate boundary (Figure 2.8).	Light to moderate earthquakes and nonexplosive volcanic eruptions
Convergent (subduction)	Ocean–continent	Oceanic plate sinks beneath continental plate.	A volcanic continental arc and deep trench are formed. Earthquakes and volcanic activity are found here.	Andes Mountains, a continental volcanic chain formed at the Nazca and South American plate boundary (Figure 2.4a).	Great earthquakes; explosive volcanic eruptions; tsunamis, flooding, mass wasting, and subsidence
Convergent (subduction)	Ocean–ocean	Older, denser, oceanic plate sinks beneath younger, less dense oceanic plate.	A volcanic island arc and deep trench is formed. Earthquakes and volcanic activity are found here.	Caribbean Islands, an ocean island arc formed at the boundary between the Caribbean and South American plates (Figure 2.4a).	Great earthquakes; explosive volcanic eruptions; tsunamis, flooding, mass wasting, and subsidence
Convergent (collision)	Continent–continent	Neither plate is dense enough to sink into the asthenosphere; compression results.	A large, high mountain chain is formed, and earthquakes are common.	Himalayan Mountains, the result of collision and shortening between the Indo-Australian and Eurasian plates (Figure 2.4a).	Major earthquakes, flooding, and mass wasting
Transform	Ocean–ocean or continent–continent	The plates slide past one another.	Earthquakes are common and may result in some topographic changes.	San Andreas fault, which defines the boundary between the North American and Pacific plates (Figure 2.11).	Strong to major earthquakes

level on continents, such as the Great Rift Valley in east Africa (see Figure 2.4a and Figure 2.8).

Convergent boundaries occur where plates move toward each other. If one of the converging plates is oceanic and the other is continental, an oceanic–continental plate collision results. The higher-density oceanic plate descends, or subducts, into the mantle beneath the leading edge of the continental plate, producing a subduction zone (Figure 2.9a). The convergence or collision of a continent with an ocean plate can result in compression. *Compression* is a type of stress, or force per unit area. When an oceanic-continental plate collision occurs, compression is exerted on the lithosphere, resulting in shortening of the surface of Earth, like pushing a tablecloth to produce folds. Shortening can cause folding, as in the tablecloth example, and faulting, or displacement of rocks along fractures to thicken the lithosphere. This process of deformation produces major mountain chains and volcanoes such as the Andes in South America and the Cascade Mountains in the U.S. Pacific Northwest.

If two oceanic lithospheric plates collide (oceanic-to-oceanic plate collision), one plate subducts beneath the other, and a subduction zone and arc-shaped chain of volcanoes known as an *island arc* are formed (Figure 2.9b) as, for example, the Aleutian Islands of the North Pacific. A **submarine trench**, a relatively narrow (usually several thousand km long and several km deep) depression on the ocean floor, is often formed as the result of the convergence of two colliding plates with subduction of one. The trench is located seaward of a subduction zone associated with an oceanic–continental plate or on the side of the downgoing plate in an oceanic–oceanic plate collision (see Figure 2.9a and b). Submarine trenches are sites of some of the deepest oceanic waters on Earth. For example, the Marianas Trench at the center edge of the Philippine plate has a depth of 11 km (~7 mi.), which is deep enough for Mount Everest to fit into. Other major trenches include the Aleutian Trench south of Alaska and the Peru-Chile Trench west of South America.

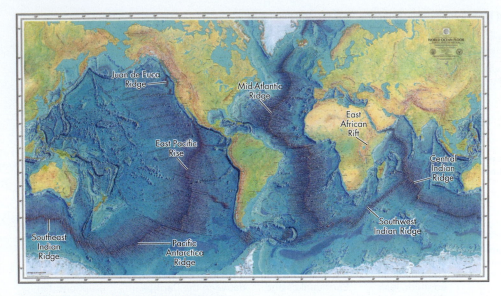

∧ FIGURE 2.8 Mid-Ocean Ridge System
Topographic map of the world's ocean basin emphasizing the tectonic significance of mid-ocean ridge divergent plate boundaries. Note that the ridges are not continuous but are composed of thousands of individual spreading centers connected by transform faults that are perpendicular to the ridges. (World Ocean Floor Panorama *by Bruce C. Hezeen and Marie Tharp. Copyright by Marie Tharp 1977/2003. Reproduced by permission of Marie Tharp Maps, LLC 8 Edward Street, Sparkill, New York 10976.)*

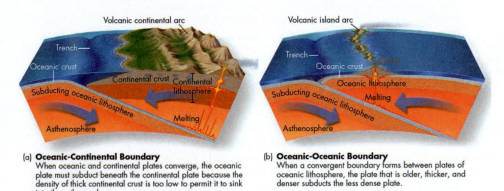

(a) Oceanic-Continental Boundary
When oceanic and continental plates converge, the oceanic plate must subduct beneath the continental plate because the density of thick continental crust is too low to permit it to sink into the asthenosphere.

(b) Oceanic-Oceanic Boundary
When a convergent boundary forms between plates of oceanic lithosphere, the plate that is older, thicker, and denser subducts the less dense plate.

(c) Continental-Continental Boundary
Subduction of ocean lithosphere inevitably leads to convergence of continents (upper panel). Because continents are too buoyant to subduct, high mountains develop as the continents are forced upwards in the collision zone (lower panel).

< FIGURE 2.9 Convergent Plate Boundaries
Idealized diagram illustrating characteristics of three types of convergent plate boundaries.

∧ FIGURE 2.10 Mountains in Italy
Mountain peaks (the Dolomites) in northern Italy are part of the Alpine mountain system formed from the collision between Africa and Europe. *(Edward A. Keller)*

If the leading edges of both plates contain relatively light, buoyant continental crust, subduction into the mantle of one of the plates is difficult. In this case, a continent–continent plate collision occurs, in which the edges of the plates collide, causing shortening and lithospheric thickening due to folding and faulting (Figure 2.9c). Where the two plates join is known as a *suture zone*. Continent–continent collision has produced some of the highest mountain systems on Earth, such as the Alpine and Himalayan mountain belts (Figure 2.10). Many older mountain belts were formed in a similar way; for example, the Appalachians in the eastern United States which formed during an ancient continent–continent plate collision 250 to 350 million years ago.

Transform boundaries, or transform faults, occur where the edges of two plates slide past each other, like the San Andreas fault in California shown in Figure 2.11. If you examine Figure 2.8, you will see that a mid-ocean ridge is not a single, continuous rift but a series of rifts that are offset from one another along connecting transform faults. Although the most common locations for transform plate boundaries are within oceanic crust, some occur within continents. A well-known continental transform boundary is the San Andreas fault in California, where the rim of the Pacific plate is sliding horizontally past the rim of the North American plate (see Figure 2.1 and Figure 2.11).

Locations where three plates border one another are known as **triple junctions**. Figure 2.11 shows several such junctions: Two examples are the meeting point of the Juan de Fuca, North American, and Pacific plates on the West Coast of North America (this is known as the Mendocino triple junction) and the junction of the spreading ridges associated with the Pacific, Cocos, and Nazca plates west of South America.

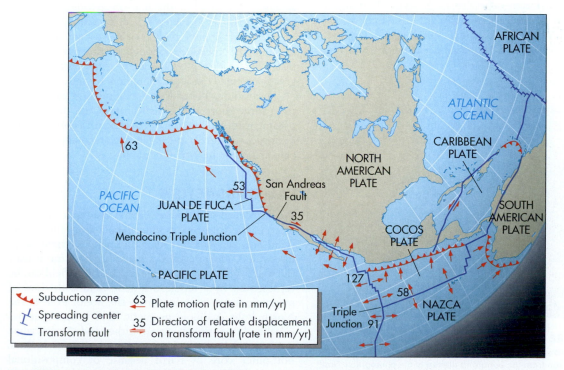

∧ FIGURE 2.11 North American Plate Boundary
Detail of boundary between the North American plate and Pacific plate. *(Based on information by Tanya Atwater)*

RATES OF PLATE MOTION

Plate Motion Is a Fast Geologic Process The directions in which plates move are shown on Figure 2.4a. In general, plates move a few centimeters per year, about as fast as some people's hair and fingernails grow. The Pacific plate moves past the North American plate along the San Andreas fault about 3.5 cm (~1.4 in.) per year, so that features such as rock units or streams are gradually displaced over time where they cross the fault (Figure 2.12). During the past 5 million years, there has been about 175 km (~110 mi.) of displacement, a distance equivalent to driving two hours at 55 mph (~90 kph) on a highway along the San Andreas fault. Although the central portions of the plates move along at a steady slow rate, plates interact at their boundaries, where collision or subduction or both occur, and movement may not be smooth or steady. The plates often get stuck together. The movement is analogous to sliding one rough wood board over another. Movement occurs when the splinters of the boards break off and the boards move quickly by each other. When rough edges along the plate move quickly, an earthquake is produced (see Chapter 3). Along the San Andreas fault, which is a transform plate boundary, the displacement is horizontal and can amount to several meters during a great earthquake. During an earthquake in 1857 on the San Andreas fault, a horse corral across the fault was reportedly changed from a circle to an "S" shape. Fortunately, such an event generally occurs at any given location only once every 100 years or so. Over long time periods, rapid displacement from periodic earthquakes and more continuous slow "creeping" displacements add together to produce the average rate of several centimeters of movement per year along the San Andreas fault.

2.3 A Detailed Look at Seafloor Spreading

When Alfred Wegener proposed the idea of continental drift in 1915, he had no solid evidence of a mechanism that could move continents. The global extent of mid-ocean ridges was discovered in the 1950s, and in 1962 geologist Harry H. Hess published a paper suggesting that continental drift was the result of the process of seafloor spreading along those ridges. New oceanic lithosphere is produced at the mid-ocean ridge spreading centers as shown in Figure 2.5 (divergent plate boundary). The lithospheric plate then moves laterally, carrying along the embedded continents in the tops of moving plates. These ideas produced a new major paradigm that greatly changed our ideas about how Earth works.[5,6,7]

The validity of seafloor spreading was established from three sources: (1) identification and mapping of oceanic ridges, (2) dating of volcanic rocks on the floor of the ocean, and (3) understanding and mapping of the paleomagnetic history of ocean basins.

PALEOMAGNETISM

We introduce and discuss Earth's magnetic field and paleomagnetic history in some detail in order to understand how seafloor spreading and plate tectonics were discovered. Earth has had a magnetic field for at least the past 3 billion years[2] (Figure 2.13). The field can be represented by a dipole magnetic field with lines of magnetic force extending from the South Pole to the North Pole. A dipole magnetic field is one that has equal and opposite charges at either end. As described in Section 2.1, convection occurs in the iron-rich, fluid, hot outer core of Earth because of compositional changes and heat at the inner–outer core boundary. As more buoyant material in the outer core rises, it starts

∧ FIGURE 2.12 The San Andreas Fault
The fault is visible as a linear depression with low hills on either side that run from the bottom to the top of the photograph. Note the prominent stream that is diverted toward the north (top of image) along the fault. *(Doc Searle/Flickr CCBYSA 2.0)*

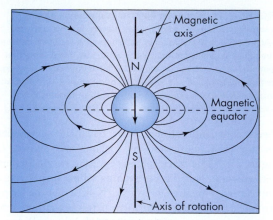

(a) Normal polarity

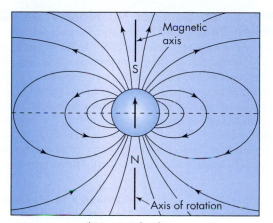

(b) Reversed polarity

∧ FIGURE 2.13 Magnetic Reversal
Idealized diagram showing the magnetic field of Earth under (a) normal polarity and (b) reversed polarity. *(From Kennett, J., Marine Geology. Englewood Cliffs, NJ: Prentice Hall, © 1982. Reprinted and electronically reproduced by permission of Pearson Education, Inc., Upper Saddle River, New Jersey.)*

the convection. The convection in the outer core, along with the rotation of Earth that causes rotation of the outer core, initiates a flow of electric current in the core. This flow of current within the core produces and sustains Earth's magnetic field.[2,7]

Earth's magnetic field is sufficient to permanently magnetize some surface rocks. For example, volcanic rock that erupts and cools at mid-ocean ridges becomes magnetized at the time it passes through a critical temperature. At that critical temperature, known as the *Curie point*, iron-bearing minerals (such as magnetite) in the volcanic rock orient themselves parallel to the magnetic field. This is a permanent magnetization known as *thermoremnant magnetization*.[7] The term **paleomagnetism** refers to the study of the magnetism of

rocks at the time their magnetic signature formed. It is used to determine the magnetic history of Earth.

The magnetic field, based on the size and conductivity of Earth's core, must be continuously generated or it would decay away in about 20,000 years. It would decay because the temperature of the core is too high to sustain permanent magnetization.[2]

Earth's Magnetic Field Periodically Reverses Before the discovery of plate tectonics, geologists working on land had already discovered that some volcanic rocks were magnetized in a direction opposite to the present-day field, suggesting that the polarity of Earth's magnetic field was reversed at the time the volcanic rocks erupted onto the seafloor and cooled (Figure 2.13b). The rocks were examined for whether their magnetic field was normal, as it is today, or reversed relative to that of today, for certain time intervals of Earth's history. A chronology for the last few million years was constructed on the basis of the dating of the "reversed" rocks. You can verify the current magnetic field of Earth by using a compass; at this point in Earth's history, the needle points to the north magnetic pole. During a period of reversed polarity, the needle would point south! The cause of **magnetic reversals** is not well known, but it is related to changes in the convective movement of the liquid material in the outer core and processes occurring in the inner core. Reversals in Earth's magnetic field are random, occurring on average every few hundred thousand years. The change in polarity of Earth's magnetic field takes a few thousand years to occur, which in geologic terms is a very short time.

What Produces Magnetic Stripes? To further explore the Earth's magnetic field, geologists towed magnetometers, instruments that measure magnetic properties of rocks, from ships and completed magnetic surveys. The paleomagnetic record of the ocean floor is easy to read because of the fortuitous occurrence of the volcanic rock basalt (see Chapter 5) that is produced at spreading centers and forms the floors of the ocean basins of Earth. The rock is fine-grained and contains sufficient iron-bearing minerals to produce a good magnetic record. The marine geologists' discoveries were not expected. The rocks on the floor of the ocean were found to have irregularities in the magnetic field. These irregular magnetic patterns were called anomalies or perturbations of Earth's magnetic field caused by local fields of magnetized rocks on the seafloor. The anomalies can be represented as stripes on maps. When mapped, the stripes form quasi-linear patterns parallel to oceanic ridges. The marine geologists found that their sequences of stripe width patterns matched the sequences established by land geologists for polarity

◀ FIGURE 2.14 Magnetic Anomalies on the Seafloor
Map showing a magnetic survey southwest of Iceland along the
Mid-Atlantic Ridge. Positive magnetic anomalies are black (normal),
and negative magnetic anomalies are white (reversed). Note that
the pattern is symmetrical on the two sides of the mid-ocean ridge.
*(Based on Heirtzler, J. R., Le Pichon, X., and Baron, J. G., "Magnetic
Anomalies Over the Reykjanes Ridge," Deep-Sea Research 13:
427–43, 1966. Reprinted by permission of Elsevier Sciences, Ltd.)*

reversals in land volcanic rocks. Magnetic survey data
for an area southwest of Iceland are shown on
Figure 2.14. The black stripes represent normally magnetized rocks and the intervening white stripes represent reversed magnetized rocks.[8] Notice that the stripes
are not evenly spaced but form patterns that are symmetrical on opposite sides of the Mid-Atlantic Ridge.

Why Is the Seafloor No More Than 200 Million Years Old?
The discovery of patterns of
magnetic stripes at various locations in ocean basins allowed geologists to infer numerical dates for the
volcanic rocks. Merging the magnetic anomalies with the
numerical ages of the rocks produced the record of seafloor spreading. The spreading of the ocean floor, beginning at a mid-ocean ridge, could explain the magnetic
stripe patterns.[9] Figure 2.15 is an idealized diagram
showing how seafloor spreading may produce the patterns of magnetic anomalies (stripes). The pattern shown

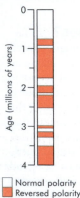

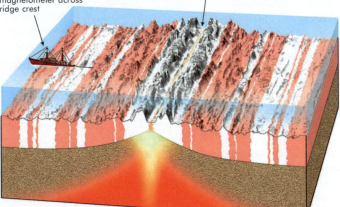

◀ FIGURE 2.15 Magnetic Reversals and Seafloor Spreading
Idealized diagram showing a
mid-ocean ridge and rising
magma in response to seafloor
spreading. As the volcanic rocks
cool, they become magnetized.
The white stripes represent
normal magnetization; the red
stripes are reversed magnetization. The record shown here was
formed over a period of several
million years. Magnetic anomalies (stripes) are a mirror image
of each other on opposite sides
of the mid-ocean ridge. Thus,
the symmetrical bands of the
normal and reversed magnetized rocks are produced by the
combined effects of the reversals and seafloor spreading.

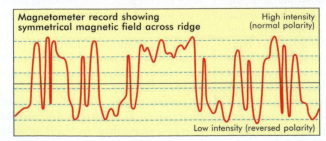

is for the past several million years, which includes several periods of normal and reversed magnetization of the volcanic rocks. White stripes represent normally magnetized rocks, and red stripes are rocks with a reversed magnetic signature. Notice that the most recent magnetic reversal occurred approximately 0.7 million years ago. The basic idea illustrated by the figure is that rising magma at the oceanic ridge is extruded, or pushed out onto the surface, through volcanic activity, and the cooling rocks become normally magnetized. When the field is reversed, the cooling rocks preserve a reverse magnetic signature (Figure 2.15). Notice that the patterns of magnetic anomalies in rocks on both sides of the ridge are mirror images of one another. The only way such a pattern might result is through the process of seafloor spreading. Thus, the pattern of magnetic reversals found on rocks of the ocean floor is strong evidence that the process of seafloor spreading is happening.

Mapping of magnetic anomalies, when combined with age-dating of the magnetic reversals in land rocks, creates a database that suggests exciting inferences; Figure 2.16 shows the age of the ocean floor as determined from this database. The pattern, showing that the youngest volcanic rocks are found along active mid-ocean ridges, is consistent with the theory of seafloor spreading. As distance from these ridges increases, the age of the ocean floor also increases, to a maximum of about 200 million years,

during the early Jurassic period. Thus, it appears that the present ocean floors of the world are no older than 200 million years. In contrast, rocks on continents are often much older than Jurassic, going back about 4 billion years, almost 20 times older than the ocean floors! We conclude that the thick continental crust, by virtue of its buoyancy, is more stable at Earth's surface than are rocks of the crust of the ocean basins. Continents form by the processes of accretion of sediments, addition of volcanic materials, and collisions of tectonic plates carrying continental landmasses. We will continue this discussion when we consider the movement of continents during the past 200 million years. However, it is important to recognize that it is the pattern of magnetic stripes that allows us to reconstruct how the plates and the continents embedded in them have moved throughout history.

HOT SPOTS

There are a number of places on Earth called **hot spots**, characterized by volcanic centers resulting from hot materials produced deep in the mantle, perhaps near the core–mantle boundary (see Figure 2.5). The partially molten materials are hot and buoyant enough to move up through the mantle and the overlying, moving tectonic plates.[5,7] An example of a continental hot spot is the volcanic region of Yellowstone National Park. Hot spots are

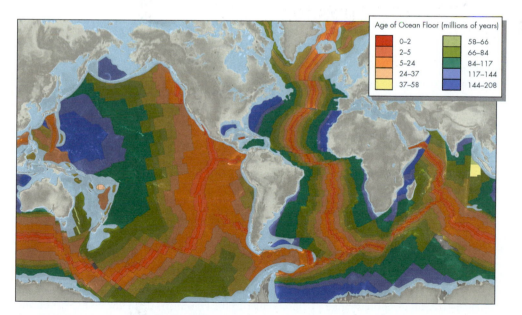

∧ FIGURE 2.16 Age of the Ocean Floor
Age of the seafloor is determined from magnetic anomalies and other methods. The youngest ocean floor (red) is located along oceanic ridge systems, and older rocks are generally farther away from the ridges. The oldest ocean floor rocks are approximately 180 million years old. (*Based on Scotese, C. R., Gahagan, L. M., and Larson, R. L., "Plate Tectonic Reconstruction of the Cretaceous and Cenozoic Ocean Basins,"* Tectonophysics *155: 27–48, 1988. Reprinted by permission of Elsevier Sciences, Ltd.*)

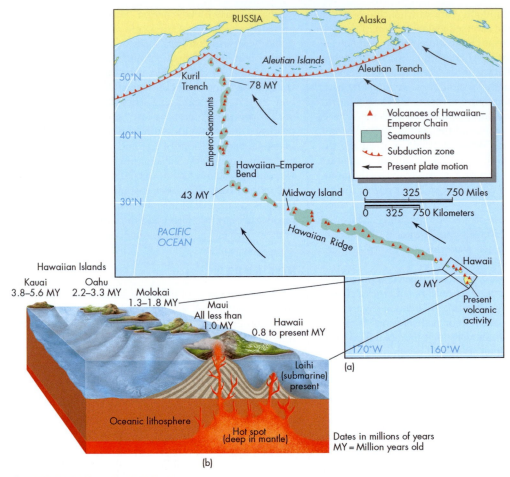

∧ FIGURE 2.17 Hawaiian Hot Spot

(a) Map showing the Hawaiian–Emperor Chain of volcanic islands and seamounts. Actually, Midway Island and the Hawaiian Islands are the only volcanoes of the chain that protrude from the ocean's surface to form islands. *(Based on Claque, D. A., Dalrymple, G. B., and Moberly, R., "Petrography and K-Ar ages of Dredged Volcanic Rocks from the Western Hawaiian Ridge and Southern Emperor Seamount Chain," Geological Society of America Bulletin 86: 991–98, 1975)* (b) Sketch map showing the Hawaiian Islands, which range in age from present volcanic activity to about 6 million years old on the island of Kauai. Notice that most of the mass of the volcanoes forming the Hawaiian Islands lies below the ocean surface. *(From Thurman, Oceanography, 5th ed. Columbus, OH: Merrill, Plate 2. Reprinted and Electronically reproduced by permission of Pearson Education, Inc., Upper Saddle River, New Jersey.)*

also found in both the Atlantic and Pacific Oceans. If the hot spot is anchored in the slow-moving deep mantle, then as the oceanic plate moves over a hot spot, a linear chain of volcanoes is produced. Perhaps the best example of this type of hot spot is the line of volcanoes forming the Hawaiian–Emperor Chain in the Pacific Ocean (Figure 2.17a). Along this chain, volcanic eruptions range in age from present-day activity on the big island of Hawaii (in the southeast) to more than 78 million years ago near the northern end of the Emperor Chain. With the exception of the Hawaiian Islands and some coral atolls (ringlike coral islands such as Midway Island), the chain primarily consists of submarine volcanoes known as *sea-mounts*. Seamounts are islands that were eroded by

waves and submarine landslides and subsequently sank beneath the ocean surface. As seamounts move farther off the fixed hot spot due to plate movement, the volcanic rocks that make up the islands cool and the oceanic crust they are on becomes denser and sinks.

Seamounts constitute impressive submarine volcanic mountains. In the Hawaiian Chain, the youngest volcano is Mount Loihi, which is still a submarine volcano, presumably directly over a hot spot, as idealized in Figure 2.17b. The ages of the Hawaiian Islands increase to the northwest, with the oldest being Kauai, about 6 million years old. Notice in Figure 2.17a that the line of seamounts makes a sharp bend at the junction of the Hawaiian and Emperor Chains. The age of the volcanic rocks at the bend is about 43 million years, and the bend is interpreted to represent a time when plate motions changed.[10] If we assume that the hot spots are fixed deep in the mantle, then the chains of volcanic islands and submarine volcanoes along the floor of the Pacific Ocean that get older farther away from the hot spot provide additional evidence to support the movement of the Pacific plate. In other words, the ages of the volcanic islands and submarine volcanoes could systematically change as they do only if the plate is moving over the hot spot.

2.3 **CHECK** your understanding

1. Describe the major process that are thought to produce Earth's magnetic field.

2. How have magnetic reversals and the study of paleomagnetism been important in understanding plate tectonics?

3. What are hot spots?

2.4 **Pangaea and Present Continents**

Movement of the lithospheric plates is responsible for the locations of mountain ranges' present shape and location of the continents. There is good evidence that the most recent global episode of continental drift, driven by seafloor spreading, started about 180 million years ago, with the breakup of a supercontinent called Pangaea (this name, meaning "all lands," was first proposed by Wegener). Pangaea (pronounced *pan-jee-ah*) was enormous, extending from pole to pole and over halfway around Earth near the equator (Figure 2.18a). Pangaea had two parts (Laurasia to the north and Gondwana to the south) and was constructed during earlier continental collisions. Figure 2.18a shows Pangaea as it was nearly 180 million years ago. Seafloor spreading over the past 200 million years separated Eurasia and North America from the

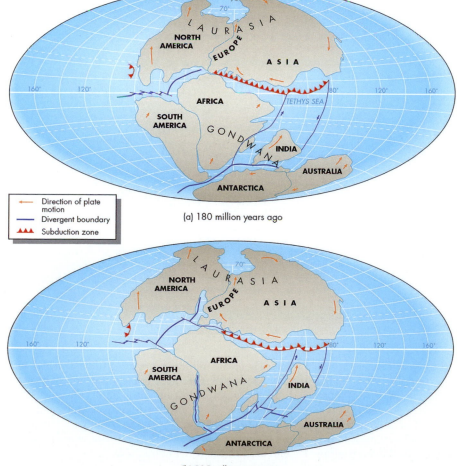

◄ **FIGURE 2.18 Two Hundred Million Years of Plate Tectonics**

(a) The proposed positions of the continents at 180 million years ago, (b) 135 million years ago, (c) 65 million years ago, and (d) at present. Arrows show directions of plate motion. See text for further explanation of the closing of the Tethys Sea, the collision of India with China, and the formation of mountain ranges. *(Based on Dietz, R. S., and Holden, J. C. 1970. Reconstruction of Pangaea: Breakup and dispersion of continents, Permian to present. Journal of Geophysical Research 75(26): 4939–56. Copyright by the American Geophysical Union. Reprinted with permission of Wiley Inc. Adaptation of block diagrams from Christopherson, R. W., Geosystems, 2nd ed. Englewood Cliffs, NJ: Macmillan/Prentice Hall, 1994. Reprinted and Electronically reproduced by permission of Pearson Education, Inc., Upper Saddle River, New Jersey.)*

- → Direction of plate motion
- — Divergent boundary
- ▲▲▲ Subduction zone

(a) 180 million years ago

(b) 135 million years ago

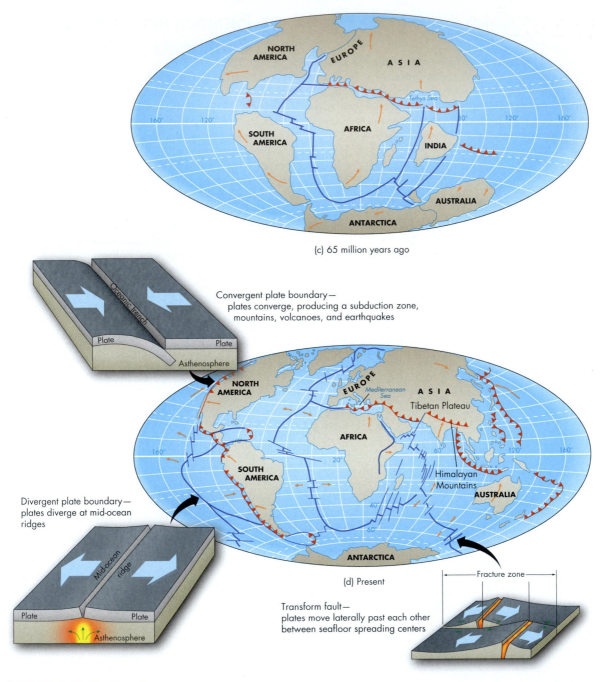

(c) 65 million years ago

Convergent plate boundary—
plates converge, producing a subduction zone,
mountains, volcanoes, and earthquakes

Divergent plate boundary—
plates diverge at mid-ocean
ridges

(d) Present

Transform fault—
plates move laterally past each other
between seafloor spreading centers

∧ FIGURE 2.18 *(Continued)*

southern landmass, Eurasia from North America, and the southern continents (South America, Africa, India, Antarctica, and Australia) from one another (Figure 2.18b–d). The Tethys Sea, between Africa and Europe–Asia (Figure 2.18a–c), closed as part of the activity that produced the Alps in Europe. A small part of this once much larger sea remains today as the Mediterranean

Sea. About 50 million years ago, India crashed into China in south Asia. That collision, which has caused India to forcefully intrude into China a distance comparable from New York to Miami, is still happening today, producing the Himalayan Mountains (the highest mountains above sea level in the world) and the Tibetan Plateau (see Figure 2.22).

Understanding Plate Tectonics Solves Long-Standing Geologic Problems Reconstruction of what the supercontinent Pangaea looked like before the most recent episode of continental drift has cleared up two interesting geologic problems:

> Occurrence of the same fossil plants and animals on different continents would be difficult to explain if they had not been joined in the past (Figure 2.19).

> Evidence of ancient glaciation on several continents, with inferred directions of ice flow, makes sense only if the continents are placed back

within Gondwanaland (southern Pangaea) near the South Pole as it was before splitting apart (Figure 2.20).

2.4 CHECK your understanding

1. What was Pangaea and when did it exist?

2. What geologic problems were solved when the existence of Pangaea was confirmed?

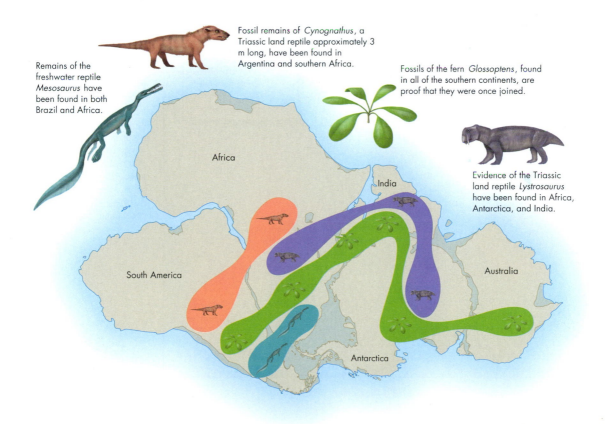

Remains of the freshwater reptile *Mesosaurus* have been found in both Brazil and Africa.

Fossil remains of *Cynognathus*, a Triassic land reptile approximately 3 m long, have been found in Argentina and southern Africa.

Fossils of the fern *Glossopteris*, found in all of the southern continents, are proof that they were once joined.

Evidence of the Triassic land reptile *Lystrosaurus* have been found in Africa, Antarctica, and India.

Africa

India

South America

Australia

Antarctica

∧ FIGURE 2.19 Paleontological Evidence for Pangaea
This map shows some of the paleontological (fossil) evidence that supports continental drift. Scientists believe that these animals and plants could not have been found on all of these continents were they not once much closer together than they are today. Major ocean basins would have been physical barriers to their distribution. *(From Hamblin, W. K., Earth's Dynamic Systems, 6th ed. New York: Macmillan, 1992. Reprinted by permission of Macmillan.)*

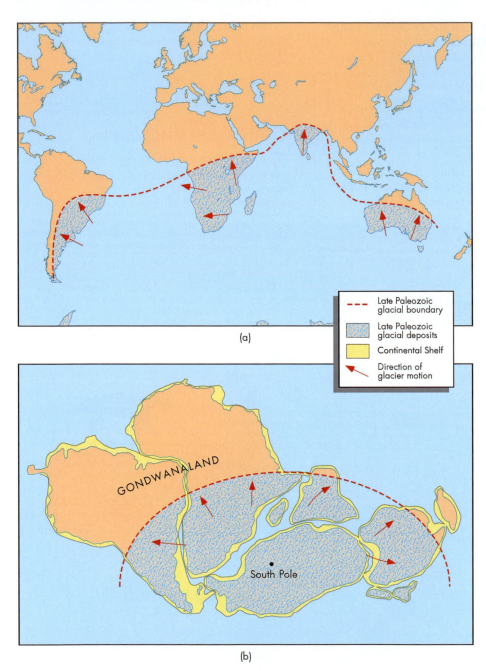

(a)

GONDWANALAND

South Pole

(b)

- - - - Late Paleozoic
glacial boundary

Late Paleozoic
glacial deposits

Continental Shelf

→ Direction of
glacier motion

> **FIGURE 2.20 Glacial Evidence for Pangaea**
(a) Map showing the distribution of evidence for late Paleozoic glaciations. The arrows indicate the direction of ice movement. Notice that the arrows are all pointing away from ocean sources. Also these areas are close to the tropics today, where glaciation would have been unlikely in the past. These Paleozoic glacial deposits were formed when Pangaea was a supercontinent before fragmentation by continental drift. (b) The continents are restored (it is thought that continents drifted north away from the South Pole). Notice that the arrows now point outward as if moving away from a central area where glacial ice was accumulating. Thus, restoring the position of the continents produces a pattern of glacial deposits that makes much more sense. *(Based on Hamblin, W. K.,* Earth's Dynamic Systems, *6th ed. New York: Macmillan, 1992)*

2.5 How Plate Tectonics Works: Putting It Together

Driving Mechanisms That Move Plates Now that we have presented the concept that new oceanic lithosphere is produced at mid-ocean ridges because of seafloor spreading and that old, cooler plates sink into the mantle at subduction zones, let us evaluate the forces that cause the lithospheric plates to actually move and subduct. Figure 2.21 is an idealized diagram illustrating the two most likely driving forces: ridge push and slab pull.

The mid-ocean ridges or spreading centers stand at elevations of about 1 to 2.5 km (~3200 to 8200 ft.) above the ocean floor as linear, gently arched uplifts with widths greater than the distance from Florida to Canada (see Figure 2.8). *Ridge push* is a gravitational

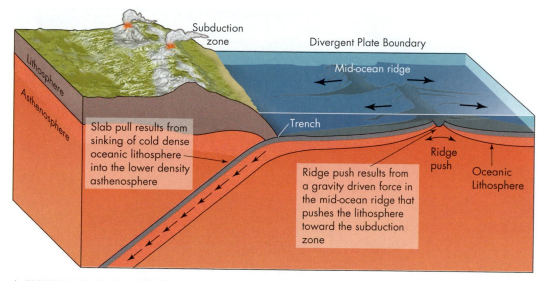

^ FIGURE 2.21 Push and Pull in Moving Plates
Idealized diagram showing concepts of ridge push and slab pull that facilitate the movement of lithospheric plates from spreading ridges to subduction zones. Both are gravity driven. The heavy lithosphere moves down the mid-ocean ridge slope and subducts down through the lighter, hotter mantle. *(Based on Cox, A., and Hart, R. B., Plate Tectonics. Boston: Blackwell Scientific Publications, 1986)*

push, like a gigantic landslide, away from the ridge crest toward the subduction zone (the lithosphere slides on the asthenosphere). *Slab pull* results when the lithospheric plate moves farther from the ridge and cools, gradually becoming denser than the asthenosphere beneath it. At a subduction zone, the plate sinks through lighter, hotter mantle below the lithosphere, and the weight of this descending plate edge pulls on the entire plate, resulting in slab pull (Figure 2.21). Which of the two processes, ridge push or slab pull, is the more influential of the driving forces? Calculations of the expected gravitational effects suggest that ridge push is of relatively low importance compared with slab pull. In addition, it is observed that plates with large subducting slabs attached and pulling on them (e.g., the subduction zones surrounding the Pacific Basin) tend to move much more rapidly than those driven primarily by ridge push alone. Thus, slab pull may be more influential than ridge push in moving plates.

2.5 CHECK your understanding

1. What is the difference between ridge push and slab pull?

2. What evidence suggests that slab pull may be the dominant driver of plate tectonics?

2.6 Plate Tectonics and Hazards

The importance of the tectonic cycle to our lives cannot be overstated. Everything living on Earth is affected by plate tectonics. As the plates slowly move a few centimeters each year, so do the continents and ocean basins, producing zones of resources (oil, gas, and minerals), as well as earthquakes and volcanoes (see Figure 2.4b). The tectonic processes occurring at plate boundaries largely determine the types and properties of the rocks upon which we depend for our land, our mineral and rock resources, and the soils in which our food is grown.

The linkage of hazardous events to plate tectonics is obvious. When plate boundaries are mapped, the boundaries are located where volcanoes and earthquakes have occurred. Beyond this, several major conclusions may be drawn:

> At divergent plate boundaries, such as along the Mid-Atlantic Ridge, the dominant hazards are earthquakes and volcanic eruption. This is especially true where the boundary briefly comes on land in Iceland. There, volcanoes that underlie glacial ice are related to both volcanic hazards and flooding (see Applying the Fundamental Concepts in Chapter 5). Subglacial lakes may form as a result of heating from below; if these suddenly burst or if a volcanic eruption occurs, flooding will occur.

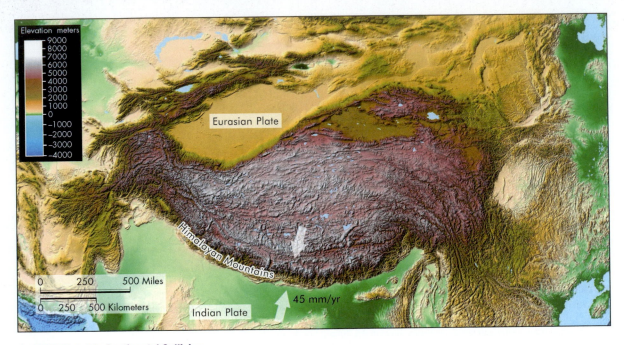

⋀ FIGURE 2.22 Continental Collision
False color topographic image showing continental collision between the Indian and Asian plates. Collision of the two plates has uplifted the Tibetan Plateau to an average of 5000 m (~16,400 ft.) and numerous peaks in the Himalayan Mountains to heights greater than 8000 m (~26,000 ft.).

> Along boundaries where one plate slides past another such as the San Andreas fault, the earthquake hazard is appreciable. There are a number of such boundaries around the world in places such as Southern California, Haiti, and New Zealand (see Figure 2.4a). Associated with these plate boundaries are often hilly or mountainous areas where rainfall may be higher and, thus, hazards such as landslides and flooding are also present.

> Convergent plate boundaries, where one plate dives beneath another or two collide, are areas particularly prone to natural hazards. Where subduction zone volcanism occurs, particularly explosive volcanoes are typical. Thus, we see the string of volcanoes in Alaska, the Pacific Northwest, New Zealand, Japan, and the western coast of South America. Where two continental landmasses collide and neither can be subducted, the highest topography in the world has been produced (Figure 2.22). Along with the earthquake hazard there is increasing precipitation and an abundance of high, steep slopes prone to landslides and flooding. Where India has collided with Asia for the past 50 million years, a high plateau (Tibetan Plateau) has formed and this high topography affects the occurrence of the summer monsoon as well as the global climate. Thus, plate tectonics can affect the occurrence of severe storms and related flooding and erosion. To a lesser extent, the same forces have produced the high plateau area of the Rocky Mountains that affects the regional climate and the monsoonal storms of the southwestern United States.

In summary, we see that plate tectonics is linked to many geologic processes operating on Earth's surface that are linked to natural hazards.

2.6 CHECK your understanding

1. Explain why convergent plate boundaries represent a greater hazard than other plate boundaries.

CONCEPTS in review

2.1 Internal Structure of Earth

LO:1 Describe the basic internal structure and processes of Earth.

- The internal structure of Earth can be divided into layers or concentric shells, based on either composition (what it's made of) or physical properties (e.g., solid or liquid, weak or strong). The compositional layers include the core, mantle, and crust.
- The uppermost physical layer of Earth is known as the lithosphere and is relatively strong and rigid compared with the soft asthenosphere underlying it.
- A convection cell is a temperature-driven circulation pattern that is assumed to operate within Earth and may be involved in driving plate tectonics.

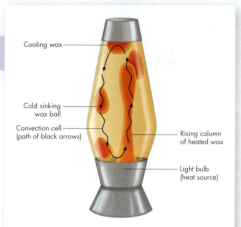

Cooling wax

Cold sinking wax ball

Convection cell (path of black arrows)

Rising column of heated wax

Light bulb (heat source)

KEY WORDS

asthenosphere (p. 40), **convection** (p. 41), **crust** (p. 40), **inner core** (p. 40), **lithosphere** (p. 40), **mantle** (p. 40), **Moho** (p. 40), **outer core** (p. 40)

2.2 Plate Tectonics

LO:2 Summarize the various lines of evidence that support the theory of plate tectonics.

LO:3 Compare and contrast the different types of plate boundaries.

- The lithosphere is broken into large pieces called tectonic plates that move relative to one another. Most earthquakes and volcanoes worldwide are associated with plate boundaries.
- The three types of plate boundaries are divergent (mid-ocean ridges, spreading centers), convergent (subduction zones and continental collisions), and transform faults. At some locations, three plates meet in areas known as triple junctions. Rates of plate movement are generally a few centimeters per year.
- Evidence supporting plate tectonic theory includes seafloor spreading, continental drift, the configurations of hot spots and chains of volcanoes, and Wadati–Benioff zones.

KEY WORDS

continental drift (p. 41), **convergent boundary** (p. 45), **divergent boundary** (p. 44), **mid-ocean ridge** (p. 41), **plate tectonics** (p. 41), **seafloor spreading** (p. 41), **spreading center** (p. 41), **subduction zone** (p. 42), **submarine trench** (p. 45), **transform boundary** (p. 47), **triple junction** (p. 47), **Wadati–Benioff zone** (p. 43)

2.3 A Detailed Look at Seafloor Spreading

LO:4 Explain the mechanisms of plate tectonics.

- Convection currents in Earth's liquid outer core generate a magnetic field that is sufficiently strong to be recorded in rocks that contain magnetic minerals. Over the past few hundred million years, periodic flips of the magnetic field between normal and reverse polarity are preserved in the rock record, the study of which is known as paleomagnetism.
- The seafloor spreading hypothesis proposed a mechanism for continental drift. The hypothesis suggested that new ocean lithosphere was created along mid-ocean ridges and moved laterally away with time, pushing the continents apart. Seafloor spreading is confirmed using the paleomagnetic signature of the seafloor centered around the mid-ocean ridges. A symmetrical pattern of magnetic anomalies, recording normal and reverse polarity, indicates that the age of the seafloor gets older away from the mid-ocean ridge, validating the seafloor spreading hypothesis.
- The age of the seafloor is everywhere younger than 200 million years old, which is 20 times younger than the age of continents. This suggests that the seafloor is recycled back into the mantle while the less dense continents are too buoyant to subduct.
- Hot spots are plumes of hot rock that rise from deep in the mantle and cause volcanoes above them at Earth's surface. Unlike the moving tectonic plates, hot spots are fixed and, therefore, as the plate moves across the hot spot a chain of volcanoes is formed, such as the Hawaiian Islands.

KEY WORDS

hot spot (p. 51), **magnetic reversal** (p. 49), **paleomagnetism** (p. 49)

2.4 Pangaea and Present Continents

LO:5 Outline how plate tectonics has changed the appearance of Earth's surface over time.

- Alfred Wegener's continental drift hypothesis suggested that all of Earth's continents were in the past assembled in a single enormous continent, known as Pangaea. We now know that this was the case about 200 million years ago.
- With the validation of the seafloor spreading, continental drift was confirmed and several long-standing geologic problems have been resolved, including (1) the existence of the fossil plants and animals on continents separated by oceans, and (2) the distribution of ancient glaciations on several continents.

2.5 How Plate Tectonics Works: Putting It Together

LO:6 Compare and contrast the two fundamental processes that drive plate tectonics.

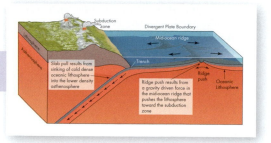

- The driving forces in plate tectonics are ridge push and slab pull. At present, we believe the process of slab pull is more significant than ridge push for moving tectonic plates from spreading centers to subduction zones.

2.6 Plate Tectonics and Hazards

LO:7 Link plate tectonic processes to natural hazards.

- Plate tectonics is extremely important in determining the occurrence and frequency of volcanic eruptions, earthquakes, and other natural hazards. Divergent plate boundaries are linked to earthquakes and volcanic eruptions but, with the exception of Iceland and East Africa, the risk associated with these hazards is low because they do not occur on land.
- Transform boundaries are linked to earthquakes and represent an appreciable risk as these faults occur on land and stretch for long distance, often through populated regions.
- Convergent plate boundaries are zones of greatest risk as these are linked to the largest recorded earthquakes, explosive volcanic eruptions, and tsunamis.
- Convergent plate boundaries also generate high mountains, which are prone to mass wasting, flooding, and severe weather and can affect global climate.

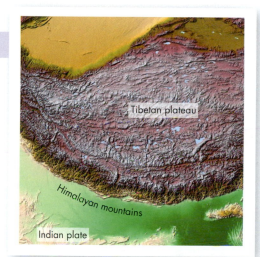

CRITICAL thinking questions

1. Assume that the supercontinent Pangaea (see Figure 2.18a) never broke up. Now deduce how Earth processes, landforms, and environments might be different from how they are today with the continents spread all over the globe. *Hint:* Think about what the breakup of the continents did in terms of building mountain ranges and producing ocean basins that affect climate and so forth.

References

1. **Wysession, M.** 1995. The inner workings of Earth. *American Scientist* 83: 134–47.
2. **Glatzmaier, G. A.** 2001. *The geodynamo.* www.es.ucsc.edu/glatz/geodynamo.html. Accessed 2/21/01.
3. **Le Pichon, X.** 1968. Sea-floor spreading and continental drift. *Journal of Geophysical Research* 73: 3661–97.
4. **Isacks, B. L., Oliver, J.,** and **Sykes, L. R.** 1968. Seismology and the new global tectonics. *Journal of Geophysical Research* 73: 5855–99.
5. **Cox, A.,** and **Hart, R. B.** 1986. *Plate tectonics.* Boston: Blackwell Scientific Publications.
6. **Dewey, J. F.** 1972. Plate tectonics. *Scientific American* 22: 56–68.

7. **Fowler, C. M. R.** 1990. *The solid Earth.* Cambridge: Cambridge University Press.

8. **Heirtzler, J. R., Le Pichon, X.,** and **Baron, J. G.** 1966. Magnetic anomalies over the Reykjanes Ridge. *Deep Sea Research* 13: 427–43.

9. **Cox, A., Dalrymple, G. B.,** and **Doell, R. R.** 1967. Reversals of Earth's magnetic field. *Scientific American* 216(2): 44–54.

10. **Claque, D. A., Dalrymple, G. B.,** and **Moberly, R.** 1975. Petrography and K-Ar ages of dredged volcanic rocks from the western Hawaiian Ridge and southern Emperor Seamount chain. *Geological Society of America Bulletin* 86: 991–98.

The 2011 Tohoku Earthquake …
released about
600 million times
more energy than the atomic bomb
dropped on Hiroshima

3

APPLYING the 5 fundamental concepts 2011 Tohoku Earthquake

①

Science helps us predict hazards.

When we return to the Tohoku earthquake at the end of this chapter you will learn about precursor events leading up to the 2011 quake, as well as the scientific data that led to preliminary forecasting and an underestimation of the maximum size of the event.

②

Knowing hazard risks can help people make decisions.

Large earthquakes are known to occur at subduction zones, the frequency of which can be predicted through scientific investigation and therefore earthquake hazard is amenable to risk analysis. However, as the 2011 Tohoku event illustrates, gaps in scientific data and invalid assumptions can lead to an underestimate of the potential earthquake risk.

Earthquakes

2011 Tohoku Earthquake

As you recall from the previous chapter, the movement of Earth's tectonic plates around the globe has reshaped the continents and ocean basins over the past few billion years. Moving at an average speed of about 5 cm (~2 in.) per year, you might think that we would be completely unaware that the plates are in motion. This would be true if movement between adjacent plates was smooth and continuous, but it is not. Rather than gradual imperceptible movement, plates move episodically in large pulses every few hundred years or so. These large episodic movements of the plates cause the violent ground shaking we know as an *earthquake*. The island country of Japan is located just 200 km (~124 mi.) west of the Japan Trench, where the Pacific plate is subducting beneath the Eurasian plate at a rate of about 9 cm (~3.5 in.) per year (Figure 3.1).[1] The convergence of these two plates means Japan experiences frequent large earthquakes.

On March 11, 2011, the strongest recorded earthquake to hit Japan, and one of the top five largest earthquakes ever documented, struck 72 km (~45 mi.) offshore of the Tohoku region of Japan's main island (Figure 3.1). Although Japan has the worlds most advanced system for detecting earthquakes,[2] and scientists had forecast a large earthquake for the Tohoku region, the 2011 quake was significantly greater than was considered possible. Fifteen times more powerful that the

◄ **Tohoku Earthquake Devastated Fukashima, Japan**
Military helicopters fly over tsunami inundated and burning coastal Japan following the 2011 earthquake. *(US Navy)*

Earthquakes are serious natural hazards that affect people across the globe, sometimes at long distances from where the quakes occur. They are especially dangerous because seismologists, the scientists who study earthquakes, cannot predict them in time for evacuations or other precautions.

After reading this chapter, you should be able to:

LO:1 Compare and contrast the different types of faulting.

LO:2 Explain the formation of seismic waves.

LO:3 Summarize the processes that lead to an earthquake and the release of seismic waves.

LO:4 Differentiate between the magnitude scales used to measure earthquakes.

LO:5 Identify global regions at most risk for earthquakes, and describe the effects of earthquakes.

LO:6 Describe how earthquakes are linked to other natural hazards.

LO:7 Discuss the important natural service functions of earthquakes.

LO:8 Explain how human beings interact with and affect the earthquake hazard.

LO:9 Discuss the United States governmental approach to minimizing the effects of earthquakes.

LO:10 Propose ways to minimize seismic risk and suggest adjustments we can make to protect ourselves.

3

Linkages exist between natural hazards.

Most of the destruction and loss of life due to the 2011 Tohoku earthquake was not due to the earthquake, but rather a tsunami triggered by the quake, which caused widespread coastal flooding of the Tohoku region (northeast) of Japan (see Chapter 4).

4

Humans can turn hazardous events into disasters.

In 2004, Indonesia was struck by a similarly large earthquake and tsunami, but while Japan's population has remained constant since 1900, Indonesia's has grown fivefold in the last century, making the 2004 Indian Ocean earthquake one of the most deadly recorded natural disaster (see Chapter 4).

5

Consequences of hazards can be minimized.

At the end of the chapter, you will learn about how widespread building and infrastructure collapse during the less powerful 1995 Kobe earthquake led to better earthquake building codes in Japan that minimized loss of life and damage during the 2011 Tohoku earthquake.

1906 earthquake that destroyed San Francisco (Table 3.1), the 2011 Tohoku earthquake released about 600 million times more energy than the atomic bomb dropped on Hiroshima at the end of World War II. In just a few minutes a 500 km (~310 mi.) long section of the Pacific plate moved as much as 40–60 m (~130–200 ft.) toward the northwest beneath Japan.[1] Shaking was felt across much of the island and lasted as long as six minutes in some locations (Figure 3.1). Had this earthquake occurred in a country where buildings are less well engineered, the loss of life due to structural collapse would have been significantly greater. In fact, the greatest loss of life was caused by another natural hazard linked to the quake—specifically, a large ocean wave triggered by the earthquake, known as a *tsunami*, which swept inland killing 93 percent of the 16,000 people who died that day (see Chapter 4).[3]

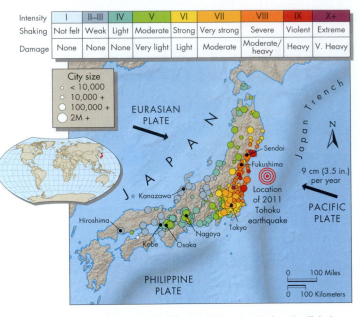

Intensity	I	II–III	IV	V	VI	VII	VIII	IX	X+
Shaking	Not felt	Weak	Light	Moderate	Strong	Very strong	Severe	Violent	Extreme
Damage	None	None	None	Very light	Light	Moderate	Moderate/heavy	Heavy	V. Heavy

∧ FIGURE 3.1 Shaking and Damage During the Tohoku Earthquake

Japan sits at the junction of three different tectonic plates. In 2011, the Pacific plate surged northwestward beneath the Eurasian plate and Japan, triggering the Tohoku earthquake, causing severe shaking and damage across most of the main island of Honshu. Intensity scale is described in Section 3.3 of this chapter. *(Based on "Poster of the Great Tohoku Earthquake (northeast Honshu, Japan) of March 11, 2011 – Magnitude 9.0," USGS, available at http://earthquake.usgs.gov/earthquakes/eqarchives/poster/2011/20110311.php)*

3.1 Introduction to Earthquakes

Worldwide, people feel an estimated 1 million earthquakes a year. However, few of these quakes are noticed very far from their source, and even fewer are considered major earthquakes. Table 3.1 lists significant earthquakes in the United States in the past 200 years. Even if you have never experienced an earthquake, you may wonder what actually happens when one occurs. To understand the effects of earthquakes and where earthquakes occur, we first need to learn about what causes earthquakes.

As discussed in Chapter 2, Earth is a dynamic, evolving system in which the movement of tectonic plates results in the formation and destruction of ocean basins, and the uplift of mountains. These processes are most active along the boundaries of lithospheric plates where *faulting* occurs (Figure 3.2).

∨ TABLE 3.1 Significant Earthquakes in the United States

Year	Locality	Damage (millions of dollars)	Number of Deaths
1811–1812	New Madrid, Missouri	Unknown	Unknown
1886	Charleston, South Carolina	23	60
1906	San Francisco, California (includes deaths from fire)	524	>3000
1925	Santa Barbara, California	8	13
1933	Long Beach, California	40	115
1940	Imperial Valley, California	6	9
1952	Kern County, California	60	12
1959	Hebgen Lake, Montana (damage to timber and roads)	11	28
1964	Prince William Sound, Alaska (includes tsunami deaths and damage near Anchorage and on U.S. West Coast)	500	125

Year	Locality	Damage (millions of dollars)	Number of Deaths
1965	SeattleTacoma, Washington	13	7
1971	Sylmar (San Fernando), California	553	65
1983	Coalinga, California	31	0
1983	Borah Peak, central Idaho	15	2
1987	Whittier Narrows, California	358	8
1989	Loma Prieta (San Francisco area), California	6000	63
1992	Landers, California	271	3
1994	Northridge, California	20,000	60
2001	Nisqually (Puget Sound), Washington	2300	1
2002	Denali Fault, southcentral Alaska	(sparsely populated area)	0
2006	Kiholo Bay, Hawaii, Hawaii	200	0
2010	Southernmost California and northern Mexico	100	several in Mexico
2011	Central Virginia	200	0
2014	South Napa Valley, California	5001	

Source: U.S. Geological Survey Earthquake Hazards Program (http://earthquakes.usgs.gov).

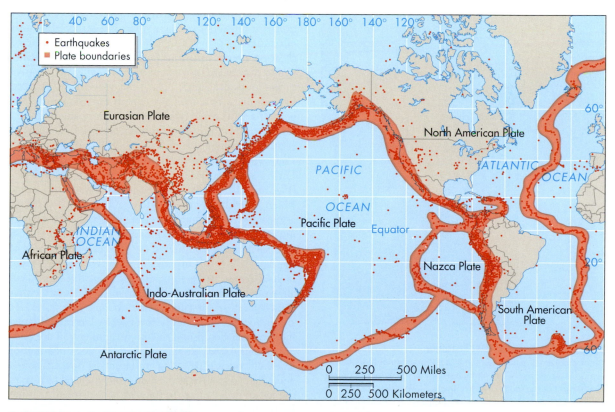

∧ **FIGURE 3.2 Earthquake Distribution**
Map showing the global distribution of moderate to large earthquakes (1963–88). Notice that most earthquakes occur along plate boundaries (shaded area), in contrast to intraplate earthquakes (isolated dots), which are infrequent. *(From the USGS National Earthquake Information Center)*

FAULTS AND FAULTING

Earthquakes occur along a plane of weakness in Earth's crust known as a **fault**, which is common along plate boundaries. A fault is a semiplanar fracture or fracture system where rocks have been displaced; that is, Earth's crust on one side of the fracture or fracture system has moved relative to the other side. Geologists use centuries-old mining terminology to describe the rocks that are displaced across faults, because many early underground mines were dug into the Earth along inclined surfaces to mine mineralized ore veins along fault zones. Specifically, the block below the fault plane where the miner would stand is known as the **footwall**, whereas the block above the fault plane where a lantern could be hung is known as the **hanging wall** (Figure 3.3).

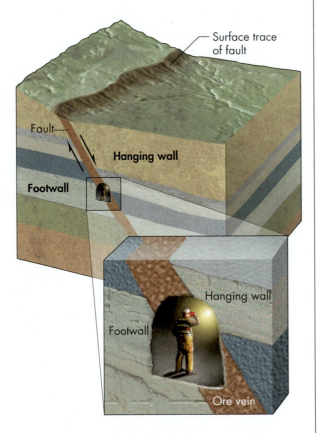

⋀ FIGURE 3.3 Basic Fault Features
Block diagram of a dip-slip fault illustrating several basic fault features. Because dip-slip faults are inclined (not vertical), rocks to the right of the fault are above the fault, whereas rocks to the left are below the fault. The hanging wall is defined as the block of rocks that lies above the fault where a miner hangs his lantern, while the block of rocks below the fault that the miner stands on is known as the footwall. *(Based on "Crustal Deformation," GeoPhysics at Ohio University)*

The process of fault rupture, or *faulting*, can be compared to sliding one rough board past another. Friction along the boundary between the boards may temporarily slow their motion, but rough edges break off and motion occurs at various places along the plane. Similarly, one lithospheric plate moving past another is slowed by friction along the fault plane that separates the plates. This "braking action" stresses the rocks along the fault, and as a result, these rocks undergo strain or deformation. When the stresses on rocks exceed their strength, that is, their ability to withstand the force, the rocks rupture and are permanently displaced along the fault plane. Put another way, **stress** is a force that results from plate tectonic movements and **strain** is the change in shape or location of the rocks due to the applied stress. As you recall from Chapter 2, there are three different types of plate boundaries (convergent, divergent, and transform); therefore, there are three different types of stress—*tensional, compressional*, and *shearing*—and three primary types of faults (Figure 3.4).

Fault Types Faults are distinguished by the direction in which rocks on either side are displaced. A *dip-slip* fault offsets rocks in a vertical motion due to compressional or tensional stresses, whereas a **strike-slip fault** offsets blocks of crust in a horizontal direction due to shearing stress. A **normal fault** is a type of fault identified by downward movement of the hanging wall relative to the footwall (Figure 3.4a). This movement is caused by tensional stress, which results in extension and thinning of Earth's crust typical of divergent plate boundaries. By contrast, a **reverse fault** is identified by an upward relative movement of the hanging wall due to compression, which results in shortening and thickening of Earth's crust typical of convergent plate boundaries (Figure 3.4b). Unlike reverse and normal dip-slip faults, which are inclined or said to have a dip, a strike-slip fault is close to vertical and, therefore, does not have a hanging wall or a footwall. But similar to a dip-slip fault, a reversal in the orientation of the stresses on either side of the fault will result in two different types of strike-slip faults. Because strike-slip faults offset Earth's crust laterally or in a horizontal motion, strike-slip faults are either right-lateral or left-lateral. The example in Figure 3.4c shows a stream that has been offset by left-lateral faulting. You can determine this by imagining you are paddling your kayak along the river—it does not matter whether you are paddling upstream or downstream—and when you arrive at the fault, you will need to turn to the left in order to get to the part of the stream that has been left-laterally offset.

Until recently it was thought that most active faults could be mapped because their most recent earthquake would cause a surface rupture. However, we now know

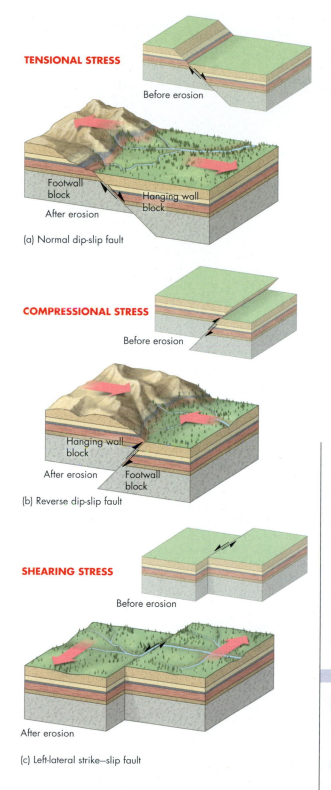

TENSIONAL STRESS

Before erosion

Footwall block

Hanging wall block

After erosion

(a) Normal dip-slip fault

COMPRESSIONAL STRESS

Before erosion

Hanging wall block

After erosion

Footwall block

(b) Reverse dip-slip fault

SHEARING STRESS

Before erosion

After erosion

(c) Left-lateral strike—slip fault

◀ **FIGURE 3.4 Types of Geologic Faults**
Three common types of faults and their effects on the land-scape. Large red arrows on the surface of the block diagrams show the stress directions, while the black arrows along the fault show relative offset (strain) across the fault. Blocks above are shown after movement on the fault with no erosion, whereas the blocks below show effects of erosion. (a) A normal dip-slip fault in which the hanging wall on the right side of the fault has dropped down. (b) A reverse dip-slip fault in which the hanging wall above the fault has moved up. (c) Strike-slip fault with horizontal displacement along the fault plane. This particular example shows left-lateral displacement.

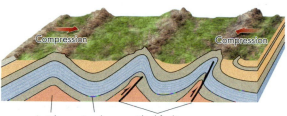

Compression   Compression

Anticline   Syncline   Blind faults

▲ **FIGURE 3.5 Blind Faults**
Blind reverse faults accommodate compressional stresses by offsetting rock layers at depth and by folding near to Earth's surface. The upward folds are called anticlines and downward folds are called synclines. (Modified after Lutgens, F., and Tarbuck, E. 1992. Essentials of geology, 4th ed. New York: Macmillan.)

that some active faults, referred to as *blind faults*, do not extend to the surface (Figure 3.5). This discovery has made it more difficult to evaluate the earthquake hazard in some areas. For example, the 1994 Northridge earthquake, which killed 60 people and caused at least $20 billion in property damages, occurred on a previously unknown active blind fault that caught the geologists working in the Los Angeles Basin completely by surprise. Since 1994, the earthquake hazard community has become focused on locating these structures in densely urbanized areas. Although blind faults do not offset rocks at the surface, geologists often locate them by identifying folds at the surface known as *anticlines and synclines* (Figure 3.5).

3.1 CHECK your understanding

1. If you were looking at the two blocks on either side of a fault, how would you differentiate the hanging wall from the footwall?

2. What types of stresses are responsible for normal, reverse, and strike-slip faulting?

3. How would you tell the difference between a strike-slip and a dip-slip fault?

3.2 The Earthquake Processes

Most geologists consider a particular fault to be an *active fault* if it has moved during the past 10,000 years of the Holocene Epoch. The Holocene, the most recent epoch of the Quaternary Period of Earth's geologic history, is preceded by the Pleistocene Epoch of the Quaternary Period. Much of our present-day landscape developed during the Quaternary Period. Faults that show evidence of movement during the Pleistocene but not the Holocene Epoch are classified as *potentially active* (Table 3.2).

Faults that have not moved during the past 2 million years are generally classified as *inactive*. However, we emphasize that it is often difficult to prove when a fault was last active, especially where no historical record of earthquakes exists. In many cases geologists must determine the **paleoseismicity** of the fault, that is, the prehistoric record of earthquakes. They do so by determining the age of the most recent faulted or folded sediments. Because the faults themselves are often buried just below our feet or are blind, determining paleoseismicity often involves excavating a paleoseismic trench across the active fault to expose the deformation (Figure 3.6).

THE EARTHQUAKE CYCLE

Observations of the 1906 San Francisco earthquake led to a hypothesis known as the **earthquake cycle**. The earthquake cycle proposes that there is a drop in elastic strain after an earthquake and an accumulation of strain before the next event.

∧ FIGURE 3.6 Paleoseismic Trench
Paleoseismic trench excavated across folded sediments above a blind fault in Southern California. *(Duane E. DeVecchio)*

As stated previously, strain is deformation resulting from stress. *Elastic strain* may be thought of as deformation that is not permanent, provided that the stress is eventually released. When the stress is released, the elastically deformed material returns to its original shape. If the stress is not released and continues to increase, the deformed material will eventually rupture, making the deformation permanent. For example, when you apply tensional stress to a rubber band—sometimes called an elastic—it stretches but will return to its original shape when the stress is released. However, if you apply too much stress, the rubber band will break and be permanently deformed. When it breaks, the broken band often

∨ TABLE 3.2 Terminology for Faults Based on Last Activity

Geologic Age				Start of Time Interval	Fault Activity
Era	Period	Epoch		(in Years before Present)	
			Historic	200	Active
		Holocene	Prehistoric	10,000[a]	
Cenozoic	Quaternary	Pleistocene		1,650,000[a]	Potentially active
	Tertiary			65,000,000	
Pre-Cenozoic time				4,600,000,000 (age of Earth)	Inactive

[a] Dates used for regulatory purposes. Actual dates for these geologic time intervals have changed (see Chapter 1).

Source: Based on California State Mining and Geology Board Classification, 1973.

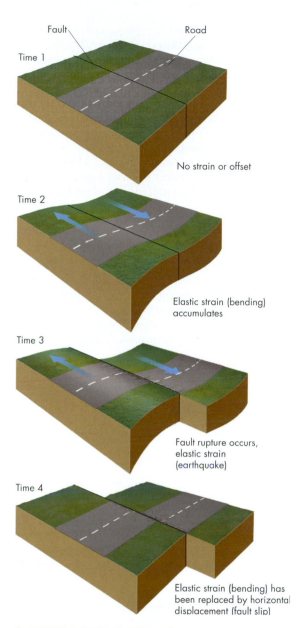

Time 1

Fault   Road

No strain or offset

Time 2

Elastic strain (bending) accumulates

Time 3

Fault rupture occurs, elastic strain (earthquake)

Time 4

Elastic strain (bending) has been replaced by horizontal displacement (fault slip)

∧ FIGURE 3.7 Elastic Rebound
Idealized diagram showing buildup and release of elastic strain on a right-lateral strike-slip fault due to continuously applied shear stress. At time 1, stress is applied, but there is no significant accumulation of strain or deformation along the fault. At time 2, accumulation of elastic strain within the rocks has caused the road to become curved. Elastic strain will continue to build until the maximum strength of the rocks along the fault is surpassed causing the fault to rupture. At time 3, fault rupture begins and elastic energy is released causing offset of the two fault blocks across the fault. The offset of the road indicates the amount of displacement. Time 4 is immediately following the fault rupture showing rebound of the elastic strain and permanent fault offset. Progression from time 1 to time 4 may take hundreds to thousands of years.

shoots across the room, releasing the pent-up energy with a "snap." Earthquakes along faults are caused by a very similar process where tectonic stresses are applied to rocks. The rocks deform elastically until a critical point is reached and the fault slips, releasing the stored elastic energy in a process know as *elastic rebound* (Figure 3.7).

Seismologists speculate that a typical earthquake cycle has three or four stages.[4] The first stage is a long period of inactivity along a segment of a geologic fault. In the second stage, accumulated elastic strain produces small earthquakes. A third stage, consisting of *foreshocks*, may occur only hours or days prior to the next large earthquake. Foreshocks are small- to moderate-sized earthquakes that occur before the main event. However, in some cases, this third stage may not occur. Finally, the fourth stage is the *mainshock*, the major earthquake, and its aftershocks.[4] An *aftershock* is a smaller earthquake occurring near the location of the mainshock anywhere from a few minutes to a year or so following the main event.

Following the mainshock of an earthquake, news reports generally give information about where the earthquake was located. This location, known as the **epicenter**, is the place on the surface of Earth above where the ruptured rocks broke to produce the earthquake. The point of initial breaking or rupturing within the Earth is known as the **focus**, or *hypocenter*, of the earthquake and is directly below the epicenter (Figure 3.8). When a rupture begins, displacement of rocks starts at the focus and

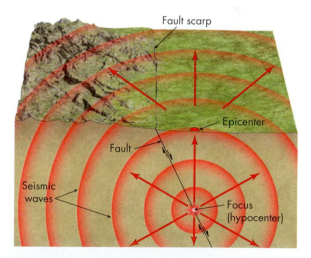

Fault scarp

Epicenter

Fault

Seismic waves

Focus (hypocenter)

∧ FIGURE 3.8 Basic Earthquake Features
The epicenter of an earthquake is the surface location directly above the focus, which is where movement on the fault begins. During large earthquakes, displacement may propagate from the focus to the surface causing surface deformation to occur, known as a fault scarp. *(Tarbuck, Edward J., Lutgens, Frederick K., and Tasa, Dennis G. Earth: An introduction to physical geology, 11th ed., © 2014, p. 363. Reprinted and electronically reproduced by permission of Pearson Education, Inc., Upper Saddle River, New Jersey.)*

then propagates up, down, and laterally along the fault plane during the earthquake. The sudden rupture of the rocks produces shock waves, called *seismic waves*, which cause ground shaking.

SEISMIC WAVES

When a fault ruptures, rocks break apart suddenly and violently, releasing stored elastic strain energy in the form of **seismic waves**. These waves radiate outward in all directions from the focus, like ripples on a pond after a pebble hits the water. It is the strong motion from seismic waves as they pass through the ground beneath our feet that we perceive as an *earthquake*, cracking the ground and damaging buildings and other structures. The term *earthquake* also refers to the fault rupture that gives rise to the waves.

Some of the seismic waves generated by the fault rupture travel within the body of the Earth, known as *body waves*, whereas others travel along the surface and are known as *surface waves*. There are two types of body waves: P waves and S waves.

P waves, also called compressional or primary waves, are the faster of the two (Figure 3.9a). They can travel through solids, liquids, and gases. P waves travel much more quickly through solids than through liquids; the average velocity for P waves through Earth's crust is 6 km (~3.7 mi.) per second, in contrast to 1.5 km (~0.9 mi.) per second through water. Interestingly, when P waves reach Earth's surface and are transmitted into the air, some people and animals may be able to hear a fraction of the seismic waves.[5] However, the sound that most people hear when an earthquake is approaching is the loud noise of objects vibrating, not the actual P wave.

S waves, also called shear or secondary waves, can travel only through solid materials (Figure 3.9b). They travel more slowly than P waves and have an average velocity through Earth's crust of 3 km (~1.9 mi.) per second. S waves produce an up-and-down motion (sideways shear) at right angles to the direction that the wave is moving. This movement is similar to the whipping back and forth of a large jump rope being held by two people on a playground. When liquids are

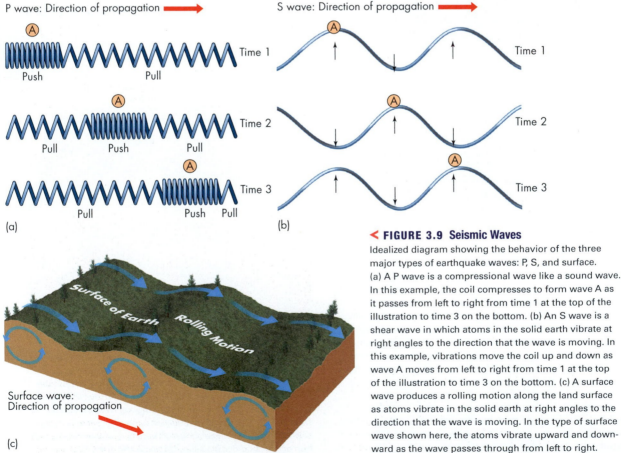

(a)

(b)

(c)

◄ FIGURE 3.9 Seismic Waves
Idealized diagram showing the behavior of the three major types of earthquake waves: P, S, and surface. (a) A P wave is a compressional wave like a sound wave. In this example, the coil compresses to form wave A as it passes from left to right from time 1 at the top of the illustration to time 3 on the bottom. (b) An S wave is a shear wave in which atoms in the solid earth vibrate at right angles to the direction that the wave is moving. In this example, vibrations move the coil up and down as wave A moves from left to right from time 1 at the top of the illustration to time 3 on the bottom. (c) A surface wave produces a rolling motion along the land surface as atoms vibrate in the solid earth at right angles to the direction that the wave is moving. In the type of surface wave shown here, the atoms vibrate upward and downward as the wave passes through from left to right.

subjected to sideways shear, they are unable to spring back, which explains why S waves can't move through liquids.

When P and S waves reach the land surface, complex *surface waves* form and move along Earth's surface. These waves travel more slowly than either P or S waves, and they cause much of the damage near the epicenter. Because surface waves have a complex horizontal and vertical ground movement or rolling motion, they may crack walls and foundations of buildings, bridges, and roads (Figure 3.9c). People caught near the epicenter of a strong earthquake have reported seeing these waves rippling across the land surface. One type of surface wave, called a *Love wave*, causes horizontal shaking that is especially damaging to building foundations.

TECTONIC CREEP AND SLOW EARTHQUAKES

Some active faults exhibit **tectonic creep**, that is, gradual movement along a fault that is not accompanied by perceptible earthquakes. For example, tectonic creep along the Cascadia subduction zone beneath southwestern British Columbia and the state of Washington produces *slow earthquakes*, which are not felt and only recently have been detected.[6] Also called *fault creep*, this process can slowly damage roads, sidewalks, building foundations, and other structures (Figure 3.10). Tectonic creep has damaged culverts under the University of California at Berkeley football stadium. Movement of 2.2 cm (~1 in.) was measured beneath the stadium in only 11 years, and periodic repairs have been necessary as the cracks developed.[7] More rapid tectonic creep has been recorded on the Calaveras Fault near Hollister, California. There, a winery on the fault is slowly being pulled apart at about 1 cm (~0.4 in.) per year.[8] Just because a fault creeps does not mean that damaging earthquakes will not occur along the fault. Often the rate of creep is a relatively small portion of the total slip rate on a fault; periodic sudden displacements that produce earthquakes can also be expected.

Slow earthquakes are similar to other earthquakes in that they are produced by fault rupture. The big difference is that the rupture, rather than being nearly instantaneous, can last from days to months. Slow earthquakes are a newly recognized fundamental Earth process. They are recognized through analysis of continuous geodetic measurement or geographic positioning systems (GPS), similar to the devices used in automobiles to identify location. These instruments can differentiate horizontal movement in the millimeter range and have been used to observe surface displacements from slow earthquakes. When slow earthquakes occur frequently,

say every year or so, their total contribution to changing Earth's surface to produce mountains may be significant over geologic time.[9]

(a)

(b)

∧ FIGURE 3.10 Tectonic Creep (a) Slow, continual movement along the San Andreas fault has split this concrete drainage ditch at Almaden Vineyards south of Hollister, California. For scale, there is a person's shoe and blue pant leg in the upperleft corner of the image. *(James A. Sugar/ National Geographic Creative)*; (b) creep along the Hayward fault (a branch of the San Andreas fault) is slowly deforming the football stadium at the University of California, Berkeley. The stadium has been redesigned to accommodate this offset along this vertical opening in the stadium wall. *(Duane E. DeVecchio)*

2015 Nepal (Gorkha) Earthquake: Forecasting a Catastrophe

Nepal is an Asian country that lies along the steep slopes of the Himalayan Mountains, which result from the collision between the Eurasian and Indian plates (Figure 3.11). The large earthquake that struck on April 25, 2015 below the Gorkha District of Nepal killed more than 9,000 people and was expected. For years, experts had been warning that the capital city of Kathmandu was overdue for a powerful earthquake that could decimate the rapidly growing city. In fact, a team of about 50 earthquake scientists and social scientists had gathered in Kathmandu just one week before the event to discuss the impact and ensuing catastrophe that would follow a large magnitude earthquake that scientists felt was imminent.[10]

The gap in historically large earthquakes west of Kathmandu between 1505 and 1866 was what concerned scientists (Figure 3.11). With a convergence rate of about 45 mm/yr, many meters of elastic strain would have accumulated over the past few hundred years (see Section 3.2). This meant that a powerful earthquake was overdue to release the accumulated strain. The 2015 earthquake caused the Eurasian plate to lurch southward over the Indian Plate by as much as 5 m (16 ft.) along 150 km (93 mi.) zone between the epicenters of the 2015 and 1866 events (Figure 3.11). The ground shaking was intense and uplifted the Kathmandu Valley by as much as 1 m (~3 ft.).[11]

With about 10 percent of the population of Nepal living within the Kathmandu Valley and the adjacent steep slopes of the highlands, the devastation of April 25 was just what the scientists had predicted.[10] Building collapse was the primary cause of death in urban environments, which was expected due to a lack of strict earthquake construction codes for homes and civil buildings. Many towns were shaken to complete destruction (Figure 3.12), whereas others were completely wiped away by massive landslides triggered by shaking of the steep slopes.[12] An avalanche triggered by the earthquake killed 21 people on Mount Everest more than 20 km (130 mi.) away, making it the deadliest day in the mountain's history.[13]

For Nepal the 2015 earthquake was catastrophic. The damage is estimated to be in excess of $10 billion or about half of Nepal's gross domestic production (GDP).[14]

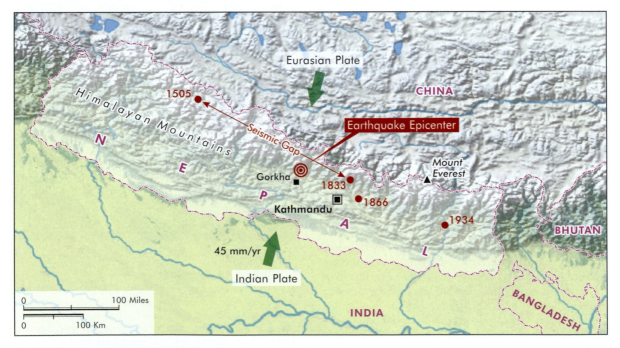

∧ FIGURE 3.11 Forecasting an Earthquake
Along one of the most active plate boundaries in the world between the Eurasian Plate and the Indian Plate, the absence of historic earthquakes (red dots) within the seismic gap northwest of Kathmandu led scientists to believe that the 2015 Nepal earthquake was eminent.

∧ FIGURE 3.12 Earthquake Flattened Town
Numerous small towns in the highlands surrounding
Kathmandu were not only shaken to the ground by the 2015
Nepal earthquake, but also bombarded by falling rock from the
slopes above. *(Courtesy of Kristen Cook)*

**∨ TABLE 3.3 Global Frequency of Earthquakes by
Descriptor Classification**

Descriptor	Average Magnitude	Annual Number of Events
Great	8 and higher	1
Major	7–7.9	14
Strong	6–6.9	146
Moderate	5–5.9	1344
Light	4–4.9	13,000 (estimated)
Minor	3–3.9	130,000 (estimated)
Very Minor	2–2.9	1,300,000 (estimated) (approx. 150 per hour)

Source: U.S. Geological Survey, "Earthquake Statistics,"
2000–15, https://earthquake.usgs.gov/earthquakes/browse/
stats.php. Accessed 7/21 /2017.

Although billions of dollars have been pledged from
around the world to help rebuild, two years after the disaster many tens of thousands of Nepalis are facing a third
monsoon season living in temporary habitations.

Although the 2015 Nepal earthquake and its effects
were anticipated, it is not possible at this time for scientists
to predict earthquakes. As you will learn in the coming
sections of this chapter, scientists often speak in probabilities of an earthquake event occurring within a given time
frame rather than prediction of a specific time and date.

3.2 CHECK your understanding

1. How are active and potentially active faults
   defined?

2. Explain the earthquake cycle.

3. Where does the energy that produces seismic
   waves come from?

4. What is the difference between the epicenter
   and focus of an earthquake?

5. What is the difference in the rates of travel of
   P, S, and surface waves?

3.3 Earthquake Shaking

Three important factors determine the shaking you will
experience during an earthquake: (1) earthquake magnitude, (2) your location in relation to the epicenter and

direction of rupture, and (3) local soil and rock conditions.
In general, strong shaking may be expected from earthquakes of moderate *magnitude* or larger (Table 3.3).

EARTHQUAKE MAGNITUDE

In addition to the epicenter location, news reports also
provide a decimal number (e.g., 6.8) that refers to the
size or magnitude of the earthquake. Immediately following an earthquake, the first magnitude estimates
were made using the **Richter scale**, developed in 1935
by the famous seismologist Charles Richter. The
Richter magnitude of an earthquake is determined by
measuring the maximum amount of ground shaking
due to the S wave. Ground motion, in either a vertical
or horizontal direction, is recorded by an instrument
known as a **seismograph** or *seismometer* (Figure 3.13a).

The Richter scale is logarithmic, which means that
the increase from one integer to the next is not linear.
For example, the ground displacement during a magnitude 6.0 earthquake is approximately 10 times as great as
displacement during a magnitude 5.0 quake. As you will
learn later in this section, the amount of ground shaking
at any one location is strongly affected by numerous
variables and, therefore, a Richter magnitude is specific
to only one location. Multiple seismographs that detect
an earthquake will record differing Richter magnitudes
depending on where they are located, giving what is
known as a *local magnitude*. In other words, Richter magnitudes are not an absolute measurement of the size of an
earthquake and, therefore, not well suited for comparing
earthquake sizes around the world.

In order to determine the absolute size of an earthquake, the **moment magnitude scale** was developed in
the 1970s. The moment magnitude is a measurement
of the actual energy released during the earthquake.

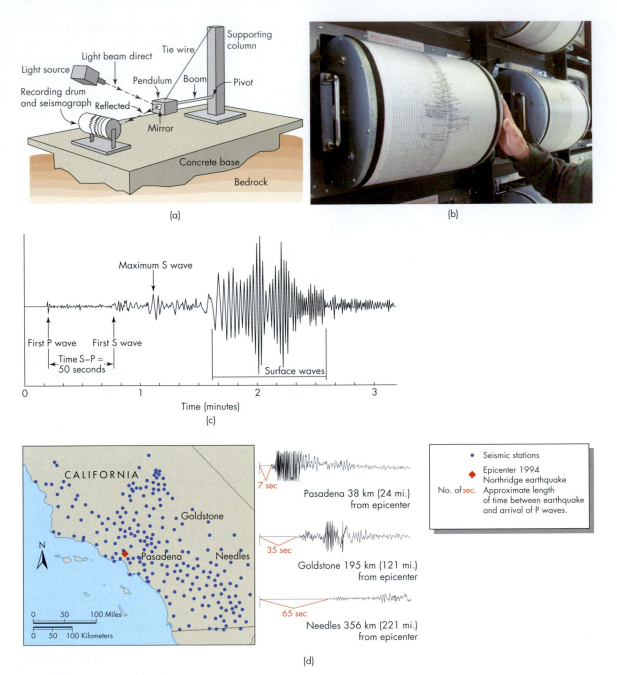

∧ FIGURE 3.13 Seismograph

(a) Simplified illustration showing how a seismograph works. *(Based on information from Southern California Earthquake Center)* (b) Modern seismograph at the Pacific Geoscience Centre near Victoria, British Columbia, showing a recording drum on which a seismogram is made. The earthquake is the February 28, 2001, Nisqually, Washington, earthquake (**M** 6.8). *(AP Photo/Victoria Times Colonist, Ian McKain)* (c) Idealized seismogram for an earthquake. The P, S, and surface waves can be identified by changes in the amplitude of the waves on the graph. (d) Differences in arrival time and amount of shaking at three seismic stations located from 38 to 356 km (~24 to 221 mi.) from the 1994 Northridge, California, earthquake. As you get farther away from the epicenter, the seismic waves take longer to reach the seismograph and the amplitude of the waves on the seismograph decreases. In general, the greater the amplitude of the waves on the seismogram, the stronger is the shaking of the ground. *(Based on information from Southern California Earthquake Center)*

The moment magnitude is determined from an estimate of the area that moved along a fault plane during the quake, the amount of movement or slippage along the fault, and the strength of the rocks near the focus of the quake. Unlike the Richter magnitude, which can be determined immediately following the quake, moment magnitude estimates can take days to months to calculate precisely. Because of this the Richter and moment magnitude scales are used in tandem, where immediately following an earthquake scientists will supply local Richter magnitudes to the media, while moment magnitude calculations are still being made. This is why news reports of the magnitude of an earthquake often change over the course of days to months.

Except for very large earthquakes, the magnitude on the Richter scale is approximately equal to moment magnitude. Because of this correlation, we will refer to the size of an earthquake simply as its magnitude, **M**, without designating Richter or moment magnitude.

It is interesting to look at how often earthquakes of various magnitudes occur. Earthquakes are given descriptive adjectives based on their magnitude (see Table 3.3). For example, most damaging earthquakes are described as major (**M** 7–7.9) or strong (**M** 6–6.9). *Major earthquakes* are capable of causing widespread and serious damage. *Strong earthquakes* can also cause considerable damage depending on their location and the nature of earth materials. Fortunately, the most powerful *great earthquakes* (**M** 8 or higher) are uncommon; the worldwide average is one per year (Table 3.3). In contrast, more than 3,500 very minor earthquakes with a magnitude of less than 3 occur each day. Most of these quakes are too small or too remote to be felt by people.

The amount of ground motion produced by an earthquake is related to its magnitude, depth, and the geologic environment in which it occurs. As earthquake magnitudes increase, the amount of ground motion changes less drastically than the amount of energy released (Table 3.4). As this table illustrates, the difference between a **M** 6 and a **M** 7 earthquake is considerable. Although the amount of ground motion, or shaking, from a **M** 7 earthquake is 10 times as great as that from a **M** 6 earthquake, the amount of energy released is 32 times as much! The energy released from a **M** 6 earthquake is about equivalent to a 30 kiloton nuclear explosion, compared to about a 1 megaton explosion for a **M** 7 earthquake. Another way of looking at this: It would take 32 **M** 6 earthquakes to release as much energy as one **M** 7 quake.

EARTHQUAKE INTENSITY

The moment magnitude scale provides a quantitative way of comparing earthquakes. In contrast, earthquake intensity is often indicated with the qualitative **Modified Mercalli Intensity scale**. The 12 categories on this scale are assigned Roman numerals (Table 3.5). Each category contains a description of how people perceived the shaking from a quake and the extent of damage to buildings and other human-made structures. For example, the 1971 Sylmar earthquake in California's San Fernando Valley had a single magnitude (**M** 6.7) but its Mercalli intensity varied from I to XI, depending on distance from the epicenter and on local geologic conditions (Figure 3.14). Because building codes are

▼ TABLE 3.4 Change in Ground Motion and Energy Released from an Incremental Change in Earthquake Magnitude

Units of Magnitude Change	Ground Motion Change[a]	Change in Amount of Energy Released
1	10 times	About 32 times
0.5	3.2 times	About 5.5 times
0.3	2 times	About 3 times
0.1	1.3 times	About 1.4 times

[a] As reflected in the maximum amplitude of S waves on a standard seismograph. Amplitude is the distance that a seismic wave is displaced from a baseline (zero line) that is established when no seismic waves are detected.

Source: U.S. Geological Survey, "Earthquake Facts and Statistics," 2010, http://earthquake.usgs.gov/earthquakes/eqarchives/year/eqstats.php. Accessed 7/12/2013.

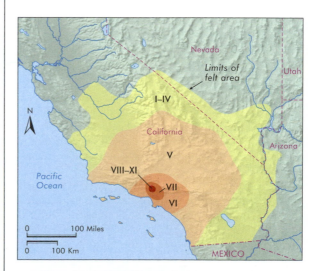

▲ FIGURE 3.14 Intensity of Shaking
Modified Mercalli Intensity map for the 1971 Sylmar, California, earthquake (**M** 6.7), determined after the earthquake. The red and orange areas experienced the most intense shaking. See Table 3.5 for an explanation of the Roman numerals. *(U.S. Geological Survey. 1974. Earthquake Information Bulletin 6[5])*

▼ TABLE 3.5 Abbreviated Modified Mercalli Intensity Scale

Intensity	Effects
I	Felt by few people under especially favorable conditions.
II	Felt by only a few persons at rest, especially on upper floors of buildings.
III	Felt quite noticeably indoors, especially on upper floors of buildings. Many people do not recognize it as an earthquake. Standing vehicles may rock slightly. Vibration feels like the passing of a truck.
IV	Felt indoors by many, outdoors by few during the day. At night some awakened. Dishes, windows, doors disturbed; walls make cracking sound. Sensation of a heavy truck striking building; standing vehicles rock noticeably.
V	Felt by nearly everyone; many awakened. Some dishes and windows broken. Unstable objects overturned.
VI	Felt by all; many frightened. Some heavy furniture moved; a few instances of fallen plaster or damaged chimneys. Damage slight.
VII	Damage negligible in buildings of good design and construction; slight to moderate in well-built ordinary structures; considerable in poorly built or badly designed structures; some chimneys broken. Noticed by vehicle drivers.
VIII	Damage slight in specially designed structures; considerable damage in ordinary substantial buildings with partial collapse; damage great in poorly built structures; fall of chimneys, factory stacks, columns, monuments, and walls. Heavy furniture overturned. Disturbs vehicle drivers.
IX	Damage considerable in specially designed structures; well-designed frame structures thrown out of plumb. Damage great in substantial buildings, with partial collapse. Buildings shifted off foundations. Underground pipes broken.
X	Some well-built wooden structures are destroyed; most masonry and frame structures with foundations destroyed; train rails bent.
XI	Few, if any, masonry structures remain standing. Bridges destroyed. Underground pipelines taken out of service. Train rails bent greatly.
XII	Damage total. Waves seen on ground surfaces. Lines of sight and level are distorted. Objects thrown into the air.

Source: Based on U.S. Geological Survey Earthquake Hazards Program https://earthquake.usgs.gov/learn/topics/mag_vs_int.php. Accessed 7/21/17.

not the same worldwide, Mercalli intensity can also vary from country to country—generally greater where human-made structures are less well constructed (see Case Study 3.2: Earthquake Catastrophes).

Earthquake intensities are commonly shown on maps. Conventional modified Mercalli intensity maps, like that shown in Figure 3.15, take days or even weeks to complete. They are based on questionnaires sent to residents near the epicenter, newspaper articles, and reports from damage assessment teams.

Recently the U.S. Geological Survey (USGS) began experimenting with the use of the Internet to collect intensity information. The USGS website "Did You Feel It?" (https://earthquake.usgs.gov/data/dyfi/) solicits email reports from people in the United States who have recently felt an earthquake (see Figure 3.25). These reports are used to prepare an online *community Internet intensity map* that is updated every few minutes after a large earthquake.

Quickly determining where the damage is most severe is one of the major challenges during a damaging earthquake. This information is now available in parts of California, the Pacific Northwest, and Utah, where there are dense networks of high-quality seismograph stations. These networks transmit direct measurements of ground motion as soon as the shaking stops. This information is known as *instrumental intensity*, and it is used to immediately produce a *shake map* that shows both intensity of shaking and potential damage. Figure 3.15 shows shake maps for the 1994 **M** 6.7 Northridge, California, earthquake and the 2001 **M** 6.8 Nisqually earthquake centered in Puget Sound, Washington. The technology to produce and distribute shake maps in the minutes following an earthquake was made available in 2002.[15] The cost of seismographs is small relative to damages from earthquake shaking, and the arrival of emergency personnel is critical in the first minutes and hours following an earthquake if people in collapsed buildings are to be rescued. The shake map is also useful for locating areas where gas lines and other utilities are likely to be damaged. Clearly, the use of this technology, especially in our urban areas vulnerable to earthquakes, is a desirable component of our preparedness for earthquakes.

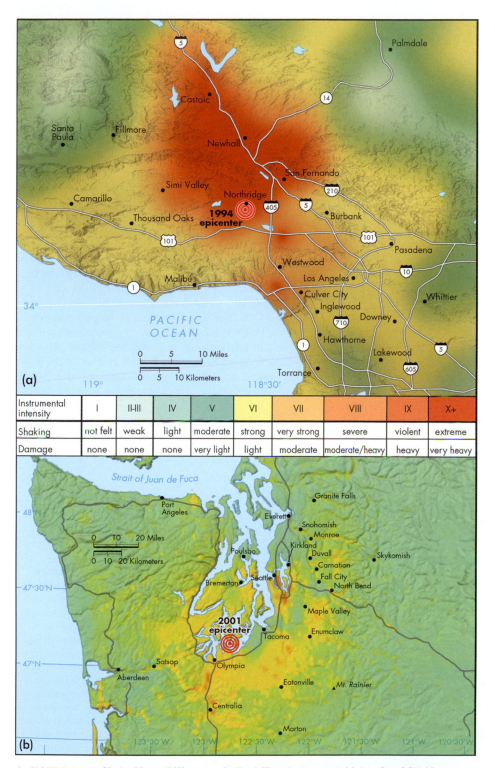

Instrumental intensity	I	II-III	IV	V	VI	VII	VIII	IX	X+
Shaking	not felt	weak	light	moderate	strong	very strong	severe	violent	extreme
Damage	none	none	none	very light	light	moderate	moderate/heavy	heavy	very heavy

∧ FIGURE 3.15 Shake Maps: Differences in Real-Time Instrumental Intensity of Shaking
(a) Instrumental Intensity map for the 1994 Northridge, California, earthquake (**M** 6.7) prepared as the earthquake occurred. *(U.S. Geological Survey and courtesy of David Wald)*; (b) The 2001 Nisqually, Washington, earthquake (**M** 6.8). *(Pacific Northwest Seismograph Network, University of Washington)*

DEPTH OF FOCUS

In addition to distance, the depth of an earthquake influences the amount of shaking. Recall that the place within Earth where the earthquake starts is the focus. In general, the deeper the focus of the earthquake, the less shaking that will occur at the surface. The seismic waves lose much of their energy before they reach the surface when the earthquake is relatively deep. This loss of energy, referred to as *attenuation*. For example, attenuation occurred in the 2001 Nisqually, Washington, earthquake (**M** 6.8) (Figure 3.15b). The Nisqually quake occurred along the Cascadia subduction zone at a depth of 52 km (~32 mi.). In comparison, there was less attenuation and more shaking in the 1994 Northridge quake (**M** 6.7), which had a shallower focus depth of 19 km (~12 mi.), but a very similar magnitude (Figure 3.15a).

DIRECTION OF RUPTURE

A third factor that influences the amount of shaking is the direction that the rupture moves along the fault during the earthquake. Although rupture may proceed in many directions from the focus, the path of greatest rupture can focus earthquake energy. This behavior, known as *directivity*, contributes to the amplification of seismic waves and thus to increased shaking. For example, in the 1994 Northridge quake, the path of greatest rupture on the fault was to the northwest (Figure 3.16). This trend caused the most intense shaking to occur northwest of the epicenter rather than directly over the focus.

DISTANCE TO THE EPICENTER

To determine the distance to the epicenter, it must first be located using information about the P and S waves detected by seismographs (see Figure 3.13a and b). The written or digital record of these waves is called a *seismogram*. On a seismogram, the P and S waves appear as an oscillating line that looks similar to an electrocardiograph (ECG) of a person's heartbeat (see Figure 3.13c). Because P waves travel faster than S waves, they will always appear first on a seismogram. Seismologists read the seismogram and use the difference between the time that the first P and S waves arrive (S–P) from a quake to determine how far away the epicenter is. For example, in Figure 3.13c, the S waves arrived 50 seconds after the P waves. This 50-second delay would occur if both types of waves traveled from an epicenter about 420 km (~261 mi.) away. Seismographs scattered across the globe will thus record different arrival times for P and S waves from an earthquake. The seismograph stations that are

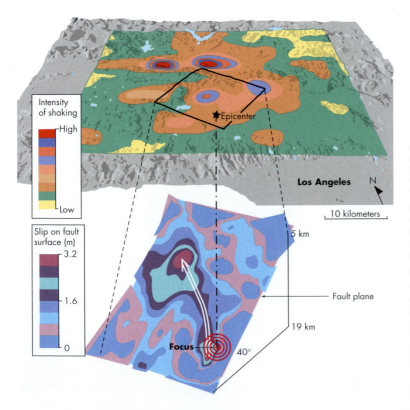

◄ FIGURE 3.16 Direction of Rupture Toward the Area of Most Intense Shaking Aerial view of the Los Angeles region from the south showing the epicenter and intensity of shaking of the **M** 6.7 1994 Northridge earthquake. The red area in the center of the bull's-eye patterns 10 km (~6 mi.) northwest of the epicenter had the most intense shaking. The bottom of the figure shows the focus and the section of the fault plane that ruptured during the quake. Colors on the fault plane show how much slippage occurred. Greatest slippage is in the reddish-purple bull's-eye pattern northwest of the focus. Rupture started at the focus and followed the direction of the white arrow to the northwest. *(U.S. Geological Survey, "USGS Response to an Urban Earthquake, Northridge '94," U.S. Geological Survey Open File Report 96–263, 1996)*

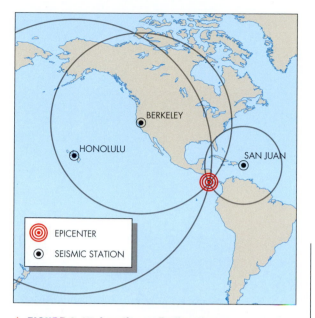

∧ FIGURE 3.17 Locating an Earthquake
Generalized concept of how the epicenter of an earthquake is located. First, the distance to the event is determined from information on seismograms from at least three seismic stations. Those distances are then used to draw circles around the seismic stations. The point where the circles come together is the epicenter, in this case the epicenter of the 2001 El Salvador earthquake. Accurate location of the epicenter is not always as simple as in this hypothetical example.

farthest from the epicenter will observe the greatest difference between P and S wave arrival times. This relationship was observed for the 1994 Northridge, California, earthquake (see Figure 3.13d).

The difference between P and S arrival times at seismographs in different locations can be used to pinpoint the epicenter of an earthquake. To accomplish this, a distance to the epicenter is calculated for each of three seismographs, and the respective values are used for the radius of a circle drawn around each seismic station. These circles will intersect in one location—the epicenter. The process of locating a feature using distances from three points is called *triangulation*. Thus, the epicenter of an earthquake in Central America can be located by triangulation from seismic stations in Honolulu, Hawaii; Berkeley, California; and San Juan, Puerto Rico (Figure 3.17).

LOCAL GEOLOGIC CONDITIONS

The nature of the local earth materials and local geologic structure strongly influences the amount of ground motion. Earth materials of various types behave differently in an earthquake. This difference is related to their degree of consolidation. Seismic waves move faster

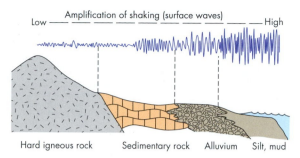

∧ FIGURE 3.18 Material Amplification of Shaking
Generalized relationship between near-surface earth material and amplification of shaking during an earthquake. The material amplification is highest in water-saturated sediment.

through consolidated bedrock than they do through unconsolidated sediment or soil. They slow even further if the unconsolidated material has high water content. For example, seismic waves typically slow down as they move from bedrock to stream deposits of sand and gravel, called *alluvium*, and then slow again as they move through coastal deposits of mud. As P and S waves slow down, the energy that was once directed forward is transferred to the vertical motion of the surface waves (Figure 3.18). This effect, known as **material amplification**, strongly influences the amount of ground motion experienced in an earthquake.

For example, in the 1989 Loma Prieta earthquake (**M** 6.9) in northern California, the most severe ground shaking was experienced along the shore of San Francisco Bay (Figure 3.19a). Intense ground motion caused the collapse of sections of the upper deck of the eastern span of the Bay Bridge and the Nimitz Freeway in Oakland, killing 41 people, and caused extensive damage in the Marina district of San Francisco (Figure 3.19b and 3.19c). The portion of the freeway that collapsed was built on bay fill, mud that had been dumped into San Francisco Bay to create new land. Likewise, the Marina district was created after the 1906 San Francisco earthquake by filling in the shoreline with debris from damaged buildings and with mud pumped from the bottom of the bay.[16] The unconsolidated nature of the bay fill and mud, combined with its high water content, caused the increased shaking. Amazingly, the effects of material amplification meant that the greatest shaking occured 65 km (~40 mi.) north of the epicenter near San Francisco (Figure 3.19a). Following the Loma Preita earthquake it was clear to geologists and engineers that the eastern span of the Bay Bridge was not earthquake safe and that a new bridge would need to be constructed. The new bridge span cost $6.5 billion and took nearly 25 years to complete, not becoming open for traffic until 2013 (Figure 3.19d).

The deaths of more than 8,000 people in the 1985 Mexico City earthquake (**M** 8.0) provide yet another

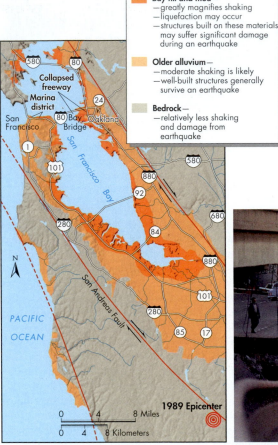

Legend:

- **Bay fill and mud—**
  —greatly magnifies shaking
  —liquefaction may occur
  —structures built on these materials may suffer significant damage during an earthquake

- **Older alluvium—**
  —moderate shaking is likely
  —well-built structures generally survive an earthquake

- **Bedrock—**
  —relatively less shaking and damage from earthquake

(a)

(b)

(c)

(d)

⋀ FIGURE 3.19 Loma Prieta Earthquake

(a) Map of earth materials in the San Francisco Bay region, showing the San Andreas fault zone and the epicenter of the 1989 **M** 6.9 earthquake (southeastern part of map). The most severe shaking was on the muddy bay shore (bright orange) where there are natural deposits of mud and the bay has been filled in to create new land. (b) Shaking caused the collapse of the Nimitz Freeway (I-880) in Oakland, which was built above bay fill and mud deposits. *(Tom Van Dyke/MCT/Newscom)* (c) Damage to buildings in the Marina district of San Francisco resulting from poorly supported first floor that has shifted due to intense shaking caused by material amplification. *(John K. Nakata/USGS)* (d) Completed in 2013 the eastern span of the Bay Bridge stands in front of the old bridge deemed not earthquake safe following the 1989 earthquake. *(Getty)*

tragic example of material amplification. This quake further demonstrated that buildings constructed on materials that are likely to accentuate and increase seismic shaking are extremely vulnerable to earthquakes, even if the event is centered several hundred kilometers away. Much of Mexico City is built on mud deposits from ancient Lake Texcoco. When seismic waves struck the unconsolidated mud, the amplitude of surface shaking appears to have increased by a factor of 4 or 5. More than 500 buildings fell down from the intense shaking.[17] In many buildings, the amplified shaking collapsed upper floors onto lower ones like a stack of pancakes.[18] Local geologic structures can also influence the amount of shaking. For example, troughlike geologic structures, such as synclines and fault-bounded sedimentary basins, can focus seismic waves the way a magnifying lens focuses sunlight. This causes severe shaking in some areas and less intense shaking in others.

3.2 CASE study

Earthquake Catastrophes: Lessons Learned

Catastrophic earthquakes are devastating events that can destroy large cities and take thousands of lives in a matter of seconds. A sixteenth-century earthquake in China reportedly claimed 850,000 lives. More recently, a 1923 earthquake near Tokyo killed 143,000 people, and a 1976 earthquake in China killed several hundred thousand. In 2005 a catastrophic earthquake struck northern Pakistan. Although the epicenter was in Pakistan, extensive damage also occurred in Kashmir and India (Figure 3.20a). More than 80,000 people were killed, and over 30,000 buildings collapsed. A catastrophic continental earthquake occurred in China's

(a)

(b)

(c)

(d)

∧ **FIGURE 3.20 Earthquake Damage**

(a) Survivors rescue an injured neighbor from a collapsed building in Balakot, Pakistan, in 2005. *(FAROOQ NAEEM/AFP/Getty Images)* (b) Collapsed buildings from 2008 earthquake in China that killed about 87,500 people. *(AP Photo/Greg Baker)* (c) Collapsed apartment building in foreground due to an earthquake in Italy, in 2009. *(Orlando/ZUMA Press/Newscom)* (d) The brown area in the center of this image is the landslide that buried more than 500 homes in the Las Colinas neighborhood of Santa Tecla, a western suburb of San Salvador, the capital of El Salvador. The landslide was triggered by a January 2001 earthquake. *(USGS)*

Sichuan Province in 2008, killing about 87,500 people. Some villages were completely buried by landslides, and more than 5 million buildings collapsed (Figure 3.20b).[19,20,21] On January 12, 2010, an earthquake struck Haiti and killed about 240,000 people (see Figure 1.1).[22] The recent earthquakes that took thousands of lives have a common dominator: They all caused tremendous loss of life because the buildings people were in collapsed. Homes, schools, hospitals, and industrial buildings were subject to collapse in strong shaking because they were not constructed with earthquakes in mind, or the required construction codes were ignored, presumably to save money. In short, most deaths were human induced. We have experience with strong shaking in California (Northridge in 1994), and more recently in Chile (February 27, 2010), where an earthquake about 500 times as strong as the Haiti earthquake only a month earlier killed not hundreds of thousands but hundreds. An earthquake several times smaller than the Northridge quake struck the town of L'Aquila in Italy in 2009, and many of the buildings there collapsed, killing about 300 people (Figure 3.20c). A similar earthquake in California probably would have caused very few deaths, if any. The Italian event was a catastrophe because the buildings were not constructed to sustain moderate seismic shaking.

Building regulations and zoning played an important role in determining the extent of damage from these earthquakes, as they did in the 2011 Tohoku earthquake (see Applying the Fundamental Concepts). Other factors, such as depth of the earthquake rupture and the nature of the soil and rock at and near the surface, were also important in explaining the pattern of damage. However, the dominant factor linked to the number of deaths was the degree to which buildings could withstand the shaking. Criminal complaints have been lodged against some construction companies when experts conclude that poor building materials were used or building codes were ignored. The pattern of damage following some earthquakes supports these complaints—many older homes were shaken, but not damaged, whereas newer homes were reduced to rubble.

Finally, a catastrophic 2001 earthquake in El Salvador illustrates that the number of casualties can be linked to where, not just how, structures were built. The Las Colinas neighborhood in Santa Tecla suffered the greatest loss of life from quake-related landslides, mainly from one giant slide (Figure 3.20d). The extensive damage appears to have resulted from building homes below a slope of loose earth material and from poor land-use practices. Balsamo Ridge, the hill above the community, is composed of unstable volcanic ash and other granular materials that are susceptible to landslides. Local authorities and environmentalists also claim that deforestation and greed were directly responsible for intensifying the disaster. Residents of Las Colinas had pleaded with the government to block

construction of mansions on the hillside above them. The residents correctly argued that fewer plants covering the bare ground would leave them vulnerable to landslides. Their pleas were ignored, the mansions went up, and during the earthquake, the slopes came down.

Although we cannot control the geologic environment or the depth of an earthquake, there is much we can do to avoid excessive damage and loss of life. The deaths in Haiti, Italy, El Salvador, and China are especially tragic because much of the devastation could have been avoided if building regulations had been enforced, and protecting lives had been more important than building cheaper structures or making more money. Sound planning techniques and earthquake engineering of buildings can prevent unnecessary damage and loss of life. For examples, the 1989 Loma Prieta earthquake near San Francisco killed about 63 people, but a similar earthquake in Pakistan, India, or Mexico could kill at least several tens of thousands of people. The difference is that homes in California are constructed to better withstand earthquake shaking.[21] The lesson here is that areas with many buildings properly designed and built to withstand shaking, as in Chile and California, result in far fewer deaths from collapsed buildings than areas where construction is substandard.[23]

3.3 **CHECK** your understanding

1. What does moment magnitude measure? How is it different from a Richter magnitude?
2. How do seismologists locate earthquakes?
3. How does the depth of an earthquake's focus relate to shaking and damage?
4. What types of earth materials amplify seismic waves?
5. How is material amplification related to earthquake damage?
6. What is the primary cause of an earthquake catastrophe due to a moderate size earthquake?

3.4 Geographic Regions at Risk from Earthquakes

Earthquakes are not randomly distributed. Most occur in well-defined zones along the boundaries of Earth's tectonic plates (see Figure 3.2). In the United States, the areas with the highest earthquake risk include the Pacific coastal areas of California, Oregon, Washington,

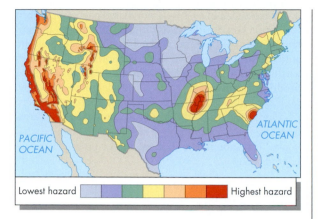

^ FIGURE 3.21 Seismic Hazard in Contiguous United States
A probabilistic approach to the seismic hazard in the contiguous United States. Colors indicate level of hazard. *(From U.S. Geological Survey Fact Sheet FS-131–02, 2002)*

Alaska, and Hawaii; an area along the California–Nevada border; and the territories of Puerto Rico and the Virgin Islands (Figure 3.21). For Canada, the risk is greatest in British Columbia and the Yukon Territory (Figure 3.22). This distribution is not surprising, since, with the exception of Hawaii, these areas are on or very close to plate boundaries. What may be more surprising are the high-risk zones in South Carolina, the central Mississippi River Valley, and the St. Lawrence River Valley. These areas are not close to present-day plate boundaries, but they have experienced large *intraplate earthquakes*.

PLATE BOUNDARY EARTHQUAKES

Earthquakes occur along all three types of plate boundaries—convergent, divergent, and transform. The world's great earthquakes with magnitude greater than **M** 9 are usually associated with subduction zones (see Applying the Fundamental Concepts). In the past 100 years, earthquakes greater than **M** 9 have been associated with subduction along what is called the "megathrust" that helps define the zone. These events are called **megathrust earthquakes**. The great Chile earthquake of 1960 was a megathrust event of **M** 9.5 that killed about 1,650 people. The earthquake occurred before the concept of subduction and the theory of plate tectonics had been proposed. The 2010 **M** 8.8 Chile earthquake, while smaller than the 1960 event, was most likely on the megathrust fault system that defines the down-going slab in the subduction zone offshore. According to the U.S. Geological Survey, the 2010 earthquake was generated at the gently sloping fault that separates the down-going Nazca plate to the east from the overriding South American plate to the west (Figure 3.23). The regional rate of convergence between the two plates is about 7 cm

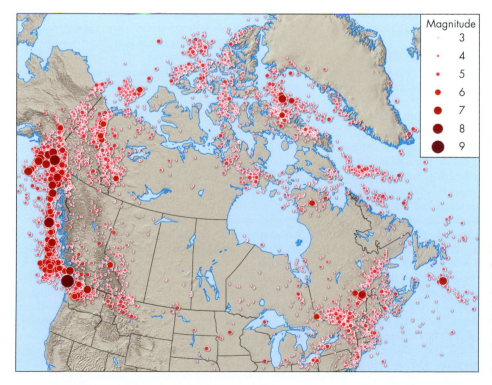

**< FIGURE 3.22
Historic Earthquakes in or near Canada**
Map of earthquake epicenters greater than **M** 3 between 1627 and 2012. *(Reproduced with the permission of Natural Resources Canada 2013, courtesy of Earthquakes Canada)*

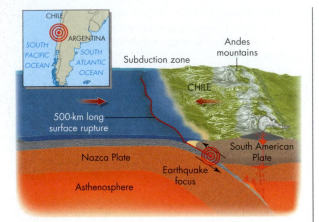

⌃ FIGURE 3.23 2010 Chile Megathrust Earthquake
The largest recorded earthquakes are known to happen along subduction zones, like the (M 9.5) 1960 and (M 8.8) 2010 Chile quakes, when the interface between the overriding plate (South American) and subducting plate (Nazca) ruptures. *(Based on Lutgens, F., and Tarbuck, E.,* Essentials of Geology. *New York: Macmillan, 1992; From the CIA World Factbook)*

(~2.7 in.) per year. The mostly offshore rupture exceeded 500 km (~310 mi.) parallel to the coast (along the subduction zone). The rupture began deep (35 km or 22 mi.) beneath the coast and spread to the west, north, and south. As the rupture propagated, the fault slip generated intense earthquake shaking. This earthquake killed about 500 people. Fault slip warped the ocean floor, generating a damaging local tsunami in Chile's coastal areas (see Case Study 3.3: Survivor Story).[24]

In the western United States, earthquakes are common along the transform San Andreas fault zone and the convergent Cascadia and Aleutian subduction zones. Whereas California and Alaska are famous for their earthquakes, Nevada, Utah, Idaho, Montana, Wyoming, Oregon, and Washington also experience quakes.

As you recall from Chapter 2, California straddles two lithospheric plates: the Pacific plate west of the San Andreas fault zone, and the North American plate to the east (see Figure 2.1). The motions of these plates in essentially opposite directions along the San Andreas and related faults result in frequent damaging earthquakes. The 1989 Loma Prieta earthquake (**M** 6.9) on the San Andreas fault system south of San Francisco caused 63 deaths, 3,757 injuries, and an estimated $5.6 billion in property damage (see Figure 3.19). Both deaths and injuries would have been significantly greater if many people had not stayed home to avoid the crowds and congestion of the third game of the World Series in Oakland. Neither the Loma Prieta earthquake nor the Northridge earthquake (**M** 6.7) was considered a great earthquake. A great earthquake (**M** 8 and higher)

occurring today in a densely populated part of Southern California could inflict $100 billion in damage and kill several thousand people.[25] Thus, neither quake was the anticipated "big one" in California. Both plate boundary earthquakes were initially puzzling because they did not produce a surface rupture. They demonstrated how much we still need to learn about earthquake processes.

This story about the Loma Prieta quake illustrates how observing a sequence of events related to an earthquake does not mean that we understand the causes behind them:

> Shortly before the quake struck, a two-year-old boy was out playing in his yard and discovered how to turn on the lawn sprinklers. His mother was not amused and took him to his room for disobeying her instructions. Shortly after she had returned to the yard, the Earth began to tremble violently. She heard her child yell and ran back into the house to find him terrified. He then remained very quiet until his father came home from work. On seeing his father, the child's first words were an emotional "Daddy, don't turn on the sprinkler!"[26]

The more we know about the probable location, magnitude, and effects of an earthquake, the better we can estimate the damage that is likely to occur and make the necessary plans for minimizing loss of life and property.

INTRAPLATE EARTHQUAKES

Although much less common than plate boundary earthquakes, intraplate earthquakes can be large and extremely damaging. Because they occur less often, there is generally a lack of preparedness, and buildings may not be able to withstand strong shaking. At least two earthquakes greater than **M** 7.5 occurred in the winter of 1811–12 in the central Mississippi Valley (Figure 3.24). The quakes nearly destroyed the town of New Madrid, Missouri, and killed an unknown number of people. They were felt in nearly every city of eastern North America from New Orleans to Quebec City in Canada, an area of more than a million square miles.[27] Seismic waves from these quakes rang church bells in Boston, more than 1600 km (~1000 mi.) away! Associated ground motion produced intense surface deformation over a wide area from Memphis, Tennessee, 230 km (~140 mi.) north of the earthquakes, to the confluence of the Mississippi and Ohio Rivers. As a result, forests were flattened, fractures in the ground opened so wide that people had to cut down trees to cross them, and land sank several meters, causing local flooding. Journal and newspaper reports also indicate that local uplift of the land surface actually caused the Mississippi River to reverse its flow for a short time.[28]

Magnitude 8.8 Earthquake and Tsunami on the Coast of Chile

Barry Keller (Figure 3.3.A) is a consulting geologist, hydrologist, and geophysicist who earned his Ph.D. at the University of California at Santa Barbara. He contributes to professional projects encompassing everything from industrial to governmental applications. He has more than 25 years of academic and professional experience in geological reconnaissance, earthquake seismology, geophysical exploration, soil and groundwater investigation and remediation, storm water management, dredge material study design, and hazardous waste planning. Dr. Keller has a home in Santa Barbara and

another in Pichilemu, a beach town in central Chile known as a surfer's paradise. It is a rare opportunity for a professional who knows the science of earthquakes to write from personal experience. The following is Dr. Keller's story.

Strong earthquakes are a recurring hazard in some parts of the world, including the rim of the Pacific Ocean. Surviving a strong earthquake entails two basic phases: The first is surviving the shaking itself, and the second is surviving the following days when basic infrastructure such as roads, water, electricity, and telephones may be out of service.

First Phase: The most important way to prepare to survive strong

shaking is to build strong buildings, as well as well-designed roads, water storage facilities, and other infrastructure. The time of day that an earthquake occurs and where people happen to be are also important factors affecting the number of fatalities and injuries.

Second Phase: Storage of emergency supplies such as food, water, candles, flashlights with batteries, and generators with fuel are very helpful in the days following an earthquake. To this might be added cash, since our ubiquitous sources of money, ATMs, do not work when the power is out.

My wife, Rosemarie, and I survived the **M** 8.8 earthquake and accompanying tsunami on

◄ FIGURE 3.3.A Survivors of the 2010 Chilean Earthquake
Barry Keller and his wife Rosie in front of the Golden Gate Bridge in San Francisco near one of Dr. Keller's field study sites. *(Edward A. Keller)*

the coast of central Chile in February 2010. The west coast of South America is above the subduction zone in which various oceanic plates are consumed beneath the South American continental plate, so very large earthquakes are a recurring fact of life. This particular part of the subduction zone had not ruptured in more than 100 years and so was identified as a "seismic gap" likely to have a "BIG ONE."

As a geophysicist who studies such risks for environmental reviews, I was well aware of the seismic gap and the potential for large earthquakes and tsunamis when we built our house in Pichilemu (Figure 3.3.B), a beach resort and surfing destination on the coast at 34.5° south latitude, which is similar to the north latitude of central California. Accordingly, the house was designed for seismic safety, built of poured concrete and concrete block with lots of rebar. It is on a hill at about 120 m (~400 ft.) elevation, well above the elevations typically reached by tsunamis.

In the dark of that very early Saturday morning, I was already awake, having gone to the bathroom and checked my watch at 3:22 A.M. The quake occurred at 3:34 A.M. The motion may have built up for a few seconds, then was VERY strong for about a minute—I was quickly thinking "This is for real!" I tried to check my watch again (as my seismology professor, Charlie Richter, taught us!), but the lights went out. Pretty strong motion was felt for about two to three more minutes. When the quake "went off," I started to regain my belief that my life expectancy was more than a few seconds and that my roof was NOT about to collapse; it was quite interesting to me as a geologist who studies such risks. However, it was quite terrifying to most other people, including my wife.

The feeling of the strong motion was similar to flying in a plane through severe turbulence but with more high-frequency jolts. I sat on the corner of the bed, holding on. The good news is that the house came through with no structural damage. However, items on shelves were strewn all over and there was lots of broken glass on the floor, so it is good advice to put on shoes before walking around in the dark in such a situation.

Within minutes, lots of people came up the hill to the roundabout at the bottom of our driveway. Many more were on the next hill over, the official evacuation spot. The lessons of the Sumatra tsunami were taken to heart by the Chilenos, and most people rapidly headed for high ground. It was a calm night with a full moon, which made evacuation easier, and people stayed until dawn. People continued to camp on both hills for several days.

That night, information was minimal; one local radio station was broadcasting, but without outside news, and an AM station from Argentina was on the air, again with minimal news. It was Sunday before we knew the extent of the damage in the rest of the country.

The tsunami struck the coast within 20 minutes of the earthquake. Damage at Pichilemu included complete destruction of a surf school (*Escuela de Surf*), restaurant, and cabañas at Punta de Lobos, a famous surf spot. In the lower downtown part of Pichilemu, there was widespread flooding, but the tsunami did not hit with enough force to destroy buildings. Only one life was lost, a child camped on the beach, so the evacuation was very effective, especially considering that this was the final weekend of

◄ **FIGURE 3.3.B The Keller Home in Pichilemu, Chile**
The home was constructed with poured concrete and rebar to strengthen the structure in case of an earthquake, and it is situated at 120 m (~400 ft.) elevation, well above the height of an earthquake-generated tsunami. (*Edward A. Keller*)

summer vacation and there were lots of people at the coast.

Farther south along the coast there was much more destruction from the tsunami, including severe damage to a Chilean Navy base at Talcahuano harbor, near Concepción.

We had no electricity for two days (until we left for Santiago—it stayed out three more days). My house has a booster pump for water, so without power we had no water pressure—although we had plenty of water in a storage tank (I have a boat bilge pump to get the water out, but it is very slow) and I had lots of bottles of water stashed, because the power goes out often. We had cell phones the first day, but then (according to the local scuttlebutt) the gasoline for the emergency generator ran out, so the phones went out.

Aftershocks were generally pretty sharp, with almost continuous motion for many hours. Daily aftershocks continued for weeks, including an **M** 6.9 very local one that cracked some plaster at our house. The aftershocks were nerve-wracking, especially for the nongeologists, and caused a great deal of mental stress for the local populace, if not much physical damage.

Most modern structures fared well in the quake, minimizing loss of life. Many older adobe structures were damaged, including historic churches, government buildings, a museum, and the funeral parlor. However, at 3:30 A.M. on a Saturday morning, most such structures were not occupied, again minimizing loss of life. The timing was also fortuitous for major roads, which suffered cracking and fallen bridges but, with little traffic at that hour, no loss of life.

—EDWARD A. KELLER

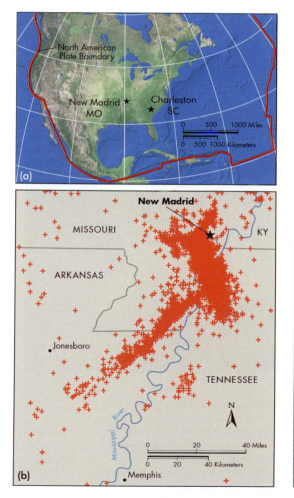

◀ **FIGURE 3.24 Intraplate Seismic Zone**
(a) The cities of New Madrid, Missouri, and Charleston, South Carolina, were the sites of several large earthquakes even though the cities are located thousands of kilometers from the nearest North American plate boundary (red lines). (b) The New Madrid seismic zone, highlighted here by earthquake epicenter locations since 1974 (red crosses), is the most earthquake prone region in the United States east of the Rocky Mountains. *(U.S. Geological Survey)*

These earthquakes occurred along the New Madrid seismic zone, a seismically active portion of a geologic structure known as the Mississippi Embayment (Figure 3.24b). The embayment is a downwarped area of Earth's crust where the lithosphere is relatively thin. At least two hypotheses exist to explain this: (1) Thinning took place near the end of the Proterozoic eon, around 600 million years ago, when a divergent plate boundary developed in the southeastern United States; and (2) the thinning is related to the passage of the North American Plate over a mantle hot spot during the Cretaceous period, around 95 million years ago. In either case, the observations of landforms and rates of uplift in the Mississippi Valley indicate that the seismic activity is recent, perhaps less than 10,000 years ago.

The *recurrence interval*, or time between events, for large earthquakes in the Mississippi Embayment is estimated to be several hundred years.[28,29] With material amplification, the New Madrid seismic zone appears to be capable of producing intensities commonly associated with great earthquakes in California. Thus, the interior of the North American Plate is far from "stable." In recognition of this, the Federal Emergency Management Agency (FEMA) and affected states and municipalities have adopted a new building code designed to mitigate major earthquake hazards.

Another large, damaging intraplate earthquake (**M** ~7.3) occurred on the night of August 31, 1886, near Charleston, South Carolina (Figure 3.24a). This earthquake killed about 60 people and damaged or destroyed most buildings in Charleston. More than 102 buildings were completely destroyed and nearly 14,000 chimneys fell, many because of poor construction following a great fire in 1838. The quake was felt from Canada to Cuba and as far west as Arkansas.[27] Effects of the earthquake were reported at distances exceeding 1000 km (~620 mi.) from the epicenter.

Intraplate earthquakes in the rest of the United States are generally more damaging and felt over a much larger area than similar magnitude earthquakes in California (Figure 3.25). Because the rocks in the eastern United States are generally stronger and less fractured, they can more efficiently transmit earthquake waves. Empirical data from the 2011 **M** 5.8 Central Virginia earthquake suggest that more than 50 million people in the eastern United States (about one-third of the country) felt the quake. The Central Virginia earthquake also triggered a landslide 245 km (~150 mi.) from the epicenter, which is four times farther away than previous studies of worldwide earthquakes of the same size. Based on what they have learned from this event, earthquake scientists at the United States Geological Survey suggest equations used to calculate ground shaking may need to be modified.[30]

3.4 CHECK your understanding

1. How do plate boundary and intraplate earthquakes differ?

2. Why are earthquakes on the East Coast felt by more people than on the West Coast?

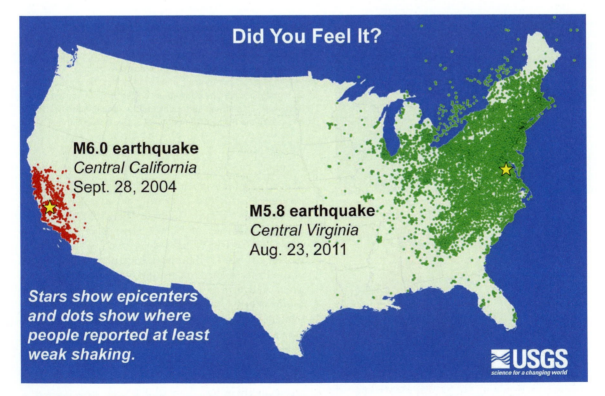

∧ FIGURE 3.25 Plate Boundary vs. Intraplate Earthquakes
U.S. Geological Survey Community Internet Intensity Map for two recent earthquakes (epicenters shown with yellow stars). Different color dots show the locations of people who submitted online responses to the "Did You Feel It?" website for either the Central Valley plate boundary earthquake (red dots) or the Central Virginia intraplate earthquake (green dots). Although the two quakes were similar in magnitude, the intraplate earthquake was felt over a much larger area than was the plate boundary quake. *(From "New Evidence Shows Power of East Coast Earthquakes: Virginia Earthquake Triggered Landslides at Great Distances," USGS 2012, www.usgs. gov/newsroom/article.asp?id=3447#.Ui9caRZyHdk)*

3.5 Effects of Earthquakes and Linkages with Other Natural Hazards

Shaking is not the only cause of death and damage during earthquakes. Many earthquakes cause other hazards and are thus an excellent example of how natural hazards are often linked. The primary effects of an earthquake are those caused directly by fault movement, such as ground shaking and its effects on people and structures, and surface rupture. Secondary effects are those that subsequently result from the faulting and shaking. These include liquefaction of the ground, regional changes in land elevation, landslides, fire, tsunamis, and disease. We will discuss the effects of tsunamis, large ocean waves that are often generated by earthquakes, in Chapter 4.

SHAKING AND GROUND RUPTURE

The immediate effects of a catastrophic earthquake can include violent ground shaking accompanied by widespread surface rupture and displacement of Earth's surface. Although most surface cracks produced by earthquakes are the result of liquefaction and landslides (discussed later), a major surface rupture may occur along the fault that was the source for the earthquake. This rupture commonly creates a low cliff, called a **fault scarp**, which may extend for kilometers along the fault (Figure 3.26). It is clear from the figure why you would not want to build your house directly along the trace of where the fault intersects Earth's surface. However, in many cases it is unavoidable, and some

human-made structures must be built across known faults. In these cases, precautions must be taken to minimize the effects of ground rupture (see Case Study 3.4: Denali Fault Earthquake).

The 1906 San Francisco earthquake (**M** 7.8) produced 6.5 m (~21 ft.) of horizontal displacement along the San Andreas fault north of San Francisco. Studies following the quake indicate that portions of the bay area reached a maximum Modified Mercalli intensity of XI.[5] At this intensity, rapid surface movements can snap and uproot large trees and knock people to the ground. This level of shaking may damage or collapse large buildings, bridges, dams, tunnels, pipelines, and other rigid structures.[31]

The great 1964 Prince William Sound, Alaska, earthquake (**M** 9.2) caused extensive damage to transportation systems, railroads, airports, and buildings. Both the 1989 Loma Prieta (**M** 6.9) and 1994 Northridge earthquakes (**M** 6.7) were much smaller than the great Prince William Sound earthquake, yet the damage they caused was far more costly. In inflation-adjusted dollars, the Loma Prieta quake was five times as costly as the Alaska quake and the Northridge quake was 20 times as costly. With losses of at least $20 billion, the Northridge earthquake was one of the most expensive hazardous events in U.S. history.[17]

The Northridge earthquake caused so much damage because there were so many human-made structures there to be damaged. Most of the destruction was in the San Fernando Valley northwest of Los Angeles (see Figure 3.15), a highly urbanized area with a high population density that experienced intense shaking.

Violent ground shaking can be especially damaging to buildings if the horizontal motion is particularly strong or if the frequency of the shaking matches the

◄ FIGURE 3.26 Fault Scarp
This small cliff, called a fault scarp, was produced by the rupturing of Earth's crust during the 1992 **M** 7.3 Landers earthquake in California. The rupture could be traced for a distance of 70 km (~43 mi.) across the Mojave Desert. *(NOAA)*

natural vibrational frequency of the building. Shaking is commonly measured as *ground acceleration* and is compared to the overall acceleration of gravity. The matching of vibrational frequencies is called **resonance** and can affect buildings a significant distance away from the epicenter. One- and two-story buildings typically have high vibrational frequencies, in contrast to tall buildings that have low vibrational frequencies. In general, high shaking frequencies damage low buildings and low shaking frequencies damage tall buildings.

Ground shaking can also be a significant hazard to dams and levees (see Applying the Fundamental Concepts). Partial failure of the Van Norman Dam in the 1971 **M** 6.7 Sylmar earthquake came perilously close to flooding more than 80,000 people downstream of the dam.[17] Parts of the dam slid into the reservoir, which fortunately at the time was only about half full of water. People living downstream were evacuated, and the water level was lowered in the reservoir as a precaution. Likewise, in northern California, ground shaking from an earthquake in the San Francisco Bay area could produce intense shaking in the Sacramento River Delta. This shaking could cause the failure of levees along the river, resulting in widespread flooding and potential disruption of the water supply to Southern California.

(a)

(b)

⋀ FIGURE 3.27 Trans-Alaska Oil Pipeline Survives M 7.9 Earthquake
(a) In its construction, the Alaskan oil pipeline was designed to withstand several meters of horizontal displacement where it crossed the Denali fault south of Fairbanks. That design was put to the test in 2002 when it shifted horizontally 4.3 m (~14 ft.) during the **M** 7.9 Denali fault earthquake. (b) Built-in bends of the pipeline on slider beams with Teflon shoes accommodated the earthquake rupture as designed. This was strong confirmation of the importance of estimating potential ground rupture and taking action to mitigate it. *(U.S. Geological Survey/U.S. Department of the Interior)*

3.4
CASE study

The Denali Fault Earthquake: Estimating Potential Ground Rupture Pays Off

On November 3, 2002, a magnitude 7.9 earthquake occurred in early afternoon on the Denali fault in a remote section of south-central Alaska (Figure 3.27a). Historical movement along the fault was primarily horizontal (right-lateral strike-slip), and the 2002 earthquake was no exception. Along the fault zone, the quake produced a surface rupture approximately 340 km (~210 mi.) in length and a horizontal displacement of up to 8 m (~26 ft.). Although the quake caused thousands of landslides, intense shaking, and many areas of liquefaction, there was little structural damage and no deaths—primarily because few people live in the affected area.[32]

The Denali fault earthquake demonstrated the value of seismic hazard assessments for geologic faults. Geologists studied the fault in the early 1970s as part of planning for the Trans-Alaskan pipeline. Where the pipeline crosses the Denali fault, the investigators determined that the fault zone is several hundred meters

wide and could experience a 6 m (~20 ft.) horizontal displacement in a **M** 8 earthquake.

These estimates were used in the engineering design of the pipeline across the fault zone. In anticipation of a future earthquake, the pipeline was elevated on long horizontal steel beams with Teflon shoes that allowed it to slide horizontally up to 6 m (~20 ft.). These slider beams were installed along a zigzag path for the pipeline for added flexibility and horizontal movement during an earthquake (Figure 3.27b). As anticipated, the 2002 earthquake occurred within the mapped fault zone and caused the pipeline to shift horizontally about 4.3 m (~14 ft.). Because of advanced engineering design, the pipeline suffered little damage and there was no oil spill from the quake.

Although the $3 million cost for the design and construction of the pipeline across the fault may have seemed excessive in 1970, in retrospect it was money well spent. Today the pipeline delivers approximately 11 percent of the domestic oil supply for the United States and transports more than half a million barrels of oil each day. A pipeline break caused by an earthquake would have a significant and costly impact on our oil supply. Current estimates are that the repair of the pipeline and cleanup of the oil spill alone would cost several million dollars.[33]

LIQUEFACTION

During earthquakes, intense shaking can cause a near-surface layer of water-saturated sand to change rapidly from a solid to a liquid. This effect, called **liquefaction**, is most common in **M** 5.5 and greater earthquakes when there is strong shaking in Holocene sediment, that is, deposits less than about 10,000 years old.[31] Layers of solid sand can liquefy into a soupy mixture of sand and water as seismic waves pass through the earth material. Vibrations from the passing seismic waves increase water pressure in the space between sand grains. When the water pressure exceeds the downward-directed pressure from the overlying material, the sand grains and other objects buried in the sand effectively "float," like pieces of vegetables suspended in a thick stew. This creates a sand slurry that flows upward, often squirting through cracks to the surface. Once the shaking slows and the water pressure decreases, the liquefied sediment re-compacts and becomes solid again. A similar change from solid to liquid and back to solid again can also occur in "quicksand" on a beach or river sand bar. Liquefaction in earthquakes commonly causes the land surface to shift or subside.

The liquefaction of sand layers about 10 m (~30 ft.) below the surface has caused four-story apartment buildings to tip over, highway bridges to collapse, and dams to fail (Figure 3.28). It has also brought empty underground tanks, pipelines, and bridge pilings floating to the surface (see Applying the Fundamental Concepts).[34] Liquefaction can cause slope failure and ground subsidence over large areas. One amazing indicator of liquefaction are "sand volcanoes," small mounds of sand formed at the surface where water from liquefied layers squirts out of the ground.

REGIONAL CHANGES IN LAND ELEVATION

Vertical deformation of the land surface is a hazard linked to some large earthquakes. This deformation includes both regional uplift and subsidence of Earth's

◄ **FIGURE 3.28 Building Collapse from Liquefaction** These apartment buildings collapsed as a result of liquefaction of sandy sediment beneath Niigata, Japan, in 1964. The **M** 7.5 earthquake that caused the liquefaction killed 26 people. *(NOAA)*

surface. These changes in elevation can cause substantial damage in coastal areas and along streams and can raise or lower the groundwater table.

The great 1964 Prince William Sound earthquake (**M** 9.2) provided a dramatic example of regional deformation in Alaska. This quake caused vertical deformation over an area of more than 250,000 km^2 (~97,000 mi.^{2}), a land surface slightly larger than the state of Vermont.[35] Within this area, uplift of as much as 10 m (~30 ft.) exposed and killed coastal marine life, lifted docks out of the water, and shifted the shoreline away from fish canneries and neighboring homes. In other areas, subsidence of as much as 2.4 m (~8 ft.) resulted in the partial flooding of several communities. Both the uplift and subsidence produced changes in the groundwater table.

LANDSLIDES

Two of the most closely linked natural hazards are earthquakes and landslides. Earthquakes are one of the most common triggers for landslides in hilly and mountainous areas. Landslides triggered by earthquakes can be extremely destructive and cause great loss of life. For example, a giant landslide triggered by the 1970 Peru earthquake buried the cities of Yungay and Ranrahirca. An estimated 20,000 of the 70,000 people who died in the quake were killed by the landslide. Both the 1964 Alaska earthquake and the 1989 Loma Prieta earthquake caused extensive landslide damage to buildings, roads, and other structures. The 1994 Northridge earthquake and associated aftershocks triggered thousands of landslides.

A giant landslide from the side of a mountain was triggered by the 2015 Gokha earthquake (Figure 3.29).

∧ FIGURE 3.29 Earthquake-Triggered Landslides
Before (above) and after (below) images of the village of Langtang, which was completely buried by a massive landslide that was triggered by the 2015 Gorkha earthquake (see Case Study 3.1). *(David Breashears/GlacierWorks)*

Thousands of other landslides were also triggered by the earthquake, most of which were on the steep slopes of the Himalayan Mountains.

The 2008 **M** 7.9 earthquake in China caused many landslides. Large landslides buried villages and blocked rivers, forming "earthquake lakes" that produced a serious flood hazard, should the landslide dam fail by overtopping or erosion. Channels were excavated to slowly drain lakes and reduce the flood hazard.[36]

As mentioned in Case Study 3.2, a large landslide associated with the January 2001 El Salvador earthquake (**M** 7.7) buried the community of Las Colinas, killing hundreds of people (see Figure 3.20d). Tragically, the landslide might have been avoided if the slope that failed had not been cleared of vegetation to construct luxury homes.

FIRES

Fire is another major hazard linked to earthquakes. Shaking of the ground and surface displacements can break electrical power and natural gas lines, thus starting fires. The threat from fire is even greater because firefighting equipment may be damaged; streets, roads, and bridges may be blocked; and essential water mains broken. In individual homes and other buildings, appliances such as gas water heaters may be knocked over, causing gas leaks that are ignited. Earthquakes in both Japan and the United States have been accompanied by devastating fires (Figure 3.30). As discussed in the introduction to Chapter 2, the San Francisco earthquake of 1906 has repeatedly been referred to as the "San Francisco Fire" because 80 percent of the damage was caused by firestorms that ravaged the city for several days. The 1989 Loma Prieta earthquake also caused large fires in the Marina district of San Francisco.

Most fires after an earthquake are not wildfires because they start in urban areas where gas and electrical lines are located. However, in a warm, dry area such as Southern California, it is possible for an urban fire to spread to wild lands, causing a wildfire.

DISEASE

Outbreaks of diseases are sometimes associated with large earthquakes. They may be caused by a loss of sanitation and housing, contaminated water supplies, disruption of public health services, and the disturbance of the natural environment. Earthquakes also rupture sewer and water lines, causing water to become polluted by disease-causing organisms.

An interesting example of how an earthquake-related disturbance can result in a disease outbreak occurred in Southern California. Desert soils in the southwestern United States and northwestern Mexico contain spores of a fungus that causes a respiratory illness known as *valley fever*. Landslides from the 1994 Northridge earthquake raised large volumes of desert dust containing these fungal spores. Winds carried the dust and spores to urban areas such as Simi Valley, where an outbreak of valley fever occurred. More than 200 cases of the disease were diagnosed within eight weeks after the earthquake, 16 times the normal number of cases. Fifty of these people were hospitalized and three died.[37]

◄ **FIGURE 3.30**
Earthquake and Fire
Fires associated with the 1995 **M** 6.9 Kobe, Japan, earthquake caused extensive damage to the city. *(The Asahi Shimbun via Getty Images)*

3.6 Natural Service Functions of Earthquakes

When we are confronted with the death and destruction caused by large earthquakes, it is difficult to imagine that there could be any benefits from such disasters. However, earthquakes, like many other natural hazards, do provide natural service functions. They contribute to the development of groundwater and energy resources, the formation and exposure of valuable mineral resources, and the development of landforms.

GROUNDWATER AND ENERGY RESOURCES

Geologic faults produced by earthquakes strongly influence the underground flow of water, oil, and natural gas (see Figure 1.15). Fault zones with coarsely broken earth materials can act as preferential paths for fluid movement. These faults may serve as conduits for the downward flow of surface water or as zones through which groundwater comes back to the surface as springs. Barton Springs in Austin, Texas, is one of many major springs that discharge from fault zones.

In other settings, faulting creates natural underground dams to slow or redirect the flow of water, oil, or natural gas. These dams develop where faulting pulverizes rock to form an impervious clay barrier and where faulting moves impervious earth materials alongside earth materials containing water, oil, or natural gas. Such subsurface dams are responsible for oases in arid areas of Southern California and for many underground accumulations of oil and gas in Texas, Oklahoma, Alberta, and elsewhere.

MINERAL RESOURCES

Faulting related to earthquakes may be responsible for the accumulation or exposure of economically valuable minerals. Mineral deposits commonly develop along cracks associated with faulting. These mineral-filled cracks, called *veins*, can be the source of precious metals such as gold, silver, and platinum (see Figure 3.3). Veins associated with large fault zones may produce enough minerals to be economically viable for extraction.

LANDFORM DEVELOPMENT

Earthquakes can form scenic landforms over long intervals of geologic time. Uplifts of earth materials along geologic faults can produce hills, rugged mountain ranges, and coastal terraces. A spectacular example of the role that earthquakes have played in creating scenic landforms is found in Grand Teton National Park in northwestern Wyoming. In the past several million years, earthquakes along faults both inside and outside the park have allowed the Teton Range to rise and Jackson Hole to subside, creating a scenic mountain front.

The fault zones themselves may become part of landforms if they contain broken rock that is more easily eroded than earth materials in surrounding areas. Easily eroded fault zones commonly become valleys cut by streams and rivers. For example, the course of the Hudson River in southeastern New York appears to be controlled by geologic faulting in the underlying bedrock, and the Rio Grande Rift Valley in New Mexico is the result of normal faulting.

FUTURE EARTHQUAKE HAZARD REDUCTION

Frequent small earthquakes may help prevent larger ones in the same area. If we assume that small earthquakes release pent-up energy, then the reduction in elastic strain may lower the chance of a devastating earthquake along a particular fault. In fact, scientists try to identify areas along active faults that have not experienced earthquakes in a long time. As discussed later, these areas, called "seismic gaps," can have the greatest potential for producing large earthquakes in the future.[4]

3.7 Human Interaction with Earthquakes

As a natural hazard, an earthquake cannot be prevented or controlled; we can only devise ways to minimize the death and destruction it causes. That does not mean, however, that we cannot cause earthquakes ourselves. In fact, several human activities are known to increase or cause earthquake activity. Damage from these earthquakes is regrettable, but the lessons learned may help control or stop large catastrophic earthquakes in the future.

EARTHQUAKES CAUSED BY HUMAN ACTIVITY

The actions of people have caused earthquakes in three ways:[39]

> Loading the Earth's crust, as in building a dam and reservoir

> Injecting liquid waste deep into the ground through disposal wells

> Creating underground nuclear explosions.

Water Reservoirs Construction of a large, deep reservoir behind a dam on a river may create, or induce, earthquakes. The huge weight of the water in a new reservoir can create or extend fractures in adjacent rock and increase water pressure in the surrounding groundwater. These effects may produce new faults or lubricate and activate existing faults. In the United States, several hundred local tremors occurred in the decade following the completion of Hoover Dam and the filling of Lake Mead in Arizona and Nevada. Most of these quakes were very small, but one was **M** 5 and two were about **M** 4.[38] In India and China, reservoir filling has triggered earthquakes with a magnitude greater than 6.0. Two of these quakes killed hundreds of people and severely damaged the reservoir dams.[31]

Deep Waste Disposal In the early 1960s, an unplanned experiment by the U.S. Army provided the first direct evidence that injecting fluids into Earth can cause earthquakes. From April 1962 to November 1965, the Denver, Colorado, area experienced several hundred earthquakes, far more than normally occur over this length of time. The largest quake (**M** 4.3) caused

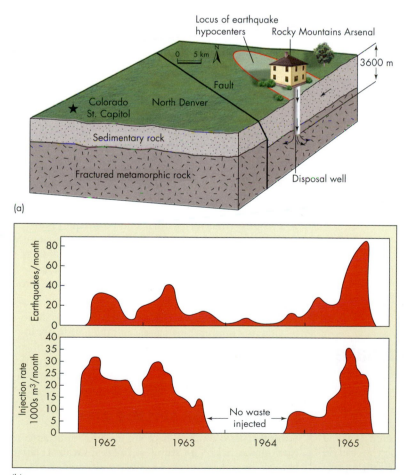

(a)

(b)

◄ FIGURE 3.31 Human-Caused Earthquakes in Colorado

(a) A red line encompasses the area where most of the earthquakes occurred and a natural fault is present about 10 km (~6.2 mi.) southwest of the disposal well. (b) Injection of waste stopped in 1966; the well was permanently sealed in 1986. Earthquake activity has settled back to the natural frequency. An earthquake of magnitude 4.2 occurred in 1982, suggesting the faulting remains active. Data from www.geosurvey.state. co.us. Rocky Mountain Arsenal—The most famous example of induced earthquakes. Accessed 12/5/10. *(Adapted from Evans, D. M., "Man-made Earthquakes in Denver," Geotimes 10: 11–18, 1966. Reprinted by permission of American Geosciences Institute.)*

sufficient shaking to knock bottles off shelves in stores. Geologist David Evans traced the source of the earthquakes to the Rocky Mountain Arsenal, a chemical weapons plant on the northeast side of Denver. Liquid waste from the plant was being injected under pressure down a deep disposal well 3600 m (~11,800 ft.) into Earth. The injected liquid increased underground fluid pressure and caused slippage of numerous fractures in metamorphic rock. Evans demonstrated a high correlation between the rate of waste injection and the location and timing of the earthquakes. When the injection of waste stopped, so did the earthquakes (Figure 3.31).[39] Deep well injection has also triggered earthquakes in Ohio and Texas.[40,41]

Recognizing that the injection of fluids into Earth could trigger an earthquake was an important development because it drew attention to the relationship between fluid pressure and earthquakes. Subsequent studies of subduction zones and active fold belts suggest that high fluid pressures are present in many areas where earthquakes occur. One hypothesis is that fluid pressure rises in an area until rocks break, triggering an earthquake and discharging fluid upward. After the event, fluid pressure begins to build up again, beginning another cycle.

Nuclear Explosions   Underground tests of nuclear weapons in Nevada have triggered numerous earthquakes with magnitudes as large as 6.3.[38] An analysis of the aftershocks suggests that these explosions may have released some natural strain within Earth. This has led to discussions by scientists as to whether nuclear explosions might be used to prevent large earthquakes by releasing strain before it reaches a critical point. Even if such techniques were feasible and legal liability and environmental issues were resolved, the explosive release of strain on one fault might trigger unanticipated earthquakes on other faults.

3.7 CHECK your understanding

1. How can humans cause earthquakes?

3.8 Minimizing the Earthquake Hazard

One important reason why earthquakes cause great damage and loss of life is that they often strike without warning. Because of this, a great deal of research is being devoted to anticipating earthquakes. The best we can do at present is to forecast the likelihood that an earthquake will occur in a particular area or on a particular segment of

a fault. These forecasts use probabilistic methods to determine the risk. A forecast will say that an earthquake of a given magnitude or intensity has a high probability of occurring in a given area or fault segment within a specified number of years. These forecasts assist planners who are considering seismic safety measures or people who are deciding where to live. However, long-term forecasts do not help residents of a seismically active area anticipate and prepare for a specific earthquake. An actual prediction specifying the time and place of an earthquake would be much more useful, but the ability to make such predictions has eluded us. Predicting earthquakes depends to a large extent on observation of natural phenomena or changes in Earth that precede an event.

THE NATIONAL EARTHQUAKE HAZARD REDUCTION PROGRAM

In the United States, the USGS as well as university and other scientists are developing a National Earthquake Hazard Reduction Program. The major goals of the program are listed here:[42]

> Develop *an understanding of the earthquake source.* This goal requires obtaining information about the physical properties and mechanical behavior of faults and developing quantitative models of the physics of the earthquake process.

> Determine *earthquake potential.* This goal requires detailed study of seismically active regions to determine their paleoseismicity, identify active faults, and determine rates of deformation (see Case Study 3.5: A Closer Look). This information is used to calculate probabilistic forecasts and to develop methods for intermediate- and short-term predictions of earthquakes.

> Predict *effects of earthquakes.* This goal requires obtaining information needed to predict ground rupture and shaking and an earthquake's effects on buildings and other structures. This information is used to evaluate losses associated with the earthquake hazard (see Case Study 3.6: Professional Profile).

> Apply *research results.* The program educates individuals, communities, states, and the nation about earthquake hazards. This goal requires better planning for earthquakes and ways to reduce loss of life and property.

ESTIMATION OF SEISMIC RISK

Several types of hazard maps are used to communicate earthquake risk. The simplest maps indicate relative hazard by showing the locations of epicenters of historic earthquakes of various magnitudes. More informative maps communicate the probability of a particular event or the amount of shaking likely to occur. For example, a seismic-shaking hazard map of California (Figure 3.32) shows areas in purple through red that have a chance of experiencing a damaging earthquake in a 30-year period. Within the colored area, the red and orange regions have the greatest seismic hazard because they are likely to experience the most intense shaking in the next 30 years. Regional earthquake-hazard maps are also helpful in establishing property zoning restrictions and determining insurance rates. Figure 3.33 shows the probability of a **M** 6.7 or larger earthquake occurring on faults in the San Francisco Bay region. There is a 62 percent probability of at least one **M** 6–7 or greater earthquake occurring before 2032.[43] The probability is based on past earthquake activity, fault displacement, and slip rate.

In addition to mapping, the state of California is now classifying faults as a method for assessing seismic risk. Geologic faults are classified based on the maximum moment magnitude that the fault can produce and the slip rate of the fault. Slip rate is determined by the amount of movement along a fault averaged over

thousands of years and numerous earthquakes. It is not actual slippage that takes place along a fault each year. Although classifying faults provides more information than simply determining if a fault has been active in the past 10,000 years, the long-term slip rates of most major faults in North America are unknown or determined only by information collected at one or two sites.[44]

SHORT-TERM PREDICTION

The short-term prediction of earthquakes is an active area of research. Predictions must rely on *precursors*, events or changes that occur before the mainshock of the earthquake. Unlike a forecast, a true earthquake prediction specifies a relatively short time period (days to weeks) in which the event is likely, an estimated magnitude for the quake, a relatively limited geographic area, and a probability of occurrence. The basic procedure for predicting earthquakes was once thought to be as easy as "one-two-three."[45] First, deploy instruments to detect potential precursors of a future earthquake; second, detect and recognize the precursors that indicate when and how big an earthquake will be; and third, after reviewing the data, publicly predict the earthquake. Unfortunately, earthquake prediction is much more complex than first thought.

Japanese seismologists made the first attempts at earthquake prediction using the frequency of small *microearthquakes* (**M** < 2), repetitive surveys to detect tilting of the land surface, and measurements of changes in the local magnetic field of Earth. They found that earthquakes in the areas they studied were nearly always accompanied by swarms of microearthquakes several months before the major shocks. Furthermore, they found that ground tilt correlated strongly with earthquake activity.[38]

Chinese scientists predicted a major earthquake (**M** 7.0) in 1975, saving thousands of lives. Although this prediction appears to have been successful by chance, because Chinese scientists had issued many unsuccessful predictions and also subsequently failed to predict major quakes, the result was beneficial.[32] The predicted earthquake destroyed or damaged about 90 percent of the buildings in Haicheng, China. Most of the town's 9000 people were saved because of a massive evacuation from potentially unsafe housing.

The short-term prediction was based primarily on a series of progressively larger foreshocks that began four days prior to the main event.

Unfortunately, foreshocks do not precede most large earthquakes, and there is no method for distinguishing a foreshock from any other earthquake. In 1976, one of the deadliest earthquakes in recorded history (**M** 7.5) struck near the mining town of Tangshan, China, killing more than 240,000 people. There were no foreshocks.

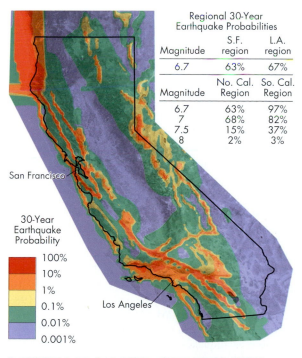

Regional 30-Year Earthquake Probabilities

Magnitude	S.F. region	L.A. region
6.7	63%	67%

Magnitude	No. Cal. Region	So. Cal. Region
6.7	63%	97%
7	68%	82%
7.5	15%	37%
8	2%	3%

30-Year Earthquake Probability

- 100%
- 10%
- 1%
- 0.1%
- 0.01%
- 0.001%

⋀ FIGURE 3.32 Probabilities of damaging (> M 6.7) Earthquakes in California Over Next 30 Years
(Field, E. H., Milner, K. R., and the 2007 Working Group on California Earthquake Probabilities. 2008. Forecasting California's earthquakes—What can we expect in the next 30 years? U.S. Geological Survey Fact Sheet. 2008–2027)

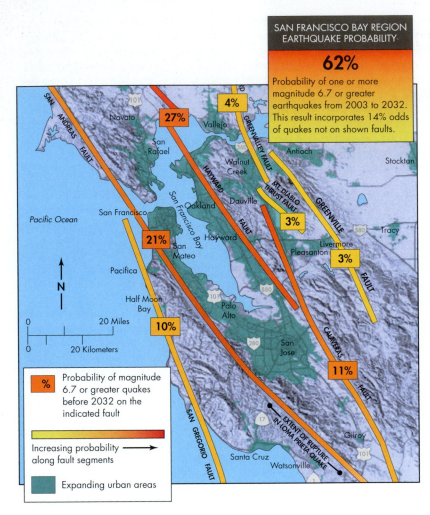

Probability of at least one **M** 6.7 or greater earthquake occurring before 2032 on any fault in the San Francisco Bay region. Small colored boxes show the probabilities of the quake occurring on each of the major faults. *(From U.S. Geological Survey Fact Sheet 039–03)*

In addition to foreshocks, many phenomena have been proposed as precursors for earthquakes. They range from lunar tides to unusual animal behavior. Reports of animal behavior have included unusual barking of dogs, chickens that refuse to lay eggs, horses or cattle that run in circles, rats perched on power lines, and snakes crawling out in the winter and freezing. To date, no scientific studies show a correlation between unusual animal behavior or lunar tides and earthquakes. Earthquake prediction is still a complex problem. Even if reliable precursors can be identified, dependable short-range predictions are many years away. Such predictions, if they are possible, will most likely be based upon precursory phenomena such as the following:

➤ Patterns and frequency of earthquakes, such as foreshocks discussed earlier with the predicted Haicheng earthquake

➤ Deformation of the ground surface

➤ Seismic gaps along faults

➤ Geophysical and geochemical changes in Earth.

Ground Deformation Changes in the land elevation, referred to as *uplift* and *subsidence*, that occur rapidly or are unusual may help predict earthquakes. For example, more than 120 km (~75 mi.) of the western Japanese coast experienced several centimeters of slow uplift in the decade prior to the 1964 Niigata earthquake (**M** 7.5).[50] A similar uplift occurred over a five-year period prior to the 1983 Sea of Japan earthquake (**M** 7.7).[46,51]

Uplifts of 1 to 2 m (~3 to 7 ft.) also preceded large Japanese earthquakes in 1793, 1802, 1872, and 1927. These uplifts were recognized by sudden withdrawals of the sea from the land. On the morning of the 1802 earthquake, the sea suddenly withdrew about 300 m (~980 ft.) in response to an uplift of about 1 m (~3 ft.). Four hours later, the earthquake struck, destroying many houses and uplifting the land another meter.[50]

Paleoseismic Earthquake Hazard Evaluation

Earthquake hazard assessment is primarily meant to safeguard against loss of life and major structural failures rather than limit damage or maintain functional aspects of society, including roads and utilities.[46] The basic philosophy behind the seismic hazard evaluation is to find what is known as a **design basis ground motion**, defined as the ground motion that has a 10 percent chance of being exceeded in a 50-year period (equivalent to 1 in 475 chance of being exceeded each year). This is determined in California by a series of maps showing strong motion PGA (peak ground acceleration) for earthquake waves of 0.3-second period for one- and two-story buildings, and 1.0-second period for taller buildings.[47] The USGS makes predictions for peak acceleration and velocity of shaking using an earthquake-planning scenario for a particular fault. Figure 3.5.A shows a scenario for a **M** 7.2 event on the San Andreas fault in San Francisco.[48]

Assessing the earthquake hazard at a particular site includes identifying the **tectonic framework** (geometry and spatial pattern of faults or seismic sources) in order to predict earthquake slip rate and ground motion (Figure 3.5.B). Predicting slip rate and ground motion can be illustrated by a hypothetical example of a dam

and reservoir site (Figure 3.5.C). The objective is to predict strong ground motion at the dam from several seismic sources (faults) in the area. The tectonic framework

shown consists of a north-dipping reverse fault and an associated fold (anticline) located north of the dam, as well as a right-lateral strike-slip fault located south of the

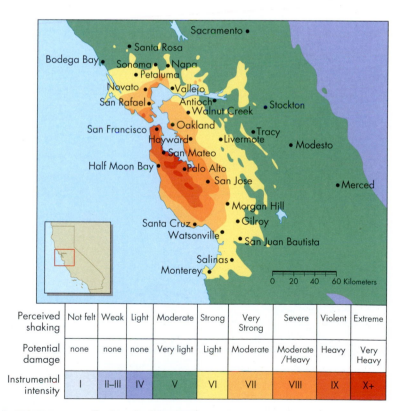

Perceived shaking	Not felt	Weak	Light	Moderate	Strong	Very Strong	Severe	Violent	Extreme
Potential damage	none	none	none	Very light	Light	Moderate	Moderate /Heavy	Heavy	Very Heavy
Instrumental intensity	I	II–III	IV	V	VI	VII	VIII	IX	X+

⋀ FIGURE 3.5.A Earthquake Planning Scenario
This scenario is for a **M** 7.2 event on the San Andreas fault in San Francisco, California. Instrumental Intensity refers to the Mercalli scale (see Table 3.5). *(From Implications for Earthquake Hazard in the San Francisco Bay Area, US Geological Survey Open File Report 03–214. By Working Group on California Earthquake Probabilities 2003. http:// seismo.berkeley. edu/~rallen/teaching/eps256-s07/WG02SumCh1Ch2.pdf)*

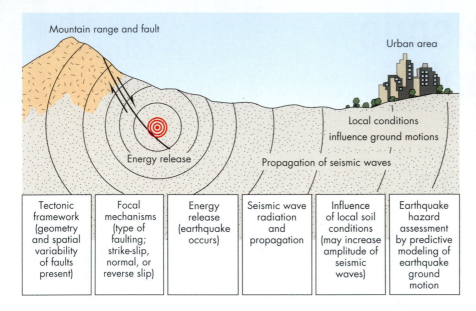

Mountain range and fault

Urban area

Local conditions
influence ground motions

Energy release

Propagation of seismic waves

Tectonic framework (geometry and spatial variability of faults present)	Focal mechanisms (type of faulting; strike-slip, normal, or reverse slip)	Energy release (earthquake occurs)	Seismic wave radiation and propagation	Influence of local soil conditions (may increase amplitude of seismic waves)	Earthquake hazard assessment by predictive modeling of earthquake ground motion

◀ FIGURE 3.5.B Assessment of Earthquake Hazard by Modeling of Ground Motion

(Based on A. Vogel and K. Brandes (eds.), Earthquake prognostics. Brannschweig/Weisbaden: Friedr. Vieweg & Sohn, 1–13)

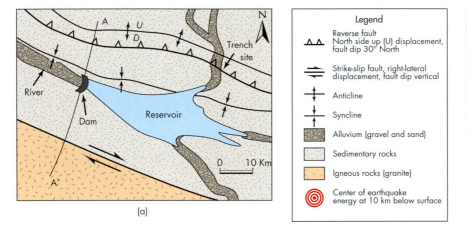

(a)

Legend

Reverse fault
North side up (U) displacement, fault dip 30° North

Strike-slip fault, right-lateral displacement, fault dip vertical

Anticline

Syncline

Alluvium (gravel and sand)

Sedimentary rocks

Igneous rocks (granite)

Center of earthquake energy at 10 km below surface

Trench site

River

Dam

Reservoir

0   10 Km

◀ FIGURE 3.5.C Tectonic Framework for a Hypothetical Dam Site

(a) Geologic map of hypothetical paleoseismic earthquake hazard evaluation site. (b) Cross-section A–A', showing seismic sources and distances of possible ruptures to the dam. See (a) for location of cross-section.

A   Dam   A'

42 km

32 km

$M_w = 6.5$ Consistent with rupture length of 30 km and slip of 1 m

$M_w = 7$ Consistent with rupture length of 50 km and slip of 2 m

(b)

dam. Figure 3.5.C is a cross-section through the dam, illustrating the geologic environment, including several different earth materials, folds, and faults. Assuming that earthquakes would occur at depths of approximately 10 km (~6 mi.), the distances from the dam to the two seismic sources (the reverse fault and the strike-slip fault) are 42 km and 32 km (~27 mi. and 20 mi.), respectively. Thus, for this area, two focal mechanisms are possible: reverse faulting and strike-slip faulting.

Another step in the process is to estimate the largest earthquakes likely to occur on these faults. Assume that fieldwork in the area revealed ground rupture and other evidence of faulting in the past, suggesting that, on the strike-slip fault, approximately 50 km (~31 mi.) of fault length might rupture in a single event with right-lateral strike-slip motion of 2 m (~6.6 ft.). The fieldwork also revealed that the largest rupture likely on the reverse fault would be 30 km (~18.6 mi.) of fault length with vertical displacement of about 1.5 m (~5 ft.). Given this information, the magnitudes of possible earthquake events can be estimated from graphs such as those shown in Figure 3.5.D.[45] Fifty kilometers (~31 mi.) of surface rupture are associated with an earthquake of approximately **M** 7. Similarly, for the reverse fault with surface rupture length of 30 km (~19 mi.) the magnitude of a possible earthquake is estimated to be **M** 6.5. Notice that in the first graph the regression line predicting the moment magnitude is for strike-slip, normal, and reverse faults. Statistical analyses have suggested that the relation between moment magnitude and length of surface rupture is not sensitive to the style of faulting.[49]

Slip rate (the vertical component) and average return period can be estimated from geologic data collected in a trench excavated across a fault. Figure 3.5.E shows the stratigraphy (trench log) for the site shown in Figure 3.5.C. The slip rate is about 1 mm/yr (1 m per 1,000 years), and the average return period is about 1,300 years.

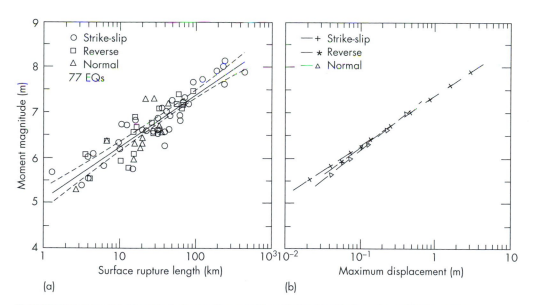

∧ FIGURE 3.5.D Relationship between Moment Magnitude of an Earthquake and Deformation
Data for 77 events were used to estimate potential earthquake magnitude based on (a) surface rupture length and (b) maximum displacement. (a) Solid line is the "best-fit" or regression line, and dashed lines are error bars at significant difference between the lines. *(Adapted from Wells, Donald L., and Coppersmith, Kevin J. New empirical relationships among magnitude, rupture length, rupture width, rupture area, and surface displacement. Bulletin of the Seismological Society of America, 84(4): 992 (figure 9), August 1994. Reprinted by permission of the authors.)*

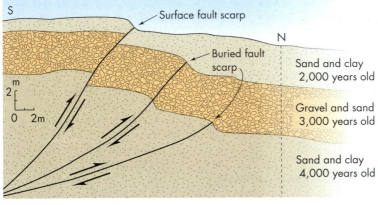

Total vertical component of slip in 4,000 yrs about 4.0 m in 4,000 yrs or
1 mm/yr (1 m/1000 years)
Most recent event about 1.5 m in 2,000 yrs or 0.8 mm/yr
Last two events about 2.8 m in 3,000 yrs or 0.9 mm/yhr

S

Surface fault scarp

N

Buried fault
scarp

Sand and clay
2,000 years old

Gravel and sand
3,000 years old

Sand and clay
4,000 years old

m
2
0   2m

Average return period 3 events in 4,000 yrs
about 1,300 year return period

**◄ FIGURE 3.5.E Paleoseismic Trench
Log**
Trench exposes sedimentary layers
that have been deformed by faulting.
Numerical dating of these sediments can
provide information about the timing
and rate of deformation. Location of the
trench site is shown on Figure 3.5.C.

Seismic Gaps Areas along an active fault zone that
are likely to produce large earthquakes but that have
not produced one recently are known as *seismic gaps*.
These areas are believed to be temporarily inactive
while they store elastic strain for future large earthquakes (see Case Study 3.1).[4] The seismic-gap concept
was the basis for what may have been the first scientific,
long-term earthquake forecast. In 1883, the multital-
ented USGS geologist G. K. Gilbert forecast a catastrophic earthquake for the Salt Lake City segment of
the Wasatch fault based on evidence of its inactivity.[32]
Since 1965, the seismic-gap concept has been used successfully in intermediate-range forecasts for at least 10
large plate-boundary earthquakes. One of these quakes
took place in Alaska, three in Mexico, one in South
America, and three in Japan. In the contiguous United
States, seismic gaps along the San Andreas fault in
California include one near Fort Tejon that last ruptured in 1857 and one along the Coachella Valley, a segment that has not produced a great earthquake for
several hundred years. Both gaps are likely to produce a
great earthquake in the next few decades.[4,52]

Geophysical and Geochemical Phenomena
Local changes in Earth's gravity, magnetic field, and ability to conduct electrical currents have been associated
with earthquakes. Changes have also been observed in
groundwater levels, temperature, and chemistry.[52] Many

of these changes may occur before an earthquake as
rocks expand and fracture and as the fractures fill with
water. For example, changes in the ability of earth materials to conduct an electrical current, referred to as *electrical resistivity*, have been reported before earthquakes in
the United States, Eastern Europe, and China.[47] Also,
significant increases in the radon content of well water
were reported in the month or so prior to the 1995
Kobe, Japan, earthquake (**M** 6.9).[53]

THE FUTURE OF EARTHQUAKE
PREDICTION

We are still a long way from a working, practical methodology to reliably predict earthquakes. However, a
good deal of information is currently being gathered
concerning possible precursor phenomena associated
with earthquakes. To date, the most useful precursor
phenomena have been patterns of earthquakes, particularly foreshocks, and recognition of seismic gaps.
Optimistic scientists around the world today believe that
we will eventually be able to make consistent long-range
forecasts (decades to centuries), intermediate-range
forecasts (months to decades), and short-range predictions (hours to weeks) for the general locations and magnitudes of large, damaging earthquakes.

Although progress on short-range earthquake prediction has not matched expectations, intermediate- to

long-range forecasting, including hazard evaluation and probabilistic analysis of areas along active faults, has progressed faster than expected (see Case Study 3.6: Professional Profile).[54] For example, the 1983 Borah Peak earthquake (**M** 6.9) on the Lost River fault in central Idaho has been lauded as a success story for intermediate-range earthquake hazard evaluation. Previous evaluation had suggested that the fault was active.[55] The earthquake killed two people and caused approximately $15 million in damages. Movement during the earthquake created a fault scarp several meters high and numerous ground fractures along a 36 km (~22 mi.) rupture zone. Investigation after the earthquake found that the new fault scarp and fractures were superimposed on previously existing fault scarps, validating the usefulness of careful mapping of scarps produced by prehistoric earthquakes. Remember—where the ground has broken before, it is likely to break again!

We hope eventually to be able to predict earthquakes. Because of this hope, nearly 30 years ago, the USGS established a plan for issuing earthquake predictions and more urgent earthquake warnings (Figure 3.34). This plan requires independent scientific review before information is transmitted to government officials and the public.

Government officials and news media that give publicity and thus credence to predictions that have not been scientifically reviewed do a great disservice to the community. This precept was demonstrated in 1990 when a pseudoscientific prediction for an earthquake in New Madrid, Missouri, was acted on by some government and business leaders and publicized by local, state, and national media—even after the prediction was rejected by the National Earthquake Prediction Evaluation Council.[56] Schools and businesses closed, public events were canceled, people evacuated their homes, and more than 30 television and radio vans converged on the predicted epicenter.[57] Widely publicizing earthquake predictions that have not been independently vetted by seismologists is the geologic equivalent of yelling "fire" in a crowded movie theater.

EARTHQUAKE WARNING SYSTEMS

Technically it is feasible to develop an earthquake warning system that would provide up to about 1 minute of warning to the Los Angeles area prior to the arrival of damaging earthquake waves from an event several hundred kilometers away. Such a system would be based on the principle that radio waves travel much faster than seismic waves. For nearly 15 years, the Japanese have had such a system that provides earthquake warnings for their high-speed trains; derailment of one of their "bullet" trains by an earthquake could kill hundreds of people (see Applying the Fundamental Concepts).

A proposed system for California involves a sophisticated network of seismometers and transmitters along the San Andreas fault. This system would first sense motion associated with a large earthquake and then send a warning to Los Angeles, which would then relay the warning to critical facilities, schools, and the general population (Figure 3.35). The warning time would range from as little as 15 seconds to as long as 1 minute depending on the location of the earthquake epicenter. This interval could be enough time for many people to shut down machinery and computers and take cover.[58] Remember that an earthquake warning system is not a prediction tool because it warns that an earthquake has already occurred.

Because the warning time is so short, some people believe that the damage to scientific credibility caused by false alarms would be far greater than the benefits of a brief warning. In the Japanese system, it is estimated that around 5 percent of the warnings turn out to be false alarms. Others have expressed concern for liability issues resulting from false alarms, warning system failures, and damage and suffering resulting from actions taken as the result of false early warnings.

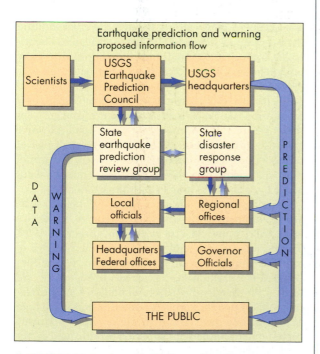

⋏ FIGURE 3.34 Issuing a Prediction or Warning
Flowchart for the U.S. plan for issuance of an earthquake prediction and warning. The plan calls for a National Earthquake Prediction Evaluation Council of seismologists from government agencies, universities, and the private sector to evaluate predictions and warnings before they are passed on to state and local officials and the public. (*From McKelvey, V. E. 1976. Earthquake prediction— opportunity to avert disaster. U.S. Geological Survey Circular 729*)

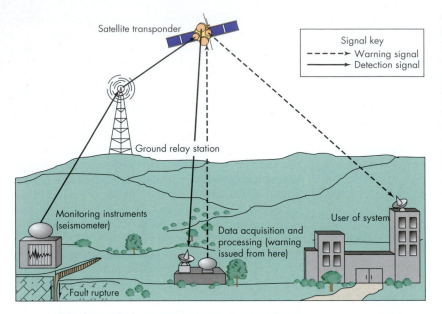

Satellite transponder

Signal key
- - - - → Warning signal
——→ Detection signal

Ground relay station

Monitoring instruments
(seismometer)

Data acquisition and
processing (warning
issued from here)

User of system

Fault rupture

◁ FIGURE 3.35 Earthquake Warning
Idealized diagram showing an earthquake warning system using radio waves, which travel faster than seismic waves. Once an earthquake is detected, a signal is sent ahead of the seismic shaking to warn people and facilities. The warning time, generally measured in seconds, depends on how far away an earthquake occurs. It could be long enough to shut down critical facilities and for people to take cover. *(Based on Holden, R., Lee, R., and Reichle, M. 1989. Technical and economic feasibility of an earthquake warning system in California, California Division of Mines and Geology Bulletin 101)*

3.8 CHECK your understanding

1. What kinds of information are useful in assessing seismic risk?
2. What kinds of phenomena may be precursors for earthquakes?
3. What is the difference between an earthquake prediction and a forecast?

3.9 Perception of and Adjustment to the Earthquake Hazard

PERCEPTION OF THE EARTHQUAKE HAZARD

The experience that Earth is not so firm in places is disconcerting to people who have felt even a moderate earthquake. The large number of people, especially children, who suffered mental distress following the San Fernando and Northridge earthquakes in California attests to the emotional and psychological effects of earthquakes. These events caused a number of families to move away from Los Angeles.

One community's experience with a large earthquake does not always translate into increased preparedness for a community in another area. For example, in the 1994 Northridge earthquake (**M** 6.7),

intense shaking caused part of the local seismograph network to malfunction. This malfunction delayed emergency response efforts because the exact location of the epicenter was not immediately available. The intense shaking also caused many unreinforced freeway bridges and buildings to collapse. A year later, the residents of Kobe, Japan, experienced nearly identical problems in a **M** 6.9 earthquake. Delays caused by communication problems and infrastructure damage kept the government from mounting a quick and effective response to the disaster. Even though Japan is one of the most earthquake-prepared countries in the world, much of the emergency relief in the Kobe quake did not arrive until about 10 hours after the earthquake!

Another example of the problems our modern society experiences with earthquakes comes from two large (**M** 7.6 and **M** 7.2) quakes in Turkey in 1999. Turkey was second only to China in the number of lethal earthquakes in the twentieth century. The first of the 1999 earthquakes occurred on August 17 and leveled thousands of concrete buildings. One-quarter of a million people were left homeless, and more than 17,000 people died. Although Turkey has a relatively high standard for new construction to withstand earthquakes, many modern buildings collapsed from the intense seismic shaking while older buildings were left standing (Figure 3.36). As with the experience of the residents of Ahmedabad, India, in the 2001 Bhuj earthquake (**M** 7.7), there is suspicion in Turkey that poor construction practices contributed to the collapse of the newer buildings. There were allegations that some of the Turkish contractors bulldozed collapsed

Andrea Donnellan, Earthquake Forecaster

Andrea Donnellan (Figure 3.6.A) talks about earthquakes the same way meteorologists talk about the weather: as dynamic, interconnected systems with their own peculiar set of rules. So it is only appropriate that she refers to her work as earthquake "forecasting."

Specifically, Donnellan and her colleagues have installed hundreds of high-precision global positioning system (GPS) receivers across Southern California, a network known as the Southern California Integrated GPS Network (SCIGN), in order to study movements in the tectonic plates. The receivers, which Donnellan says are more highly sophisticated than commercial GPS devices such as those used in cell phones, can measure millimeter-scale slips for faults and give scientists valuable data for understanding how and why earthquakes take place. Donnellan's work goes far beyond studying just the earthquakes themselves. In fact, most of the work occurs in between all the moving and shaking. "I look at the quiet part of the earthquake cycle," Donnellan said. "Each earthquake changes where the next earthquake will be."

Like meteorologists, Donnellan and her colleagues have reproduced earthquake systems using supercomputers in order to understand how the dynamic systems change over time. "My focus is on modeling earthquake systems," she said. "We want to treat it like weather, where the system is always changing."

But Donnellan's research consists of far more than just sitting in front of a computer all day. In fact, her studies have taken her to places around the globe, from Antarctica to Mongolia to Bolivia, to name a few. Her current research has taken her across much of Southern California, where she and her team are studying the San Andreas fault.

Back in the lab, Donnellan is also involved in an ambitious project called QuakeSim, in which scientists are developing sophisticated, state-of-the-art computer models of earthquake systems. She said the results of the project will eventually be accessible to anyone and that schools are already using some of the software for educational purposes.

In terms of practical implications, Donnellan's research is already giving scientists a much clearer idea of where earthquakes may next strike, allowing communities to prepare much further in advance. Although, like the weather, earthquakes will probably never be totally predictable, a close study of the ever-shifting activity beneath the ground by Donnellan and her colleagues is making the picture a lot clearer.

— CHRIS WILSON

◄ FIGURE 3.6.A Geophysicist Develops Earthquake Forecasting Method

Dr. Andrea Donnellan, deputy manager of NASA's Jet Propulsion Laboratory Earth and Space Sciences Division, leads a team of scientists working on earthquake forecasting. She was co-winner of Women in Aerospace's 2003 Outstanding Achievement Award and twice a finalist in the NASA astronaut selection process. *(NASA)*

⋀ FIGURE 3.36 Collapse of Buildings in Turkey
Damage to the town of Golcuk in western Turkey from the
M 7.6 Izmit earthquake in August 1999. The very old mosque
on the left remains standing, whereas many modern buildings
collapsed. *(AP Photo/Enric Marti)*

buildings soon after the earthquake to remove evidence of shoddy construction. If that was true, those contractors also tied up bulldozers that could have been used to help rescue people trapped in other collapsed buildings.

The lessons learned from Northridge, Kobe, and Turkey are bitter ones. Our modern society is vulnerable to catastrophic loss from large earthquakes. Older, unreinforced concrete buildings or buildings not designed to withstand strong ground motion are most vulnerable. In Kobe, reinforced concrete buildings constructed with improved seismic building codes experienced little damage compared to those constructed earlier. Minimizing the hazard requires new thinking about it.

COMMUNITY ADJUSTMENTS TO THE EARTHQUAKE HAZARD

Since it is clearly impossible to avoid all human habitation in earthquake-prone areas, certain steps must be taken by countries, states, communities, and individuals to adjust to the earthquake hazard. These steps include careful location of critical facilities, structural protection, education, and increased insurance and relief measures. Individuals can also take steps to protect themselves. The extent to which these adjustments occur depends in part on people's perception of the hazard.

Location of Critical Facilities Buildings and other structures, often referred to as *facilities*, that are critical to the community must be located as safely as possible. These structures include hospitals, schools, utility plants, communication systems, and police and fire stations. Selecting safe locations first requires site-specific investigation of earthquake hazards such as the liquefaction, ground motion, and landslide potential. This involves *microzonation*, the detailed location and land-use planning of areas having various earthquake hazards. Microzonation is necessary because the ground's response to seismic shaking can vary greatly within a small area. In urban areas, where individual property values may exceed millions of U.S. dollars, detailed maps of ground response are required to accomplish microzonation. These maps help engineers and architects design buildings and other structures that can better withstand seismic shaking. Clearly, microzonation requires a significant investment of time and money; however, it provides the basic information needed to adequately predict the ground motion at a specific site. Although we may never be able to prevent injury and death from earthquakes, we can reduce the number of casualties through the safe location of critical facilities.

Structural Protection The often-repeated statement "earthquakes don't kill people, buildings kill people" succinctly expresses the importance of building design and construction in reducing the hazard of earthquakes. In regions where there is a significant earthquake hazard, buildings, bridges, pipelines, and other structures must be constructed to withstand at least moderate shaking. To accomplish this, communities must adopt and enforce building codes with earthquake sections similar to those in the International Building Code and International Residential Code. In establishing seismic provisions in building codes, engineers try to balance reducing risk to human life with the high construction costs of earthquake-resistant design.[44]

Of equal importance to seismic provisions in building codes is the inspection and strengthening of existing structures to withstand the ground motion of large earthquakes. Making engineering changes to existing structures, referred to as *retrofitting*, is a costly but extremely important activity (Figure 3.37). For example, a recent study showed that more than half of the hospitals in Los Angeles County could collapse in a large earthquake.[44] The cost of retrofitting these buildings is likely to be more than $8 billion. Retrofitting is especially critical in developing countries and in U.S.

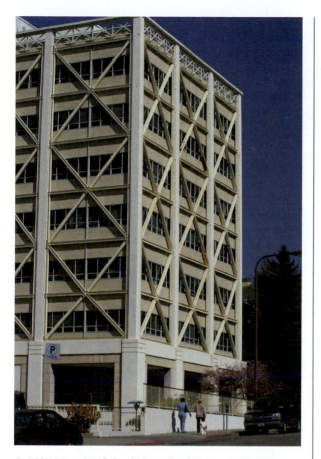

∧ FIGURE 3.37 Seismic Retrofit of University Building
The cross braces and horizontal bracing on the building exterior
were added to University Hall on the University of California,
Berkeley, campus to help the structure withstand shaking in the
next major earthquake. *(Robert H. Blodgett)*

communities such as Memphis, Tennessee, that have
only recently included earthquake-resistant design in
their building codes.

Earthquake-resistant design and retrofitting of
buildings, which first began in the 1920s, has generally
been successful in the United States. The effect of
these improvements in construction can be appreciated by comparing two events of approximately equal
magnitude: the 1988 Armenia (**M** 6.8) and the 1994
Northridge (**M** 6.7) earthquakes. The loss of life and
destruction in Armenia was staggering (at least 25,000
people were killed, compared with 60 in California)
and some towns near the epicenter were almost totally
destroyed. Most buildings in Armenia were constructed of unreinforced concrete and instantly crum-
bled into rubble, crushing or trapping their occupants.

This is not to say that the Northridge quake was not a
catastrophe—it certainly was: 25,000 people were left
homeless, several freeway overpasses collapsed, at least
8,000 people were injured, and many billions of dollars
in damages to buildings and other structures occurred.
However, since most buildings in the Los Angeles
Basin are constructed with wood frames or reinforced
concrete, thousands of deaths in Northridge were
avoided.

Education   As with most other natural hazards, education is an important component of preparedness at
the community level. This educational effort could include the public distribution of pamphlets and videos;
workshops and training sessions for engineers, architects, geologists, and community planners; and the
availability of information on the Internet. In areas of
high seismic or tsunami risk, education also takes place
through earthquake and tsunami drills in schools and
earthquake disaster exercises for government officials
and emergency responders.

An excellent example of educating those at great risk
from earthquake hazards is Mexico's annual preparedness exercise. The mock earthquake drill is conducted
each year on September 19, which is the anniversary
day of the devastating 1985 Mexico city earthquake that
killed more than 8,000 people (see Section 3.1). The
drill is designed to teach participants proper building
evacuation procedures given an assumed 40 second
warning period prior to hypothetical ground shaking
(Figure 3.38a). Coincidentally, just 2 hours after the
2017 annual earthquake preparedness drill, earthquake
warning sirens went off a second time. Fifteen seconds
later strong ground shaking rocked Mexico City, triggered by an **M** 7.1 earthquake with an epicenter 120 km
(~75 mi.) to the south. Although collapsing buildings
claimed the lives of 370 people (Figure 3.38b), countless citizen reports credit the mock earthquake drills for
helping them to safely evacuate following the second
round of sirens that day.

Increased Insurance and Relief Measures   Insurance and relief measures are vital to help a community, state, or country recover from an earthquake.
Losses from a major earthquake can be huge. This fact
was graphically demonstrated in the 1906 San
Francisco earthquake (**M** 7.8) when only 6 of the 65
insurance companies were able to pay their liabilities
in full.[5] Although risk assessment and the insurance
industry have greatly improved in the past century, the
potential losses from a large earthquake in a densely
urbanized area such as Los Angeles, the San Francisco

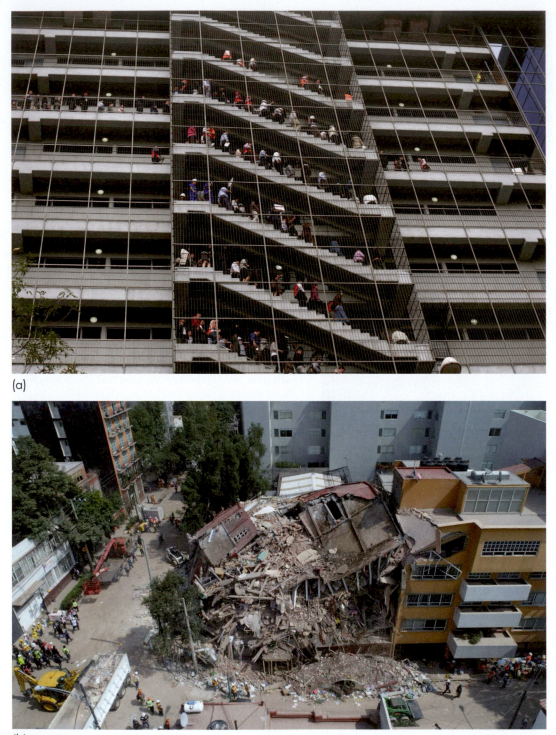

(a)

(b)

˄ FIGURE 3.38 Mexico Remembrance and Preparadness
Every year the country of Mexico conducts earthquake drills on September 19, the anniversary day of the 1985 Mexico City earthquake that killed thousands. (a) Civilians conducting building evacuation preparedness drills at 11 A.M. (b) Structural collapse of a building 2 hours after the earthquake drill due to the 2017 Central Mexico earthquake.

Bay area, Seattle, New York City, or Boston are enormous. The $15.3 billion in insurance claims paid in the 1994 Northridge quake (**M** 6.7) does not come close to the $200 billion estimate of losses in the 1995 Kobe quake (**M** 6.9).[17] The state of California has partially addressed this problem with state-subsidized earthquake insurance through the nonprofit California Earthquake Authority. Even with state subsidies, barely 25 percent of California residents have earthquake insurance, in part because of the high deductible.[17] Unlike flood insurance and crop insurance, there is no federally subsidized earthquake insurance. Federal officials and the insurance industry have yet to agree on a national strategy to deal with catastrophic earthquake loss.

Development of methods to estimate the potential impact of earthquakes is important to insurance companies, government agencies, and homeowners.[59] Estimates of losses from earthquakes and other hazards can now be accomplished with "Hazards U.S." (HAZUS), a FEMA computer program. This free software can be downloaded from the FEMA website for use on a personal computer.

PERSONAL ADJUSTMENTS: BEFORE, DURING, AND AFTER AN EARTHQUAKE

Individual actions before, during, and after a major earthquake can reduce casualties and property damage. Close to 150 million people in 39 U.S. states live in seismically active areas. In damage alone, billions of dollars could be saved if our buildings and contents were better secured to withstand shaking from earthquakes.

If you live in a seismically active area, a home safety check could increase the probability that you and your property survive a large earthquake. This inspection should include checking chimneys and foundations for reinforcement and the security of large objects, such as water heaters, that might fall over in an earthquake.[60] Probably the most important thing is to plan exactly what you will do should a large earthquake occur. This plan might include teaching your family to "drop, cover, and hold on," which means to "drop" to the ground; take "cover" under a sturdy desk, table, or bed away from falling objects; and "hold on" to something until the shaking stops. Contrary to some suggestions, standing in most doorways is not a good option.[5] Preparations should also include maintaining an adequate stock of bottled water, food, medical supplies, batteries, and spare cash.[60]

During the quake, remember what you have learned in this chapter. P waves will arrive first, causing initial vibrations followed by heavy shaking from the S and surface waves. If there has been at least 20 seconds between the P and S waves, then the quake is from a distant source. The noise of the ground vibrating will be deep and loud and may begin suddenly like a loud clap of thunder if you are close to the epicenter.[52] Books, dishes, glass, furniture, and other objects may come crashing down. Car alarms will go off, and there may be explosive flashes from electrical transformers and falling power lines. The length of shaking will vary with the magnitude. For example, shaking during the **M** 6.7 Northridge quake lasted around 15 seconds, whereas shaking during the **M** 7.8 San Francisco quake lasted nearly 2 minutes.

After the shaking stops, you may feel dizzy and sick.[61] Take several deep breaths, look around, and then leave the building only after the shaking stops, carefully watching for fallen and falling objects. More injuries occur when people attempt to move to a different location inside a building or try to leave. Turn off the main gas line and do not light matches or lighters. Move to an open area, away from fallen power lines and buildings or trees that might fall during aftershocks. The number and intensity of aftershocks are also related to the magnitude of the quake. For example, in the **M** 6.9 Loma Prieta quake, there were 2 **M** 5, 4 **M** 4, and 65 **M** 3 aftershocks.[62] The largest aftershock is typically at least one magnitude value lower than the mainshock. Overall, the most hazardous period for aftershocks is in the minutes, hours, and days following the mainshock. Both the number and the intensity of aftershocks will generally decrease in the days and weeks following the mainshock.

The good news is that in most developed countries, large earthquakes are survivable. In areas of greatest hazard, such as California, buildings are generally constructed to withstand earthquake shaking, and wood-frame houses seldom collapse. In the United States, earthquake-proof construction is progressively less common as one moves from west to east. Many brick houses, buildings, and other structures in New England, New York City, and Washington, DC, will not withstand shaking in a large earthquake.[26] Therefore, it is important to be well informed and prepared for earthquakes.

3.9 CHECK your understanding

1. What kinds of adjustments can a community make to the earthquake hazard?
2. What is retrofitting?
3. List personal adjustments you can make to prepare for an earthquake.

APPLYING the ⑤ Fundamental Concepts

2011 Tohoku Earthquake

❶ Science helps us predict hazards.

❷ Knowing hazard risks can help people make decisions.

❸ Linkages exist between natural hazards.

❹ Humans can turn disastrous events into catastrophes.

❺ Consequences of hazards can be minimized.

Japan, which is about the size of California, and the United States both experience about 1,000 moderate (>**M** 5) earthquakes each year. However, over the past two decades Japan has been hit by 19 major quakes (**M** 7–7.9), whereas the United States has experienced only three during this same interval, all in sparsely populated Alaska. Although less frequent, Japan also experiences **M** 8 earthquakes every few years. With the regular occurrence of such significant earthquakes, and Japan's technological capability and wealth, it is no surprise that many consider Japan to be the world's best-equipped country in earthquake preparedness.[2] Modern structures built in Japan are extremely well engineered to resist ground shaking, and the country hosts the world's densest seismometer network, which is used for forecasting, locating, and studying earthquakes. In addition, the seismometer network is linked to an advance earthquake-warning system, and Japan's people are rigorously drilled on what to do when an earthquake warning is given. Japan's high level of earthquake preparedness represents an evolving process that has grown out of decades of lessons learned from devastating earthquakes, with the 2011 Tohoku earthquake providing the most recent lesson.

Although relatively small compared to the Tohoku earthquake, the **M** 6.8 Kobe earthquake that struck Japan in 1995 killed about 6,500 people, about 90 percent of whom perished beneath collapsed buildings. Even though new seismic building codes had been introduced in 1981, homes, buildings, freeways, and bridges constructed before that time were not required to upgrade to the new code. This led to widespread collapse and toppling of critical roadways, which greatly inhibited the damage control and search-and-rescue operations following the quake (Figure 3.39). Japan's realization that it was not as prepared as it or the world

⋀ FIGURE 3.39 Damage from 1995 Kobe Earthquake
Eighteen 3.1-m (~10 ft.) diameter columns fail due to intense ground shaking during the Kobe Earthquake causing 630 m (~0.4 mi.) of the Hanshin elevated expressway to topple, scattering automobiles into the city below. Failure of critical infrastructure such as this severely limited the rescue effort. (*Yoshiaki Nagashima/Pacific Press Service/Alamy*)

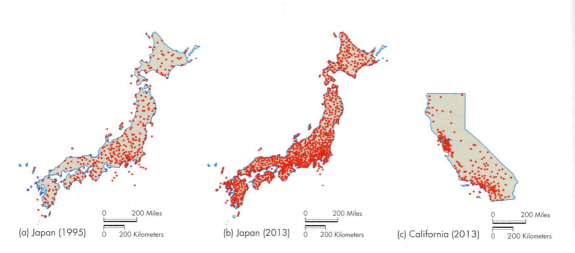

(a) Japan (1995)

0   200 Miles
0   200 Kilometers

(b) Japan (2013)

0   200 Miles
0   200 Kilometers

(c) California (2013)

0   200 Miles
0   200 Kilometers

∧ FIGURE 3.40 Seismometer Networks

Following the Kobe earthquake, (a) the deficient seismometer network emplace during the 1995 quake network was significantly upgraded *(Okada, Y., et al. "Recent Progress of Seismic Observation Networks in Japan: Hi-net, F-net, K-net and KiK-net," Earth Planets Space 56: 15–28, 2004. Reprinted by permission of Terrapub.)* (b) *(From "Next Steps: Improving the Sensor Network," USGS, http://earthquake.usgs.gov/research/earlywarning/nextsteps.php)* (c) Seismometer network in California, the most dense network in the United States, for comparison. *(From "Next Steps: Improving the Sensor Network," USGS, http://earthquake.usgs.gov/research/earlywarning/nextsteps.php)*

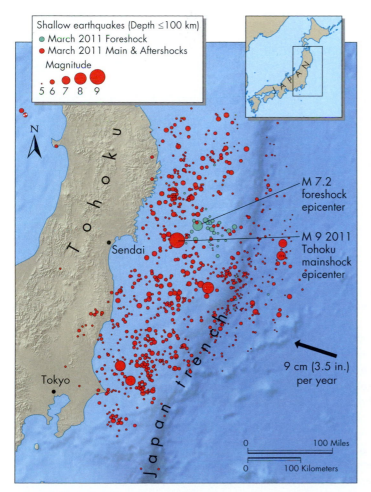

Shallow earthquakes (Depth ≤100 km)
● March 2011 Foreshock
● March 2011 Main & Aftershocks
Magnitude
5 6 7 8 9

M 7.2 foreshock epicenter

M 9 2011 Tohoku mainshock epicenter

9 cm (3.5 in.) per year

0   100 Miles
0   100 Kilometers

believed led to a more proactive approach to minimizing the effects of earthquakes. This included significant spending increases to retrofit older buildings and infrastructure throughout the country and an investment of $1 billion in research and development for an early warning system, which included an overhaul of Japan's seismometer network (Figure 3.40). Fifteen years later Japan would be tested again— this time by a much larger quake.

Two days before the Tohoku earthquake on March 9, 2011, the seismic network detected a **M** 7.2 earthquake east of the Tohoku region of Japan (Figure 3.41). Because the frequency and size of earthquakes for this region over the past 1,100 years was relatively well known, the timing and size of this earthquake came as no surprise to the

< FIGURE 3.41 2011 Tohoku Earthquake Sequence

Epicenteral foreshock and aftershock locations surrounding the mainshock of the 2011 Tohoku earthquake with respect to the big island of Japan. *(Based on Ammon, C. J., Lay, T., Kanamori, H., and Cleveland, M., "A Rupture Model of the 2011 of the Pacific Coast of Tohoku Earthquake," Earth Planets Space 63: 693–96, 2011)*

113

seismologists.[63] In fact, earthquake forecasting along this part of the subduction zone estimated a 60–70 percent chance of an **M** 7 quake occurring in the next 20 years; in earthquake forecasting, this is like saying an earthquake could happen any day.[64] Therefore, there was no reason to suspect that this earthquake was a foreshock to an imminent quake that would be 500 times more powerful than the earthquake history suggested. For the next 50 hours small foreshocks continued, until March 11 at 2:46 P.M. Japan time, when the frictional forces keeping the Pacific plate from sliding under the Eurasian plate were overcome. In a geological instant the Pacific plate lurched northwestward beneath Japan as much as 40–60 m (~130–200 ft.), releasing the stored elastic strain energy that had likely accumulated over the past 1,000–1,200 years and producing the **M** 9.0 Tohoku mainshock (Figure 3.41).[63] Not only was the size of the earthquake significantly larger than the maximum size that seismologists had predicted—it also produced a larger than predicted tsunami (see Chapter 4: Applying the Fundamental Concepts). Although seismologists significantly underestimated the maximum potential earthquake magnitude for the region, the lessons learned from the 1995 Kobe earthquake and development of the early warning system was about to save thousands of lives.[2]

Just 8 seconds after the first P waves from the Tohoku earthquake were confirmed by the seismometer network, an automated alert went out to 124 television stations and 52 million telephones. Across the country critical infrastructure systems such as power plants and Japan's famous bullet trains automatically shut down upon receiving the alert. For the cities closest to the Tohoku epicenter, like Sendai (Figure 3.41), the alert provided about 10 seconds of early warning before the slower, more destructive S waves arrived. This was enough time for the earthquake-prepared citizens to take immediate shelter as they had practiced. The 9 million people living in the capitol city of Tokyo 373 km (~232 mi.) to the south received about 80 seconds of early warning before very strong ground motion was felt.[2]

Since the Tohoku earthquake in 2011 about 10,000 aftershocks were recorded, with about 2,000 occurring within seven days of the mainshock. Some of the aftershocks were even large enough to trigger the early warning system again, with 39 aftershocks having magnitudes greater than **M** 6, three of which were greater than **M** 7 (Figure 3.41).

Buildings, bridges, and other critical facilities built or retrofitted since 1995 withstood the ground shaking extremely well with few buildings having collapsed as a direct result of shaking. However superficial damage and minor structural damage were widespread. Even though many of the buildings with structural damage will need to be demolished, the fact that they did not completely collapse allowed occupants within to escape with their lives unlike 15 years earlier during the Kobe earthquake.

∧ FIGURE 3.42 Damage Due to Liquefaction
Manholes to underground sewer system forced through overlying sidewalks due to ground surface subsidence (lowering) caused by liquefaction. *(Koki Nagahama/Getty Images)*

In fact, the overall resilience of the structures is particularly impressive considering that the earthquake building codes for the region near the epicenter were based on the predicted maximum earthquake magnitude, which had been grossly underestimated.

Outside of tsunami damage, liquefaction and landsliding triggered by ground shaking were responsible for a significant portion of the damage to buildings, roads, and subsurface utilities such as water, gas, and sewage pipes. Subsidence of the ground surface due to liquefaction was particularly destructive (Figure 3.42) and widespread, with about 6,000 homes needing to be demolished due to liquefaction damage, some as far away as Tokyo. The extent of the liquefaction damage surprised many earthquake hazard experts worldwide, raising questions about whether building codes in Japan, as well as the United States, adequately account for this hazard in urban areas.[65] The Tohoku earthquake also triggered about 4000 landslides along the east coast of Japan, most in sparsely populated mountainous regions.[66] However, about 20 people were killed by landslides, some of whom died when a landslide caused an earthen dam to collapse unleashing a flood on the community below.

The 2011 Tohoku earthquake and lessons learned preceding and following the quake are excellent examples of our unending struggle to understand, predict, prepare, and survive Earth's natural hazards. Over the past 20 years Japan has invested tens of billions of dollars in this struggle. The cost of the 2011 Tohoku earthquake and tsunami is estimated to be more than $200 billion, making it the most expensive natural disaster in history. Did Japan's investment pay off? Could the effects have been better minimized? The answer to both these questions is clearly *yes*! Like the lessons learned from the Kobe earthquake, which clearly saved thousands of lives in 2011, lessons from the Tohoku quake will undoubtedly save countless future lives, not only in Japan but also worldwide. The data recorded by Japan's seismometer network have provided the first detailed record of a megathrust earthquake. From this record, seismologists have already learned more about why both the earthquake and the tsunami were larger than predicted and expect that the data will greatly improve global earthquake forecasting.

Following the success of Japan's early warning system in 2011, United States experts have begun an aggressive push to acquire funds to implement a similar system along the West Coast. On March 11, 2013, a pilot early warning system in California detected a **M** 4.7 earthquake 160 km (~100 mi.) south of Pasadena (near Los Angeles) and alerted Caltech seismologists in Pasadena 30 seconds before minor shaking was felt. A robust early warning system for the entire state of California would cost about $80 million and could be operating within five years, but no substantial funds for this program have yet been approved.

APPLYING THE 5 FUNDAMENTAL CONCEPTS

1. Discuss Japan's successes and shortcomings in forecasting the 2011 Tohoku earthquake, using the earthquake history of the region and the application of modern tools used to detect precursory events.

2. Compare and contrast the seismic hazard to the citizens of the United States living on the west Coast of California to those living in Japan.

3. Explain the links between the Tohoku earthquake and other natural hazards. How do the effects of the earthquake and linked natural hazards compare?

4. Similar in size to Japan, California is growing at a rate higher than the national average and by 2100 its population could be about the size of Japan's population today. Discuss how rapid population growth in California combined with its current level of earthquake hazard preparedness could lead to a catastrophe if an event similar to the Tohoku earthquake struck the United States West Coast.

5. Summarize and evaluate Japan's strategy for minimizing the effects of earthquakes following the 1999 Kobe earthquake, which has earned the country the title "best equipped country in earthquake preparedness."

CONCEPTS in review

3.1 Introduction to Earthquakes

LO:1 Compare and contrast the different types of faulting.

- Earthquakes are common along tectonic plate boundaries where faulting is common.
- Faults are fractures where rocks on one side of the fracture have been offset with respect to rocks on the other side.
- Displacement is caused by compressional, tensional, or shearing stresses. Displacement can be mainly horizontal, as along a strike-slip fault, or mainly vertical, as on a dip-slip fault.

KEY WORDS
fault (p. 68), **footwall** (p. 68), **hanging wall** (p. 68), **normal fault** (p. 68), **reverse fault** (p. 68), **strain** (p. 68), **stress** (p. 68), **strike-slip fault** (p. 68)

3.2 The Earthquake Processes

LO:2 Explain the formation of seismic waves.

- A fault is usually considered active if it has moved during the past 10,000 years and potentially active if it has moved during the past 2 million years.
- Before an earthquake, elastic strain builds up in the rocks on either side of a fault as the sides pull in different directions. When the stress exceeds the strength of the rocks, they rupture, producing elastic rebound.
- Released elastic strain energy radiates outward in all directions from the ruptured surface of the fault in the form of seismic waves.
- Seismic waves are vibrations that compress (P) or shear (S) the body of Earth or travel across the ground as surface waves. Although P waves travel the fastest, the S and surface waves cause most of the shaking and damage. Some faults exhibit tectonic creep, a slow displacement not accompanied by felt earthquakes.

KEY WORDS
earthquake cycle (p. 70), **epicenter** (p. 71), **focus** (p. 71), **P waves** (p. 72), **paleoseismicity** (p. 70), **S waves** (p. 72), **seismic waves** (p. 72), **tectonic creep** (p. 73)

3.3 Earthquake Shaking

LO:3 Summarize the processes that lead to an earthquake and the release of seismic waves.

LO:4 Differentiate between the magnitude scales used to measure earthquakes.

- Large earthquakes release a tremendous amount of energy. Seismologists measure this energy on a magnitude (**M**) scale. On this scale, an increase from one whole number to the next represents a tenfold increase in the amount of shaking and a 32-fold increase in the amount of energy released.

- Earthquake intensity varies with the severity of shaking and is affected by proximity to the epicenter, the local geological environment, and the engineering of structures.
- Buildings highly subject to damage are those that (1) are constructed on unconsolidated sediment, artificially filled land, or water-saturated sediment, all of which tend to amplify shaking; (2) are not designed to withstand significant horizontal acceleration of the ground; or (3) have natural vibrational frequencies that match the frequencies of the seismic waves.

KEY WORDS
material amplification (p. 81), **Modified Mercalli Intensity scale** (p. 77), **moment magnitude scale** (p. 75), **Richter scale** (p. 75), **seismograph** (p. 75)

3.4 Geographic Regions at Risk from Earthquakes

LO:5 Identify global regions at most risk for earthquakes, and describe the effects of earthquakes.

- Most earthquakes occur on faults near tectonic plate boundaries, such as the San Andreas fault in California, the Cascadia subduction zone in the Pacific Northwest, and the Aleutian subduction zone in Alaska.
- Intraplate earthquakes are also common in Hawaii, the western United States, the southern Appalachians and South Carolina, and in the northeastern United States.
- Some of the largest historic earthquakes in North America occurred within the plate in the central Mississippi Valley in the early 1800s.

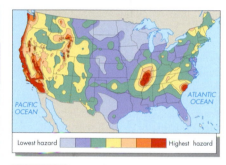

Lowest hazard Highest hazard

KEY WORD
megathrust earthquake (p. 85)

3.5 and 3.6 Effects of Earthquakes and Linkages with Other Natural Hazards Natural Service Functions of Earthquakes

LO:6 Describe how earthquakes are linked to other natural hazards.

LO:7 Discuss the important natural service functions of earthquakes.

- The primary effect of an earthquake is violent ground motion accompanied by fracturing, which may shear or collapse large buildings, bridges, dams, tunnels, pipelines, levees, and other structures.
- Other effects include liquefaction, regional subsidence, uplift of the land, landslides, fires, tsunamis, and disease.
- Earthquakes provide natural service functions such as enhancing groundwater and energy resources and exposing or contributing to the formation of valuable minerals deposits.

KEY WORDS
fault scarp (p. 91), **liquefaction** (p. 93), **resonance** (p. 92)

3.7 Human Interaction with Earthquakes

LO:8   Explain how human beings interact with and affect the earthquake hazard.

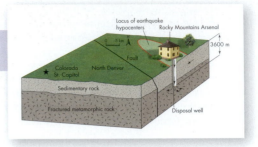

- Human activity has locally increased earthquake activity by fracturing rock and increasing water pressure underground below large water reservoirs, by raising fluid pressure in faults and fractures through the deep-well disposal of liquid waste, and by setting off underground nuclear explosions.
- Understanding how we have caused earthquakes may eventually help us control or stop large natural earthquakes.

3.8 Minimizing the Earthquake Hazard

LO:9   Discuss the United States governmental approach to minimizing the effects of earthquakes.

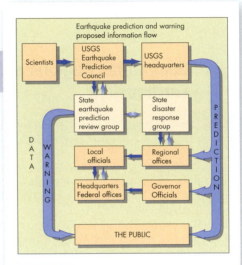

- Reducing earthquake hazards requires detailed mapping of geologic faults, the cutting of trenches to determine earthquake frequency, and detailed mapping and analysis of earth materials sensitive to shaking.
- Adjustments to earthquake hazards include improving structural design to better withstand shaking, retrofitting existing structures, microzonation of areas of seismic risk, and updating and enforcing building codes.
- To date, scientists have been able to make long- and intermediate-term forecasts for earthquakes using probabilistic methods but not consistent, accurate short-term predictions. A potential problem of predicting earthquakes is that their pattern of occurrence is often variable with clusters of events separated by longer periods of time with reduced earthquake activity.
- Early warning systems have been shown to be effective in Japan, but no such system exists in the United States or Canada.

KEY WORD
tectonic framework (p. 101)

3.9 Perception of and Adjustment to the Earthquake Hazard

LO:10   Propose ways to minimize seismic risk and suggest adjustments we can make to protect ourselves.

- Warning systems and earthquake prevention are not yet reliable alternatives to earthquake preparedness. More communities must develop emergency plans to respond to a predicted or unexpected catastrophic earthquake. Such plans should include earthquake education, disaster response, drills, and improved insurance coverage.
- At a personal level, individuals who live in or visit areas of seismic risk can learn how to "duck, cover, and hold on" before the next large earthquake. Preparation both reduces the earthquake hazard and eases recovery.

CRITICAL thinking questions

1. You live in an area that has a significant earthquake hazard. Public officials, the news media, and citizens are debating whether an earthquake warning system should be developed. Some people are worried that false alarms will cause a lot of problems, and others point out that the response time may be very long. What are your views on this issue? Should public funds be used to finance an earthquake warning system, assuming such a system is feasible? What are potential implications if a warning system is not developed and a large earthquake results in damage that could have been partially avoided with a warning system in place?

2. You are considering buying a home on the California coast. You know that earthquakes are common in the area. What questions would you ask before purchasing the home? For example, consider the effects of earthquakes, the relationship of shaking to earth material, and the age of the structure. What might you do to protect yourself (both financially and physically) if you decide to buy the house?

3. You are working for the Peace Corps in a developing country where most of the homes are built out of unreinforced bricks. There has not been a large damaging earthquake in the area for several hundred years, but earlier there were several earthquakes that killed thousands of people. How would you present the earthquake hazard to the people living where you are working? What steps might be taken to reduce the hazard?

References

1. **Ammon, C. J., Lay, T., Kanamori, H.,** and **Cleveland, M.** 2011. A rupture model of the 2011 off the Pacific coast of Tohoku Earthquake. *Earth Planets Space* 63: 693–96.

2. **Talbot, D.** 2011. 80 Seconds of Warning for Tokyo: Earthquake-detection investment pays off for Japan. *Massachusetts Institute of Technology (MIT) Review*, March 11, 2011, 2 p.

3. Learning from Earthquakes. The March 11, 2011, Great East Japan (Tohoku) Earthquake and Tsunami: Societal Dimensions. Earthquake Engineering Research Institute Special Earthquake Report, August, 2011, 23 p.

4. **Hanks, T. C.** 1985. The National Earthquake Hazards Reduction Program: Scientific status. U.S. Geological Survey Bulletin 1659.

5. **Bolt, B. A.** 2006. *Earthquakes: 2006 Centennial update*, 5th ed. San Francisco: W. H. Freeman.

6. **Cervelli, P.** 2004. The threat of silent earthquakes. *Scientific American* 290: 86–91.

7. **Radbruch, D. H.,** and **Lennert, B. J.** 1966. Damage to culvert under Memorial Stadium University of California. In *Tectonic creep in the Hayward fault zone California*, pp. 3–6. U.S. Geological Survey Circular 525.

8. **Steinbrugge, K. V.,** and **Zacher, E. G.** 1960. Creep on the San Andreas fault. In *Focus on environmental geology*, ed. R. W. Tank, pp. 132–37. New York: Oxford University Press.

9. **Melbourne, T. I.,** and **Webb, F. H.** 2003. Slow but not quite silent. *Science* 300: 1886.

10. **Chicago Tribune.** *2015 Experts gathered in Nepal a week ago to ready for earthquake.* www.chicagotribune.com/news/nationworld/chi-nepal-earthquake-experts-20150425-story.html. Accessed 5/19/17.

11. **Wang., K.,** and **Fialko, Y.** 2015. *Slip Model of the 2015 Mw 7.8 Gorkha (Nepal) earthquake from inversions of ALOS-2 and GPS Data.* American Geophysical Union.

12. **Kargel, J. S., Leonard, G. J., Shugar, D. H., Haritashya, U. K., Bevington, A., Fielding, E. J., Fujita, K., Geertsema, M.,** and **Miles, E.S.** 2016. *Geomorphic and geologic controls of geohazards induced by Nepal's 2015 Gorkha earthquake.* Interdisciplinary Arts andSciences Publications. Paper 342.

13. **New York Times.** Everest Climbers are Killed as Nepal Quake sets off Avalanche. www.nytimes.com/2015/04/26/world/asia/everest-climbers-killed-as-nepal-quake-sets-off-avalanche.html?_r=0. Accessed 5/19/17.

14. **Bloomberg Media.** 2015. Nepal says earthquake rebuilding cost to exceed $10 Billion. www.bloomberg.com/news/articles/2015-04–28/nepal-rebuilding-cost-to-exceed-10-billion-finance-chief-says. Accessed 5/19/17.

15. **Wald, D., Wald, L., Worden, B.,** and **Goltz, J.** 2003. *Shake Map—A tool for earthquake response.* U.S. Geological Survey Fact Sheet FS-087-03.

16. **Hough, S. E., Friberg, P. A., Busby, R., Field, E. F., Jacob, K. H.,** and **Borcherdt, R. D.** 1989. Did mud cause freeway collapse? *EOS, Transactions, American Geophysical Union* 70(47): 1497, 1504.

17. **Yeats, R. S.** 2001. *Living with earthquakes in California: A survivor's guide.* Corvallis: Oregon State University Press.

18. **Jones, R. A.** 1986. New lessons from quake in Mexico. *Los Angeles Times,* September 26.

19. **U.S. Geological Survey.** 2006. *Magnitude 7.6—Pakistan.* U.S. Geological Survey Earthquake Hazards Program. http://earthquake.usgs.gov/eqcenter/eqinthenews/2005/usdyae/. Accessed 6/9/07.

20. **Achenbach, J.** 2006. The next big one. *National Geographic.* 209(4): 120–47.

21. **U.S. Geological Survey.** 1996. *USGS response to an urban earthquake, Northridge '94.* U.S. Geological Survey Open File Report 96–263.

22. **USGS.** 2010. Magnitude 7.0 Haiti region. http://earthquake.usgs.gov. Accessed 3/15/10.

23. **Eberhard, M. O., Baldridge, S., Marshall, J., Mooney, W.,** and **Rix, G. J.** 2010. The M_w 7.0 of January 12, 2010: USGS/EERI Advance reconnaissance team report. U.S. Geological Survey Open-File Report 2010–1048. Executive Summary. Washington, DC.

24. **U.S. Geological Survey.** 2010. *Magnitude 8.8— offshore Maule Chile.* http://earthquake.usgs.gov/earthquakes/eqinthenews/2010/us2010tfan/. Accessed 8/11/10.

25. **Advisory Committee on the International Decade for Natural Hazard Reduction.** 1989. *Reducing disaster's toll.* National Research Council. Washington, DC: National Academy Press.

26. **Hough, S.** 2002. *Earthshaking science: What we know (and don't know) about earthquakes.* Princeton, NJ: Princeton University Press.

27. **Sieh, K.,** and **LeVay, S.** 1998. *The Earth in turmoil: Earthquakes, volcanoes and their impact on humankind.* New York: W. H. Freeman.

28. **Hamilton, R. M.** 1980. Quakes along the Mississippi. *Natural History* 89: 70–75.

29. **Mueller, K., Champion, J., Guccione, E. M.,** and **Kelson, K.** 1999. Fault slip rates in the modern New Madrid Seismic Zone. *Science* 286: 1140–38.

30. **U.S. Geological Survey Newsroom.** 2012. New evidence shows power of East Coast earthquakes. www.usgs.gov/newsroom/article.asp?ID=3447. Accessed 7/29/13.

31. **Yeats, R. S., Sieh, K.,** and **Allen, C. R.** 1997. *The geology of earthquakes.* New York: Oxford University Press.

32. **Eberhart-Phillips, D., Haeussler, P. J., Frey-mueller, J. T., Frankel, A. D., Rubin, C. M., Craw, P., Ratchkovsaki, N. A., Anderson, G., Carver, G. A., Crone, A. J., Dawson, T. E., Fletcher, H., Hansen, R., Harp, E. L., Harris, R. A., Hill, D. P., Hreinsdóttir, S., Jibson, R. W., Jones, L. M., Kayen, R., Keffer, D. K., Larsen, C. F., Moran, S. C., Personius, S. F., Plafker, G., Sherrod, B., Sieh, K., Sitar, N.,** and **Wallace, W. K.** 2003. The 2002 Denali fault earthquake, Alaska: A large magnitude, slip-partitioned event. *Science* 350: 1113–18.

33. **Fuis, G. S.,** and **Wald, L. A.** 2003. *Rupture in South-Central Alaska—the Denali earthquake of 2002.* U.S. Geological Survey Fact Sheet 014–03.

34. **Youd, T. L., Nichols, D. R., Helley, E. J.,** and **Lajoie, K. R.** 1975. Liquefaction potential. In *Studies for seismic zonation of the San Francisco Bay region,* ed. R. D. Borcherdt, pp. 68–74. U.S. Geological Survey Professional Paper 941A.

35. **Hansen, W. R.** 1965. *The Alaskan earthquake, March 32, 1964: Effects on communities.* U.S. Geological Survey Professional Paper 542A.

36. **Liu, J. G.,** and **Kusky, T.** 2008. After the quake. *Earth* 53(10): 48–51.

37. **U.S. Geological Survey.** 1996. *USGS response to an urban earthquake; Northridge '94.* U.S. Geological Survey Open-File Report 96–263.

38. **Pakiser, L. C., Eaton, J. P., Healy, J. H.,** and **Raleigh, C. B.** 1969. Earthquake prediction and control. *Science* 166: 1467–74.

39. **Evans, D. M.** 1966. Man-made earthquakes in Denver. *Geotimes* 10(9): 11–18.

40. **Reed, C.** 2002. Triggering quakes with waste. *Geotimes* 47(3): 7.

41. **Frohlich, C.,** and **Davis, S. D.** 2002. *Texas earthquakes.* Austin: University of Texas Press.

42. **Page, R. A., Boore, D. M., Bucknam, R. C.,** and **Thatcher, W. R.** 1992. *Goals, opportunities, and priorities for the USGS Earthquake Hazards Reduction Program.* U.S. Geological Survey Circular 1079.

43. **Field, E. H., Milner, K. R.,** and **the 2007 Working Group on California Earthquake Probabilities.** 2008. Forecasting California's earthquakes—What can we expect in the next 30 years? *U.S. Geological Survey Fact Sheet* 2008–2027.

44. **Committee on the Science of Earthquakes.** 2003. *Living on an active Earth: Perspectives on earthquake science.* National Research Council. Washington, DC: National Academies Press.

45. **Scholz, C.** 1997. Whatever happened to earthquake prediction? *Geotimes* 42(3): 16–19.

46. **State of California Uniform Building Code.** 1997. Chapter 16.

47. **State of California.** 2008. California Geological Survey—Probabilistic seismic hazards assessment— Peak ground acceleration. www.conservation.ca.gov. Accessed 11/30/08.

48. **U.S. Geological Survey.** 2008. Deterministic and scenario ground-motion maps. www.earthquake.usgs.gov/research/hazmaps/scenario. Accessed 8/11/10.

49. **Wells, D. L.,** and **Coppersmith, K. J.** 1994. New empirical relationships among magnitude, rupture length, rupture width, rupture area, and surface displacement. *Bulletin of the Seismological Society of America* 84: 974–1002.

50. **Press, F.** 1975. Earthquake prediction. *Scientific American* 237: 14–23.
51. **Scholz, C. H.** 2002. *The mechanics of earthquakes and faulting*, 2nd ed. New York: Cambridge University Press.
52. **Rikitakr, T.** 1983. *Earthquake forecasting and warning*. London: D. Reidel.
53. **Silver, P. G.,** and **Wakita, H.** 1996. A search for earthquake precursors. *Science* 273: 77–78.
54. **Allen, C. R.** 1983. Earthquake prediction. *Geology* 11: 682.
55. **Hait, M. H.** 1978. Holocene faulting, Lost River Range, Idaho. *Geological Society of America Abstracts with Programs* 10(5): 217.
56. **Gori, P. L.** 1993. The social dynamics of a false earthquake prediction and the response by the public sector. *Bulletin of the Seismological Society of America* 83: 963–80.
57. **Yeats, R. S.** 2004. *Living with earthquakes in the Pacific Northwest*, 2nd ed. Corvallis: Oregon State University Press.
58. **Holden, R., Lee, R.,** and **Reichle, M.** 1989. *Technical and economic feasibility of an earthquake warning system in California*. California Division of Mines and Geology Special Publication 101.
59. **Whitman, R. V., Anagon, T., Kircher, C. A., Lagurio, H. J., Lawson, R. S.,** and **Schneider, P.** 1997. Development of a national earthquake loss estimation methodology. *Earthquake Spectra* 13(4): 643–61.
60. **Southern California Earthquake Center.** 2007. *Putting down roots in earthquake country.* Los Angeles: University of Southern California.
61. **Coburn, A.,** and **Spence, R.** 1992. *Earthquake protection.* New York: John Wiley & Sons.
62. **Stein, W.,** and **Wysession, M.** 2003. *An introduction to seismology, earthquakes, and earth structure.* Malden, MA: Blackwell.
63. **Choi, C. Q.** 2011. Japan quake's size surprised scientists. Live Science. www.livescience.com/13744-japan-earthquake-size-suprising.html Accessed 7/24/13.
64. **Cyranoski, D.** Systems for forecasting, early warning and tsunami protection all fell short on 11 March, 2011, *Nature* 471: 556–57.
65. **Ashford, S. A., Boulanger, R. W., Donahue, J. L.,** and **Stewart, J. P.** 2011. Geotechnical Quick Report on the Kanto Plain Region during the March 11, 2011, Off Pacific Coast of Tohoku Earthquake, Japan. Geotechnical Extreme Events Reconnaissance (GEER).
66. **Wartman, J., Dunham, L., Tiwari, B.,** and **Pradel, D.** 2013. Landslides in Eastern Honshu induced by the 2011 Tohoku earthquake. *Bulletin of the Seismological Society of America* 103(2B): 1503–21.

Cooling systems failed as pumps used **to circulate cooling water** in three nuclear reactors were **flooded by the tsunami**

4

APPLYING the **5** fundamental concepts Japan 2011 Tsunami and Nuclear Disaster

❶ Science helps us predict hazards.

Tsunamis result from large vertical movement of the ocean floor. Such movement is typical of large subduction zone earthquakes, like the 2011 Tohoku earthquake. By understanding this relationship, our ability to predict regions susceptible to the effects of tsunamis is greatly improved.

❷ Knowing hazard risks can help people make decisions.

Following the 2011 tsunami, Japan has reevaluated the risk of damage to land and property from future large tsunamis. Continued risk analysis provides an important example of how future natural hazards need to be viewed from a risk management perspective and how risk is linked to numerous changing factors, such as population growth.

Tsunamis

Japan 2011: Tsunami and Nuclear Disaster

The Pacific Ocean bottom 130 km (~80 mi.) offshore of Fukushima, Japan was vertically displaced by as much as 9 m (~30 ft.) due to the 2011 magnitude 9.0 Tohoku earthquake. This displacement of the ocean floor generated a large set of tsunami waves up to about 40 m (~130 ft.) high at the shore. These waves quickly reached the coastline of northern Japan where millions of people live and several nuclear power plants were located.

Within seconds to 3 minutes of the earthquake, an automated tsunami warning system alerted the people of coastal communities to evacuate to high ground as they had been taught, and the nuclear power plants immediately shut down as they were designed to do. Although the warning system greatly minimized the loss of life, the nuclear reactor shutdown did not prevent a disaster that was ultimately more serious than any other nuclear accident of the last several decades. Like many of the sea walls that protect 40 percent of Japan's coastline from tsunami waves, the 6 m (~20 ft.) high sea wall designed to keep the power plant from flooding was quickly overtopped. Cooling systems failed when back-up diesel generators powering the circulation pumps were inundated by seawater,

◁ Tsunami Overriding Seawall Defenses

About 1 hour and 30 minutes after the Tohoku earthquake, the 6 m (~20 ft.) high sea wall protecting this seaside village on the east coast of Japan is overtopped by the first tsunami wave to hit the coast. In this photograph, the seawall is completely obscured by the sea water cascading over it onto the road below. *(Toru Yamanaka/AFP/Getty Images)*

The tsunami is one of Earth's most destructive natural hazards. These ocean waves are both fascinating in their behavior and awesome in their power. Tsunamis are common in some coastal regions and rare in others. Although they have long been known to cause disasters and catastrophes, the hazard posed by tsunamis has generally been underestimated. Translating the increased hazard awareness into improved warning, preparedness, and mitigation is proceeding at an excruciatingly slow pace. This chapter will explain the natural tsunami process and assess the hazard that these waves pose to people.

After reading this chapter, you should be able to:

LO:1 Explain the process of tsunami formation and development.

LO:2 Locate on a map the geographic regions that are at risk for tsunamis.

LO:3 Synthesize the effects of tsunamis and the hazards they pose to coastal regions.

LO:4 Summarize the linkages between tsunamis and other natural hazards.

LO:5 Understand that tsunamis are not caused by or affected by human activities, but damages are compounded as coastal populations increase.

LO:6 Discuss what nations, communities, and individuals can do to minimize the tsunami hazard.

LO:7 Identify the actions you should take and not take if a tsunami warning is issued.

3

Linkages exist between natural hazards.

Flooding and coastal erosion due to earthquake-generated tsunami waves serves as an exceptional example of how different natural hazards are linked.

4

Humans can turn disastrous events into catastrophes.

Nuclear reactor meltdown following tsunami inundation triggered the world's second worst nuclear disaster. Although no deaths are attributed directly to the meltdown, evacuation of the contaminated region doubled the loss of life in the region.

5

Consequences of hazards can be minimized.

The high level of Japanese tsunami preparedness, including an advanced early warning system, elaborate tsunami seawall defense, and a well-informed, vigilant population, resulted in a 96 percent survival rate for those living in the areas inundated by the tsunami.

resulting in a nuclear meltdown (overheating of the reactor core causes nuclear fuel to melt). The disaster was compounded by explosions of hydrogen gas that built up in the reactor core containment buildings due to the meltdown, causing a significant release of radiation into the atmosphere[1-5] and the need to evacuate nearly 200,000 people living in the region. Six years after the disaster (March 2017), the Japanese are still pumping several hundred tons of water through the buried hot nuclear fuel in an attempt to cool it. The water becomes contaminated with radiation and is collected and stored in over 1,000 large tanks. The Japanese are struggling with how the radioactive water might be treated and disposed of when storage space for more new tanks runs out.

Due to the 2011 and nuclear incident, Japan initiated new research regarding its tsunami hazard, and other countries around the world have begun a new discussion about the safety of nuclear power.

4.1 Introduction to Tsunamis

Tsunamis (the Japanese word for "large harbor waves") are produced by the sudden vertical displacement of ocean water.[6] These waves are a serious natural hazard that can cause a catastrophe thousands of kilometers from where they originate. They may be triggered by several types of events, including a large earthquake that causes a rapid uplift or subsidence of the seafloor; an underwater landslide that may be triggered by an earthquake; the collapse of part of a volcano that slides into the sea; a submarine volcanic explosion; and an impact in the ocean of an extraterrestrial object, such as an asteroid or comet. Asteroid impact can produce a "mega" tsunami, a wave that is about 100 times higher than the largest tsunami produced by an earthquake, that could put hundreds of millions of people at risk.[6] Fortunately, the frequency of large asteroid impact is low. Of the previously mentioned potential causes, tsunamis produced by earthquakes are by far the most common.

Damaging tsunamis in historic times have been relatively frequent, mostly in the Pacific Basin, and include the following:[7,8]

> The 1755 (~**M** 9) Lisbon, Portugal, earthquake produced a tsunami that, along with the earthquake and resulting fire, killed an estimated 20,000 people. Tsunami waves that crossed the Atlantic Ocean amplified to heights of 7 m (~23 ft.) or more in the West Indies.

> The 1883 violent explosion of Krakatoa volcano in the Sundra Strait between Java and Sumatra

caused the top of the volcano to collapse into the ocean. This sudden collapse produced a giant tsunami more than 35 m (~115 ft.) high that destroyed 165 villages and killed more than 36,000 people.

> The 1946 (**M** 8.1) Aleutians (Alaska) earthquake produced a tsunami in the Hawaiian Islands that killed about 160 people.

> The 1960 (**M** 9.5) Chile earthquake triggered a tsunami that killed 61 people in Hawaii after traveling for 15 hours across the Pacific Ocean.

> The 1964 (**M** 9.2) Alaska earthquake generated a deadly tsunami that killed about 130 people in Alaska and California.

> The 1993 (**M** 7.8) earthquake in the Sea of Japan caused a tsunami that killed 120 people on Okushiri Island, Japan.

> The 1998 (**M** 7.1) Papua New Guinea earthquake triggered a submarine landslide that produced a tsunami that killed more than 2,100 people.

> The 2004 (**M** 9.1) Sumatra earthquake generated a tsunami that killed about 230,000 people (see Case Study 4.2: Indonesian Tsunami).

> The 2009 (**M** 8.1) Samoa earthquake generated a tsunami that killed about 200 people.

> The 2010 (**M** 8.8) Chile earthquake generated a tsunami that killed about 700 people in coastal towns.

> The 2011 (**M** 9.1) Japan earthquake generated a tsunami that killed over 20,000 people (see Applying the Fundamental Concepts).

HOW DO EARTHQUAKES CAUSE A TSUNAMI?

An earthquake can cause a tsunami by causing the seafloor to move or by triggering a landslide. Seafloor movement is probably the more common of these two mechanisms. This movement occurs when the seafloor sits on a block of Earth's crust that shifts up or down during a quake. In general, it takes an earthquake of **M** 7.5 or greater to create enough displacement of the seafloor to generate a damaging tsunami. The upward or downward movement of the seafloor displaces the entire mass of water from the sea bottom to the ocean surface. This starts a four-stage process that eventually leads to landfall of tsunami waves on the shore (Figure 4.1):

1. For example, if an earthquake rupture uplifts the seafloor, the water surface above the uplift initially forms an elongated dome parallel to the geologic

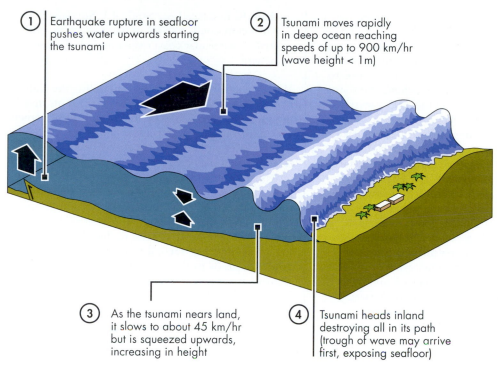

① Earthquake rupture in seafloor pushes water upwards starting the tsunami

② Tsunami moves rapidly in deep ocean reaching speeds of up to 900 km/hr (wave height < 1m)

③ As the tsunami nears land, it slows to about 45 km/hr but is squeezed upwards, increasing in height

④ Tsunami heads inland destroying all in its path (trough of wave may arrive first, exposing seafloor)

∧ FIGURE 4.1 Tsunami
Idealized diagram showing the process of how a tsunami is produced by an earthquake. *(Modified after the United Kingdom Hydrographic Office)*

fault. That dome collapses and generates a tsunami wave. Oscillations of the water surface and, in some cases, aftershocks along the fault produce additional waves. These waves radiate outward like the pattern made by a pebble thrown into a pond of water.

2. In the deep ocean, the tsunami waves move very rapidly and are spaced long distances apart. Their velocity is equal to the square root of the product of the acceleration of gravity and the water depth. The acceleration of gravity is approximately 10 m/sec² and average water depth in deep ocean is 4000 m (~13,100 ft.). Thus, when we do the math, by taking the product of 10 m/sec² and 4000 m and then taking the square root of that number, we arrive at a velocity of 200 m/sec. Converting 200 m/sec to km/hour, we find that tsunamis travel at 720 km (~450 mi.) per hour, close to the average airspeed of a jetliner, but they can reach up to 900 km/hr. In the deep ocean, the spacing between the crests of tsunami waves may be more than 100 km (~60 mi.), and the height of the waves is generally less than 1 m (~3 ft.). Consequently, sailors rarely notice a passing tsunami in the deep ocean.

3. As the tsunami nears land, the water depth decreases, so the velocity of the tsunami also decreases. Near land, the forward speed of a tsunami may be about 45 km (~28 mi.) per hour—too fast to outrun but not nearly as fast as in the open ocean. This decrease in velocity also decreases the spacing between wave crests, that is, the wavelength. As the water slows down and piles up, the height of the waves increases.

4. When the first tsunami wave reaches the shore and moves inland, it may be several meters to several tens of meters high and destroys everything in its path. Sometimes the trough of the wave may arrive first, which exposes the seafloor. When tsunamis strike the coast, they generally do not come in as a giant breaking wave as you might see on the surfing beaches of the north shore of Oahu or the California coast. Instead, when the wave arrives, it is more like a very strong and fast-rising increase in sea level. On the rare occasions when tsunamis do break, they may appear as a vertical wall of turbulent water. The movement of a tsunami inland is called the **runup** of the wave. Runup refers to the furthest horizontal and vertical distance that the largest wave of a tsunami

moves inland. Runup from tsunamis varies considerably with the shape of the seafloor immediately offshore and with the type of topography and vegetation landward of the beach. Both the offshore and shoreline topography can greatly accentuate tsunami wave height. Figure 4.2 shows the runup (~wave height at coast) from the 2011 tsunami. Notice that the greatest runup of about 25 m+ occurred directly offshore from the earthquake. Near the nuclear power plant (red circle), it was about 12 m (~40 ft.). Other coastal areas received runup of less than 5m (16ft). The 2011 tsunami moved inland from 1 to 8 km (0.6 to 5 mi.), depending on wave height and topography. Once a wave has moved to its farthest extent inland, most of the water then returns back to the open ocean in a strong and often turbulent flow. A tsunami can also generate other types of waves, known as *edge waves*, that travel back and forth parallel to the shore. The interaction between edge waves and additional incoming tsunami waves can be complex. As a result of this interaction, wave amplification may occur, causing the second or third tsunami wave to be even larger than the first (see Case Study 4.1: Survivor Story). Most commonly, a series of tsunami waves will strike a particular coast over a period of several hours.[9]

When an earthquake ruptures and uplifts the seafloor close to land, both distant and local tsunamis may be produced (Figure 4.3). The initial fault displacement and uplift raise the water above mean sea level. This uplift creates potential energy that drives the horizontal movement of the waves. The uplifted dome of water collapses downward and expands radially, splitting into two waves. One wave, known as a **distant tsunami**, travels out across the deep ocean at high speed. Distant tsunamis can travel thousands of kilometers across the ocean to strike remote shorelines with little loss of energy. The second wave, known as the **local tsunami**, heads in the opposite direction toward the nearby land. A local tsunami can arrive quickly following an earthquake, giving coastal residents and visitors little warning time. When the initial tsunami wave is split, each (distant and local) tsunami has a wave height about one-half of that of the original dome of water.[9]

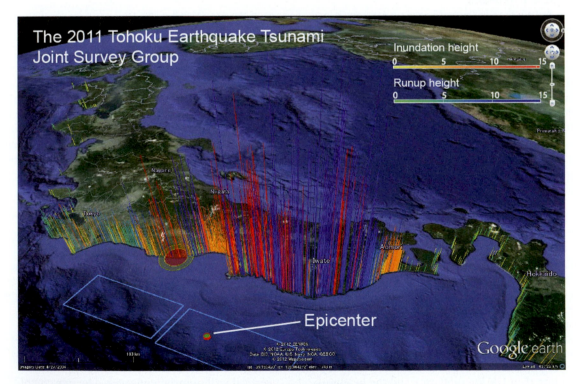

∧ FIGURE 4.2 Tsunamis runup height and inundation from the 2011 Tsunami
The epicenter of the earthquake is 130 km offshore. The red circle is the location of a nuclear power plant. Notice the height is extremely variable, depending on location. *(The 2011 Tokohu Earthquake Tsunami Joint Joint Survey Group, 2012)*

Tsunami in the Lowest Country on Earth

Boxing Day, December 26, 2004, was a hot sunny day in the Maldives, a beautiful group of atolls in the Indian Ocean west of India (Figure 4.1.A). The Republic of Maldives is the lowest country on Earth; its highest point is only 2.4 m (~8 ft.) above sea level. Dave Lowe was working in his office at a resort on South Ari Atoll when he heard a strange bump on the door,

and outside people were screaming "the children." He opened the door, and to his horror, the ocean was level with the island, and a boiling, frothing wall of water was bearing straight down on him: "There was a strange mist that looked like a fog, so I stopped breathing and tried to decide where to run."

The resort had no two-story buildings, and he was just a meter

(~3 ft.) above sea level! He ran to the reception area where there were pillars to hold on to. Others were there screaming as the first waves began to hit: "Three glass windows exploded, and, within seconds, the water was up to my waist. I couldn't tell if the island was sinking or if the sea was rising."

Dave made his way to the reception counter, grabbing two children

˄ FIGURE 4.1.A Resort Employees Climb Down from Tsunami Refuge
The roof of this South Ari Atoll resort in the Maldives Islands was the highest place to take refuge from the multiple tsunami waves that flooded the island. On islands closer to the earthquake epicenter, this building would have been destroyed by the waves. *(SIPA Press)*

who were being swept out to sea, and threw them up onto the counter. Wave after wave hit, and water passed over his head: "I knew I was going to die and blacked out from fear. I became conscious when a receptionist was yelling, 'What is happening?'"

As quickly as the water came up, it was gone, leaving fish flopping on the lobby floor and seaweed draped everywhere. The resort guests regrouped. Then Dave saw a second wave coming. It was worse than the first. People desperately tried to hang onto anything they could. Wave after wave came for the next 6 hours, as the sea gradually calmed down. Guests suffered in the hot sun, and the resort staff built a shelter for 15 children whose parents were missing. When darkness came, people stood watch, looking for more waves under a full moon. No one slept, and in the morning, debris was everywhere: "It looked like *Titanic*, *Lost*, *Lord of the Flies*, and *Survivor* all rolled into one." Dave and other employees were evacuated by seaplane two days later. As they flew over the islands, they saw the unbelievable devastation, and several staff broke down crying.

They had survived, but they didn't know how

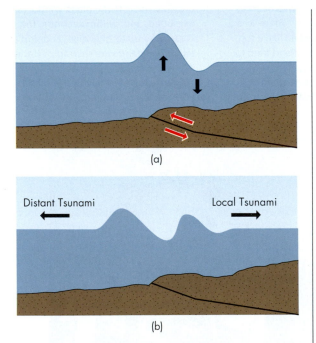

(a)

(b)

▲ FIGURE 4.3 Distant and Local Tsunamis
(a) Fault displacement lifts the water above mean sea level, which creates potential energy that drives the horizontal propagation of the waves. In these diagrams, wave height is greatly exaggerated compared to the depth of the water. Actual height of the waves is at most a meter or so, but it is spread out to tens or several hundreds of kilometers in length. (b) The initial wave is split into a tsunami that travels out across the deep ocean (distant tsunami) and another tsunami that travels toward nearby land (local tsunami). The wave heights are also split, and each (distant and local) tsunami has a height about one-half the original wave in (a). *(Based on U.S. Geological Survey; http://walrus.wr.usgs.gov/tsunami/basics.html)*

An example of a local tsunami occurred on the island of Okushiri in western Japan. On July 12, 1993, a **M** 7.8 earthquake in the Sea of Japan produced a local tsunami that extensively damaged Anoae, a small town on the southern tip of the island. The town was struck several times by 15 to 30 m (~50 to 100 ft.) waves coming around both sides of the island. There was virtually no warning because the earthquake epicenter was very close to the island. The huge waves arrived only 2 to 5 minutes after the earthquake, killing 120 people and causing $600 million in property damage.[10]

HOW DO LANDSLIDES CAUSE A TSUNAMI?

Although most large tsunamis are produced by fault rupture along subduction zones at tectonic plate boundaries, landslides have both generated and contributed to huge tsunamis.[11,12]

These landslides can take place underwater, where they are referred to as *submarine landslides*, or they can be large rock avalanches that fall from mountains into the sea. In most cases, the landslides are triggered by an earthquake. For example, in 1998, a **M** 7.1 earthquake occurred off the north shore of the island of New Guinea. The earthquake was felt at Sissano Lagoon, and shortly thereafter a tsunami arrived with little warning. In less than an hour, coastal villages were swept away, leaving 12,000 people homeless and more than 2,000 dead. The tsunami waves, which reached heights of 15 m (~50 ft.), appeared to have resulted primarily from a submarine landslide triggered by the earthquake.[13] The epicenter of the earthquake was only about 50 km (~30 mi.) offshore, so there was little, if any, warning time for coastal residents. This event emphasized the potentially devastating damage that can occur from a large tsunami produced by a nearby earthquake and submarine landslide.[8,13] The earthquake itself probably would not have generated a large tsunami had it not been combined with a landslide that displaced water at the bottom of the sea.

A submarine landslide may have significantly contributed to the 2011 Japan tsunami,[11] and recent studies of seafloor images have apparently solved the 1964 mystery of the tsunami associated with the Great Alaskan Earthquake that hit the Alaskan village of Chenega in the Prince William Sound. The tsunami destroyed all but two of the buildings and killed 23 of the 75 people living there. The origin of the tsunami was uncertain; was the tsunami due to the earthquake alone or to the combination of the earthquake and landslides? The discovery of a large underwater landslide deeper than the area studied in 1964 suggests the tsunami that destroyed the village was largely due to submarine landsliding.[12]

Perhaps the most famous landslide-induced giant wave occurred at Lituya Bay, Alaska, in 1958. The landslide was set in motion by a **M** 7.7 earthquake on a nearby fault. Approximately 30.5 million cubic meters (~46 million cubic yards) of rock fell from a cliff into the bay, instantly displacing a huge volume of seawater.[7] That volume of earth material would fill the National Mall in Washington, DC, to a height of about 25 m (~80 ft.). The huge mass of broken rock caused waters in the bay to surge upward to an elevation of about 524 m (~1,720 ft.) above the normal water level.[7] This splash of water was so tall that it would have washed over the Sears Tower in Chicago with 82 m (~269 ft.) to spare! Although the wave that swept out of the bay into the ocean wasn't nearly as large, it did pick up a fishing boat with two people onboard. The boat's captain estimated that the tsunami wave was 30 m (~100 ft.) high when it lifted the boat over the sand spit at the mouth of the bay. Two other fishing boats in the bay disappeared after being caught in the wave.[14] A panoramic view of the bay taken a few days after the tsunami shows that the erosive force of the wave stripped the shoreline bare of trees (Figure 4.4). In the area beyond the direct splash effect of the fallen rock, the high waterline above the barren shoreline indicates that the local tsunami had a runup height of 20 to 70 m (~65 to 230 ft.).[7]

The tsunami risk of high, unstable rock slopes associated with fjords (long, narrow, deep, cliff-sided inlets along the coast), such as in Alaska and Norway, that collapse by rockfall landslides into the sea is now recognized.[15] Norway experienced three tsunami disasters due to subareal rockslides in the twenty-first century. One in 1934 was up to 60 m (~200 ft) high and killed 40 people. A potential tsunami from one 900 m high mountain looming over a fjord with unstable slopes (large cracks have been mapped) could generate a giant landslide with local a tsunami up to 80 m (~260 ft) high. As a result, Norway is working on hazard risk evaluation from tsunamis. Included are hazard risk maps and land-use planning, including hazard zoning to restrict

Location of landslide

Trees removed by wave

∧ FIGURE 4.4 Landslide-Produced Giant Wave Lituya Bay, Alaska, following local tsunami. Denuded shoreline of the bay was produced as trees were removed by the erosive force of the wave. *(USGS)*

building in hazardous areas. The most recent event in Nowway was in 2014 when an apparent submarine rock slide produced a tsunami as much as 15 m high that damaged or destroyed homes, boats, and vehicles. So, both subareial and submarine landslides can produce a serious to catastrophic tsunami risk.[15]

4.1 CHECK your understanding

1. Define *tsunami* and identify several tsunami trigger events.
2. Describe how earthquakes produce a tsunami.
3. What is the difference between a local tsunami and a distant tsunami?
4. Define tsunami *runup*.

4.2 Geographic Regions at Risk from Tsunamis

Although all ocean and some lake shorelines are at risk from tsunamis, certain coasts are more at risk than others. The heightened risk comes from the geographic location of a coast in relation to potential tsunami sources, such as earthquakes, landslides, and volcanoes. Coasts in near proximity to a major subduction zone, or directly across the ocean basin from a major subduction zone, are at greatest risk. For the purpose of this discussion, a major subduction zone is considered to be capable of periodically generating a major earthquake of

M 9 or greater. Ruptures produced by these great earthquakes may extend for 1000 km (~625 mi.) or more along the subduction zone and produce significant uplift of the seafloor (see Case Study 4.2: Indonesian Tsunami). Earthquake and subduction zone processes that can lead to a large tsunami are illustrated in Figure 4.5 and discussed in the following paragraph.

As you recall from Chapter 3, the interval between large earthquake-generating fault slip events along any one particular plate boundary segment may be hundreds of years. In subduction zones, this is due to an area of strong coupling between the overriding and subducting plates, where no fault slip occurs. Such a fault segment is referred to as a *locked fault* (Figure 4.5a). Because there is no displacement along the interface between the two plates, the plates must deform elastically. As Figure 4.5b shows, as the subducting plate drags the leading edge of the overriding plate downward, subsidence (sinking) occurs offshore while onshore uplift occurs. Between earthquakes, there is distortion, deformation, or bulging (uplift) and leading-edge subsidence (sinking). When the fault finally slips and a tsunami-generating earthquake occurs, the overriding plate rebounds elastically. This causes the ocean floor to uplift, triggering a tsunami, and the on-land bulge to collapse and coastal areas

to subside (Figure 4.5c,d). Sometimes, forests near a shoreline may sink several meters (up to 10 feet) or so, leaving a drowned forest (called a ghost forest; see Figure 4.6).[16] The 1964 great Alaskan earthquake caused coastal forests to be inundated by salt marshes as the land subsided. These drowned, dead trees may persist long enough in the landscape to be recognized, while others are buried by coastal sediment to become part of the recent geologic record.

The greatest tsunami hazard, with return periods of several hundred years, is adjacent to those major subduction zones with a convergence of a few centimeters per year. These include the Cascadia subduction zone in the Pacific Northwest, the Chilean trench along southwestern South America, and the subduction zones off the coast of Japan.

Tsunamis can range in height from a few centimeters to 30 m (~100 ft.) or more. Those with a runup height of at least 5 m (~16 ft.) are considered significant tsunamis and are commonly produced by high-magnitude earthquakes and associated submarine slides. One assessment of global tsunami hazard ranks the hazard from relatively low to greatest based on the return period of a significant tsunami (at least 5 m [~16 ft.] runup) (Figure 4.7). However, not enough is known about

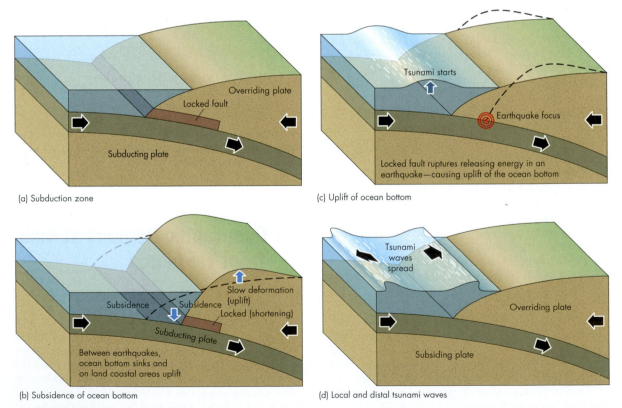

(a) Subduction zone

(c) Uplift of ocean bottom

(b) Subsidence of ocean bottom

(d) Local and distal tsunami waves

∧ FIGURE 4.5

Idealized diagram of subduction zone cycle that produces large earthquakes and tsunamis.
(Based on U.S. Geological Survey Circular 1187)

◄ FIGURE 4.6 Ghost Forest
Drowned forest in Washington produced by subsidence during an earthquake about 300 years ago. *(USGS/http://pubs. usgs. gov/pp/pp1707/)*

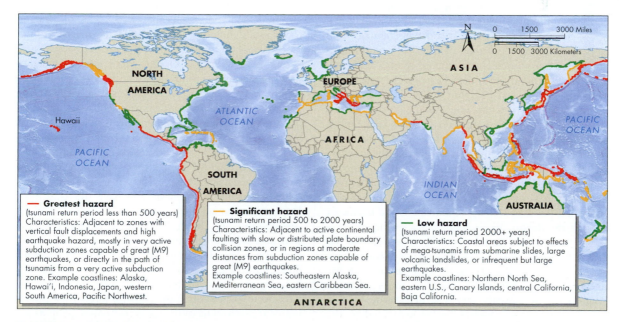

— Greatest hazard
(tsunami return period less than 500 years)
Characteristics: Adjacent to zones with vertical fault displacements and high earthquake hazard, mostly in very active subduction zones capable of great (M9) earthquakes, or directly in the path of tsunamis from a very active subduction zone. Example coastlines: Alaska, Hawai'i, Indonesia, Japan, western South America, Pacific Northwest.

— Significant hazard
(tsunami return period 500 to 2000 years)
Characteristics: Adjacent to active continental faulting with slow or distributed plate boundary collision zones, or in regions at moderate distances from subduction zones capable of great (M9) earthquakes.
Example coastlines: Southeastern Alaska, Mediterranean Sea, eastern Caribbean Sea.

— Low hazard
(tsunami return period 2000+ years)
Characteristics: Coastal areas subject to effects of mega-tsunamis from submarine slides, large volcanic landslides, or infrequent but large earthquakes.
Example coastlines: Northern North Sea, eastern U.S., Canary Islands, central California, Baja California.

⋀ FIGURE 4.7 Global Tsunami Hazard
Map of the relative hazard of coastlines to experience a tsunami that is at least 5 m (~16 ft.) high. Fjords in Norway (not shown) also have a very significant hazard from landslide generated local tsunamis.
(Modified from Risk Management Solutions. 2006. 2004 Indian Ocean Tsunami report. Newark, CA: Risk Management Solutions, Inc.)

global seismicity to accurately predict return periods of earthquakes of **M** 9 or above—those quakes most likely to produce a significant tsunami. Examination of the map reveals that most zones of greatest hazard surround the Pacific Ocean. This is no surprise, given that most major subduction zones are found on the margins of the Pacific. Other regions judged to have the greatest hazard are parts of the Mediterranean, as well as the northeastern side of the Indian Ocean. The good news about the risk from significant tsunamis is that almost all of them will cause substantial damage only within a couple of hundred kilometers from their source. However, the very largest subduction zone earthquakes may cause substantial damage several thousands of kilometers from the source. That certainly was the case with the 2004 Indonesian tsunami, which raced across the Indian Ocean causing death and destruction as far away as the east coast of Africa. However, most of the death and destruction occurred much closer to the earthquake epicenter, as described in Case Study 4.2: Indonesian Tsunami.

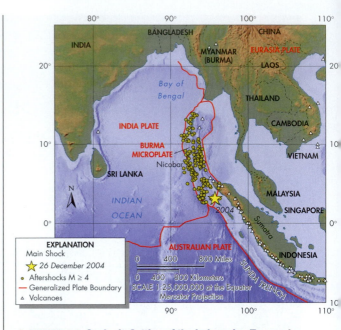

∧ FIGURE 4.8 Geologic Setting of the Indonesian Tsunami The gold star locates the epicenter of the mainshock of the **M** 9.1 earthquake on December 26, 2004. *(Based on U.S. Geological Survey)*

4.2 CASE study

Indonesian Tsunami

Prior to December 26, 2004, few people knew what the Japanese word *tsunami* meant. That changed in the span of only a few hours, as close to 230,000 people were killed, many hundreds of thousands injured, and millions displaced in more than a dozen countries surrounding the Indian Ocean. With no warning system in place, residents of coastal area after coastal area around the Indian Ocean were struck by a series of tsunami waves without notice.

The source of this tsunami was the world's largest earthquake in the past four decades that struck on the morning of December 26, just off the west coast of the Indonesian island of Sumatra (Figure 4.8). Now assigned a magnitude of at least 9.1, this earthquake caused the most lethal tsunami in recorded history.[17] The earthquake occurred where the giant Indian and Australian plates are being subducted to the northeast beneath the Burma microplate. In this earthquake, the Burma microplate moved about 20 m (~65 ft.) to the west-southwest along a gently inclined subduction zone (Figure 4.9). Because of the large amount of displacement along the thrust faults in the subduction zone, geologists classify this type of earthquake as a "mega-thrust event." Seismic waves generated by the rupture caused several minutes of shaking on nearby islands. The total length of the rupture along the subduction zone was more than 1500 km (~930 mi.), about

the length of the state of California. Not only did the seafloor shift about 20 m (~65 ft.) horizontally, but also it rose several meters vertically.[14] Displacement of the sea bottom disturbed and displaced the overlying waters of the Indian Ocean, and a series of tsunami waves was generated. Waves radiated outward at high speed across the Indian Ocean (Figure 4.10).

Unlike the Pacific Ocean, there was no tsunami warning system for the Indian Ocean, and people were mostly caught by surprise. Scientists on duty at the Pacific Tsunami Warning Center in Hawaii recognized that the earthquake could potentially produce a tsunami, but the geophysical techniques to calculate the size of an earthquake of such magnitude were not immediately available to them.[18] Once they determined that a tsunami was likely, the scientists managed to contact other scientists in Indonesia and have the U.S. State Department relay its concerns to some nations surrounding the Indian Ocean. Unfortunately, the warnings did not reach authorities in time to take action, and even if they had, there was no system in place for directly notifying coastal residents. If the warnings had been received in time, tens of thousands of lives could have been saved because the tsunami waves took several hours to reach some of the coastlines where people died (Figure 4.10). Even a warning of half an hour or so before the tsunami hit would have given sufficient time to move many people from the low-lying coastal areas.[6]

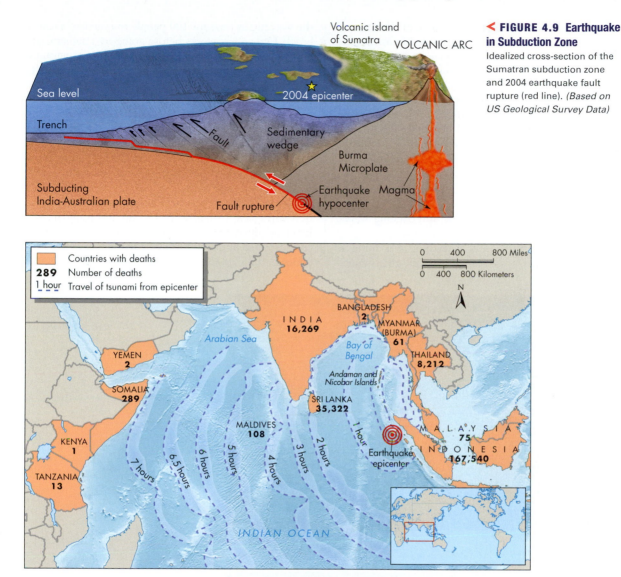

< FIGURE 4.9 Earthquake in Subduction Zone
Idealized cross-section of the Sumatran subduction zone and 2004 earthquake fault rupture (red line). *(Based on US Geological Survey Data)*

∧ FIGURE 4.10 2004 Tsunami Killed People on Both Sides of Indian Ocean
Map showing the path of the tsunami produced by a **M** 9.1 earthquake off the west coast of Sumatra, Indonesia, on December 26, 2004. Shown with blue dashed lines is the location of the lead tsunami wave each hour after the earthquake. Note that the tsunami waves took approximately 7 hours to reach Somalia on the east coast of Africa. Most deaths were in Indonesia, where the first local tsunami wave arrived an hour or less after the earthquake. The number of deaths shown in each country is the total number of people dead or still missing. *(Data from Causalities summarized in Telford, J., and Cosgrave, J., Joint Evaluation of the International Response to the Indian Ocean Tsunami: Synthesis Report. London: Tsunami Evaluation Coalition; Tsunami travel time data from NOAA, 2006)*

More than three-quarters of the deaths were in Indonesia, which experienced both the intense shaking from the earthquake and the inundation by tsunami waves within less than an hour. In contrast, the first tsunami wave took between 90 minutes and 2 hours to reach Sri Lanka and India, where many thousands died, and more than 7 hours to reach Somalia on the east coast of Africa (see Figure 4.10). The first of many tsunami waves reached other locations earlier or later, depending on their distance from the rupture in the subduction zone.

At the northern tip of the island of Sumatra, the Indonesian provincial capital of Banda Aceh was nearly destroyed (Figure 4.11). Destruction was caused by shaking from the earthquake, the force of the tsunami waves, and flooding as the land subsided following the earthquake. Coastal subsidence of a tectonic plate is common in megathrust events along a subduction zone (see Chapter 3).

Tourist areas in the region were hard hit, especially in Thailand where several thousand tourists from Europe, the United States, and elsewhere were killed. Most visitors, and many first-generation residents, were unfamiliar with tsunamis. They did not know how to recognize that a tsunami was about to take place or what to do if one occurred. This was true for many people in Phuket, Thailand (Figure 4.12). Some people seemed to be mesmerized by the approaching waves, while others ran in panic.

On some coastlines, not everything was quite as bleak. In Thailand, a 10-year-old British girl sounded the warning in time for 100 people to evacuate a resort beach. Only a few weeks before going to Thailand, she had received a lesson in school about plate tectonics, earthquakes, and tsunamis. As part of that lesson, she learned that sometimes the sea recedes prior to the arrival of a tsunami wave. That is precisely what she observed, and her screaming to get off the beach eventually convinced her mother and others to take action. Thanks to her persistence, the beach was successfully evacuated. Her mother later stated she did not even know what a tsunami was but that her daughter's school lesson had saved their lives.

On another beach in Sri Lanka, a scientist on vacation witnessed a small wave rise up and inundate the hotel swimming pool. In the next 20 minutes, the sea level dropped by around 7 m (~23 ft.). The scientist recognized these events as a sign that a big wave was

0 100m

0 300 ft.

(a)

(b)

◄ FIGURE 4.11 Nearly Complete Destruction Associated with Indonesian Tsunami of 2004
(a) Satellite view of Banda Aceh, the Indonesian provincial capital on the northern end of the island of Sumatra, before the earthquake and tsunami of December 26, 2004. *(AFP/GETTY IMAGES/Newscom)* (b) Two days after the tsunami, it is apparent that nearly all the development has been damaged or destroyed. Note that the shoreline at the top of the photograph has been extensively eroded, leaving behind only a few small islands. Large parts of the city flooded because of subsidence caused by the earthquake that combined with inundation from the tsunami. *(AFP/GETTY IMAGES/Newscom)*

^ FIGURE 4.12 Tourists Running for Their Lives
Man in the foreground is looking back at a brown wave of the
2004 Indonesian tsunami rushing toward him. The wave is
higher than the building behind him at this Phuket, Thailand,
resort. Many people living there, as well as the tourists, did not
initially think the wave would inundate the area where they
were, and when the waves arrived, they thought they would
be able to outrun the rising water. In some cases, people did
escape, but all too often they drowned. *(John Russell/The Age/
ZUMAPRESS/Newscom)*

coming and sounded the alarm. A hotel employee then
used a megaphone to warn people to get off the beach.
Many people had gone down to the beach out of curiosity to see the exposed seafloor. Shortly thereafter, the
sea rose 7 m (~23 ft.) above its normal level. Fortunately,
most people were either near hotel stairs or had escaped to higher floors. None of the staff or guests
drowned, but several people on the ground floor were
swept out and survived only by clinging to palm trees in
the hotel garden.[19]

In the Andaman and Nicobar Islands in the northeastern Indian Ocean, as well as in parts of Indonesia,
some native people have a collective memory of tsunamis. When the earthquake occurred, they applied that
knowledge and moved to high ground, saving entire
small tribes on some islands.

In Khao Lak, Thailand, elephants rather than people sounded the warning and saved lives.[20] The elephants started trumpeting about the time the **M** 9.1
earthquake occurred more than 600 km (~370 mi.) to
the west. The elephants then became agitated again

about an hour later. Those elephants not taking tourists for rides broke loose from their strong chains and
headed inland. Other elephants carrying tourists on
their backs ignored their handlers and climbed a hill
behind the beach resort, saving the tourists from the
fate that fell upon about 4,000 people who were killed
by the tsunami at the resort. When handlers recognized the advancing tsunami, they got other elephants
to lift tourists onto their backs with their trunks and
proceed inland. The elephants did so even though
they were accustomed only to people mounting them
from a wooden platform. Tsunami waves then surged
about 1 km (~0.5 mi.) inland. The elephants stopped
just beyond where the waves ended their destructive
path.

Did the elephants know something that people did
not? Animals have sensory ability that differs from humans. It's possible the elephants heard the earthquake,
because earthquakes produce sound waves with low
tones referred to as *infrasonic sound*. Some people can
sense infrasonic sound, but they don't generally perceive it as a hazard. The elephants may also have sensed
the motion as the land vibrated from the earthquake. In
either case, the elephants fled inland away from the advancing tsunami. Although the link between the elephants' sensory ability and their behavior is still
speculative, the actions of the elephants nevertheless
saved at least a dozen lives.[20]

Numerous reports received after the Indonesian
tsunami indicate the vital role of education in tsunami
preparedness. Thousands of lives would have been
saved had more people recognized from Earth's behavior that a tsunami was likely. Those people who felt the
large earthquake would have known that a tsunami
might be coming and could have evacuated to higher
ground. Even thousands of miles away where the seismic waves were not felt, there were still signals of what
was about to happen. As the British schoolgirl knew, if
the water suddenly recedes from the shore and exposes
the sea bottom, you can expect it to come back as a
tsunami wave. This is the signal to move to higher
ground. People should also be informed that tsunamis
are seldom one wave but are, in fact, a series of waves,
with later ones many times larger and more damaging
than earlier ones. Educating people close to where a
tsunami may originate is particularly important, as
waves may arrive within 10 to 15 minutes following an
earthquake. Geologists have warned that it is likely another large tsunami will be generated by earthquakes
offshore of Indonesia in the next few decades.[21,22] In
fact, during the past 2,800 years, there have been four
tsunamis identified by shallow excavations (Figure 4.13).
The most recent tsunami before the 2004 event occurred 500 to 700 years ago.[23]

∧ FIGURE 4.13 Shallow Excavation Following the 2004 Tsunami
The top layer of light sand is the 2004 tsunami deposit. Three older light-colored tsunami deposits, all less than 2,800 years old, are shown here. *(Brian Atwater/USGS)*

Since 2004, an ocean-bottom tsunami warning system consisting of sensors that recognize tsunami waves has been put in place in the Indian Ocean, and sirens are now in place in some coastal areas around the Indian Ocean to warn of an approaching tsunami. In addition, much more has been done to educate people about the tsunami hazard. A tsunami in 2009 generated by a **M** 8.1 earthquake struck American Samoa and other islands in the region. The number of lives lost was about 200, fewer than might have occurred without the program of tsunami awareness and education. People got the word out about the tsunami hazard and what to look for, which resulted in evacuations that saved lives. However, with even more education and improved warning systems linked to evacuation processes, more lives could have been saved.[24]

TSUNAMI THREE CENTURIES AGO

An interesting example of damage from a distant earthquake comes from three centuries ago. In the year 1700, a tsunami generated by an estimated **M** 9 earthquake on the Cascadia subduction zone reached the shores of Japan approximately 12 hours later (Figure 4.14). You might wonder how we are able to determine the magnitude of the earthquake and know that it created a tsunami in Japan. This detective story started with geologic investigations in North America that found logs and soil that were buried below a tsunami deposit sometime after 1660. The logs and other plant material in the soil were radiocarbon dated to determine when they were last alive, and the

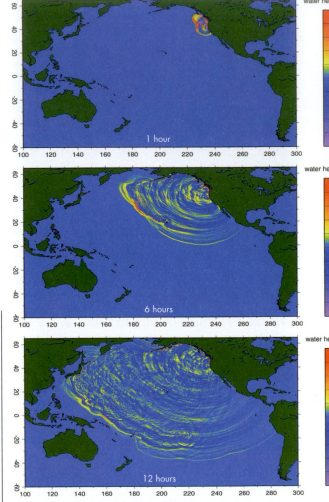

∧ FIGURE 4.14 Tsunami of 1700
This output from a computer model of the January 1700 tsunami is based on arrival times at several sites in Japan. The maps show wave height and location 1 hour, 6 hours, and 12 hours after an ~**M** 9 earthquake occurred on the Cascadia subduction zone. Such an earthquake would produce a fault displacement of around 19 m (~60 ft.) over a distance of about 1100 km (~680 mi.) from British Columbia south to northern California. *(Used by permission of Kenji Satake)*

deposits showed evidence of subsidence at the time of the earthquake. Based on the geologic evidence, the length of rupture was estimated to be about 1000 km (~620 mi.).[25,26,27] The evidence from Japan consists of historical descriptions of a tsunami with a 1 to 5 m (~3 to 16 ft.) runup height that took place in 1700. Knowing the time of arrival of the tsunami in Japan, scientists have inferred that the tsunami originated at the Cascadia subduction zone about 9:00 P.M. PST on January 26, 1700.[25]

4.3 Effects of Tsunamis and Linkages with Other Natural Hazards

The effects of tsunamis are both primary and secondary. Primary effects are related to the inundation of the water and the resulting flooding and erosion. Virtually nothing can stand in the path of a truly high-magnitude tsunami. The wave energy is sufficient to tear up beaches and most coastal vegetation, as well as all homes and buildings in its path. These effects diminish with distance from the coast. Much of the damage both to the landscape and to human structures results from the tremendous amount of debris carried by the water as it moves inland and then back out again to the ocean. What is left behind is often bare, eroded ground and areas covered with all sorts of human and natural debris (Figure 4.15).

Secondary effects of tsunamis occur in the hours, days, and weeks following the event. Immediately following the tsunami, fires may start in urban areas from ruptured natural gas lines or from the ignition of flammable chemicals released from damaged tanks. Water supplies may become polluted from floodwaters, damaged wastewater treatment systems, and rotting animal carcasses and plants. Disease outbreaks may occur, as people surviving the tsunami come in contact with polluted water and soil. Following the 2004 Indonesian tsunami, public health officials were initially concerned that there would be outbreaks of waterborne illnesses, such as malaria and cholera. Fortunately, this did not become a serious problem because of the quick action of relief agencies and the destruction of mosquito breeding grounds by saltwater flooding. What did occur is a pneumonia-like disease known as "tsunami lung," a condition that developed in people who inhaled bacteria in muddy salt water. Because the disease was not initially recognized and there were few antibiotics or medical personnel available for treatment, many of the patients developed advance lung infections that resulted in paralysis. Fortunately, treatment with antibiotics eventually brought the infections under control.[28]

(a)

(b)

∧ **FIGURE 4.15 Damage and Debris from 2004 Indonesian Tsunami**
(a) In some areas, the tsunami removed all but the most sturdy buildings, such as this mosque in Aceh Province, Indonesia. *(Spencer Platt/Getty Images)* (b) In other areas, such as this part of the Indonesian resort town of Pangandaran, the tsunami piled up huge amounts of human and natural debris. This made it difficult for these soldiers to locate victims. *(Dimas Ardian/Getty Images)*

Several linkages exist between tsunamis and other natural hazards. Tsunamis are obviously closely linked to submarine and coastal earthquakes and landslides, as well as island volcanic explosions and oceanic impacts of asteroids and comets. For earthquake-generated tsunamis, coastal communities near the epicenter experience casualties and property damage from both the ground shaking of the earthquake (see Chapter 3) and the inundation by the tsunami. Powerful tsunami waves interact with coastal processes to change the coastline through erosion and deposition of sediment. There was dramatic evidence of this interaction following the 2004 tsunami in the area near Banda Aceh, Indonesia (see Figure 4.11). A combination of erosion caused by the tsunami and subsidence caused by the earthquake has altered the coastal area to the point that it scarcely resembles what it was prior to the event.

4.3 CHECK your understanding

1. What are the primary and secondary effects of tsunamis?
2. List the ways tsunamis are linked to other hazards.

4.4 Natural Service Functions of Tsunamis

Large tsunamis are capable of causing catastrophic death and destruction. It is hard to imagine much in the way of benefits from such disasters. The movement of vast amounts of seawater on land undoubtedly brings many chemicals from the ocean to the land, which may have long-term effects on ecosystems that might otherwise be deprived of nutrients. Tsunamis also bring ashore a large volume of sediment that, over a long period, contributes to the general development of the landscape. In the coming years, studies of landscape and ecosystem changes following the 2004 Indonesian tsunami may reveal other natural service functions of tsunamis.

4.4 CHECK your understanding

1. What are potential natural service functions of tsunamis?

4.5 Human Interaction with Tsunamis

We cannot prevent tsunamis, and few, if any, tsunamis have been influenced by human activities. The magnitude and frequency of these events are not in any way tied to human activity. As with hurricanes and rising sea level, people who move to coasts that have an elevated risk for tsunamis are increasing the likelihood that they will be affected by this natural process. As populations of countries around the globe increase and more people inhabit coastal areas, the number of people affected by a given tsunami event will likely rise.

Damage from tsunamis is regrettable, but there are some lessons from past tsunamis that can be applied to reduce the damage from future events. For example, we can plant buffer zones of trees along a coast that will absorb some of the impact of incoming tsunami waves. We might also consider constructing our buildings to withstand the onslaught of moderate tsunamis and moving other structures inland where they are less likely to be damaged.

4.5 CHECK your understanding

1. How do growing populations and changing land uses influence the incidence and consequences of tsunamis?

4.6 Minimizing the Tsunami Hazard

A number of strategies are available for minimizing the tsunami hazard, including detection and warning, structural control, construction of tsunami runup maps, land-use planning, probability analysis, education, and implementation of tsunami-ready status.

DETECTION AND WARNING

Since nearly all large tsunamis are associated with giant earthquakes, our first warning comes from an earthquake in an offshore area that is large enough to produce a tsunami. We have the capability to detect distant tsunamis in the open ocean and accurately estimate their arrival time to within a few minutes. This information has been used to create a successful tsunami warning system in the Pacific Ocean.

A tsunami warning system has three components: a network of seismographs to accurately locate and determine the depth and magnitude of submarine and coastal earthquakes, automated tidal gauges to measure unusual rises and falls of sea level, and a network of sensors connected to floating buoys (Figure 4.16). Surface buoys with a bottom sensor, known as a *tsunameter*, detect small changes in the pressure exerted by the increased volume of water as a tsunami passes overhead (Figure 4.16a). This information is relayed by satellite to a warning center and is combined with tidal gauge information to predict tsunami arrival times. For example, an underwater earthquake in Hawaii could produce a tsunami that would radiate outward across the entire Pacific Ocean and arrive at different times in California, Alaska, Japan, and Papua New Guinea (Figure 4.16b). Following the Indonesian tsunami, similar systems are being created in the Indian Ocean and

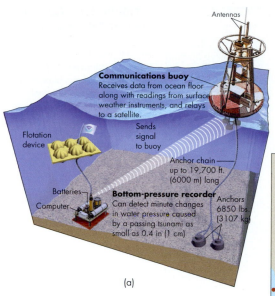

◄ **FIGURE 4.16 Tsunami Warning System**

(a) A surface buoy and bottom sensor to detect a tsunami. *(From "Dart Mooring System," NOAA, http://nctr.pmel.noaa.gov/Dart/ dart_ms1.html)* (b) Travel time (each band is 1 hour) for a tsunami generated in Hawaii. The wave arrives in Los Angeles in about 5 hours. It takes about 12 hours for the waves to reach South America. Locations of 30 DART tsunameter buoys are shown with red dots. At least 14 additional DART tsunameter buoys are planned for the Pacific Ocean and Atlantic Ocean. *(From "Dart Locations," NOAA Magazine, www. noaanews.noaa.gov/stories2007/s2904.htm)*

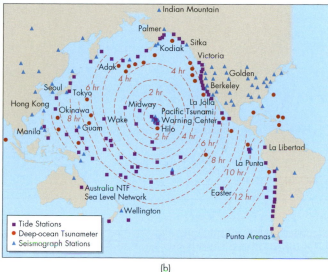

(b)

Atlantic Ocean, including warning sensors for Puerto Rico and the east coast of the United States and Canada.

For a local tsunami that strikes land very close to the source of the earthquake, there may be little warning time. People close to the source, though, will probably feel the earthquake and can immediately move inland to higher ground. Certainly, if one observes the water receding, it is a sign to run inland. Some coastal communities in Hawaii, Alaska, the Pacific Northwest, British Columbia, Japan, and elsewhere also have warning sirens to alert people that a tsunami may soon arrive.

STRUCTURAL CONTROL

Tsunamis that are even a meter or two high have such power that houses and small buildings are unable to withstand their impact.[7] However, building designs for larger structures, such as high-rise hotels and critical facilities, can be engineered in such a way as to greatly reduce or minimize the destructive effects of a tsunami. For example, the city of Honolulu, Hawaii, has special requirements for construction of buildings in areas with a tsunami hazard. However, for most hazard areas, the current building codes and guidelines do not adequately address the effect of a tsunami on buildings and other structures.[29]

Building seawalls is another type of structural control. The goal is to have walls that will not allow tsunamis to move from the coast inland. The hard part is deciding how high walls should be. Walls that are not sufficiently high to protect a city can lead to a false sense of security.

TSUNAMI RUNUP MAPS

Following a tsunami, it is a fairly straightforward procedure to produce a **tsunami runup map** that shows the level to which the water traveled inland. Such a map was created for the island of Oahu, Hawaii, following the 1946 tsunami that originated from an earthquake in Alaska's Aleutian Islands (Figure 4.17). That map illustrates the tremendous variability of runup height from a typical tsunami. The runup from the 1946 tsunami varied from about 0.6 m (~2 ft.) in Honolulu to 11 m (~36 ft.) at Kaena Point and Makapuu Point on opposite ends of the island.

Before a tsunami strikes, a community can produce a hazard map that shows the area that is likely to be inundated by a given height. Huntington Beach, California, one of a number of American and Canadian communities that has a significant tsunami hazard, has prepared such a map (Figure 4.18a). Although the probability of a large tsunami in southern California is relatively low, the consequences of such an event could be catastrophic. For example, the entire city of Huntington Beach has an elevation of less than 33 m (~110 ft.), and three-quarters of the city is less than

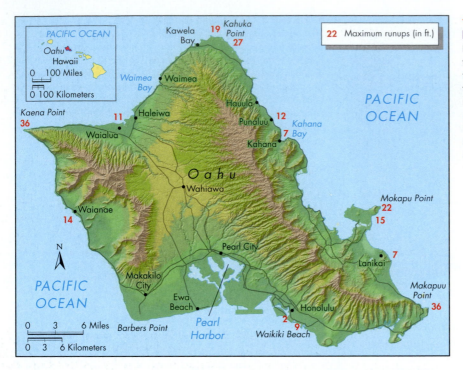

◄ FIGURE 4.17 Tsunami Runup on Oahu, Hawaii
Map of Oahu showing runup of the 1946 tsunami that originated from an earthquake in the Aleutian Islands, Alaska. Maximum runups are indicated in feet. *(Based on Walker, D., "Tsunami Facts,"* SOEST Technical Report *94–03. School of Ocean and Earth Science and Technology, 1994)*

(a)

(b)

◄ FIGURE 4.18 Huntington Beach, California, Tsunami Runup Map
(a) Predicted runup map showing areas likely to be inundated by tsunami runup in red. These areas are more susceptible to runup due to their lower elevations with respect to yellow regions. *(From the City of Huntington Beach)* (b) The popular beach with pier and area to the southeast. *(Arthur Eugene Preston/Shutterstock)*

8 m (~25 ft.) above sea level. Huntington Beach itself faces to the southwest and, as such, is vulnerable to a tsunami from either the south or west. Such an event, for example, might originate in Alaska, Japan, or the South Pacific. The beach attracts up to 100,000 people or more each day during the summer, so the city is giving serious consideration to the possibility of a tsunami. Many other cities and areas have produced tsunami runup maps, and this trend will undoubtedly continue in the future.

January 10, 2003   (a)   December 29, 2004   (b)

∧ FIGURE 4.19 **Tsunami Damage to Trees**
(a) Before and (b) after 2004 Indonesian tsunami removed many trees. (© *Getty Images/DigitalGlobe/ScapeWare3d/Contributor*)

LAND USE

In the aftermath of the 2004 Indonesian tsunami, scientists discovered that tropical ecology played a role in determining tsunami damage. Where the largest waves struck the shore, there was inevitably massive destruction (Figure 4.19). In other areas, the damage was not uniform. Some coastal villages were destroyed, whereas others were less damaged. Those villages that were spared destruction were partly protected from the energy of the tsunami by either a coastal mangrove forest or several rows of plantation trees that reduced the velocity of incoming water.[30] Coastal land-use and land-cover studies have documented the advantage of locating villages inland behind a coastal forest (Figure 4.20). It has been suggested that places where the forest grows naturally may be somewhat protected from wave attack by the nearshore and land topography. Thus, development in areas with a tsunami hazard might look to land-use planning that includes identifying where serious tsunami damage has occurred in the past and avoiding the most hazardous places. That is using the principle of uniformitarianism discussed in Chapter 1.

With rapid coastal development for a variety of purposes, including tourism, coastal mangroves are often removed and replaced by homes, high-rise hotels, and other buildings. These structures are often located close to the beach and, thus, are vulnerable to tsunami attack. Although it may not be practical to move tourist areas inland, planting or retaining native vegetation could provide a partial buffer from a small-to-moderate tsunami attack. However, large tsunami waves can erode large trees from the soil, and their trunks and branches can then move inland with the water, producing a serious hazard to people and structures. The best and safest approach to lessen the tsunami hazard is always quick evacuation.

PROBABILITY ANALYSIS

The risk of a particular event may be defined as the product of the probability of that event occurring and the consequences should it occur. Thus, determining the likelihood or probability of a tsunami is an important component in analyzing the potential hazard. For the most part, the hazard analysis for tsunamis has relied on evidence from past tsunamis to determine the hazard rather than to attempt to calculate the probability of a future tsunami. The more deterministic approach is simply to derive a tsunami inundation map, such as that shown in Figure 4.18a, and use that to develop evacuation maps. In contrast, analysis of the probability of a tsunami hazard provides information not only on the likelihood of the event but also on other aspects as well, such as location of the event, the extent of the runup, and the possible severity of damage. The approach taken in developing a probabilistic analysis of the tsunami hazard should include the following steps:

> Identify and specify the potential earthquake sources and their associated uncertainties.

> Specify relationships that will either attenuate or reduce tsunami waves as they travel from the source area.

> Apply probabilistic analysis to the tsunami hazard similar to what is currently being done for earthquake hazard analysis. However, with tsunamis,

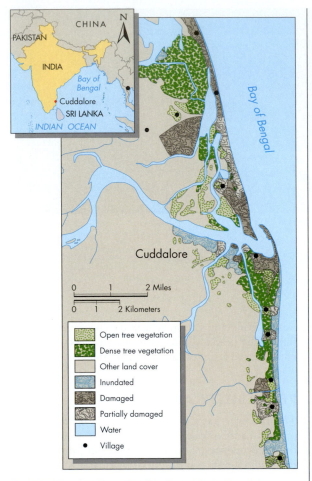

∧ FIGURE 4.20 Trees Provide Some Protection from Tsunami Damage

Tree cover existing pre-tsunami with post-tsunami damages to the land and villages in the Cuddalore District of India. *(Based on Danielsen, F., Serensen, M. K., Olwig, M. F., Selvam, V., Parish, F., Burgess, N. D., Hiraishi, T., Karunagaran, V. M., Rasmussen, M. S., Hansein, L. B., Quarto, A., and Suryadiputra. N. 2005. "The Asian Tsunami: A Protective Role for Coastal Vegetation,"* Science 310: 643)

the analysis must consider the fact that tsunamis can originate from multiple sources, including ones that are far away from the location where the hazard is present.

This probabilistic approach to tsunami hazard assessment is still being developed. One difficulty is that tsunamis at a particular location are generally rare events. If there is not sufficient activity in the past to develop the probability of future events, then the analysis depends on a statistical technique known as Monte Carlo simulation. This technique has been used to simulate the behavior of earthquakes and tsunamis. The overall objective is to determine tsunami return periods and probabilities for both distant and local sources. To apply this technique, the *Monte Carlo simulation* selects a random sample of earthquakes of various magnitudes and determines the tsunamis that would be propagated from these quakes. Using these tsunamis, a mathematical model is then constructed to estimate the tsunami amplitude or runup along a particular coast. This must be done for each of the potential seismic sources for a particular coastline.[31]

EDUCATION

Education concerning the tsunami hazard is critical to minimizing risk (see Case Study 4.3: Professional Profile). This was shown dramatically during the 2004 Indonesian tsunami when numerous lives were saved because people recognized that the receding seawater was a warning sign of an impending tsunami. Likewise, it is important to teach coastal residents and visitors the difference between a **tsunami watch**, which is notification that an earthquake that can cause a tsunami has occurred, and an actual **tsunami warning** that a tsunami has been detected and is spreading across the ocean toward their area. For a distant tsunami, there will likely be several hours before the waves arrive after a warning has been issued. In a local tsunami, there may be little warning time; therefore, more attention must be given to nature's warning systems, such as earthquake shaking or water receding from a shoreline. People must also be taught that tsunamis come in a series of waves and that the second and third waves may be larger than the first one. Finally, people must be informed that the water returning to the ocean once a wave has run up is just as dangerous as the incoming water.

TSUNAMI-READY STATUS

In order for a community to be prepared, or **tsunami ready**, it must follow these steps:

> Establish an emergency operation center with 24-hour capability.

> Have ways to receive tsunami warnings from the National Weather Service, Canadian Meteorological Centre, Coast Guard, or other agencies.

> Have ways to alert the public.

> Develop a tsunami preparedness plan with emergency drills.

> Promote a community awareness program to educate people concerning a tsunami hazard.

These guidelines for tsunami-readiness status are being applied to a number of communities in California and other areas. For example, the University of California,

Santa Barbara, and the city of Huntington Beach, California, have been certified as tsunami ready (see Figure 4.18).

The educational component is of particular importance. Most people may not know what to do if a tsunami watch or warning is issued. For example, in 2005, there was an earthquake far away from the city of Santa Barbara in the Pacific. As it turned out, a tsunami did not occur, but a tsunami watch was placed up and down the California coastline. Since nothing was said about the size of the possible tsunami, some people, on hearing the notice, drove to the top of a nearby mountain pass thousands of feet above sea level. Those locations were great for views, but people certainly did not need to drive that far or that high to evacuate the potential danger zone. There was no plan in place to directly observe the tsunami, and the news media reported that some people perched on a sea cliff at night, whereas others climbed palm trees to see if the wave was coming. This experience in Santa Barbara, and other West Coast communities, suggests that even many "tsunami-ready" communities are not adequately prepared for the tsunami that will eventually strike their coasts.

4.6 CHECK your understanding

1. How are tsunamis detected in the open sea?
2. What are structural control options to reduce tsunami damage?
3. How can producing a tsunami runup map help reduce potential tsunami damage?
4. How can vegetation help minimize tsunami damage?
5. What are the three parts of developing a probabilistic approach to analyze future tsunami damages?
6. What is the role of education in minimizing the consequences of a tsunami?
7. What factors are necessary for obtaining tsunami-ready status?

4.7 Perception and Personal Adjustment to Tsunami Hazard

The preceding discussion suggests that many people do not know the signs of an approaching tsunami or what to do if a watch or warning is issued. From a personal perspective, if a warning is issued, you can take the following actions and follow tsunami safety rules:[32,33]

> Because large earthquakes have the potential to cause tsunamis, if you feel a strong earthquake and are at the beach, leave the beach and low-lying coastal area immediately.

> A small tsunami at one beach can be a giant a few miles away. Don't let the modest size of a tsunami in one location make you lose respect for what may strike your location.

> If the trough of a tsunami wave arrives first, the ocean will recede, exposing the seafloor. This is one of nature's warning signs that a wave is on the way and that you should run from the beach for higher ground.

> Although a tsunami may be relatively small at one location, it may be much larger nearby. Therefore, do not assume that all locations are safe because of an absence of dangerous waves elsewhere.

> Homes and other buildings located in low-lying coastal areas are not safe from tsunamis. Do not remain in such buildings if there is a tsunami warning. Upper floors or the roof of high, multistory, reinforced concrete buildings can provide refuge from a tsunami. Use them if there is no time to quickly move inland or to higher ground. If all else fails, climb a tall, strong tree.

> If you are on a boat and weather and time-permit, move your boat to deeper water at least 50 m (150 ft.). If there is concurrent severe weather, it may be safer to leave the boat at the pier and physically move to higher ground.

> Unpredictable currents from tsunamis can affect harbor conditions for a period of time after the tsunami's arrival. Be sure conditions are safe before you return your boat to the harbor.

> Approaching large tsunamis are usually accompanied by a loud roar that may sound like a train. If a tsunami arrives at night when the ocean cannot be seen, this is one of nature's tsunami warnings and should be heeded.

> As coastal communities gain tsunami-readiness status, they will have warning sirens; if you hear a siren, move away from the beach to higher ground (at least 20 m [~60 ft.] above sea level) and listen for emergency information.

> If you are aware that a tsunami watch or warning has been issued, do not go down to the beach to watch the tsunami. If you can actually see the wave, you are probably too close to escape. These waves look small out at sea because of the vast distance to the horizon. Remember that these waves move fast and can be deadly. A 2 m (~6 ft.) tall person is very small compared to a 15 m (~50 ft.) tsunami

4.3 Professional Profile

Jose Borrero— Tsunami Scientist

Tsunami scientist Jose Borrero has witnessed first-hand the combined powers of land and sea unleashed (Figure 4.3.A). As a researcher with the University of Southern California Tsunami Research Center, he has traveled to areas hit by some of the most massive natural disasters in recent decades.

When a 10 m (~30 ft.) tsunami killed 2,100 people in New Guinea in 1998, Borrero and his colleagues were there just a week later to study the extent of the wave damage. "It looked like a hurricane came through. We saw dead bodies on the beach, ghost towns that looked completely bombed out—it was shocking," Borrero said.

The researchers traveled up and down the coast, looking for flattened trees, measuring water-lines on house walls, and asking local people what they remembered about the timing and size of the waves. Back home, they used their wave damage data, seismometer measurements, and computer models to reconstruct the events leading up to the tsunami. Their work revealed that an underwater landslide triggered by a relatively small earthquake created the large and powerful tsunami.

After the 2004 Indonesian earthquake and tsunami, Borrero and a team of *National Geographic* filmmakers were among the first outsiders to reach one of the hardest-hit areas—the city of Banda Aceh, Indonesia. What Borrero saw horrified him: "It was the worst of the worst of what I saw in New Guinea, except that, instead of being confined to just one town, it went on for 200 miles." In some areas, 10 m (~30 ft.) waves had pushed boats onto balconies and stripped the trees off the side of a mountain 24 m (~80 ft) up.

Back home, Borrero and colleagues used the data to develop a report about the risk Sumatra faces from future earthquakes and tsunamis along the next segment of

⋀ FIGURE 4.3.A Dr. Jose Borrero
A research professor at the University of Southern California Department of Civil Engineering Tsunami Research Group, Dr. Borrero was part of an international damage assessment team that studied the effects of the 2004 Indonesian tsunami. Dr. Borrero is shown here with boats that were destroyed by the tsunami in Banda Aceh, Indonesia. *(Dr. Jose Borrero/USC Tsunami Research Group)*

the earthquake fault. "If you know what's possible, you can make a plan. We can give that information directly to cities and towns so they can make evacuation routes and public awareness programs," he said.

Tsunami education, according to Borrero, is critical for coastal areas located directly on top of earthquake faults that produce tsunamis.

"Tsunami warning systems only work for areas that are more than two hours away from the area of ground shaking. The only way to alert people is through education. If you feel an earthquake, and you're near the coast, don't sit and wait— head for high ground."

Growing up surfing in California, Borrero always wanted a career in which he could work directly with

the ocean. "There's this primal fear people have of giant walls of water. Many people have told me they have this recurring nightmare where they're trapped in a tsunami and can't escape. I've never had dreams like that; maybe understanding it keeps me from worrying too much."

— KATHLEEN WONG

wave (Figure 4.21). Tsunamis can move faster than a person can run.

> If trapped in the water and swept away, look for something that floats to use as a raft.

> Tsunamis generally consist of a series of waves with possibly up to an hour between waves. Therefore, stay out of dangerous areas until a notice that all is clear is given by the proper authorities.

> During a tsunami watch or warning, stay tuned to your local radio, marine radio, NOAA Weather Radio, or television stations during a tsunami emergency – bulletins issued through your local emergency management

office and National Weather Service offices can save your life.

4.7 CHECK your understanding

1. What are two warnings from nature that a tsunami may be coming?

2. What should you do if a tsunami warning is issued?

3. What can you do if you have had little warning and tsunami waves will arrive before you can evacuate?

∧ FIGURE 4.21 A Person Versus a Tsunami
Tsunami height can be huge compared to the waves we normally see on a beach.

15 m (45 ft)

Beach   (6 ft)

Japan 2011: Tsunami and Nuclear Disaster

1 Science helps us predict hazards.

2 Knowing hazard risks can help people make decisions.

3 Linkages exist between natural hazards.

4 Humans can turn disastrous events into catastrophes.

5 Consequences of hazards can be minimized.

The Japanese people have studied earthquakes and tsunamis for decades and have been keeping records of their experiences with both for hundreds of years. People in Japan are well aware that large earthquakes in the region and resulting tsunamis are likely in the future. Major earthquake faults lie about 100 km (~65 mi.) offshore of the east coast of Japan where the Pacific Plate is forced beneath the Eurasian Plate. These faults have been studied for decades and are known to be capable of producing **M** 7 to **M** 8 tsunami-generating earthquakes (see Chapter 3: Applying the Fundamental Concepts).

With more than 12,500 km (~7750 mi.) of tsunami seawalls protecting about 43 percent of Japan's coastline, the country has invested billions of dollars on these concrete barriers to minimize the tsunami risk. For example, the deepest tsunami defensive barrier ever constructed was completed in 2009 outside the historic town of Kamaishi at a cost of $1.6 billion. The Kamaishi wall is anchored to the sea floor at a depth of 65 m (~215 ft.), and is about 2000 m (~6400 ft.) long, 60 m (~207 ft.) wide, and stands 6 m (~20 ft.) above sea level. Like most of Japan's seawalls, however, the Kamaishi wall was designed to minimize the effects tsunami waves with heights up to 5 m (~16 ft.). This design specification means that most of Japan's seawalls are capable of protecting the coastline from typical tsunami waves that result from **M** 8 or less subduction zone earthquakes, which occur every 100 to 150 years.

The unexpectedly large magnitude (**M** 9) of the earthquake that struck the Tohoku (northeastern) region of Japan on March 11, 2011, was a surprise to most, as was the height of the tsunami. Following the earthquake, tsunami warnings were issued, giving coastal communities nearest the epicenter about 12 to 15 minutes to evacuate. Although the Japanese population is well educated and trained on how to survive an earthquake, many people underestimated the tsunami risk to their communities because of a false sense of security provided by the extensive seawall network. With waves approximately 10 m (~30 ft.) high and some locally exceeding 40 m (~130 ft.), seawalls were quickly overtopped and destroyed by the tsunami waves (Figure 4.22), including the Kamaishi seawall. The deluge of water caused extensive coastal erosion and destroyed almost everything in its path, carrying debris such as cars, houses, and large sea-going vessels as far as 10 km (~6 mi.) inland. Of the approximately 16,000 people killed by the 2011 Tohoku earthquake and tsunami, most died as a result of flooding by the tsunami wave and not the earthquake.

A decade earlier, a Japanese scientist had suggested that larger than expected tsunamis were possible and that existing tsunami protection barriers may not be large enough. The existence of very large tsunamis in the region with long return periods of about a thousand years was discussed as early as 2001. Buried tsunami deposits were found inland about as far as the 2011 tsunami inundated the land. Unfortunately, these scientific findings were either not believed or not taken seriously. In other words, warnings from scientists were largely ignored, except in scientific circles.[1,2] Interestingly,

◄ **FIGURE 4.22**
Tsunami Wave Reaches the Coast of Japan
Over-topping 4 m high sea walls, the first tsunami wave inundates Sendai, one of the first coastal communities to be flooded by the 2011 Tohoku tsunami. *(z03/ ZUMA Press/Newscom)*

the small village of Fudai survived the tsunami because in the late 1960s the mayor insisted on building a much higher wall—about 16 m (~52 ft.), which at the time people in the village thought to be foolish and too expensive (Figure 4.23). But the mayor remembered burying people after the 1933 tsunami and had knowledge of a large killer tsunami that had occurred in 1896. He persisted in building the higher wall, and the village survived the 2011 tsunami, while others in the region were completely destroyed.

ᐱ **FIGURE 4.23 Village of Fudai Saved by Seawall**
Fudai's current mayor, Hiroshi Fukawatari, near the 16 m (~52-ft) seawall (not shown) that protected homes from the tsunami waves. The seawall and floodgates were a project begun by former mayor Kotaku Wamura, who remembered the devastation of a tsunami in 1933. *(AP Photo/Hiro Komae)*

Tsunami warnings were also issued across the Pacific as distant tsunamis raced across the ocean, reaching Hawaii in about 7 hours and California in about 10 hours (Figure 4.24). Tsunami waves of 1–2 m (~3–6 ft.) inundated California harbors, costing millions of dollars in damages.

About the time the distant tsunami was arriving on the shores of Hawaii, an evacuation order was issued for all persons within a 3 km (~1.9 mi.) radius of the Fukushima Daiichi nuclear power station. The Fukushima power station is one of 23 reactor complexes in the country and one of six located on Japan's east coast impacted by the 2011 tsunami. Although backup power generators for the reactor cooling system were functioning within two minutes of the automated earthquake shutdown procedure, less than an hour later tsunami waves breached the power station's seawalls, flooding the generators. With no way to cool the reactor cores, it was only a matter of time before the nuclear fuel in the three active reactors at the Fukushima complex would melt down. Over the next few days, several massive hydrogen gas explosions within the containment buildings led to release of radiation into the atmosphere.

At first, people outside the 3 km (~1.9 mi.) evacuation zone were advised to stay in their homes and keep their doors and windows closed to minimize protect themselves from the

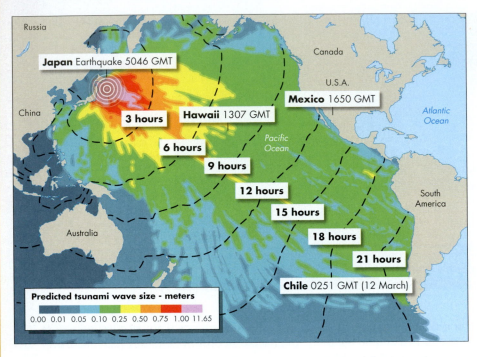

Russia

Japan Earthquake 5046 GMT

China

3 hours

Canada

U.S.A.

Mexico 1650 GMT

Hawaii 1307 GMT

Atlantic
Ocean

Pacific
Ocean

6 hours

9 hours

12 hours

South
America

15 hours

Australia

18 hours

21 hours

Chile 0251 GMT (12 March)

Predicted tsunami wave size - meters

0.00 0.01 0.05 0.10 0.25 0.50 0.75 1.00 11.65

< **FIGURE 4.24 Tsunami Prediction**
Predicted arrival times and average wave heights of tsunami waves from the March 11, 2011 event are shown. Maximum observed wave heights in Hawaii and California were several times the predicted average. *(NOAA)*

radiation. However, authorities soon realized that the hydrogen explosions had released particles of radioactive cesium dust, which had blanketed the landscape and was being remobilized every time the wind blew.[3] Twenty-eight hours after the earthquake and four hours after the first hydrogen explosion, the mandatory evacuation radius around the Fukushima Daiichi power station was expanded to 20 km (~12.5 mi.), displacing about 200,000 people.

When a catastrophe occurs, government and people sometimes "freeze" at first, which inhibits proper response. Nearly 25 percent of the people in Japan are over age 65, compared to about 12 percent in the United States. Many of the elderly were stranded for several days as they tried to survive in freezing weather. Imagine being an elderly patient at a hospital near the damaged power plants. Now imagine being ordered to evacuate, but no vehicles are available that could take you, along with a hundred other elderly patients, to safety. These patients had to wait in the freezing cold with no electricity until help finally came. When these elderly patients were finally evacuated to another location (a temporary move to a high school), many died in transit, and the new site lacked running water and had little medicine or food.

Although evacuation conditions improved over the next few years, much of the 20 km evacuation zone is still off limits, and the town of Tomioka, near the failed reactor, is not expected to be habitable until 2017. Current death toll estimates of refugees from the Fukushima nuclear meltdown due to poor evacuation conditions is about 1,600, which exceeds the number killed by the earthquake and tsunami for the region.

Leaks of radioactive material into the environment were thought to have been contained within the first few months following the meltdown. However, in the summer of 2013, it was discovered that the plant was actively leaking radioactive water into the Pacific Ocean and had likely been leaking since the incident began in 2011. As of late 2013, it is estimated that about 120 tons of radioactive water have been leaked into the world's oceans. No clear solution for stopping the leak had yet been devised.

The 2011 Tohoku tsunami and nuclear disaster has invigorated a global discussion about defending against infrequent, larger-than-expected hazardous natural events and how humans can compound their destructive effects. For example, following several months of anti-nuclear protests in Germany following the events of March 2011, Germany's government announced it would close all nuclear power plants by 2022, in large part due to the perception that the safety of nuclear power generation could not be ensured.

Although about 180 km (~112 mi.) of the 300 km (~186 mi.) of seawalls impacted by the 2011 tsunami were destroyed and the seawall network largely failed in stopping coastal flooding, studies suggest that some seawalls delayed coastal inundation by the first tsunami wave by several minutes, saving countless lives. Japan is already reconstructing many destroyed seawalls, but can they be built high enough to protect against the combined effects of sea-level rise due to global warming and long frequency extreme events like the one the struck in 2011? In the United States, we should learn from Japan's catastrophe and ensure that we are ready when a large tsunami occurs. Along the Pacific coast, from northern California to Seattle, Washington, large subduction zones have periodically produced giant earthquakes that presumably could produce large damaging tsunamis every few hundred years.[34]

APPLYING THE ⑤ FUNDAMENTAL CONCEPTS

1. Evaluate Japan's overall level of preparedness for the tsunami hazard of the region and discuss the successes and failures of the country to predict the 2011 Tohoku tsunami (see also Chapter 3, Applying the 5 fundamental concepts: 2011 Tohoku Earthquake).

2. Discuss the factors that led the scientific community and the general population of Japan to underestimate the tsunami risk presented by an event like the one that struck in 2011.

3. Describe how the 2011 tsunami is linked to other hazards. What are the temporal and spatial aspects of these hazards?

4. Explain how and why the 2011 tsunami became a much worse catastrophe due to human error before and following the event.

5. Discuss Japan's successes and failures to minimize the effects of the 2011 tsunami. How might Japan and its population better minimize the consequences of future events?

CONCEPTS in review

4.1 Introduction to Tsunamis

LO:1 Explain the process of tsunami formation and development.

- A tsunami is produced by the sudden vertical displacement of ocean water. Several possible processes can produce tsunamis, including underwater landslides, submarine volcanic explosions, and impact of extraterrestrial objects.
- The major source of large damaging tsunamis over the past few millennia has been giant earthquakes associated with the major subduction zones on Earth. These tsunamis have formed where geologic faulting ruptures the seafloor and displaces the overlying water.
- When seafloor earthquakes occur, both distant and local tsunamis may be produced. Distant tsunamis can travel thousands of kilometers across the ocean to strike a remote shoreline. On the other hand, local tsunamis head toward a nearby coast and can strike with little advance warning.

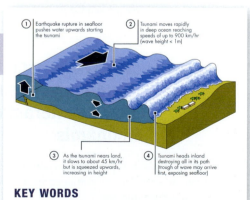

① Earthquake rupture in seafloor pushes water upwards starting the tsunami

② Tsunami moves rapidly in deep ocean reaching speeds of up to 900 km/hr (wave height < 1m)

③ As the tsunami nears land, it slows to about 45 km/hr but is squeezed upwards, increasing in height

④ Tsunami heads inland destroying all in its path (trough of wave may arrive first, exposing seafloor)

KEY WORDS

distant tsunami (p. 126), **local tsunami** (p. 126), **runup** (p. 125), **tsunamis** (p. 124)

4.2   Geographic Regions at Risk from Tsunamis

LO:2   Locate on a map the geographic regions that are at risk for tsunamis.

- The largest tsunamis generated by earthquakes are produced at subduction zones.

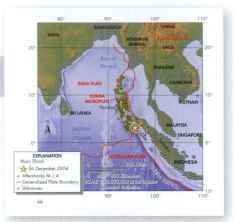

(a)

4.3   Effects of Tsunamis and Linkages with Other Natural Hazards

LO:3   Synthesize the effects of tsunamis and the hazards they pose to coastal regions.

LO:4   Summarize the linkages between tsunamis and other natural hazards.

- Effects of a tsunami are both primary and secondary. The primary effects are related to the powerful water from the tsunami runup that results in flooding and erosion. Virtually nothing can stand in the path of a large-magnitude tsunami. In 2004 in Indonesia, huge concrete barriers were moved by the force of the waves. Secondary effects of a tsunami include a potential for water pollution, fires in urban areas, and disease to people surviving the event.
- Tsunamis are linked to other natural hazards. They are obviously tightly linked to the earthquakes that cause them; thus, their effects are often combined with the ground shaking, fires, and subsidence associated with the quakes. Tsunami waves also interact with coastal processes to change the coastline through erosion and deposition of sediment. Following an earthquake and tsunami, a coastal area may scarcely resemble what it was prior to the event.

4.4 and 4.5   Natural Service Functions of Tsunamis Human Interaction with Tsunamis

LO:5   Understand that tsunamis are not caused by or affected by human activities, but damages are compounded as coastal populations increase.

- Tsunamis can move sediment and nutrients inland with long-term ecosystem benefits.
- Occurrences of tsunamis are not influenced by human activities, but we can take actions to reduce their potential impacts.

4.6   Minimizing the Tsunami Hazard

LO:6   Discuss what nations, communities, and individuals can do to minimize the tsunami hazard.

- A number of strategies are available to minimize the tsunami hazard, including detection and warning, structural control, construction of tsunami runup maps, land-use practices, probability analysis, education, and achieving tsunami-ready status. Of these, the detection and warning system is of paramount importance.
- We can detect distant tsunamis in open ocean and accurately estimate their arrival time to within a few minutes. It is more difficult to provide adequate warning for local tsunamis, as tsunami wave formation quickly follows an earthquake or earthquake-produced landslide. In this case, warning sirens can alert people that a tsunami may soon arrive.
- Without adequate education, watches and warnings are often ineffective because many people do not know how to recognize a tsunami or take appropriate action to save themselves and others. People can learn the natural warning signs of an approaching tsunami, which may include the earthquake shaking and withdrawal of seawater prior to arrival of the large wave. People must understand that tsunamis come in a series of waves and that the second or third wave may be even larger than the first wave. In addition, the water returning to the ocean following tsunami inundation can cause as much damage as the runup of the incoming water.
- Along coasts with great or significant tsunami hazard, most communities have not adequately prepared for this underestimated natural hazard. Adequate preparation includes improved perception of the hazard, development of ways to alert the public, preparation and implementation of a tsunami-readiness plan, and promotion of community awareness and education concerning the hazard.

KEY WORDS

tsunami ready (p. 142), **tsunami runup map** (p. 139), **tsunami warning** (p. 142), **tsunami watch** (p. 142)

4.7   Perception and Personal Adjustment to Tsunami Hazard

LO:7   Identify the actions you should take and not take if a tsunami warning is issued.

There are several personal adjustments to minimize your tsunami hazard.

- Most important is to heed tsunami warnings.
- Local tsunamis may arrive quickly following a tsunami-generating event; leave the beach if you feel a strong earthquake or if the water suddenly withdraws.
- Remember that tsunami waves often arrive as a series of waves separated by a variable time period.

CRITICAL thinking questions

1. You are placed in charge of developing an educational program with the objective of raising a community's understanding of tsunamis. What sort of program would you develop and on what would it be based?
2. What do you think the role of the media should be in helping to make people more aware of the tsunami hazard? How should scientists be involved in increasing the perception of this hazard?
3. You live in a coastal area that is subject to large but infrequent tsunamis. You are working with the planning department of the community to develop tsunami-ready status. What issues do you think are most important in obtaining this status and how could you convince the community that it is necessary or in everyone's best interest to achieve tsunami-ready status?

References

1. **Normile, D.** 2011. Scientific consensus on great quake came too late. *Science* 322: 22–23.
2. **Fackler, M.** 2012. Nuclear disaster in Japan was avoidable, critics contend. *New York Times*, Asia Pacific edition, March 9.
3. **Gupta, T.** 2011. Exploring the nuclear accidents of Japan: A tremor, a tsunami, and a dark cloud left behind. *US Science Review*. www.scf.usc. edu/~uscience/radiation_nuclear_accidents.html. Accessed 5/21/12.
4. **Meyers, C.** 2012. Japan earthquake anniversary: Nuclear evacuees scarred by disaster one year later. www.huffingtonpost.com/2012/02/13/japan-earthquake-anniversary_n_1272447.html#254326. Accessed 5/21/12.
5. **Author unknown.** 2011. Hot spots and blind spots. *The Economist*, October 8.
6. **U.S. Geological Survey.** 2005. *Magnitude 9.1—off the west coast of northern Sumatra*. U.S. Geological Survey Earthquake Hazards Program.
7. **Bryant, E.** 2001. *Tsunami: The underrated hazard*. New York: Cambridge University Press.
8. **Bolt, B. A.** 2006. *Earthquakes*, 5th ed.; 2006 Centennial update. New York: W. H. Freeman.
9. **U.S. Geological Survey.** 2005. *Life of a tsunami*. Western Coastal and Marine Geology Program. http://walrus.wr.usgs.gov/tsunami/basics.html. Accessed 5/25/07.
10. **Hokkaido Tsunami Research Group.** 1993. Tsunami devastates Japanese coastal region. *EOS, Transactions American Geophysical Union* 74(37): 417–32.
11. **Tappin, D. R. and 9 others.** 2014. Did a submarine landslide contribute to the 2011 Tokohu tsunami ? *Marine Geology* 357: 344–61.
12. **Brothers, D. S. and 6 others.** 2016. A submarine landslide source for the devastating 1964 Chenega tsunami, southern Alaska. *Earth and Planetary Science Letters* 438: 112–21.
13. **Tappin, D. R., Watts, P., McMurtry, G. M., Lafoy, Y.,** and **Matsumoto, T.** 2001. The Sissano, Papau New Guinea tsunami of July 1998—offshore evidence of the source mechanism. *Marine Geology* 175: 1–23.
14. **Stover, C. W.,** and **Coffman, J. L.** 1993. *Seismicity of the United States, 1958–1989* (revised). U.S. Geological Survey Professional Paper 1527.
15. **Harbitz, C. B. and 5 others.** 2014. Rockslide tsunamis in complex fjords: From an unstable rock slope at Akerneset to tsunami risk in western Norway. *Coastal Engineering* 88: 101–22.
16. **Sataki, K.,** and **Atwater, B. F.** 2007. Long-term perspectives on giant tsunamis at subduction zones. *Annual Review of Earth and Planetary Sciences* 35: 349–74.
17. **Subarya, C., Chlieh, M., Prawirodirdjo, L., Avouac, J.-P., Bock, Y., Sieh, K., Meltzner, A. J., Natawidjaja, D. H.,** and **McCaffrey, R.** 2006. Plate-boundary deformation associated with the great Sumatra-Andaman earthquake. *Nature* 440: 46–51.
18. **Kerr, R. A.** 2005. Failure to gauge the quake crippled the warning effort. *Science* 307: 201.
19. **Chapman, C.** 2005. The Asian tsunami in Sri Lanka: A personal experience. *EOS, Transactions, American Geophysical Union* 86(2): 13–14.
20. **Bendeich, M.** 2005. *Elephants saved tourists from tsunami*. Reuters. http://savetheelephants.org. Accessed 5/25/07.
21. **Achenbach, J.** 2006. The next big one. *National Geographic* 209(4): 120–47.
22. **Sieh, K.** 2006. Sumatran megathrust earthquakes: From science to saving lives. *Philosophical Transactions Royal Society* 364: 1947–63.
23. **Jankew, K. et al.** Medieval forewarning of the 2004 Indian Ocean tsunami in Thailand. *Nature* 455: 1232–34.
24. **Jaffe, B. E., Gelfenbaum, G., Buckley, M. L., Watt, S., Apostos, A., Stevens, A. W.,** and

Richmond, B. M. 2010. *The limit of inundation of the September 29, 2009 tsunami in Tutuila, American Samoa.* U.S. Geological Survey Open File Report 2010–1018.

25. **Satake, K., Wang, K.,** and **Atwater, B. F.** 2003. Fault slip and seismic moment of the 1700 Cascadia earthquake inferred from Japanese tsunami descriptions. *Journal of Geophysical Research* 108(B11): 148–227.

26. **Nelson, A. R., Atwater, B. F., Bobrowsky, P. T., Bradley, L.-A., Clague, J. J., Carver, G. A., Darienzo, M. E., Grant, W. C., Krueger, H. W., Sparks, R., Stafford Jr., T. W.,** and **Stuiver, M.** 1995. Radiocarbon evidence for extensive plate-boundary rupture about 300 years ago at the Cascadia subduction zone. *Nature* 378: 371–74.

27. **Atwater, B. F.** 1992. Geologic evidence for earthquakes during the past 2000 years along the Copalis River, southern coastal Washington. *Journal of Geophysical Research* 97(B2): 1901–19.

28. **Potera, C.** 2005. In disasters wake: Tsunami lung. *Environmental Health Perspectives* 113(11): A734.

29. **California Seismic Safety Commission.** 2005. The tsunami threat to California: Findings and recommendations on tsunami hazards and risks. Report CSSC 05–03.

30. **Danielsen, F., Serensen, M. K., Olwig, M. F., Selvam, V., Parish, F., Burgess, N. D., Hiraishi, T., Karunagaran, V. M., Rasmussen, M. S., Hansen, L. B., Quarto, A.,** and **Suryadiputra**, N. 2005. The Asian tsunami: A protective role for coastal vegetation. *Science* 310: 643.

31. **Geist, E. L.,** and **Parsons, T.** 2006. Probabilistic analysis of tsunami hazards. *Natural Hazards* 37: 277–314.

32. **Atwater, B. F.** 1999, updated 2012. Surviving a tsunami—Lessons from Chili, Hawaii, and Japan. *U.S. Geological Survey Circular 1187.*

33. Tsunami Safety Rules. 2017. National Ssunami Warning Center. ntwc@noaa.gov.

34. **Valentine, D. W., Keller, E. A., Carver, G., Li, W.-H., Manhart, C.,** and **Simms, A. R.** 2012. Paleoseismicity of the southern end of the Cascadia Subduction Zone, Northwestern California. *Bulletin of the Seismological Society of America* 102(3): 1059–2078.

An ash plume rising 9.5 km into the atmosphere ... caused the **shutdown of airspace above northern Europe**

5

APPLYING the **5** fundamental concepts
Eyjafjallajökull 2010 Eruption

① Science helps us predict hazards.

By understanding the relationship between seismic activity and volcanic activity, scientists were able to forecast the 2010 eruption of Eyjafjallajökull volcano.

② Knowing hazard risks can help people make decisions.

Although volcanoes are present on all continents, in the oceans, and on islands like Iceland, their locations are not random but are expected in particular geologic settings. If you live in a volcanically active setting, a certain probability of experiencing an eruption exists depending on the frequency with which the volcanoes erupt.

Volcanoes

Eyjafjallajökull 2010 Eruption

Located in the middle of the Atlantic Ocean, Iceland—an island one-quarter the size of California—is home to more than 30 active volcanoes. Even though volcanic eruptions light up the Icelandic skies every three to four years,[1] the effects of the eruptions are usually limited to those living on the island. In April 2010, however, an ash plume rising 9.5 km (~6 mi.) into the atmosphere above the erupting Eyjafjallajökull volcano caused the shutdown of airspace above northern Europe, the United Kingdom, Ireland, Sweden, and Norway, which lie more than 1000 km (~620 mi.) to the southeast (Figure 5.1). This volcanic activity caused the largest aerial closure in Europe since World War II—about 95,000 flights were canceled, stranding hundreds of thousands of travelers and costing the airline industry more than a billion dollars during a one-week period.[2]

Increased seismic activity around Eyjafjallajökull in December 2009 signaled the awakening of the volcano, which had not erupted since 1821. Although the 2010 eruption was not completely unexpected, the height of the plume of ash ejected from the volcano was surprising since all previous eruptions in the past 1,500 years had been comparatively small.[1] Such large volcanic ash plumes are unusual for Icelandic volcanic eruptions, with the last occurring nearly 100 years ago, long before airplanes filled the skies like today.

‹ 2010 ICELANDIC VOLCANIC ERUPTION

Ash, steam, and glowing molten rock erupt skyward from beneath the ice-covered Eyjafjallajokull volcano, in southern Iceland. *(Orvar Porgiersson / Barcroft Media / Getty Images)*

LEARNING Objectives

Of the approximately 1,500 active volcanoes on Earth, nearly 400 have erupted in the past century. In fact, while you are reading this paragraph, at least 20 volcanoes are erupting on our planet. Volcanoes occur on all seven continents as well as in the middle of the ocean. When human beings live in the path of an active volcano, the results can be devastating.

After reading this chapter, you should be able to:

LO:1 Explain the relationship of volcanoes to plate tectonics.

LO:2 Identify the different types of volcanoes and their associated features.

LO:3 Locate on a map the geographic regions at risk from volcanoes.

LO:4 Describe the effects of volcanoes and how they are linked to other natural hazards.

LO:5 List the potential benefits of volcanic eruptions.

LO:6 Discuss how humans can minimize the volcanic hazard.

LO:7 Recommend adjustments we can make to avoid death and damage from volcanoes.

3

Linkages exist between natural hazards.

The confluence of "fire and ice" on the island of Iceland is an exceptional example of how hazards can be linked to one another, and often it is the linked hazards that are the most damaging.

4

Humans can turn disastrous events into catastrophes.

As the 2010 airline shutdowns showed, in our highly connected modern world, a perceived local natural hazard can have global impacts.

5

Consequences of hazards can be minimized.

Volcanic and hydrologic monitoring around Eyjafjallajökull's volcano enabled early warnings to be issued and successful evacuation of the Icelanders living in harm's way.

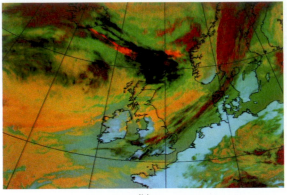

(a)   (b)

∧ FIGURE 5.1 Volcanic Ash Shuts Down Air Travel
(a) Volcanic ash plume rising from Eyjafjallajökull volcano reaches an elevation of about 9.5 km (~6 mi.). *(Lee Frost/Robert Harding World Imagery/Alamy)* (b) Enhanced color satellite image highlighting a volcanic ash plume (black colour) moving from Iceland toward the United Kingdom on April 15, 2010; all flights in and out of Britain's airports were grounded. *(Matt Cardy/Getty Images News/Getty Images)*

< FIGURE 5.2 Fire and Ice
A massive volume of water cascades down the mountain, flooding the valleys below, as the Icelandic ice cap partially melts due to the erupting volcano. Such volcanically induced floods are known by the Icelandic name, jökulhlaup. *(UPPA/ZUMAPRESS/Newscom)*

as locally. For many Icelanders living around Eyjafjallajökull, the eruption was a small taste of what might lie ahead if the volcano's more explosive neighbor Katla were to erupt. Since 2010, Katla has shown signs of unrest, and Iceland's president has issued this warning: "The time for Katla to erupt is coming close ... we have prepared ... it is high time for European governments and airline authorities all over Europe and the world to start planning."[3]

5.1 Introduction to Volcanism

Volcanic activity, or volcanism, is typically related to plate tectonics, with most volcanoes being located near active plate boundaries. Approximately two-thirds of all active volcanoes above sea level on Earth are located within the Pacific "Ring of Fire" (Figure 5.3).[4] This relationship exists because spreading or sinking plates at plate boundaries interact with other earth materials to produce molten rock, called **magma** within Earth, or **lava** when it erupts onto the surface. Volcanoes are constructed around an opening, called a **volcanic vent**, through which lava and other volcanic materials are extruded onto the surface. Vents may be roughly circular, or they may be elongated cracks called *fissures*.

Many of Iceland's large volcanoes lie beneath thick glacial ice caps. In fact, Eyjafjallajökull means "island mountain glacier," and it is this glacial ice cap from which the volcano takes its name. During volcanic eruptions, escaping heat from Earth's interior can rapidly melt large volumes of ice and trigger massive flooding, which is one of the most significant natural hazards on Iceland (Figure 5.2). Because such glacial floods are common in Iceland, scientists often use the Icelandic word *jökulhlaups* for these megafloods. To mitigate the effects of jökulhlaups, a network of hydrological stations monitors water flow downstream of glacially capped volcanoes; when flooding occurs, Icelanders living in hazardous regions are immediately notified.

Although the 2010 eruption was relatively small—about one-tenth the size of the 1980 Mount St. Helens eruption in Washington state (see Case Study 5.3: Mount St. Helens)—the effects were felt globally as well

∧ FIGURE 5.3 The "Ring of Fire"
The thick orange line outlines the Ring of Fire surrounding the Pacific plate. Although no fire is present, the Ring of Fire contains most of the world's explosive volcanoes. *(From E. J. Tarbuck and L. K. Lutgens,* Earth, *11e.* © *2014 Pearson Education, Inc.)*

Because the processes that result in volcano formation are different for different tectonic settings, volcanoes are not all the same. By understanding the various ways that magma forms and the plate tectonic settings of volcanically active regions, you will be able to understand the different types, shapes, and eruption behavior of volcanoes around the world.

HOW AND WHERE MAGMA FORMS

It is a common misconception that the plates move around on an ocean of molten rock. This is not true. With the exception of the outer core, deep within Earth's interior, the planet is composed of solid rock (see Section 2.2). So then, where does the magma that forms Earth's volcanoes come from?

Most magma comes from the asthenosphere—the weak layer within the mantle that allows the overlying rigid lithospheric plates to move around the globe (see Section 2.1). The asthenosphere is weak and able to flow because it is close to its melting temperature. This means that the asthenosphere is more prone to melting than the lithosphere above or lower mantle below.[5] Figure 5.4 illustrates the different geologic settings

where melting occurs and the three principal processes that generate magma—decompression melting, addition of volatiles, and addition of heat.

1. *Decompression melting* occurs when the overlying pressure exerted on hot rock within the asthenosphere is decreased. The melting temperature for mantle rocks at Earth's surface is approximately 1200°C (~2200°F). Although this temperature is exceeded within the upper 50 km (~30 mi.) of Earth's surface in most tectonic settings, the mantle is a solid.[5] Like temperature, pressure also increases with depth. This great pressure, generated by the weight of overlying rock, keeps Earth's mantle in a solid state far above the surface melting temperature of the rock. In other words, for a given increase in depth, an additional amount of heat is required to melt the rock. Conversely, if rocks are close to their melting temperature and the pressure from above is decreased, **decompression melting** occurs and magma will begin to form even though the temperature stays the same.

   Decompression melting predominates at locations where Earth's lithosphere is being stretched and thinned. About three-quarters of all

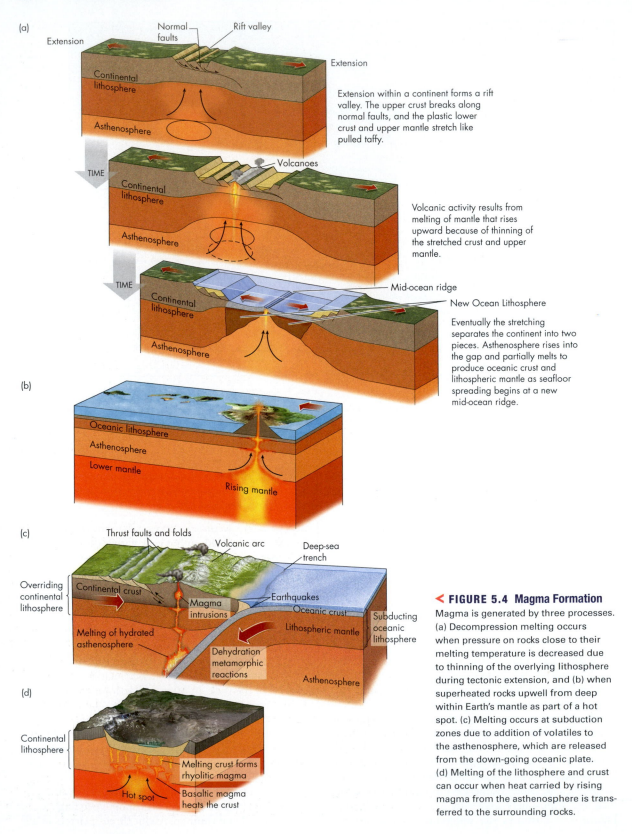

(a)

Normal faults

Rift valley

Extension

Extension

Continental lithosphere

Asthenosphere

Extension within a continent forms a rift valley. The upper crust breaks along normal faults, and the plastic lower crust and upper mantle stretch like pulled taffy.

TIME

Volcanoes

Continental lithosphere

Asthenosphere

Volcanic activity results from melting of mantle that rises upward because of thinning of the stretched crust and upper mantle.

TIME

Mid-ocean ridge

Continental lithosphere

New Ocean Lithosphere

Asthenosphere

Eventually the stretching separates the continent into two pieces. Asthenosphere rises into the gap and partially melts to produce oceanic crust and lithospheric mantle as seafloor spreading begins at a new mid-ocean ridge.

(b)

Oceanic lithosphere

Asthenosphere

Lower mantle

Rising mantle

(c)

Thrust faults and folds

Volcanic arc

Deep-sea trench

Overriding continental lithosphere

Continental crust

Earthquakes

Magma intrusions

Oceanic crust

Subducting oceanic lithosphere

Melting of hydrated asthenosphere

Lithospheric mantle

Dehydration metamorphic reactions

Asthenosphere

(d)

Continental lithosphere

Melting crust forms rhyolitic magma

Hot spot

Basaltic magma heats the crust

< FIGURE 5.4 Magma Formation
Magma is generated by three processes. (a) Decompression melting occurs when pressure on rocks close to their melting temperature is decreased due to thinning of the overlying lithosphere during tectonic extension, and (b) when superheated rocks upwell from deep within Earth's mantle as part of a hot spot. (c) Melting occurs at subduction zones due to addition of volatiles to the asthenosphere, which are released from the down-going oceanic plate. (d) Melting of the lithosphere and crust can occur when heat carried by rising magma from the asthenosphere is transferred to the surrounding rocks.

lava erupted on Earth is extruded from undersea mid-ocean ridges, which are divergent plate boundaries.[6] Recall from Chapter 2 that a mid-ocean ridge forms when a continent is pulled apart and a rift valley develops due to extension and normal faulting. Continued stretching causes thinning of the entire lithosphere and the asthenosphere must upwell toward the surface to fill the space (Figure 5.4a). Now nearer to Earth's surface, the asthenosphere is under less pressure than before and decompression melting occurs. Some of the magma at this point may rise rapidly to the surface and erupt onto the continent to produce a volcano. If the stretching is great enough and continues for long enough (tens to hundreds of million years), the region will subside below sea level and a mid-ocean ridge will develop (Figure 5.4a). Cooling of the voluminous amounts of magma generated by decompression melting at mid-ocean ridges has created all of Earth's oceanic lithosphere over the course of the past 200 million years.

Decompression melting also occurs at "hot spots," where a plume of superheated rock rises from deep within the mantle toward the surface (Figure 5.4b). Compared to magma formation at mid-ocean ridges, hot spots are responsible for a relatively small amount of decompression melting on Earth. However, melting at any one hot spot can be extensive due to the great depth from which the mantle upwells.

2. *Addition of volatiles* lowers the melting temperature of rocks by helping to break chemical bonds within minerals. **Volatiles** are chemical compounds, such as water (H_2O), carbon dioxide (CO_2), and sulfur dioxide (SO_2), that evaporate easily and exist in a gaseous state at Earth's surface. However, volatiles are present deep in Earth as solid compounds that are incorporated into minerals. When volatiles, such as water, are added as a fluid to rocks that are close to their melting temperature, the rocks will begin to melt. For example, at a depth of 100 km (~62 mi.) the mantle has a dry (volatile-free) melting temperature of about 1500°C (~2700°F), but in the presence of a small percentage of water, the mantle melting temperature is about 800°C (~1500°F).[7] This process is analogous to using salt on sidewalks in cold climates to melt ice—salt lowers the melting temperature of water from 0°C (~32°F) to –8°C (~17°F); therefore, salted ice will melt even though it is below freezing outside.

Magma formation due to the addition of volatiles occurs at subduction zones and is responsible for the "Ring of Fire" that surrounds the Pacific Ocean, which represents more than half of Earth's active volcanoes above sea level (see Figure 5.3).[8] This ring has formed above the subduction zones where the Pacific, Nazca, Cocos, Philippine, and Juan de Fuca plates are sinking below adjacent plates. Volatiles trapped within the sediments on the seafloor and in minerals that compose the oceanic lithosphere are released as the subducting oceanic plate is forced downward into the hotter mantle (Figure 5.4c). The volatiles are then free to rise upward into the formerly dry (volatile-free) asthenosphere causing magma to form due to a lowering of the mantle melting temperature. The volatiles become dissolved into the rising magma and play an important role in the eruptive behavior of the volcano.

3. *Addition of heat* to rocks will induce melting if the temperature exceeds the melting temperature of the rocks at that depth. The best way to move heat rapidly from areas of higher temperature to areas of lower temperature is within large bodies of magma. As the magma rises within the mantle and ultimately through the crust, heat from deep within Earth is carried toward the surface and along the way is transferred to the surrounding rocks. This heat may cause the cooler, shallower rocks to melt and this newly formed magma may become mixed with the rising magma. The degree to which this mixing occurs depends on whether the surrounding rocks are already close to their melting temperature and the amount and temperature of the rising magma.

Magma formation due to the addition of heat is likely widespread wherever magma rises and comes into contact with cooler rocks, but it is more difficult to quantify. However, where hot spots occur beneath continents, pooling of basaltic magma near the base of Earth's crust can cause extremely large magma chambers to develop (Figure 5.4d). Perhaps more important than the volume of magma at continental hot spots is the effect of crustal melting and magma mixing—*crustal assimilation*—on the chemical composition of the magma at these locations. As you will see in the next section, magma composition has a strong effect on the type of volcano and the eruptive behavior of the different volcano types, which is why continental hot spot volcanoes are some of the largest and most dangerous on Earth.

MAGMA PROPERTIES

Magma is composed of melted silicate rocks and dissolved gases. Silica (SiO_2), the most abundant chemical compound in silicate minerals, is the primary constituent of magma. The three major types of magma—*basaltic*, *andesitic*, and *rhyolitic*—vary in their silica content (Figure 5.5). All magmas that form by melting of the asthenosphere are basaltic. Because most magmas form by melting the asthenosphere, as you learned in the previous section, basaltic magma is the most common magma type. Basaltic magmas have a relatively low silica content (45–55 percent) and are referred to as *mafic* in composition. Rhyolitic magma contains more than 65 percent silica and is referred to as *felsic*, whereas andesitic magma is *intermediate* in composition (55–65 percent silica). When these magmas erupt onto Earth's surface, each lava type solidifies to produce a different volcanic rock—*basalt*, *andesite*, and *rhyolite* (Figure 5.5).

Magma is less dense than the surrounding rock; therefore, it rises buoyantly toward the surface and sometimes accumulates near Earth's surface in large pools, known as **magma chambers**. While pooled in a magma chamber or along its path toward the surface, chemical changes due to cooling and crustal assimilation change the composition of the melt and it becomes more felsic (increasing silica content) over time. The change in magma composition from primitive mafic basalt toward felsic magma as it ascends toward the surface is known as *magma evolution*. The degree of magma evolution depends on how fast the magma rises through the lithosphere and the amount of volatiles dissolved into the melt—greater volatile content increases the rate of magma evolution. When the magma reaches the surface,

the type of volcano formed and its eruptive behavior are controlled not only by the composition but also by the viscosity and volatile content of the melt.

Viscosity   Greater amounts of silica make it more difficult for magma to flow. Resistance to flow in fluids is called *viscosity*. Fluids that do not flow easily, such as molasses or refrigerated honey, are like silica-rich felsic magma—they have a high viscosity. Other fluids that flow readily, such as water, warm honey, and low-silica mafic magma, have a low viscosity. As the example of honey suggests, viscosity is typically influenced by temperature as well as composition (see Figure 5.5). For example, a low-silica content, basaltic lava flow may have a low viscosity when it erupts onto Earth's surface, but as it flows down the flank of the volcano and cools, its viscosity will increase, causing the flow to move more slowly and change in form (see Section 5.4).

The variability in magma viscosity strongly influences both the mobility of the magma under the surface and the velocity and form it takes if it reaches the surface as a lava flow. Rhyolitic lava flows have high viscosity, move slowly, may be a hundred feet thick, are generally restricted to the vent region, and form steep-sided domes. In contrast, basaltic lava flows can move rapidly, are often thin (< 3 m or 10 ft.), and may travel tens of kilometers from the vent. These differences in flow characteristics affect the shape of the volcano formed by different lava types. More

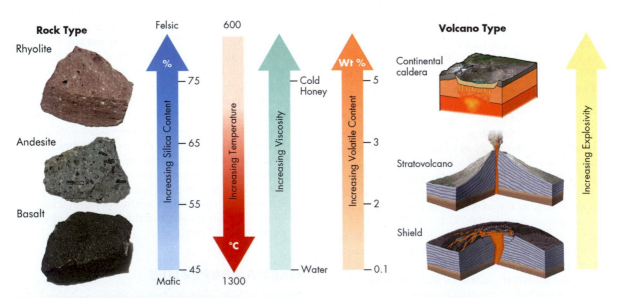

⋀ FIGURE 5.5 Volcano Characteristics
The eruptive character, shape, and type of volcano that forms are directly related to the silica content of the magma, which defines the rock type and also affects the viscosity and volatile content of the melt.

importantly, the viscosity is often correlated to the volatile content of the magma, which is responsible for whether or not volcanic eruptions are peaceful or explosive (Figure 5.5).

Volatile Content Uncorking a bottle of champagne and opening a shaken carbonated beverage are good analogies to explain why volcanoes erupt explosively. When the champagne is depressurized by uncorking, the dissolved volatiles (CO_2) come out of solution from the liquid, forming bubbles. The rapid expansion of the bubbles causes the liquid to vigorously squirt from the bottle. Similarly, a high concentration of dissolved volatiles within the magma will cause an explosive eruption when the magma chamber is decompressed. In fact, the characteristic semicircular depression at the top of many volcanoes, called a volcanic **crater**, is caused by explosive extrusion of volcanic material during an eruption.

In general, volatile content increases with increasing silica content (Figure 5.5). Andesitic-to-rhyolitic magma has more dissolved gas (2–5 weight percent) than basaltic magma (< 1 wt. percent). Thus, volcanoes with andesitic-to-rhyolitic magma are more prone to explosive eruptions than volcanoes with basaltic lava flows.

Eruption of volatile-poor basaltic magma results in **effusive eruptions**, which are typified by passive outpouring of low-viscosity lava onto Earth's surface (Figure 5.6a). By contrast, the rapid formation and expansion of bubbles within volatile-rich, high-viscosity magma causes explosive eruptions, in which the molten material fragments into fine-grained particles that are propelled high into the atmosphere (Figure 5.6b). Volcanic materials, such as ash, that are explosively ejected from a volcano are collectively called **pyroclastic debris** (see Section 5.4).

5.1 CHECK your understanding

1. What is magma and where does it come from?
2. Describe the physical and chemical changes that cause rocks to melt.
3. What is the relationship between plate tectonics and the location of volcanoes?
4. What is viscosity and what determines magma viscosity?
5. Explain the relationship between magma composition, viscosity, and gas content.

(a)

(b)

⋀ FIGURE 5.6 Effusive vs. Explosive Eruptions
The degree to which the direct effects of a volcanic eruption are hazardous is related to the viscosity of the magma. (a) Low-viscosity lava shoots 30 m (~100 ft.) from Pu'u 'Ō'ō crater on Hawaii in 1983, creating a spectacular fire fountain display, and then flows from the vent downhill like a river. *(J. D. Griggs/USGS)* (b) Rapid expansion of volatiles trapped in high-viscosity magma explosively propels ash more than 8 km (~5 mi.) into the atmosphere during the 2008 eruption of Chaitén volcano in Chile. *(Oliver Wien/epa European Pressphoto Agency creative account/Alamy)*

5.2 Volcano Types, Formation, and Eruptive Behavior

Volcanoes are landforms that vary greatly in size, shape, composition, and eruptive behavior (Table 5.1). Some volcano types form during a single eruption, while others may be the results of hundreds or thousands of eruptions over millions of years. The type of volcano that forms is the result of how and where the magma formed, the amount of magma evolution (silica enrichment) that occurred on the way to the surface, and the volatile content of the melt. Although every volcanic eruption is unique, and even eruptions from the same volcano differ, volatile content and viscosity are the primary controls on whether a volcano erupts effusively or explosively. The eruptive behavior of specific volcano types can be broadly generalized and predicted. A somewhat more quantitative means of evaluating a volcano's eruption is the **volcanic explosivity index (VEI)**. The VEI is a relative scale by which different eruptions can be compared based on quantitative and qualitative observations of explosivity. Explosivity is based on the height of the eruption plume and the volume of ejected material, as well as the overall violence of the eruption (Table 5.2). With the exception of VEI 0 through VEI 2, the scale is logarithmic, with each interval representing a tenfold increase in the volume of ejected material.

Stratovolcanoes Known for their beautiful conical shapes (Figure 5.7), Mount Fuji in Japan and Mount St. Helens, Mount Rainer, and Mount Shasta in the United States are examples of **stratovolcanoes**. Over time, stratovolcanoes will erupt a variety of different lava compositions from basalt to rhyolite, but intermediate to felsic compositions dominate (see Table 5.1). Because of their high viscosity, these lavas do not travel far from the vent and therefore pile up, giving stratovolcanoes their steep sides. Stratovolcanoes have high water content because they form above subduction zones where magma formation is caused by addition of volatiles (see Section 5.1). The high-volatile content of these volcanoes means that eruptions can be extremely explosive. In fact, stratovolcanoes are a composite of a series of layered pyroclastic debris that rains down after an explosive eruption and effusive lava flows (Figure 5.8a). As a result, these volcanoes are also called *composite cones*.

Stratovolcanoes are common within the Ring of Fire, where subduction is the dominant tectonic process, such as the Cascadia subduction zone in the Pacific Northwest and southwestern Canada (see Figure 5.3). As the volatile-rich basaltic melt rises upward from the asthenosphere through the overriding plate, the magma evolves, becoming more felsic. The volcanoes formed by these magmas are commonly andesitic to rhyolitic in composition and their silica content lies somewhere between basaltic oceanic crust and more silica-rich

▼ TABLE 5.1 Volcano Characteristics

Volcano Type	Composition	Volatile Content	Shape	Eruption Type	Volcano Examples
Stratovolcano, or composite cone	Andesite	High	Cone shaped with steep sides; built of lava flows and pyroclastic deposits	Combination of lava flows and pyroclastic activity	Mount Shasta, California (Figure 5.7); Mount St. Helens, Washington (Figure 5.34); Eyjafjallajökull, Iceland (chapter-opener photograph)
Lava dome	Andesite to rhyolite	Low to moderate	Dome shaped; steep sided	Mostly effusive with lavas piling up near the vent, but can be explosive.	Mount Lassen, California; Mt. Unzen dome, Japan (Figure 5.9)
Shield volcano	Basalt	Low	Gentle arch or shield shape, with shallow slopes; built up of many lava flows	Far-traveling lava flows	Mauna Loa, Hawaii (Figure 5.10); Kīlauea, Hawaii
Cinder cone	Basalt	Low to moderate	Cone shaped with steep sides and summit crater	Mostly tephra ejection (nut to fist sized) lava flows	SP Crater, Arizona (Figure 5.14); Parícutin, Mexico (Figure 5.15); Eldfell, Iceland (Figure 5.44)
Continental calderas	Rhyolite	High	Broad uplift with large summit depression	Tephra	Yellowstone caldera; Mount Mazama (Figure 5.17)

Source: Data partially derived from Ollier, C., *Volcanoes*. Cambridge, MA: MIT Press, 1969.

continental crust. The higher silica content and higher volatile content make these volcanoes explosively unpredictable (Table 5.2). Stratovolcanoes are responsible for more than 80 percent of the volcanic eruptions in historic times and for most of the death and destruction caused by volcanoes throughout history.[9]

Explosive eruptions of stratovolcanoes are known as *Plinian-type eruptions*—named after the 17-year-old Roman boy Pliny the Younger, who first described this type of eruption at Mount Vesuvius in Italy in A.D. 79. During a Plinian-type eruption, such as Chaitén, a Chilean volcano shown in Figure 5.6b, tens of cubic kilometers of rock and pyroclastic debris may be ejected (Table 5.2). The pyroclastic material is propelled skyward by decompressed hot volatilies, forming a plume that may climb to more than 30 km (~18.6 mi.) into the atmosphere (Figure 5.8a). Although much of the pyroclastic debris will fall near the volcano, as discussed in the next section, fallout from an eruption plume is one of the effects of volcanoes that can have global impacts.

Lava Domes Similar to stratovolcanoes, **lava domes** are characterized by highly viscous felsic magma and are common along the Ring of Fire. Unlike stratovolcanoes, however, lava domes are relatively small and have low to moderate volatile content. Nevertheless, the high viscosity of the magma means that these volcanoes can explode and, therefore, present a serious hazard to anyone living in the area.

Lava domes often form in the vent of a stratovolcano days to years after an explosive eruption. Having exhausted much of the volatile content of the magma during the preceding eruption, the degassed high-viscosity lava slowly oozes out and cools, forming a steep-sided plug in the vent that may be only a few hundred meters high and a kilometer across (Figure 5.8b and Figure 5.9). A lava dome may grow in a single event immediately following an eruption or may grow episodically over decades, as did the Mount St. Helens dome following the 1980 eruption (see Case Study 5.3: Mount St. Helens). Once the vent is plugged, volatile content begins to build again

▼ **TABLE 5.2 Volcanic Explosivity Index (VEI)**

VEI	Eruption Type	Typical Volcano Type	Eruption Behavior	Plume Height	Ejecta Volume	Frequency	Eruption Examples
0	Hawaiian	Shield	Effusive, not explosive	<100 m	<10,000 m^3	Constant	Kilauea (1983–present)
1	Hawaiian/ Strombolian	Shield, Cinder cone	Effusive, mildly explosive	100–1000 m	<10,000 m^3	Daily	Stromboli, Eldfell (1973)
2	Strombolian/ Vulcanian	Cinder cone, Stratovolcano	Mild	1–5 km	>1,000,000 m^3	Weekly	Galeras (1993), Mount Sinabung (2010)
3	Vulcanian/ Peléan	Stratovolcano, Lava dome	Severe	3–15 km	>10,000,000 m^3	Few months	Mount Unzen (1995), Nevado del Ruiz (1985), Soufriere Hills (1995)
4	Peléan/Plinian	Stratovolcano	Cataclysmic	10–25 km	>0.1 km^3	≥1 yr	Mount Pelée (1902), Eyjafjallajökull (2010)
5	Plinian	Stratovolcano	Cataclysmic	20–35 km	>1 km^3	≥10 yrs	Mount Vesuvius (A.D. 79), Mount St. Helens (1980), Chaitén (2008; Figure 5.6)
6	Plinian/Ultra-Plinian	Stratovolcano, Continental caldera	Colossal	> 30 km	>10 km^3	≥100 yrs	Krakatoa (1883), Mount Pinatubo (1991)
7	Ultra-Plinian	Continental caldera	Super-colossal	> 40 km	>100 km^3	≥1,000 yrs	Mazama (5600 B.C.), Tambora (1815)
8	Supervolcanic	Continental caldera	Mega-colossal	> 50 km	>1,000 km^3	≥10,000 yrs	Yellowstone (640,000 B.C.), Toba (74,000 B.C.)

Source: Based on "Volcanic Explosivity Index," Wikipedia, http://en.wikipedia.org/wiki/Volcanic_Explosivity_Index.

∧ **FIGURE 5.7 Stratovolcano**
Mount Shasta, in northern California, is the most voluminous volcano in the Cascade Range (85 cubic miles)
and reaches an elevation of 4,322 m (~14,179 ft.). *(Duane E. DeVecchio)*

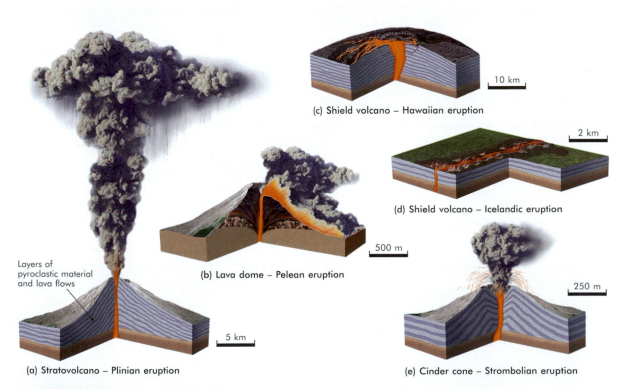

Layers of
pyroclastic material
and lava flows

(a) Stratovolcano – Plinian eruption

(b) Lava dome – Pelean eruption

(c) Shield volcano – Hawaiian eruption

(d) Shield volcano – Icelandic eruption

(e) Cinder cone – Strombolian eruption

∧ **FIGURE 5.8 Volcanoes and Eruption Types**
Most volcanoes will erupt numerous times over their lifetime, and although the eruption type is not unique to
volcano type (see Table 5.2), characteristic eruption types for each of the different volcanoes is provided here.
(Based on "Icelandic eruption: types of volcanic eruption," Britannica Kids, http://kids.britannica.com/comp-
tons/art-156012/The-six-major-types-of-volcanic-eruptions-differin-their_Britannica)

< FIGURE 5.9 Lava Dome
Glowing high-viscosity lava builds up in the vent of Japan's Mount Unzen. The large blocks of cooled magma periodically collapse, releasing trapped volatiles and causing a Peléan eruption. *(Stocktrek Images/SuperStock)*

< FIGURE 5.10 Shield Volcano
Peering above the clouds view, Mauna Loa, the largest shield volcano on Earth is located on the big island Hawaii. Notice the gently sloping profile of the volcano, similar to that of a warrior's shield lying on the ground. Dark patches are less vegetated regions that mark the locations where more recent low-viscosity basaltic lavas have flowed down the flank of the mountain. Note reddish basaltic cinder cones in the foreground. *(Duane E. DeVecchio)*

and periodic eruptions may occur during lava dome growth and collapse (Figure 5.8b). Collapse of a growing lava dome can cause explosive release of trapped volatiles and large destructive volcanic avalanches, such as those during eruptions of Mount Unzen in Japan between 1991 and 1995 (see Case Study 5.1: Mount Unzen). Mount Lassen in California similarly experienced a series of eruptions from 1914 to 1917, which included a tremendous horizontal (lateral) blast that affected a large area. These eruptions are known as *Peléan-type*, and have VEIs of 3 to 4 (see Table 5.2).

Shield Volcanoes The largest volcanoes in the world are **shield volcanoes**. They are common in the Hawaiian Islands, Iceland, and in the East African Rift (see Figure 5.3). Shield volcanoes are constructed from basaltic lavas and in profile appear as broad arc with gentle slopes (Figure 5.10). Because of the low viscosity of the lava, when a shield volcano erupts the lava can flow great distances away from the vent down the flanks of the volcano. An accumulation of thousands of thin basaltic lava flows over a large region gives these volcanoes their gentle slopes (Figure 5.10). Although shield volcanoes are much wider than they are tall, they are

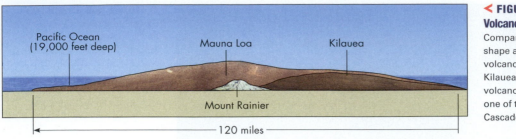

◄ FIGURE 5.11
Volcano Dimensions
Comparison between the shape and size of a shield volcano (Mauna Loa and Kilauea) and a strato-volcano (Mount Rainier, one of the largest in the Cascade Range).

still among the tallest mountains on Earth when measured from their bases, which are often located on the ocean floor. By contrast, a stratovolcano constructed of viscous lava, such as Mount Rainier in Washington State, has a height more similar to its width (Figure 5.11).

Shield volcanoes are most common at hot spots in the oceanic lithosphere and at divergent plate boundaries and continental rifts, where decompression melting occurs (see Figure 5.4a). Because the lithosphere is thin in these environments, basaltic magma is able to rise rapidly to Earth's surface with little time to chemically evolve. The low-silica content and the absence of volatiles during magma formation mean these volcanoes erupt effusively rather than explosively. Consequently, rather than having an explosive crater near the vent, shield volcanoes develop a **caldera**, which forms by the inward collapse of the volcano top due to removal of magma from the subsurface (Figure 5.8c).

Effusive eruptions such as the ones that build shield volcanoes are known as *Hawaiian-type* eruptions, after the island of Hawaii, which is the most active and largest shield volcano on Earth (see Figure 5.10). Although

these eruptions are not explosive, during the initial phase of a Hawaiian-type eruption, lava may shoot up more than 300 m (~1000 ft.), forming a "fire fountain" like the one that formed at Pu'u 'Ō'ō' crater on Hawaii in 1983 (see Figure 5.6a). The passive behavior of these types of eruptions encourages large groups of geotourists to flock to Hawaii or Iceland with the hope of seeing lava flowing across the surface. By contrast, when a stratovolcano like Mount St. Helens or Pinatubo in the Philippines erupts, large areas must be evacuated (see Figure 5.6b).

Another type of effusive eruption, known as an *Icelandic-type* eruption, occurs when basaltic lava erupts from an elongate fissure rather than from a central vent, producing what is called a "curtain of fire." The absence of a central vent means that shield volcanoes constructed from Icelandic-type eruptions are typically smaller and characterized by a broad and flat upland surface (Figure 5.8d). Fissure eruptions, similar to Icelandic-type eruptions, also occur on the flanks of shield volcanoes, such as the 2011 Kamoamoa eruption on Kilauea shield volcano in Hawaii (Figure 5.12).

◄ FIGURE 5.12 **Fissure Eruption**
Lava fountaining from the 2.3 km (~1.4 mi.) long fissure during the 2011 Kamoamoa eruption on Kilauea, a shield volcano. *(Jay Robinson/NPS)*

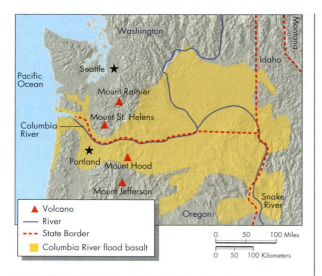

^ FIGURE 5.13 Columbia River Flood Basalts
From approximately 17 to 15.5 million years ago, thousands of basalt flows poured out onto the surface of the Pacific Northwest, blanketing the states of Washington, Oregon, and Idaho in volcanic rock that locally reaches a thickness of 3500 m (~11,000 ft).

Although the preceding paragraphs have emphasized that eruptions of low-volatile content basaltic magma are typically nonexplosive with VEIs of 0 to 1 (Table 5.2), don't be tricked into believing that there is no hazard associated with basaltic eruptions. Some of the largest and most catastrophic eruptions in Earth's history have been Icelandic-type eruptions of basaltic magma. Many of these massive outpourings of lava, called *flood basalts*, are correlated to sudden loss of large numbers of plant and animal species on Earth (see Chapter 14). The best-known accumulation of flood basalts in the United States is the Columbia Plateau in Washington, Oregon, and the Snake River Plain in Idaho (Figure 5.13). These flood basalts cover about 130,000 km^2 (~50,000 mi.^{2}), an area slightly larger than the state of Mississippi.

Basaltic eruptions can also be explosive when magma comes into contact with groundwater, snow, or ice, which can cause large steam blasts creating pyroclastic debris that may rise 10 km (~6.2 mi.) into the atmosphere (see Applying the 5 Fundamental Concepts: Eyjafjallajökull 2010 Eruption). These types of eruptions are known as *phreatomagmatic* or *Vulcanian-type* eruptions and can have a VEI as high as 3.

Cinder Cones   Built up by the accumulation of volcanic rock near a volcanic vent, **cinder cones** are relatively small and composed of nut- to fist-sized pieces of vesicular red or black lava (Figure 5.14).[10] Scoria, the name for vesicular mafic volcanic rock, forms when basaltic magma with intermediate amounts of volatile

^ FIGURE 5.14 Cinder Cone
SP Crater, a cinder cone with a small crater at the top in the San Francisco Peaks volcanic field near Gray Mountain, Arizona. The lava flow in the upper part of the photograph originated from the base of the cinder cone and extends about 6 km (~4 mi.). *(Michael Collier)*

content is explosively ejected, cools, and falls to the ground (Figure 5.8e). The conical shape of the volcano comes from the accumulation of loose scoria around the vent; therefore, these volcanoes are also called *scoria cones*. Scoria formed at cinder cones is the "lava rock" used widely in commercial landscaping, and the vesicles, which are small holes formed by bubbles in the magma, indicate that this mafic lava had intermediate volatile contents.

Basaltic eruptions with intermediate explosivity are known as *Strombolian-type* eruptions—named for the volcano Stromboli offshore southwestern Italy. Strombolian-type eruptions are more explosive than Hawaiian-type eruptions due to a higher volatile content (Table 5.2) but are similarly associated with lava flows. Because basaltic magma is the most common type of magma formed in most tectonic settings, it is common for small amounts to reach the surface forming a small cinder cone. As such, cinder cones are common on the flanks of larger volcanoes, along normal faults, and along cracks or fissures.

The Parícutin cinder cone in the Itzicuaro Valley of central Mexico, about 320 km (~200 mi.) west of Mexico City, offered a rare opportunity to observe the birth and rapid growth of a volcano at a location where none had existed before (Figure 5.15). On February 20, 1943, following several weeks of

∧ FIGURE 5.15 Rapid Growth of a Volcano
In this 1943 photograph, Parícutin cinder cone in central Mexico is erupting a cloud of volcanic ash and gases (background). A basaltic lava flow has nearly buried the village of San Juan Parangaricutiro, leaving only the church steeple exposed.
(© Getty Images/Evans/Stringer)

earthquakes and sounds like thunder coming from beneath Earth's surface, an astounding event occurred. As Dionisio Pulido was preparing his cornfield for planting, he noticed that a hole he had been trying to fill for years had opened in the ground at the base of a knoll. As Señor Pulido watched, the surrounding ground swelled, rising more than 2 meters (more than 6 ft.), while sulfurous gases and ash began billowing from the hole. By that night, the hole was ejecting glowing-red rock fragments high into the air. By the next day, the cinder cone had grown to 10 m (~33 ft.) high as rocks and ash continued to be blown into the sky by the eruption. After only five days, the cinder cone had grown to more than 100 m (~330 ft.) high. In June 1944, a fissure opened in the base of the cone that was now 400 m (~1300 ft.) high. The fissure produced a basaltic lava flow that overran the nearby village of San Juan Parangaricutiro, leaving little but the church steeple exposed. No one was killed in these eruptions, and within a decade, Parícutin cinder cone became a dormant volcano. Nevertheless, over nine years more than a billion cubic meters (~1.3 billion cubic yards) of ash and 700 million m³ (~900 billion yd.³) of lava were erupted from Señor Pulido's cornfield. Crops failed, especially when buried by ash faster than they could grow, and livestock sickened and died. Although several villages were relocated to other areas, some residents moved back to the vicinity of Parícutin. Locating property boundaries has been difficult because markers are generally covered by ash and lava, resulting in many land ownership disputes.[10]

Continental Caldera The largest and most violent eruptions occur when calderas form due to collapse of the land surface or volcanic edifice following partial emptying of the magma chamber (Figure 5.16). Unlike the passive formation of a caldera on a shield volcano, this inward collapse causes an explosive release of the magma and the volatiles that still remain in the magma chamber. Fortunately, these eruptions are rare and none have occurred in the United States in the past few thousand years. However, at least 10 such eruptions have occurred in the past million years, three of which were in North America. Large caldera-forming eruptions, known as *ultra-Plinian* type, may explosively extrude more than 100 km³ (~62 mi.³) of pyroclastic debris consisting mostly of ash (see Table 5.2). A quantity this large would cover the island of Manhattan to a height of about 1.6 km (~1 mi.)—four times the height of the Empire State Building. This volume is approximately 1,000 times the amount of ash ejected by the 1980 eruption of Mount St. Helens! Ash deposits from such an eruption could be 100 m (~300 ft.) thick near the crater's

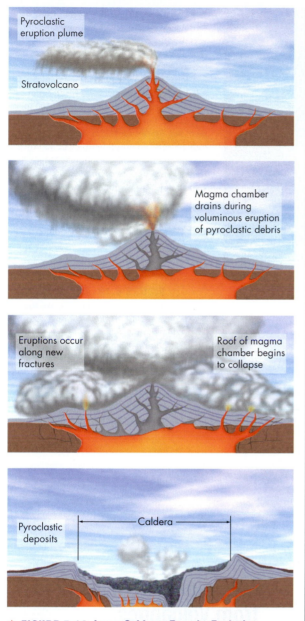

Pyroclastic
eruption plume

Stratovolcano

Magma chamber
drains during
voluminous eruption
of pyroclastic debris

Eruptions occur
along new
fractures

Roof of magma
chamber begins
to collapse

Pyroclastic
deposits

Caldera

^ FIGURE 5.16 Large Calderas Form by Explosion and Collapse

Large calderas, such as the one illustrated here, generally form by explosion and subsequent collapse of the magma chamber below a composite volcano. The resulting depression may be 20 km (~12 mi.) or more in diameter. This process may occur once, like Crater Lake in Oregon, or multiple times, like the formation of resurgent calderas in Yellowstone National Park, Wyoming. *(Based on Smith, G. A., and Pun, A. How does Earth Work? Physical Geology and the Process of Science. © 2006 Upper Saddle River, NJ: Pearson Prentice Hall)*

rim and a meter (~3 ft.) or so thick 100 km (~60 mi.) away from the source.[11] Crater Lake in Oregon fills the caldera that formed following the 5677 B.C. eruption of Mount Mazama (Figure 5.17). Other notable historic ultra-Plinian eruptions include the 1815 Tambora and 1883 Krakatau explosions in Indonesia.

The largest caldera eruptions are known as the *supervolcanic*-type (see Table 5.2). These eruptions are associated with rhyolitic magmas having very high volatile content. The high silica and volatile content suggests that significant amounts of continental crust have been melted and mixed with the magma. This mixing occurs in areas where large volumes of magma are generated and must rise through thick felsic crust, such as when a hot spot is located beneath continental lithosphere (see Figure 5.4d). The most recent incident occurred 27,000 years ago in New Zealand. Today this caldera is the site of Lake Taupo, a famous trout fishery on North Island. The eruption ejected between 1000 and 2000 km^3 (~240–480 $mi.^3$) of volcanic ash and created a caldera that has an area of 616 km^2 (~238 $mi.^2$). The best-known caldera-forming eruptions in North America occurred about 640,000 years ago at Yellowstone National Park in Wyoming and 780,000 years ago in Long Valley, California. These eruptions that blanketed large parts of North America in ash (Figure 5.18) would devastate the U.S. economy and food production capabilities, not to mention the negative impacts on global climate change, if they happened today. Although the most recent volcanic eruptions at Long Valley occurred about 400 years ago, potential hazards from a future volcanic eruption do still exist. In the early 1980s, measurable uplift of the land accompanied by swarms of earthquakes up to **M** 6 suggested that magma was moving upward at Long Valley. This prompted the U.S. Geological Survey (USGS) to issue a potential volcanic hazard warning that was subsequently lifted. Although activity at Long Valley has subsided, the future remains uncertain.

The main events in a caldera-producing eruption can occur quickly—in a few days to a few weeks—but intermittent, lesser-magnitude volcanic activity can linger on for a million years. Caldera-producing eruptions at Yellowstone have left us hot springs and geysers, including Old Faithful, whereas the Long Valley eruption has left us a potential volcanic hazard. In fact, both sites are still capable of producing volcanic activity because magma is present at shallow depths beneath the caldera floors. Both are considered *resurgent calderas* because their floors have slowly domed upward since the explosive eruptions that formed them.

◀ FIGURE 5.17 Crater Lake Caldera
Aerial view of Crater Lake, Oregon, which was produced by an explosive eruption of Mount Mazama around 7,700 years ago. Crater Lake is 8 by 10 km (~5 by 6 mi.) across and almost 600 m (~2000 ft.) deep, making it the second deepest lake in North America. Originally a stratovolcano around 1200 m (~4000 ft.) higher than it is today, the explosion and subsequent collapse of the caldera took place in an eruption that was 42 times greater than the 1980 eruption of Mount St. Helens. Each year close to 500,000 people visit Crater Lake National Park. *(USGS)*

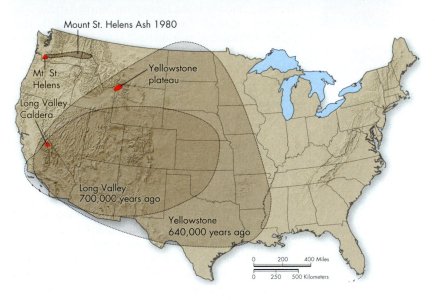

◀ FIGURE 5.18 Ash Fall Hazard
Caldera-forming eruptions, like the most recent eruptions at Yellowstone and Long Valley, can affect a very large area. Distribution of heavy ash fallout from the 1980 Plinian eruption of Mount St. Helens is shown for comparison.

5.2 CHECK your understanding

1. How is a volcanic crater different from a caldera?
2. Describe the major types of volcanoes and their composition.
3. What characteristic of a volcano controls its shape, and how?
4. How are Icelandic-type and Hawaiian-type eruptions different and the same?
5. Both stratovolcanoes and lava domes form in the same tectonic setting and have similar composition, but why do they have different explosive behavior?
6. Explain how large caldera eruptions occur and why they are so dangerous.

5.3 Geographic Regions at Risk from Volcanoes

Like earthquakes, volcanoes are intimately related to plate tectonics and most occur along the Ring of Fire surrounding the Pacific Ocean basin (see Figure 5.3). Areas outside the Ring of Fire can also experience volcanoes because of hot spot activity (Yellowstone) or because they are located on a portion of the mid-ocean ridge (Iceland). East Africa has volcanism related to the rifting, or pulling away from one another, of three tectonic plates. Also large, infrequent eruptions at Long Valley and those centered in Yellowstone National Park in North America indicate that these areas may be at risk in the future. The area at highest risk of local volcanic activity in the contiguous United States and Canada

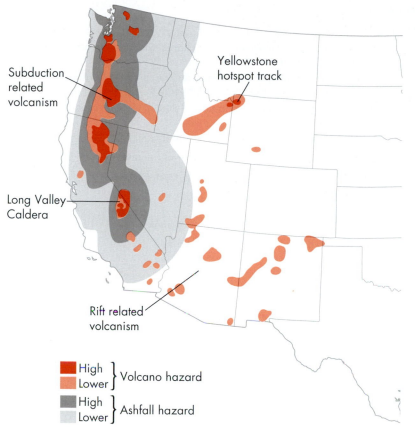

Volcanic hazards for the contiguous United States based on activity during the past 15,000 years. Red colors show high and lower risk of local volcanic activity, whereas gray colors show regions at risk of receiving 5 cm (~2 in.) or more of ash fall from large explosive eruptions. The hazard belt in Washington State continues north into southern British Columbia. *(U.S. Geological Survey)*

is in the mountainous regions of the Pacific Coast and at Yellowstone (Figure 5.19). Other isolated areas of the U.S. Southwest are at risk, whereas the eastern two-thirds of the United States and Canada is risk free of local volcanic activity. However, the effects of a large caldera explosion in the western United States or Canada would likely be felt far from the source in the form of ash fall and ash clouds in the atmosphere.

5.3 CHECK your understanding

1. What part of the United States has the greatest volcanic hazard, and why?

2. Why is there no volcanic hazard on the East Coast of the United States and Canada?

5.4 Effects of Volcanoes

Worldwide, 50 to 60 volcanoes erupt each year.[12] With the exception of the Kilauea volcano on the big island of Hawaii, which has been erupting continuously since 1983, eruptions in the United States occur two or three times per year, mostly in Alaska. Eruptions are often in sparsely populated areas of the state, causing little, if any, loss of life or economic damage. However, when an eruption takes place near a densely populated area, the effects can be catastrophic (Table 5.3).[13] Approximately 500 million people on Earth live close to volcanoes, and as the human population grows, more and more people are living on the flanks, or sides, of active or potentially active volcanoes. In the past 100 years, nearly 100,000 people have been killed by volcanic eruptions—approximately 28,500 lives were lost in the decade of the 1980s alone.[12,13] Densely populated countries with many active volcanoes, such as Japan, Mexico (especially near Mexico City), the Philippines, and Indonesia, are particularly vulnerable.[14] Several active or potentially active volcanoes in the western United States are near cities, such as Seattle, with populations of more than 720,000 people (Figure 5.20).

Volcanic hazards include the *primary effects* of volcanic activity, which are the direct results of an eruption, and the *secondary effects*, which may be caused by the primary effects. Lava flows, pyroclastic activity such as ash fall, pyroclastic flows and lateral blasts, and release of volcanic gases are primary effects. Secondary effects include debris flows, mudflows, landslides or debris

▼ **TABLE 5.3** Selected Historic Volcanic Events

Volcano or City	Year	Effect
Mount Vesuvius, Italy	A.D. 79	Destroyed Pompeii and killed 16,000 people. City was buried by volcanic activity and rediscovered in 1595.
Skaptar Jokull, Iceland	1783	Killed 10,000 people (many died from famine) and most of the island's livestock. Also killed some crops as far away as Scotland.
Tambora, Indonesia	1815	Global cooling; killed 10,000 people and 80,000 starved; produced "year without a summer."
Krakatau, Indonesia	1883	Tremendous explosion; more than 36,000 deaths from tsunami.
Mount Pelée, Martinique	1902	Ash flow killed 30,000 people in a matter of minutes.
La Soufrière, St. Vincent	1902	Killed 2,000 people and caused the extinction of the Carib Indians.
Mount Lamington, Papua New Guinea	1951	Killed 6,000 people.
Villarica, Chile	1963–64	Forced 30,000 people to evacuate their homes.
Mount Helgafell, Heimaey Island, Iceland	1973	Forced 5,200 people to evacuate their homes.
Mount St. Helens, Washington, United States	1980	Debris avalanche, lateral blast, and mudflows; killed 57 people, destroyed more than 100 homes.
Nevado del Ruiz, Colombia	1985	Eruption generated mudflows that killed at least 22,000 people.
Mount Unzen, Japan	1991	Ash flows and other activity killed 41 people and burned more than 125 homes. More than 10,000 people evacuated.
Mount Pinatubo, Philippines	1991	Tremendous explosions, ash flows, and mudflows combined with a typhoon killed more than 740 people; several thousand people evacuated.
Montserrat, Caribbean	1995	Explosive eruptions and pyroclastic flows; south side of island evacuated, including capital city of Plymouth; several hundred homes destroyed.
Chaitén, Chile	2008	Explosive eruptions and pyroclastic flows: 5,000 people evacuated and disrupted aviation in South America for weeks.
Mount Merapi, Indonesia	2010	Multiple explosions and dome collapses; 320,000 people displaced, and more than 300 killed in Indonesia.
Eyjafjallajökull, Iceland	2010	Large ash emission: disrupted air travel in the United Kingdom and northern Europe for several weeks.

Source: Based on USGS Volcanic Activity Alert-Notification System and U.S. Geological Survey Fact Sheet 2006–3139.

avalanches, floods, fires, and tsunamis (discussed in Chapter 4). At the planetary level, global cooling of the atmosphere for a year or so is a secondary effect of a large eruption.[11]

LAVA FLOWS

A **lava flow** is one of the most familiar products of volcanic activity. Lava flows result when magma reaches the surface and overflows the central crater or erupts from a volcanic vent along the flank of the volcano. The three major types of lava take their names from the volcanic rocks they form: basaltic—by far the most abundant of the three—andesitic, and rhyolitic.

Lava flows can be quite fluid and move rapidly or be relatively viscous and move slowly. Basaltic lavas, which have lower viscosity and higher eruptive temperatures,

are the fastest and can move 15–35 km (~10–20 mi.) per hour near the vent. Called *pahoehoe* (pronounced pa-hoy-hoy), these lavas have a smooth, sometimes ropey surface texture when they harden (Figure 5.21a). As the lava cools, it becomes more viscous and may move only a few meters (around 6 to 10 ft.) per day. Called *aa* (pronounced ah-ah), these flows have a blocky surface texture after cooling and hardening (Figure 5.21b). With the exception of some flows on steep slopes or near to the vent, most lava flows are slow enough for people to easily move out of the way as they approach.[15]

In addition to flowing down the sides of a volcano, lava can move away from its source in other ways. Magma can move for many kilometers just below the surface in *lava tubes* (Figure 5.22a). The rock walls of these tubes insulate the magma, keeping it hot and fluid for great distances. When an eruption ends the lava will

∧ FIGURE 5.20 Locations of Volcanoes in the United States
Index maps show locations of active and potentially active volcanoes and nearby population centers (not all labeled) of the United States. There are at least 11 active and potentially active volcanoes in British Columbia and the Yukon Territory. *(From Wright, T. L., and Pierson, T. C., U.S. Geological Survey Circular 1073, 1992)*

drain from the tube, leaving behind a sinuous cavern system (Figure 5.22b). These form natural pipes for movement of groundwater and may cause engineering problems when encountered during construction projects.

Lava flows from rift eruptions on the flank of Kilauea in Hawaii began in 1983 with the eruption of Pu'u 'Ō'ō' crater (see Figure 5.6a), which has become the longest and largest eruption of Kilauea in history.[4] Most of the time, tubes carry the lava for more than 9.5 km (~6 mi.) from the crater to the coast where it enters the sea, but periodic eruptions from the tubes

and overland flow have caused significant damage. By 2005, more than 50 structures in the village of Kalapana were destroyed by lava flows, including the National Park Visitors Center. Lava flowed across part of the famous Kaimu Black Sand Beach and into the ocean. The village of Kalapana has virtually disappeared, and it will be many decades before much of the land is productive again. On the other hand, the eruptions, in concert with beach processes, have produced new black sand beaches. Black sand is produced when the molten lava enters the relatively cold ocean water and shatters into sand-sized particles.

(a)

(b)

(a)

(b)

⋀ FIGURE 5.22 Lava Tubes
Magma can move many kilometers below ground in tunnels called lava tubes that form below cooled crusts of basaltic lava. Both active and drained tubes can be hazards when roofs collapse. (a) Holes, called skylights, develop in the roof of some active tubes such as this one on the Kilauea volcano in Hawaii. *(Jeffrey B. Judd/USGS)* (b) Older drained tubes, such as the Thurston (Nahuku) lava tube shown here near the summit of Kilauea, are attractions for tourists. *(Simone Genoese/Alamy Stock Photo)*

⋀ FIGURE 5.21 Lava Flows
(a) Pahoehoe Lava Flow. The basaltic lava flow surrounding and destroying this home at Kalapana, Hawaii, in 1990 has the characteristic smooth surface texture of most pahoehoe. Such flows destroyed more than 100 structures, including the National Park Service Visitors Center. Some pahoehoe flows are referred to as ropey lava because the surface texture can look like a piece of rope lying on its side as if it were coiled. *(Bettmann/Corbis)*. (b) Aa Lava Flow. This active aa lava flow is moving over an earlier flow of solidified pahoehoe lava. Both aa and pahoehoe are common types of basaltic lava; aa has a blocky surface that develops from cooler, thicker, slower-moving, and less fluid lava. *(Corbis)*

PYROCLASTIC ACTIVITY

Pyroclastic activity refers to explosive volcanism in which magma and the rocks that compose the volcano are physically blasted from a volcanic vent into the atmosphere forming *pyroclastic debris* (Greek *pyro*, "fire," and *klastos*, "broken"). Pyroclastic debris is known as **tephra** and can range from fine dust to sand-sized ash

less than 2 mm (~0.1 in.), to small gravel-sized *lapilli* 2 to 64 mm (~0.1 to 2.5 in.), to large angular *blocks* and smooth-surfaced *bombs* greater than 64 mm (~2.5 in.). Tephra entrained in the hot, low-density, rising gases will form an *eruption column* that is carried skyward until it cannot rise any longer or is blown by prevailing winds, forming an eruption cloud (Figure 5.23). Like the old saying, "What goes up must come down," tephra may come down cool and lightly like falling snow, or hot, fast, and heavy like a freight train. Accumulation of tephra forms a **pyroclastic deposit** that when deposited hot can be fused to form a *pyroclastic rock*.

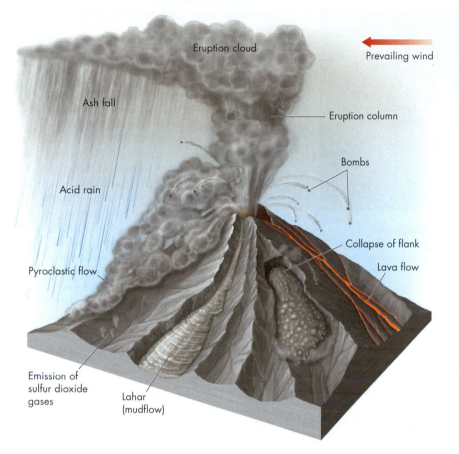

Eruption cloud

Prevailing wind

Ash fall

Eruption column

Bombs

Acid rain

Collapse of flank

Lava flow

Pyroclastic flow

Emission of
sulfur dioxide
gases

Lahar
(mudflow)

Fall Volcanic ash develops due to explosive fragmentation of magma during an eruption. This ash can cover hundreds or even thousands of square kilometers and if it reaches high enough it can encircle the globe making it the farthest reaching volcanic hazard. **Ash fall** occurs downwind of the volcano as the fine particles carried aloft cool and fall back toward Earth's surface (Figure 5.23). Towns and agricultural lands upwind of the volcano will likely be unaffected by ash fall during the eruption. The thickness of the ash fall in any one location will be controlled by the amount and duration of ash ejection, local and regional wind patterns, and distance from the volcano—decreasing thickness with increasing distance. Ash fallout creates several direct hazards:

> Vegetation, including crops and trees, may be destroyed.
> Surface water may be contaminated by sediment and toxic compounds. The very fine particles clog the gills of fish and kill other aquatic life. Chemical coatings on the ash can cause a temporary increase in the acidity of the water, which may last for several hours after an eruption ceases.

> Structural damage to buildings may occur as ash piles up on roofs (Figure 5.24). As little as 1 centimeter (~0.4 inch) of ash can place an extra 2.5 tons of weight on an average house with a 140 m^3 (~1500 ft.^{3}) roof.
> Health hazards, such as irritation of the respiratory system and eyes, are caused by contact with volcanic ash and associated caustic fumes.[13]
> Engines of jet aircraft may "flame out" as melted silica-rich ash forms a thin coating of volcanic glass in the engines. In 1989, a KLM 747 jet on the way to Japan flew though a cloud of volcanic ash from Redoubt volcano in Alaska. Power to all four of its jet engines was lost and the plane began a silent, 4270 m (~14,000 ft.) fall before the engines could be restarted.[16] Since this event, Federal Aviation Administration (FAA) guidelines limit air traffic near volcanic ash clouds (see Applying 5 Fundamental Concepts: Eyjafjallajökull 2010 Eruption).

Pyroclastic Flows One of the most lethal aspects of volcanic eruptions is **pyroclastic flow** (Figure 5.23). Unlike ash fallout, these flows of pyroclastic

FIGURE 5.24 Volcanic Tephra on Buildings
Fall out of ash and cinders may completely bury houses and increase the load on walls and roofs causing structural collapse. *(PhotoAlto sas/Alamy)*

debris are hot and race down the side of the volcano, attaining speeds in excess of 400 km/hr (~250 mph).[12] The gases within the flow can reach temperatures of 1000°C (~1830°F), which is hot enough to incinerate any organic matter in its path. Pyroclastic flows are also known as ash flows, hot avalanches, or *nuée ardentes* (French for "glowing clouds"). As the pyroclastic flows move down the side of the mountain, hot expanding gases carry low-density ash upward, forming a billowing gray cloud that rises above the higher-density base of the flow. In addition to ash, the base of the flow contains larger debris such as pumice, rock fragments, and whatever else the flow sweeps up off the surface on its way down the mountain. Pyroclastic flows are associated with Peléan-, Plinian-, Ultra-Plinian- and supervolcanic-type eruptions (see Table 5.2). Overall, these flows have killed more people than any other volcanic hazard in the past 2,000 years.[17]

Pyroclastic flows can be catastrophic if a populated area is in their path. A tragic example occurred in 1902 on the Caribbean island of Martinique. On the morning of May 8, a flow of hot, glowing ash, steam, and other gases roared down Mount Peleé and through the town of St. Pierre, killing an estimated 30,000 people. A jailed prisoner was one of only two survivors, and he was severely burned and horribly scarred. Reportedly, he spent the rest of his life touring circus sideshows as the "Prisoner of St. Pierre." The second survivor was a shoemaker who ran inside a building. Although he was burned, he escaped being suffocated by the ash that killed nearly everyone else. Flows like these have occurred on volcanoes of the Pacific Northwest, Alaska, and Japan in the past and also can be expected to occur in the future.

Pyroclastic flows form by three different volcanic processes:

1. During large ash-generating eruptions (VEI > 3), the eruption column is loaded with pyroclastic debris. Although this debris is denser than the surrounding atmosphere, it is carried upward by the hot expanding gases erupting from the volcanic crater. However, an increase in the density of the pyroclastic debris aloft or a decrease in the upward thrust of the gases can cause parts of the eruption column to collapse, forming a flow of hot pyroclastic material down the eruption column and the flank of the volcano (Figures 5.23 and 5.25). Multiple collapses of the eruption column during the A.D. 79 eruption of Mount Vesuvius killed more than 16,000 people living in the Roman towns of Pompeii and Herculaneum. Buried beneath as much as 20 m (~65 ft.) of pyroclastic debris, the remains of the towns and many of the residents who died there lay undisturbed for about 1,500 years before being unearthed (Figure 5.26).

2. Pyroclastic flows known as **lateral blasts** form when an explosion destroys part of the volcano, ejecting hot gases and pyroclastic debris horizontally from the side of the mountain rather than vertically. These are the fastest-moving pyroclastic flows that can mow down entire forests in their path (see Case Study 5.3: Mount St. Helens).

3. Lava dome collapse is the most common volcanic process that causes pyroclastic flows. Recall that lava domes are made of extremely viscous

FIGURE 5.25 Pyroclastic Flow
A dense flow of superheated ash, pumice, and rock fragments cascade down the slopes of Mayon Volcano in the Philippines during the 2006 eruption. The pyroclastic flow was caused by partial collapse of the eruption column. *(AP Photo/Bullit Marquez)*

⋏ FIGURE 5.26 Plaster Casts of Volcano Victims
In A.D. 79, pyroclastic flows from Mount Vesuvius destroyed the Roman towns of Pompeii and Herculaneum, killing more than 16,000 people. Archaeologists have excavated the remains of some 2,000 people asphyxiated in the eruption. Plaster casts of molds of the victims reveal adults, children, and dogs in their death positions. Pompeii was buried by up to 3 m (~10 ft.) of ash and pumice, and Herculaneum was excavated from beneath 20 m (~65 ft.) of volcanic debris. *(Jim Zuckerman/Alamy Stock Photo)*

magma that often forms a plug in a volcanic vent, causing volatiles to build within the magma (see Figure 5.9). Over years to centuries, periodic growth of the lava dome may cause the steep flanks of the dome to collapse, decompressing the volatiles that lie beneath. The avalanche of volatile-rich boulders and fresh magma continues to explosively decompress forming the characteristic gray billowing ash cloud as the hot material makes its way down the mountain, as happened at Mount Unzen in 1991 (see Case Study 5.1: Mount Unzen).

5.1 CASE study Mount Unzen

Just over 220 years ago, Mount Unzen in southwestern Japan erupted, triggering both a volcanic avalanche that devastated the center of Shimabara City on its way to the sea and a tsunami that killed an estimated 15,000 people. The volcano then lay dormant until November 1989 when a swarm of earthquakes was detected 20 km (~12 mi.) beneath the volcano, signaling an awakening of the mountain. Over the next 12 months earthquake foci became shallower and migrated toward the summit of the mountain, indicating movement of molten rock toward the summit. Explosive eruptions began in November 1990 and continued

episodically for the next five years. During this period, Mount Unzen produced more than 9,000 pyroclastic flows due to lava dome growth and collapse.[18] The great number and regularity of pyroclastic flows produced at Unzen attracted numerous volcanologists interested in studying these flows to Japan (Figure 5.27).[12,14] The numerous pyroclastic flows also resulted in large accumulations of volcanic debris on the flanks of the volcano, which could spell disaster during the rainy season. Given Shimabara's

⋏ FIGURE 5.27 Pyroclastic Flow from Mount Unzen
The gray cloud on the forested slope of Mount Unzen at the top of the photo is moving rapidly downslope toward a village in the June 1991 eruption. The cloud is a pyroclastic flow of extremely hot gases, volcanic ash, and large rocks. Both the firefighters in the truck and the individual are running for their lives. Fortunately, the flow stopped before reaching both the firefighters and the village. *(© Getty Images/Mike Lyvers)*

history with Mount Unzen, specially designed debris flow channels were constructed through the center of town to divert volcanic mudflows. Unfortunately, frequent debris flows caused overtopping of the levees and destroyed more than 1,700 buildings (Figure 5.28).

In May 1991, monitoring of the volcano indicated the potential for increased pyroclastic activity so more than 12,000 people were evacuated from the area. But before evacuation was complete, on June 3, 1991, a landslide triggered by an earthquake caused a devastating pyroclastic flow that killed 43 people, including three volcanologists. By the end of 1993, Mount Unzen had produced about 0.2 km^3 (~0.05 mi.^{3}) of lava.

⋀ FIGURE 5.28 Mount Unzen
The large mountain in the top center is Mount Unzen, one of Japan's 19 active volcanoes. Dormant for more than two centuries until 1991, the volcano threatens parts of Fukae Town and Shimabara City in the foreground. In the 1991 eruption, searing hot pyroclastic flows killed 43 people and burned a school and more than 600 homes. This 1993 aerial photograph shows a channel constructed to divert volcanic mudflows to the sea. *(Michael S. Yamashita/Corbis)*

VOLCANIC GASES

A number of gases, including water vapor (H_2O), carbon dioxide (CO_2), carbon monoxide (CO), sulfur dioxide (SO_2), and hydrogen sulfide (H_2S), are emitted during volcanic activity. Water and carbon dioxide make up more than 90 percent of all emitted gases. Toxic concentrations of hazardous volcanic gases rarely reach populated areas. A notable tragic exception occurred at Lake Nyos, a deep crater lake on a dormant volcano in the Cameroon highlands of West Africa. Over a period of time, carbon dioxide escaping from the crater floor accumulated in the sediment and the bottom waters of the lake. On an August night in 1986, with little warning other than a loud rumbling, Lake Nyos released a misty cloud of dense, mainly carbon dioxide gas. Nearly odorless, the gas cloud flowed from the volcano into valleys below, displacing the air. As the cloud lost its water droplets, it became invisible, spreading silently through five villages. The cloud traveled 23 km (~14 mi.) from Lake Nyos, suffocating 1,742 people, an estimated 3,000 cattle, and numerous other animals (Figure 5.29a and Figure 5.29b).[19]

Since 1986, Lake Nyos has continued to accumulate carbon dioxide in the bottom of the lake, and another release could occur at any time.[19] Although the lake area was to remain closed to all but scientists studying the hazard, thousands of people are returning to farm the land. Scientists have installed an alarm system at the lake that will sound if carbon dioxide levels become high. They have also installed a pipe from the lake bottom to a degassing fountain on the surface of the lake that allows the carbon dioxide gas to escape slowly into the atmosphere (Figure 5.29c). The hazard is being slowly reduced as the single degassing fountain is now releasing a little more carbon dioxide gas than is naturally seeping into the lake. Additional pipes with degassing fountains will be necessary to adequately reduce the hazard.

Sulfur dioxide, a gas that smells like gunpowder, is often released from volcanoes. It can be a direct hazard to people, and evacuations or area closures are ordered if large amounts are released. For example, in 2007 part of Crater Rim Drive at the Kilauea volcano in Hawaii was closed because of high levels of the gas. Sulfur dioxide can also react in the atmosphere to produce *acid rain* downwind of an eruption. Toxic concentrations of some chemicals emitted as gases may be adsorbed by volcanic ash that falls onto the land. In some cases, these chemicals have contaminated the soil and have been absorbed by plants that are eaten by people and livestock. For example, in 1783 a "dry fog" of sulfur dioxide gas from the eruption of the Skaptar Jokull volcano in Iceland blanketed much of Europe. The cloud

(a)

(c)

(b)

∧ FIGURE 5.29 Poisonous Gas from Dormant Volcano
(a) Water in Lake Nyos, a crater lake in Cameroon, Africa, accumulated and then released immense volumes of carbon dioxide in 1986. A landslide into the lake triggered the carbon dioxide eruption and is responsible for the muddy water in this photograph. *(Thierry Orban/Sygma/Corbis)* (b) Aerial view of some of the 3,000 cattle that died by asphyxiation from the carbon dioxide. *(Peter Turnley/Corbis)* (c) Gas is being released from the bottom waters of Lake Nyos with a degassing fountain in 2001. *(Louise Gubb/Corbis)*

contained fluorine in the form of fine droplets of acidic hydrogen fluoride. These droplets contaminated pastures with fluorine and caused the death of grazing cattle in as little as two days.[20]

In Japan, volcanoes are monitored to detect the release of poisonous gases such as hydrogen sulfide. When releases are detected, sirens are sounded to advise people to evacuate to high ground to escape the gas.

Volcanoes can also create air pollution over larger areas. This pollution can take the form of volcanic smog, known as *vog* (volcanic material, "v," and fog, "og"), or a haze from steam plumes created by lava flowing into the ocean. For example, the eruptions of Kilauea in Hawaii since 1983 have at times produced vog mixed with acid rain downwind of the volcano. This combination has covered the southeastern part of the Big Island with a thick, blue acidic haze that far

exceeds air quality standards for sulfur dioxide. Public health warnings have been issued because small, acidic aerosol particulates and sulfur dioxide concentrations can induce asthma attacks and cause other respiratory problems. Residents and visitors have reported breathing difficulties, headaches, sore throats, watery eyes, and flulike symptoms when exposed to vog. In addition, acid rain has made the water in some shallow wells and household rainwater-collection systems undrinkable. The acidic rainwater has dissolved lead from metal roofing and water pipes and may have caused elevated lead levels in the blood of some residents.[21] These eruptions have also produced nearly continuous lava flows into the ocean, resulting in steam explosions (Figure 5.30). Haze from the steam explosions contains corrosive hydrogen chloride gas from the vaporized seawater and tiny glass fragments that can irritate and sometimes damage eyes.[22]

∧ FIGURE 5.30 Lava Creates Corrosive Steam Cloud
Volcanoes produce a wide range of hazards including toxic
gases and corrosive clouds of steam. In this picture, lava from
Hawaii's Kilauea volcano vaporizes seawater at temperatures
of more than 1000°C (~1800°F). Flowing from the end of a lava
tube, the molten rock creates a cloud of steam mixed with corrosive hydrogen chloride gas before it cools in the Pacific Ocean.
(Brocken Inaglory/Wikimeda)

∧ FIGURE 5.31 Volcanic Mudflow Catastrophe
Armero, Columbia was almost completely buried by a volcanic
mudflow when pyroclastic fallout caused rapid melting of
mountain snowpack during the 1985 eruption of Nevado del Ruiz
stratovolcano. *(STf/AFP/Getty Images)*

DEBRIS FLOWS, MUDFLOWS, AND VOLCANIC LANDSLIDES

The most serious secondary effects of volcanic
activity are **debris flows** and **mudflows**, known
collectively by their Indonesian name, *lahar*. Lahars
are produced when large amounts of loose volcanic
ash and other tephra are saturated with water, become unstable, and suddenly move downslope. Debris
flows differ from mudflows in that they are coarser;
more than half of their particles are larger than sand
grains. Unlike pyroclastic flows, lahars can occur without an eruption and are generally low-temperature
flows.

Debris Flows
Even relatively small eruptions of
hot volcanic material may quickly melt large volumes
of snow and ice on a volcano. This rapid melting produces a flood of meltwater that erodes the slope of
the volcano to create a debris flow. Volcanic debris
flows are fast-moving mixtures of fine sediment and
large rocks that have a consistency similar to wet
concrete. Debris flows can travel many tens of kilometers down valleys on the flanks of the volcano
where they formed, which makes them particularly
hazardous, like the one that destroyed Armero,
Colombia, in 1985.

Nevado del Ruiz in northwestern Colombia is a
stratovolcano that developed above the Andean subduction zone, which runs the length of the western
coast of South America (see Figure 5.3). In November
1985 the volcano erupted, sending a Plinian-type

eruption column about 30 km (~19 mi.) into the atmosphere. Successive collapse of the eruption column
triggered pyroclastic flows that melted summit glaciers and unleashed a torrent of water. Floodwaters
mixed with pyroclastic debris, forming a dense slurry
that moved down the river valleys at 100 km (~60 mi.)
per hour, sweeping up everything along the way. By
the time the debris flows reached the mouth of the
valley, 45 km (~28 mi.) downslope from where Armero
was located, the volume of the flows had increased
fourfold and had attained a maximum width of 50 m
(~200 ft.). Armero was almost completely wiped out,
killing three-quarters of the population, or about
21,000 people (Figure 5.31).

Mudflows
Gigantic mudflows have originated on
the flanks of volcanoes in the Pacific Northwest in
both historic and prehistoric times. Two of these, the
Osceola mudflow and the Electron mudflow, started
on Mount Rainier (Figure 5.32). Approximately 5,000
years ago, the Osceola mudflow moved 1.9 km³
(~0.5 mi.³) of sediment a distance of more than 80 km
(~50 mi.) from the volcano. This volume would fill the
Mall between the U.S. Capitol and the Washington
Monument with a pile of debris more than a mile high
(10.5 times the height of the Washington Monument).
Deposits of the younger, 500-year-old Electron mudflow extended about 56 km (~35 mi.) from the
volcano.

Hundreds of thousands of people now live on the
area covered by these old flows with no guarantee that
similar flows will not occur again. Many areas southeast of Puget Sound are at potential risk for debris,

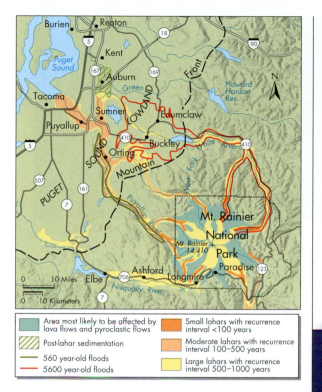

<figure>
∧ FIGURE 5.32 Mount Rainier Volcanic Hazard Map
Map of Mount Rainier and vicinity showing the extent of the
5,600-year-old Osceola mudflow in the White River Valley
and the 560-year-old Electron mudflow in the Puyallup River
Valley. Shaded areas show the modern potential hazards
from lahars, lava flows, and pyroclastic flows. The Seattle/
Tacoma suburbs of Puyallup, Sumner, Orting, and Auburn
are all in potential hazard areas. *(Based on Crandell, D.
R., and Mullineaux, D. R., U.S. Geological Survey Bulletin 1238, and Hoblitt and others, 1998, USGS Open-File
Report 98–428)*
</figure>

Legend:
- Area most likely to be affected by lava flows and pyroclastic flows
- Post-lahar sedimentation
- 560 year-old floods
- 5600 year-old floods
- Small lahars with recurrence interval <100 years
- Moderate lahars with recurrence interval 100–500 years
- Large lahars with recurrence interval 500–1000 years

mud, lava, or pyroclastic flows from Mount Rainier
(Figure 5.32). An observer in the valley would see a debris or mudflow as a wall of mud the height of a ranch
house moving toward them at close to 30 km (~20 mi.)
per hour. With the flow moving at 8.3 m (nearly 30 ft.)
per second, the observer would need a car headed in
the right direction toward high ground to escape being
buried alive.[15]

The USGS has recently developed an automated,
solar-powered lahar detection system for several volcanoes in the United States (e.g., Mount Rainier),
Indonesia, the Philippines, Mexico, and Japan. These
systems have acoustic-flow monitors that sense ground
vibrations from a moving lahar and can warn people
that a flow is moving down the valley. This system replaces visual sightings and cameras that are unreliable

in bad weather or at night and require continual maintenance. Dangerous lahars can occur quickly and with
little warning, so people in their path must be alerted
immediately. Alarms from these USGS lahar detection
systems should give people sufficient time to evacuate
to the safety of higher ground.[23]

Landslides Volcanic landslides are another secondary effect of volcanic activity. Like lahars, they may be
triggered by events other than an eruption. Large volcanic landslides may affect areas far from their source.
For example, massive landslide deposits on the seafloor off the coasts of Hawaii and the Canary Islands
are likely to have generated huge tsunamis (see Case
Study 5.2: Volcanic Landslides and Tsunamis).[24]
These large waves would cause catastrophic damage
far from the islands where they originated (see
Chapter 4).

5.2 Volcanic Landslides and Tsunamis

CASE study

What may be the largest active landslides
on Earth are located on Hawaii. They are
up to 100 km (~60 mi.) wide, 10 km (~6
mi.) thick, and 20 km (~12 mi.) long and
extend from a volcanic rift zone on land
to an endpoint beneath the sea. Currently
these landslides creep along at around
10 cm (~4 in.) per year and contain blocks of rock
the size of Manhattan Island. The fear is that they
might again become a giant, fast-moving submarine
debris avalanche. This avalanche would generate a
huge tsunami capable of lifting marine debris hundreds of meters above sea level onto nearby islands,
causing catastrophic damage around the Pacific
Basin. Fortunately, such high-magnitude events are
rare; the average recurrence interval seems to be
about 100,000 years or so.[4]

Other huge volcano-related landslides or debris avalanches have been documented in the Canary Islands,
located in the Atlantic Ocean off the western coast of
Africa. On Tenerife, the largest island, six huge landslides have occurred during the past several million
years (Figure 5.33), the most recent of which took place
less than 150,000 years ago. The seafloor just north of
Tenerife is covered by 5500 km^2 (~2150 mi.^{2}) of landslide deposits, an area nearly as large as the state of
Delaware and more than twice the land surface of the
island.[24]

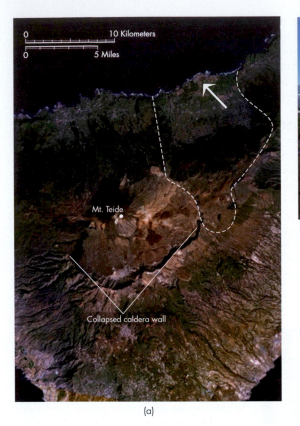

(a)

(b)

∧ FIGURE 5.33 Giant Landslide on Volcano
(a) Aerial view of part of the island of Tenerife, the largest island of the Canary Islands in the Atlantic Ocean. Labeled are the collapsed La Cañadas caldera and the Mount Teide volcano (elevation 3.7 km [~12,100 ft.]). The white dashed line outlines the extent of the Orotova landslide, one of many large landslides on the island. The white arrow in both photographs (a) and (b) points to the seaward end of this landslide. *(NASA)* (b) Seaward end of the Orotova landslide is outlined by the white dashed line. Snow-capped Mount Teide is in the distance. *(Neirfy/shutterstock)*

5.3 CASE study

Mount St. Helens 1980–2010: From Lateral Blasts to Lava Flows

The eruption of Mount St. Helens on May 18, 1980, in the southwestern corner of Washington State exemplifies the many types of volcanic events expected from a Cascade volcano (Figure 5.34). The eruption, like many other natural events, was unique and complex, making generalizations somewhat difficult. Nevertheless, we have learned a great deal from Mount St. Helens, and the story is not yet complete.

Mount St. Helens awoke in March 1980, after 120 years of dormancy, with seismic activity and small explosions created by the boiling of groundwater as it came in contact with hot rock. By May 1, a prominent bulge on the northern flank of the mountain could clearly be observed, and it grew at a rate of about 1.5 m (~5 ft.) per day (Figure 5.35a). Only a few weeks later, at 8:32 A.M.

on May 18, 1980, a magnitude 5.1 earthquake centered below the volcano triggered a large landslide/debris avalanche of approximately 2.3 km^3 (~0.6 mi.^{3}) of earth material.[25] The avalanche, which involved the entire bulge area (Figure 5.35b), shot down the north flank of the mountain, displacing water in nearby Spirit Lake. It then struck and overrode a ridge 8 km (~5 mi.) to the north and made an abrupt turn, moving 18 km (~11 mi.) down the North Fork of the Toutle River (Figure 5.36a).

Seconds after the failure of the bulge, Mount St. Helens erupted with a lateral blast directly from the area that the bulge had occupied (Figure 5.35c). The blast moved at speeds of more than 480 km (~300 mi.) per hour—greater than the velocity of the fastest bullet train. Effects of the blast were felt nearly 30 km (~19 mi.) from its source and devastated about 600 km^2 (~230 mi.^{2}).[26] Debris-avalanche deposits, blasted-down timber, scorched timber, pyroclastic flows, and mudflows from the eruption covered large areas north of the volcano (Figure 5.36a).

The first of several mudflows consisting of water, volcanic ash, rock, and organic debris such as logs

(a)   (b)

∧ FIGURE 5.34 Mount St. Helens, Before and After
(a) Mount St. Helens before the May 18, 1980, eruption. *(Dennis Hallinan/Alamy)* (b) Mount St. Helens after the 1980 eruption. During the eruption, much of the northern side of the composite volcano was blown away, and the altitude of the summit was reduced by approximately 400 m (~1314 ft.). The lateral blast originated in the amphitheater-like area in the top center of the photograph and knocked down 600 km² (~230 mi.²) of trees, some of which can be seen in the foreground. *(Danita Delimont/Alamy)*

occurred minutes after the start of the eruption. These flows and accompanying floods raced down the valleys of both forks of the Toutle River at estimated speeds of 29 to 55 km (~18 to 34 mi.) per hour, threatening the lives of people camped along the river.

Two young people on a fishing trip on the North Fork of the Toutle River about 36 km (~22 mi.) downstream from Spirit Lake were awakened that morning by a loud rumbling noise from the river, which was covered by felled trees. The campers attempted to run to their car, but water from the rising river poured over the road, preventing their escape. A mass of mud then crashed through the forest toward their car, and the couple climbed on its roof to escape the mud. They were safe only momentarily, however, as the mud pushed the vehicle over the bank and into the river. Leaping off the roof, they fell into the river, which was by now a rolling mass of mud, logs, collapsed train trestles, and other debris. They disappeared several times beneath the surface of the flow but were lucky enough to emerge again. The two were carried downstream for approximately 1.5 km (~0.9 mi.) before another family of campers spotted and rescued them.

Within an hour after the lateral blast, a large vertical cloud had quickly risen to an altitude of approximately 19 km (~12 mi.), extending more than 4 km (~2.5 mi.) into the stratosphere (Figure 5.35d). Eruption of the vertical column continued for more than nine hours, and large volumes of volcanic ash fell on a wide area of

Washington, northern Idaho, and western and central Montana. During the nine-hour eruption, a number of pyroclastic flows swept down the northern slope of the volcano (Figure 5.36a). The total amount of volcanic ash ejected was about 1 km³ (~0.26 mi.³), and a large cloud of ash moved over the United States, reaching as far east as New England (Figure 5.36b). In less than three weeks, the ash cloud had circled the Earth.

The entire northern slope of the volcano, which is the upper part of the North Fork of the Toutle River watershed, was devastated. Forested slopes were transformed into a gray landscape of mounded volcanic ash, rocks, blocks of melting glacial ice, narrow gullies, and hot steaming pits (Figure 5.37).[25]

When the volcano could be viewed again following the eruption, the top of the mountain had been reduced by about 400 m (~1314 ft.). What was originally a symmetrical volcano was now a huge, steep-walled amphitheater facing northward, and what was originally Spirit Lake (Figure 5.34a) was now filled with sediment (Figure 5.34b). The debris avalanche, horizontal blast, pyroclastic flows, and mudflows devastated an area larger than Chicago, Illinois. The eruption killed 57 people and associated flooding destroyed more than 100 homes. Approximately 800 million board feet of timber were flattened by the blast—enough wood to build 50,000 homes. Total damage from the eruption exceeded $1 billion.

Both volcanic and seismic activity has continued since the deadly 1980 eruption. During the first six years

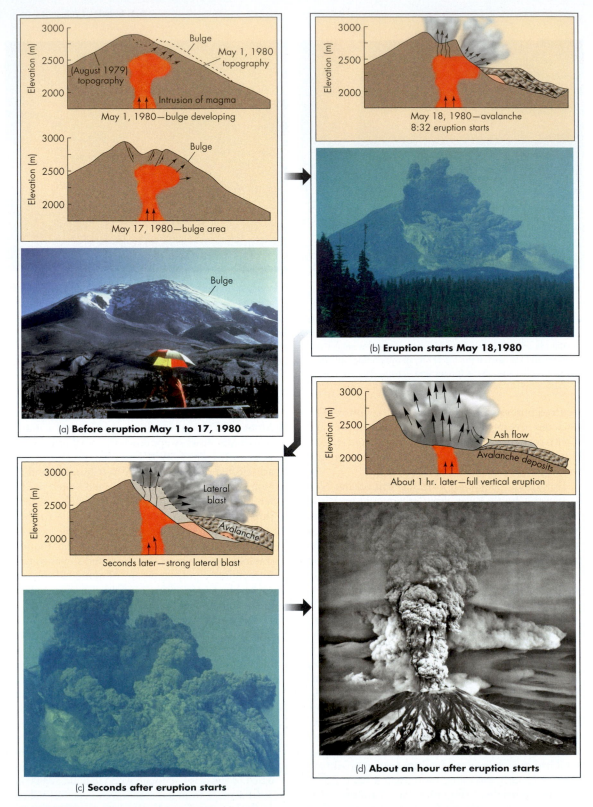

^ FIGURE 5.35 Mount St. Helens Erupts
Diagrams and photographs showing the sequence of events for the May 18, 1980, eruption of Mount St. Helens. *(A-Pf/Alamy)* Photographs (b) and (c) of the lateral blast were taken less than 10 seconds apart. *(Keith Ronnholm)* (d) *(R.M. Krimme/USGS)*

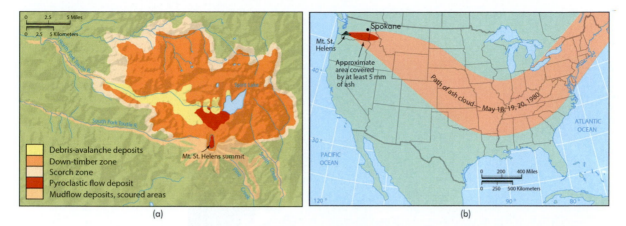

⋀ FIGURE 5.36 Debris Avalanche and Ash Cloud
(a) Debris-avalanche deposits, zones of trees blown down or scorched, mudflows, and pyroclastic flow deposits associated with the May 18, 1980, eruption of Mount St. Helens. *(Based on information from the USGS)* (b) Path of the ash cloud (orange) from the 1980 eruption. Area covered by at least 5 mm (~0.2 in.) is shown in red in the upper-left corner. *(Data from various U.S. Geological Survey publications)*

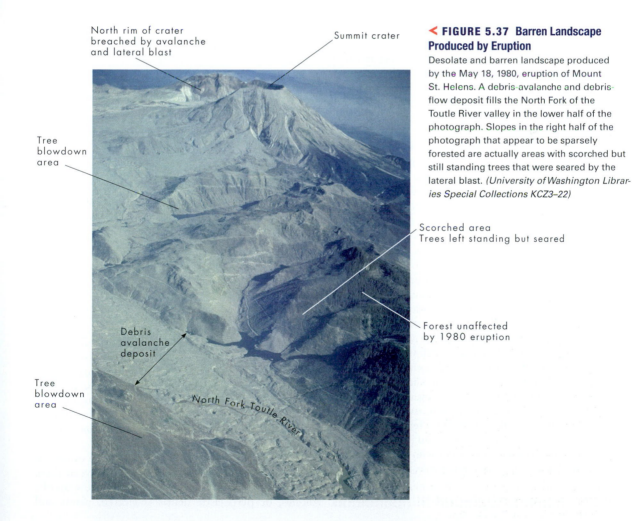

◄ FIGURE 5.37 Barren Landscape Produced by Eruption
Desolate and barren landscape produced by the May 18, 1980, eruption of Mount St. Helens. A debris-avalanche and debris-flow deposit fills the North Fork of the Toutle River valley in the lower half of the photograph. Slopes in the right half of the photograph that appear to be sparsely forested are actually areas with scorched but still standing trees that were seared by the lateral blast. *(University of Washington Libraries Special Collections KCZ3–22)*

◄ **FIGURE 5.38 Continued Growth of Dome in Crater** Steam and volcanic gases rise from cooling lava being added to a dome in Mount St. Helens's crater. Growth of this dome occurred from 1980 to 1986 and resumed in 2004. Heat from the new growth on the dome melts snowfall. The opening in the crater wall on the lower right is where the avalanche and lateral blast occurred in 1980 (Figure 5.35c). *(USGS)*

following the eruption, 19 smaller eruptions occurred. Lava flows from these eruptions began building a lava dome above the 1980 crater floor (Figure 5.38). Between 1989 and 2004, earthquake activity increased, and the quakes were sometimes accompanied by small explosions, ash flows, and mudflows.

On September 23, 2004, Mount St. Helens came back to life when magma began moving up toward the crater floor. Construction of the lava dome resumed and continued until January 2008. An estimated 0.5 m³ (~0.6 yd.³) of lava, about the volume of a five-drawer file cabinet, was being added to the dome every second.[26] The lava dome now rises 350 m (~1155 ft.) above the crater floor, close to the height of the Empire State Building.[27,28]

Scientists at the USGS Cascade Volcano Observatory in Vancouver, Washington, monitor Mount St. Helens at all times. This monitoring includes a network of automated seismographs and global positioning system (GPS) satellite receivers to detect earthquakes and ground deformation, aerial gas sampling, acoustic monitors for mudflows, and surveillance with webcams.[26]

By 1998, 18 years following the main eruption, life had returned to the mountain and its surrounding area, making many places green once more. However, the mounds of landslide deposits are still prominent, a reminder of the catastrophic event of 1980. Mount St. Helens National Monument now has two visitor centers and has attracted more than 3 million visitors since 1983, who have come to the enjoy the beauty and power of the volcano.

5.4 CHECK your understanding

1. Describe the primary and secondary effects of volcanic eruptions.

2. Compare and contrast ash falls and pyroclastic flows.

3. What are the major gases emitted in a volcanic eruption? How can they be hazardous?

4. Explain how volcanoes can produce gigantic debris flows or mudflows.

5.5 Linkages between Volcanoes and Other Natural Hazards

Volcanoes are intimately linked to their physical environment as well as to several other natural hazards. We have stressed the relationship between volcanoes and plate tectonics, the connection between plate tectonic setting and the type of magma, and the influence that magma type has on the nature of volcanic eruptions. Volcanoes are also directly related to other natural hazards such as fire, earthquakes, landslides, and climate change.

Although volcanoes are not a major cause of fires, it isn't difficult to imagine that a linkage between these hazards exists. As molten lava pours down the sides of a volcano, or pyroclastic debris is ejected, plants and

∧ FIGURE 5.39 Lava Ignites Fires in Africa
A lava flow erupting from the Nyiragongo volcano in the Democratic Republic of Congo ignited this fire in January 2002. More than 400,000 people were displaced by fires and lava flows in the city of Goma. *(PEDRO UGARTE/AFP/Getty Images)*

∧ FIGURE 5.40 Volcano Temporarily Cools Global Climate
An eruption of Mount Pinatubo in 1991, shown here, ejected vast amounts of volcanic ash and sulfur dioxide up to about 30 km (~19 mi.) into the atmosphere. Extremely fine particles, called aerosols, from the eruption remained in the upper atmosphere and circled Earth for more than a year. This aerosol cloud temporarily lowered the average global temperature. The Mount Pinatubo eruption was the second largest in the twentieth century. *(D. Harlow/USGS)*

manufactured structures commonly catch fire. In fact, the Hawaiian Volcano Observatory on the Big Island of Hawaii warns tourists of the fires caused by Kilauea's active lava flows. These fires can start from explosions of methane gas that form as vegetation is intensely heated by the lava. In January 2002, Africa's most destructive volcanic eruption in 25 years sent residents of the Democratic Republic of Congo fleeing raging fires ignited by lava flows (Figure 5.39). Sadly, the lava also sparked an explosion, killing 60 people.

Earthquakes commonly accompany or precede volcanic eruptions as magma rises through Earth's crust. For example, weeks of earthquakes preceded the appearance of the Parícutin volcano 320 km (~200 mi.) west of Mexico City. Some earthquakes may be large enough to do damage independent of the volcano. Similar examples are also discussed in the Mount Unzen and Eyjafjallajökull case studies.

Landslides are possibly the most common side effect of volcanic activity. These mass movements can be part of the slope or flank of the volcano (see Case Study 5.2: Volcanic Landslides and Tsunamis) or lahars, the volcanic debris flows and mudflows discussed earlier. Both types of mass wasting have the potential to do great damage and take many lives.

Finally, volcanic eruptions may affect our global climate. The best-known example of climate change from a volcanic eruption occurred following the great eruption of the Tambora volcano in Indonesia in 1815. The eruption created a global cloud of sulfuric acid droplets, referred to as aerosols, in the stratosphere, which were later detected in glacial ice formed that year in both Greenland and Antarctica.[29,30] This cloud caused regional cooling of up to 1°C (~2°F) and resulted in 1816 being called the "year without a summer" in New England. Climate change from the eruption caused major hardships in both North America and Europe, including crop failures, famine, and disease.[29] A smaller but still significant climatic cooling occurred after the 1991 eruptions of Mount Pinatubo in the Philippines. The cloud of ash and sulfuric acid aerosol from the Pinatubo eruption remained in the atmosphere for more than a year (Figure 5.40). Ash particles and aerosol droplets scattered incoming sunlight and slightly cooled the global climate for two years following the eruptions.[31]

5.5 **CHECK** your understanding

1. How can volcanic eruptions affect the global climate?

2. Why are volcanoes commonly linked with earthquake activity?

5.6 Natural Service Functions of Volcanoes

Although volcanoes pose a serious threat to those who live in their paths, like most other hazards, they also provide important natural service functions. Perhaps their greatest gift to us occurred billions of years ago when gases and water vapor released from volcanoes contributed to our atmospheric and hydrologic systems, allowing life, as we know it, to evolve. Additionally, volcanoes provide us with fertile soils, a source of power, mineral resources, and recreational opportunities, as well as the creation of new land.

VOLCANIC SOILS

From an agricultural perspective, volcanic eruptions are quite valuable, providing an excellent growth medium for plants. The nutrients produced by the weathering of volcanic rocks allow crops such as coffee, maize, pineapple, sugarcane, and grapes to thrive in volcanic soils. However, rich, fertile soils produced by volcanoes encourage people to live in hazardous areas. So although volcanic soils provide an important resource, nearby volcanic activity can make it difficult to use that resource safely.

GEOTHERMAL POWER

Another benefit provided by volcanoes is their potential for geothermal power. The internal heat associated with volcanoes may be used to create power for nearby urban areas. In fact, volcanic energy is being harnessed as geothermal power in Kilauea, Hawaii; Santa Rosa, California; and Long Valley, California. An important benefit of geothermal energy is that it can be a renewable resource. Unlike fossil fuels, it can be used at a rate that doesn't outpace its replenishment. However, care must be taken so that the heat and/or steam driving the system is not removed faster than it can be restored naturally or it will become depleted, at least temporarily. In addition to using hot geothermal water to make electricity, the heat can be used for heating and industrial processes. The hot water can be pumped directly through a building or utilized with a recirculating heat-exchange system. Reykjavik, Iceland; Paris, France; and Klamath Falls, Oregon, are just a few of the many locations that use geothermal water for heating.[32]

MINERAL RESOURCES

Volcanism is the source and volcanic rocks are the host for many mineral resources. These include economic concentrations of metals, such as gold, silver, platinum, copper, nickel, lead, and zinc, and nonmetallic resources, such as pumice, tuff, perlite, scoria, basalt, and volcanic clays. Although these mineral deposits form today in volcanic belts around the world, including underwater at mid-ocean ridges, most of those that we use formed in the geologic past. In the case of metallic mineral resources, most are found in much older Precambrian rocks. These older deposits formed at a time when the composition of some magma differed from today's volcanoes. Examples include gold accumulations in Precambrian volcanic rocks in Western Australia and Ontario, Canada. In addition to metals, volcanic rocks are used in a wide range of commercial products, including soap, building stone, aggregate for roads and railroads, oil and gas drilling mud, landscaping gravel, ceiling tile, cement, plaster, and cat litter.[33]

RECREATION

Besides being an energy source, the heat associated with volcanoes provides recreational opportunities. Many health spas and hot springs are developed in volcanic areas. Volcanoes also provide opportunities for hiking, snow sports, and education. More than 1 million tourists visit the Kilauea volcano each year, many of whom come to observe the volcano during eruption (Figure 5.41).

CREATION OF NEW LAND

In our discussion of the benefits of volcanoes, we should not forget to mention that they are responsible for creating much of the land we inhabit. For the past several decades, the residents of Hawaii are reminded of this fact as lava flows from Kilauea build deltas of land into the Pacific Ocean (see Figure 5.30 and

∧ FIGURE 5.41 Erupting Volcanoes as Tourist Attractions Tourists on the cliff to the right view a lava flow from Kilauea volcano on the coast of Hawaii. *(Paul A. Souders/CORBIS)*

Figure 5.41). Not only are volcanic processes the major force that builds continents, but also oceanic islands such as Hawaii and Iceland would not exist without volcanoes!

5.6 CHECK your understanding

1. What contribution to human and animal food supply do volcanoes provide?
2. What is geothermal energy and what is it's primary advantage over fossil fuels?

5.7 Human Interactions with Volcanoes

Unlike earthquakes, volcanoes do not lend themselves to human tinkering. That is to say, there is little we can do to affect the timing and severity of volcanic eruptions. Whereas deep-well disposal may increase the number of earthquakes in an area, and clearing the land for agriculture may contribute to flooding, there does not appear to be any human activity that affects volcanoes. They truly are a hazard that is beyond our control, and the best we can do is to attempt to minimize the loss of life and property associated with eruptions.

5.8 Minimizing the Volcanic Hazard

Forecasting volcanic eruptions is a major component of the goal to reduce volcanic hazards. A "forecast" for a volcanic eruption is a probabilistic statement describing the time, place, and character of an eruption. It is analogous to forecasting the weather and is not as precise a statement as a prediction.[4]

FORECASTING

It is unlikely that we will be able to accurately forecast the majority of volcanic activity in the near future, but valuable information is being gathered about phenomena that occur prior to eruptions. One problem is that most forecasting techniques require experience with actual eruptions before the mechanism is understood. Thus, we are better able to predict eruptions in the Hawaiian Islands because we have had so much experience there.

Forecasting volcanic eruptions uses information gained by

> monitoring seismic activity
> monitoring thermal, magnetic, and hydrologic conditions
> monitoring the land surface to detect tilting or swelling of the volcano
> monitoring volcanic gas emissions
> studying the geologic history of a particular volcano or volcanic center.[13,30]

Seismic Activity Monitoring Our experience with volcanoes, such as Mount St. Helens and those on the Big Island of Hawaii, suggests that earthquakes may provide the earliest warning of an impending volcanic eruption. Shallow earthquakes and tremors are produced below a volcano as upward-moving magma fractures the surrounding rock and gas bubbles in the magma form and burst. In the case of Mount St. Helens, earthquake activity started in mid-March before the eruption in May. Activity in March began suddenly with nearly continuous, shallow swarms of earthquakes. Unfortunately, there was no additional increase in earthquakes immediately preceding the catastrophic eruption on May 18. In Hawaii, earthquakes have been used to monitor the movement of magma as it approaches the surface.

Several months prior to the 1991 Mount Pinatubo eruptions, small steam explosions and earthquakes began.[4] Unlike Mount St. Helens, Pinatubo was an eroded ridge that did not have the classic shape of a volcano. Furthermore, it hadn't erupted in 500 years, so the majority of people living near it did not even know it was a volcano! After the initial steam explosions, scientists began monitoring earthquakes on the volcano and studying past volcanic activity, which was determined to have been explosive. Earthquakes increased in number and magnitude prior to the catastrophic eruption, and foci migrated from deep beneath the volcano to shallow depths beneath the summit.[4]

Recently, geophysicists have proposed a generalized model for seismic activity that may help in predicting eruptions.[34] This model is used for explosive stratovolcanoes, such as those in the Cascade Mountains, which may awaken after an extended period of inactivity (Figure 5.42). In a reawakening volcano, the magma must fracture and break previously solidified igneous rock above the magma chamber to work its way to the surface. Several weeks before reawakening, increasing pressure creates numerous fractures in the plugged volcanic conduit above the chamber. At first, the increase in seismic events will be gradual and a seismologist may need 10 days or so to confidently recognize an accelerating trend

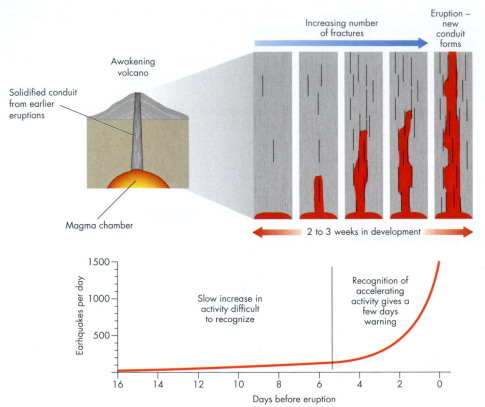

◄ FIGURE 5.42 A Volcano Reawakens Increased seismic activity is a good indicator of a forthcoming volcanic eruption. As a dormant volcano reawakens, rising magma fractures rock above. At first, the fracturing slowly increases the rate of seismic activity; then both the fracturing and seismic activity accelerate a few days prior to an eruption. *(Adapted from Kilburn, C. R. J., and Sammonds, P. R., "Maximum Warning Times for Imminent Volcanic Eruptions," Geophysical Research Letters 32: L24313, 2005. Copyright © 2005. Reprinted with permission of Wiley Inc.)*

toward an eruption. However, once the trend has been recognized, there will still be several days before the eruption occurs. Unfortunately, this short warning time may be insufficient for a large-scale evacuation. Thus, to forecast eruptions it may be best to use seismic activity in concert with other eruption precursors discussed later. It is fortunate that, in contrast to earthquakes, volcanic eruptions always provide warning signs.[34]

Thermal, Magnetic, and Hydrologic Monitoring

Prior to a volcanic eruption, a large volume of magma accumulates in a holding reservoir beneath the volcano. The hot material changes the local magnetic, thermal, hydrologic, and geochemical conditions. As the surrounding rocks heat up, the rise in temperature of the rocks at the surface may be detected by satellite remote sensing or infrared aerial photography. Increased heat may melt snowfields or glaciers; thus, periodic remote sensing of a volcanic chain may detect new hot places related to volcanic activity. This method was used with some success at Mount St. Helens prior to the main eruption in 1980.

When older volcanic rocks are heated by new magma, magnetic properties, originally imprinted when the older rocks cooled and crystallized, may change. These changes can be detailed by ground or aerial magnetic surveys.[13,35]

Land Surface Monitoring Monitoring changes in the land surface and seismic activity of volcanoes has been useful in forecasting some volcanic eruptions. Hawaiian volcanoes, especially Kilauea, have supplied most of the data. The summit of Kilauea tilts and swells prior to an eruption and subsides during the actual outbreak (Figure 5.43). Kilauea also experiences swarms of small earthquakes from the underground movement of magma shortly before an eruption. The tilting of the summit, in conjunction with earthquake swarms, was used to predict a volcanic eruption in the vicinity of the farming community of Kapoho on the flank of the volcano 45 km (~28 mi.) from the summit. As a result, the inhabitants were evacuated before the event, in which lava overran and eventually destroyed most of the village.[36] Because of the characteristic swelling and earthquake activity before eruptions, scientists expect the Hawaiian volcanoes to continue to be more predictable than others.

Monitoring ground movements such as tilting, swelling, and opening of cracks, or changes in the water level of lakes on or near a volcano, can identify movement that might indicate a forthcoming eruption.[13] Today, satellite-based interfrometric synthetic aperature radar (InSAR), and a network of GPS satellite receivers can be used to monitor change in volcanoes, including surface deformation, without sending people into a hazardous

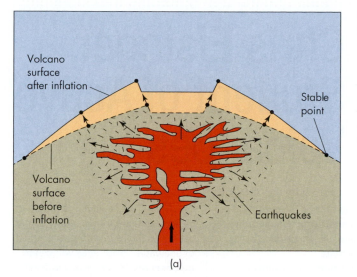

(a)

◀ **FIGURE 5.43 Inflation and Tilting Before Eruption**
(a) Idealized diagram of the Kilauea volcano in Hawaii illustrating inflation and surface tilting as magma moves up. The red area is the underground magma chamber that fills before an eruption. *(Solarfilma ehf)* (b) Graphs showing the tilting of the surface of Kilauea in two directions, the east–west component and the north–south component, from 1964 to 1966. Notice the slow increase in ground tilt before an eruption and rapid subsidence, or lowering of the land surface, during eruption. *(From Fiske, R. S. and Koyanagi R. Y., U.S. Geological Survey Professional Paper 607, 1968)*

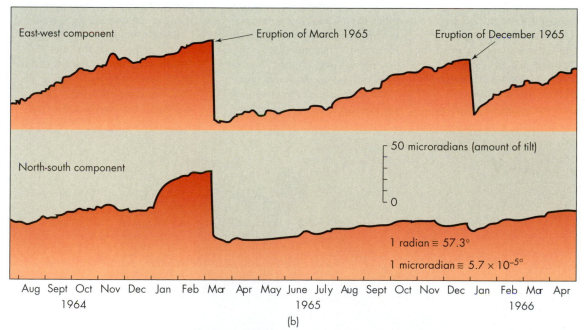

(b)

area.[37] These tools are capable of measuring millimeter- to centimeter-scale (0.4–4 in.) changes in Earth's surface over spans of days to years.

Monitoring Volcanic Gas Emissions

The primary objective of monitoring volcanic gas emissions is to recognize changes in chemical composition (see Case Study 5.4: Professional Profile). Changes in the relative amounts of carbon dioxide and sulfur dioxide, as well as changes in gas emission rates, are thought to correlate with subsurface volcanic processes. These changes may indicate movement of magma toward the

surface. This technique was useful in studying eruptions at Mount St. Helens and Mount Pinatubo. The volume of sulfur dioxide emitted from Pinatubo increased more than a million times two weeks before the explosive eruptions.

Geologic History

An understanding of the geologic history of a volcano or volcanic system is useful in predicting the types of future eruptions. The primary tool for reconstructing geologic history is geologic mapping of volcanic rocks and deposits. Scientists attempt to determine the age of past lava flows and pyroclastic

5.4 Professional Profile

Chris Eisinger, Student of Active Volcanoes

Chris Eisinger (Figure 5.4.A) describes volcanoes the same way he might discuss old, temperamental friends. "For the most part volcanoes are very approachable," he says, "as long as you know their cycles." As a graduate student in the Department of Geological Sciences at Arizona State University, Eisinger has 15 years of firsthand experience with volcanoes. He began as an undergraduate and spent time at the Hawaiian Volcano Observatory and in Indonesia, a region rife with volcanic activity.

But the summit of an active volcano is anything but a breezy affair. On the contrary, volcanoes emit malodorous gases, as Eisinger can attest. "The most distinct things you notice are the fumes," he says. "You get a strong sulfur smell."

If the volcano is near enough to the ocean for lava to flow into the sea, Eisinger also says the runoff creates a steam that is highly acidic, thanks to the high chlorine content of seawater. "Your eyes will water," he observes. "It's typically difficult to breathe and you end up coughing if you're not wearing a gas mask."

Rainstorms can further cloud the air, creating steam when the falling water strikes the hot surface of the lava. At times, visibility for volcanologists studying at the summit is reduced to a few feet.

∧ FIGURE 5.4.A Volcanologist Sampling Molten Sulfur
Volcanologist Chris Eisinger uses a crowbar to hit some molten sulfur on Kawah Ijen volcano in Indonesia. The 200°C (~390°F) molten sulfur is still red hot and will cool to a yellow or green color. *(Chris Eisinger)*

The heavy influx of gases doesn't affect only the eyes and nose. The taste, if one isn't wearing a mask, can be "pretty nasty," Eisinger says, adding that the smell adheres to clothing as well.

For volcanoes in a constant state of eruption, Eisinger notes that there are distinct patterns in the emission cycles. Volcanoes he has visited in Indonesia, for example, would emit a stream of gaseous material every 20 minutes to an hour.

Although most volcanologists are able to remain safe by paying close attention to eruption cycles, fatalities have happened. In August 2000, two volcanologists died at the summit of Semeru on the island of Java in Indonesia when the volcano erupted with no warning.

Active volcanoes also emit low, rumbling sounds, Eisinger says, due largely to subterranean explosions that tend to be muffled. Hawaiian volcanoes are unique in that they also emit an intense hissing sound, "like a jet engine," from the expulsion of gases. Eruptions are often foreshadowed by seismic activity that contributes to the rumble of the explosions. Although Eisinger himself has never witnessed a volcanic eruption of historic order, he noted that accounts of such colossal eruptions as that of Krakatau in 1883 produced reports of a boom that could be heard thousands of miles away.

— CHRIS WILSON

deposits. Geologic mapping, in conjunction with the dating of volcanic deposits at Kilauea, led to the discovery that more than 90 percent of the land surface of the volcano has been covered by lava in only the past 1,500 years. The town of Kalapana, destroyed by lava flows in 1990, might never have been built if this information had been known prior to development because the risk might have been considered too great. The real value of geologic mapping and dating of volcanic events is that it allows hazard maps to be prepared to assist in land-use planning and disaster preparedness.[13] Such maps are now available for a number of volcanoes around the world.

VOLCANIC ALERT OR WARNING

At what point should the public be alerted or warned that a volcanic eruption may occur? This question has been partially addressed by volcanologists and policy makers. The USGS recently established an alert notification system for volcanic activity for use by its five volcanic observatories.[38] This system has two components: ground-based volcano alert levels and aviation-based color code levels (Table 5.4). Each component has four levels and, for most monitored volcanoes and eruptions, the alert and aviation code will be at the

▼ **TABLE 5.4 U.S. Geological Survey Volcanic Alert Levels and Aviation Color Codes[a]**

Ground Alert Level	Volcanic Condition	Aviation Color Code
NORMAL	1. Typical background, noneruptive state.	GREEN
	Or	
	2. If downgraded from higher alert level, activity has ceased and returned to a background, noneruptive state.	
ADVISORY	1. Elevated unrest above known background level.	YELLOW
	Or	
	2. If downgraded from a higher alert level, activity has decreased significantly with close monitoring for possible renewed increase.	
WATCH	1. Heightened or escalating unrest with increased potential of eruption, time frame uncertain.	ORANGE
	Or	
	2. Eruption underway with limited hazards; for aviation—no or minor volcanic ash emissions.	
WARNING	Hazardous eruption imminent, underway, or suspected; for aviation—significant emission of volcanic ash into atmosphere.	RED

[a] Note: For most eruptions, the ground alert level and aviation color code will be at the same levels; however, for some eruptions there will be a greater hazard on the ground or to aviation, and different levels will be assigned for the two environments.

Source: Modified from http://volcanoes.usgs.gov/activity/alertsystem/icons.php and U.S. Geological Survey Fact Sheet 2006–3139.

same level. For some eruptions, however, the hazard posed to either those on the ground or those in the air will differ and higher alerts or codes will be issued. For example, if a lava flow from Kilauea volcano threatens a subdivision, the Hawaiian Volcano Observatory may issue a warning because a hazardous eruption is underway but only issue an orange aviation code because there are minor volcanic-ash emissions. Likewise, if a volcano erupts on an isolated island in the Aleutians, the Alaska Volcano Observatory may issue a watch because the eruption poses limited hazards on the ground and a red aviation code because of significant emissions of volcanic ash.

Although this system is a good start, the hard questions remain: When should evacuation begin? When is it safe for people to return? These are questions that public officials will have to answer when the USGS issues volcanic watches or warnings. The USGS volcanic observatories monitor approximately 50 of an estimated 170 active volcanoes in Alaska, Hawaii, the western contiguous United States, and the Northern Mariana Islands.[38]

5.8 CHECK your understanding

1. Describe how seismic activity can indicate a forthcoming volcanic eruption.

2. How is land surface monitoring used to detect volcanic activity?

3. Explain the USGS alert notification system for volcanic eruptions.

5.9 Perception of and Adjustment to the Volcanic Hazard

PERCEPTION OF VOLCANIC HAZARDS

Information concerning how people perceive a volcanic hazard is limited. People live near volcanoes for a variety of reasons: (1) They were born there and in the case of some islands, such as the Canary Islands, all land is volcanic; (2) the land is fertile and good for farming; (3) people are optimistic and believe an eruption is unlikely; and (4) they cannot choose where they live—for example, they may be limited by economics. One study of people's perception in Hawaii found that a person's age and length of residence near a volcanic hazard are significant factors in his or her knowledge of the hazard and possible adjustments to it.[39] One

reason the evacuation of 60,000 people prior to the 1991 eruption of Mount Pinatubo was successful was that the Philippine government had educated people about the dangers of violent ash eruptions with debris flows. A video depicting these events was widely shown before the eruption, and it helped convince local officials and residents that they faced a real and immediate threat.[4]

The science of volcanoes is becoming well known. However, good science is not sufficient (see Case Study 5.5: Survivor Story). Probably the greatest hazard reduction will come from an increased understanding of human and societal issues that arise during an emerging **volcanic crisis**. A volcanic crisis can develop when scientists predict that an eruption is likely in the near future. In such a crisis, improved communication among scientists, emergency managers, educators, media, and private citizens is particularly important. The goal is to prevent a volcanic crisis from becoming a disaster or catastrophe.[39]

ADJUSTMENTS TO VOLCANIC HAZARDS

Apart from the psychological adjustment to losses, the primary human adjustment to volcanic activity is evacuation. Mount Vesuvius, which devastated the cities of Pompeii and Herculaneum in A.D. 79 (see Figure 5.26), is considered by some to be one of the most hazardous volcanoes on Earth with nearly 3 million people living in the surrounding area. In an attempt to minimize the potential for a future catastrophe if another large eruption were to occur, the Italian government is offering 30,000 euros to anyone living in the 18 towns within the immediate area who is willing to relocate.[40]

ATTEMPTS TO CONTROL LAVA FLOWS

Several methods, such as *hydraulic chilling* and wall construction, have been employed to deflect lava flows away from populated or otherwise valuable areas. These methods have had mixed success. They cannot be expected to modify large flows, and their effectiveness with smaller flows requires further evaluation.

The world's most ambitious hydraulic chilling effort was initiated in January 1973 on the Icelandic island of Heimaey. Basaltic lava flows from Eldfell volcano nearly closed the harbor of Vestmannaeyjar, the island's main town and Iceland's main fishing port. The situation prompted immediate action.

Three favorable conditions existed: (1) Slow movement of the flows allowed the time needed to initiate a control effort; (2) transport by sea and roads allowed the delivery of pipes, pumps, and heavy equipment; and

A Close Call with Mount St. Helens

For someone who makes his living as a reporter, Don Hamilton was awfully lucky to be nowhere near a telephone on the evening of May 17, 1980 (Figure 5.5.A).

The next day, at 8:32 A.M., Mount St. Helens would explode in one of the largest eruptions in North American history, spewing monstrous proportions of ash that would eventually circle the entire planet and cover the surrounding region in hundreds of feet of debris.

But that evening, Hamilton was sitting on the porch of a lodge on Spirit Lake, no more than 16 km (~10 mi.) from the mountain's peak, visiting the lodge's resident, a local eccentric named Harry Truman. By this point, smaller earthquakes and seismic observations had geologists clamoring about the imminent eruption, and Truman, then 84 years old, had become a minor celebrity for his stalwart refusal to evacuate.

Hamilton was working for the now-defunct *Oregon Journal*, and that night he might have stayed with Truman if he had had a telephone. But Hamilton needed some way to file that day's story.

As it turned out, leaving was a wise decision. When the volcano erupted the very next morning, with Hamilton safely in bed in Portland, it buried Truman and his lodge in a massive landslide of debris and pyroclastic flow.

In the days and weeks leading up to the eruption, Hamilton had witnessed several violent earthquakes while visiting Truman. "I was up there a few times in his lodge, up on his front porch, when some pretty good quakes hit," he said. "I could see the road rippling. It was the weirdest thing I'd ever seen."

By this time the media had already caught up with Truman, who had appeared on the *Today* show and in the *New York Times*. But for all his gusto, Hamilton said, there were cracks in Truman's resolve. "I saw some genuine fear in his face," he recalled. "He looked drawn. His eyes were bugging out a little bit. It made him stop and pay attention."

But in spite of all the warnings, Hamilton thought most people weren't prepared for the sheer magnitude of the eruption. "The geologists were certainly warning us that this was a very dangerous and unstable situation," he said. "But I don't think anybody really anticipated the enormity of what happened. There wasn't a lot of documented history of this kind of thing."

◀ **FIGURE 5.5.A Oregon Reporter who Escaped Death on Mount St. Helens**
A fateful decision on May 17, 1980, to leave Harry Truman's lodge on the slopes of Mount St. Helens and return to Portland saved Don Hamilton's life. Don recounted his story nearly 25 years later when Mount St. Helens erupted again in March 2005. *(Don Hamilton)*

(3) water was readily available. Initially, the edges and surface of the flow were cooled with water from numerous fire hoses (Figure 5.44).

Then bulldozers were moved up on the slowly advancing flow to make a path for a large water pipe. The plastic pipe did not melt as long as water was flowing in it, and small holes in the pipe allowed the cooling of hot spots along parts of the flow. Watering had little effect the first day, but then the back edge flow began to slow down and in some cases stopped.

(a)

(b)

(c)

◀ **FIGURE 5.44 Fighting Lava Flows**
Eruptions of Eldfell volcano on the island of Heimaey, Iceland. (a) Lava fountain buries homes in volcanic cinders. Numerous homes were destroyed by structural collapse or fire. *(Bettmann/CORBIS)* (b) White steam pours from the front of the lava flow as emergency crews use fire hoses to slow the advance of the flow on the town and the precious harbor. *(S Jonasson/FLPA Image Broker/Newscom)* (c) Aerial view taken several months after the eruption shows the town covered in black ash and the harbor that was kept free of lava, partly due to human efforts. *(Scholz/picture-alliance/dpa/AP Images)*

These actions undoubtedly had an important effect on the lava flows from Eldfell volcano. They restricted lava movement, reduced property damage, and allowed the harbor to remain open. When the eruption stopped five months later, the harbor was still usable.[41] In fact, by fortuitous circumstances, the shape of the harbor was actually improved because the cooled lava provided additional protection from the sea. More recent efforts to chill and deflect lava flows on the slope of Mount Etna in Sicily have also had some success.

5.9 CHECK your understanding

1. What methods have been used to control lava flows?

2. Why was the evacuation prior to the Mount Pinatubo eruptions so successful?

3. How has the risk from an eruption of Mount Vesuvius changed in the past two 2,000 years?

APPLYING the **5** Fundamental Concepts

Eyjafjallajökull 2010 Eruption

1 **Science helps us predict hazards.**

2 **Knowing hazard risks can help people make decisions.**

3 **Linkages exist between natural hazards.**

4 **Humans can turn disastrous events into catastrophes.**

5 **Consequences of hazards can be minimized.**

The large number of active volcanoes on Iceland is not surprising given its location on the Mid-Atlantic Ridge. Iceland is literally being ripped apart at a rate of 20 cm (~8 in.) per year (Figure 5.45a)![42] The island also sits atop a hot spot, enhancing magma formation and volcanic activity at the surface—Iceland experiences a volcanic eruption once every three to four years.[1] Fortunately, the process by which magma forms in this setting means that most Icelandic eruptions are generally effusive and have a VEI of no more than 1 (see Table 5.2). This means that the effects of Icelandic volcanic eruptions are typically limited to the island and often go unnoticed by the rest of the world. This all changed with the eruption of Eyjafjallajökull volcano in 2010.

Frequent eruptions means that Iceland's people and government agencies must be constantly prepared. As recently as 2006, the government and citizens of southern Iceland had taken part

in a full-scale test of the evacuation plan for an eruption of Katla[43]—Eyjafjallajökull's larger and more active neighbor to the east (Figure 5.45b). The Icelandic Meteorological Office (IMO) also operates a complex array of monitoring and early warning equipment to minimize the effects of volcanic eruptions, including 56 seismic stations and 70 continuous GPS stations that measure earthquake activity and ground displacement indicative of magma movement. An automated network of hydrological stations also provides early warning for jökulhlaups by measuring the amount and temperature of water flowing in rivers downstream of glacially covered volcanoes (Figure 5.45b).[1]

For almost 20 years prior to the 2010 eruption, intermittent deep earthquake activity was recorded throughout the region surrounding the Eyjafjallajökull and Katla volcanoes.[1] By January 2010, however, just three months before the eruption, seismic activity became localized about 5 km (~3.1 mi) east of Eyjafjallajökull's summit crater and earthquake depths were becoming more shallow with time. Concurrently, GPS stations on the southeast flank of the volcano were detecting uplift and tilting (Figure 5.46). Slow at first, by early March the rate of deformation was as much as 5 mm (~0.2 in.) per day. These data left no doubt that magma in the subsurface was rising up from great depths and pooling within the upper 4 km (~2.5 mi.) of the surface. For the next two weeks the volcano

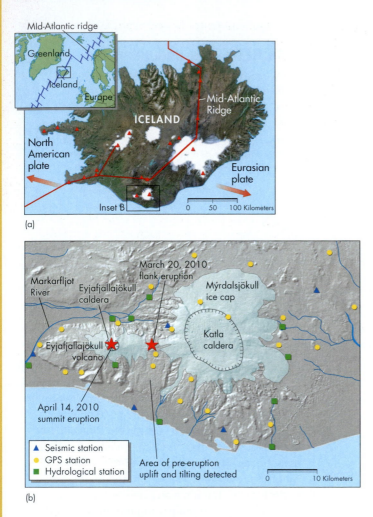

(a)

(b)

∧ FIGURE 5.45 Icelandic Volcano Monitoring
(a) Map showing the location of Iceland's major volcanoes and ice caps. *(Based on Sigmundson, F. et al., "Intrusion triggering of the 2010 Eyjafjal-lajökull explosive eruption,"* Nature Letters *468: 426–32, 2010)* (b) Monitoring equipment surrounding Eyjafjallajökull and Katla volcanoes on the southern tip of Iceland. See (a) for location. *(Based on Erik Sturkell et al., "Katla and Eyjafjallajökull Volcanoes,"* Developments in Quaternary Science 13: 521, 2010)

was suspended after 24 hours of no increased stream flow detected by hydrologic stations.

The effusive phase continued for three weeks and generated a spectacular fire curtain display and lava falls cascading more than 300 m (~985 ft.) down to the local river valleys. Without any obvious hazard, more than 25,000 geotourists visited the volcano to watch the eruption (Figure 5.47).[44] Some visitors actually witnessed the formation of a second fissure, which fortunately opened 200 m (~650 ft.) northwest of the first and not where the crowds had gathered.

After erupting for three straight weeks, 10–20 m (~30–60 ft.) of lava had accumulated around the vent covering 1.5 km² (~1 mi².). Yet despite this large volume of extruded lava, GPS data showed no evidence for volcano deflation during this phase (Figure 5.46). The lack of deflation indicated a steady inflow of magma, suggesting that the eruption might not end anytime soon. However, on April 12 the lava stopped flowing from the fissures.

For the next 48 hours the seismicity increased beneath Eyjafjallajökull's summit caldera indicating that the magma was working its way to the surface along another path. Then, early in the morning on April 13, an explosive eruption occurred from beneath the 200 m (~650 ft.) thick ice cap (Figure 5.45b). Emergency response plans were enacted and residents on the south side of the volcano were being evacuated by 1:00 A.M. At about 7:00 A.M., hydrologic stations on both the north and the south sides of the volcano detected jökulhlaups, forcing the evacuation of another 800 people in nearby river valleys.[43] Over the next three days several jökulhlaups would be unleashed down the Markarfljót River valley, the largest occurring shortly after the explosive phase began (Figure 5.48a). Fortunately, the floods were relatively small (< 3000 m/s) and flood control structures along the river were able protect some of the roads and agricultural lands in the Valleys (Figure 5.48b). But because the initial floods were charged with debris from the eruption

continued to inflate and seismic activity steadily increased with as many as 1,000 earthquakes occurring around March 4 (Figure 5.46).[42]

Eyjafjallajökull's eruption occurred in two phases—an effusive, Icelandic-type eruption phase followed by an explosive phase. The effusive phase began on March 20 with the opening of a 500 m (~1,640 ft.) long fissure on the flank of the volcano 10 km (~6.2 mi.) east of Eyjafjallajökull's caldera (Figure 5.45b). About 500 farm families were immediately evacuated from the area. Fortunately, the fissure opened between the Eyjafjallajökull and Mýrdalsjökull ice caps and the evacuation

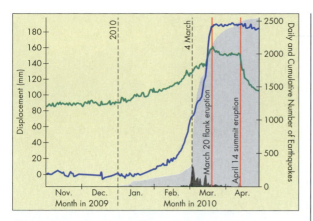

∧ FIGURE 5.46 GPS and Seismic Data
Seismic and GPS monitoring data showing earthquake activity and surface displacement measurements before and during the 2010 eruption of Eyjafjallajökull volcano. Displacement of two GPS stations (blue and green lines show inflation and deflation of the mountain (left axis), while seismic stations the number of daily (dark gray) and cumulative (light gray) earthquakes since the beginning of 2010. *(Based on Sigmundson, F. et al., "Intrusion triggering of the 2010 Eyjafjallajökull explosive eruption," Nature Letters 468: 426–32, 2010)*

and icebergs from the melting glaciers, some local infrastructure and farms sustained damage (Figure 5.48c).

Unlike the effusive eruption phase, a towering eruption column of ash could be seen rising

9 km (~5.9 mi.) into the atmosphere above Eyjafjallajökull within 24 hours of the eruption (see Figure 5.1a).[45] By April 15, unusually stable high-altitude winds (see jet streams, Chapter 9, section 9.4) blowing toward the southeast carried the ash cloud straight toward Europe (Figure 5.49a). Because volcanic ash can stall and destroy aircraft engines in flight, the Civil Aviation Authority declared a "no fly zone" for any airspace containing more than 4 milligrams per cubic meter of ash. By April 18 most air travel in and out of Europe had been suspended (Figure 5.49b). For more than a week northern Europe's airspace remained closed, stranding countless travelers and costing businesses relying on air travel billions of dollars. This situation persisted until wind patterns changed. Over the next five weeks, the height of the eruption column and the path of the ash varied widely due to changing wind velocities and direction, which meant that temporary airport closures throughout Europe continued through the rest of April and May.

The southerly winds carrying the ash out to sea and causing Europe's woes spared much of Iceland from being blanketed in ash. However, the island's southeastern corner, within about

∧ FIGURE 5.47 Geotourists Witness Fissure Eruption
Eyjafjallajökull volcano draws a crowd during the effusive first phase of the eruption. *(PaKos Photography/Alamy)*

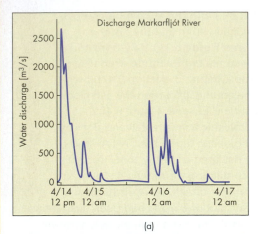

(a)

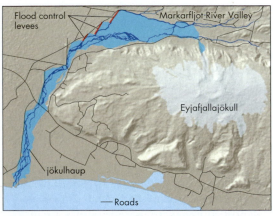

(b)

(c)

˄ FIGURE 5.48 Markarfljót River Valley Floods
(a) Hydrograph on the Markarfljót River showing multiple jökulhlaups over a four-day period of the eruption, with the largest occurring on 14 April. *(From Oddur Sigursson et al., "Flood warning
system and jökulhlaups-Eyjafjallajökull" (4/7/13), Iceland Meteorological office, http://en.vedur.is/hydrology/articles/nr/2097)*
(b) Markarfljót River Valley, showing the area covered by the 14
April jökulhlaup (shaded blue) and the effectiveness of flood control structures (red lines) in minimizing the effects of the torrent
of water. (c) Roadway across the Markarfljót River destroyed by
jökulhlaup. *(Caters News/ZUMAPRESS/Newscom)*

(a)

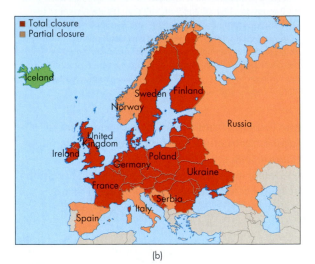

(b)

˄ FIGURE 5.49 Ash Shuts Down Air Travel in Europe
(a) Satellite image showing ash cloud from Eyjafjallajökull being carried southward toward Europe. *(NASA)* (b) For 10 days,
beginning with the second phase of the eruption on 14 April, the
airspace above most of Europe was shut down either partially
or completely, affecting an estimated 10 million travelers. *(From
"Travel crisis in Europe 2010," http://en.wikipedia.org/wiki/
File:Travel_crisis_in_Europe_2010.png)*

20 km (~12.5 mi.) of Eyjafjallajökull, was covered
by several centimeters (1–2 inches) of ash, and
from April 17 until early April 18 the region was
plunged into total darkness by the ash fallout
(Figure 5.50).[45] Volcanic lightning within the
eruption column during this period was also
particularly vigorous with as many as 15 strikes
an hour being recorded (see chapter-opener
photograph), but it was the accumulation of ash
that caused problems for local farmers. Toxic
levels of water-soluble fluoride in the ash could

sicken or kill livestock if animals digested or inhaled the fine ash or drunk from contaminated streams. Farmers were therefore told to keep themselves and their animals inside, but this was problematic due to calving season and also confusing because these instructions contradicted the recently practiced Katla evacuation plan, which stated that animals should be released during an eruption.

The unexpectedly large ash column produced during the second phase of the eruption was, in part, due to the interaction of water melted from the ice cap and the erupting lava. Rapid cooling of the hot lava caused explosive fragmenting of the melt during steam explosions, forming fine ash that was carried upward into the eruption column. In addition, the magma erupting from Eyjafjallajökull's caldera had a higher silica content and viscosity than the lava that erupted from the fissures on the flank of the volcano. The 2010 eruptions of Eyjafjallajökull was not declared over until six months later in October. During the volcano's explosive phase, it was estimated that 0.27 km^3 (~0.05 mi.^{3}) of magma erupted, and together with the ash plume height, the explosive power of the eruption ranked as a 3 on the VEI (see Table 5.2).[45] No deaths were directly attributed to the eruption, but two people did die in their travels to witness the effusive phase of the eruption.

Eyjafjallajökull had only erupted twice before in the past thousand years, and eruptions were relatively small.[46] Katla, however, had erupted 20 times over this same interval and the eruptions often included large ash plume. This history combined with a larger ice cap means Katla has always been considered the greater threat. Although most scientists do not believe that the volcanic magma pluming systems of these two volcanoes are linked, the previous two eruptions of Eyjafjallajökull occurred in tandem with Katla eruptions. Therefore, seismic activity surrounding the Katla volcano was closely monitored throughout Eyjafjallajökull's 2010 eruption. Since the 2010 eruption, an increase in seismic activity beneath Katla's caldera indicates that magma is moving within the volcano, leading to concerns about an impending Katla eruption.

APPLYING THE 5 FUNDAMENTAL CONCEPTS

1. What kind of monitoring is done to predict Eyjafjallajökull and Katla eruptions? Figure 5.46 shows some monitoring station data. Discuss how these data changed from January through April and how these changes are related to forecasting and monitoring changes within the volcano.

2. How does the risk presented by Eyjafjallajökull compare to that of Katla? Prior to 2010, the global risk of volcanic eruptions on Iceland was thought to be very low. Explain how and why the 2010 eruption has changed the perceived global risk for future Icelandic eruptions.

3. In addition to the direct effects of the volcanic eruption, such as ash fall and lava flows, several other natural hazards were linked to the eruption. Explain these linkages and discuss the geographic regions that could be affected by the linked hazards.

4. Iceland's very low population density and the volcano's rural setting limited the potentially disastrous effects of the 2010 eruption. How might the local effects of the eruption have been different if the city where you live now was located at the foot of Eyjafjallajökull in April 2010?

5. Discuss the preparedness of Iceland's government and people for volcanic eruptions and the effectiveness of the scientific equipment, evacuation planning, and engineering structures used to mitigate the effects of the eruption (refer to Figures 5.45 and 5.48).

∧ **FIGURE 5.50 Ash Fallout Endangers Animals**
Farmers rescue cattle from toxic volcanic ash fallout southeast of Eyjafjallajökull on 17 April. *(Brynjar Gauti/AP Images)*

CONCEPTS in review

5.1 Introduction to Volcanism

LO:1 Explain the relationship of volcanoes to plate tectonics.

- Volcanic activity is directly related to plate tectonics. Most volcanoes are located at plate boundaries, where magma is produced in the spreading or sinking of lithospheric plates. Two-thirds of the world's volcanoes are associated with the sinking of lithospheric plates along the Ring of Fire surrounding most of the Pacific Ocean.
- Lava is magma that has been extruded from a volcano. Its viscosity, a characteristic related to the temperature and silica content, and volatile content are important in determining the eruptive style of the different types of volcanoes.
- Magma comes from melting of the asthenosphere. Three ways to melt the asthenosphere are by decompression, addition of volatiles, and addition of heat.
- Features of volcanoes include ven**hield volcano**ts, craters, and calderas.

KEY WORDS

crater (p. 161), **decompression melting** (p. 157), **effusive eruption** (p. 161), **lava** (p. 156), **magma** (p. 156), **magma chamber** (p. 160), **pyroclastic debris** (p. 161), **volatiles** (p. 159), **volcanic vent** (p. 156)

5.2 Volcano Types, Formation, and Eruptive Behavior

LO:2 Identify the different types of volcanoes and their associated features.

- Most explosive volcanic eruptions are from the classic, cone-shaped stratovolcanoes that occur above subduction zones, particularly around the Pacific Rim. These volcanoes are characterized by explosive eruptions and are composed primarily of silica-rich lavas, such as andesite, and pyroclastic deposits.
- Volcanic domes are smaller than stratovolcanoes but develop similarly above subduction zones and are composed of viscous magma. Domes can be explosive or nonexplosive.
- The largest volcanoes, shield volcanoes, are common at mid-ocean ridges such as Iceland, and over mid-plate hot spots such as the Hawaiian Islands. They are characterized by relatively nonexplosive lava flows of basalt.
- Large calderas are created by infrequent, huge, violent eruptions. Following an explosive beginning, they often resurge and may present a volcanic hazard for a million years or longer.

KEY WORDS

caldera (p. 166), **cinder cone** (p. 167), **lava dome** (p. 163), **shield volcano** (p. 165), **stratovolcano** (p. 162), **volcanic explosivity index (VEI)** (p. 162)

5.3   Geographic Regions at Risk from Volcanoes

LO:3   Locate on a map the geographic regions at risk from volcanoes.

- Specific geographic regions of North America at risk from volcanoes include the northwestern coast of California; the western coasts of Oregon and Washington and parts of British Columbia and Alaska, Long Valley, and the Yellowstone National Park area.

5.4   Effects of Volcanoes

LO:4   Describe the effects of volcanoes and how they are linked to other natural hazards.

- Primary effects of volcanic activity include lava flows, pyroclastic hazards, and, occasionally, the emission of poisonous gases.
- Lava flows, which can move as fast as 35 mph, but often move much slower, do not represent a serious hazard for loss of life because they can often be avoided. However, they are not easily diverted so loss of property and land can result from lava flows.
- Pyroclastic hazards include volcanic ash falls, which may cover large areas with carpets of cool ash that can destroy some structures and ruin agricultural land, but loss of life is less common. By contrast, pyroclastic flows travel at very high velocities, are extremely hot, and destroy and kill everything in their path.
- Secondary effects of volcanic activity include debris flows and mudflows, generated when melting snow and ice or precipitation mix with volcanic ash. These flows can devastate an area many kilometers from the volcano. All these effects have occurred in the recent history of the Cascade Range of the Pacific Northwest and will occur there in the future.

KEY WORDS

ash fall (p. 175), **debris flows** (p. 180), **lateral blast** (p. 176), **lava flow** (p. 172), **mudflows** (p. 180), **pyroclastic deposit** (p. 174), **pyroclastic flow** (p. 175), **tephra** (p. 174)

5.5 and 5.6   Linkages between Volcanoes and Other Natural Hazards Natural Service Functions of Volcanoes

LO:5   List the potential benefits of volcanic eruptions.

LO:6   Discuss how humans can minimize the volcanic hazard.

- Volcanoes are linked to other natural hazards such as fire, earthquakes, landslides, and climate change.
- Volcanoes provide fertile soils, a source of power, mineral resources, recreational opportunities, as well as newly created land.

5.7, 5.8, and 5.9 Human Interactions with Volcanoes
Minimizing the Volcanic Hazard
Perception of and Adjustment to the Volcanic Hazard

LO:7 Recommend adjustments we can make to avoid death and damage from volcanoes.

- Sufficient monitoring of seismic activity; thermal, magnetic, and hydrologic properties; and changes in the land surface, combined with knowledge of the recent geologic history of volcanoes, may eventually result in reliable forecasting of volcanic activity.
- Forecasts of eruptions have been successful, particularly for Hawaiian volcanoes and Mount Pinatubo in the Philippines.
- The U.S. Geological Survey has developed an alert notification system for volcanic activity that has four levels for hazards on land and four color codes for aviation hazards.
- Efforts to meet the goal of reducing volcanic hazards are focusing on human and societal issues of communication; the objective is to prevent a volcanic crisis from becoming a disaster or catastrophe. Worldwide, however, it is unlikely that we will be able to accurately forecast most volcanic activity in the near future.
- Perception of the volcanic hazard is apparently a function of a person's age and length of residency near the hazard. Community-based education plays an important role in informing people about the hazards of volcanoes. Apart from psychological adjustment to losses, the primary human adjustment to volcanic activity is evacuation. Some attempts to control lava flows once an eruption has begun have been successful.

KEY WORD
volcanic crisis (p. 194)

CRITICAL thinking questions

1. While looking through some old boxes in your grandparents' home, you find a sample of volcanic rock collected by your great-grandfather. No one knows where it was collected. You take it to school, and your geology professor tells you that it is a sample of andesite. What might you tell your grandparents about the type of volcano from which it probably came, its geologic environment, and the type of volcanic activity that likely produced it?

2. In our discussion of perception of and adjustment to volcanic hazards, we established that people's perceptions and what they will do in case of an eruption are associated with both their proximity to the hazard and their knowledge of volcanic processes and necessary adjustments. With this association in mind, develop a public relations program that could alert people to a potential volcanic hazard. Keep in mind that the tragedy associated with the eruption of Nevado del Ruiz was in part due to political and economic factors that influenced the apathetic attitude toward the hazard map prepared for that area. Some people were afraid that the hazard map would lower property values in high-risk areas.

3. You are going to take a scout troop to Hawaii to see the Kilauea volcano. Some children have seen

documentaries of the 1980 eruption of Mount St. Helens and are afraid; others are fearless and want to try cooking on a lava flow. What will you tell the fearful scouts? Describe the safety precautions you will follow for all the scouts.

4. Given what you know about the effects of a volcanic eruption and the linked natural hazards, where

might you build your house if you were living near a composite volcano such as Mount Rainer in Washington?

5. Do you think our current ability to monitor precursor events and predict an eminent eruption is sufficient for large populations to live on the flank of an active volcano? Why or why not?

References

1. **Gudmundsson, M. T.,** and **Pedersen, R.** 2010. Eruption of Eyjafjallajökull Volcano, Iceland. *Eos News*, 91(2).

2. **Mazzocchi, M., Hansstein, F.,** and **Ragona, M.** 2010. The 2010 volcanic ash cloud and its financial impact on the European airline industry. CESifo Forum/A joint initiative of Ludwig-Maximilians-Universitat and the Ifo Institute for Economic Research 1(1): 92–100.

3. **BBC Newsnight.** 2010. Interview with President Grimsson of Iceland, April 20, 2010. http://news.bbc.co.uk/2/hi/programmes/newsnight/8631343.stm. Accessed 5/17/13.

4. **Decker, R.,** and **Decker, B.** 2006. *Volcanoes*, 4th ed. New York: W. H. Freeman.

5. **Smith, G. A.,** and **Pun, A.** 2010. *How does Earth work*, 2nd ed. Upper Saddle River, NJ: Pearson Prentice Hall.

6. **Kious, W. J.,** and **Tilling, R. I.** 1996. *This dynamic Earth: The story of plate tectonics*. Washington, DC: U.S. Geological Survey.

7. **T. L. Grove, N. Chatterjee, S. W. Parman,** and **E. Medard.** 2006. The influence of H2O on mantle wedge melting. *Earth and Planetary Science Letters* 249: 74–89.

8. **Brantley, S. R.** 1994. Volcanoes of the United States: USGS General Interest Publication.

9. **Schmincke, H.** 2004. *Volcanism*. New York: Springer-Verlag.

10. **Fisher, R. V., Heiken, G.,** and **Hulen, J. B.** 1997. *Volcanoes*. Princeton, NJ: Princeton University Press.

11. **Francis, P.** 1983. Giant volcanic calderas. *Scientific American* 248(6): 60–70.

12. **Wright, T. L.,** and **Pierson, T. C.** 1992. *Living with volcanoes*. U.S. Geological Survey Circular 1073.

13. **IAVCEE Subcommittee on Decade Volcanoes.** 1994. Research at decade volcanoes aimed at disaster prevention. *EOS, Transactions, American Geophysical Union* 75(30): 340, 350.

14. **Pendick, D.** 1994. Under the volcano. *Earth* 3(3): 34–39.

15. **Crandell, D. R.,** and **Waldron, H. H.** 1969. Volcanic hazards in the Cascade Range. In *Geologic hazards and public problems*, conference proceedings, eds. R. Olsen and M. Wallace, pp. 5–18. Office of Emergency Preparedness Region 7.

16. **Neal, C. A., Casadevall, T. J., Miller, T. P., Hendley, J. W. II,** and **Stauffer, P. H.** 1998. *Volcanic ash—Danger to aircraft in the North Pacific*. U.S. Geological Survey Fact Sheet 030–97.

17. **Nakada, S.** 2000. Hazards from pyroclastic flows and surges. In *Encyclopedia of volcanoes*, eds. H. R. Sigurdsson, B. F. Houghton, S. R. McNutt, H. Rymer, and J. Stix, pp. 945–55. San Diego, CA: Academic Press.

18. **Nakada, S., Shimizu, H.,** and **Ohta, K.,** 1999. Overview of 1990–1995 eruption at Unzen volcano. *J. Volcanol. Geotherm. Res.* 89: 1–22.

19. **Holloway, M.** 2000. The killing lakes. *Scientific American* 286(3): 90–99.

20. **Simkin, T., Siebert, L.,** and **Blong, R.** 2001. Volcano fatalities—lesson from the historical record. *Science* 291:255.

21. **Scarth, A.** 1999. *Vulcan's fury: Man against the volcano*. New Haven, CT: Yale University Press.

22. **Sutton, J., Elias, T., Hendley II, J. W.,** and **Stauffer, P. H.** 2000. *Volcanic air pollution—A hazard in Hawai'i*. U.S. Geological Survey Fact Sheet 169–97, version 1.1.

23. **U.S. Geological Survey.** 1999. *Pilot project, Mount Rainier volcano lahar warning system*. http://volcanoes.usgs.gov/About/Highlights/RainierPilot/Pilot_highlight.html. Accessed 6/17/07.

24. **Watts, A. B.,** and **Masson, D. G.** 1995. A giant landslide on the north flank of Tenerife, Canary Islands. *Journal of Geophysical Research* 100: 24487–98.

25. **Hammond, P. E.** 1980. Mt. St. Helens blasts 400 meters off its peak. *Geotimes* 25(8): 14–15.

26. **U.S. Geological Survey.** 2007. *Thoughts on the 27th anniversary of catastrophic eruption of Mount St. Helens and on ongoing activity*. Vancouver, WA: USGS Cascade Volcano Observatory, http://vulcan.wr.usgs.gov/Volcanoes/MSH/Eruption04/MediaInfo/May07/talking_points_may2007.pdf. Accessed 6/18/07.

27. **Gardner, C.** 2005. Monitoring a restless volcano: The 2004 eruption of Mount St. Helens. *Geotimes* 50(3): 24–29.

28. Pendick, D. 1995. Return to Mount St. Helens. *Earth* 4(2): 24–33.

29. Self, S. 2005. Effects of volcanic eruptions on the atmosphere and climate. In *Volcanoes and the environment*, eds. J. Martí, and G. G. J. Ernst, pp. 152–74. New York: Cambridge University Press.

30. Oppenheimer, C. 2003. Climatic, environmental and human consequences of the largest known historic eruption: Tambora volcano (Indonesia) 1815. *Progress in Physical Geography* 27(2): 230–59.

31. Anonymous. 1991. Pinatubo cloud measured. *EOS, Transactions, American Geophysical Union* 72(29): 305–06.

32. Duffield, W. A. 2005. Volcanoes, geothermal energy, and the environment. In *Volcanoes and the environment*, eds. J. Martí, and G. G. J. Ernst, pp. 304–32. New York: Cambridge University Press.

33. Heiken, G. 2005. Industrial uses of volcanic materials. In *Volcanoes and the environment*, eds. J. Martí, and G. G. J. Ernst, pp. 387–403. New York: Cambridge University Press.

34. Kilburn, C. R. J., and **Sammonds, P. R.** 2005. Maximum warning times for imminent volcanic eruptions. *Geophysical Research Letters* 32: L24313.

35. Francis, P. 1976. *Volcanoes.* New York: Pelican Books.

36. Richter, D. H., Eaton, J. P., Murata, K. J., Ault, W. U., and **Krivoy, H. L.** 1970. *Chronological narrative of the 1959–60 eruption of Kilauea Volcano, Hawaii.* U.S. Geological Survey Professional Paper 537E.

37. Tilling, R. I. 2000. Volcano notes. *Geotimes* 45(4): 19.

38. Gardner, C. A., and **Guffanti, M. C.** 2006. *U.S. Geological Survey's alert notification system for volcanic activity.* U.S. Geological Survey Fact Sheet 2006–3139.

39. Murton, B. J., and **Shimabukuro, S.** 1974. Human adjustment to volcanic hazard in Puna District, Hawaii. In *Natural hazards: Local, national, global*, ed. G. F. White, pp. 151–59. New York: Oxford University Press.

40. *Geotimes.* 2005. Vesuvius' next eruption, April 50(4).

41. Williams Jr., R. S., and **Moore, J. G.** 1973. Iceland chills a lava flow. *Geotimes* 18(8): 14–17.

42. Sigmundson, F. et al. 2010. Intrusion triggering of the 2010 Eyjafjallajökull explosive eruption. *Nature Letters* 468: 426–32.

43. Bird, D. K., and **Gisladottir, G.** 2012. Residents' attitudes and behaviour before and after the 2010 Eyjafjallajökull eruptions—a case study from southern Iceland. *Bulletin of Volcanology* 74: 1263–79.

44. Global Volcanism Program, Eyjafjallajökull Volcano. www.volcano.si.edu/index.cfm. Accessed 5/10/2013.

45. Gudmundsson, M. T., Thordarson, T., Höskuldsson, A. et al. 2010. Ash generation and distribution from the April–May 2010 eruption of Eyjafjallajökull, Iceland. *Scientific Reports* 2(572): 20.

46. Sturkell, E., Einarsson, P., Sigmundsson, F. et al. 2010. Katla and Eyjafjallajökull Volcanoes. *Developments in Quaternary Science* 13: 521.

Many small government **and** **private boats** rescued people from flooded homes

6

APPLYING the **5** fundamental concepts Houston Floods of 2017

1

Science helps us predict hazards.

Flooding is a serious and costly natural hazard, one that more people experience than any other natural hazard. Fortunately, flooding is usually linked to meteorological phenomena that can be used to provide some advance warning about the potential for flooding, as was the case in 2017.

2

Knowing hazard risks can help people make decisions.

Following the 2017 Houston floods, the risk of damage to land and property from the flood is being studied in the context of population increase, land-use planning and long-term trends and climate change. However, for Houston and many other regions knowing the risk may not prevent people from living and working in flood-prone areas.

Flooding

Houston Flood of 2017

In a period spanning a few days, as much as 125 cm (~50 in.) of rain from Hurricane Harvey fell on the city of Houston, Texas, producing the largest and most serious urban flood in U.S. history. Total damages will probably approach $200 billion. The flood caused immense human suffering to the urban population (see opening photograph). Houston is the fourth largest city in the U.S., with about 6.5 million people spread over 1500 square kilometers (6000 square miles, two-thirds the area of Maryland). The city is sufficiently inland from the Gulf Coast to avoid storm surge from hurricanes, but it is sited on low-lying land with numerous streams (some, called bayous, are mostly very slow-moving water courses with abundant plants) which flow south through Houston to the sea. The result of the unprecedented amount of rain was catastrophic flooding which inundated much of the city and surrounding areas with several feet of water. Freeways and streets became rivers (Figure 6.1). The floodwaters displaced about 1 million people and damaged about 200,000 homes and several hundred thousand automobiles. Many thousands of people were rescued by boat and helicopter (Figure 6.2) and spent time in emergency shelters. Evacuation of the city was not an option because massive traffic jams would have trapped people in cars on flooding streets and perhaps have caused many more deaths than those which occurred to people sheltering in their homes.

Flooding is the most universally experienced natural hazard. Floodwaters have killed more than 10,000 people in the United States since 1900. For the past decade, property damage from flooding has averaged more than $4 billion per year. Flooding is a natural process that will remain a major hazard as long as people live and work in flood-prone areas.

After reading this chapter, you should be able to:

LO:1 Explain basic river processes.

LO:2 Compare and contrast flash floods, downstream floods, and megafloods and know why the differences are important.

LO:3 Summarize regions at risk from flooding and the global impact.

LO:4 Differentiate between the effects of flooding and the linkages with other natural hazards.

LO:5 Describe the benefits of periodic flooding.

LO:6 Discuss how human activity can significantly affect river processes and flooding.

LO:7 Summarize why being proactive with flood hazards is, in the long term, more significant in reducing the flood hazard than being reactive after a flood.

LO:8 Evaluate the adjustments we can make to minimize flood deaths and damage.

LO:9 Summarize how people perceive flooding.

3

Linkages exist between natural hazards.

Flooding is linked to several other natural hazards; in the case of Pakistan, it was linked to severe weather and climate, as well as waterborne disease.

4

Humans can turn disastrous events into catastrophes.

With a population that has grown dramatically in recent decades, the 2017 Houston flood illustrates that densely populated flood-prone areas are becoming more vulnerable to large events with the potential to cause catastrophes.

5

Consequences of hazards can be minimized.

When we return to the 2017 Houston flood, you will learn about the challenges faced by governments with minimal resources to minimize future flood hazards through prediction, land-use planning, and engineering.

∧ **FIGURE 6.1 Aerial View of 2017 Houston Flood**
Thousands of homes were damaged and roads were covered by
floodwaters. *(David J. Phillip/AP)*

∧ **FIGURE 6.2 Flood Victims Were Rescued by Boats**
Private and government boats of many types joined together to
rescue people from floodwaters. *(ABC News.Go.Com chron.com)*

Years before the Houston flood of 2017, concerns
and warnings were voiced about the risk and impact of
flooding as a result of the rapid increase in Houston's
urban population (increasing exposure to flooding),
drainage problems (increasing vulnerability of flooding)
as the land was paved over, and poor land-use planning
(increasing exposure to flooding) that allowed increased
development with more people living in areas that were
becoming more susceptible to flooding over time. After
the flood, people reported that this was the first flood
they had experienced, and few had flood insurance.
Concern was also raised about emerging long-term
trends linked to climate change which, with increase in
intensity of storms, would likely increase the risk and
damage from flooding.

The response to the flood (people helping people) as
it emerged over a week of prolonged, intense rain was a
heartwarming aspect of the catastrophic urban flood. Private citizens from far and near in all sorts of small boats,

along with military support, moved through flooded
areas, rescuing people from homes and stranded cars
and moving them to shelters. Around the country, millions of dollars were donated from fellow Americans to
assist those in need. Houston-area deaths from the flood
numbered about 30 people. Swift responses and rescues
saved lives. While Houston was being flooded, unusually
strong seasonal (monsoon) rains in India, Nepal, and
Bangladesh (South Asia) flooded cities and surrounding
areas, killing over 1,000 people (about 40 million people
were affected by the flooding). As in Houston, the city of
Mumbai, India's richest city with an urban population of
about 22 million, was brought to a standstill by flooding
(Figure 6.3). The geography, topography, and climate is
very different in Houston compared to Mumbai, but the
risk of urban flooding in both cities is high due to similar
circumstances (high population, potential of prolonged
intense rainfall, and poor land-use planning). Flooding
is the most widespread and common natural hazard in
the world today. Major U.S. cities that have a significant
urban flood hazard include New York, Boston, New
Orleans, Miami, and Los Angles.

Flood disasters are becoming more common in large
urban areas, suggesting that a new approach to mitigating floods, such as reducing exposure and vulnerability
and, thus, risk (see the discussion of risk in Chapter 1)
through restricting development on floodplains and better control, using engineering of floodwaters, is necessary. An urban flood control dam failed to control runoff
in Houston, requiring release of water that flooded
downstream areas. Links to climate change must also
be carefully considered in future flood hazard evaluation
as storms become more intense. We will return to the
Houston flood in more detail at the end of the chapter
when we discuss the event with respect to the fundamental concepts of this book.

∧ **FIGURE 6.3 People Wading through Floodwater**
While Houston was being flooded, so was the city of Mumbai,
India's richest city with an urban population of about 22 million.
(newGettyimages 840507052)

6.1 **An Introduction to Rivers**

The range of drainage area, sediment size and volume, water, climate, and human use make each river and each flood unique. Nevertheless, an understanding of basic river processes greatly improves the understanding of floods, which are the most commonly experienced natural hazard, whether you live in a rural area, small town, or large urban area. Streams and rivers are part of the water or hydrologic cycle, and *hydrology* is the study of this cycle (see Chapter 1). In the hydrologic cycle, water evaporates from Earth's surface (primarily from the oceans) into the atmosphere, then precipitates back onto Earth's surface, and eventually returns to the oceans by flowing underground and across the land surface. Water that falls on the land as rain and snow will infiltrate into the ground, evaporate from the land surface, or drain off the land, following a course determined by the local topography.

Surface drainage, referred to as *runoff*, finds its way to small streams, which may merge as *tributaries* to form a larger stream or **river**. Streams and rivers differ only in size; that is, streams are small rivers. Local usage varies as to what constitutes a creek, brook, and river. Geologists, however, commonly use the term *stream* for any body of

water that flows in a channel. The region drained by a single stream or river is variously called a **drainage basin**, *watershed*, river basin, or catchment (Figure 6.4a). Thus, each stream has its own drainage basin or watershed that collects rain and other precipitation. A large river basin, such as the Susquehanna River basin in the northeastern United States, is made up of hundreds of small drainage basins, commonly called watersheds, which catch the precipitation that drains into the smaller tributary streams and rivers. Except for the very few who live right next to the river, most people who live on land that drains into the Susquehanna live in a watershed of one of its tributaries. In fact, virtually everyone lives in a watershed that is a part of a larger drainage basin.

One important characteristic of a river is the slope of the land over which it flows. Referred to as the *gradient*, this slope is determined by calculating the vertical drop in elevation of the channel over some horizontal distance. Gradient is commonly given in meters per kilometer or feet per mile. Also used is meter (drop) per meters in channel length or foot (drop) per feet of channel length. If a river falls 2 feet per mile, the slope is 2/5280 or 0.0004 (0.04 percent). The gradient of a river is steepest at higher elevations and levels off as the

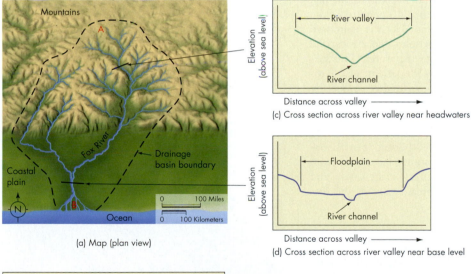

(a) Map (plan view)

(c) Cross section across river valley near headwaters

(d) Cross section across river valley near base level

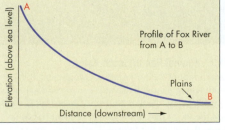

(b) Longitudinal profile

◄ **FIGURE 6.4 Drainage Basin and River Profile**
Idealized diagram of the Fox River showing (a) its drainage basin outlined by the black dashed line, that is, the area where surface runoff drains into the river or its tributaries. (b) The longitudinal profile from point A at the river's head to point B at the river's mouth; note that the vertical scale on this diagram is greatly exaggerated. (c) The V-shaped cross-section of the river valley near its headwaters where the valley floor is mainly the channel. (d) Cross-sectional profile across the Fox River valley on the coastal plain where the valley floor is mainly the floodplain.

river approaches its base level. The *base level* of a river is the lowest elevation to which it may erode. Most often, this elevation is at or close to sea level, although a river may have a temporary base level such as a lake. Rivers, thus, flow downhill to their base level, and a graph showing the downstream changes in a river's elevation is the *longitudinal profile* (Figure 6.4b).

A river usually has a steeper-sided and deeper valley at high elevations near its *headwaters* (Figure 6.4c) than closer to its base level where a wide floodplain may be present (Figure 6.4d). This happens because, at higher elevations, the steeper river gradient increases flow velocity, which, in turn, increases erosional down-cutting by the river.

EARTH MATERIAL TRANSPORTED BY RIVERS

Rivers not only move water but also transport a tremendous amount of visible and invisible material. The quantity of this material, called the *total load*, is generally subdivided into bed load, suspended load, and dissolved load, based on how the river carries the material. The bed load of most rivers consists of sand and gravel particles that slide, roll, and bounce along the channel bottom in rapidly moving water. The bed load commonly makes up less than 10 percent of the total load. In contrast, the suspended load is composed mainly of small silt and clay particles that are carried above the streambed by the flowing water. The suspended load accounts for nearly 90 percent of the total load and makes rivers look muddy, especially during a flood. Finally, the dissolved load consists of electrically charged atoms or molecules, called *ions*, which are carried in chemical solution. Most dissolved load is derived from chemical weathering of earth materials in the drainage basin. In some places, ions from discharging underground springs, wastewater, and chemical pollution are a significant part of the dissolved load.

RIVER VELOCITY, DISCHARGE, EROSION, AND SEDIMENT DEPOSITION

Rivers are the basic transportation system of that part of the rock cycle involving erosion and deposition of sediment. They are a primary erosion agent in the sculpture of our landscape, removing and carrying sediment from one site and depositing it in another. The velocity, or speed, of water in a river varies along its course, affecting both erosion and deposition of sediment. Hydrologists combine measurements of flow velocity and channel area (capacity) to determine discharge, a more useful indicator of stream flow.

Discharge (Q) is the volume of water moving through a cross-section of a river per unit time. In this case, a cross-section is the side-to-side view of the river that you would have if you took a huge knife and sliced across the valley from top to bottom at right angles to each riverbank. Once the topographic cross-section is completed with survey equipment, a hydrologist can determine the cross-sectional area that is occupied by water by simply measuring the water depth. Mean velocity is measured with a current meter. Discharge is then calculated by multiplying the cross-sectional area of the water in the channel (A) and the velocity of flow (V) and is reported in either cubic meters per second (cms) or cubic feet per second (cfs).

If there were no additions or deletions of flow along a given length of river, then its discharge would not change. Where the cross-sectional area of flow decreases, the velocity of the water must increase for discharge to remain constant. You can prove this to yourself with a garden hose. Turn on the water and observe the velocity of the water as it exits the hose. Then put your thumb partly over the end of the hose, reducing the area where the water flows out of the hose, and observe the increase in the velocity. This explains why velocity is higher where a river flows through a narrow channel in a canyon than where it spreads out and expands into a larger area downstream of the narrow channel.

Stream flow expands and slows down as a river goes from mountains onto plains or from a channel into an ocean, lake, or pond. At these points, the river commonly forms a fan-shaped deposit on land known as an *alluvial fan* (Figure 6.5) or a triangular or irregularly

⋀ FIGURE 6.5 Alluvial Fan in Death Valley
Along the western foot of California's Black Mountains, this alluvial fan formed where a stream leaves a canyon and expands outward into Death Valley. The bent line in the lower left is a road that cuts across the lower part of the fan. Infrequent floods on this fan drain into the white salt flats of Death Valley, seen in the upper and lower left. *(Michael Collier)*

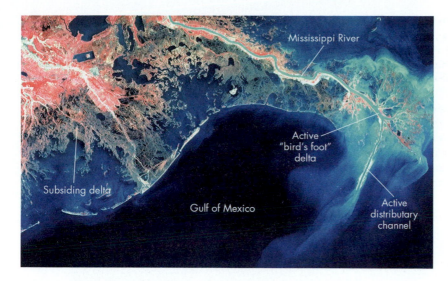

◄ FIGURE 6.6 Mississippi Delta In this false-color satellite image of the Mississippi River Delta, living vegetation appears red, sediment-laden waters are white or light blue, and deeper water is darker blue because it contains less suspended sediment. The main active distributary is the light-blue channel that starts in the top-left center and slopes toward the middle right. In the far-right center, the main distributary branches into several channels that make the shape of a "bird's foot." This shape indicates that river flow is stronger than the wave and current action in the Gulf of Mexico. Distance across the image is about 180 km (~112 mi.). *(University of Washington Libraries Special Collections SL-23–40-H)*

shaped deposit known as a *delta* if it extends into a larger body of water (Figure 6.6).

River flow on alluvial fans and deltas is different from flow in a river valley and its adjacent area. Rivers reaching base level and entering an alluvial fan or delta often split into a system of *distributary channels*. That is, the main river divides into several channels that carry floodwaters and sediment to different parts of the fan or delta. Furthermore, these channels may change position rapidly during a high-flow event or from one flow event to the next, creating a system and channels that are difficult to predict.

CHANNEL PATTERNS AND FLOODPLAIN FORMATION

Most physical features of rivers and floodplains result from the interaction of flowing water and moving sediment. Such features include various forms of sediment accumulation and the river channel itself.

A bird's-eye view of a stream channel as seen from an airplane reveals that streams and rivers develop distinctive **channel patterns**. The most common patterns are *meandering*, a single-thread channel similar to the curves of a moving snake, and *braided*, a multithread channel where there are two or more channels that unite and divide as the river flows downstream.

Meandering Pattern Many rivers have curving channel bends called *meanders* that migrate back and forth across the **floodplain** (relatively flat land adjacent to a river that is produced by river processes). Floodplains are a major river landform that develop over a period of years to decades and much longer (Figure 6.7). It is interesting that although the

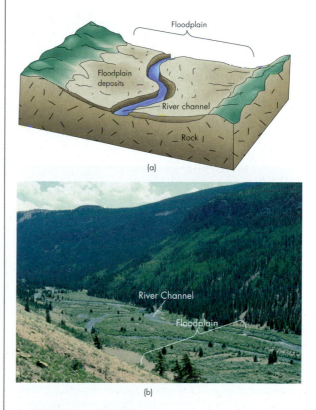

▲ FIGURE 6.7 Floodplain (a) Diagram illustrating the location of a river's floodplain. Most floodplains have deposits that are finer grained than those found in, and immediately adjacent to, the channel. (b) Floodplain of the Rio Grande near its headwaters in Colorado. *(Edward A. Keller)*

meandering behavior of rivers has been studied for nearly a century by many hydrologists, geologists, engineers, and physicists, including Albert Einstein, we don't know for sure why rivers meander. We do, however, know a great deal about how water flows in a meandering stream.

Canoeists have long known that water moves faster along the outside of a meander bend during low to relatively high flows. The fast-moving water erodes the riverbank on the outside of the bend to form a steep or near-vertical slope known as a *cutbank* (Figure 6.8). In contrast, slower water on the inside of

a meander bend deposits sand and sometimes gravel to form a *point bar*. Continual erosion of the cutbank and deposition on the point bar cause each meander bend to migrate laterally in a different direction. This process is important in constructing and maintaining some floodplains (see Figure 6.7). Floodplains are also built during *overbank flow*, a condition that develops when rising water spills over the riverbank onto the floodplain. Flow that expands out of its channel is slowed and then deposits fine sediment, such as very fine sand, silt, and clay, which builds up the floodplain. Much of the sediment transported in rivers is

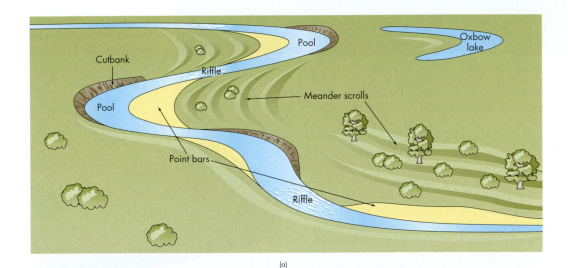

(a)

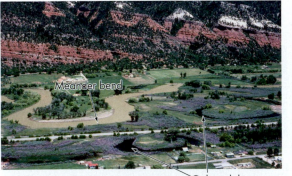

(b)

(c)

⋀ FIGURE 6.8 Characteristics of a Meandering River

(a) Idealized diagram of a meandering stream showing important features. Historical migration of meander bends is commonly indicated by low, curving, vegetation-covered ridges called meander scrolls. The scrolls on the point bar in the left center of the diagram indicate that it migrated from right to left. (b) Animas River north of Durango, Colorado, at high flow with brown muddy water filling the channel. The dark-blue abandoned channels that form oxbow lakes in the lower right indicate that the river has meandered back and forth across the valley to build its floodplain. In contrast to the river, the lakes have little suspended sediment in the water. *(Robert H. Blodgett)* (c) The sandy point bar on the left side of (b) exposed at low flow. A farmer has placed junked cars on the cutbank of a meander bend in the right center to slow the erosion of his pasture. *(Robert H. Blodgett)*

periodically stored by deposition in the channel and on the adjacent floodplain.

Meandering channels often contain a series of regularly spaced pools and riffles (Figure 6.9). *Pools* are deep areas produced by scour, or erosion, at high flow, and *riffles* are shallow areas formed by sediment deposited at high flow. At low flow, pools have relatively deep, slow-moving water while riffles have shallow, fast-moving water. These changes in water depth and velocity along a stream create different *habitats*, that is, environmental conditions in which organisms live.

The variety of environmental conditions found in these habitats increases the diversity of aquatic life. For example, fish may feed in riffles and seek shelter in pools. Also, pools have different types of insects than are found in riffles.

Braided Pattern Overall, braided channels (Figure 6.10) tend to be wide and shallow compared to meandering channels. During a raft trip down the braided Chilkat River in Alaska, for example, the raft may frequently run aground on gravel bars despite the

◄ FIGURE 6.9 Pools and Riffles
Well-developed pool and riffle sequence in Sims Creek near Blowing Rock, North Carolina. Deep pools lie under the smooth, reflective water surface in the center and lower right; shallow riffles lie under the rough, nonreflective water in the far distance and left foreground. *(Edward A. Keller)*

(a)   (b)

∧ FIGURE 6.10 Braided Rivers
(a) The braided pattern of the North Saskatchewan River, Alberta, Canada, is formed by shallow channels flowing around and across numerous sand and gravel bars and islands. Coarse bed-load sediment of the river comes from melting glaciers in the Canadian Rocky Mountains. *(University of Washington Libraries Special Collections KC9330)* (b) Surface view of a braided channel near Granada, southern Spain, with subdividing channels, a steep gradient, and coarse gravel. Distance near the channel in the foreground is about 7 m (~23 ft.). *(Edward A. Keller)*

guide's best efforts. When that occurs, the guide may yell "shuffle," and the passengers bounce up and down until the raft floats back into the main channel. Rivers similar to the Chilkat are likely to have a braided pattern where the stream has a steep gradient and abundant, coarse bed-load sediment. These conditions are often found in areas where tectonic processes are rapidly uplifting the land surface and where rivers receive water and sediment from melting glaciers.

River Systems A drainage basin with the sediment and water it transports makes up a **river system** that consists of three distinct zones, each with different channel characteristics[1] (Figure 6.11):

> Zone 1: The zone of water and sediment production, also called the zone of production. This is generally in the upper parts of the system where topography is steeper and more precipitation falls. Water velocity is fast, and downcutting and erosion occur. Channels may be in steep-sided, V-shaped valleys with waterfalls and rapids controlled by hard rocks.

> Zone 2: The zone of transport, where water and sediment are conveyed by a river with a broad valley and floodplain, created as the river moves laterally rather than downcutting a steep valley. Sediment is frequently deposited in river bars or on the floodplain, but such deposition is temporary in the history of a river. The channel pattern may be braided or meandering or some combination of both.

> Zone 3: The zone of deposition, where water velocity slows near base level and sediment is deposited. This may be an alluvial fan, lake, or, more commonly, a delta.

Understanding the characteristics of these three zones is basic to evaluating a river system. Life is very different in each of the zones; where you live does make a difference.

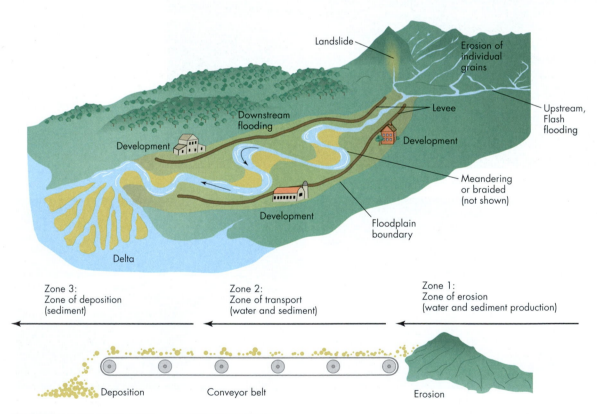

∧ FIGURE 6.11 Idealized Diagram of the River System
A river system consists of three zones: water and sediment production, transport of water and sediment, and deposition of sediment. *(Schumm, S. A. 1977. The fluvial system. New York: John Wiley & Sons; Kondolf, M. G. 1994. "Geomorphic and Environmental Effects of Instream Gravel Mining,"* Landscape and Urban Planning *28: 225–43)*

Stream channels flowing through sediment, such as sand, gravel, or mud, rarely stay fixed in the same place for a long period of time. Either erosion or deposition causes the stream channel to shift position. These shifts can be continual or they can be abrupt, such as when a channel shifts to an entirely new location during a flood. In abrupt changes, a stream may abandon part of its existing channel and form a new one nearby. This process, known as *avulsion*, can have unexpected consequences for people living in a stream valley.

Having presented some of the characteristics and processes of the flow of water and sediment in rivers, we turn to the process of flooding in greater detail.

(a)

6.1 CHECK your understanding

1. What is a drainage basin?
2. What is the difference between dissolved load, suspended load, and bed load?
3. Define alluvial fan and delta.
4. What are the two main channel patterns? How and why do these differ?
5. Describe the three zones of a river system.

6.2 Flooding

INTRODUCTION

The natural process of overbank flow is termed **flooding** (Figure 6.12). Most river flooding is related to the amount and distribution of precipitation in the drainage basin, the rate at which the precipitation infiltrates (soaks) into the earth, and how quickly surface runoff from that precipitation reaches the river.

The amount of moisture in the soil at the time the precipitation starts also plays an important role in flooding. Water-saturated soil is similar to a wet sponge that cannot hold any additional moisture. If significant precipitation falls on a saturated drainage basin, flooding will occur. If the same amount of precipitation falls on a dry basin, the soil may be able to absorb considerable moisture and, thus, help prevent flooding.

In mountainous areas (zone 1; see Figure 6.11) and in locations at higher latitudes, such as Alaska, Canada, and the northern contiguous United States, floods are common during early spring and midwinter thaws. Rain falling on frozen ground or accumulated snow, or

(b)

∧ FIGURE 6.12 Floodplain Flooding from Snowmelt Gaylor Creek, Yosemite National Park, California, during spring snowmelt. (a) In the morning, water stays within the stream channel. *(Edward A. Keller)* (b) In the afternoon, during daily peak snowmelt, overbank flow covers the floodplain. *(Edward A. Keller)*

the rapid melting of ice and snow, can cause flooding. In these areas, large masses of floating ice can create ice jams across rivers that temporarily dam their flow. Floodwaters can back up behind the ice jams or the jams can break apart abruptly, causing flooding downstream. For example, an ice jam on the Winooski River in Vermont in 1992 threatened to flood Montpelier, the state capital, which was located upstream of the jam. Flooding can also occur in zones 2 or 3 (see Figure 6.11), downstream of zone 1. Large downstream floods may last weeks to months and be regional in extent.

A flood can be characterized in several ways. One is the *flood discharge*, defined as the discharge of the stream

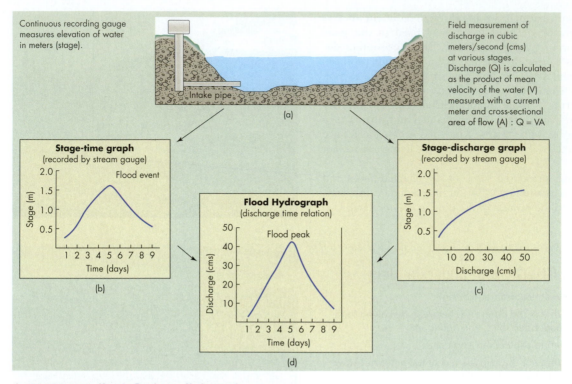

∧ FIGURE 6.13 How to Produce a Hydrograph
Field data (a) consists of a continuous recording of the water level, or stage, which is used to produce a stage-time graph (b). Field measurements at various flows also produce a stage-discharge (c). Graphs (b) and (c) are combined to produce the final hydrograph (d).

at the point where water overflows the channel banks. Floods can also be defined by the height of water in the river, referred to as the *stage* of the river. A graph showing changes in stream discharge, water depth, or stage over time is called a *hydrograph*.

The term *flood stage* is frequently used to indicate that the elevation of the water surface has reached a level likely to cause damage to personal property. This definition is based on human perception, so the elevation that is considered flood stage depends on human use of the floodplain.[2] Therefore, the size of a flood (total discharge) may or may not coincide with the extent of property damage. The relationships among the stage, discharge, and recurrence interval of floods are described and illustrated in the following discussion.

MAGNITUDE AND FREQUENCY OF FLOODS

Flooding is intimately related to the amount and intensity of precipitation and runoff. Catastrophic floods reported on television and in newspapers often are produced by infrequent, large, intense storms. Smaller

floods or flows may be produced by less intense storms that occur more frequently. All flow events can be measured or estimated from stream-gauging stations to produce hydrographs (Figure 6.13) and arranged in order of their magnitude of discharge, generally measured in cubic meters per second (Figure 6.14). This list of annual peak (largest) flows can be plotted on a discharge-frequency curve by deriving the recurrence interval R for each flow from the relationship.

$$R = (N + 1) \div M$$

where R is a **recurrence interval** in years, N is the number of years of record, and M is the rank of the individual flow within the recorded years (Figure 6.14).[3] For example, the highest flow for 9 years of data for the Patrick River is approximately 280 m³ (~9900 ft.³) per second, and that flow has a rank $M = 1$.[4] The recurrence interval of this flood is

$$R = (N + 1) \div M = (9 + 1) \div 1 = 10,$$

which means that a flood with a magnitude equal to or exceeding 280 m³ (~9900 ft.³) per second can be

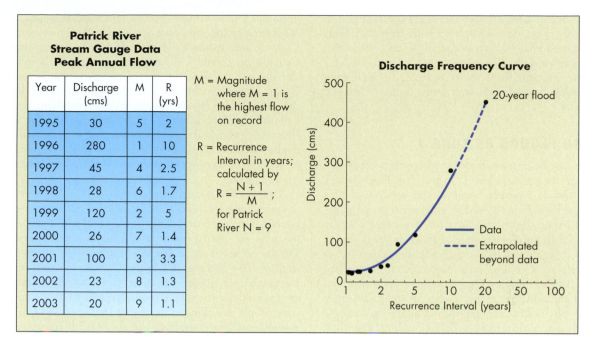

| Patrick River Stream Gauge Data Peak Annual Flow | | | |
Year	Discharge (cms)	M	R (yrs)
1995	30	5	2
1996	280	1	10
1997	45	4	2.5
1998	28	6	1.7
1999	120	2	5
2000	26	7	1.4
2001	100	3	3.3
2002	23	8	1.3
2003	20	9	1.1

M = Magnitude where M = 1 is the highest flow on record

R = Recurrence Interval in years; calculated by $R = \dfrac{N+1}{M}$; for Patrick River N = 9

^ FIGURE 6.14 Example of a Discharge-Frequency Curve
To make a discharge-frequency graph for the Patrick River, nine years of discharge measurements in cubic meters per second (cms) were ranked (1 though 9) on the basis of the size (magnitude) of the largest flow each year. The recurrence interval, or frequency, of the largest annual flow was then calculated using the formula shown in the text. Data from the table on the left were then plotted on the graph on the right. The curve was then extended (extrapolated) to estimate the discharge of the 20-year flood, which was 450 m^3 (~15,900 ft.^{3}) per second. *(Based on Leopold, L. B., U.S. Geological Survey Circular 554, 1968)*

expected about every 10 years; we call this a 10-year flood. The probability that the 10-year flood will occur in any one year is 1 ÷ 10 or 0.1 (10 percent). Likewise, the probability that the 100-year flood will occur in any year is 1 ÷ 100 = 0.01, or 1 percent.

Extending, that is, extrapolating, the discharge-frequency curve is risky. The curve shouldn't be extended much beyond twice the number of years for which there are discharge records. Studies of many streams and rivers show that channels are formed and maintained by *bankfull discharge*, defined as a flow with a recurrence interval of 1.5 to 2 years. Applying this concept to the Patrick River (Figure 6.14), the bankfull discharge with an interval of 1.5 years was a discharge of 27 m^3 (~950 ft.^{3}) per second. Bankfull is the flow that just fills the channel. Therefore, we can expect a stream to emerge from its banks and cover part of the floodplain with water and sediment once every year or so.

As longer flow records are collected, we can more accurately predict floods. However, designing structures for a 10-year, 25-year, 50-year, or even 100-year flood, or, in fact, any flow, is a calculated risk, since

predicting such floods is based on a statistical probability. For many streams, the flow record is far too short to accurately predict the magnitude and frequency of large floods. In the long term, a 25-year flood happens on the average of once every 25 years, but two 25-year floods could occur in any given year, as could two 100-year floods![5] Finally, very large but not unprecedented floods may not have occurred during the period of stream flow record. As a result, the flow frequency analysis may not be of much use in dealing with the "so called extreme events." To consider these flows we need to look to the historic and prehistoric record that is often present in old news stories and the sediment record the Earth provides. The flood record in the sediment in a valley or on a floodplain is the real deal – never fake news. The Earth does not lie. What is needed is better science to incorporate extreme floods (too often neglected in flood hazard analysis as unprecedented) into our flood hazard analysis. The 2017 Houston flood that opened this chapter was considered unprecedented with a recurrence interval of the precipitation of hundreds of years or more. It was not unprecedented, similar rainfall events have occurred in Texas several times in past 100 years, often associated with tropical storms (hurricanes) that stall and dump 100 cm (~40 in.) of rain in a few days. What was unprecedented in 2017 was the horrific damages as a result of increasing population, urbanization, and poor land-use practices.

FLASH FLOODS OF ZONE 1

Floods can be characterized by where they occur in a drainage basin (see zones 1, 2, and 3 in Figure 6.11). **Flash floods** typically occur in the upper part of a drainage basin (zone 1) and in some small drainage basins of tributaries to a larger river. They are generally produced by intense rainfall of short duration over a relatively small area. If these floods are sudden and of relatively great volume, peak discharge can be reached in less than 10 minutes. Flash flooding is most common in arid and semiarid environments, in areas with steep topography or little vegetation, and following breaks of dams, levees, and ice jams (see Case Study 6.1: Survivor Story).

Although flash floods do not generally cause flooding in the larger streams, they can join downstream and may be quite severe locally. For example, a high-magnitude flash flood occurred in July 1976 in the Front Range of the Rocky Mountains in Colorado. This flood was caused by a complex system of thunderstorms that swept through several canyons west of Loveland and delivered up to 25 cm (~10 in.) of rain in a few hours. The flash flooding killed 139 people and inflicted more than $35 million in damages to

highways, roads, bridges, homes, and small businesses. Most damage and loss of life occurred in Big Thompson Canyon, where hundreds of residents, campers, and tourists were caught with little or no warning. Although the storms and flood were rare events in the Front Range canyons, comparable floods have occurred in the past.[6–8] Flash floods returned to the Front Range of Colorado in September 2013. Record rainfall over a short time produced record floods in many streams, including Boulder Creek in Boulder, Colorado. The area of flooding covered several thousand square kilometers (square miles).

Over 1,500 homes were destroyed (17,000 were damaged), many bridges were washed out, and cars were washed into creeks and rivers (Figure 6.15). Many people who die during flash floods are in automobiles. Deaths occur when people attempt to drive through shallow, fast-moving floodwater. A combination of buoyancy and the strong lateral force of the water sweeps automobiles off the road into deeper water, trapping people in sinking or overturned vehicles. Most automobiles will be carried away by 0.6 m (~2 ft.) of water. More than ten thousand people were evacuated during the 2013 Colorado flash floods, and about 1,000 were airlifted to safety. Eight people were killed as a direct result of flooding, and damages will likely exceed $100 million.

DOWNSTREAM FLOODS OF ZONE 2

It is the large **downstream floods** in zone 2 that often make national television and newspaper headlines. In 2004, heavy rains from a series of hurricanes and

(a)

(b)

⌃ FIGURE 6.15 Heavy Rain and Flooding in Colorado
(a) Three vehicles crashed into a creek after the road washed out from beneath them in Broomfield, Colorado, September 12, 2013, in heavy flooding. *(Andy Cross/The Denver Post via Getty Images)* (b) A road crew works on a stretch of highway washed out by flooding along the South Platte River in Weld County, Colorado, near Greeley, Saturday, September 14, 2013. *(AP Photo/John Wark)*

Flash Flood

Jason Lange and his four companions were all set for a leisurely canoe trip down Santa Elena Canyon in Big Bend National Park in southwestern Texas. And there wasn't any reason to suspect it would be anything but that. The five University of Wisconsin–Whitewater students were in Big Bend during their spring break, and as Lange says, "it was supposed to be the dry season."

They set off on their planned two-day trip on March 20, and for the first day the waters were quiet, as is typical for early spring. Oftentimes, the water was so shallow that they were forced to carry their boats, and Lange estimated that it never got much deeper than 1 m (~3 ft.).

They had stopped for lunch on the second day about a quarter of a mile from the entrance to Santa Elena Canyon when, without warning, the group saw a "wall of water coming down the river" more than 2 m (~7 ft.) high.

"It literally sounded like a train coming," Lange said. "It immediately made the river explode. The river was casually running along, and all of the sudden it was whitewater."

The area where the group stopped was quickly being consumed by the rising water level, and they had no choice but to get in their canoes and enter the canyon (Figure 6.1.A).

The students would later learn that the flash flood that caught them unprepared had originated in Mexico and caused a 6 m (~20 ft.) depth gauge to overflow, leading park officials to estimate that it was more than 7 m (~23 ft.) tall and had boasted Class V rapids. "The canoes we were in were not set up for those kinds of rapids whatsoever," Lange said.

Once in the water, the group quickly ran into trouble. Almost immediately one of Lange's companions, Nick Gomez, capsized in his one-person canoe. After joining Lange and one of their other companions, the five set off again down the rapids in a pair of two-person canoes.

After a second boat flipped on an unexpected drop, the group stopped at a gravel bar to regroup. Repose, however, was brief; the very land beneath their feet quickly disappeared under the rising water. Within minutes, both remaining boats had capsized, and the five students were scattered.

Amazingly, Lange's original partner in his canoe, Lisa Chowdhury, managed to cling to one canoe all the way down the river. The remaining four were split into two groups, battling hypothermia through the night. A hiker finally spotted one party the next morning, and park rangers rescued the group, using spotter planes and jet boats.

Once all the students had learned that their companions were safe, Lange said the entire experience was thrilling beyond belief. "Once everyone got rescued, it was the most insane rush," he said. And, what's more, they want to go back.

— CHRIS WILSON

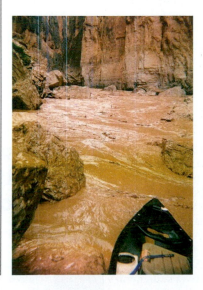

◄ **FIGURE 6.1.A Flash Flood on the Rio Grande**
Lange was able to snap a quick photograph as he and his companions were tossed into the rockslide rapids. The rocks pictured and his canoe were submerged within 2 minutes. This is a photograph of the Rio Grande in Santa Elena Canyon, Big Bend National Park, Texas-Mexico border. *(Jason Lange/Big Bend National Park)*

Λ FIGURE 6.16 Downstream Flooding on the Ohio River
Marietta, Ohio, experienced its worst flooding in 40 years from the Ohio River because of heavy rains from the remnants of Hurricane Ivan in 2004. The town used snowplows to clear the mud deposited by the floodwaters. *(AP Photo/The Marietta Times, Mitch Casey)*

tropical storms in the eastern United States caused record and near-record flooding. Major downstream flooding occurred because the ground remained saturated as one storm after another crossed the same drainage basins. In Pennsylvania, the Susquehanna River crested 2.5 m (~8 ft.) above flood stage for one of the five greatest floods in history. Downstream flooding on the Ohio River in Marietta, Ohio, was the worst in 40 years (Figure 6.16), and flooding in Atlanta, Georgia, set all-time records.

Downstream floods of zone 2 are often associated with inundation of the floodplain. Zone 2 floods may cover a wide area and are usually produced by storms of long duration that saturate the soil and produce increased runoff. Although flooding in small tributary drainage basins is generally limited, the combined runoff from thousands of tributary basins produces a large flood downstream. This kind of flood is characterized by the downstream movement of the flood crest (the peak height of the floodwaters) with a large rise and fall of discharge at a particular location.[9] An example of a downstream flood comes from the Chattooga and Savannah Rivers of Georgia and South Carolina (Figure 6.17a). As a flood crest on this river system migrated 257 km (~160 mi.) downstream, it took a

progressively longer time for the water to rise and fall (Figure 6.17b). The hydrograph of the flood at the downstream gauging station in Clyo, Georgia, shows that the floodwaters took five days to reach their peak of more than 1700 m^3 (~60,000 ft.^{3}) per second.[10] Another way of looking at a flood is to examine the volume of discharge per unit area of the drainage basin (Figure 6.17c). This approach eliminates the effect of downstream increases in discharge and better illustrates the shape and form of the flood peak as it moves downstream.[11]

6.2

CASE study

Mississippi River Floods of 1973–2008: Zone 2 Floods

In 1973, spring flooding of the Mississippi River caused the evacuation of tens of thousands of people, as thousands of square kilometers of farmland were inundated throughout the Mississippi River Valley.

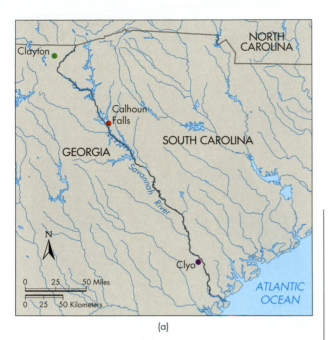

(a)

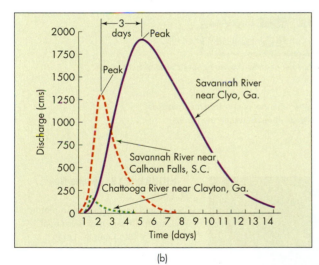

(b)

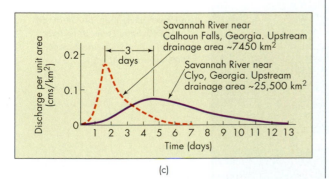

(c)

< FIGURE 6.17 Downstream Movement of a Flood Crest
Floodwaters, as they moved downstream along the Savannah River on the state line between South Carolina and Georgia. The distance from Clayton to Clyo is 257 km (~160 mi.). (a) Map of the area. (b) Volume of water passing Clayton, Calhoun Falls, and Clyo each second; the discharge increases as tributaries add more water. (c) Volume of water per unit area at the same points; flooding at Clyo lasts much longer than flooding in Clayton because of the addition of water from thousands of small tributaries. *(Based on data from Hoyt, W. G. and Langbein, W. B., Floods, Princeton, NJ: Princeton University Press, 1955)*

Fortunately, there were few deaths, but the flooding resulted in approximately $1.2 billion in property damage.[12] The 1973 floods occurred despite a tremendous investment in upstream flood-control dams on the Missouri River, the largest tributary to the Mississippi. Reservoirs behind these dams inundated some of the most valuable farmland in the Dakotas, and despite these structures, the flood near St. Louis was record breaking.[13] As impressive as this flood was at the time, it did not compare—either in magnitude or in the suffering it caused—with the flooding that occurred 20 and 35 years later.

During the summers of 1993 and 2008, the land around the Mississippi River and its tributaries experienced two of the largest floods in the last hundred years. There was more water, in both 1993 and 2008, than during the 1973 flood, which had a magnitude of flow expected on average once in 100 years.

The 1993 flood resulted from rainstorms from April through July in the region of Cedar Rapids, Iowa, where low-pressure areas produced about 90 cm (~35 in.) of rain. Flooding lasted from late June to early August and caused 50 deaths and more than $15 billion in property damage. In all, about 55,000 km² (~21,000 mi.²), including numerous towns and farmlands, were inundated with water.[14–16]

The 2008 floods resulted from winter to late spring storms that in June dumped as much as 25.5 cm (~10 in.) of rain in a 24-hour period in Indiana. Flood defenses again failed, about 24 people died, and the damage totaled around $9 billion.

The 1993 floodwaters were high for a prolonged time, putting pressure on the flood defenses of the Mississippi River, particularly **levees**, which are earthen embankments constructed parallel to the river to contain floodwaters and reduce flooding (Figure 6.18). Levees in the St. Louis region on the Mississippi, Missouri, and Illinois Rivers are shown in Figure 6.19. Notice that levees completely circle some areas on the floodplains.

Before construction of the levees, the Mississippi River's floodplain was much wider and contained extensive wetlands. Since the first levees were built in 1718, approximately 60 percent of the wetlands in Wisconsin, Illinois, Iowa, Missouri, and Minnesota—all hard hit by the flooding in 1993—have been lost. In some urban areas, such as St. Louis, Missouri, levees have been replaced with flood walls designed to protect the city against high-magnitude floods. The effects of these flood walls can be seen in a satellite image taken in mid-July 1993 (Figure 6.20). This image shows that the river is narrow at St. Louis, where it is contained by the flood walls, and broad upstream near Alton, Illinois, where extensive flooding occurred.

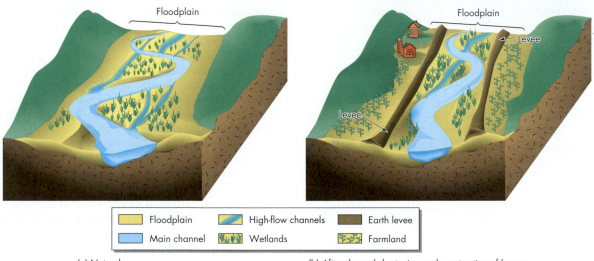

Floodplain	High-flow channels	Earth levee
Main channel	Wetlands	Farmland

(a) Natural

(b) After channel shortening and construction of levees

⋀ FIGURE 6.18 Floodplain With and Without Levees

Idealized diagram of (a) natural floodplain with wetlands. (b) Floodplain after channel is shortened and levees are constructed. Land behind levees is farmed, and wetlands are generally confined between the levees. *(Edward A. Keller)*

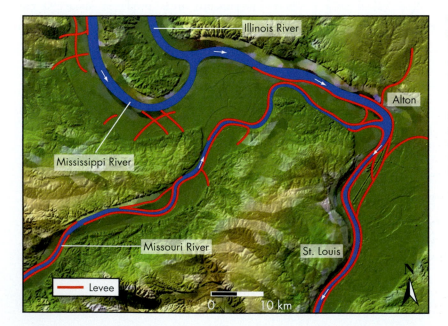

◄ FIGURE 6.19 Levees Near St. Louis

Levees (red lines) almost completely enclose the Mississippi River near the cities of Alton and St. Louis, Missouri. *(Data from R. E. Criss and T. M. Kusky (eds.), Finding a Balance between Floods, Flood Protection and River Navigation, pp. 34–40. St. Louis, MO: Saint Louis University Center for Environmental Science. 2009. www.ces.slu.edu)*

The flood walls produce a bottleneck effect—they force floodwater through a narrow channel between the walls and cause it to back up while waiting to get through. This effect contributed to the 1993 flooding upstream of St. Louis.

Despite the high walls constructed to prevent flooding, the rising flood peak came to within about 0.6 m (~2 ft.) of overtopping the flood walls at St. Louis. As the floodwaters in 1993 rose to record levels, these flood walls began to leak and required round-the-clock efforts to keep them from failing. Failure of levees downstream from St. Louis partially relieved the pressure, possibly saving the city from flooding. Levee failures were common during the flood event (Figure 6.21). In fact, the vast majority of the private levees, that is, levees built by farmers and homeowners, along the Mississippi River and its tributaries failed.[15–18] However, most of the levees built by the federal government survived the flooding and undoubtedly saved lives and property. Unfortunately, there is no uniform building code for levees, so some areas have levees that are higher or lower than others. Failures occurred as a result of overtopping and breaching, or rupturing, resulting in massive flooding of farmlands and towns (Figure 6.21).[15,17]

In addition to flood walls, other engineering structures designed to control the Mississippi have actually increased the long-term flood hazard.[19] For example, river-control structures designed to optimize the movement of loaded riverboats and barges during normal flow conditions can increase flooding at high flow. When floods overtop these structures, the floodwaters slow down and begin to back up, increasing flood height in a manner similar to the bottleneck effect created by the flood walls in St. Louis.[20]

Of course, building houses, industrial plants, public buildings, and farms on a floodplain invites disaster; too many floodplain residents have refused to recognize the natural floodway of the river for what it is—part of the natural river system. The floodplain is, in fact, produced by the process of flooding (see Figure 6.7). If the floodplain and its relation to the river are not recognized, flood control and drainage of wetlands, including floodplains, become prime concerns.

An important lesson learned from the 1993 and 2008 floods is that construction of levees provides a false sense of security. Construction of a levee does not remove land from the floodplain, nor does it eliminate the flood risk. It is difficult to design levees to withstand extremely high-magnitude floods for a long period of time. Furthermore, by constructing a levee, the result is less floodplain space to "soak up" the floodwaters.[18] The 1993 floods caused extensive damage and loss of property; in 2008, floodwaters of the Mississippi River system inundated floodplain communities once again. National trends show that flood-control expenditures and damages have increased since the mid-twentieth century. Total flood losses have dramatically increased, in spite of the

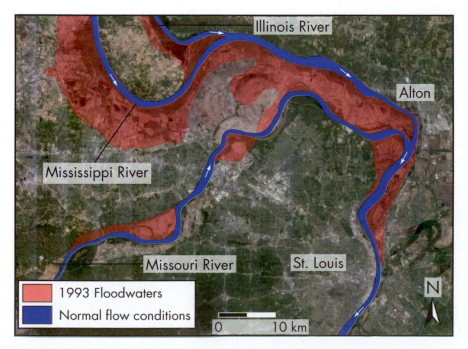

< FIGURE 6.20 Mississippi River Flood of 1993

Satellite view of the extent of flooding in 1993 where the Illinois and Missouri Rivers join the Mississippi. The 1993 floodwaters are shown in light red, and dark blue shows normal river flow. In the lower right, a series of flood walls built to protect St. Louis constricts the Mississippi. This bottleneck to the flow caused widespread flooding upstream, including in the town of Alton, Illinois, on the east bank of the river near the right center edge of the image. Alton has a notorious history of flooding.

◄ **FIGURE 6.21 Levee Failure**
(a) A breach of this levee in Illinois during the 1993 floods of the Mississippi River caused flooding of the town of Valmeyer. Rapidly flowing whitewater can be seen rushing through the breach in the lower half of this grass-covered earthen levee. (©*Getty Images/Chicago Tribune/Contributor*) (b) Damage to farmlands during the peak of the 1993 flood of the Mississippi River. (© *Getty Images/Win McNamee/Staff*)

ever-increasing cost of flood-control structures.[21] The increase is partly due to changing frequency of high-intensity, high-volume storms. Since about 1950, the extreme participation index, defined as a two-day precipitation event (total participation) that is exceeded only once in a five-year period, in the U.S. (not including Hawaii and Alaska) has increased by about 40 percent.[22] The occurrence of extreme participation events are becoming more common. The increase is consistent with global warming, insofar as warming oceans impart more energy into the atmosphere, increasing storm intensity.

In the first six to seven years after the 1993 flood, the federal government bought out many homeowners in the floodplain upstream from St. Louis.[22] Now, real estate interests have reversed the trend, and local governments are allowing floodplain development behind levees.[22] Weak floodplain regulations and government subsidies have allowed the construction of more than 25,000 new homes and other buildings on the floodplain behind new, higher levees.[23] Fortunately, not all communities along the river are shortsighted and have reevaluated their strategies concerning the flood hazard—they have moved to higher ground! Of course, this is exactly the adjustment that is appropriate.

DOWNSTREAM FLOODS OF ZONE 3: ALLUVIAL FANS AND DELTAS

Floods on alluvial fans and deltas (zone 3) are particularly hazardous because of uncertain and changing flow paths.[24] The main channel of a fan or delta during one flood may not be the same channel in a subsequent flood. Fans and deltas have distributary channels that developed as the fan or delta formed. For example, the Mississippi River Delta has had numerous distributary channels over the past several thousand years (see Figure 6.6). The present channel passes New Orleans, giving much of the charm to the riverfront city. Upstream, a potential distributary channel is closer to the Gulf of Mexico, and only a system of barriers keeps the river from moving there and inundating New Orleans. One day, the barriers may fail.

Flooding on the Delta of the Ventura River

In 1905, the philosopher George Santayana said, "Those who cannot remember the past are condemned to repeat it." Scholars may debate the age-old question of whether or not cycles in human history repeat themselves, but the repetitive nature of natural hazards, such as floods, is undisputed.[25,26]

Better understanding of the historical behavior of a river is, therefore, valuable in estimating its present and future flood hazards. Consider the February 1992 Ventura River Flood in southern California. The flood severely damaged the Ventura Beach Recreational Vehicle (RV) Resort, which had been constructed a few years earlier on an active distributary channel of the Ventura River Delta (Figure 6.22). Although the recurrence interval for the 1992 flood is approximately 22 years, earlier engineering studies had suggested that the area of the RV park had a 100-year flood recurrence interval.[27] What had gone wrong?

> Planners did not recognize that the RV park was constructed on a historically active distributary channel of the Ventura River Delta. In fact, early reports did not even mention a delta.

> Engineering models that predict flood inundation are inaccurate when evaluating distributary channels on river deltas where extensive channel filling and scouring, as well as lateral movement of the channel, are likely to occur.

> Historical documents, such as maps dating back to 1855, and more recent aerial photographs showed that the channels were apparently not evaluated. Maps rendered from these documents suggest that the distributary channel passing through the RV park was, in fact, present in 1855 (Figure 6.23).

Clearly, the historic behavior of the river was not evaluated as part of the flood-hazard assessment. If it had been, the site would have been recognized as unacceptable for development, given that a historically active channel was present. Nevertheless, necessary permits were issued for development of the park, and, in fact, the park was rebuilt after the flood.

Records show that the distributary channel carried water during 1969, 1978, and 1982. Following the 1992 flood event, the channel carried floodwaters in the winters of 1993, 1995, and 1998, again flooding the RV park. During the 1992 floods, the discharge increased from less than 25 m^3 per second (~880 ft.^{3} per second) to a peak of 1,322 m^3 per second (~46,670 ft.^{3} per second) in only about 4 hours! This rate of flow is approximately twice as much as the daily high discharge of the Colorado River through the Grand Canyon in the summer, when river rafters navigate it. This volume is an immense discharge for a relatively small river with a drainage area of only about 585 km^2 (~226 mi.^{2}). The flood occurred during daylight, and one person was killed. However, if the flood had occurred at night, the death toll would likely have been much higher.[27]

A warning system developed for the park has, so far, been effective in providing early warning of an

RV park

<FIGURE 6.22 Flooding of California's Ventura Beach RV Resort in February 1992

The RV park was built directly across a historically active distributary channel of the Ventura River Delta. The recurrence interval of this flood is approximately 22 years. A similar flood occurred again in 1995. On the left, U.S. 101, the Pacific Coast Highway, is completely closed by the flood event. *(AP Photo/Mark J. Terrell)*

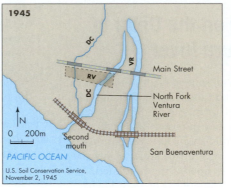

(a)

(b)

(c)

< **FIGURE 6.23 Historical Maps of the Ventura River Delta**
Distributary channel and location of the RV park are shown. (a) In 1855, a
small distributary channel (DC) was flowing through the future site of the RV
park (dashed box) and joined the main Ventura River channel (VR) near the
river's mouth. (b) By 1945, the 1855 channel had widened to flow directly to
the ocean as the North Fork of the Ventura River. A second distributary was
flowing through the western part of the future RV park and joined the North
Fork upstream of its mouth. (c) In 1989, around the time the RV park was built,
the original 1855 channel was still active as one of the major distributaries. A
levee had been built on the east side of the delta to protect San Buenaventura.
In the 1992 flood (arrows), the levee and earthen fill elevating the railroad and
U.S. 101 acted as dams to widen the extent of flooding on the delta. *(Based on
Keller, E. A., and Capelli, M. H., "Ventura River Flood of February, 1992: A Lesson
Ignored?" Water Resources Bulletin 25(5): 813–31, 1992)*

impending flood. The park, with or without the RVs
and people, is a "sitting duck." This situation was dramatically illustrated in 1995 and 1998, when winter
floods again swept through the park.[27] Although the
warning system worked and the park was successfully
evacuated, the facility was again severely damaged. A
move is now afoot to purchase the park and restore the
land to a more natural delta environment—a good
move but, as of 2017, not yet a reality.

MEGAFLOODS

Disasters resulting from natural hazards occur frequently over the surface of the Earth. Often, these disasters are local to regional in extent and cause up to
several billions of dollars of property loss, as well as loss
of human life. Less frequently, **megafloods** may cause
tens to hundreds of billions of dollars of damage and
loss of human life that goes into the thousands or even
the hundreds of thousands. Southern California from
Los Angeles to Santa Barbara receives on average

38–51 cm (~15–20 in.) of rain annually, with more to
the north, which could be plus or minus about 400 percent for a given year. Much of the rainfall in a given
winter often comes in a few storms where it might rain
13 to 18 cm (~5 to 7 in.) over several days. Less frequently, it may rain for a week at a time. The so-called
"Pineapple Express," moist air coming across the
Pacific from the Hawaiian Islands, would occasionally
make landfall in southern California, causing intense
and sometimes prolonged rains (see Chapter 10). In the
late 1990s, scientists discovered atmospheric rivers that
were, in fact, discrete zones of high moisture that could
move, as do rivers, for thousands of kilometers (Figure
6.24).[28] Recently, it was learned that the amount of
moisture is truly astounding, as some of the larger atmospheric rivers contain as much water as the Amazon
or 10 to 20 times as much water as carried by the
Mississippi River. Furthermore, these atmospheric rivers carry high winds, and when they reach high mountains, the storms are forced upward, increasing
precipitation even more.[28]

Several atmospheric rivers arrive in southern
California each year; most are moderate in extent and
bring welcome precipitation. However, we have

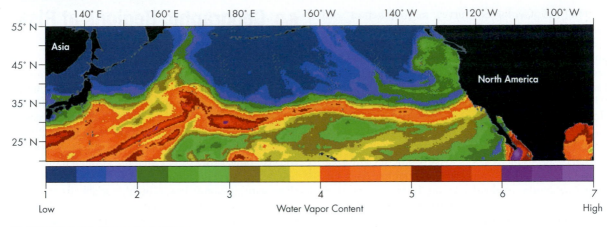

∧ **FIGURE 6.24 Atmospheric River**
A relatively narrow band of concentrated moisture in the atmosphere forms a conceptual river of water flowing
from western Pacific to the West Coast of North America bringing heavy precipitation. *(NOAA)*

recently learned that sometimes a series of atmospheric rivers will arrive over a period of a month or longer, bringing prolonged moderate to intense rainfall that can cause catastrophic flooding. When this occurs, the term that meteorologists and hydrologists now use is *megaflood*.

As we learn more about meteorology and what drives climate and link the science to the geologic record, we find that what were considered extremely rare, one-time-only events are, in fact, not nearly as singular as we thought. Megafloods, for example, occur in southern California every 200 years or so, and the last one struck during the winter of 1861–62. Stories about these floods are almost mythical in nature.[29] These atmospheric rivers and resulting storms and flooding caused havoc from San Diego to well north of San Francisco. Rainfall in the foothills of the Sierra Nevada Mountains and the Sacramento area was intense and prolonged, lasting over one month. Sacramento was flooded, and the state capital had to be moved to San Francisco for several months while Sacramento dried out. What happened in the Great Valley of California is difficult to imagine. Because the precipitation from the Sierra Nevada Mountains and the runoff were very high, part of the valley literally became a lake that extended for as much as 480 km (~300 mi.) along the valley, with widths exceeding 32–97 km (~20–60 mi.).[28]

Extensive flooding in 1861–62 occurred along the Ventura River, and much of the town of Ventura was flooded. Ventura in 1881 was strung out along the river delta area, which today is protected by a levee. However, a repeat of the 1861–62 event would probably threaten that levee, as well as other flood protection normally in place for large floods.[29]

The looming question becomes how likely is it that we will see a repeat of the 1861–62 floods. The more we learn about natural hazards and their long-term return period, the more we realize that these events, while rare, are not unexpected. This holds true for earthquakes, tsunamis, landslides, and other natural hazards. What do we know now about megafloods that could strike California and other regions in the future?

The Santa Barbara Channel is a natural collecting ground for sediment from runoff from streams and rivers. That sediment has been extensively studied for various reasons, including trying to discern past climate. A recent study on megaflood deposits in the channel found that, since approximately 107 B.C. to the present, there has been periodic deposition of massive flood deposits in the channel. The three most recent occurred in 1532, 1761, and 1861 (the flood for which we have most knowledge). The only time there were fewer floods was during the medieval warming period from about A.D. 950 to 1250, when Earth's surface was considerably warmer than what climatologists today call normal.[28]

Reading personal accounts of the megaflood in northern California, along with extensive references to southern California as well, provides sobering thoughts about what might occur in the future. The problem is that our flood programs simply are not designed for the type of flood that occurred in 1861–62. Such a flood was considered an unimaginable event that is unlikely to occur again. Unfortunately, it is likely that such floods will occur, even though the return period between events may be several hundred years.[29]

The U.S. Geological Survey, in conjunction with many other agencies and people, is working on storm scenarios for events such as the one in 1861–62. This approach allows us to work through a trial event and test our emergency response. Some people say that it is impossible to plan for a megaflood that might have total damages in excess of $500 billion. The actual megaflood scenario run for planning purposes is quite a bit smaller than the 1861–62 event. The scenario is based on a 23-day storm period in which two major storms caused by atmospheric rivers strike southern California: a 10-day storm, then a reprieve for about five days, followed by another eight-day storm. Keep in mind that, while two-thirds of the water resources in California are in the northern part of the state, three-fourths of the people live in southern California. The scenario suggests that initial damages directly to property would be on the order of $400 billion, of which less than 10 percent could be recovered by public and commercial insurance companies. The report states that if business interruptions, evacuations, and so on are taken into account, total damages could be on the order of $700 billion, about three times the predicted losses for a large earthquake on the San Andreas fault that impacts the Los Angeles area.

So, what does all this doom-and-gloom news mean? It means that extreme events are likely rather than unlikely—only the time frame may be long between such events. Realizing this, it would be prudent to think about the consequences of a future megaflood. In many cases, an option is to look more closely at flood-proofing (discussed in more detail in Section 6.7). Flood walls can be decorative and built around critical facilities such as airports, train stations, and other vulnerable places. Individual companies can also consider flood-proofing buildings with specially designed flood doors that can be closed to keep floodwaters out for a period of time. So far, scientists think that the number of megafloods will not increase significantly due to climate change, but the intensity may increase because warmer seawater would mean more water vapor, which can be translated into more rainfall.

6.2 CHECK your understanding

1. Define flooding.
2. What are flash floods and where are they likely to occur?
3. Why are downstream zone 2 floods potentially so dangerous?
4. Contrast the alluvial fans and delta flows of zone 3 with zone 2 river flows.
5. What are atmospheric rivers and megafloods?

6.3 Geographic Regions at Risk for River Flooding

Water covers about 70 percent of Earth's surface and is critical to supporting life on the planet. However, water can also cause a significant hazard to human life and property in certain situations, such as a flood. Analysis of the global impact of flooding from 1980 to 2012 suggests annual losses worldwide of about 6,000 lives lost and economic losses of $23 billion.[30] Although the number of floods and the number of people killed or affected, along with economic loss, is highly variable from region to region (e.g., Asia compared to the Americas, Europe, or Africa), the greatest losses of life and property, as well as people affected, is in Asia. This number reflects the very large populations in Asia that live along rivers in areas with a climate characterized by intense precipitation. As a result, both exposure (people and property at risk) and vulnerability (degree of loss due to exposure) are high, resulting in a relatively high risk.

At the global level, the vulnerability from 1990 to 2010 to flooding declined. This is in spite of the fact that total global impacts of flooding remain very significant and, in many cases, is rising in particular areas. Rising incomes worldwide in the past several decades has reduced vulnerability to flooding, resulting in fewer deaths and direct economic losses. Vulnerability (recall that vulnerability is the degree of loss due to exposure) has been historically high in developing countries, compared to developed countries, but that is changing. Vulnerability of developing countries compared to developed countries (high income countries) is converging (especially since about 2000) as a result of developing countries (low income countries) successfully lowering their vulnerability through a mixture of structural and nonstructural flood control and other adaption measures. The number of lives lost as a percent of exposed population in low income countries compared to high income countries has converged since about 1995. The rate of deaths in 1995 was much higher in developing countries than in developed countries. Death rates remain higher today in developing countries, but differences are much less.[30]

Floods in the United States were the number one disaster during the twentieth century; on average, 80 lives are lost each year from river flooding. Tragically, developing countries suffer much greater losses because of a lack of monitoring facilities, warning systems, adequate infrastructure and transportation systems, and effective disaster relief.

Virtually all areas of the United States are vulnerable to floods. A single flood can cause billions of dollars of property damage and hundreds of lives to be lost (Table 6.1). Not listed in Table 6.1 are the hundreds of

▼ **TABLE 6.1 Selected River Floods in the United States**

Year	Month	Location	Number of Lives Lost	Property Damage (Millions of Dollars)
1937	Jan.–Feb.	Ohio and lower Mississippi River basins	137	418
1938	March	Southern California	79	25
1940	August	Southern Virginia and Carolinas and eastern Tennessee	40	12
1947	May–July	Lower Missouri and middle Mississippi River basins	29	235
1951	June–July	Kansas and Missouri	28	923
1955	December	West Coast	61	155
1963	March	Ohio River basin	26	98
1964	June	Montana	31	54
1964	December	California and Oregon	40	416
1965	June	Sanderson, Texas (flash flood)	26	3
1969	Jan.–Feb.	California	60	399
1969	August	James River basin, Virginia (flash flood)	154	116
1971	August	New Jersey	3	139
1972	June	Rapid City, South Dakota (flash flood)	242	163
1972	June	Eastern United States	113	3,000
1973	March–June	Mississippi River	0	1,200
1976	July	Big Thompson River, Colorado (flash flood)	143	35
1977	July	Johnstown, Pennsylvania	76	330
1977	September	Kansas City, Missouri, and Kansas	25	80
1979	April	Mississippi and Alabama	10	500
1983	September	Arizona	13	416
1986	Winter	Western states, especially California	17	270
1990	Jan.–May	Trinity River, Texas	0	1,000
1990	June	Eastern Ohio (flash flood)	21	Several
1993	June–Aug.	Mississippi River and tributaries	50	15,000
1997	January	Sierra Nevada, Central Valley, California	23	Several hundred
2001	June	Houston, Texas, Buffalo Bayou (coastal river)	22	2,000
2004	Aug.–Sept.	Georgia to New York and the Appalachian Mountains	13	400
2006	June–July	Mid-Atlantic (Virginia to New York)	16	1,000
2008	June	Mississippi River System	24	9,000
2010	May	Cumberland and other rivers (Nashville, Tennessee region)	24	1,500
2013	September	Colorado Front Range (flash flood)	8	Several hundred
2015	Late May	Houston, Texas area (flash flood)	6	45
2015	September	Utah-Arizona border area (flash flood)	20	Not evaluated
2016	August	Louisiana	13	10,000
2017	August	Houston, Texas area (long duration high intensity rainfall)	~50	100,000
2018	May	Ellicot City Maryland (flash flood)	2	Several million
2018	August	Hawaii Islands (flooding Hurricane Lane)	4	Prop damage >3 billion (3,000 million)

small drainage basins that experienced flooding during the time period covered by the table. For example, central Texas had six major episodes of flash flooding during the five-year period, with damage totals exceeding $10 million.[31] By realizing how many more areas are at risk, we can recognize how serious the flood hazard can be and how the hazard might be minimized.

6.3 CHECK your understanding

1. Why are flood losses so great in Asia?
2. Examine Table 6.1. Which flood caused the most deaths? Which caused the greatest economic loss? Can you make any generalizations about zone 1 and zone 2 floods?

6.4 Effects of Flooding and Linkages between Floods and Other Hazards

The effects of flooding may be primary, that is, directly caused by the flood, or secondary, caused by disruption and malfunction of services and systems because of the flood.[11] Primary effects include injury, loss of life, and damage caused by swift currents, debris, and sediment to farms, homes, buildings, railroads, bridges, roads, and communication systems. Erosion and deposition of sediment during a flood may also involve considerable loss of soil and vegetation.

Secondary effects may include short-term pollution of rivers, hunger and disease, and displacement of persons who have lost their homes. Failure of wastewater ponds, treatment plants, sanitary sewers, and septic systems often contaminate floodwaters with disease-causing microorganisms. For instance, in June 1998, flooding from record rainfall in southern New England caused partially treated sewage to float into Boston Harbor, and many areas of Rhode Island's Narragansett Bay were closed to swimming and shell fishing because of sewage-contaminated floodwaters.

Several factors affect the damage caused by floods:

> Land use on the floodplain
> Depth and velocity of floodwaters
> Rate of rise and duration of flooding
> Season of the year in which flooding takes place
> Quantity and type of sediment transported and deposited by floodwaters
> Effectiveness of forecasting, warning, and evacuation.

In general, commercial and residential properties may experience more damage than land used for farming, ranching, or recreation. On the other hand, long-duration flooding during the growing season may completely destroy crops, whereas the same flooding of cropland during winter months may be less damaging. For downstream floods, accurate flood forecasts by the National Weather Service sometimes provide the warning time needed to build temporary levees, remove property, or evacuate residents from the hazard area.

Floods can be a primary effect of hurricanes (Chapter 10) and a secondary effect of earthquakes (Chapter 3) and landslides (Chapter 7). Although it may seem counterintuitive, floods may also cause fires in urban areas. Floodwaters may produce shorts in electric circuits and erode and break natural gas mains, resulting in dangerous fires.[11] For example, the 1997 flood in Grand Forks, North Dakota, caused a fire that burned part of the city center. Floods may also contribute to coastal erosion, especially in estuaries and deltas where high water levels may allow storm waves to reach further inland (Chapter 11). River processes in general may also lead to landslides where stream banks are eroded.

6.4 CHECK your understanding

1. What are the major factors that influence the damage a flood causes?
2. Differentiate, with examples, the primary and secondary effects of flooding.

6.5 Natural Service Functions of Floods

Although flooding is considered to be a natural hazard and sometimes a disaster, it is important to remember that "flooding" is no more than the natural process of overbank flow. It becomes a hazard only when people live or build structures on the floodplain or try to cross a flooding river. In fact, there are many benefits to periodic flooding. Floods provide water resources, transport fertile sediment for farming, benefit aquatic ecosystems, and, in some cases, help keep land above sea level.

WATER RESOURCES AND FLOODING

Water security for humans and ecosystems is one of the major environmental concerns of the twenty-first century. Surface water in the form of rainfall and snow supplies much of the runoff that fills our reservoirs and infiltrates the land to renew groundwater resources. Runoff (especially in the southwestern USA) often comes from high-intensity rainfall that produces high-flow events (floods) or snowmelt that also produces high amounts of spring and summer runoff. Thus, the process of flooding upstream of reservoirs is an important source of water that is controlled and released for agriculture and of water for human consumption. Water from reservoirs may be released to flow down river valleys to support fisheries and augment groundwater resources as it seeps into the river channel. Water supply is intimately related to the concept of the **Water–Energy–Food Nexus** that consists of the complex linkages between water, energy, and agriculture (Figure 6.25). Water supply from runoff to rivers is stored in reservoirs where it provides a number of services, such as helping provide the water supply for human consumption, energy for power, and water to irrigate crops. Increasing human population in a changing climate is raising concerns about water, energy, and food security. The water–energy–food nexus is proving to be a useful concept in the discussion of resource use from the regional to global scale. The nexus is linked to the flood hazard because as we plan the use of water for energy production, urban water supply, and agriculture, the level of flood control from reservoirs is constantly changing. If a reservoir is managed primarily for flood control, then the water for other uses such as agriculture will be impacted as lake level is kept low to accommodate storage space for unexpected upstream floodwater or change in snow runoff. If the reservoir is managed for agriculture, then storage space may be reduced early in the season to provide late season water for irrigation, which lowers the flood protection aspect.

FERTILE LANDS

Floodplains are built by floods. As the river overflows its banks, the velocity of flow decreases, and fine sand, silt, clay, and organic matter are deposited on the floodplain. These periodic deposits explain why floodplains are some of the most fertile and productive agricultural areas in the world. Recognizing the value of floods, the ancient Egyptians planned their farming around regular flooding of the Nile. They learned that the higher the flood, the better that year's harvest would be, and even referred to flooding as "The Gift of the Nile." Unfortunately, with completion of the Aswan Dam in 1970, the river's annual floods in Egypt have effectively been stopped. Now farmers must use fertilizer to successfully grow crops on what was once naturally fertile land.

RIPARIAN AND AQUATIC ECOSYSTEMS

Riparian (of the river) refers to the banks and other landforms that are heavily influenced by a river. The riparian area along a stream or river has different vegetation and animals than adjacent upland areas. Floods form and maintain stream banks and, thus, the riparian zone. Floods help flush out stream channels, scouring pools and debris, such as boulders, logs, and tree branches that may have accumulated. Such events generally have a positive effect on stream and riparian ecosystems. Floods also sweep nutrients and other food supplies downstream, potentially improving the environment for downstream aquatic organisms.

SEDIMENT SUPPLY

In some cases, flooding is needed simply to keep the elevation of a landmass above sea level. For example, the Mississippi River Delta in southeastern Louisiana is built

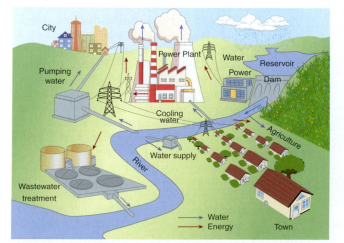

◄ FIGURE 6.25 Water–Energy–Food Nexus
Linkages between stored river water and service function that help produce water, energy and food security. *(Modified from US DOE)*

up of sediment deposited as the Mississippi River repeatedly overflowed its banks through time. Construction of levees along the Mississippi has all but eliminated flood sedimentation, with the result that much of the delta is now slowly subsiding (see Chapter 8). Some areas, such as the city of New Orleans, are beneath sea level and are regularly threatened by hurricanes that cause both coastal and river flooding (see Chapter 10).

6.4

CASE study

Experimental Floods on the Colorado River

The Grand Canyon of the Colorado River (Figure 6.26) provides a good example of the natural service functions provided by periodic flooding and delivery of sediment. In 1963, the Glen Canyon Dam was built upstream from the Grand Canyon (Figure 6.27). Construction of the dam drastically altered both the natural pattern of flow and channel processes downstream—from a hydrologic viewpoint, the Colorado River was tamed.

∧ FIGURE 6.26 The Colorado River in the Grand Canyon at Marble Canyon *(National Park Service)*

Before the Glen Canyon Dam was built, the river reached maximum flow in May or June during the spring snowmelt, and flow receded during the remainder of the year, except for occasional flash floods caused by upstream rainstorms. During periods of high discharge, the river had a tremendous capacity to transport sediment and vigorously scour its channel. This competence was evident when the river moved large boulders from the rapids. Shallow areas in the river, which become rapids, develop where the river flows over an alluvial fan or debris flow deposits delivered from tributary canyons. As the summer low flow approached, the river was able to carry less sediment and deposited sand and gravel along the channel to form large bars and terraces, known as *beaches* to people who raft the river.

After the dam was built, the mean annual flood, determined by averaging the highest flow each year, was reduced by about 66 percent, and the 10-year flood was reduced by about 75 percent. The dam also trapped sediment, greatly reducing the sediment load immediately downstream from the dam. The sediment load farther downstream is less reduced because tributary channels continue to add sediment to the channel.[32]

The Colorado River's changed flow conditions in the Grand Canyon have greatly altered both the channels and the banks. Rapids may be becoming more dangerous because large floods are no longer moving many of the boulders from shallow areas of the channel. In addition, some of the large sandbars are disappearing because the river is deficient in sediment below the dam. The river is eroding this valuable habitat.

Changes in the river flow, mainly the loss of high flows, have also shifted vegetation. Before the dam was built, three nearly parallel belts of vegetation were present on the slopes above the river. Adjacent to the river and on sandbars, plant growth was scoured by yearly spring floods. Above the high waterline of the annual floods were clumps of thorned trees, such as mesquite and acacia, mixed with cactus and Apache plume. Higher yet was a belt of widely spaced brittlebush and barrel cactus. For the first 20 years following its completion, the dam significantly reduced the spring floods. The absence of high, scouring flows allowed plants not formerly found in the canyon, including tamarisk (salt cedar) and indigenous willow, to become established in a new belt along the riverbanks.

In June 1993, a record snowmelt in the Rocky Mountains forced the release of about three times the normal amount of water from the dam. This release was about the same volume as an average spring flood before the dam was built. The resulting flood scoured sediment from the riverbed and riverbanks, which replenished the sandbars, and it scoured out or broke off some of the tamarisk and willow trees. This large

◄ FIGURE 6.27 The Colorado River Basin
Glen Canyon Dam, just south of the Arizona-Utah state line, divides the Colorado River basin (shaded areas) in half for management purposes. Flaming Gorge Reservoir in Wyoming and Lake Powell in Utah store runoff in the Upper Basin, and Lake Mead on the Nevada–Arizona state line stores runoff in the Lower Basin. The delta, at the head of the Gulf of California, was once a large wetland area in Mexico. Today, it is severely degraded because of the diversion of Colorado River water for other uses in the United States.

release of water was, thus, beneficial to the river, and it emphasizes the importance of large floods in maintaining the system in a more natural state.

Three years later, a "test flood" was released from the dam as an experiment to redistribute the sand supply. The experimental flood formed 55 new beaches and added sand to 75 percent of the existing beaches. It also helped rejuvenate marshes and backwaters, which are important habitats for native fish and some endangered species. The experimental flood was hailed a success, although a significant part of the new sand deposits subsequently eroded away.

Although the 1996 test flood distributed sand from the channel bottom and banks to sandbars, tributary streams added little new sand, since they were not in flood. The sand scoured from the river below the dam is a limited, nonrenewable source that cannot resupply sandbars on a sustainable basis. Two additional experimental floods were run down the canyon in 2004 and 2008 to learn more about river processes and sand supply. Evaluation of the hydrology of the Colorado River and tributaries suggests that the opportunity to replenish sand on the beaches occurs periodically, but high flow releases must be carefully timed to restore or recreate river flow and sediment transport conditions which are as close as possible to those that existed prior to the construction of Glen Canyon Dam.[33,34]

One final effect of the Glen Canyon Dam has been to increase the number of people rafting through the Grand Canyon. Although rafting is now limited to about 20,000 people annually, the rafters' long-range impact on canyon resources is bound to be appreciable. Prior to 1950, fewer than 100 explorers and river runners had made the trip through the canyon.

We must concede that the Colorado River is a changed river. Despite the three experimental floods that reversed some of the changes, river restoration efforts cannot be expected to return the river to what it was before construction of the dam.[33–35] On the other hand, better management of the flows and sediment transport will improve and help maintain the river ecosystem.

6.5 CHECK your understanding

1. What are the main natural service functions of flooding?

2. River flow contributes to our water resources, but also causes damaging flooding. Discuss the implications of this dichotomy.

3. What is the water–energy–food nexus, and why is it important to understanding the flood hazard?

4. What were the main purposes of the three experimental floods through the Grand Canyon?

6.6 Human Interaction with Flooding

Unlike some other natural hazards, human activity can significantly affect river processes, including the magnitude and frequency of flooding. Land-use change and dam construction may affect a stream's sediment supply, which can, in turn, alter its gradient and the shape of its channel. Urbanization, with its addition of paved areas, buildings, and storm sewers, greatly affects the flood hazard of an area.

LAND USE CHANGES

Streams and rivers are open systems that generally maintain a rough *dynamic equilibrium*, that is, an overall balance between the work that the river does transporting sediment and the load that it receives. Sediment is supplied from tributaries and from earth material that moves down hillsides to the stream. A stream tends to have the gradient and cross-sectional shape that provide the flow velocity needed to move its sediment load.[35]

An increase or decrease in the amount of water or sediment received by a stream usually brings about changes in its gradient or cross-sectional shape, effectively changing the velocity of the water. The change in velocity may, in turn, increase or decrease the amount of sediment carried in the system. Therefore, land-use changes that affect a stream's sediment or water volume may set into motion a series of events that results in a new dynamic equilibrium.

Consider, for example, a land-use change from forest to an agricultural row crop, such as corn. Because croplands have higher erosion rates, they provide more sediment to the stream. At first, the stream will be unable to transport the entire load and will deposit sediment, increasing the channel gradient. The new, steeper slope of the channel will increase the velocity of water and allow the stream to move more sediment. If we assume that the base level remains constant, this process will continue until the stream is flowing fast enough to carry the new load. A new dynamic equilibrium may be reached, provided the rate of sediment accumulation levels off and the channel gradient and shape can adjust before another land-use change occurs. Suppose the reverse situation now occurs; that is, farmland is converted to forest. The sediment supply to the stream will decrease, and less sediment will be deposited in the channel. Erosion of the channel will eventually lower the gradient, which, in turn, will lower the velocity of the water. The predominance of erosion over deposition will continue until equilibrium is again achieved between sediment supply and work done.

The sequence of change just described occurred in parts of the southeastern United States during the past two and a half centuries. On the Piedmont, an area of rolling hills between the Appalachian Mountains and the Atlantic Coastal Plain, most forests were cleared for farming by the 1800s. The land-use change from forest to farming accelerated soil erosion and subsequent deposition of sediment in local streams. This caused the channel that existed before farming to begin to fill with sediment. After 1930, the land reverted to pine forests, and this change, in conjunction with soil conservation measures, reduced the quantity of sediment delivered to streams. Thus, by 1969, formerly muddy streams choked with sediment had cleared and eroded their channels (Figure 6.28).

DAM CONSTRUCTION

Dams and their accompanying reservoirs often have multiple purposes, including water for urban and agricultural areas, recreation, generation of electricity, and flood control. It is difficult to maximize different uses at the same time. Reservoir storage is important for flood control. A full reservoir provides no flood control. A reservoir that lowers storage for flood control water may not be optimal for agriculture. There are about 8,000 large dams in the United States; California alone has several hundred large dams. Almost all dams in the United States are over 50 years old, and, thus, dams are part of the aging, deteriorating infrastructure of the United States. Dam failures are rare events that are often linked to the geology of the dam site. Examples are

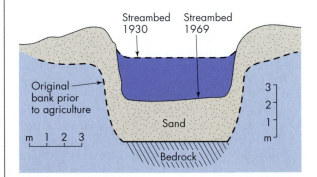

∧ FIGURE 6.28 Changes in a Streambed from Changes in Land Use

Cross-section of a stream at the Mauldin Millsite, Georgia, in 1969 showing changes in channel position through time. Land use changed from natural forest to agriculture, which increased stream sedimentation until 1930, and then back to woodland, which increased stream erosion. *(Trimble, S. W.,* Culturally Accelerated Sedimentation on the Middle Georgia Piedmont. *Master's thesis, Athens, Georgia: University of Georgia, 1969. Reproduced by permission of author.)*

the St. Francis Dam failure in 1928, 50 km (30 mi.) northwest of Los Angeles in San Francisquito Canyon, a tributary of the Santa Clara River. The floodwaters rushed down the Santa Clara River, killing about 500 people and washing many out to the ocean. Constructed as part of the water supply from the Sierra Nevada to Los Angeles, the concrete dam was about 60 m (200 ft. high). The geology of the dam site included landslide-prone schist and sedimentary rocks that could dissolve when wet, both separated by a fault zone with water-soluble minerals. Another example is the Teton Dam, which failed in 1976, just as it was nearly filled. The earthfill dam, consisting of layers of compacted soil, was about 90 m (300 ft. high), and failure was linked to the geology of the foundation which consisted of fractured volcanic rock that had the potential to transmit water beneath the dam. There were significant geologic studies of the site prior to construction. The St. Francis Dam failure was remembered, and detailed geologic evaluations were completed. Some adverse geologic conditions were recognized before and during construction, but the decision to proceed to build the dam was given. Failure, when it occurred, was swift; the dam collapsed, and a wall of water rushed down the Teton River Valley, killing 11 people and causing about $2 billion in damages.

Significant safety issues of dams (under normal operation) are water containment and control of water that is released. Dams have gates and spillways to lower lake levels in times of upstream runoff (floods) that can quickly fill some reservoirs. Dams also have emergency features (secondary spillways) to safely lower reservoir water levels when necessary. When a lot of water has to be released, downstream flooding may occur. Unfortunately, many cities are sited downstream of dams, presenting a potential flood hazard. The worst dam failure in U.S. history occurred in 1889 when floodwaters from a collapsed dam roared through Johnstown, Pennsylvania, killing about 2,000 people. The private earthfill dam had an inadequate, partly plugged emergency spillway that could not handle the heavy runoff that fell prior to the failure. The dam filled, and water overtopped the structure, quickly eroding it and causing the dam to collapse. Alarms had been sounded about the safety of the dam, and an inspection that revealed major flaws was completed. Unfortunately, the results of the inspection were ignored! This is an all too common decision that was repeated in 2017 when a major flood disaster was narrowly avoided in Oroville, California. The Feather River runs through the Oroville area, and there has been major development downstream of the Oroville Dam (Figure 6.29), an aging structure constructed in 1968. The dam is an earthfill structure and is the highest dam (235 m; 770 ft.) in the United States. The Oroville Dam is designed for water supply, agriculture, flood control, and electrical power. A multiple-year drought ended for Northern California in the winter of 2016–17. Intense, long-duration storms from the Pacific Ocean filled reservoirs. When Lake Oroville is nearly filled, water flows down a concrete spillway, lowering the lake level. Rock strength and the state of erosion were not evaluated when the spillway was constructed decades go, and the concrete and, in large areas under the spillway, highly weathered and fractured metamorphic rock was not anchored to fresh, strong

◀ **FIGURE 6.29 Oroville, California Dam Site**
The spillway is the white bar just northwest of the dam. Notice the extensive development along the Feather River downstream of the dam. *(2016 Google Earth Image)*

rock. Fresh rock is intact and relatively unweathered; it is the desired substrata for dams and spillways. If weathered rock is present, it must be treated or scraped off, exposing strong, competent rock. Fortunately, this was done for the dam foundation. There is also an emergency spillway with a 10 m (30 ft.) high concrete wall at the top (to protect the spillway). The emergency spillway is not lined and flows over the same highly fractured, weathered weak rock that is under the main concrete-lined spillway (Figure 6.30). Concern over the viability of the emergency spillway was raised in 2005, but no action was taken (the same all too common decision that led to the Johnstown flood). Suddenly, in early 2017, the main spillway failed when a growing hole in the concrete opened and began eroding. A large portion of the spillway failed, and a decision was made to use the emergency spillway. The main dam was not threatened. When the emergency spillway began eroding much faster than expected, it was feared that if the protective wall at the top eroded, a large volume of water would be released downstream, causing serious flooding. Urgent repairs to the emergency spillway were attempted by dropping large rocks below the concrete wall at the top of the emergency spillway. Fearing that a large, damaging flood might be imminent, 200,000 people were quickly evacuated downstream of the dam. A disaster was avoided, but, perhaps, by a narrow margin. Following the emergency, repairs have started on both spillways. If proper repairs had been taken, both the Johnstown flood and the Oroville emergency could have been avoided. Repairs as of 2018 are expected to exceed $1 billion.

URBANIZATION AND FLOODING

Human activities increase both the magnitude and the frequency of floods in small urban drainage basins of a few square kilometers. The rate of increase is determined by the percentage of the land that is covered with roofs, pavement, and cement, referred to as *impervious cover* (Figure 6.31), and the percentage of the area served by storm sewers. In most urban areas, storm sewers start at drains along the sides of streets and carry runoff to stream channels much more quickly than in natural settings. Therefore, impervious cover and storm sewers are collectively a measure of the degree of urbanization. An urban area with 40 percent impervious cover and 40 percent of its area served by storm sewers can expect to have about three times more floods of a given magnitude than before urbanization (Figure 6.32). This ratio applies to floods of small and intermediate frequency. However, as the size of the drainage basin increases, large floods with frequencies of approximately 50 years are less affected by urbanization.

Floods are a function of the relationship between rainfall and runoff, which is significantly changed by urbanization. One study showed that urban runoff from larger storms is nearly five times that of pre-urban conditions.[36] However, the extent of urban flooding is related not only to the peak discharge of a flood but also to the condition of the drainage system. For example, long periods of only moderate precipitation can also cause flooding if storm drains become blocked with sediment and storm debris. In this case, water begins to pond behind a debris dam, causing flooding in low areas. An analogy is water rising in a bathtub when the drain becomes partly blocked by soap.

ᐱ FIGURE 6.30 Orville Dam Main Spillway Damaged by Erosion
Large holes were eroded in the concrete spillway. The spillway was poorly bonded to the blue and rust-red colored metamorphic rock when the spillway was constructed decades ago. The rust-red rock is more highly weathered and weaker than the blue rock. Both are highly fractured. *(California Department of Water Resource)*

ᐱ FIGURE 6.31 Urbanization Increases Impervious Cover
Aerial view of Santa Barbara, California, which, like most other U.S. cities, has much of its land surface covered by paved streets, sidewalks, parking lots, and buildings. This impervious land cover blocks the infiltration of water and increases surface runoff. *(Edward A. Keller)*

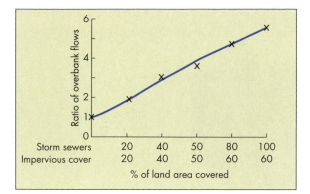

∧ FIGURE 6.32 Overbank Flooding Before and After Urbanization
Relationship between the ratio of overbank flows (after urbanization compared to before urbanization) and the primary measures of urbanization: the percentage of area with impervious cover and storm sewers. For example, a ratio of 3 to 1, or simply 3, means that after urbanization there are three floods for every one that took place before urbanization, or that flooding is three times as common after urbanization. Note that the percentage of the land area with impervious cover typically reaches its maximum of 60 percent when about 80 percent of the urban area has storm sewers. The percentage of impervious cover then stays the same as the storm sewer system is completed for the community. This graph shows that as the degree of urbanization increases; the number of overbank flows per year also increases. *(Based on Leopold, L. B., "Hydrology for Urban Landplanning—A Guidebook on the Hydrologic Effects of Urban Land Use," U.S. Geological Survey Circular 554, 1968)*

Besides increasing runoff and flooding frequency, urbanization affects how rapidly floods develop. Before urbanization, a considerable delay, or *lag time*, existed between when most rainfall occurred and the subsequent flooding (Figure 6.33a). A comparison of hydrographs before and after urbanization shows that there is a significant reduction in lag time after urbanization (Figure 6.33b). Short lag times, referred to as *flashy discharge*, are characterized by rapid rise and fall of floodwater. Urbanization also affects stream discharge by greatly reducing stream flow during the dry season. Normally, urban streams continue to flow during dry periods because groundwater seeps into the channel. However, because urbanization significantly reduces infiltration into the drainage basin, less groundwater is available to seep into streams. This reduced flow affects both water quality and the appearance of a stream. Very low discharge effectively concentrates pollutants in the water.[4] Some of these pollutants, such as nitrogen and phosphorus from fertilizers, can cause the growth of large amounts of algae and harm aquatic life.

During periods of surface runoff and flooding, pollutants from urban lands may enter streams as fine

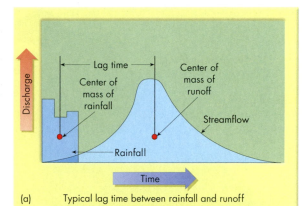

(a)   Typical lag time between rainfall and runoff

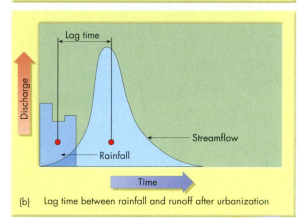

(b)   Lag time between rainfall and runoff after urbanization

∧ FIGURE 6.33 Urbanization Shortens Lag Time
Generalized hydrographs. (a) Hydrograph shows the typical lag between the time when most of the rainfall occurs and the time when the stream floods. (b) Here, the hydrograph shows the decrease in lag time and an increase in discharge because of urbanization. *(Based on Leopold, L. B., U.S. Geological Survey Circular 559, 1968; modifed after Tarbuck, E. J., and Lutgens, F. K., Earth: An introduction to physical geology, 8th ed. Upper Saddle River, NJ: Pearson Prentice Hall, 2005)*

sediment, and wastewater treatment plants may be overwhelmed and fail (Figure 6.34). Impervious cover and storm sewers are not the only forms of construction that can increase flooding. Some flash floods occur because bridges built across small streams block the passage of floating debris, which then forms a temporary dam. When the debris breaks loose, a wave of water moves downstream. This sequence of events has been repeated in many flash floods around the world. All too often, the supports for bridges are too close together or drainage pipes under roads, referred to as culverts, are too small to allow large debris to pass through; instead, they become temporary dams that either cause flooding upstream or, when they fail, cause flash flooding downstream.

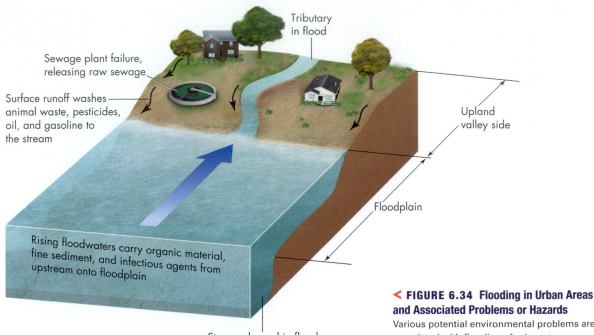

Sewage plant failure, releasing raw sewage

Surface runoff washes animal waste, pesticides, oil, and gasoline to the stream

Tributary in flood

Upland valley side

Floodplain

Rising floodwaters carry organic material, fine sediment, and infectious agents from upstream onto floodplain

Stream channel in flood

◀ **FIGURE 6.34 Flooding in Urban Areas and Associated Problems or Hazards** Various potential environmental problems are associated with flooding of urban streams.

Flooding in streams and rivers flowing through large cities in poor countries, where high concentrations of people are living in substandard housing (shanties) and encroaching on a floodplain, can result in a catastrophe. The combination of population pressure and poor land-use choices in Manila, the capital of the Philippines, resulted in a catastrophic event in late September 2009 when a tropical storm produced about 44 cm (~18 in.) of rain in a 12-hour period.[37] Swirling floodwaters inundated poor urban areas. Flooding also occurred in middle-class areas, as well as in some commercial areas. The population of Manila had grown to more than 11 million by the time the flooding occurred, with about one-fifth of the people living in crowded shantytowns on a wide floodplain with little or no infrastructure. They had no protection when the Marikina River flooded. Hundreds of people were killed, drowned in their houses and cars and in shopping malls (Figure 6.35).

▶ **FIGURE 6.35 Flood Damage** Urban flooding east of Manila in the Philippines was devastated by floodwaters in late September 2009. The combination of a high-magnitude storm in a densely populated city with many thousands of people living on a floodplain produced a catastrophe. *(Pat Roque/AP Images)*

6.6 CHECK your understanding

1. How does changing land use from forest cover to cropland affect stream channel sediment balance?
2. Why might a stream channel erode below a dam?
3. Why do impervious covers and storm sewers define the degree of urbanization?
4. What are the main effects of the urbanization of small drainage basins on flooding?

6.7 Minimizing the Flood Hazard: Structural Control to Floodplain Management

∧ FIGURE 6.36 Mississippi River Levee at Baton Rouge Bike Path
Levee tops are sometimes used as bike paths. Levee is protecting development. (*U.S. Army Corps of Engineers Vicksburg District Levee Safety Program*)

Historically, particularly in the nineteenth century, humans have responded to floods by attempting to prevent them by modifying streams and rivers. Physical barriers, such as dams and levees, have been created, or the shape of the stream has been changed by widening, deepening, or straightening the channel. The channel shape is changed so that a stream will drain the land more efficiently. Every new flood-control project has the effect of luring more people to the floodplain in the false hope that the flood hazard is no longer significant. We have yet to build a dam or channel capable of controlling the heaviest runoff, and, when the water finally exceeds the capacity of the structure, flooding may be extensive.[11,38]

THE STRUCTURAL APPROACH

Physical Barriers Measures to prevent flooding include construction of physical barriers, such as earthen *levees* and concrete flood walls (Figure 6.36). Unfortunately, the potential benefits of these physical barriers are often lost because of increased development on the floodplains that they are supposed to protect. Also, older levees are often in poor condition and subject to failure during floods. Levee failures can damage agricultural land by creating high-energy flows that erode topsoil and deposit a layer of sand and gravel on otherwise fertile soil in the floodplain.

Capture and control of storm water runoff is an important strategy for reducing urban flooding. For example, specially designed parking lots can temporarily store runoff that is released slowly, reducing the rate of runoff to storm sewers and urban streams. Storm water retention basins of variable size (Figure 6.37) may be constructed that temporarily store urban runoff that is slowly released to reduce flooding. In addition, ground level planters may be constructed and planted with trees and other plants that capture runoff to be released slowly to streams, thus reducing flooding. Concrete can be made pervious, so water will infiltrate the ground rather than runoff. Rooftops of urban areas may have plants and soil that capture and delays runoff.[39]

Channelization Straightening, deepening, widening, clearing, and/or lining existing stream channels are all methods of **channelization**. Objectives of this engineering technique include controlling floods and erosion, draining wetlands, and improving navigation. Thousands of kilometers of streams in the United States have been modified without adequate consideration of the adverse effects of channelization.

Opponents of modifying natural streams emphasize that the practice is antithetical to the production of fish and wetland wildlife and causes extensive aesthetic degradation. Not all channelization causes serious environmental degradation; in many cases, drainage projects are beneficial. Benefits are probably best observed in urban areas subject to flooding and in rural areas where previous land use has caused drainage problems. In other areas, channel modification has improved navigation or reduced flooding and has not caused environmental disruption.

Engineering solutions to reduce flooding will remain an important aspect of flood control in areas where previous high-value development is present. However, engineering is not the only option. Alternatives include flood insurance and controlling the land use on floodplains. Planners, policy makers, and hydrologists generally agree that no one adjustment is best in all cases. Rather, an integrated approach

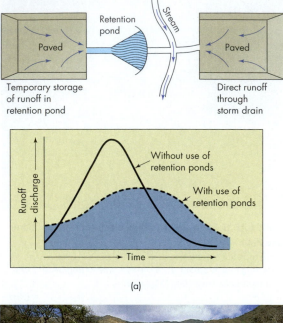

(a)

(b)

⋀ FIGURE 6.37 Retention Ponds Reduce Flood Discharge
(a) Comparison of runoff from a paved area, which goes directly through a storm drain to a stream, with runoff that is stored temporarily in a retention pond before draining to a stream. The graph shows that the use of a retention pond reduces peak discharge and the likelihood that the runoff will contribute to flooding of the stream. (b) A nearly dry retention pond near Santa Barbara, California; retention ponds also capture sediment that reduces pollution levels and sedimentation in streams. *(Edward A. Keller)*

to minimizing the flood hazard is more effective, especially if it incorporates adjustments that are appropriate for a particular situation.

Flood Insurance

In 1968, the federal government took over the **flood insurance** business when private companies became reluctant to continue to offer policies. Congress established the U.S. National Flood Insurance Program to make flood insurance available at subsidized rates. Administered by the Federal Emergency Management Agency (FEMA), this program requires the mapping of special flood hazard areas, defined as those areas that would be inundated by a 100-year flood. Flood hazard areas are designated along streams, rivers, lakes, alluvial fans, and deltas and along low-lying coastal areas susceptible to flooding during storms or very high tides.

New property owners in flood hazard areas must buy insurance at rates determined by the risk. Basic risk evaluation depends on identifying the area that would be inundated by the 100-year flood. The insurance program is intended to provide short-term financial aid to victims of floods and to establish long-term land-use regulations for the nation's floodplains. As part of this program, building codes have been revised to limit new construction of nonresidential buildings in a flood hazard area (floodplain) to flood-proofed buildings. The floodway is defined as the channel and parts of the adjacent floodplain that are reasonably necessary to carry the floodwater of a stream or river. No development is allowed in the floodway. Figure 6.38 shows an idealized FEMA Map (Zone AE is the 100-year floodplain). For a community to join the National Flood Insurance Program, it must have FEMA-prepared maps of the 100-year floodplain. The 100-year flood is the flood that has a 1 percent chance of being equaled or exceeded in a given year. Using probability rules, the chance of the 100-year flood occurring once in the 30-year period of a mortgage is 1 in 4 (25 percent). Communities must also adopt minimum standards of land-use regulation within the flood hazard areas. Nearly all U.S. communities with a significant flood risk have basic flood hazard maps and have initiated some form of floodplain regulation. Several million property owners in the United States presently have flood insurance policies.

A problem with flood insurance using the 100-year flood is that it is often based on precipitation and stream flow data that may only be available for 20 years or so and the data may be from years ago. Many people were flooded during the 2017 Houston floods that were well outside of the mapped 100-year flood. The scientific problem is to somehow plan for extreme events that are often not included in flood hazard mapping but are not unprecedented. When we apply standard engineering flood frequency probability equations and modeling to extreme events with long return periods we get answers but those answers are often wrong.

By the early 1990s, policy makers and flood-control professionals recognized that the insurance program

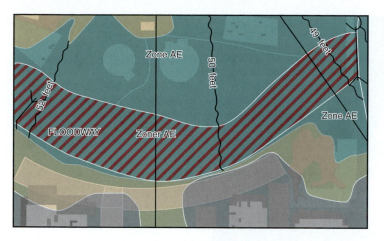

◀ FIGURE 6.38 Idealized FEMA Flood Hazard Map
The floodway and Zone AE are defined as areas subject to 1 percent annual chance flood (100-year flood). *(FEMA Flood Map Map Service Center* https://msc.fema.gov*)*

was in need of reform. This prompted Congress to pass the National Flood Insurance Reform Act of 1994. Provisions of the act encourage additional opportunities to mitigate flood hazards, such as flood-proofing, relocations, and buyouts of properties likely to be frequently flooded.[40]

Flood-Proofing This is the addition of structural changes to buildings or around buildings that are designed to reduce flooding. Several methods of **flood-proofing** are currently available:

> Raising the foundation of a building above the flood hazard level, using piles or columns or by extending foundation walls or earth fill.

> Constructing flood walls or earthen mounds around buildings to isolate them from floodwaters.

> Using waterproofing construction, such as water-proofed doors, basement walls, and windows.

> Installing improved drains with pumps to remove incoming floodwaters.

Other modifications to buildings are designed to minimize flood damage while allowing floodwaters to enter a building. For example, ground floors along expensive riverfront properties in some communities in Germany are designed so that they are not seriously damaged by floodwaters and may easily be cleaned for reuse following a flood.[43]

Floodplain Regulation From an environmental point of view, the best adjustment to the flood hazard in urban areas is **floodplain regulation**. The objective of floodplain regulation is to obtain the most beneficial use of floodplains while minimizing flood damage and the cost of flood protection. This approach is a compromise between the indiscriminate use of floodplains, which

results in loss of life and tremendous property damage from floods, and the complete abandonment of floodplains, which means giving up a valuable natural resource.

There are circumstances, however, when physical barriers, reservoirs, and channelization works are required. Structural controls may be necessary to protect lives and property in areas where there is extensive development on floodplains. We need to recognize, however, that the floodplain belongs to the river system, and any encroachment that reduces the cross-sectional area of the floodplain increases flooding. An ideal solution would be to discontinue floodplain development that necessitates new physical barriers. In other words, the ideal is to "design with nature." Realistically, the most effective and practical solution is a combination of physical barriers and floodplain regulations that results in less physical modification of the river system. For example, reasonable floodplain zoning in conjunction with a floodwater diversion channel or upstream reservoir may result in a smaller channel or reservoir than would be necessary without the zoning regulations.

A preliminary step toward floodplain regulation is flood hazard mapping, which is a means of providing information about the floodplain for land-use planning.[40] Flood hazard maps may delineate past floods or floods of a particular frequency, for example, the 100-year flood. Such maps are useful in regulating private development, purchasing land for public use as parks and recreational facilities, and creating guidelines for future land use on floodplains.

Flood hazard evaluation may be accomplished in a general way by direct observation and measurement of physical parameters. For example, the extensive flooding of the Mississippi River Valley during the summer of 1993 was mapped in detail, using imagery from

◄ FIGURE 6.39 Scour Marks Indicate Flood Height
Floodwaters of the San Gabriel River in Los Angeles County, California, removed bark from the roots and lower trunks of these trees. Hydrologists measured the elevation of the top of the scour marks to determine the height of the 2005 flood on this segment of the river. *(Theresa Modrick, Ph.D., P.E./Hydrologic Research Center)*

satellite and aircraft (see Figure 6.20). Flooding can also be estimated from the field measurement of high waterlines, flood deposits, scour marks (Figure 6.39), and the distribution of trapped debris in the channel banks or on the floodplain after water has receded.[41] Once flood hazard areas have been established, planners can modify zoning maps, regulations, and building codes.

Relocating People from Floodplains: Examples from North Carolina and North Dakota

For several years, local, state, and federal governments have been selectively purchasing and removing homes damaged by floodwaters in order to reduce future losses. In September 1999, nearly 50 cm (~19 in.) of rain from Hurricane Floyd flooded many areas of North Carolina. The flooding damaged approximately 700 homes in Rocky Mount, North Carolina, population 60,000. The state and federal governments subsequently decided to spend nearly $50 million to remove 430 of these homes, the largest single-home buyout ever approved. Following the purchase, the homes were demolished, and the land was preserved as open space. Following Hurricane Matthew in 2016, a buyout program is being considered for about 50 homes damaged by flooding associated with the hurricane.

At Churchs Ferry, North Dakota, a wet cycle since 1993 caused nearby Devils Lake to rise approximately 8 m (~26 ft.). With no natural outlet and flat land surrounding its shore, the lake more than doubled in area. The swollen lake inundated the land around Churchs Ferry, and by late June 2000, the town was all but deserted; the population had shrunk from approximately 100 to 7 people. Most residents took advantage of a voluntary federal buyout plan and moved to higher ground, many to the town of Leeds, approximately 24 km (~15 mi.) away. The empty houses left behind will be demolished or moved to safer ground.

This lucrative $3.5 million buyout plan seemed to be assured. Those people who participated were given the appraised value of their homes plus an incentive; most considered the offer too good to turn down. There was also recognition that the town would eventually have come to an end as a result of flooding. Nevertheless, there was some bitterness among the town's population, and not everyone participated. Among the seven people who decided to stay were the mayor and fire chief. The buyout program for Churchs Ferry demonstrated that the process is an emotional one; it is difficult for some people to make the decision to leave their homes, even though they know the homes are likely to be destroyed by floodwaters in the relatively near future. The story of Churches Ferry is full of anxiety, fear, and anger, as well as hope. The details of people lives, chronicled in three articles in *The Atlantic* are well worth a read.[42–44]

Forest and Flood Management

Forested land may serve as a giant sponge that traps precipitation and releases it slowly into streams and rivers, reducing flooding and augmenting low stream flow during dry times. Thus, maintaining and restoring forests may be considered a flood control strategy. This strategy is more attractive to developing countries that may lack funding for expensive flood control programs. Forested lands are most effective for flood control of small to moderate-sized drainage basins. Large catastrophic floods resulting from high intensity (extreme event) storms in large drainage basins may not be much affected by the presence of forested land. This is because high volumes of rainfall in a short period of time may overwhelm other variables such as forest cover. Nevertheless, deforestation of as little as 10 percent of native forest increases the frequency of floods by 4–28 percent and lengthens flood duration by 4–8 percent. Thus, protecting and restoring forests can help reduce both the frequency and the duration of flooding.[45]

SOCIAL MEDIA AND FLOOD DAMAGE MEDIATION

The 2011 Bangkok flood was Thailand's worst flood in 50 years. The flooding came in three stages over several months of prolonged rainfall. Flooding occurred in spite of two dams that were thought to be able to mitigate severe flooding. Over the greater Bangkok Metropolitan Area, deaths during the first stage totaled about 60, and several hundred thousand homes and businesses were damaged. The relatively low number of deaths was due, in part, to the fact that people in flood-prone areas of Bangkok have adjusted to floods for centuries. Homes and other buildings are often constructed on stilts or they have two-story housing so they can move possessions to the second story during floods. Flood risk in Thailand has increased in recent decades as a result of expansion of agriculture that caused deforestation and increase in population that exposes more people and urban resources to flooding. Furthermore, urban growth has reduced the ability of the land's drainage systems to remove floodwaters. Past flood damage mitigation has focused on early warning to alert people of eminent flooding. The 2011 Bangkok flood was the first flood in a major metropolitan area in Southern Asia where many people had smartphones and were connected to the internet. The flood was slow in developing, so people had time to communicate and alert one another to the danger.[46]

Facebook worldwide has about 2 billion users, and Twitter has about 300 million. People are hooked on communicating via such means, and these communications can help reduce flood losses. For example, social media during the 2011 Bangkok flood reduced losses by an average of 37 percent. This even though many poorer people did not have smartphones (this has changed rapidly following the flood, and many more people now have smartphones). Many people in Bangkok were informed in real time of the dynamic nature of the coming flood. They learned what the depth of flooding was going to be, and many moved property to a safer location (often upstairs). The data from social media was often more reliable than from other sources and often offered information not otherwise available. In the United States, the Federal Emergency Management Agency is testing the use of social media for distributing disaster information. In developing countries, expanded access to mobile networks and broadband is justified, based on people being better informed during unfolding disasters. Today, there is rapid expansion of the use of smartphones that will be used during disasters to better inform people in harms way about urban floods.[47] Social media was also important during the 2017 Houston flood. People called and sent text messages to those who responded with water rescues.

6.7 CHECK your understanding

1. Why might levees, flood walls, and dams produce a false sense of security?

2. What is channelization and how has it been used in flood control?

3. Describe the main adjustments to the flood hazard that do not involve engineered structures.

4. Google FEMA Flood Map Service Center, enter your home address, and examine your personal flood hazard.

5. What is the potential role of social media in reducing the flood hazard?

6.8 Perception of the Flood Hazard

Whereas most governmental agencies, planners, and policy makers have an adequate perception and understanding of flooding, many individuals do not. The public knowledge of flooding, anticipation of future flooding, and willingness to accept adjustments caused by the hazard are highly variable.

Progress at the institutional level includes the preparation of thousands of maps of flood-prone areas. The maps show areas susceptible to flooding along streams, lakes, and coastlines; areas with a flash-flood potential downstream from dams; and areas where urbanization is likely to cause problems in the near future. In addition, the federal government has encouraged states and local communities to adopt floodplain management plans.[44] Still, the idea of restricting or prohibiting development on floodplains or of relocating present development to locations off the floodplain raises its own problems; it needs further community discussion before the general population will accept it.

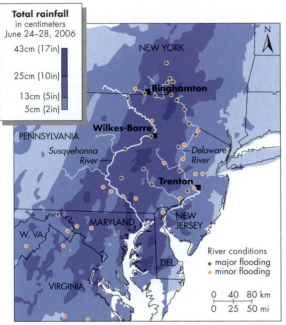

(b)

∧ FIGURE 6.40 Mid-Atlantic Floods of June and July 2006
(a) Map of major and minor flooding. *(Based on the* New York Times *with data from National Weather Service)* (b) Collecting mail from flooded home in Wilkes-Barre, Pennsylvania. *(Matt Rourke/AP Images)*

This need was tragically evident in the 2006 floods in the mid-Atlantic states, when severe river flooding impacted the region from Virginia to New York (Figure 6.40). More than 200,000 floodplain residents were evacuated in Wilkes-Barre, Pennsylvania, alone, and damages of approximately a billion dollars were incurred. About 16 people lost their lives as cars were swept away by floodwaters and people drowned in flood-swollen creeks and rivers. At least 70 people were rescued from rooftops. In Conklin, New York, near Binghamton, nearly three-fourths of the town was flooded as the Susquehanna River crested about 4.2 m (~14 ft.) above flood level.

Increasing public awareness of the dangers of flash flooding could save many lives each year (see Case Study 6.5: Professional Profile). Public safety campaigns by the National Weather Service and local governments, including the city of Boulder, Colorado; Clark County Regional Flood Control District, Nevada; and the city of Austin, Texas, have been especially successful in increasing public awareness about flash floods. With at least half of all flood deaths in automobiles, the need to educate drivers is especially critical. The National Weather Service slogan "Turn Around, Don't Drown" summarizes a life-saving safety precaution for all motorists not to drive on flooded roads. For hilly or mountainous terrain, road and trail signs can warn people to climb to higher ground if floodwaters begin to rise. In a narrow stream valley, this can mean abandoning your vehicle for the safety of higher ground if floodwaters are rising rapidly ahead of you or if you hear a roaring sound coming from upstream.

PERSONAL ADJUSTMENT: WHAT TO DO AND WHAT NOT TO DO

Flooding is the most commonly experienced natural hazard. Although we can't prevent floods, individuals can be better prepared by learning what to do and what not to do before, during, and after a flood (Table 6.2).

6.8 CHECK your understanding

1. What is being done to raise public awareness of the flood hazard?

2. Summarize the personal adjustments you can make to reduce your flood hazard.

Nicholas Pinter, Southern Illinois University then U.C. Davis in California

Nicholas Pinter (Figure 6.5.A) was a doctoral student at the University of California, Santa Barbara, where he studied the high desert of the Owens Valley and in the adjacent Sierra Nevada and White Mountains of California. After doing postdoctoral research at Yale University, Nick went to Southern Illinois University at Carbondale. His research focuses on "quantifying human influences on the geology of the earth's surface." Recent research has shown that modern humans move more material each year than all the wind, running water, waves, and glacial ice on Earth. "It's both easy and dangerous," Nick says, "to underestimate how much people have influenced the planet, both locally and at the global scale."

Southern Illinois University at Carbondale is located near where the Mississippi, Ohio, Missouri, Illinois, and other rivers all come together. This central location kindled his interest in river processes and issues of flood hazard in particular. "Some research questions are purely theoretical, but as soon as you start looking at big rivers, the science becomes relevant to people's lives and government policy really fast," says Nick. He realized early on that some of the structures being built by the U.S.

Army Corps of Engineers to facilitate shipping at low flows were behaving as large roughness elements and actually increasing the flood hazard during large floods. Over time, squeezing the river between levees and slowing down flows with structures built in the channel were leading to increasing flood stages. This means that, say 50 years ago, a flood of a certain discharge would inundate the banks at some depth. Today, the same discharge is associated with water many feet deeper and, hence, a greater flood hazard. Although a number of outside scientists have reached similar conclusions, the U.S. Army Corps of Engineers did not receive the news that its own construction projects were actually making the flooding much worse.

Nick Pinter and his students have worked diligently to counter the political resistance with rock-solid science: quantifying past increases in flood stages and risk and confirming what caused those increases with hydrologic, statistical, geospatial, and modeling analyses. Nick points out that it is difficult to achieve two goals at the same time. That is, if you want to increase a river's ability to transport goods through navigation at low flows, you may have one strategy. However, that strategy may not be the optimum solution for passage of a 100-year flood safely through the same stretch of river. "We live in the twenty-first century," says Nick, "in which robust science is vital to sustainable management of natural systems that are stretched to their limits in many places by the demands of modern society."

◀ FIGURE 6.5.A Professor Nicholas Pinter
Professor Pinter is at the University of California, Davis. He is an expert on flooding of the Mississippi River and the role of human intervention through building flood defenses and structures to improve navigation. Nick holds an Endowed chair and I am very proud of him. He was one of my Ph.D. students at UCSB. *(Nicholas Pinter)*

▼ **TABLE 6.2 What to Do and What Not to Do Before and After a Flood**

Preparing for a Flood	
What to Do	• Check with your local flood-control agency to see if your property is at risk from flooding. • If your property is at risk, purchase flood insurance if you can and be sure that you know how to file a claim. • Buy sandbags or flood boards to block doors. • Make up a flood kit, including a flashlight, blankets, rain gear, battery-powered radio, first-aid kit, rubber gloves, and key personal documents. Keep it upstairs if possible. • Find out where to turn off your gas and electricity. If you are not sure, ask the person who checks your meter when he or she next visits. • Talk about possible flooding with your family or housemates. Consider writing a flood plan, and store these notes with your flood kit.
What Not to Do	• Underestimate the damage a flood can do.

When You Learn a Flood Warning Has Been Issued	
What to Do	• Be prepared to evacuate. • Observe water levels and stay tuned to radio and television news and weather reports. • Move people, pets, and valuables upstairs or to higher ground. • Move your car to higher ground. It takes only 0.6 m (~2 ft.) of fast-flowing water to wash your car away. • Check on your neighbors. Do they need help? They may not be able to escape upstairs or may need help moving furniture. • Do as much as you can in daylight. If the electricity fails, it will be hard to do anything. • Keep warm and dry. A flood can last longer than you think, and it can get cold. Take warm clothes, blankets, a Thermos, and food supplies.
What Not to Do	• Walk in floodwater above knee level: It can easily knock you off your feet. Utility access holes, road works, and other hazards may be hidden beneath the water.

After a Flood	
What to Do	• Check home for damage; photograph any damage. • If insured, file a claim for damages. • Obtain professional help in removing or drying carpets and furniture as well as cleaning walls and floors. • Contact gas, electricity, and water companies. You will need to have your supplies checked before you turn them back on. • Open doors and windows to ventilate your home. • Wash water taps and run them for a few minutes before use. Your water supply may be contaminated; check with your water supplier if you are concerned.
What Not to Do	• Touch items that have been in contact with the water. Floodwater may be contaminated and could contain sewage. Disinfect and clean thoroughly everything that got wet.

Source: Based on Environment Agency, United Kingdom, Floodline: Prepare for Flooding, 2004.

APPLYING the **5** Fundamental Concepts

Houston Flood of 2017

1 **Science helps us predict hazards.**

2 **Knowing hazard risks can help people make decisions.**

3 **Linkages exist between natural hazards.**

4 **Humans can turn disastrous events into catastrophes.**

5 **Consequences of hazards can be minimized.**

The 2017 flood in Houston is the latest and most severe flood event in a region that has a serious flood hazard. As recently as 2016, a flood killed about eight people and flooded about 1,000 homes. A serious flood on Memorial Day in 2015 caused extensive damage, albeit less than in 2001 when floods inflicted about $9 billion in damages, destroying about 14,000 homes and claiming 23 lives. These floods occur when remnants of hurricanes stall over Houston, unfortunately a more common occurrence than previously thought. The 2017 flood was a 1,000-year event, by far the largest in Houston's history, but it occurred due to the same set of circumstances surrounding earlier floods. Since 1989, Houston has experienced 26 major floods; of these, six (25 percent) were associated with the remnants of hurricanes.

The 2007 flood map, showing major channels and the 100- and 500-year flood plains, are shown on Figure 6.41. The maps are considered too outdated and inadequate for the future. The maps are expensive to update, do not adequately consider climate change, and are subject to manipulation by local government. More development is allowed if flood maps shrink the flood zone. Nearly half of those flooded in 2017 were outside of mapped flood hazard areas. Of course, the flood greatly exceeded even the 500-year flood.

Houston is a huge metropolitan area with up to 40 percent of the area rendered impermeable to water from roofs, streets, freeways, and parking lots. The many trillions of gallons of rainfall that caused the 2017 flood was caused by the remnant of Hurricane Harvey moving very slowly and even doubling back for a second hit. More than the annual rainfall of the city fell in a couple of days. Water ran off quickly from urban areas and piled up in the normally sluggish stream channels and bayous.

Complicating factors make up a familiar list: Rapid population and development in recent decades; poor land-use planning that allowed development in flood hazard areas; inadequate flood maps that provided a false sense of security; loss of control of runoff from flood control dams; poor understanding of the total risk, given the recent flood history; and climate change. Houston is far from alone in terms of these complicating factors. Many U.S. cities from coast to coast, from New York to Los Angles, are vulnerable to catastrophic flooding.

As population continues to grow, we need to reduce the flood hazard in ways that do not require massive response but are proactive. We need plans for future flood hazard reduction that do not require massive evacuation from flood-prone areas but that avoid the hazard through better land-use practices, such as avoiding development in floodplains. That is a proactive solution for minimizing the hazard of future flooding in the United States. As we learned in Chapter 1, you can pay now in preparation or pay many times more if not prepared. Total flood losses are the sum of direct losses (Houston will flood from storms) and losses related to human actions that encourage poor land-use planning (development in flood risk areas); inadequate engineering structures, such as levees, flood walls, and flood control dams; and lack of flood insurance. Sometimes people and government are greedy, incompetent, or just lazy and not doing the right thing. Paying after a disaster is many times more expensive than

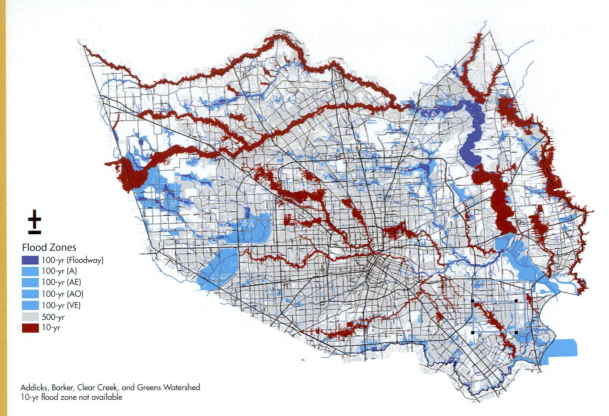

Flood Zones
- 100-yr (Floodway)
- 100-yr (A)
- 100-yr (AE)
- 100-yr (AO)
- 100-yr (VE)
- 500-yr
- 10-yr

Addicks, Barker, Clear Creek, and Greens Watershed
10-yr flood zone not available

∧ FIGURE 6.41 Flood Map for Houston Texas
Boundries for the 100-year flood (probability of 0.01) and 500-year flood (probability of 0.002) events. These FEMA maps are outdated (probably 2007 maps) and did not consider development on the past decade or impact of climate that has produced more intense storms. Many homes outside of the 500-year event boundary flooded. This suggests there existed a false sense of security that probably led many to not purchase flood insurance. (*Modified from FEMA flood hazard maps and Houston Flood Agency*)

prevention. Flood-proofing a building may cost a few thousand dollars; replacing the building after flood damages could cost thousands of times more. For example the 2001 flood in Houston greatly damaged the large Houston Medical Center. Following the flood the Center developed an ambitious flood-proof project that included 23 large flood doors and other barriers. Just prior to the 2017 flood the doors were closed and the Center was not damaged, saving massive flood damages and loss of medical care for people. Figure 6.42 shows the floodwaters surrounding but not entering the Medical Center.

Encouraging property owners to play an active part in limiting their potential losses to their property (e.g., flood-proofing) can save property owners from wading through polluted floodwater that enters their buildings. Government can and often does take on big projects to help reduce the flood hazard, and this is very important and often most important. In the end though, it is the property owner of a site who is damaged.

APPLYING THE 5 FUNDAMENTAL CONCEPTS

1. Flooding is usually a predictable event in countries where accurate weather forecasts and stream monitoring are performed. Could the 2017 floods in Houston have been predicted? Why or why not?

2. Discuss the flood risk in Houston before and after the 2017 flood, based on what is known and unknown about the probability of flooding in the future.

< FIGURE 6.42 Flood Proofing of Texas Medical Center in Houston
Twenty-three large flood doors and flood barriers kept floodwaters out saving many millions of dollars in potential damages. *(floodbreak. com)*

3. What other hazards were linked to the 2017 flood? Specifically, what flood processes were responsible and why?

4. What human decisions before and during the long 2017 flood made the consequences of the flood much worse?

5. What specific recommendation can you make to help minimize future consequences of flooding in Houston?

CONCEPTS in review

6.1 An Introduction to Rivers

LO:1 Explain basic river processes.

■ Streams and rivers form a basic transport system of the rock cycle and are a primary erosion agent in shaping the landscape. The region drained by a stream system is called a drainage basin or watershed. Rivers carry chemicals in their dissolved load and sediment in their suspended and bed loads. Discharge refers to the volume of water moving past a particular location in a river per unit of time.

■ Sediment deposited by lateral migration of meanders in a stream and by the periodic overflow of the stream banks during a flood forms a floodplain. The configuration of the stream channel as seen in an aerial view is called the channel pattern. A pattern can be braided or meandering or both characteristics may appear in the same river.

Coarse gravel

KEY WORDS
channel patterns (p. 213), **discharge** (p. 212), **drainage basin** (p. 211), **floodplain** (p. 213), **river** (p. 211), **river system** (p. 216)

6.2 Flooding

LO:2 Compare and contrast flash floods, downstream floods, and megafloods and know why the differences are important.

- The natural process of overbank flow is termed *flooding*. Flash floods in small drainage basins can be produced by intense, brief rainfall over a small area. Downstream floods in major rivers are produced by storms of long duration over a large area that saturate the soil, increasing runoff from thousands of tributary basins.

- Flooding magnitude and frequency are difficult to predict on many streams because of changes in land use and limited historical records. This difficulty is especially pronounced for extreme events, such as 100-year floods. The probability that a 100-year flood or greater will take place each year is the same regardless of when the last 100-year flood occurred.

KEY WORDS

downstream floods (p. 220), **flash floods** (p. 220), **flooding** (p. 217), **levee** (p. 223), **megafloods** (p. 228), **recurrence interval** (p. 218)

6.3, 6.4, and 6.5   Geographic Regions at Risk for Flooding Effects of Flooding, Linkages between Floods and Other Hazards Natural Service Functions of Floods

LO:3 Summarize regions at risk from flooding and the global impact.

LO:4 Differentiate between the effects of flooding and the linkages with other natural hazards.

LO:5 Describe the benefits of periodic flooding.

- River flooding is the most universally experienced natural hazard. Floods can occur just about anywhere there is water, and much of the United States faces the possibility of flooding.

- Annual losses from flooding worldwide are about 6,000 lives lost and economic losses of $23 billion. The number of floods and the number of people killed or affected, along with economic loss, is highly variable from region to region, reflecting the population of people living along rivers and exposed to flooding.

- Floods produce primary effects such as loss of life and damage to structures by floodwaters. Secondary effects include water pollution, disease, displacement of people, and fire.

- Although flooding causes many deaths and much damage, it does provide natural service functions such as the production of fertile lands, benefits to aquatic ecosystems, and maintenance of ample sediment supplies to naturally subsiding deltas such as the Mississippi.

- Water supply is intimately related to the concept of the **Water–Energy–Food Nexus** that consists of the complex linkages between water, energy, and agriculture. Water supply from runoff to rivers is stored in reservoirs where it provides a number of services, such as helping provide the water supply for human consumption, energy for power, and water to irrigate crops.

6.6   Human Interaction with Flooding

LO:6   Discuss how human activity can significantly affect river processes and flooding.

- Land-use changes, especially urbanization, have increased flooding in small drainage basins by covering much of the ground with impervious surfaces, such as buildings, parking lots, and roads, thereby increasing the runoff of storm water. Construction of a dam can significantly change the hydrology and flood frequency and magnitude downstream.

6.7   Minimizing the Flood Hazard: Structural Control to Floodplain Management

LO:7   Summarize why being proactive with flood hazards is, in the long term, more significant in reducing the flood hazard than being reactive after a flood.

LO:8   Evaluate the adjustments we can make to minimize flood deaths and damage.

- Engineering structures will still be needed to protect existing development in highly urbanized areas. These structures include physical barriers, such as levees and flood walls, and structures that regulate the release of water, such as dams and reservoirs.
- Channelization is the straightening, deepening, widening, cleaning, or lining of existing streams, usually with the goal of controlling floods or improving drainage. Channelization has often caused environmental degradation, so new projects must be closely evaluated.
- Preferred environmental adjustments to the flood hazard include flood insurance, flood-proofing, and floodplain regulation. Floodplain regulation is critical because engineered structures tend to encourage further development of floodplains by producing a false sense of security. The first step in floodplain regulation is flood hazard mapping, which can be difficult and expensive. Planners use flood hazard maps to zone flood-prone areas for appropriate uses. In some cases, homes in flood-prone areas have been purchased and demolished by the government and their owners relocated to safe ground.
- Social media is becoming a tool in minimizing flood damages by providing real time data to people during a flood.

KEY WORDS

channelization (p. 241), **flood-proofing** (p. 243), **flood insurance** (p. 242), **floodplain regulation** (p. 243)

6.8   Perception of the Flood Hazard

LO:9   Summarize how people perceive flooding.

- An adequate perception of flood hazards exists at the institutional level; however, on the individual level, more public awareness programs are needed to help people clearly perceive the hazard of living in flood-prone areas.

Scour mark

CRITICAL thinking questions

1. You are a planner working for a community that is expanding into the headwater portions of drainage basins. You are aware of the effects of urbanization on flooding and want to make recommendations to avoid some of these effects. Outline a plan of action.

2. You are working for a regional flood-control agency that has been channelizing streams for many years. Although bulldozers are usually used to straighten and widen the channel, the agency has been criticized for causing extensive environmental damage. You have been asked to develop new plans to minimize the damage to streams from channelization. What would you recommend?

3. Does the community you live in have a flood hazard? If not, why not? If there is a hazard, what has been done or is being done to reduce the hazard? What more could be done?

4. Your parents have recently heard about atmospheric rivers and past megafloods on a documentary film shown on television. They ask you what you think, having taken a course on natural hazards. How do you answer their question?

5. How might we better integrate the use of social media into flood hazard rediction?

References

1. **Schumm, S. A.** 1977. *The fluvial system*. New York: John Wiley & Sons.

2. **Beyer, J. L.** 1974. Global response to natural hazards: Floods. In *Natural hazards: Local, national, global*, ed. G. F. White, pp. 265–74. New York: Oxford University Press.

3. **Linsley Jr., R. K., Kohler, M. A.,** and **Paulhus, J. L.** 1958. *Hydrology for engineers*. New York: McGraw-Hill.

4. **Leopold, L. B.** 1968. *Hydrology for urban land planning*. U.S. Geological Survey Circular 554.

5. **Seaburn, G. E.** 1969. *Effects of urban development on direct runoff to East Meadow Brook, Nassau County, Long Island, New York*. U.S. Geological Survey Professional Paper 627B.

6. **McCain, J. F., Hoxit, L. R., Maddox, R. A., Chappell, C. F.,** and **Caracena, F.** 1979. Meteorology and hydrology in Big Thompson River and Cache la Poudre River Basins. In *Storm and flood of July 31–August 1, 1976, in the Big Thompson River and Cache la Poudre River Basins, Larimer and Weld Counties, Colorado*. U.S. Geological Survey Professional Paper 1115A.

7. **Shroba, R. R., Schmidt, P. W., Crosby, E. J.,** and **Hansen, W. R.** 1979. Geologic and geomorphic effects in the Big Thompson Canyon area, Larimer County. In *Storm and flood of July 31–August 1, 1976, in the Big Thompson River and Cache la Poudre River Basins, Larimer and Weld Counties, Colorado*. U.S. Geological Survey Professional Paper 1115B.

8. **Bradley, W. C.,** and **Mears, A. I.** 1980. Calculations of flows needed to transport coarse fraction of Boulder Creek alluvium at Boulder, Colorado. *Geological Society of America Bulletin* Part II, 91: 1057–90.

9. Agricultural Research Service. 1969. *Water intake by soils*. Miscellaneous Publication No. 925. U.S. Department of Agriculture.

10. **Strahler, A. N.,** and **Strahler, A. H.** 1973. *Environmental geoscience*. Santa Barbara, CA: Hamilton Publishing.

11. **Office of Emergency Preparedness.** 1972. *Report to Congress*, Vol. 3: *Disaster preparedness*. Washington, DC: U.S. Government Printing Office.

12. **Rahn, P. H.** 1984. Flood-plain management program in Rapid City, South Dakota. *Geological Society of America Bulletin* 95: 838–43.

13. **U.S. Department of Commerce.** 1973. *Climatological data, national summary* 24(13).

14. **Anonymous.** 1993. The flood of '93. *Earth Observation Magazine*, September: 22–23.

15. **Mairson, A.** 1994. The great flood of 1993. *National Geographic* 185(1): 42–81.

16. **Bell, G. D.** 1993. The great midwestern flood of 1993. *EOS, Transactions, American Geophysical Union* 74(43): 60–61.

17. **NOAA.** 2008. *Climate of 2008. Midwestern U.S. flood overview*. www.ncdc. noaa.gov. Accessed 4/22/10.

18. **Anonymous.** 1993. Flood rebuilding prompts new wetlands debate. *U.S. Water News*, November: 10.

19. **Pinter, N., Thomas, R.,** and **Wollsinski, J. H.** 2001. Assessing flood hazard on dynamic rivers. *EOS, Transactions, American Geophysical Union* 82: 333–39.

20. **Pinter, N.** 2009. Non-stationary flood occurrence on the upper Mississippi River–lower Missouri River system: Review and current status. In *Finding a balance between floods, flood protection and river navigation*, eds. R. E. Criss and T. M. Kusky. pp. 34–40. St. Louis, MO: Saint Louis University Center for Environmental Science. www.ces.slu.edu.

21. **Hey, D., Kostel, J.,** and **Montgomery, D.** 2009. An ecological solution to the flood damage problem. In *Finding a balance between floods, flood protection and river navigation*, eds. R. E. Criss and T. M. Kusky. pp. 72–79. St. Louis, MO: Saint Louis University Center for Environmental Science. www.ces.slu.edu.

22. **U.S. Global Change Research Program.** 2017. *Our changing planet*.

23. **Saulny, S.** 2007. Development rises on St. Louis area floodplains. *New York Times*, May 15.

24. **Pinter, N.** 2005. One step forward, two steps back on U.S. floodplains. *Science* 308: 207–08.

25. **Committee on Alluvial Fan Flooding.** 1996. *Alluvial fan flooding*. National Research Council. Washington, DC: National Academy Press.

26. **Edelen Jr., G. W.** 1981. Hazards from floods. In *Facing geological and hydrologic hazards, earth-science considerations*, ed. W. W. Hays, pp. 39–52. U.S. Geological Survey Professional Paper 1240–B.

27. **Keller, E. A.,** and **Capelli, M. H.** 1992. Ventura River flood of February 1992: A lesson ignored? *Water Resources Bulletin* 28(5): 813–31.

28. **Dettinger, M. D.,** and **Ingram, B. L.** 2013. The coming megafloods. *Scientific American* 308(1): 64–71.

29. **Taylor, W. L.,** and **Taylor, R. W.** 2006. The Great California Flood of 1862. The Fortnightly Club. Redlands, CA. www.redlandsfortnightly.org/papers/taylor06.htm.

30. **Jongman, B.** and 6 others. 2015. Declining vulnerability to river floods and the global benefits of adaption. *PNAS* 112: E2271–80.

31. **Slade Jr., R. M.,** and **Patton, J.** 2003. *Major and catastrophic storms and floods in Texas*. U.S. Geological Survey Open-File Report 03–0193.

32. **Dolan, R., Howard, A.,** and **Gallenson, A.** 1974. Man's impact on the Colorado River and the Grand Canyon. *American Scientist* 62: 392–401.

33. **Grams, P. E.** and 5 others. 2015. Building sandbars in the Grand Canyon. *EOS Earth and Space Science News* 96: 12–16.

34. **Schmidt, J. C.** and **Grams, P. E.** 2011. The high flows-physical science results. In *Effects of three high-flow experiments on the Colorado River ecosystem downstream from Glen Canyon Dam, Arizona*, ed. T. S. Melis (ed.), pp. 53–91. Geological Survey Circular 1366.

35. **Mackin, J. H.** 1948. Concept of the graded river. *Geological Society of America Bulletin* 59: 463–512.

36. **Terstriep, M. L., Voorhees, M. L.,** and **Bender, G. M.** 1976. *Conventional urbanization and its effect on storm runoff*. Illinois State Water Survey Publication, ISWS CR-177.

37. **Lagmay, A. M. F., Rodolfo, R. S.,** and **Bato, M. G.** 2010. The perfect storm: Floods devastate Manila. *Earth* 55(4): 50–55.

38. **Mount, J. F.** 1997. *California rivers and streams*. Berkeley: University of California Press.

39. **Gaines, J. M.** 2016. Water potential. *Nature* 531: 554–55.

40. **Smith, K.,** and **Ward, R.** 1998. *Floods*. New York: John Wiley and Sons.

41. **Baker, V. R.** 1976. Hydrogeomorphic methods for the regional evaluation of flood hazards. *Environmental Geology* 1: 261–81.

42. **Hamilton, L. M.** 2011. Where the roads end in water: the lake that won't stop rising. *The Atlantic*, May 13.

43. **Hamilton, L. M.** 2011. Flooded lives: The fight to survive Devils lake. *The Atlantic*, May 24.

44. **Hamilton, L. M.** 2011. Spirit Lake rising: Living with a never ending flood. *The Atlantic*, May 31.

45. **Laurance, W. F.** 2007. Forest and Floods. *Environmental Science* 449: 09.

46. **Nabangchang, O.** 2014. Economic costs incurred by households in the 2011 Greater Bangkok flood. *Water Resources Research* 51: 58–77.

47. **Allaire, M. C.** 2016. Disaster loss and social media: can online information increase flood resilience? *Water Resources Research* 52: 7408–23.

400,000 tons of water-saturated soil and rock tore through **the upper part of the community**

7

APPLYING the **5** fundamental concepts
La Conchita, Southern California: Landslide Disaster

① Science helps us predict hazards.

Geologic studies illustrate that regular landsliding has occurred over a protracted period in this region. La Conchita and other communities along this part of the California coast, therefore, have a high probability of experiencing mass wasting events.

② Knowing hazard risks can help people make decisions.

Many of the residents of La Conchita consider the high mass wasting risk acceptable, being outweighed by the opportunity to live in this idyllic beach community. In contrast, other residents stay because they are unable to sell their homes due the high risk of mass wasting.

Mass Wasting

La Conchita, Southern California: Landslide Disaster

Monday, January 10, 2005, was a disaster for the small beachside community of La Conchita, California, 80 km (~50 mi.) northwest of Los Angeles. Ten people lost their lives, and about 30 homes were destroyed or damaged when 400,000 tons of water-saturated soil and rock tore through the upper part of the community (see chapter-opener photograph).

Although it had been raining for days prior to the 2005 slide and a similar but larger landslide had occurred in the same location in 1995 (Figure 7.1),[1] neither residents nor local officials recognized that another slide was imminent. As a result, some people were trapped in their homes and others ran for their lives.

Following the 1995 landslide, a geologic investigation was conducted to evaluate the risk of future events as well as to estimate the cost of stabilizing the hillslope behind the community. Although the report showed that landslides were common along the steep sea cliff behind La Conchita, the findings indicated that the slope could be completely stabilized for a cost of about $150 million,[2] or

< **Reactivated Landslide Turns Deadly**
Search and rescue personnel attempt to stabilize collapsed home before attempting to enter and search for survivors buried in the debris and mud generated by the 2005 La Conchita landslide. *(John Shea/Fema.)*

Landslides, the movement of materials down a slope, constitute a serious natural hazard in many parts of North America and the world. Landslides are often linked to other hazards such as earthquakes and volcanoes. Most landslides are small and slow, but a few are large and fast. Both may cause significant loss of life and damage to property, particularly in urban areas.

After reading this chapter, you should be able to:

LO:1 Describe slope processes and the different types of landslides.

LO:2 Analyze the forces that act on slopes and how they affect the stability of a slope.

LO:3 Evaluate what geographic regions are at risk from landslides.

LO:4 Compare different mass wasting processes and their linkages with other natural hazards.

LO:5 Describe the ways in which people can affect the landslide hazard.

LO:6 Explain the adjustments people can make to avoid death and damage caused by landslides.

LO:7 Summarize the ways in which you can protect yourself and property from the processes of mass wasting.

LEARNING Objectives

3 Linkages exist between natural hazards.

The 2005 slide was clearly linked to the heavy rain that preceded the event. When we return to La Conchita at the end of this chapter, you will learn about several other natural hazards that contribute to mass wasting along this section of California's scenic coastline.

4 Humans can turn disastrous events into catastrophes.

According to the 2010 U.S. census data, Southern California is in the top 10 states for population growth. Homes in coastal communities with similar geologic settings and natural hazards as La Conchita are highly desirable, leading to more structures being built in hazard zones.

5 Consequences of hazards can be minimized.

The 1995 and 2005 landslides in La Conchita exemplify the challenges facing scientists, government agencies, and individuals seeking to minimize the effects of mass wasting.

(a)   (b)

∧ FIGURE 7.1 La Conchita 1995 and 2005
(a) Aerial view of the 1995 La Conchita landslide with part of the community of La Conchita in the foreground.
The cliff face contains mostly older slide deposits and sediments near the top of the slope, a large slump block
partially covered with vegetation on the side of the slope, and earthflow deposits at the base of the slope. This
slide buried, collapsed, or damaged houses at the base of the slope. *(USGS)* (b) Aerial view of the 2005 reactivation of the same landslide. For reference, the white arrow on the left side of each photograph points to the
same building. To the right of this building the partially buried and destroyed retaining wall built following the
1995 slide can be seen. At the base of the slope, the landslide has formed two lobe-shaped earthflows, a large
one on the right and a much smaller one in the center of the photograph. Mud and debris have covered parts of
four streets and damaged or buried about 30 houses. *(AP Photo/Kevork Djansezian)*

about $1 million for every home in the community.
Due to the prohibitive cost of complete slope stabilization, several less costly strategies, such as building a
retaining wall and installing a limited drainage system,
were employed to minimize the potential hazard and
to mitigate the effects of a future slide. However, the
2005 event clearly showed that this approach was inef-
fectual.

    Although property values in La Conchita plummeted
following each slide, within just a few years some housing prices rebounded and the community remains a
highly desirable place to live along the rapidly urbaniz-
ing coast of California.

7.1 An Introduction to Landslides

Mass wasting is a comprehensive term for any type of
downslope movement of earth materials. In its more
restricted sense, mass wasting refers to a rapid
downslope movement of rock or soil as a more or less
coherent mass. In this chapter, we consider landslides in
this restricted sense. We will also discuss the related
phenomena of earthflows and debris flows, rockfalls,
and avalanches. For the sake of convenience, we sometimes refer to all of these as **landslides**.

SLOPE PROCESSES

Slopes are the most common landforms on Earth. Although most slopes appear stable and static, they are actually dynamic, evolving systems. The processes that are active in these systems generally do not produce uniform slopes. Rather, most slopes are composed of several segments that are either straight or curved (Figure 7.2).

A spectacular example of these segments can be seen on the north side of Yosemite Valley in Yosemite National Park, California (Figure 7.2a). The prominent landform there, El Capitan, has a high cliff face or *free face* of hard granite that forms a straight, nearly vertical slope segment. The free face and adjacent valley wall regularly shed pieces of rock that accumulate at the base of the cliff to form a **talus** slope (Figure 7.2a and Figure 7.3). Both the free face and the talus slope are segments of the overall slope. Frequent rockfalls keep soil from developing on the free face and much of the talus slope.

In contrast to the strong granite of Yosemite Valley, gentler hillslopes develop on other types of rocks. For example, on Santa Cruz Island, California, slopes are developed on weaker metamorphic rocks and lack a free face. Instead, the slopes have three segments: an upper convex slope, a straight slope on the hillside, and a lower concave slope (Figure 7.2b).

As illustrated in Yosemite Valley and Santa Cruz Island, slopes are usually composed of different slope segments. The five segment types in the foregoing examples—free face, talus, upper convex, straight, and lower concave slopes—are sufficient to describe most slopes encountered in nature. Which slope segments are present on a particular slope depends on the rock type and the climate of the area. Free face development is more common on strong, hard rocks or in arid environments where there is little vegetation. Convex and concave slopes are more common on softer rocks or in

(a)

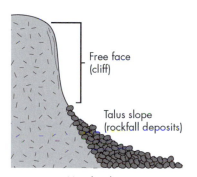

(b)

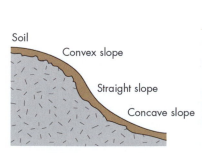
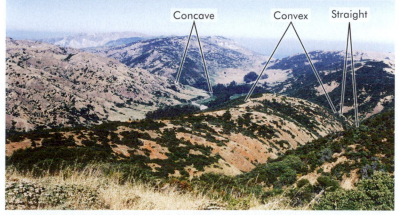

⋀ FIGURE 7.2 Slope Segments
(a) Slope on hard granite in Yosemite National Park, California. The cliff on the left side of the image is the several-thousand-foot-high free face of El Capitan. Below the cliff is a talus slope. *(Edward A. Keller)* (b) These slopes on Santa Cruz Island, California, formed on relatively weak metamorphic rock with convex, straight, and concave slope segments. *(Edward A. Keller)*

◀ FIGURE 7.3 Rockfall and Talus Pile
View of a huge rockfall in 2009 near Half Dome in Yosemite National Park, California. The rockfall originated from 540 m (~1800 ft.) above the valley floor (out of this photo), knocking down hundreds of trees and burying a part of the popular Mirror Lake trail. The impact of the rock on the valley floor registered on seismographs around California as a magnitude 2.5 earthquake. Broken rock from the large rockfall forms the light-colored talus pile covering the ground where the trees have been bowled over. *(Duane E. DeVecchio)*

humid environments where thick soil and vegetation are present. However, there are many exceptions to these general rules, depending on local conditions. For example, the gentle, convex, red-colored slopes in the foreground of Figure 7.2b have formed on weak, easily eroded metamorphic rock (schist) in a semiarid climate on Santa Cruz Island, California.

Material on most slopes is constantly moving down the slope at rates that vary from an imperceptible creep of soil and rock to thundering avalanches and rockfalls that move at velocities of 160 km/hr (~100 mph) or more. Through time, this downslope movement is important in eroding valley walls. Slope processes are one reason that valleys are usually much wider than the streams they contain.

TYPES OF LANDSLIDES

Earth materials on slopes may fail and move or deform in several ways (Table 7.1). **Falling** refers to the free fall of earth material, as from the free face of a cliff (Figure 7.4a). **Sliding** is the downslope movement of a

▼ TABLE 7.1 Common Types of Landslides and Other Downslope Movements

Mechanism	Type of Mass Movement	Characteristics
Falling	Rockfall	Individual rocks fall through the air and may accumulate as talus.
Sliding	Slump	Cohesive blocks of soft earth material slide on a curved surface; also called a rotational landslide.
	Soil slip	Soil and other weathered earth material slide on a tilted surface of bedrock or cohesive sediment; also called a debris slide or earth slide.
	Rock slide	Large blocks of bedrock slide on a planar surface, such as layering in sedimentary or metamorphic rocks.
Flowing	Avalanche	Granular flow of various combinations of snow, ice, organic debris, loose rocks, or soil that moves rapidly downslope.
	Creep	Very slow, downslope movement of rocks and soil.
	Earthflow	Wet, partially cohesive, and internally deformed mass of soil and weathered rock.
	Debris flow	Fluid mixture of rocks, sand, mud, and water that is intermediate between a landslide and a water flood; includes mudflows and lahars.
More than one	Complex	A combination of two or more types of sliding, flowage, and occasionally falls; forms where one type of landslide changes into another as it moves downslope.

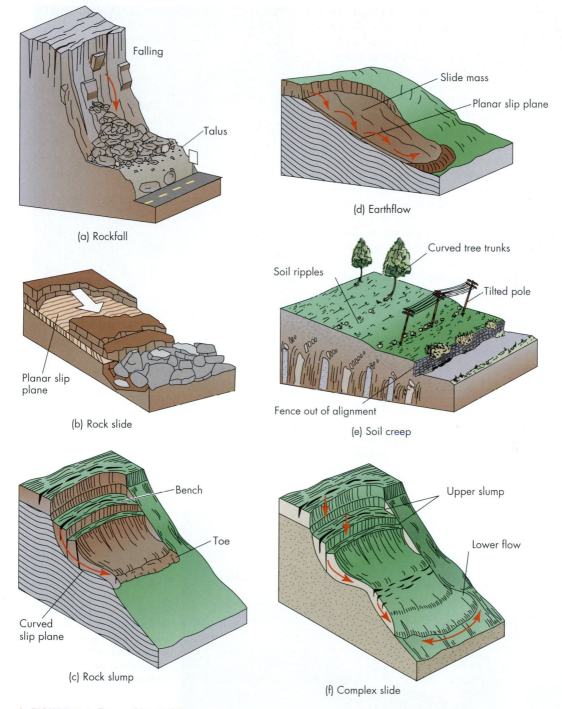

∧ FIGURE 7.4 Types of Landslides

(a) Rockfall, in which blocks of bedrock fall through air and accumulate at the base of the cliff as a talus slope. (b) Rock slide, in which blocks of bedrock slide down a bedding plane in sedimentary rock. Soil slips are similar movements except soil, rather than rock, slides down a bedrock surface. (c) Slump, in which cohesive blocks of soft earth material slide down a curved slip surface. (d) Earthflow, in which mud, sand, rock, and other material are mixed with water and become partially liquified. (e) Soil creep, causing telephone poles to tilt and trees trunks to become curved. (f) Complex landslide, consisting of upper slump and lower flow.

coherent block of earth material along a planar *slip plane* (Figure 7.4b). **Slumping** of rock or soil is sliding along a curved slip plane producing *slump blocks* (Figure 7.4c). *Flowage*, or **flow**, is the downslope movement of unconsolidated material in which particles move about and mix within the mass, such as earthflows, debris flows, or avalanches (Figure 7.4d). Very slow flowage, called **creep**, progressively tilts telephone poles, fences, and tree trunks (Figure 7.4e). Creep occurs at such a low rate that the progressive upward growth of trees outpaces the tilting, resulting in trees trunks that are curved skyward (Figure 7.5).

Many landslides are complex combinations of sliding and flowage (Figure 7.4f). For example, a landslide may start out as a *slump*, then pick up water as it moves downslope, and transform into an earthflow in the lower part of the slide. Some complex landslides may form when water-saturated earth materials flow from the lower part of a slope, undermining the upper part and causing slump blocks to form (Figure 7.4f and Figure 7.6).

Downslope movements are classified according to four important variables: (1) the mechanism of movement (slide, fall, flow, or complex movement), (2) type of earth material (e.g., solid rock, soft consolidated sediment, or loose unconsolidated material), (3) amount of water present, and (4) rate of movement. In general, movement is considered rapid if it can be discerned with the naked eye; otherwise, it is classified as slow. Actual rates vary from a slow creep of a few millimeters or centimeters per year, to a rapid 1.5 m (~5 ft.) per day, to an extremely rapid 30 m (~100 ft.) or more per second.[3]

⋀ FIGURE 7.5 Trees Affected by Soil Creep
Curved tree trunks in Lassen Volcanic National Park are due to soil creep. Soil creep is likely the result of construction of the road (lower left corner of the photo), which steepened the hillside after the older trees were firmly anchored into the bedrock, which is why only the younger trees exhibit highly curved trunks. *(Duane E. DeVecchio)*

⋖ FIGURE 7.6 Houses Buried by Complex Landslide
In the distance is the amphitheater-like scar in the hillslope that was the source of the 1995 La Conchita landslide. The slide started out as a slump and then liquefied into an earthflow as it moved downhill. Several homes, including the three-storied home behind the orange plastic fence, were destroyed. *(Edward A. Keller)*

FORCES ON SLOPES

To determine the causes of landslides, we must examine the forces that influence slope stability. The stability of a slope can be assessed by determining the relationship between **driving forces** that move earth materials down a slope and **resisting forces** that oppose such movement. The most common driving force is the downslope component of the weight of the slope material. That weight can be from anything superimposed or otherwise placed on the slope, such as vegetation, fill material, or buildings. The most common resisting force is the **shear strength** of the slope material, that is, its resistance to failure by sliding or flowing along potential slip planes (see Case Study 7.1: A Closer Look). Potential *slip planes* are geologic surfaces of weakness in the slope material, such as bedding planes in sedimentary rocks and fractures in all types of rock.

Slope stability is evaluated by computing a **safety factor** (SF), defined as the ratio of the resisting forces to the driving forces. If the safety factor is greater than 1, the resisting forces exceed the driving forces and the slope is considered stable. If the safety factor is less than 1, the driving forces exceed the resisting forces and a slope failure can be expected. Driving and resisting forces are not static; as local conditions change, these forces may change, increasing or decreasing the safety factor.

Driving and resisting forces on slopes are determined by the interrelationships of the following variables:

> type of earth materials
> slope angle and topography
> climate
> vegetation
> water
> time.

The Role of Earth Material Type The material composing a slope can affect both the type and the frequency of downslope movement. Important material characteristics include mineral composition, degree of cementation or consolidation, the presence of zones of weakness, and the ability of the earth material to transmit water. These weaknesses may be natural breaks in consistency of the earth materials, such as sedimentary and metamorphic layering, or they may be zones along which the earth has moved before, such as an old landslide slip surface or geologic fault. Weak zones can be especially hazardous if the zone or plane of weakness intersects or parallels the slope of a hill or mountain. Where the planes of weakness are rock bedding planes, they are referred to as *daylighting beds* (Figure 7.7).

∧ FIGURE 7.7 Daylighting Beds
(a) Bedding planes that intersect the surface of the land on a slope are said to "daylight." Such beds are potential slip planes. (b) This slide occurred in late 2003. Failure was along a daylighting bedding plane. Slide deposits cover part of the beach. *(Edward A. Keller)*

Forces on Slopes

The fundamental driving force of mass wasting processes is gravity. Over time, the ever-constant downward pull of gravity exceeds resisting forces and causes slopes to fail, rocks to fall, soils to creep, and water-saturated earth materials to flow. Over geologic time scales, gravity-driven erosional processes have sculpted Earth's surface and worn away countless ancient mountain ranges and washed them out to the sea.

The downward pull of gravity exerts a quantifiable force toward the center of Earth, giving weight to potential slide masses (Figure 7.1.A). Yet, because earth materials cannot be pulled straight down through the underlying rocks, the gravitational pull is divided between a downslope force and the normal force, which are at right angles to each other. The downslope force or driving force (D) is parallel to the slope of the potential slip plane, whereas the normal force (N) is perpendicular (Figure 7.1.A). The shear strength of the potential slip plane (resisting force) is proportional to the magnitude of the normal force, which adds frictional resistance onto the surface of the plane. The relative magnitudes of the normal and driving forces are directly related to the angle of the slope. Specifically, as the slope angle (θ) increases, the normal force decreases and the driving force increases proportionally; the opposite is true as θ decreases. It follows then that the safety factor decreases with increasing slope angle, because the safety factor is the magnitudinal ratio of resisting forces over the driving forces.

As local conditions change naturally or as a result of human interaction with the landscape, driving and resisting forces will

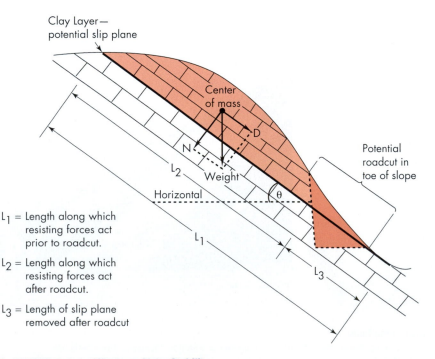

L_1 = Length along which resisting forces act prior to roadcut.

L_2 = Length along which resisting forces act after roadcut.

L_3 = Length of slip plane removed after roadcut

∧ FIGURE 7.1.A Effects on Slope Stability
Cross-section of a slope showing the principal forces acting on a potential slide mass and the effects of undermining a slope due to road construction.

change. These changes may lead to an increase or decrease in the safety factor. For example, consider the construction of a road in which part of a hillside must be excavated to provide a suitable surface for automobile traffic. Removal of the material at the toe (base) of the slope exposes a potential slip plane, a daylighting bed in this case (see Figure 7.1.A and Figure 7.8). Although the roadcut reduces the driving forces on the slope by removing some of the weight of the potential slide mass, the cut also decreases the length of the slip plane, thereby reducing the resisting force (shear strength). If you examine Figure 7.1.A, you can see that only a small portion of the potential slide mass has been removed, whereas a relatively large portion of the length of the slip plane has been removed (L1 versus L2). Therefore, the overall effect of the roadcut is to *lower the safety factor* because the reduction of the driving force is small compared to the reduction of the resisting force.

For slides, the shape of the slip surface is strongly controlled by the type of earth material. Slides have two basic patterns of movement, *rotational* and *translational*. Rotational slides, or slumps, have curved slip surfaces (Figure 7.4c), whereas translational slides generally have planar slip surfaces (Figure 7.4b).

In slumps, the rotational sliding of slump blocks tends to produce small topographic benches, which may rotate and tilt in an upslope direction (Figure 7.4c). Slumps are most common in unconsolidated earth material and in mudstone, shale, and other weak rock types. In translational slides, material moves along inclined slip planes within and parallel to a slope (Figure 7.7). The zones of weakness for these slip planes include fractures in all rock types, surfaces between layers in sedimentary rocks referred to as *bedding planes*, weak clay layers, and surfaces between layers in metamorphic rocks referred to as *foliation planes*. Generally, once a slip surface is established for a landslide, that plane of weakness will continue to exist within the earth material. The material that moves along these planes can be large blocks of bedrock or unconsolidated and weathered earth material.

A common type of translational side is a *soil slip*, a shallow slide of unconsolidated material over bedrock (Figure 7.8). The plane of weakness for soil slips is usually in **colluvium**, a mixture of weathered rock and other debris that is above the bedrock and below the soil.

Material type is also a factor in rockfalls. Where resistant rock overlies weaker rock, rapid erosion of the underlying weaker rock may cause a slab failure and a subsequent rockfall (Figure 7.9).

The nature of the earth material composing a slope can greatly influence the type of slope failure. For example, on slopes of shale or weak volcanic pyroclastic material, failure commonly occurs as creep, earthflows, debris flows, or slumps. Slopes formed in resistant rock such as well-cemented sandstone, limestone, or granite are much less likely to experience the same problems. Thus, builders should assess the landslide hazard before starting construction on shale or other types of weak rock.

The ability of the earth material to transmit water, referred to as *permeability*, determines how easily soil, sediment, or rock absorbs water (see Appendix 2). As discussed later, water is involved in nearly all landslides, and permeability barriers can allow the buildup of water. For example, soil slips commonly develop because of contrasts between more permeable soil and less permeable, unweathered bedrock below the soil. Downward percolating rainwater builds up in the soil because it cannot easily infiltrate the underlying bedrock. The accumulation of water can break down cohesion between soil particles and facilitate slippage of the soil, especially on steep hillsides.

The Role of Slope and Topography Two factors are important with slope and topography: steepness of the slope and amount of topographic relief. By *slope*, we mean the slant or incline of the land surface. In general, the steeper the slope, the greater the driving force. For example, a study of landslides that occurred during two rainy seasons in California's San Francisco Bay area established that 75 percent to 85 percent of landslide activity is closely associated with urban areas on steep slopes.[4]

Topographic *relief* refers to the height of the hill or mountain above the land below. Areas of high relief are hilly or mountainous, have tens or hundreds of meters of relief, and are generally more prone to mass wasting. Within the contiguous United States and Canada, the coastal mountains of California, Oregon, Washington, and British Columbia; the Rocky Mountains; the Appalachian Mountains; and coastal cliffs and bluffs have the greatest frequency of landslides. All types of downslope movement occur on steep slopes within these areas.

Steep slopes are also associated with rockfalls, avalanches, and soil slips. In Southern California, shallow soil slips are common on steep slopes when a hillside

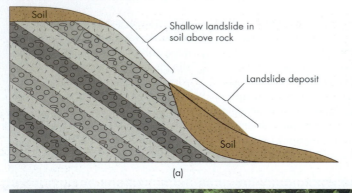

Soil

Shallow landslide in
soil above rock

Landslide deposit

Soil

(a)

Vegetation and thin veneer of soil
removed by slide exposing bedrock
that is parallel to the slope

Retaining wall –
to keep landslide
deposits off the road

(b)

◀ FIGURE 7.8 Soil Slips

(a) Cross-sectional diagram of a shallow soil slip
in which the gray bands are sedimentary layers
parallel to the slope. Soil slides off the slope and
piles up at its base. (b) This shallow soil slip in
North Carolina has removed slope vegetation.
Highway engineers installed the white barrier
at the base of the slide to keep debris off of the
road. (Edward A. Keller)

◀ FIGURE 7.9 Rockfall Damages House

The large boulder embedded in the right
end of this house fell 120 m (400 ft.)
from a ledge of resistant sandstone in
the early morning of October 18, 2001, in
Rockville, Utah. Weighing an estimated
200 to 300 tons and measuring approximately 5 m (~16 ft.) by 5 m (~16 ft.) by
3.6 m (~12 ft.), the boulder destroyed the
living room, bathroom, and service area.
It then entered a bedroom, narrowly
missing the sleeping homeowner. The
specific cause of the rockfall is unknown,
but there are numerous large boulders
in the area from earlier rockfalls. (Utah
Geological Survey)

becomes saturated with water. Once soil slips move
downslope, they often become earthflows or debris
flows, which can be extremely hazardous (Figure 7.10).
In Southern California, earthflows are common on

moderate slopes, and creep can be observed on gentle
slopes.

Debris flows are thick mixtures of mud, debris,
and water. They range in consistency from thick mud

∧ FIGURE 7.10 Shallow Soil Slips Can Kill
Following several days of rain in 2011, mass wasting events, like this soil slip, killed at least 600 people in the Rio De Janeiro area. *(AP Photo/Felipe Dana)*

soups to wet concrete and are capable of carrying house-size boulders. Debris flows can move either slowly or rapidly depending on conditions. They can move down established stream valleys, flow from channels filled with colluvium, or take long narrow tracks or chutes on steep hillsides. These flows can be relatively small to moderate events, transporting a few hundred to hundreds of thousands of cubic meters of debris down a single valley, or they can be huge events transporting cubic kilometers of material from the failure of an entire flank of a mountain (see volcanic mudflows and debris flows discussed in Chapter 5).

The Role of Climate *Climate* can be defined as the weather that is typical in a place or region over a period of years or decades. A description of climate is more than just the average air temperature and amount of precipitation. It also includes the type of precipitation and its seasonal patterns. In North America, examples of type and pattern are winter rains along the West Coast of the United States, summer thunderstorms in the southwestern United States, winter blizzards in the Arctic and Great Plains, heavy snows on the eastern shores of the Great Lakes, and late summer and fall hurricane activity on the Mexican and southeastern U.S. coasts. For landslides, the type of climate influences the amount and timing of water that infiltrates or erodes a hillslope and the type and abundance of hillslope vegetation.

In arid and semiarid climates, vegetation tends to be sparse, soils are thin, and bare rock is exposed. Free faces and talus slopes are common where erosion-resistant bedrock is present (Figure 7.11a). Resistant rock forms steep slopes (cliffs), and less resistant rock forms more gentle slopes. Rockfalls, debris flows, and shallow soil slips are common types of landslides in these climatic regions.

In subhumid to humid regions of the world, abundant vegetation and thick soils cover most slopes (Figure 7.11b). These two characteristics cause the widespread development of convex and concave slope segments (see Figure 7.2b). Landslide activity in these areas includes deep complex landslides, earthflows, and soil creep.

The Role of Vegetation Vegetation has a complex role in the development of landslides and related phenomena. The nature of the vegetation in an area is a function of the climate, soil type, topography, and fire history, each of which also independently influences

(a)   (b)

∧ FIGURE 7.11 Climate and Mass Wasting
Different climate zones around the world play an important role in not only the appearance of the landscape but also in the processes that sculpt it. (a) Semiarid landscape in Boysen State Park, Wyoming. *(Jack Jeffers/Super-Stock)* (b) Humid landscape in Taiwan in Southeast Asia. *(Sean Sprague/The Image Works)*

what happens on slopes. Vegetation is a significant factor in slope stability for three reasons:

1. Vegetation provides a protective cover that cushions the impact of falling rain. This cushion allows the water more time to infiltrate the slope rather than running downhill, retarding surface erosion.

2. Plant roots add strength and cohesion to slope materials. They act like steel rebar reinforcements in concrete, anchoring soil and unconsolidated material, and increasing the resistance of a slope to landsliding.[5]

3. Vegetation also adds weight to a slope.

In some cases, this additional weight increases the probability of a landslide, especially with shallow soils on steep slopes. Such soil slips are common along the California coast where ice plant, an invasive plant imported from South Africa in the early 1900s, covers steep-cut slopes (Figure 7.12). During especially wet winter months, these shallow-rooted plants take up water and store it in their leaves. This stored water adds considerable weight to steep slopes, thereby increasing the driving forces. These plants also cause increased water infiltration into the slope, which decreases the resisting forces. When failure occurs, the plants and several centimeters of roots and soil slide to the base of the slope.

In Southern California, soil slips are also a serious problem on steep slopes covered with chaparral, a dense growth of native shrubs and small trees (Figure 7.13). Chaparral facilitates water infiltration into a slope and lowers its safety factor.[6]

∧ FIGURE 7.13 Multiple Shallow Slides
Shallow soil slips on steep slopes in Southern California are common due to low, brushy vegetation that does not root deeply, known as chaparral. *(Edward A. Keller)*

The Role of Water Water is almost always directly or indirectly involved with landslides, so its role is particularly important.[7] When studying a landslide, scientists first examine what water is doing both on and within the slope. Water affects slope stability in three basic ways:

1. Many landslides, such as shallow soil slips and debris flows, develop during rainstorms when slopes become saturated.

2. Other landslides, such as slumps, develop months or even years following the deep infiltration of water into a slope.

(a)   (b)

∧ FIGURE 7.12 Ice Plants on Slopes Are Often Unstable
Ice plant was first introduced in California to protect slopes along railroads and was later widely used along highways. Instead of stabilizing slopes, ice plant often contributes to landslides because of its shallow roots and the added weight of water stored in its leaves. The plant caused these soil slips near Santa Barbara, California, (a) on a steep roadside embankment *(Edward A. Keller)* and (b) at a home site. A black plastic sheet was placed over the upper part of the slide to reduce the infiltration of rain water. *(Edward A. Keller)*

3. Water erosion of the base or toe of a slope decreases its stability.

Erosion by flowing water along a stream bank or by wave action on the coast can remove earth material from the base of a hill or cliff. This loss of material steepens the slope, reduces the safety factor, and increases the likelihood of a landslide (Figure 7.14). Erosion is especially problematic if it removes the toe of an old landslide, thereby increasing the potential for

the landslide to move again (Figure 7.15). Therefore, it is important for planners and engineers to identify old landslides before cutting into hillslopes for the construction of roads and buildings. Excavations of hillslopes must be carefully planned so that potential problems can be isolated and corrected prior to construction.

Another way that water can cause landslides is by contributing to the spontaneous liquefaction of quick or expansive clay. These clays are fine-grained earth

(a)

(b)

◄ **FIGURE 7.14 Water Eroding the Toe of a Slope Causes Instability**
(a) Stream-bank erosion seen in the lower left corner of this image has caused slope failure and damage to a road in the San Gabriel Mountains, California. *(Edward A. Keller)* (b) Wave erosion was a major cause of this landslide in Cove Beach, Oregon. Further movement of the slide threatens the homes above. *(Gary Braasch/ZUMAPRESS/Newscom)*

∧ FIGURE 7.15 Reactivation of a Slide
(a) Part of the beach (end of thin arrow) in Santa Barbara, California, has been buried by reactivation of an older landslide. Reactivation of the slide may have resulted due to erosion of the old slide toe by ocean waves. *(Donald W. Weaver)* (b) Close-up of the head of this slide where it destroyed two homes. The thick, black arrow in (a) points to the location of this picture. *(Donald W. Weaver)*

material that lose their shear strength and flow as a liquid when disturbed (see Case Study 7.2: Portuguese Bend, California). The shaking of quick clay beneath Anchorage, Alaska, during the 1964 earthquake produced this effect and caused three extremely destructive landslides. In Quebec, Canada, several large slides associated with quick clays destroyed numerous homes and killed more than 70 people. These slides occurred on river valley slopes when initially solid material was converted into a liquid mud as the sliding movement began.[8] The slides in Quebec are especially interesting because the liquefaction of clays occurred without earthquake shaking. In many cases, the slides were triggered by river erosion at the toe of a slope. Liquefaction first started in a small area and spread to a much larger area. Since these movements have often involved the reactivation of older slides, future problems may be avoided by mapping existing slides and restricting development on them.

The Role of Time The forces acting on slopes often change with time. For example, both driving and resisting forces may change seasonally with fluctuations in the moisture content of the slope or with changes in the position of the water table. Much of the chemical weathering of rocks, which slowly reduces their strength, is caused by the chemical action of water in contact with soil and rock near Earth's surface. Soil water is often acidic because it reacts with carbon dioxide in the atmosphere and soil to produce weak carbonic acid that can dissolve rock. This type of chemical weathering is significant in areas underlain by limestone, a rock type that is easily dissolved by carbonic acid. Changes are greater in especially wet years, as reflected by the increased frequency of landslides during or following these years. In other slopes, resisting forces may continuously diminish through time from weathering, which reduces the cohesion in slope materials, or from an increase in underground water pressure within the slope. A slope that is becoming less stable with time may have an increasing rate of creep until failure occurs.

7.2   Portuguese Bend, California

CASE study

The Portuguese Bend landslide along the Southern California coast near Los Angeles is a famous example of how people can increase the landslide hazard. This landslide, which destroyed more than 150 homes, is part of a larger ancient slide (Figure 7.16). Road building and

changes in subsurface drainage associated with urban development reactivated the ancient slide.

Aerial photography of the lower part of the reactivated landslide shows evidence of its movement (Figure 7.17). This evidence includes bare ground west of the highway, where recent movement destroyed homes and roads, and a kink in the pier caused by landsliding into the ocean. Eventually the slow-moving landslide destroyed the pier and the adjacent swim club. Recent movement of the slide started in 1956 during construction of a county road. Fill dirt placed over the upper part of the ancient slide increased its instability. During subsequent litigation, Los Angeles County was found responsible for the landslide.

From 1956 to 1978, the slide moved continually at an average rate of 0.3 to 1.3 cm (~0.1 to 0.5 in.) per day. Several years of above-normal precipitation accelerated the movement to more than 2.5 cm (~1 in.) per day in the late 1970s and early 1980s. Since 1956, the total displacement of the slide near the coast has been more than 200 m (~660 ft.). The above-normal rainfall reactivated a second part of the ancient slide to form the Abalone Cove landslide to the north (Figure 7.16). The second slide prompted additional geologic investigation, and a successful landslide-control program was initiated. Wells were drilled into the wet rocks of the Portuguese Bend landslide in 1980 to remove groundwater from the slide mass. The pumping was an attempt to dry out and stabilize the rocks in the slide. By 1985, the slide had apparently been stabilized. However, precipitation and groundwater conditions in the future will determine the fate of the "stabilized" slides.[9]

During the two-decade period of activity, homes on the Portuguese Bend landslide continued to move. One home constantly shifted position as it moved more than 25 m (~80 ft.). Other homes were not in constant motion but still moved up to 50 m (~160 ft.) in the same time period. Homes remaining on Portuguese Bend during the slide's active period were adjusted every year or so with hydraulic jacks. Utility lines were placed on the surface to avoid breakage as the ground shifted. With one exception, no new homes have been constructed since the landslide began to move. The remaining occupants have elected to adjust to the landslide rather than bear the total loss of their property (Figure 7.18). Nevertheless, few geologists would choose to live there now.

The story of the Portuguese Bend landslides emphasizes that science can be used to understand landslides, sometimes allowing us to at least temporarily stop their movement. However, recognizing landslides and using land-use planning to avoid building on active

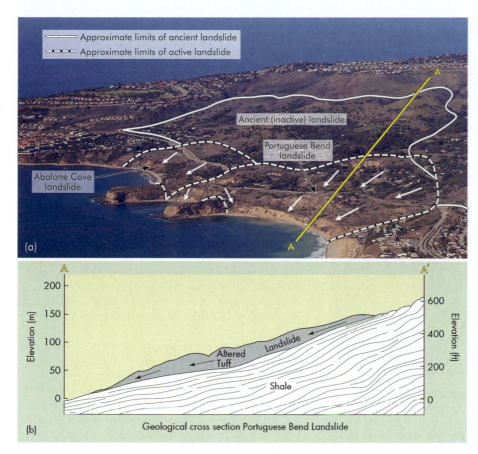

(a)

(b) Geological cross section Portuguese Bend Landslide

< FIGURE 7.16 Ancient Landslides Can Be Reactivated

(a) Extent of ancient inactive landslide, part of which was reactivated to form the Portuguese Bend slide in the 1950s and Abalone Cove slide in the 1970s. Arrows show the direction of landslide movement toward the Pacific Ocean in the foreground. *(California State University)* (b) Cross-section through the right side of the Portuguese Bend slide. The landslide is underlain by a consolidated pyroclasitc volcanic tuff that has been altered to bentonite, a type of expansive clay that deforms readily when a force is applied to it. *(Photograph and cross-section courtesy of Los Angeles County, Department of County Engineer)*

< FIGURE 7.17 Homes Destroyed

The kink in the pier in the upper-left portion of this image shows initial damage from the Portuguese Bend landslide. Eventually most of the homes seen here, as well as the swim club and pier, were destroyed by the slow-moving landslide. *(University of Washington Library Special Collections KC5892)*

∧ FIGURE 7.18 Living with Landslides
Homes, roads, and streets were built on Portuguese Bend southwest of Los Angeles in spite of geologic maps published in the 1940s that revealed most of the area as a landslide. Palos Verdes Drive South, shown here, crosses the landslide and requires constant roadwork as it moves several feet each year toward the ocean. *(Chris Cantelmo)*

slides is preferable to reacting to landslide movement after homes have been constructed.

SNOW AVALANCHES

A **snow avalanche** is the rapid downslope movement of snow and ice, sometimes with the addition of rock, soil, and vegetation. Thousands of avalanches occur every year in the western United States and Canada. As more people venture into avalanche-prone areas and more development takes place, the loss of life and property from avalanches increases. Avalanche accidents in the winter of 2005–06 killed 32 people in North America.[10] Most avalanches that kill people are triggered by the victims themselves or by members of their party.[11]

Three variables interact to create unstable conditions for snow avalanches: steepness of the slope, stability of the snowpack, and the weather. Snow avalanches generally occur on slopes steeper than about 25°.[11] The steepest angle at which snow, or any loose material, is stable is its *angle of repose*. For snow, this angle is affected by temperature, wetness, and shape of the snow grains.[12] Most snow avalanches occur on slopes between 35° and 40°.[12] On slopes steeper than 50° to 60°, the snow tends to continually fall off the slope. Snow-covered slopes may become unstable when the wind piles up snow on the leeward or downwind side of a ridge or hillcrest, when rapid precipitation adds weight to the slope, or when temperatures rapidly warm to make the snow very wet.

There are two common types of snow avalanches: loose-snow and slab. *Loose-snow avalanches* typically start at a point and widen as they move downslope. *Slab avalanches* start as cohesive blocks of snow and ice that move downslope. The latter type of avalanche is far more dangerous and damaging. Slab avalanches typically are triggered by the overloading of a slope or the development of zones of weakness in the snowpack. Millions of tons of snow and ice then move rapidly downslope at velocities of up to 100 km (~60 mi.) per hour.

Avalanches tend to move down tracks, called *chutes*, which have previously produced avalanches (Figure 7.19). As a result, it is easier to prepare maps showing potentially hazardous areas. Avoiding these areas is obviously the preferable and least expensive adjustment to avalanches. Other adjustments include clearing excess snow with carefully placed explosives, constructing buildings and structures to divert or retard avalanches, or planting trees on slopes in avalanche-prone areas to better anchor the snow.

7.1 CHECK your understanding

1. What are slope segments, and how do the common types of slope segments differ?

2. What are the three main ways that materials on a slope may fail?

3. What is the safety factor, and how is it defined?

4. How do slumps (rotational slides) differ from soil slips and rock slides (translational slides)?

5. How and where do debris flows occur?

6. Name the five factors that affect driving and resisting forces on slopes, and explain how these are interrelated.

7. What are the three ways that vegetation affects slope stability?

8. What is the relationship between the downslope force and normal force?

9. What are the two types of snow avalanches, and how do they differ?

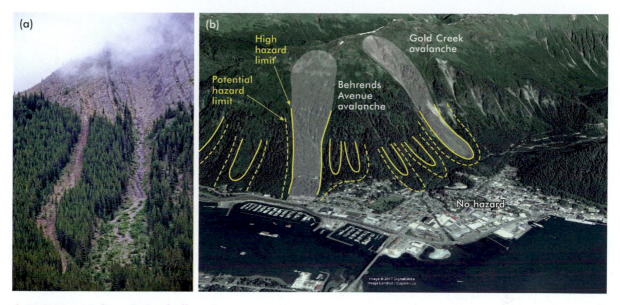

∧ FIGURE 7.19 Snow Avalanche Hazard
(a) Avalanche chutes can be clearly identified in forested areas by an absence of large trees. Only young saplings with flexible trunks survive in areas that experience regular avalanches. *(Robert McGouey/Alamy)* (b) Map of part of Juneau, Alaska, showing areas at highest risk for snow avalanches. Most of the downtown business district is in the "no hazard" area in the lower right. *(Robert McGouey/Alamy)*

7.2 Geographic Regions at Risk from Landslides

As you may imagine, landslides occur everywhere that significant slopes exist, and mountainous areas have a higher risk for landslides than most areas of low relief. The latter generalization is supported by a map of major landslide areas in the contiguous United States (Figure 7.20). On this map, the areas shown in red, which experience frequent landslide activity, are concentrated in the mountainous areas of the West Coast, the Rocky Mountains, and the Appalachian Mountains. The Plains states are flatter and, therefore, relatively safe from landslides. The Rocky Mountains and coastal mountains of the East Coast and West Coast are at particularly high risk. Parts of central Canada, such as Saskatchewan, Manitoba, and eastern Ontario, have a relatively lower risk.

Three factors are expected to increase worldwide landslide activity in the decades to come:

1. Urbanization and development will expand in landslide-prone areas.

2. Tree cutting will continue in landslide-prone areas.

3. Changing global climate patterns will result in regional increases in precipitation.[13]

The consequences of these human activities alter the variables that affect driving and resisting forces on slopes as explained in Section 7.5.

7.2 CHECK your understanding

1. Using Figure 7.20, explain the general distribution of regions at risk from mass wasting.

2. Why is there little risk of mass wasting in the Midwest and Great Plains region of the United States?

3. How might processes involved in urbanization increase or decrease the stability of slopes?

7.3 Effects of Landslides and Linkages with Other Natural Hazards

EFFECTS OF LANDSLIDES

Landslides and related phenomena have the capacity to cause substantial damage and loss of life. In the United States alone, on average 25 people are killed each year

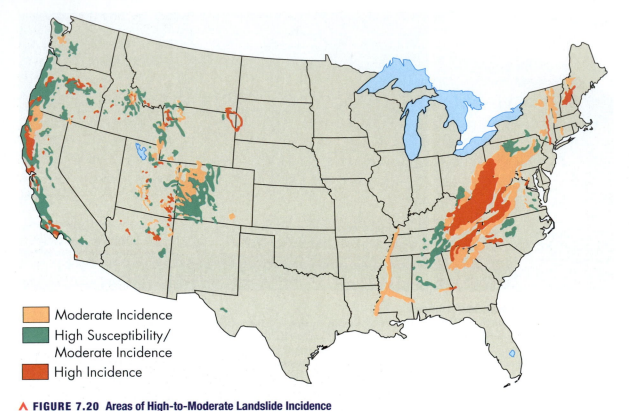

∧ FIGURE 7.20 Areas of High-to-Moderate Landslide Incidence
On this map of landslide areas in the contiguous United States, the colored areas either have a high-to-moderate number of landslides or they have a moderate number of landslides and are highly susceptible to additional landslides. *(U.S. Geological Survey)*

Legend:
- Moderate Incidence
- High Susceptibility/ Moderate Incidence
- High Incidence

by landslides, and this number increases to between 100 and 150 if collapses of trenches and other excavations are included. The total annual cost of damages exceeds $2.8 billion and may be as high as $8.5 billion.[14]

Direct effects of landslides on people and property include being hit with or buried in falling debris (see Case Study 7.4: Survivor Story). Landslides may also damage homes, roads, and utilities that have been constructed on the top or side of a hill. They regularly block roads and railroads, delaying travel for days or weeks. One massive landslide, the 1983 Thistle slide discussed later, blocked commerce on major highways and a transcontinental rail line for months. Rerouting of traffic significantly increased transportation costs in south-central Utah and resulted in the temporary or permanent closure of coal mines, oil companies, motels, and other businesses. Landslides may even block shipping lanes. In 1980, a debris flow from the volcano Mount St. Helens (see Chapter 5) filled the Columbia River with more than 34 million m³ (~44 million yd.³) of sediment. This sediment stopped cargo ships from reaching Portland, Oregon, until dredging was completed.

Indirect effects include flooding upstream from a landslide that blocks a river and the transmission of disease from fungal spores disturbed by landslides, as happened following the 1994 Northridge earthquake (see Chapter 3).

LINKAGES BETWEEN LANDSLIDES AND OTHER NATURAL HAZARDS

Landslides and other types of mass movement are intimately linked to just about every other natural hazard. Earthquakes, volcanoes, storms, and fires all have the potential to cause landslides (see Case Study 7.5: Professional Profile). These relationships are discussed in their respective chapters. As described earlier, landslides themselves may be responsible for flooding if they form an earthen dam across a river. A large landslide may also cause a tsunami or widespread flooding if it displaces water out of a lake or bay. In 1963, more than 2.5 km² (~1 mi.²) of a hillside slid into the Vaiont reservoir in northeastern Italy. The landslide and resulting flood killed more than 2,600 people during this tragedy.

7.3 **CHECK** your understanding

1. Explain how large landslides are related to flooding and tsunamis.

2. How does the annual cost of damage due to landslides compare to other natural hazards?

7.3

CASE study

2014 Oso Landslide, Washington

The deadliest landslide in U.S. history occurred on March 22, 2014, along the North Fork of the Stillaguamish River valley east of Oso, Washington. Killing 43 and destroying 49 homes and other structures in the Steelhead Haven neighborhood, the Oso landslide buried a 2.5 km² (~1 mi²) area several meters deep (Figure 7.21). Although the Director of the County's Department of Emergency Management stated that the slide was "completely unforeseen," engineers, geologists, and local residence for decades had referred to the hillside above the community as "Slide Hill."[15] Moreover, a 2010 study warned that the hillslope was one of the most dangerous in the county.[15]

Historic landslides and evidence of mass wasting all along the Stillaguamish River valley are visible in the landscape (Figure 7.22).[16] In fact, much of the 2014 landslide mass was composed of material that had slumped off of the Whitman Bench in the 2006 Hazel Landslide (Figure 7.22). Including the Hazel slide, the runouts—distance from landslide headscarp to toe—of the three previous landslides, were only 100–200 m (325–650 ft.).[17] Therefore, some experts did not predict that landslide debris from Whitman bench could flow more than 1000 m (~3,300 ft.).[17] However, that is just what happened in 2014.

∧ **FIGURE 7.21** **Deadliest Landslide in U.S. History**
Near complete destruction of the Steelhead Haven neighborhood (shaded area) occurred when millions of tons of earth and rock of the 2014 Oso landslide raced down the steep slope and crossed the North Fork of the Stillaguamish River—killing 43 people.

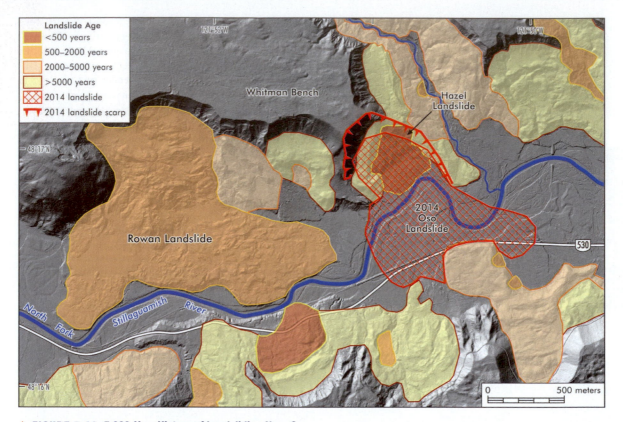

∧ FIGURE 7.22 5,000-Year History of Landsliding Near Oso
Shadded-relief map of the North Fork of the Stillaguamish River highlighting the location of numerous historic
landslides (colored by age) in the area surrounding the location of the 2014 Oso landslide.

For the 20 days preceding the Oso landslide, the region experienced more than 200 percent the normal precipitation for that time of year. It is likely that this high rainfall played a fundamental role in not only triggering the landslide, but also the great mobility of the slide material. The Oso landslide can best be described as a complex slide (see Figure 7.4f). Figure 7.21 clearly shows a single large slump block which detached from the Whitman bench and moved downslope. However the majority of the slide mass, saturated with water, became completely fluidized and flowed across the Stillaguamish River, the Steelhead Haven neighborhood, and State Route 530 beyond (Figure 7.21)

The Oso landslide mass blocked the Stillaguamish River for 10 days following the event causing upstream inundation of the flood plain, and further blocking State Route 530 (Figure 7.21). Concerns about the mud and debris damn failing and unleashing a torrent of water lead the National Weather Service (NWS) to issue a flash flood watch for communities downstream. Although, on April 2 the river began flowing in a new channel across the slide mass, the NWS continued to issue flood warnings for the Stillaguamish Valley for another month until the water-level lowered beyond concern (Figure 7.21).

Some experts have speculated that forestry practices on Whitman Bench over the past 50 years has contributed to the instability of the high banks of the Stillaguamish River.[17] A triangular region that was cleared of all trees in 2004 immediately upslope of the Oso Landslide head scarp can be seen clearly in Figure 7.21. Such forestry practices lead to a loss of sediment cohesion, as well as increased surface and subsurface water runoff all, of which are known to cause instability of steep slopes (see Section 7.1). Consequently, the State of Washington and the forestry company responsible for logging the Whitman Bench paid a $60 million settlement in 2016 to the Oso Landslide survivors and victim's families.[18]

Landslide in Colorado

Danny Ogg Thought He Was "a Goner" When the First Boulder Hit His Truck

Danny Ogg was goin' round the mountain, as they say, on Interstate 70 west of Denver, when a giant granite boulder crashed across his windshield, leaving him with only one thought: "I'm going to be buried alive."

The boulder was one of about 20 or 30 that rocketed down the steep slopes onto the interstate highway, which winds through the Rocky Mountains, early on the morning of April 8, 2004 (Figure 7.4.A). The largest of the rocks weighed 30 tons. Surprisingly, though, Ogg, a 50-year-old driver from Tennessee, said he had no indication that a landslide was occurring until his semitrailer hit the first rock. After

that, it was "bang bang bang, crunch crunch," as he put it, as his truck collided with the falling debris and eventually hit the median. The vehicle's fuel tank was torn open in the process, spilling hundreds of gallons of diesel fuel.

Ogg's sense of time was skewed in the terrifying moments of the landslide. "After I hit the first boulder, it took two, three minutes, I guess. It was fast but it wasn't," he said.

Luckily, Ogg suffered only an injured back and some torn cartilage and ligaments, compared to the much more serious injuries he could have incurred. "I thought I was a goner," he said. "I don't know how long I was in there. It idled me.

I was silly as a goose, as we say in Tennessee."

His sense of location was also askew: "You know you ain't dead yet, but you don't know where you are," he said. "I didn't know if I was hanging off the mountain or what."

Ogg said some of the boulders left potholes in the road a foot or two deep, and a *Rocky Mountain News* article reported that some of them bounced as high as 20 feet after impact.

Although landslides are difficult to predict and avoid, Ogg said he would be more vigilant now when driving in the mountains. "When you go around the mountain now, you look up," he said.

— CHRIS WILSON

◄ FIGURE 7.4.A Boulders on I-70
These granite boulders, some as large as 4 m^3 (~140 ft.^{3}), fell onto I-70 near Glenwood Springs, Colorado, on the morning of May 9, 2003. Heavy rain, combined with ice wedging in cracks within the granite, dislodged one or more of these large rocks from a free face more than 100 m (~330 ft.) above the interstate. One car-sized boulder bounced across the highway and landed in the nearby Colorado River. A highway crew had to drill or blast apart the largest boulders and push them over to the shoulder so traffic could resume. The low retaining wall behind the boulders in this image was shattered in several places. *(USGS)*

7.5 Professional Profile

Bob Rasely, Mass Wasting Specialist

Bob Rasely's line of work may be the very definition of "down and dirty" (Figure 7.5.A). A Utah-based geologist with the Natural Resources Conservation Service, Rasely specializes in predicting the likelihood of mass wasting in the wake of a wildfire. "You don't see the dirt you got on you until you take your shirt off," Rasely said.

While most of the situations Rasely encounters are not emergencies, there are certainly times when the possibility of mass wasting, especially debris flows, poses an immediate threat to populated areas. "If there's a town down

below, we have to rush to get ahead of the next storm," Rasely said. "In the worst-case scenario, we plant someone halfway up the mountain with a radio." In such cases, Rasely said, they often set up evacuation plans that can remain in effect for long periods of time.

The probability and scale of a potential disaster depend on many factors. Rasely classifies most wildfires or "burns" as either light or heavy. Light burns destroy only the surface growth and have a recovery span of about one year, whereas heavy burns destroy the roots and require three

to five years for recovery. "First, we want to know the condition of the burn, if the roots are intact," he said. Light burns increase the erosion rate on a given surface by 3 to 5 times, whereas heavy burns increase the rate 6 to 10 times.

Other factors include the steepness and length of the slope, as well as whether established paths for water to flow already exist on the surface—a bad sign, since this means the water will run off the mountain at a much faster rate. Pine needles and oaks also leave a waxy residue when they burn, which can further expedite the flow, Rasely said.

◄ **FIGURE 7.5.A Geologist Examines Recent Landslide**
Bob Rasely, retired Utah State geologist for the Natural Resources Conservation Service, stands on the uneven, hummocky surface of the Thistle landslide shown in Figure 7.32. Irregular land surfaces such as this are common on landslides. *(Bob Rasely)*

Once the general risk of mass wasting has been assessed, Rasely and his team have a variety of tricks they can use to minimize the potential damage, including erecting fencing made out of mesh material called "geofabric," as well as digging what they call a "debris basin" to catch falling material. "If there's any kind of little terrace, you can stop a lot of stuff," he said.

Rasely's job has taken him to all but two of Utah's mountain ranges, and he frequently works in the many state and national parks in Utah. And although the severity of the situation may vary widely, time is always important in this line of work. "It's by the seat of the pants, live time," Rasely said.

— CHRIS WILSON

7.4 Natural Service Functions of Landslides

Although most hazardous natural processes provide important natural service functions, it is difficult to imagine any good coming from landslides except the creation of new habitats in forests and aquatic ecosystems.

Landslides, like fire, are a major source of ecological disturbance in forests. For some old-growth forests, this disturbance can be beneficial by increasing both plant and animal diversity.[19] In aquatic environments, landslide-dammed lakes create new habitat for fish and other organisms.[20]

Mass wasting can produce deposits that become mineral resources. Weathering frees mineral grains from rocks, and mass wasting transports these minerals downslope. Heavier minerals, particularly gold and diamonds, can be concentrated at the base of the slope and in adjacent streams. Although not as abundant as in stream deposits, gold and diamonds have been mined from colluvium and debris flow deposits.

7.4 CHECK your understanding

1. List the different natural service functions of landslides.
2. What role do mass wasting and erosion play in the mining of precious earth materials, such as gold and diamonds?

7.5 Human Interaction with Landslides

Landslides and other types of ground failure are natural phenomena that would occur with or without human activity. However, in many instances, the expansion of urban areas, transportation networks, and natural resource use has increased the number and frequency of landslides. In other settings where the potential for landslides has been recognized, the number of events has decreased because of preventive measures. For example, grading of land surfaces for housing developments may initiate landslides on previously stable hillsides; on the other hand, building stabilizing structures and improving drainage may reduce the number of landslides on naturally sensitive slopes.

The effect of human endeavors on the magnitude and frequency of landslides varies from nearly insignificant to very significant. In cases where human activities rarely cause landslides, we may still need to learn all we can about this natural phenomenon to avoid development in hazardous areas and to minimize damage. When human activities such as road construction (see Case Study 7.1: A Closer Look) increase the number and severity of landslides, we need to learn how our practices cause slides and how we can control or minimize their occurrence. Following is a description of several human activities that interact with landslides.

TIMBER HARVESTING AND LANDSLIDES

The possible cause-and-effect relationship between timber harvesting and erosion is a major environmental and economic issue in northern California, Oregon, and Washington. Two controversial practices are *clear-cutting*, the harvesting of all trees from large tracts of land, and road building, the construction of an extensive network of logging roads used to remove cut timber from the forest. Landslides, especially shallow soil slips, debris avalanches, and more deeply seated earthflows are responsible for much of the erosion in these areas.

One 20-year study in the Western Cascade Range of Oregon found that shallow slides are the dominant form of erosion. This study also found that timber-harvesting activities, such as clear-cutting and road building, did not significantly increase landslide-related erosion on geologically stable land; however, logging on weak, unstable slopes did increase landslide erosion by several times compared to slopes that had not been logged.[21]

Road construction in areas to be logged can be an especially serious problem because roads can interrupt surface drainage, alter subsurface movement of water, and adversely change the distribution of earth materials on a slope by cut-and-fill or grading operations.[21] As we learn more about erosional processes in forested areas, soil scientists, geologists, and foresters are developing improved management practices to minimize the adverse effects of logging. Nevertheless, landslide erosion problems continue to be associated with timber harvesting (see Case Study 7.3).

URBANIZATION AND LANDSLIDES

Human activities are most likely to cause landslides in urban areas where there are high densities of people, roads, and buildings. Examples from Los Angeles, California, Oso, Washington, and Rio de Janeiro, Brazil illustrate this situation (see Case Studies 7.2, 7.3, and 7.6). Los Angeles in particular (and Southern California in general) has also experienced a high frequency of landslides associated with hillside development. Large variations in topography, rock and soil types, climate, and vegetation make interactions with the natural environment complex and notoriously unpredictable. For this reason, the area has the sometimes dubious honor of showing the ever-increasing value of studying urban geology.[22]

It took natural processes many thousands, and perhaps millions, of years to produce valleys, ridges, and hills. In a little over a century, humans have developed the machines to grade them. Nearly 40 years ago, F. B. Leighton, a geological consultant in Southern California, wrote: "With modern engineering and grading practices and appropriate financial incentive, no hillside appears too rugged for future development."[22] Thus, human activity has become a geological agent that is capable of carving out the landscape at a much faster pace than glaciers and rivers. Almost overnight, we can convert steep hills into a series of flat lots and roads. The grading process, in which benches, referred to as pads, are cut into slopes for home sites, has been responsible for many landslides. Because of the extent of grading, Los Angeles has led the nation in developing codes concerning grading for development.

Grading codes that minimize the landslide hazard have been in effect in the Los Angeles area since 1963. These codes were adopted in the aftermath of destructive and deadly landslides in the 1950s and 1960s (see Case Study 7.2: Portuguese Bend, California). Since these codes have been in effect and detailed engineering geology studies have been required, the percentage of hillside homes damaged by landslides and floods has been greatly reduced. Although initial building costs are greater because of the strict codes, they are more

than balanced by the reduction of losses in subsequent wet years.

Oversteepened slopes, increased water from lawn irrigation and septic systems, and the additional weight of fill material and buildings make formerly stable slopes unstable (Figure 7.23). As a rule, any project that steepens or saturates a slope, increases its height, or places an extra load on it may cause a landslide.[23]

Landslides related to urbanization have also been a problem in the eastern United States. In Cincinnati and surrounding Hamilton County, Ohio, most slides have taken place in colluvium, although clay-rich glacial deposits and soil developed on shale have also slipped.[23] At an average cost of more than $14 million per year (estimates adjusted for 2013), these slides are a serious problem.[14]

In Pittsburgh and surrounding Allegheny County, Pennsylvania, urban construction is estimated to be responsible for 90 percent of the landslides.[24] Most of these slides are slow moving and take place on hillslopes of weathered mudstone or shale. In one deadly exception in an adjacent county, a rockfall crushed a bus and killed 22 passengers. Most of the landslides in Allegheny County are caused by placing fill dirt or buildings on the top of a slope, by cuts in the toe of a slope, or by the alteration of water conditions on or beneath a slope.[20] Damages from these slides average $4 million per year.

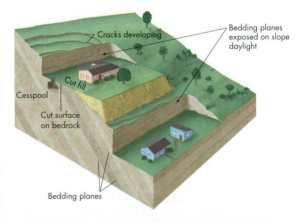

∧ FIGURE 7.23 Urbanization and Landslide Potential
This diagram shows how building on hillslopes can contribute to slope instability. The diagonal lines in the gray side of the diagram are bedding planes in sedimentary rock. Cuts into the hillside behind the houses have removed support from these bedding surfaces. The yellow fill material used to extend the flat pad for building adds weight to the slope. Cracks shown in the upper part of the diagram are an early sign that a landslide is likely to occur soon. Leaking wastewater from "cesspools" and landscape irrigation can add lubricating water to the hillslope. *(Based on F. Beach Leighton, Landslides and Urban Development, 1966)*

Rio de Janeiro, Brazil

Rio de Janeiro, with a population of more than 6.3 million, may have more slope-stability problems than any other city of its size.[25] Several factors contribute to a serious landslide problem in the city and surrounding area: (1) The beautiful granite peaks that spectacularly frame the city (Figure 7.24) have steep slopes of fractured rock that are covered with thin soil, (2) the area is periodically inundated by torrential rains, (3) cut-and-fill construction has seriously destabilized many slopes, and (4) vegetation has been progressively removed from the slopes.

The landslide problem started early in the city's history when many of the slopes were logged for lumber and fuel and to clear land for agriculture. Landslides associated with heavy rainfall followed the early logging activity. More recently, a lack of building sites on flat ground has led to increased urban development on slopes. The removal of additional vegetative cover and the construction of roads have led to building sites at progressively higher elevations. Excavations have cut the base of many slopes and severed the soil cover at critical points. In addition, placing fill material on slopes to expand the size of building sites has increased the load on already unstable land. Since this area periodically experiences tremendous rainstorms, it becomes apparent that Rio de Janeiro has a serious problem.

In February 1988, an intense rainstorm dumped more than 12 cm (~5 in.) of rain on Rio de Janeiro in 4 hours. The storm caused flooding and debris flows that killed at least 90 people and left 3,000 people homeless. Most of the deaths occurred from debris flows in hill-hugging shantytowns where housing is precarious and control of storm water runoff nonexistent. However, more affluent mountainside areas were not spared from destruction. In one area, a landslide destroyed a more affluent nursing home and killed 25 patients and staff. Restoration costs for the entire city have exceeded $100 million. A similar story played out in April 2010, when heavy rains caused numerous landslides that destroyed 60 homes and killed more than 200 people. Much of the damage and loss of life occurred in low-income slums, where, in addition to poor construction practices, the homes were built on the steep slopes that surround the city (Figure 7.24b). If future disasters are to be avoided, Rio de Janeiro must take extensive and decisive measures to control storm runoff and increase slope stability.

▼ FIGURE 7.24 Landslides are Common in the Rio De Janeiro Area
(a) Panoramic view of Rio de Janeiro, Brazil, showing the steep "sugarloaf" mountains and hills (right center of the image). A combination of steep slopes, fractured rock, shallow soils, and intense rainfall contributes to the landslide problem, as do human activities such as urbanization, logging, and agriculture. Virtually all the bare rock slopes were at one time vegetated, and that vegetation has been removed by landslides and other erosional processes. *(B. Martinez/Roger Viollet/Getty Images)* (b) Aerial view of the Morro dos Prazeres slum, where several homes were destroyed by landsliding in Rio de Janeiro on April 8, 2010. *(Antonio Scorza/AFP/Getty Images/Newscom)*

(a)

(b)

7.5 **CHECK** your understanding

1. Explain how timber harvesting, specifically clear-cutting, affects the occurrence of landslides.

2. How does road construction increase the occurrence of landslides?

3. Explain how human activity in Rio de Janeiro has resulted in devastating landslides.

7.6 **Minimizing the Landslide Hazard**

To minimize the landslide hazard, it is necessary to identify areas in which landslides are likely to occur, design slopes or engineering structures to prevent them, warn people of impending slides, and control slides after they have started moving. The most preferable and least expensive option to minimize the landslide hazard is to avoid development on sites where landslides are occurring or are likely to occur.

IDENTIFYING POTENTIAL LANDSLIDES

Recognizing areas with a high potential for landslides is the first step in minimizing this hazard. These areas can be identified where slopes are underlain by sensitive earth materials, such as clay or shale, and by a variety of surface features:

> Crescent-shaped cracks or terraces on a hillside (see Figure 7.23)

> A tongue-shaped area of bare soil or rock on a hillside (Figure 7.25)

> Large boulders or talus piles at the base of a cliff (see Figure 7.2a)

> A linear path of cleared or disturbed vegetation extending down a hillslope (Figure 7.19)

> Exposed bedrock with layering that is parallel to the slope (see Figure 7.8b)

> Tongue-shaped masses of sediment, especially gravel, at the base of a slope (Figure 7.25) or at the mouth of a valley

> **FIGURE 7.25 Evidence of a Mass Wasting**
The long, narrow area of bare soil on this Northern California hillslope was created by a debris flow. Trees and sediment carried by the flow are in a tongue-shaped pile at the bottom of the hill along the bank of the Klamath River. A logging road at the head of the flow (end of the arrow) may have helped destabilize the slope. (*Edward A. Keller*)

> An irregular, often referred to as *hummocky*, land surface at the base of a slope (Figure 7.23).

Geologists look for these features in the field and on aerial photographs. This information is used to assess the hazard and to produce several types of maps.

The first type of map is the direct result of the landslide inventory just described. It may be a reconnaissance map showing areas that have experienced slope failure or a more detailed map showing landslide deposits in terms of their relative activity (Figure 7.26a). Information concerning past landslides may be combined with land-use considerations to develop a *slope stability map* for engineering geologists or a *landslide hazard map* with recommended land uses (Figure 7.26b) for planners. These maps do not take the place of detailed evaluation of a specific site. Preparing a *landslide risk map* is more complicated because it involves evaluating the probability of a landside occurring and an assessment of potential losses.[26]

PREVENTING LANDSLIDES

Prevention of large, natural landslides is difficult, but common sense and good engineering practices can help minimize the hazard. For example, loading the top of slopes, cutting into sensitive slopes, placing fill material on slopes, or changing water conditions on slopes should be avoided or done with caution.[19] Common engineering techniques for landslide prevention include surface and subsurface drainage, removal of

Head of flow

Track of flow

Flow deposits

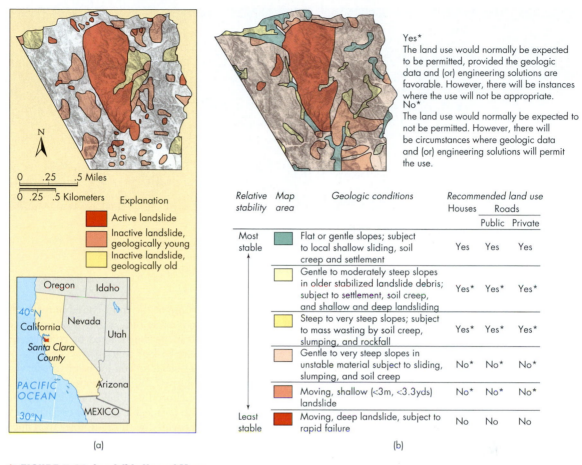

Yes*
The land use would normally be expected to be permitted, provided the geologic data and (or) engineering solutions are favorable. However, there will be instances where the use will not be appropriate.
No*
The land use would normally be expected to not be permitted. However, there will be circumstances where geologic data and (or) engineering solutions will permit the use.

Relative stability	Map area	Geologic conditions	Recommended land use		
			Houses	Roads	
				Public	Private
Most stable		Flat or gentle slopes; subject to local shallow sliding, soil creep and settlement	Yes	Yes	Yes
		Gentle to moderately steep slopes in older stabilized landslide debris; subject to settlement, soil creep, and shallow and deep landsliding	Yes*	Yes*	Yes*
		Steep to very steep slopes; subject to mass wasting by soil creep, slumping, and rockfall	Yes*	Yes*	Yes*
		Gentle to very steep slopes in unstable material subject to sliding, slumping, and soil creep	No*	No*	No*
		Moving, shallow (<3m, <3.3yds) landslide	No*	No*	No*
Least stable		Moving, deep landslide, subject to rapid failure	No	No	No

(a)   (b)

Λ FIGURE 7.26 Landslide Hazard Maps
(a) Landslide inventory map and (b) a landslide risk and land-use map for part of Santa Clara County, California. Goals and tasks of the landslide part of a ground-failure hazards reduction program. *(Based on U.S. Geological Survey, "Goals and Tasks of the Landslide Part of a Ground-failure Hazards Reduction Program," U.S. Geological Survey Circular 880, 1982)*

unstable slope materials, construction of retaining walls or other supporting structures, or some combination of these techniques.[3]

Drainage Control Surface and subsurface drainage control is usually effective in stabilizing a slope. The objective is to divert water to keep it from running across or infiltrating the slope. Surface runoff can be diverted around the slope by a series of surface drains (Figure 7.27a). The amount of water infiltrating a slope can also be controlled by covering the slope with an impermeable layer, such as soil-cement, asphalt, or even plastic (Figure 7.27b). Underground water can be inhibited from entering a slope by subsurface drains. To construct a drain, a drainpipe with holes along its length is surrounded with permeable gravel or crushed rock and is positioned underground to intercept and divert water away from a potentially unstable slope.[3]

Grading Although grading of slopes for development has increased the landslide hazard in many areas, carefully planned grading can increase slope stability. In a single cut-and-fill operation, material from the upper part of a slope is removed and placed near the base. The overall gradient is thus reduced, and material is removed from the upper slope, where it contributes to the driving force, and placed at the toe of the slope, where it increases the resisting force. However, this method is not practical on a high, steep slope. Instead, the slope may be cut into a series of benches or steps, each of which contains surface drains to divert runoff. The benches reduce the overall slope and are good collection sites for falling rock and small slides (Figure 7.28).[3]

∧ **FIGURE 7.27 Two Ways to Increase Slope Stability**
(a) Concrete drain on a roadcut that removes surface water runoff before it can infiltrate the slope. *(Edward A. Keller)* (b) Workers are covering this slope in Greece with soil-cement to reduce the infiltration of water and provide strength. Soil-cement is a blend of pulverized soil, Portland cement, and water. *(Edward A. Keller)*

∧ **FIGURE 7.28 Benches on a Highway Roadcut**
The stepped surfaces, or benches, seen in the upper-right part of this image reduce the overall steepness of the slope and provide better drainage. Benches along roadcuts into bedrock can catch rockfalls before they reach a highway. *(Edward A. Keller)*

Slope Supports One of the most common methods of slope stabilization is a retaining wall that supports the slope. These walls can be constructed of concrete or brick; stone-filled wire baskets called *gabions* or series of piles of long concrete, steel, or wooden beams are driven into the ground (Figure 7.29). To function effectively, the walls must be anchored well below the base of the slope, back-filled with permeable gravel or crushed rock (Figure 7.30), and provided with drain holes to reduce water pressure in the slope (Figure 7.29). With the addition of plants, these walls can be aesthetically pleasing or blend in with the natural hillside (Figure 7.31).

Preventing landslides can be expensive, but the rewards can be well worth the effort. Estimates of the benefit-to-cost ratio for landslide prevention range from 10 to 2,000. That is, for every dollar spent on

< **FIGURE 7.29 How to Support a Slope**
The types of slope support shown in this illustration include a roadside concrete retaining wall that is deeply anchored into the slope, concrete or steel piles sunk into stable rock, and subsurface drains that reduce water pressure within the slope.

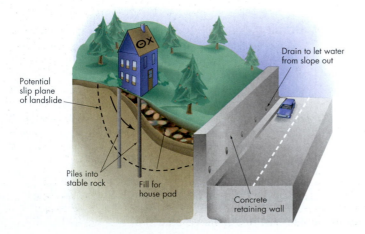

∧ FIGURE 7.30 Retaining Wall
This retaining wall of concrete cribbing was installed and then backfilled to help stabilize the roadcut. *(Edward A. Keller)*

landslide prevention, the savings will range from $10 to $2,000.[27] The cost of *not* preventing a slide is illustrated by the massive Thistle landslide southeast of Salt Lake City in April 1983. This slide moved down a mountain slope and across a canyon to form a natural dam about 60 m (~200 ft.) high. This dam created a lake that flooded the community of Thistle, the Denver–Rio Grande Railroad switchyard and tracks, and two major U.S. highways (Figure 7.32).[26] The total costs (direct and indirect) from the landslide and associated flooding exceeded $400 million.[28]

The Thistle slide was a reactivated older slide, which had been known for many years to be occasionally active in response to high precipitation. It came as no surprise that those extremely high rainfall totals were produced by a climatic event called El Niño (see Chapter 12). In fact, a review of the landslide history suggests that this slide was recognizable, predictable, and preventable! A network of subsurface and surface drains would have prevented failure. The cost of preventing the landslide was between $300,000 and $500,000, a small amount compared to the damages caused by the slide.[27] Because the benefit-to-cost ratio in landslide prevention is so favorable, it seems prudent to evaluate active and potentially active landslides in areas where considerable damage may be expected and possibly prevented.

LANDSLIDE WARNING SYSTEMS

Landslide warning systems do not prevent landslides, but they can provide time to evacuate people and their possessions and stop trains or reroute traffic. Surveillance provides the simplest type of warning. Hazardous areas

(a)

(b)

(c)

∧ FIGURE 7.31 Steps in Making a Retaining Wall
(a) This shallow slide occurred in the early 1990s. *(Edward A. Keller)* (b) A retaining wall of cemented stone was constructed in 1999 to correct the problem. *(Edward A. Keller)* (c) Benches behind the retaining walls were landscaped as shown by this 2001 photograph. *(Edward A. Keller)*

Slide blocks canyon

∧ **FIGURE 7.32 Landslide Blocks a Canyon**
The costliest landslide in U.S. history, this 1983 slide at Thistle, Utah, was a reactivation of an older slide. Debris from the slide blocked the Spanish Fork River and created a natural dam. This dam produced a lake that flooded the community of Thistle, the Denver–Rio Grande Railroad, and two major U.S. highways. Nearly the entire extent of the landslide moved again in 1999. *(Michael Collier)*

can be visually inspected for apparent changes, and small rockfalls on roads and other areas can be noted for quick removal of debris. Human monitoring has the advantages of reliability and flexibility but becomes difficult during adverse weather and in hazardous locations.[29] For example, snow avalanche conditions in the United States have been monitored for decades by park rangers and volunteers of the National Ski Patrol.

Other warning methods include electrical systems, tiltmeters, and geophones that pick up vibrations from moving rocks. Many U.S. and Canadian railroads have slide fences on slopes above their tracks that are tied to signal systems. When a large rock hits the fence, a signal is sent to stop trains before they are in danger. This and other mitigation measures have significantly reduced the number of railroad accidents, injuries, and fatalities.[29]

In western Washington State, the U.S. Geological Survey and Pierce County joined forces in 1995 to develop and deploy acoustic flow monitors to detect debris flows (lahars) from Mt. Rainier.[30] By detecting the vibrations caused by a large debris flow, the automated detection and warning system is capable of giving communities in danger between 40 minutes and three hours to evacuate. With no warning a large debris flow from the mountain, similar to one that occurred about 500 years ago, would devastate more than a dozen communities and kill thousands or possibly tens of thousands of people.

For existing slides, shallow wells can be drilled into a slope and monitored to signal when the slide contains a dangerous amount of water. In some regions, a rain-gauge network is useful for warning when a precipitation threshold has been exceeded and shallow soil slips are probable.

7.6 CHECK your understanding

1. What types of surface features are associated with landslides?

2. What are the main steps that can be taken to prevent landslides?

3. Explain how a cut-and-fill operation can increase the safety factor of a particular slope.

7.7 Perception of and Adjustment to the Landslide Hazard

PERCEPTION OF THE LANDSLIDE HAZARD

Homeowners talking about landslides typically say, "It could happen on other hillsides, but never this one."[25] Just as flood hazard mapping does not prevent development in flood-prone areas, landslide hazard maps will not prevent many people from moving into hazardous areas. Prospective hillside occupants who are initially unaware of the hazard may not be swayed by technical information. The infrequency of large slides reduces awareness of the hazard, especially where evidence of past events is not readily visible. Unfortunately, it often takes catastrophic events to bring the problem to people's attention. In the meantime, people in many parts of the Rocky Mountains, Appalachian Mountains, California, and elsewhere continue to build homes in areas subject to future landslides.

ADJUSTMENT TO THE LANDSLIDE HAZARD

Although the most reasonable adjustment to the landslide hazard is simply not to build in landslide-prone areas, many people still go ahead. As long as people continue to build and buy expensive homes "with a view," we will need to find other adjustments to avoid death and damage from landslides. Such adjustments can include locating critical facilities away from landslide-prone areas and landslide correction.

Siting of Critical Facilities As with earthquakes (Chapter 3), safely siting critical facilities such as hospitals, schools, and police stations is crucial. Ensuring that these buildings are not located on or directly below hillsides is a simple way to guarantee that they will remain functional in the event of a landslide. In urban areas that have been built up against mountains, such as Juneau, Alaska, such selective siting can be difficult.

Landslide Correction After a slide has begun, the best way to stop it is to attack the process that started the slide. In most cases, the cause is an increase in underground water pressure within and below the slide. The pressure may be reduced by an effective drainage program. This program often includes surface drains at the head of the slide to reduce surface water infiltration and subsurface drainpipes or wells to remove groundwater and lower the water pressure. Draining tends to increase the resisting force of the slope material, thereby stabilizing the slope.[7]

PERSONAL ADJUSTMENTS: WHAT YOU CAN DO TO MINIMIZE YOUR LANDSLIDE HAZARD

Consider the following advice if purchasing property on a slope:

> Landslides often develop in areas of complex geology, and a geologic evaluation by a professional geologist is recommended for any property on a slope.

> Avoid homes at the mouth of a valley or canyon, even a small one, where debris flows may originate from upstream slopes and travel down the channel.

> Consult local agencies such as city or county engineering departments that may be aware of landslides in your area.

> Watch out for "little landslides" in the corner of the property—they usually get larger with time.

> If purchasing a home, look for cracks in house walls and look for retaining walls that lean or are cracked. Be wary of doors or windows that stick or floors that are uneven. Foundations should be checked for cracks or tilting. If cracks in walls of the house or foundation can be followed to the land outside the house, be especially concerned that a landslide may be present.

> Be wary of leaks in a swimming pool or septic tank, trees or fences tilted downslope, or utility wires that are taut or sagging.

> Be wary if small springs are present because landslides tend to leak water. Look for especially "green" areas, other than a septic system leach field, where more subsurface water is present.

> Walk the property and surrounding property, if possible, looking for linear or curved cracks (even small ones) that might indicate instability of the land surface.

> Look for the surface features listed earlier that geologists use to identify potential landslides.

> Although correcting a potential landslide problem is often cost-effective, it can still be expensive; much of the fix is below ground where you will never see the improvement. Overall it is better not to purchase land with a potential landslide hazard.

> The presence of one or more of these features does not prove that a landslide is present or will occur. For example, cracks in walls and foundations, doors or windows that stick, or floors that are uneven can be caused by soils that shrink and swell. However, further investigation is warranted if these features are present.

If you enjoy snow sports, hike, climb, live, or travel in areas that have snow avalanches, you should be aware of the following:

> Most avalanches that involve people are triggered by the victims themselves or by others in their party.[11]

> Obtain forecasts from the nearest avalanche center. Remember that changing weather conditions can affect the avalanche hazard.

> Most people who survive avalanches are rescued by the other members in their party.[12]

> Learn avalanche safety procedures and how to evaluate snow conditions before traveling in an area that has avalanches.

7.7 CHECK your understanding

1. List five signs you would look for as evidence of potential landslides if you were inspecting a piece of property and house for purchase.

2. With the exception of buying property on a low-relief surface, what is the best thing you could do to minimize the probability of a landslide occurring on your land? Explain.

APPLYING the ⑤ Fundamental Concepts

La Conchita, Southern California: Landslide Disaster

① **Science helps us predict hazards.**

② **Knowing hazard risks can help people make decisions.**

③ **Linkages exist between natural hazards.**

④ **Humans can turn disastrous events into catastrophes.**

⑤ **Consequences of hazards can be minimized.**

La Conchita is one of only three small communities occupying a 25 km (~16 mi.) long stretch of land along California's scenic coastline between Ventura and Carpinteria. Development along this section of the coast is fairly limited because there are only a few patches between the mountains and the Pacific Ocean wide enough for development. In most places, the coastal strip is barely wide enough for U.S. Highway 101 to navigate the rugged coastline (Figure 7.33). To most people driving up the coast, with little or no education in the geological sciences or natural hazards, La Conchita would appear to be an idyllic location to own a home.

The high topography along this stretch of the coastline is indicative of active tectonic uplift of the region. One of the highest peaks in this part of Southern California is Rincon Mountain, which is uplift on the Red Mountain reverse fault east of La Conchita (Figure 7.33). The Red Mountain fault has extremely high rates of deformation and is also responsible for the steep 180 m (~800 ft.) high sea cliff that looms above the community (Figure 7.33 and Figure 7.34a). Exposed in the sea cliff about 250 m (~820 ft.) above La Conchita are unconsolidated beach sand and shell material that were deposited at sea level 40,000 years ago and have since been uplifted by the Red Mountain and other parallel faults in the area. This rapid rate of uplift is similar to that of the Himalayan

Mountains! In addition to the obvious earthquake hazard presented by the presence of such an active faulting so close to La Conchita, movement along the fault during seismic events has caused pervasive fracturing of the older rocks that underlie the marine deposits in the sea cliff.

Even if no knowledge of the fault-related damage of the bedrock was known before the 1995 event, the potential for a major landslide of the sea cliff behind La Conchita was clear. While driving U.S. 101, landslide after landslide in the slopes above can be seen all along this section of the coastline. Some of these slides are recent, while others are probably thousands of years old, including the prehistoric slide within which the 1995 and 2005 events were located (Figure 7.34a). Furthermore, every few years, including in 1995 and 2005, Highway 101 is closed for a day or so due to mudflows covering the roadway (Figure 7.34b).

During the summer of 1994, about six months before the 1995 landslide, cracking in the upper part of the slope was observed above La Conchita.[31] Cracking was a clear sign of movement, and these cracks continued to grow into the rainy season. By December 1994 open cracks had started to channel water into the subsurface beneath the slope. January 1995 was a particularly wet month for the region with about 62.5 cm (~24.6 in.) of rain, which is more than the total annual average rainfall and six times as much as usually falls in January. Although February was a relatively dry month, heavy rains fell again on March 2–3. On March 4, 1.3 million m^3 (~1.7 million $yd.^3$) of rock and soil started to move downslope toward La Conchita.

It took the landslide several minutes to move about 100 m (~330 ft.) down the slope and into the community. The moving mass formed a complex landslide with most of the material

< FIGURE 7.33 Geology and Topography of La Conchita
False color hillshade image of part of the California coast showing the location of La Conchita with respect to important geologic and topographic features. The dashed white line shows the approximate boundaries of the proposed Rincon Mountain megaslide.

(a)

(b)

< FIGURE 7.34 Mass Wasting Activity Near La Conchita
(a) Annotated photograph of La Conchita illustrating the locations of the ancient and active landslides near the 1995 and 2005 La Conchita slides. *(Modified after Gurrola, L. D., DeVecchio, D. E., Keller, E. A., 2010, Rincon Mountain Megaslide, La Conchita, Ventura County, California,* Journal of Geomorphology *114: 311–18)* (b) Debris flows are common along the La Conchita coast, such as this one that occurred on the same day as the deadly 2005 La Conchita mudflow. The debris flow covered about 0.5 km (~0.3 mi.) of U.S. 101 just north of La Conchita. The flow broke through concrete barriers and continued onto the beach on the left. *(©Getty Images/Mathew Imaging/Contributor)*

moving as a slump block, while small regions lost cohesion and moved as an earthflow down the slope (see Figure 7.7). It is thought that the two days of heavy rain preceding the slide caused the slope failure by adding to the already elevated groundwater level that had built up due to January's extraordinarily high rainfall.[1] When the mass finally came to rest, it was 120 m (~400 ft.) long, covered about 10 acres, and had destroyed or severely damaged nine houses (see Figure 7.1a). Fortunately, the slide moved slowly enough that residents of the community were able to leave their homes and get out of the way. La Conchita was immediately declared a geologic hazard area by Ventura County, and restrictions on new construction were imposed.

Following the 1995 landslide, the county commissioned a study to evaluate the risk of future events and to estimate the cost of stabilizing the sea cliff behind La Conchita. One of the key findings of the study showed that landslides are a common feature along this part of the coast and that the land beneath the homes is composed of about 25 m (~80 ft.) of landslide deposits. Age dating of these deposits showed that these sediments had all accumulated in the last 6,000 years.[32] A more recent study suggests that a large part of Rincon Mountain is part of a mega-landslide, similar to Portuguese Bend (see Case Study 7.2: Portuguese Bend, California), that occurred about 20,000 years ago.[33] Ironically, it appears as though La Conchita owes its existence to landslides, which have provided the sediment over the past several thousand years to build the broad strip of land on which the community is constructed.

The study also proposed constructing benches to reduce the slope of the sea cliff and installing an elaborate dewatering system, measures estimated to total about $150 million. However, due to the prohibitive cost of engineering such a structure for this small community, a less aggressive approach was undertaken, which included a limited drainage system and a modest wall designed to prevent mud from small slides and hillslope erosion from covering the road (Figure 7.35). This wall was not designed to stop another slide, and to make things worse, part of the landslide toe had to be removed to construct the wall, which likely weakened the slope.

Between 1995 and 2005, residents were repeatedly warned of the landslide hazard. Some people moved away following the 1995 slide; others, however, decided to stay or had no choice but to stay due to decreased property values following the slide. Others felt that the area was safer following the slide, and some people bought new property in the community. These were personal decisions based on individual perceptions of the hazard. However, the question was never whether another slide would occur, but when!

(a)   (b)

⋀ FIGURE 7.35   Retaining Wall Built Following the 1995 Landslide
(a) This retaining wall was constructed following the 1995 landslide and was installed to stop loose material from covering the road but was not designed to prevent another landslide. *(R. L. Schuster/USGS)* (b) The retaining wall was partially destroyed and locally overtopped during the 2005 mudflow (see also Figure 7.1b). *(USGS)*

◄ **FIGURE 7.36 Earthflow Buries People, Houses, and Cars** The toe of the 2005 La Conchita landslide shown here buried at least 10 people and damaged or destroyed about 30 houses. When this mass of mud and debris flowed into and over these homes, it had the consistency of thick concrete pouring out of a cement truck. *(Edward A. Keller)*

The 2005 La Conchita landslide occurred on January 10 following 15 days of near-record rainfall levels in many parts of Southern California. U.S. Highway 101 had been shut down earlier that day due to a debris flow from one of the canyons northwest of La Conchita that had buried the roadway (Figure 7.34b). News crews were on site filming the aftermath of the debris flow and highway closure when at 12:30 P.M. the sea cliff above La Conchita started to move again. Unlike the 1995 landslide, the 2005 slide moved rapidly down the slope as an incohesive mass of flowing debris, which was actually caught on film by news crews. The earthflow was composed almost entirely of remobilized material from the southeastern portion of the 1995 landslide (Figure 7.34a). Although only about one-fifth the volume of the 1995 slide, the fluid flow of the 2005 landslide knocked down and locally overtopped the retaining wall built after the 1995 slide and moved into La Conchita, causing widespread damage, destroying 13 homes and severely damaging 23 others (Figure 7.36 and see Figure 7.1b). It is estimated that the flow moved at about 5 m (~15 ft.) per second through La Conchita.[1] Those who were outside and saw the flow coming ran for their lives, while those indoors had no time to escape; unfortunately, 10 people were killed when their homes were buried in the muddy debris.

The residents of La Conchita organized after the 2005 mudflow to coordinate with government officials on the best way to protect the community from yet another future landslide. Govenor Arnold Schwarzenegger allocated $667,000 toward scientific research into what could be done to stop mass wasting of the sea cliff behind La Conchita in March 2006. A lawsuit filed by residents of the community against the La Conchita Ranch Co., which operated an avocado ranch above the community (Figure 7.34a), found that the ranch was 50 percent negligent due to a lack of adequate drainage during the heavy rains that preceded the 2005 slide. Although the courts found human negligence to be partly to blame for the La Conchita disaster, it is clear now that the nature of the location, the original site selection for the community, and the residents' decision to stay were to blame for the damage and loss of life.

A 2010 report suggests that most of the homes could be protected from future events for a cost of only $50 million.[34] However, to date very little has been done to stabilize the slope. Moreover, the fear of triggering yet another slide by removing debris from the 2005 slide means the residents of La Conchita live with a daily reminder of the mass wasting hazard they face, and those who choose to live there do so at their own risk.

APPLYING THE 5 FUNDAMENTAL CONCEPTS

1. Discuss the evidence of the landslide hazard that was known before the landslide disasters in La Conchita. What precursor events occurred before the 1995 and 2005 landslides?

2. What steps could have been taken to reduce the landslide risk in La Conchita following the 1995 landslide, and what factors prevented these steps from being implemented?

3. Describe the linkages between the landslide hazard in La Conchita and other natural hazards.

4. Explain how and why human error made the disaster at La Conchita even worse.

5. How were the residents of La Conchita and Ventura County and state government agencies involved in attempting to minimize the consequences of landslides in the community?

CONCEPTS in review

7.1   An Introduction to Landslides

LO:1   Describe slope processes and the different types of landslides.

LO:2   Analyze the forces that act on slopes and how they affect the stability of a slope.

- The most common landforms are slopes—dynamic, evolving systems in which surface material is constantly moving downslope in the process of mass wasting at rates ranging from imperceptible creep to thundering avalanches.
- Slope failure may involve flowage, sliding, or falling of earth materials; landslides are often complex combinations of sliding and flowage.
- The forces that produce landslides are determined by the interactions of several variables: the type of earth material on the slope, topography and slope angle, climate, vegetation, water, and time.
- Determining the cause of most landslides can be accomplished by examining the relations between forces that tend to make earth materials slide, the driving forces, and forces that tend to oppose movement, the resisting forces.
- The most common driving force is the weight of the slope materials, while the most common resisting force is the shear strength of the slope materials. Geologists and engineers determine the safety factor of a slope by calculating the ratio of resisting forces to driving forces; a ratio greater than 1 means that the slope is stable; a ratio less than 1 indicates potential slope failure. The type of rock or soil on a slope influences both the type and the frequency of landslides.
- Water has an especially significant role in producing landslides. Moving water in streams, lakes, or oceans erodes the base of slopes, increasing the driving forces. Excess water within a slope increases both the weight and the underground water pressure of the earth material, which in turn decreases the resisting forces on the slope.
- Snow avalanches present a serious hazard on snow-covered, steep slopes. Loss of human life from avalanches is increasing as more people venture into mountain areas for winter recreation.

KEY WORDS

colluvium (p. 265), **creep** (p. 262), **debris flows** (p. 266), **driving forces** (p. 263), **falling** (p. 260), **flow** (p. 262), **landslides** (p. 258), **mass wasting** (p. 258), **resisting forces** (p. 263), **safety factor** (p. 263), **shear strength** (p. 263), **sliding** (p. 260), **slumping** (p. 262), **snow avalanche** (p. 272), **talus** (p. 259)

7.2, 7.3, and 7.4   Geographic Regions at Risk from Landslides Effects of Landslides and Linkages with Other Natural Hazards Natural Service Functions of Landslides

LO:3   Evaluate what geographic regions are at risk from landslides.

LO:4   Compare different mass wasting processes and their linkages with other natural hazards.

- Landslides may occur just about anywhere that slopes exist. The areas of greatest hazard in the United States include the mountainous areas of the West Coast and Alaska, the Rocky Mountains, and the Appalachian Mountains.
- Landslides are also linked to other hazards, especially floods, earthquakes, and wildfires.
- Landslides are locally responsible for creation of new habitats in forest and aquatic ecosystems and the generation of new land along the continental coasts.
- Landslides also expose deeply buried rocks and may expose and move precious minerals and elements so they may be concentrated at the base of the slope or in river deposits to be mined.

7.5   Human Interaction with Landslides

LO:5   Describe the ways in which people can affect the landslide hazard.

- The effects of land use on the magnitude and frequency of landslides range from insignificant to very significant. Where landslides occur independent of human activity, we need to avoid development or provide protective measures. In other cases, when land use has increased the number and severity of landslides, we need to learn how to minimize their recurrence.
- In some cases, filling large water reservoirs has altered groundwater conditions along their shores and caused slope failure. Logging operations on weak, unstable slopes have increased landslide erosion. Grading of slopes for development has created or increased landslide problems in many urbanized areas of the world.

7.6   Minimizing the Landslide Hazard

LO:6   Explain the adjustments people can make to avoid death and damage caused by landslides.

- To minimize the landslide hazard, it is necessary to establish identification, prevention, and correction procedures. Monitoring and mapping techniques help identify hazardous sites.
- Identification of potential landslides has been used to establish grading codes that have reduced landslide damage.
- Preventing large natural slides is difficult, but careful engineering practices can minimize the hazard where they cannot be avoided. Engineering techniques for landslide prevention include drainage control, proper grading, and construction of supports such as retaining walls.
- Efforts to stop or slow existing landslides must attack the processes that started the slide—usually by initiating a drainage program that lowers water pressure in the slope. Even with these improvements in recognizing, predicting, and mitigating landslides, the incidence of landslides is expected to increase throughout this century.

7.7   Perception of and Adjustment to the Landslide Hazard

LO:7   Summarize the ways in which you can protect yourself and property from the processes of mass wasting.

- Most people perceive the landslide hazard as minimal unless they have had prior experience. Furthermore, hillside residents, like floodplain occupants, are not easily swayed by technical information. Nevertheless, the wise person will have a geologist inspect property on a slope before purchasing.

CRITICAL thinking questions

1. Your consulting company is hired by the national park department in your region to estimate the future risk from landsliding. Develop a plan of attack that outlines what must be done to achieve this objective.
2. Why do you think that few people are easily swayed by technical information concerning hazards such as landslides? Assume you have been hired by a municipality to make its citizens more aware of the landslide hazard on the steep slopes in the community. Outline a plan of action and defend it.
3. The Wasatch Front in central Utah frequently experiences wildfires followed by debris flows that exit mountain canyons and flood parts of communities built next to the mountain front. You have been hired by the state emergency management office to establish a warning system for subdivisions, businesses, and highways in this area. How would you design a warning system that will alert citizens to evacuate hazardous areas?

References

1. **Jibson, R.** 2005. *Landslide hazard at La Conchita, California*. U.S. Geological Survey, Open-File Report 2005–1067, 12 p.

2. **Griggs, G.** 2006. State to study La Conchita's slide problem. *Los Angeles Times*. http://articles.latimes.com/2006/mar/31/local/me-laconchita31. Accessed 9/15/13.

3. **Pestrong, R.** 1974. *Slope stability*. American Geological Institute. New York: McGraw-Hill.

4. **Nilsen, T. H., Taylor, F. A.,** and **Dean, R. M.** 1976. *Natural conditions that control landsliding in the San Francisco Bay Region*. U.S. Geological Survey Bulletin 1424.

5. **Burroughs, E. R. Jr.,** and **Thomas, B. R.** 1977. *Declining root strength in Douglas fir after felling as a factor in slope stability*. USDA Forest Service Research Paper INT-190.

6. **Campbell, R. H.** 1975. *Soil slips, debris flows, and rainstorms in the Santa Monica Mountains and vicinity, southern California*. U.S. Geological Survey Professional Paper 851.

7. **Terzaghi, K.** 1950. Mechanism of landslides. In *Application of geology to engineering practice*, ed. S. Paige, pp. 83–123. Geological Society of America Berkeley Volume.

8. **Leggett, R. F.** 1973. *Cities and geology*. New York: McGraw-Hill.

9. **Ehley, P. L.** 1986. The Portuguese Bend landslide: Its mechanics and a plan for its stabilization. In *Landslides and landslide mitigation in southern California*, ed. P. L. Ehley, pp. 181–90. Guidebook for fieldtrip, Cordilleran Section of the Geological Society of America meeting, Los Angeles, California.

10. **WestWide Avalanche Network.** 2007. *US & Canada, avalanche fatalities & close calls*. Alta, UT: Center for Snow Science. www.avalanche.org. Accessed 6/30/07.

11. **Abromeit, D., Deveraux, A. M.,** and **Overby, B.** 2004. *Avalanche basics*. U.S. Forest Service National Avalanche Center. www.fsavalanche.org/basics/basic_index.html. Accessed 6/29/07.

12. **Perla, R. I.,** and **Martinelli, M. Jr.** 1975. *Avalanche handbook*. U.S. Department of Agriculture, Agricultural Handbook 489.

13. **Schuster, R. L.** 1996. Socioeconomic significance of landslides. In *Landslides investigation and mitigation*, ed. A. K. Turner and R. L. Schuster, pp. 12–35. Transportation Research Board Special Report 247, National Research Council. Washington, DC: National Academy Press.

14. **Flemming, R. W.,** and **Taylor, F. A.** 1980. *Estimating the cost of landslide damage in the United States*. U.S. Geological Survey Circular 832.

15. *Seattle Times*. 2014. Risk of slide "unforeseen"? Warnings go back Decades. March 24. http://old.seattletimes.com/html/localnews/2023218573_mudslidewarningsxml.html. Accessed 6/23/17.

16. **LaHusen, S. R., Duvall, R, A., Booth, A. M.,** and **Montgovery, D. R.** 2016. Surface roughness dating of long-runout landslides near Oso, Washington (USA), reveals persistent postglacial hillslope instability. *Geology* 44(2): 111–14.

17. **Geotechnical Extreme Events Reconnaissance Report.** *The 22 March 2014 Oso Landslide, Snohomish County, Washington*. https://snohomishcountywa.gov/DocumentCenter/View/18180. Accessed 6/23/17.

18. **CBS News.** 2016. Families reach $10 million settlement with Grandy Lake Forest Associates over deadly 2014 landslide. October 10. www.cbsnews.com/news/families-10-million-settlement-grandy-lake-forest-associates-deadly-2014-landslide/. Accessed 10/10/16.

19. **University of California, SNEP Science Team and Special Consultants.** 1996. Sierra Nevada ecosystems. In *Status of the Sierra Nevada*, Sierra Nevada Ecosystem Project. Final report to Congress, Berkeley, CA, v. 1. http://ceres.ca.gov/snep/pubs/web/PDF/v1_ch01.pdf. Accessed 7/14/07.

20. **Kattleman, R.** 1996. Impacts of floods and avalanches. In *Status of the Sierra Nevada*, Sierra Nevada Ecosystem Project. Final report to Congress, Berkeley, CA, v. 2. http://ceres.ca.gov/snep/pubs/web/PDF/vII_C49.PDF. Accessed 7/14/07.

21. **Swanson, F. J.,** and **Dryness, C. T.** 1975. Impact of clear-cutting and road construction on soil erosion by landslides in the Western Cascade Range, Oregon. *Geology* 7: 393–96.

22. **Leighton, F. B.** 1966. Landslides and urban development. In *Engineering geology in southern California*, ed. R. Lung and R. Proctor, pp. 149–97, Special Publication, Los Angeles Section, Association of Engineering Geologists.

23. **Committee on the Review of the National Landslide Hazards Mitigation Strategy.** 2004. *Partnerships for reducing landslide risk*. National Research Council. Washington, DC: The National Academies Press.

24. **Pennsylvania Department of Conservation and Natural Resources.** 2001. Landslides in Pennsylvania. Educational Series No. 9. www.dcnr.state.pa.us/cs/groups/public/documents/document/dcnr_014592.pdf.

25. **Jones, F. O.** 1973. Landslides of Rio de Janeiro and the Sierra das Araras Escarpment, Brazil. U.S. Geological Survey Professional Paper 697.

26. **Jones, D. K. C.** 1992. Landslide hazard assessment in the context of development. In *Geohazards*, ed. G. J. McCall, D. J. Laming, and S. C. Scott, pp. 117–41. New York: Chapman & Hall.

27. **Slosson, J. E., Yoakum, D. E.,** and **Shuiran, G.** 1986. Thistle, Utah, landslide: Could it have been prevented? In *Proceedings of the 22nd Symposium on Engineering Geology and Soils Engineering*, ed. S. H. Wood, pp. 281–303. Boise, ID: 22nd Symposium on Engineering Geology and Soils Engineering.

28. **Spiker, E. C.,** and **Gori, P. L.** 2003. *National landslides mitigation strategy—A framework for loss reduction*. U.S. Geological Survey Circular 1244.

29. **Piteau, D. R.,** and **Peckover, F. L.** 1978. Engineering of rock slopes. In *Landslides*, ed. R. Schuster and R. J. Krizek, pp. 192–228. Transportation Research

Board Special Report 176, National Research Council. Washington, DC: National Academy Press.

30. **U.S. Geological Survey, Volcano Hazards Program, Cascades Volcano Observatory.** 2013. Monitoring Lahars at Mount Rainier. http://volcanoes.usgs.gov/volcanoes/mount_rainier/mount_rainier_monitoring_99.html. Accessed 9/21/13.

31. **O'Tousa, J.** 1995. La Conchita landslide, Ventura County, California. *Association of Engineering Geologists AEG News* 38(4): 22–24.

32. **Stoney, G. F.,** and **Miller, M. J.** 1996. Summary geotechnical evaluation of the La Conchita landslide—March 4, 1995. Stoney Miller Consultants, Inc., Ventura County.

33. **Gurrola, L. D., DeVecchio, D. E.,** and **Keller, E. A.** 2010. Rincon Mountain megaslide, La Conchita, Ventura County, California. *Journal of Geomorphology* 114: 311–22.

34. **Wilson, K.** 2010. Study finds $50 million grading project would make La Conchita safe. www.vcstar.com/news/2010/aug/16/study-finds-50-million-grading-project-would-la-/#ixzz2enoqdJHh. Accessed 9/13/13.

Natural and human-induced **processes are** **contributing** to the subsidence of **Venice**

8

APPLYING the **5** fundamental concepts Venice is Sinking

1

Science helps us predict hazards.

Short-term annual subsidence estimates form satellite data and longer-term estimates from scientific investigation of historical records and anthropological studies provide valuable data from which to understand the causes (natural and human-induced) of subsidence and enable preventative action to be taken.

2

Knowing hazard risks can help people make decisions.

Italy has evaluated the subsidence of Venice, as well as the risk of damage to land and property from future sub-sidence, for many decades. However the city is a global tourist attraction, and residents and tourists consider the risk to be acceptable—for now.

Subsidence and Soils

Venice is Sinking

Italy's beautiful and famous city of Venice faces a serious geologic problem; the city is sinking, or subsiding, at more that 1.5 mm (~0.06 in.) per year in some areas.[1] Venice is built on 118 small islands in a coastal lagoon. The islands are separated by waterways and connected by more than 400 bridges. Although its coastal location and numerous canals are part of Venice's attraction as a tourist destination, the presence of so much water surrounding a subsiding city makes Venice extremely prone to flooding. Today the city uses a network of hydrologic stations, capable of predicting high-water incursions 48 hours in advance, linked to sirens (and ringing of church bells) to warn the residents and tourists of impending flood events.

Historic records indicate that Venice has subsided by about 25 cm (~10 in.) since the late 1800s, leaving much of the land on which the city is built only a few centimeters above sea level.[1] Although subsidence (vertical deformation, in this case sinking) has been occurring naturally for millions of years, over pumping of groundwater from the 1930s to the 1960s significantly increased the historic rate of subsidence.[2] This subsidence has resulted in numerous floods from the sea due to such things as minor as high tides (see chapter-opener

◀ **Flooding in St. Mark's Square, Venice, Italy**
An exceptional high tide (137 cm (~57 in.)) on October 17, 2015, caused the flooding of 80 percent of Venice in northern Italy. Flooding is becoming more common as Venice subsides and sea-level rises. *(Panoramio (grumpylumixuser))*

Subsidence, the sinking of the land, and expansion and contraction of the soil are important geologic processes capable of causing extensive damage in some areas of the world.

After reading this chapter, you should be able to:

LO:1 Describe what a soil is and the processes that form and maintain soils.

LO:2 Explain why soil erosion might threaten our society.

LO:3 Summarize the causes and effects of subsidence and volume changes in the soil.

LO:4 Locate on a map the geographic regions at risk for subsidence and volume changes in the soil.

LO:5 Discuss the hazards associated with karst regions.

LO:6 Summarize the effects of subsidence and soil volume changes.

LO:7 Identify linkages between subsidence, soil expansion and contraction, and other hazards, as well as natural service functions of karst.

LO:8 Explain how humans interact with subsidence and soil hazards.

LO:9 Evaluate what can be done to minimize the hazard from subsidence and volume changes in the soil.

LEARNING Objectives

3

Linkages exist between natural hazards.

Subsidence of Venice is linked to flooding, which is increasing due to global climate change and consequent sea-level rise.

4

Humans can turn disastrous events into catastrophes.

Ground water pumping during the mid-twentieth century has led to increased total and rates of subsidence over the past 100 years, exacerbating the flood hazard to this densely populated and highy visited city.

5

Consequences of hazards can be minimized.

When we return to Venice at the end of the chapter you will learn about the city's strategy for minimizing the effects of subsidence-driven flooding caused by high tides and storm surges beyond their early warning system.

photograph). The human response to this flooding has been to raise buildings and streets above the flood level, but the former ground floors of many old houses are uninhabitable. Both natural- and human-induced subsidence is still occurring and together with sea-level rise over the past century the frequency of floods has increased. As recently as February 2013, floodwaters entered historic buildings and shops in Venice, forcing some Venetians to wade in waist-deep water.

8.1 Soil and Hazards

Soil develops in the upper few meters (~10 ft.) of Earth's surface where rock, water, atmospheric gasses, and life interact for a prolonged period of time. Soil is composed of a mixture of rock and mineral fragments, organic material, liquids, gasses, and innumerable organisms that together support life. The actual definition of soil, however, may be defined in different ways. To geologists, it is rocky earth material that has been altered by physical and chemical processes, and is a medium for water transport and storage. To biologists, it is the top layer of Earth in which rooted plants grow, and serves as a habitat for thousands of species. To civil engineers, on the other hand, soil is any solid earth material that can be manipulated or removed without blasting. Each of these definitions highlight different soil properties, all of which are important to understanding soil-related earth processes as hazards.

Understanding soil development, in addition to being fundamental to soil-hazard science, is useful for evaluating other natural hazards, such as floods, landslides, earthquakes, and climate change. For example, the study of soils has been a powerful tool in establishing the local age of Earth's surface deformed by both faulting and landsliding, which has led to better site calculations of event frequency and thus assist in minimizing future impacts. Soil is also intimately involved in all components of the geologic cycles (see Section 1.3), and as Earth's largest terrestrial carbon reservoir, global soils are intergral to the planet's global climate system and may be instrumental in efforts to slow the rapid rise of atmospheric CO_2.

The development of a soil from inorganic and organic materials is a slow and complex process. The first step in this process is the physical breakdown and/or chemical decomposition of rock and mineral material into *sediment*—a process known as **weathering**. In addition to liberating precious nutrients nessessary for life, chemical weathering creates clay minerals which are intimately involved in many soil hazards (see Section 8.2). Further weathering of sediment and soil is enhanced by the activity of organisms that reside in the soil, as well as reactions with gasses and liquids trapped in or passed through the pore spaces (the spaces between soil particles).

A soil can be considered an open system that interacts with all components of the geologic cycle. Although the more insoluble weathered material may remain in place, some weathered material will be transported by water, wind, or glaciers and deposited far away from the rock source from which it was weathered. Different soil characteristics and properties are the result of local environmental factors, including: *climate* (temperature, precipitation), *topography* (slope, elevation), *parent material* (composition of the rock), *age* (time to develop), and *biota* (organisms) where the soils form.

SOIL HORIZONS

Vertical and horizontal movements of material in a soil system over time creates a *soil profile*—a collection of distinct soil layers parallel to the surface. The layers are called zones or **soil horizons**, and can be distinguished based on soil color, texture and composition. Our discussion of soil profiles will mention only the horizons most commonly present in soils. Additional information is available from detailed soils texts.[3,4]

Figure 8.1a shows the common master (prominent) soil horizons. The O horizon and A horizon contain highly concentrated organic material; the differences between these two layers reflect the amount of organic material present. Generally, the O horizon consists entirely of plant litter and other organic material, whereas the underlying A horizon contains a good deal of both organic and mineral material. Below the O or A horizon, some soils have an E horizon, or zone of leaching, a layer that is leached of iron-bearing components and is light in color because it contains less organic material than either the O or A horizon and little inorganic coloring material such as iron oxides.

The B horizon, or zone of accumulation, underlies the O, A, or E horizon and consists of a variety of materials translocated downward from the overlying horizons. Several types of B horizon have been recognized. Probably the most important type is the argillic B, or B_t horizon. A B_t horizon is enriched with clay minerals that have been translocated downward by soil-forming processes. Another type of B horizon of interest to environmental geologists is the B_k horizon, characterized by the accumulation of calcium carbonate which whitens the colors the soil horizon (Figure 8.1b). The carbonate coats individual soil particles in the soils and may fill some pore spaces, but it does not dominate the morphology (structure) of the horizon. A soil horizon that is so impregnated with calcium carbonate that its morphology is dominated by the carbonate is designated a K horizon. Carbonate completely fills the pore

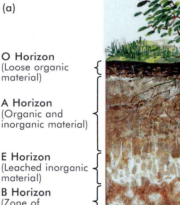

O Horizon
(Loose organic
material)

A Horizon
(Organic and
inorganic material)

E Horizon
(Leached inorganic
material)

B Horizon
(Zone of
accumulation)

C Horizon
(Broken and
chemically altered
parent material)

R Horizon
(Unweathered
parent material)

~3m

(a)

(b)

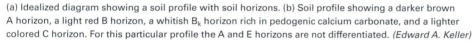

∧ FIGURE 8.1 Soil Profiles

(a) Idealized diagram showing a soil profile with soil horizons. (b) Soil profile showing a darker brown
A horizon, a light red B horizon, a whitish B_k horizon rich in pedogenic calcium carbonate, and a lighter
colored C horizon. For this particular profile the A and E horizons are not differentiated. *(Edward A. Keller)*

spaces in K horizons, and the carbonate often forms in layers parallel to the surface. The term *caliche* is often used for irregular accumulation or layers of calcium carbonate in soils.

The C horizon lies directly over the unaltered parent material and consists of parent material partially altered by weathering processes. The R horizon, or unaltered parent material, is the consolidated bedrock that underlies the soil. However, some of the fractures and other pore spaces in the bedrock may contain clay that has been translocated downward.[3]

The term *hardpan* is often used in the literature to describe a dense soil layer. Hardpans are often composed of compacted clay within the B_t horizon, or a well-cemented carbonate layers in the B_k horizon that are typical of soils in arid and semi-arid landscapes. Hardpan horizons are nearly impermeable and, thus, restrict the downward movement of soil water. Such hardpans, or impermeable soil horizons, lead to increased runoff during heavy rain fall events and, therefore, are linked to downstream and flash flooding events.

SOIL COLOR

One of the first things we notice about a soil is its color, or the colors of its horizons. The color of a particular horizon is a function, in part, of its chemical and physical composition reflecting different soil processes within the profile (Table 8.1). The O and A horizons tend to be dark because of their abundant organic material. The E horizon, if present, may be almost white due to the leaching of iron and aluminum oxides. The B horizon shows the most dramatic differences in color, varying from yellow-brown to light red-brown to dark red, depending on the presence of clay minerals and iron oxides. The B_k horizons may be light-colored due to their carbonates, but they are sometimes reddish as a result of iron oxide accumulation. If a true K horizon has developed, it may be almost white because of its accumulation of calcium carbonate. Although soil color can be an important diagnostic tool for analyzing a soil profile, one must be cautious about calling a red layer a B horizon. The original parent material, if rich in iron,

▼ TABLE 8.1 Soil Profile Descriptions by Horizon

Soil Horizon	Horizon Description
O	Horizon is composed mostly of organic materials, including decomposed or decomposing leaves, twigs, etc. The color of the horizon is often dark brown or black.
A	Horizon is composed of both mineral and organic materials. The color is often light black to brown. Leaching, defined as the process of dissolving, washing, or draining earth materials by percolation of groundwater or other liquids, occurs in the A horizon and moves clay and other material such as iron and calcium to the B horizon.
E	Horizon is composed of light-colored materials resulting from leaching of clay, calcium, magnesium, and iron to lower horizons. The A and E horizons together comprise the zone of leaching.
B	Horizon is enriched in clay, iron oxides, silica, carbonate, or other material leached from overlying horizons. This horizon is known as the zone of accumulation.
C	Horizon is composed of partially altered (weathered) parent material; rock is shown in Figure 8.1a but the material could also be alluvial in nature, such as river gravels in other environments. The horizon may be stained red with iron oxides.
R	Unweathered (unaltered) parent material.

may produce a very red soil, even with relatively little soil profile development.

Soil color can be an important indicator of how well drained a soil is. Well-drained soils are well aerated (oxidizing conditions), and iron oxidizes form giving these soils a red color. Poorly drained soils are wet, and iron in it is reduced rather than oxidized. The color of such a soil is often yellow. This distinction is important because poorly drained soils are associated with environmental problems such as lower slope instability. These soils are also a poor choice for siting waste disposal functions such as household sewage systems (septic tanks and leach fields).

SOIL TEXTURE

The texture of a soil is defined by the relative proportions of sand-, silt-, and clay-sized particles (Figure 8.2). *Clay particles* have a diameter of less than 0.004 mm, *silt particles* have diameters ranging from 0.004 to 0.074 mm, and *sand particles* are 0.074 to 2.0 mm in diameter. Earth materials with particles larger than 2.0 mm in diameter are called *gravel*, *cobbles*, or *boulders*, depending on the particle size. Note that the sizes of particles given here are for engineering classification and are slightly different from those used by the U.S. Department of Agriculture for soil classification.

Soil texture is commonly estimated in the field and then refined in the laboratory by separating the sand, silt, and clay and determining their proportions. A useful field technique for estimating the size of sand-sized or smaller soil particles is as follows: It is sand if you can see individual grains, silt if you can see the grains with a 10× hand lens, and clay if you cannot see grains with such a hand lens. Another method is to feel the soil:

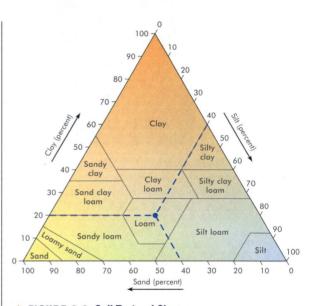

▲ FIGURE 8.2 Soil Textural Classes
The classes are defined according to the percentage of clay-, silt-, and sand-sized particles in the soil sample. The point connected by dashed lines represents a soil composed of 40 percent sand, 40 percent silt, and 20 percent clay, which is classified as loam. (*U.S. Department of Agriculture standard textural triangle*)

Sand is gritty (crunches between the teeth), silt feels like baking flour, and clay is cohesive. When mixed with water, smeared on the back of the hand, and allowed to dry, clay cannot be dusted off easily, whereas silt or sand can.

Soil particles often cling together in aggregates, called *peds*, that are classified according to shape into several types. Figure 8.3 shows some of the common structures of peds found in soils. The type of structure

Soil Structure Type		Typical size range	Horizon usually found in	Comments
Granular		1–10 mm	A	Can also be found in B and C horizons
Blocky		5–50 mm	B_t	Are usually designated as angular or subangular
Prismatic		10–100 mm	B_t	If columns have rounded tops, structure is called *columnar*
Platy		1–10 mm	E	May also occur in some B horizons

∧ FIGURE 8.3 Chart of Different Soil Structures (Peds)

present is related to soil-forming processes, but some of these processes are poorly understood.[3] For example, *granular structure* is fairly common in A horizons, whereas *blocky* and *prismatic structures* are most likely to be found in B horizons. Soil structure is an important diagnostic tool in helping to evaluate the development and approximate age of soil profiles. In general, as the profile develops with time, structure becomes more complex and may go from granular to blocky to prismatic as the clay content in the B horizon increases.

RELATIVE SOIL PROFILE DEVELOPMENT

Most geologists will not need to make detailed soil descriptions and analyses of soil data. However, it is important for geologists to recognize differences among weakly developed, moderately developed, and well-developed soils, that is, to recognize their relative profile development. These distinctions are useful in the preliminary evaluation of soil properties and help determine whether the opinion of a soil scientist is necessary in a particular project:

> *A weakly developed soil profile* is generally characterized by an A horizon directly over a C horizon (there is no B horizon or it is weakly developed). The C horizon may be oxidized. Such soils tend to be only a few hundred years old in most areas, but they may be several thousand years old.

> *A moderately developed soil profile* may consist of an A horizon overlying an argillic B_t horizon that overlies the C horizon. A carbonate B_k horizon may also be present but is not necessary for a soil to be considered moderately developed. These soils have a B horizon with translocated changes, a better-developed texture, and redder colors than those that are weakly developed. Moderately developed soils often date from at least the Pleistocene (more than 10,000 years old).

> *A well-developed soil profile* is characterized by redder colors in the B_t horizon, more translocation of clay to the B_t horizon, and stronger structure. A K horizon may also be present but is not necessary for a soil to be considered strongly developed. Well-developed soils vary widely in age, with typical ranges between 40,000 and several hundred thousand years and older.

A *soil chronosequence* is a series of soils arranged from youngest to oldest on the basis of their relative profile development. Such a sequence is valuable in hazards work because it provides information about the recent history of a landscape, allowing us to evaluate site stability when locating such critical facilities as a waste disposal operation or a large power plant. A chronosequence combined with numerical dating (applying a variety of dating techniques, such as radiocarbon[14]C, to obtain a date for the soil in years before the present time) may provide the data necessary to make such inferential statements as, "There is no evidence of ground rupture due to earthquakes in the past 1000 years," or "The last mudflow was at least 30,000 years ago." It takes a lot of work to establish a chronosequence in soils in a particular area. However, once such a chronosequence is developed and dated, it may be applied to a specific problem.

Consider, for example, the landscape shown in Figure 8.4, an offset alluvial fan along the San Andreas fault in the Indio Hills of Southern California. The fan is offset about 0.6 km, which is equal to 60,000 cm (~0.4 mi.). Soil age estimates from pits excavated in the alluvial fan suggest that it is at least 20,000 years old but younger than 45,000 years. The age was estimated based on correlation with a soil chronosequence in the nearby Mojave Desert, where numerical dates for similar soils are available. Soil development on the offset alluvial fan allowed the age of the fan to be estimated. Thus, the slip rate (the amount of offset of the fan divided by the age of the fan—that is, 60,000 cm divided by 20,000 years) for this part of the San Andreas fault could be estimated at about 3 cm (~1 in.) annually.[5,6] The slip rate for this segment of the fault was not previously known. The rate is significant because it is a necessary ingredient in the eventual estimation of the probability and the recurrence interval of large, damaging earthquakes (see Chapter 3).

More recent work, using a numerical dating technique known as exposure dating, suggests the age of 35,000 +/− 2,500 years and a total displacement of about 570 m. (~1870 ft.). This new and more accurate estimation of age provides a maximum slip rate of about 1.7 cm (~0.7 in.) per year. Thus, the earlier soil date has been improved by more recent technology. As numerical dating has improved, the use of soil development as a dating tool has decreased.

WATER IN SOILS

If you analyze a block of soil, you will find it is composed of bits of solid inorganic and organic matter with pore spaces between particles. The pore spaces are filled with gases (mostly air) or liquids (mostly water). If all the pore spaces in a block of soil are completely filled with water, the soil is said to be in a *saturated condition*; otherwise, it is said to be *unsaturated*. Soils in swampy areas may be saturated year-round, whereas soils in arid regions may be saturated only occasionally (Figure 8.5).

The amount of water in a soil, called its *water content* or its *moisture content*, can be important in determining engineering properties such as the strength of a soil and its potential to shrink and swell. If you have ever built a sand castle at the beach, you know that dry sand is impossible to work with but that moist sand will stand vertically, producing walls for your castle. Differences between wet and dry soils are also apparent to anyone who lives in or has visited areas with dirt

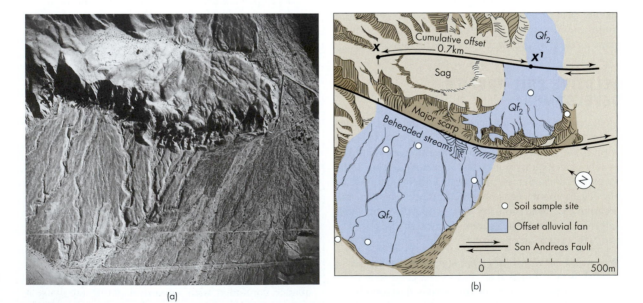

(a) (b)

∧ FIGURE 8.4 Offset Alluvial Fan Along the San Andreas Fault Near Indio, California
(a) Aerial photograph of an alluvial fan offset across the San Andreas fault and (b) sketch map of the important tectonic features used to estimate offset. *(Woodward-Clyde Consultants)*

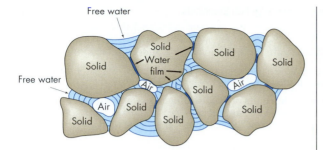

⋀ FIGURE 8.5 Partly Saturated Soil Showing Particle–Water–Air Relationships
Particle size is greatly magnified. Attraction between the water and soil particles (surface tension) develops a stress that holds the grains together. This apparent cohesion is destroyed if the soil dries out or becomes completely saturated. *(Based on R. Pestrong,* Slope Stability. *American Geological Institute, 1974)*

roads that cross clay-rich soils. When the soil is dry, driving conditions are excellent but, following a rain storm, the same roads become mud pits and are nearly impassable.[7]

Water in soils may flow laterally or vertically through soil pores, which are the void spaces between grains, or in fractures produced as a result of soil structure. The flow is termed *saturated flow* if all the pores are filled with water and *unsaturated flow* when, as is more common, only part of the pores are filled with water.

In unsaturated flow, movement of water is related to processes such as thinning or thickening of films of water in pores and on the surrounding soil grains.[8] The water molecules closest to the surface of a soil particle are held most tightly, and as the films thicken, the water content increases and the outer layers of water may begin to move. Flow is, therefore, fastest in the center of pores and slowest near the edges.

Soil moisture relations and movement of water and other liquids in soils, along with ways to monitor movement of liquids, are important research topics. Water in the soil is related to many processes, including slope stability and subsidence of earth materials.

CLASSIFYING SOILS

Both the terminology and the classification of soils present a problem in environmental studies because we are often interested in both soil processes and the human use of soil. A *taxonomy* (classification system) that includes engineering as well as physical and chemical properties would be most appropriate, but none exists. We must, therefore, be familiar with two separate systems of soil classification: *soil taxonomy*,

used by soil scientists, and the *engineering classification*, which groups soils by material types and engineering properties.

Soil scientists have developed a comprehensive and systematic classification of soils known as *soil taxonomy*, which emphasizes the physical and chemical properties of the soil profile. This classification is a sixfold hierarchy that groups soils into several levels: Orders, Suborders, Great Groups, Subgroups, Families, and Series. The eleven Orders (Table 8.2) are mostly based on gross soil morphology (number and types of horizons present), nutrient status, organic content (plant debris, etc.), color (red, yellow, brown, white, etc.), and general climatic considerations (amount of precipitation, average temperature, etc.). With each step down the hierarchy, more information about a specific soil becomes known.

Soil taxonomy is especially useful for agricultural and related land-use purposes. It has been criticized for being too complex and for lacking sufficient textural and engineering information to be of optimal use in site evaluation for engineering purposes. Nevertheless, the serious earth scientist must know and understand this classification because it is commonly used by soil scientists and Quaternary geologists who study earth materials and the processes of recent earth history (past 1.8 million years).

The *unified soil classification system*, widely used in engineering practice, along with the evaluation of hazards is shown in Table 8.3. Because all natural soils are mixtures of coarse particles (gravel and sand), fine particles (silt and clay), and organic material, the major divisions of this system are *coarse-grained soils*, *fine-grained soils*, and *organic soils*. Each group is based on the predominant particle size or the abundance of organic material. Coarse soils are those in which more than 50 percent of the particles (by weight) are larger than 0.074 mm in diameter. In fine soils, less than 50 percent of the particles are greater than 0.074 mm in diameter.[7] Organic soils have a high organic content and are identified by their black or gray color and sometimes by an odor of hydrogen sulfide, which smells like rotten eggs.

SOIL EROSION AS A HAZARD

A big question with soil as it relates to our society and our future prosperity is whether the agricultural systems that we depend on to feed a growing human population can maintain and improve soil fertility while minimizing soil erosion.[9] It appears that many of our agricultural practices are mining the soil; that is, soils are being eroded faster than they are being produced by soil-forming processes (Figure 8.6). Eventually,

▼ **TABLE 8.2** **General Properties of Soil Order Used with Soil Taxonomy by Soil Scientists**

Order	General Properties
Entisols	No horizon development; many are recent alluvium; synthetic soils are included; often young soils.
Vertisols	Include swelling clays (greater than 35 percent) that expand and contract with changing moisture content. Generally form in regions with pronounced wet and dry seasons.
Inceptisols	One or more horizons have developed quickly; horizons are often difficult to differentiate; most often found in young but not recent land surfaces; have appreciable accumulation of organic material. Most common in humid climates but range from the Arctic to the tropics; native vegetation is most often forest.
Aridisols	Desert soils; soils of dry places; low organic accumulation; have subsoil horizon where gypsum, caliche (calcium carbonate), salt, or other materials may accumulate.
Mollisols	Soils characterized by black, organic-rich A horizon (prairie soils); surface horizons are also rich in bases. Commonly found in semiarid or subhumid regions.
Andisols	Soils derived primarily from volcanic materials; relatively rich in chemically active minerals that rapidly take up important biologic elements such as carbon and phosphorus.
Spodosols	Soils characterized by ash-colored sands over subsoil, accumulations of amorphous iron-aluminum sesquioxides and humus. They are acid soils that commonly form in sandy parent materials; found principally under forests in humid regions.
Alfisols	Soils characterized by a brown or gray-brown surface horizon and an argillic (clay-rich) subsoil accumulation with an intermediate to high base saturation (greater than 35 percent as measured by the sum of cations, such as calcium, sodium, magnesium, etc.). Commonly form under forests in humid regions of the midlatitudes.
Ulfisols	Soils characterized by an argillic horizon with low base saturation (less than 35 percent as measured by the sum of cations); often have a reddish-yellow or reddish-brown color. Restricted to humid climates and generally form on older landforms or younger, highly weathered parent materials.
Oxisols	Relatively featureless, often deep soils, leached of bases, hydrated, containing oxides of iron and aluminum (laterite) as well as kaolinite clay. Primarily restricted to tropical and subtropical regions.
Histosols	Organic soils (peat, muck, bog).

Source: Based on *Soil Survey Staff, Keys to Soil Taxonomy*, 6th ed. Soil Conservation Service, U.S. Department of Agriculture, 1994.

▼ **TABLE 8.3** **Unified Soil Classification System Used by Engineers**

Major Divisions				Group Symbols	Soil Group Names
COARSE-GRAINED SOILS (Over half of material larger than 0.074 mm)	Gravels	Clean Gravels	Less than 5% fines	GW	Well-graded gravel
				GP	Poorly-graded gravel
				GM	Silty gravel
		Dirty Gravels	More than 12% fines	GC	Clayey gravel
	Sands	Clean Sands	Less than 5% fines	SW	Well-graded sand
				SP	Poorly-graded sand
				SM	Silty sand
		Dirty Sands	More than 12% fines	SC	Clayey sand
FINE-GRAINED SOILS (Over half of material smaller than 0.074 mm)	Silts Non-plastic			ML	Silt
				MH	Micaceous silt
				OL	Organic silt
	Clays Plastic			CL	Silty clay
				CH	High plastic clay
				OH	Organic clay
		Predominantly Organics		PT	Peat and muck

∧ FIGURE 8.6 Soil Erosion in Eastern Washington State
An example of serious soil erosion, with development of shallow gullies, in a wheat field. *(Jack Dykinga/USDA Forest Service)*

agricultural soil loss could erode the foundation of our civilization, as has happened to other civilizations throughout history.

Soil erosion may be defined in three ways: (1) the grain-by-grain removal of mineral and organic material in a soil by wind and/or water; (2) removal of soil material at an unacceptable rate; and (3) removal of soil material at a rate faster than it is being produced. Soil erosion has multiple effects, including loss of soil resources as well as degradation of the quality of the water due to sedimentation. Soil erosion is a problem in urban environments, where vegetation is often removed prior to development. Although we have many safeguards in effect to minimize soil erosion resulting from agriculture and other land-use change, the problem persists in many parts of the United States and is severe in many parts of the world where protection of soil resources may not be a high priority.[9,10]

CASE study

Haiti and Soil Erosion

An example of extreme soil erosion comes from the Caribbean. Haiti and the Dominican Republic share the island of Hispaniola, but the two countries are very different. In Haiti, the median age of the population is 18, compared to 24 for the Dominican Republic. Life expectancy for Haitians is 60 years, compared to 72 for Dominicans. Haiti's per capita income is a fraction of that of the Dominican Republic. Another significant difference is the amount of forest cover: 3.8 percent in Haiti compared to 28.4 percent in the Dominican Republic (Figure 8.7). The percentage of forest cover is much lower for Haiti compared to the other Caribbean islands as well. Haiti is distinguished by its extensive deforestation, accompanied by a serious soil erosion problem, both of which are linked to an inadequate and unsafe water supply for the population (Figure 8.7).[11,12]

Historians and other scientists have studied the differences between Haiti and the Dominican Republic. There is general agreement that many of these differences result from their different colonial histories. The people of Haiti went through a difficult, violent revolution, gaining their freedom from France in 1804. During French rule, Haiti was prosperous with a high population density. Following the revolution, the Haitian people mistrusted Europeans to such an extent that trade

∧ FIGURE 8.7 Clearly Visible Boundary between Haiti and Dominican Republic
Deforestation in Haiti has resulted in severe soil erosion, threatening slope stability and water quality and affecting the country's ability to feed its people. The boundary between Haiti and the Dominican Republic is clearly visible by the absence of vegetation in Haiti. *(NASA)*

with other countries was much reduced. Haiti was relatively wealthy following independence and did well for a while until political, social, and economic forces eventually led to a downward spiral that resulted in Haiti becoming the poorest country in the Western Hemisphere. The path to independence was different for the Dominican Republic. Spain was wealthy in terms of her colonies, and the Dominican people were left with the Spanish language, a dominant language of the time,[11] as well as good contacts with Europe, and so were able to effectively develop trade and commerce with the rest of the world.

Later in their respective histories, absolute dictators ruled both countries—but one of the Dominican Republic's dictators was interested in preserving forests for sustainable timber harvesting in order to increase the money in his own accounts. For this and other reasons, the Dominican Republic today has retained much of its forest cover and has a good deal of land in national parks. In Haiti, however, the French began the deforestation process by taking back a lot of timber to Europe and, following independence, deforestation continued at accelerating rates. Trees were unable to regenerate effectively, due to Haiti's lesser amounts of rainfall and the extensive soil erosion that occurred. Today, Haiti is virtually denuded of forests and has a seriously high erosion rate that favors landsliding and other problems related to tropical storms in the Caribbean.[11,12]

The population of Haiti has grown, as has that of the Dominican Republic. However, in Haiti the high density of people living in degraded land areas has led to a situation in which Haitians cannot adequately support themselves on their own land and, as a result, must depend on imported food and other commodities necessary for life. A devastating earthquake in 2010 only worsened Haiti's already precarious economy.

In summary, Haiti is an example of what can happen when environmental degradation linked to high population density, unprotected water and soil resources, and inadequate infrastructure results in extreme poverty.

Rates of Soil Erosion   The depth of the soil at a particular site represents a balance between weathering of fresh parent material (rock or sediment) beneath the soil and erosion processes that removes soil. We are most concerned with top, organically rich soil that is best for growing crops. It takes about 500 to 1,000 years to form 50 mm (~2 in.) of soil. Although rates are highly variable, depending on climate, topography, and parent material, the rate of soil formation for agricultural land is 0.1 to 0.05 mm (~0.004–0.002 in.) per year. Accelerated erosion due to poor soil conservation can remove centuries of accumulation in less than a decade.

Fortunately, agricultural practices such as no-till agriculture that minimize soil disturbance, along with other known conservation measures, can significantly reduce erosion rates.[10]

Rates of soil erosion are measured as a volume, mass, or weight of soil that is removed from a location within a specified time and area, for example, kilograms per year per hectare. Soil erosion rates vary with the engineering properties of the soil, land use, topography, and climate. The natural rate of soil erosion is about 0.5 to 1.0 ton per hectare per year (tons/ha/yr). The erosion rate of 6 tons per hectare will decrease soil depth by about 1 mm. That doesn't sound like much, but the process of soil loss is generally slow and difficult to recognize unless the erosion is concentrated in easily visible gullies. However, soil erosion rates in North, Central, and South America average 10–20 tons/ha/yr, and in Asia, the average is about 30 tons/ha/yr. These rates are many times the natural rate, suggesting that soil erosion in many agricultural regions is excessive and is depleting the soil resource and land productivity along with it.[13,14]

There are several approaches to measuring rates of soil erosion. The most direct method is to make actual measurements on slopes over a period of at least several years and use these values as representative of what is happening over a wider area and longer time span. This approach is rarely used, however, because data from individual slopes and drainage basins are very difficult to obtain. A second approach is to use data obtained from resurveying reservoirs to calculate the change in the reservoirs' storage capacity of water; the depletion of storage capacity is equivalent to the volume of sediment eroded from upstream soils. A third approach is to use an equation to calculate rates of sediment eroded from a particular site. One of the most common is the Universal Soil Loss Equation.[15] This equation uses data on rainfall, runoff, the size and shape of the slope, the soil cover, and erosion control practices to predict the amount of soil moved from its original position.[16]

With this introduction to soils behind us, we will turn to more specific aspects of subsidence.

8.1 CHECK your understanding

1. Define soil.
2. Decribe the main soil horizons.
3. What is relative soil profile development, and how is it useful in evaluating hazards?
4. What is the role of water in soils?
5. Explain the two primary ways that soils are classified.
6. Why is soil erosion a serious problem?
7. How are soil erosion rates determined?

8.2 Introduction to Subsidence and Soil Volume Change

In contrast to mass wasting (Chapter 7), which originates on slopes, subsidence and changes in the volume of the soil occur on both slopes and flat ground. **Subsidence** is a type of ground failure characterized by nearly vertical deformation, or the downward sinking of earth materials. It often produces circular surface pits, but it may produce linear or irregular patterns of failure. Volume changes in the soil result from various natural processes. These processes include shrinking and swelling caused by changes in the water content of the soil and frost heaving related to the freezing of water in the soil. With volume changes of 5–15 percent, the effects of changes of shrink/swell and freeze/thaw movement in soil are generally not as dramatic as some subsidence—you are unlikely to see a headline "Dozens Die in Denver as Soil Expands." However, volume changes in the soil are one of the most widespread and costliest of natural hazards.

Subsidence is commonly associated with the dissolution of soluble rocks, such as limestone, beneath the surface. The resultant landscape has closed depressions and is known as **karst topography**. Other major causes of subsidence include the thawing of frozen ground and compaction of recently deposited sediment. To a lesser degree, earthquakes and the underground drainage of magma are also responsible for causing subsidence. Human-induced subsidence, discussed in Section 8.7, can result from the withdrawal of underground fluids, from the collapse of soil and rock over underground mines, and from the draining of marshes and swamps.

KARST FORMATION

Although the term *karst* is not in most people's vocabulary, it is the name given to a common type of landscape in the United States and in many other parts of the world.[17] In this type of topography, subsidence results from chemical weathering—the dissolution of rocks beneath the land surface. Dissolution occurs as surface water or groundwater percolates through rock that is easily dissolved. Four common sedimentary rocks—rock salt and rock gypsum, which are evaporites; limestone and dolostone, which are carbonates—and one common metamorphic rock, marble, are easily dissolved. Rock salt and rock gypsum dissolve rapidly as fresh surface water flows through holes in the rock, whereas limestone, dolostone, and marble will dissolve if the percolating water is acidic. Although rock salt is approximately 7,500 times more soluble and rock gypsum 150 times more soluble than limestone, most karst topography is underlain by limestone because of its greater abundance in near-surface rocks.[18]

Percolating water may become acidic when carbon dioxide is dissolved in it. This acidification generally occurs in the soil where carbon dioxide is produced by bacterial decomposition of dead plants and animals. Most soil bacteria "breathe" like humans; that is, their respiration takes in oxygen and releases carbon dioxide. Dissolving carbon dioxide in water produces carbonic acid, the same weak acid that is present in carbonated soft drinks such as Coke and Pepsi. Where acidified water comes in contact with limestone, it dissolves the bedrock and the land surface is lowered at an average rate of 10 cm (~4 in.) per century.[19] Carbonic acid dissolves limestone more readily than dolostone; thus, land underlain by limestone is more susceptible to dissolution and more likely to become karst topography (Figure 8.8).

Areas underlain by dense, thin-bedded, fractured, or well-jointed crystalline limestone are especially vulnerable to dissolution. In such areas, surface waters are easily diverted to subterranean routes along irregular fractures or to planar cracks between sedimentary layers. If the percolating surface water has become acidic, it can enlarge these openings in the rock to produce underground holes, or voids, of various shapes and sizes. Where the void space is relatively close to the surface, pits known as **sinkholes** may develop (see Figure 8.8).

Sinkholes may exist individually or develop in large numbers to form a pockmarked surface known as a *karst plain*. The limestone plateau in southern Indiana and central Kentucky and Tennessee is an example of a karst plain (Figure 8.9).

In addition to sinkholes, many karst areas have other features developed by the chemical weathering of bedrock. In humid temperate climates, karst landscapes are characterized by beautiful rolling hills with alternating areas of subsidence and undisturbed land. These areas may be underlain by large natural caves, commonly referred to as *caverns*, if the voids excavated by dissolution are at sufficient depth below the surface (see Figure 8.8). Cave openings can be the site of disappearing streams, where surface water goes underground, or the place where groundwater comes out at the surface to form springs. In humid tropical regions, extensive dissolution removes most of the soluble bedrock, leaving behind a landscape of steep hills known as *tower karst*.

Sinkholes In karst areas, sinkholes vary in size from one to several hundred meters in diameter and can open up extremely rapidly. There are two basic types of sinkholes (Figure 8.10):

1. **Solutional sinkholes:** Solutional sinkholes form by dissolution on the top of a buried bedrock surface and are the more common type of

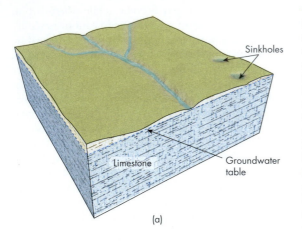

(a)

◄ **FIGURE 8.8 Kart Topography Formation**
(a) In the early stage of karst formation in a limestone terrain, water from the surface seeps through fractures and along layering in the soluble rock. Weak acid in the water then dissolves the rock. (b) As a river erodes more deeply into the land surface, the groundwater table drops. Caves begin to form and collapse to become sinkholes. Some surface streams disappear underground to become groundwater. (c) In later stages of karst development, a downward-eroding river continues to lower the groundwater table. Large caverns and sinkholes develop and eventually merge to form solution valleys that form without surface streams. In humid tropical climates, intense dissolution removes nearly all of the rock, leaving behind pillars of limestone referred to as *karst towers. (Tarbuck, Edward J., Lutgens, Frederick K., and Tasa, Dennis G., Earth: An Introduction to Physical Geology, 10th ed. Upper Saddle River, NJ: Pearson, © 2011, p. 484. Reprinted and Electronically reproduced by permission of Pearson Education, Inc., Upper Saddle River, New Jersey.)*

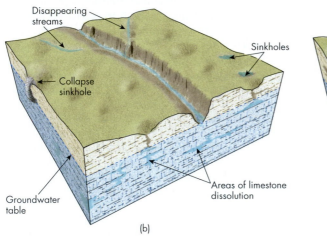

(b)

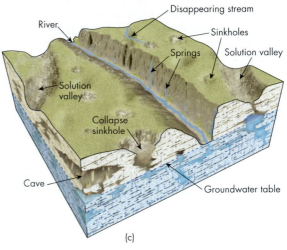

(c)

sinkhole. Dissolution occurs where the downward infiltration of acidic groundwater becomes concentrated in holes created by joints and fractures. In the formation of these sinkholes, groundwater is typically drawn into a cone above a hole in the limestone, like water being drawn into a sink drain.

2. **Collapse sinkholes:** Collapse sinkholes develop when surface or near-surface material collapses into an underground cavern. As subsidence features, these sinkholes can develop into spectacular collapse structures (Figure 8.11).

Some sinkholes have openings into subterranean passages that allow water to escape during a rainstorm. Most, however, are filled with rubble that blocks any underground passages. Blocked sinkholes usually fill up with water, forming small lakes. Most of these lakes eventually drain when the water is able to filter through

the debris.[20] Similar drainage problems can result when artificial ponds and lakes are constructed over sinkholes (see Case Study 8.2: Survivor Story).

Cave Systems As solutional pits enlarge and water moves downward through limestone, a series of caves or larger caverns may be produced. Mammoth Cave, Kentucky, and Carlsbad Caverns, New Mexico, are two of the famous caverns, or **cave systems**, in the United States. The primary mechanism for forming caves is groundwater moving through rock (see Figure 8.10a). Cave systems tend to develop at or near the present groundwater table. With respect to karst, the groundwater table refers to the level of groundwater below which rock fractures and caverns are filled with water (saturated). At and near the groundwater table there is a continuous replenishment of water that is not saturated with the weathering products of the limestone. Many cave systems have passages and

⋀ FIGURE 8.9 Karst Topography
This rolling landscape in west-central Kentucky is typical of karst topography in a humid temperate climate.
(Tony Waltham/Robert Harding Picture Library Ltd/Alamy)

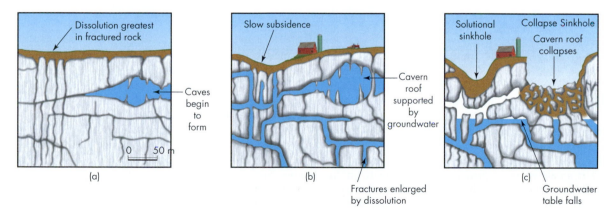

⋀ FIGURE 8.10 Formation of Sinkholes
(a) Initial dissolution of soluble bedrock takes place along vertical and horizontal fractures, and some fractures become small caves. (b) Over geologic time, dissolution continues to enlarge fractures and some caves become caverns with roofs partly supported by groundwater. (c) Subsequent lowering of the groundwater table removes the cavern roof support, and it falls into the cave to form a *collapse sinkhole*. Collapse may occur only a few years after the groundwater table has lowered. The heavily fractured area of bedrock is dissolved to create a *solutional sinkhole*.

∧ FIGURE 8.11 Giant Sinkhole
Swallowing a 3-story building, this 18 m (60 ft.) diameter, 100 m (300 ft.) deep sinkhole opened in Guatemala
City in 2010. Although the sinkhole was triggered by torrential rain during tropical storm Agatha, it appears a
broken sewer pipe helped hollow out the subsurface cavity.

underground rooms on a number of levels. In these
systems, each level may represent cave formation at
different times that are in response to a fluctuating
groundwater table. Fundamentally, caves are enlarged
as groundwater moves through limestone along fractures or along planes between sedimentary layers,
eventually forming a cavern. Later, if the groundwater
table moves to a lower level, water seeping into the
cave from above will deposit calcium carbonate on the
sides, floor, and ceiling. These deposits form beautiful
cave formations such as *flowstone*, *stalagmites*, and *stalactites* (Figure 8.12).

Tower Karst Large, steep limestone "towers" rising
above the surrounding landscape are known as *tower
karst*. Most common in humid tropical regions, karst
towers are residual landforms of a highly eroded karst
landscape. Cuba and Puerto Rico exhibit tower karst, as
does much of Southeast Asia. The most spectacular

examples come from Guilin, China, where they have
been painted by artists since the first century
(Figure 8.13).

Disappearing Streams Karst regions are often underlain by a complex groundwater system that occasionally intersects with the land surface. In these areas,
surface streams may suddenly disappear into cave openings. Such *disappearing streams* do not actually disappear; rather, they flow directly into the groundwater
system and continue along a subterranean route (see
Figure 8.8b).

Springs Areas where groundwater naturally discharges at the land surface are known as *springs*. Most
springs in karst areas are highly productive, especially
during rainy periods. Karst springs are an important
resource, but many are drying up as a result of overpumping of groundwater.

∧ FIGURE 8.12 Cave Formations
Carlsbad Caverns, New Mexico, contains stalactites, which hang
from the ceiling; stalagmites, which grow up from the cave floor;
and flowstone, which forms as water flows slowly down the
walls or across an incline. Stalactite hangs "tight" to the ceiling
and a stalagmite "might" grow up from the cave floor to meet
the stalactite. *(Bruce Roberts/Science Source)*

∧ FIGURE 8.13 Tower Karst
Tower karst south of Guilin in South China's Guangxi Zhuang
Autonomous Region. *(Tony Waltham/Robert Harding)*

THERMOKARST FORMATION

Some karst terrains develop from processes other than
the groundwater dissolution of rock. In polar regions,
and at high elevations, much of the soil and underlying
sediment remain frozen throughout the year. This natural condition, called **permafrost**, may be continuous
throughout the soil or, in slightly warmer climates, may
exist as discontinuous patches or thin layers. Permafrost
consists of particles of soil or sediment that remain cemented with ice for at least two years. The melting of
this frozen earth material can produce subsidence, especially if it contains a large amount of ice. Land subsidence of several meters or more is possible through
the melting of permafrost.[21]

   If undisturbed by human activity, the thawing of
permafrost is generally limited to the upper few meters
of earth material. This forms an active layer directly below the land surface, in which the permafrost thaws
during summer months and the water in the soil refreezes in the winter. However, more extensive thawing
of permafrost can produce an irregular land surface
known as *thermokarst*. Climatic warming over the past
five decades has thawed large areas of Arctic permafrost
and formed thermokarst. In some areas, the uppermost
layers of permafrost are thawing at rates of close to
20 cm (~8 in.) per year.[22]

SEDIMENT AND SOIL COMPACTION

Rapidly deposited fine sediment, soil and sediment
cemented with soluble minerals, and organic-rich
soil are all susceptible to subsidence. This subsidence
may occur as the rapidly deposited sediment compacts; as sediment and soil collapse when the binding
material is loosened; or, in organic soils, as water is
drained from the soil. Compaction of sediment and
soil can occur naturally or as the result of human
activities.

Fine Sediment Fine sand and mud that have been
transported and deposited in a relatively short amount
of time contain a great many water-filled spaces between the sediment particles. With time, the amount of
water between the particles is reduced, and the sediment compacts. Rapid deposition and compaction are
especially common in river deltas. In deltas, natural
subsidence has to be balanced by additional sedimentation to keep the land surface of the delta, called the *delta
plain*, from sinking below sea level. Episodic events
such as floods and earthquakes can cause deltaic sediment to remobilize and subside. Historically, this subsidence has submerged coastal cities, including Eastern
Canopus, Herakleion, and part of Alexandria on the
Nile Delta in Egypt.[23]

Sinkhole Drains Lake

The shores of Scott Lake, in central Florida, were a coveted address for residents of the area. Influential executives built multimillion-dollar mansions alongside the 285-acre private waterway. Scott Lake's pristine waters attracted birds, otters, alligators, and people who fished for the teeming populations of largemouth bass, grass carp, bluegill, and catfish.

Then, within the space of a week, virtually the entire lake disappeared.

The story began on June 13, 2006, when the lake received more than 8 inches of rain. But instead of rising, the water level of the lake began dropping. Over the next few hours, the water began to drain in earnest. Residents on the lake's southeastern shore watched the lake's normally placid waters begin to bubble, then whirlpool around a single point. Boathouses collapsed. Docks across the lake were left standing high and dry (Figure 8.2.A).

Dave Curry, who has lived on the lakeshore for more than 30 years, guessed the cause immediately: a sinkhole.

As in most of Florida, the soil around Scott Lake is shallow, just a few tens of feet deep. Beneath the soil is a porous layer of limestone and the water-saturated rock of the great Floridan aquifer. Several seasons of drought had lowered the water level in the aquifer, undermining support for the soil above. With the sudden weight of so much rain, the lake bottom collapsed into a hole more than 1.5 m (~5 ft.) in diameter and 16 m (~52 ft.) deep.

"This happens all the time in Florida," Curry says. He had watched the lake drain once before, in 1969, likely due to a previous sinkhole.

This time, Scott Lake lost more than a billion gallons of water, more than 95 percent of the lake's volume. The event was catastrophic for lake wildlife. "I counted seven alligators down in that one hole. They couldn't get out; they ended up being washed into the Floridan aquifer along with the fish and the turtles," Curry says.

Of the estimated 85,000 pounds of fish in the lake, those not sucked into the sinkhole were left gasping on the muddy bottom. "There were huge dead fish laying all around the banks. But the smell didn't last too long because thousands of buzzards and all kinds of other bird life I'd never seen before came in and cleaned it up in a few days. Otters were catching some of the live fish in distress," Curry says.

On shore, houses nearest the sinkhole fared the worst. The foundations of two houses cracked, windows shattered, and fissures opened up in lawns and swimming pools. The collapse of the soil had steepened the shoreline, causing the houses to slip. "It looked like a California mudslide," Curry says.

Since then, Scott Lake has steadily begun to refill. Mud and other debris seem to have formed a natural plug, and the lake, about one-third of its original size, attracts birds and other wildlife. The lake—and housing values—may recover, at least until Mother Nature strikes again.

— KATHLEEN WONG

⋀ FIGURE 8.2.A Dave Curry Viewing Drop in Lake Level
Property owner Dave Curry walks out on a boat dock that once served Scott Lake in south-central Florida. The lake level was lowered drastically when a sinkhole opened in the lake bottom. *(Dave Curry/Curry Controls Company)*

The natural tendency for deltas to subside can be amplified if sedimentation on the delta plain is slowed or stopped. Sedimentation has been sharply reduced on both the Mississippi and the Nile Deltas during the past 125 years.[23] On the Mississippi, sedimentation on the delta plain was slowed and stopped by levees built on both sides of the river. These levees have protected communities such as New Orleans from river flooding, but they have also kept riverbanks from eroding and new sediment from being added to the delta plain. In the case of the Nile, construction of the Aswan dams upstream and the diversion of two-thirds of the river water into canals have stopped sediment from reaching much of the delta plain.[24] Part of the subsidence in both the Mississippi and the Nile Deltas is also the result of compaction of organic soils. We will now explore the subsidence of the Mississippi Delta in more detail.

▼ FIGURE 8.14 Mississippi Delta Built in Stages

During the past 7,500 years, the Mississippi and Atchafalaya Rivers have built the Mississippi Delta plain in a series of stages. In each stage, the rivers built a complex lobe of sand and mud into the Gulf of Mexico. Each lobe became inactive when the river breeched its levee during a flood and shifted its flow to a shorter channel to the sea. *(Based on Roberts, H. H., "Dynamic Changes of the Holocene Mississippi River Delta Plain: The Delta Cycle,"* Journal of Coastal Research *13(3): 605–27, 1997)*

8.3 CASE study — Mississippi River Delta Subsidence

Home to nearly 40 percent of the coastal marsh and swamp wetlands in the United States, the Mississippi Delta is the major land feature in southern Louisiana. Geologically young, the present coastline of the delta was built by sedimentation from both the Mississippi and the Atchafalaya Rivers during the past 7,500 years.[25] As these rivers reach the sea, their channels subdivide into multiple distributaries. Actively flowing distributary channels build out a lobe-shaped pile of sand and mud into the ocean. This outbuilding continues until floodwaters break through a channel bank and the river finds a shortcut to the sea. If the break is permanent, then the river abandons its old distributaries and forms a new lobe of deltaic sediment. For the Mississippi Delta, this has occurred, on average, once each millennium (Figure 8.14).

As the river shifts its distributaries, two natural processes keep the land surface—the delta plain—above sea level: the deposition of sand and mud by distributary floodwaters, and the accumulation of organic soil

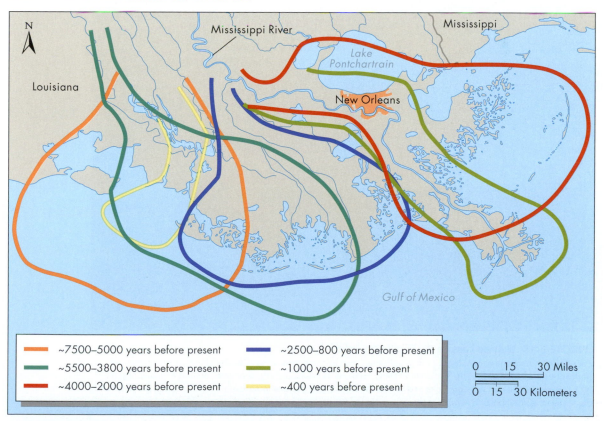

━━ ~7500–5000 years before present	━━ ~2500–800 years before present
━━ ~5500–3800 years before present	━━ ~1000 years before present
━━ ~4000–2000 years before present	━━ ~400 years before present

0 15 30 Miles
0 15 30 Kilometers

from dead marsh and swamp vegetation. The resulting sedimentary deposit is loosely packed together and saturated with water, much like a wet sponge. Over time, this deposit naturally compacts, and the land surface will subside unless new sediment is added to the delta plain.

Nearly three centuries ago, French settlers in New Orleans began to alter the natural balance between sedimentation and subsidence with the construction of levees along the Mississippi. By the early twentieth century, the U.S. Army Corps of Engineers had completed levees on both sides of the river and essentially stopped the Mississippi from flooding the delta plain. Without river flooding, the delta plain lost its regular supply of freshwater, nutrients, and sediment, and near-surface subsidence increased.

Starving the delta plain of new sediment and nutrients has not been the only cause of subsidence. Withdrawal of large amounts of groundwater, oil, and natural gas from below the delta surface appears to have significantly contributed to subsidence and a loss of wetlands.[26] Other scientists have hypothesized that tectonic subsidence, the sinking of Earth's crust below the delta, has contributed to lowering the delta surface. Investigations testing this hypothesis have yielded conflicting results, and further study will have to be done.[27,28]

A major consequence of the increased subsidence of the Mississippi Delta plain has been a significant loss of wetlands. Marshes and swamps in coastal Louisiana are disappearing at an average rate of a football-field-sized area every 30 minutes, which translates to the disappearance of 50 to 89 km^2 (~20 to 35 mi.^{2}) of land each year.[29] Wetland loss makes the coastline, including New Orleans, even more vulnerable to hurricane storm surges (see Chapter 10) and has resulted in major economic losses for fisheries.

In addition to subsidence, rising sea levels and the construction of numerous navigation and drainage canals have also contributed significantly to the loss of wetlands. The canals, built primarily for oil and gas operations, have exposed the delta plain to erosion and brought in saltwater that has killed wetland plants (Figure 8.15).[30]

An estimated 25 percent of the original coastal wetlands in Louisiana disappeared in the past century

Oil drilling rig

^ FIGURE 8.15 Wetland Loss on the Mississippi Delta
Excavation of numerous canals for oil and gas drilling, such as these near Leesville, Louisiana, have contributed significantly to wetland loss in the Mississippi Delta. Each canal drains the adjacent wetland and exposes organic soil to decomposition and erosion. Loss of this soil causes the land surface to subside. In many areas, the only land remaining is the spoil piles of sediment that were dredged to make the canals. *(Philip Gould/Corbis)*

◄ FIGURE 8.16 Cutting a Levee to Restore a Wetland The floating dredge *California* in the lower center of the photograph is cutting a channel through the levee along the west bank of the Mississippi River south of New Orleans. The channel, part of the $22 million U.S. Army Corps of Engineers West Bay Sediment Diversion Project, was completed in 2003. This project was designed to divert 1400 m³ (~50,000 ft.³) of river water per second to build a small, deltaic wetland in an adjacent bay. *(U.S. Army Corps of Engineers, New Orleans District)*

through subsidence, vegetation loss, and erosion. What will happen to the remaining 14,000 km² (~5500 mi.²) of wetlands? In the late 1990s scientists, engineers, government officials, and business leaders in Louisiana determined that a massive wetland and barrier island restoration effort was needed to combat the loss of coastal marshes and swamps (Figure 8.16).[29] Although this effort is projected to cost more than $14 billion over the next 40 to 50 years, it is substantially less than the $135 billion in property damage from Hurricane Katrina (see Chapter 10) and other large consumer and federal expenditures. Dealing with coastal land loss, such as that on the Mississippi Delta, will be one of the major environmental and disaster-planning challenges for the United States in the twenty-first century.

Collapsible Soils Some windblown dust deposits, referred to as *loess*, and stream deposits in arid regions are loosely bound with clay particles and water-soluble minerals. These deposits dry out rapidly after they form and may remain essentially dry for a long time. This dry condition can change if water ponds on the surface long enough for it to percolate downward through the entire deposit. The percolating water weakens the bonds of clay particles and dissolves minerals that hold the soil together. The entire deposit can then collapse and lower the land surface, sometimes by more than a meter (~3 ft.).[31] Soil and sediment that are prone to this behavior are referred to as **collapsible soil**.[32]

Organic Soils Some wetland soils forming in marshes, bogs, and swamps contain large amounts of organic matter. Called *organic soils*, these earth materials consist of partially decayed leaves, stems, roots, and, in colder regions, moss, which soak up water like a sponge. When water is drained from these soils, they dry out, compact, and are exposed to processes that cause them to disappear.

Bacterial decomposition is the primary process causing subsidence in drained organic soil. This process converts organic carbon compounds to carbon dioxide gas and water. Other destructive processes include water and wind erosion and the combustion of peat in prescribed burns and wildfires. The decomposition, erosion, and burning of organic soils cause the irreversible subsidence of drained wetlands. Part of the subsidence of the city of New Orleans can be attributed to compaction of organic soil.[33]

One of the most dramatic examples of subsidence in organic soils has occurred in the Florida Everglades. Land drainage, primarily for agriculture and urban development, has combined with droughts to cause more than half of the freshwater Everglades to subside from 0.3 to 3 m (~1 to 9 ft.) during the twentieth century.[34,35]

EARTHQUAKES

Although we commonly think of earthquakes as associated with uplift of the ground surface, they may also cause subsidence. In the Pacific Northwest, the geologic record contains evidence of repeated episodes of subsidence along the coasts of British Columbia, Washington, and Oregon. See Chapter 4 and Figures 4.5 and 4.6 for a discussion of subsidence with earthquakes.

UNDERGROUND DRAINAGE OF MAGMA

Subsidence can also occur from volcanic activity, related to either the central magma chamber of the volcano or shallow tunnels on the flank of the volcano. As magma moves upward into underground chambers below a volcano, the surface of the volcano may be forced upward. When it erupts, the volume of magma in the underground chamber is reduced and the land initially uplifted by the magma will subside. As mentioned in Chapter 5, cycles of uplift and subsidence are useful in predicting volcanic eruptions.

When basaltic lava flows down the flank of a volcano, the surface of the flow may cool as molten lava continues to move underground through a tunnel called a *lava tube* (see Figure 5.22). Lava eventually drains from the tube, leaving a void near the surface that is susceptible to collapse. Unfortunately, some unsuspecting home buyers on the Big Island of Hawaii have discovered too late that it is not advisable to build on a lava tube.

EXPANSIVE SOILS

Changes in moisture conditions can produce substantial movement in some clay-rich soils. Referred to as **expansive soils**, these soils shrink significantly during dry periods and expand or swell during wet periods. Most of the swelling is caused by the chemical attraction of water molecules to the surface of very fine particles of clay (Figure 8.17).[36] Swelling can also be caused by the chemical attraction of water molecules to layers within the crystal structure of some clay minerals.

The *smectite* group of clay minerals, including the mineral montmorillonite, typically has the smallest clay crystals and thus, in bulk, has the greatest surface area to attract water molecules. Smectites are abundant in many clay and shale deposits derived from the chemical weathering of volcanic rock, and they are the primary mineral in bentonite, a rock that forms from the alteration of volcanic ash. Clay, shale, and clay-rich soil containing smectite have the greatest potential for shrinking and swelling.

The presence of *shrink-swell clays* such as smectites can often be recognized from features on the land surface. These features include deep cracks produced as these soils dry out (Figure 8.18a), a popcorn-like weathering texture on bare patches of clay, an alternating pattern of small mounds and depressions in the land surface, a series of wavy bumps in asphalt pavement (Figure 8.18b), the tilting and cracking of blocks of concrete and bricks in sidewalks and foundations (Figure 8.18c), and the random tilting of utility poles and gravestones.

Shrinking and swelling of expansive soils cause significant structural damage to building foundations and the shifting and breaking of pavement and utility lines. This decrease or increase in the overall volume of the soil results from changes in its moisture content. Factors that affect the moisture content include climate, vegetation, topography, and drainage.[37] Regions that have a pronounced wet season followed by a dry season often have shrink-swell soils. These regions, such as the southwestern United States, are more likely to experience problems with expansive soils than do regions where precipitation is more evenly distributed throughout the year. Vegetation can also cause changes in the moisture content of a soil. Large trees draw and use a lot of local soil moisture, especially during a dry season. This withdrawal of water may produce soil shrinkage (Figure 8.17b).

Expansive soils swell when they become wet and can produce an expansion pressure in excess of 500 kPa (~10,000 lb per ft.[2]).[38] This intense pressure is strong enough to move the foundation and break open walls of multistory buildings, cause building support pilings to rise up out of the ground, and rupture major underground pipelines.

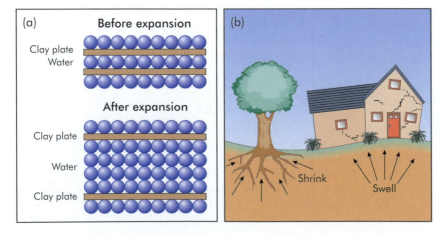

◄ FIGURE 8.17 Expansive Soils
(a) Smectite clay expands as water molecules are added onto and within the clay particles. (b) Effects of soil's shrinking and swelling at a home site. *(Based on Mathewson, C. C., and J. P. Castleberry, II,* Expansive Soils: Their Engineering Geology. *College Station, TX: Texas A&M University)*

(a)

(b)

(c)

∧ FIGURE 8.18 Damage from Expansive Soil
(a) Drying of an expansive soil produces this popcorn-like surface and a network of polygonal desiccation cracks. Pen in lower left for scale. *(USGS)* (b) Uneven shrinking and swelling of expansive clays in layers of steeply dipping bedrock produced the rolling surface of this road and sidewalk in Colorado. *(David C. Noe/Colorado Geological Survey)* (c) Uneven shrinking and swelling of expansive soil cracked the concrete in this driveway. *(Edward A. Keller)*

FROST-SUSCEPTIBLE SOILS

Freezing temperatures over a sufficient period of time can produce a great deal of expansion in silty soil. Silt is gritty sediment that is finer than sand but coarser than clay. Soils that have the potential to move and expand when they freeze are referred to as *frost-susceptible soils*. These soils expand because of the 9 percent volume increase that occurs when water changes to ice. This volume increase causes an upward movement of soil particles and the land surface in a process known as *frost heaving*. Upward movement is generally much greater than 9 percent because additional water is drawn to the zone of freezing, thereby increasing the amount of ice that is formed. The additional ice accumulates to form discrete pods or lenses of ice in the earth material (Figure 8.19).

> FIGURE 8.19 Ice Formation in Soil Causes Frost Heaving
In cold regions, pods or lenses of ice can form in the soil and cause the land surface to heave upward. This photo of Arctic tundra in northern Alaska shows ice lenses below mounds of heaved soil. *(John A. Kelley, USDA Natural Resources Conservation Service)*

Three conditions are necessary for frost heaving to occur in a soil: (1) Temperatures in the soil must remain below freezing long enough for ice to form, (2) water must be present to create a substantial amount of ice, and (3) there must be a significant proportion of silt-size particles to allow water to be drawn through and retained in the soil.

8.2 CHECK your understanding

1. Define subsidence.

2. How is karst topography formed? Give some examples of karst features.

3. Compare the different types of sinkholes.

4. How do limestone cave systems form?

5. Why is the Mississippi River Delta subsiding?

6. How are thermokarst, sediment and soil compaction, earthquakes, and drainage of magma linked to subsidence?

7. What is expansive soil?

8.3 Geographic Regions at Risk for Subsidence and Soil Volume Change

Dissolution of soluble rocks, the presence of permafrost, and compaction of sediment and soil are the most common causes of subsidence. In particular, the dissolution of limestone produces karst topography, which is estimated to compose up to 10 percent of Earth's surface in all parts of the world. In the contiguous United States, approximately 25 percent of the land surface is underlain by limestone or evaporites that have some degree of karst development (Figure 8.20). Major belts of karst topography include (1) a region extending through Tennessee, Virginia, Maryland, and Pennsylvania; (2) south-central Indiana and west-central Kentucky; (3) the Salem-Springfield plateaus of Missouri and northernmost Arkansas; (4) the Edwards Plateau of central Texas; (5) most of central Florida; and (6) Puerto Rico. Subsidence and other karst-related phenomena are a major problem in these areas. The most extensive development of karst in evaporite rocks in the United States is in the Permian basin of eastern New Mexico and western Texas.[39]

Permafrost covers more than 20 percent of the world's land surface.[22] Most of Alaska and more than half of Canada and Russia are underlain by permafrost. Hazard maps show that many towns and smaller settlements are threatened by thawing permafrost, including Barrow, Alaska; Inuvik, Northwest Territories, Canada; and Yakutsk, Norilsk, and Vorkuta in Russia.[22]

Subsidence caused by the compaction of sediment is most pronounced in areas where the sediment was rapidly deposited or where it contains abundant organic matter. These areas include many of the world's marine deltas, such as those of the Mississippi River, the Sacramento–San Joaquin Rivers in California, and the Nile River in Egypt.

Organic-rich soils susceptible to subsidence are common in cold-region wetlands of New England, New

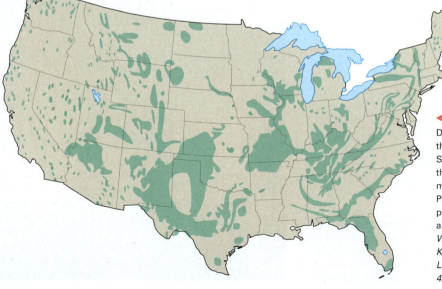

◀ FIGURE 8.20 Karst Map
Distribution of karst topography in the contiguous states of the United States. Approximately 40 percent of the land in the eastern half of this map is karst topography. Alaska, Puerto Rico, Hawaii, and Canadian provinces and territories also have areas of karst. *(Based on White, W. B., Culver, D. C., Herman, J. S., Kane, T. C., and Mylroie, J. E., "Karst Lands," American Scientist 83: 450–59, 1995)*

York State, the Upper Midwest, Washington State, Alaska, and parts of Canada, where these wetlands are variously called "bogs," "fens," "moors," and "muskeg." Coastal wetlands underlain by peat or muck deposits are also susceptible to subsidence in the Florida Everglades, the Sacramento–San Joaquin Delta of California, and coastal Louisiana and North Carolina.

In North America, expansive soils are a problem primarily in the western United States and Canada. Soils with the highest swelling potential are found in the blackland prairies from Texas to Alabama and in parts of the High Plains from eastern Colorado north to Saskatchewan and Alberta, Canada, although swelling clays are found in soils in all 50 U.S. states as well as Puerto Rico, the Virgin Islands, and in most Canadian provinces and territories.[40]

Frost heaving occurs in regions underlain by permafrost, as well as in colder parts of the United States and Canada that do not have permanently frozen ground. Frost-susceptible soils occur throughout Canada, Alaska, the northern contiguous United States, and at higher altitudes in the Appalachians, Rocky Mountains, Sierra Nevada, and Cascades.

Large-scale, seismic-related subsidence is a risk for areas of Alaska, Hawaii, and the Pacific Northwest in both Canada and the United States. Deflation of a magma chamber can cause subsidence in any volcanic area, and subsidence over lava tubes occurs in the Hawaiian Islands and in some lava flows of western North America.

8.3 CHECK your understanding

1. List some areas in the United States that are particularly vulnerable to subsidence, and explain why.

8.4 Effects of Subsidence and Soil Volume Change

Both karst formation and soil expansion and contraction cause significant economic damage each year. Karst regions are home to a host of problems such as sinkhole formation, groundwater pollution, and variable water supply. Heaving in expansive and frost-susceptible soils often damages highways, buildings, pipelines, and other structures. Additional damage from subsidence occurs in marine deltas, drained wetlands, areas of collapsible soils, and in many areas underlain by permafrost.

SINKHOLE DAMAGE

Sinkholes have caused considerable damage to highways, homes, sewage facilities, and other structures. Both natural and artificial fluctuations in the water table are probably the trigger mechanism for the formation of collapse sinkholes. High groundwater table conditions enlarge caverns closer to the surface of Earth by dissolving the roof and side of the caves. As long as a cavern remains filled with water, the buoyancy of the water helps support the weight of the overlying earth material, but lowering of the groundwater table eliminates some of the buoyant support, and the cave roof may collapse. This situation was dramatically illustrated in Winter Park, Florida, on May 8, 1981, when a large collapse sinkhole began developing. The sinkhole grew rapidly and within 24 hours had swallowed a house, part of a community swimming pool, half of a four-lane street, parts of three businesses, and parking lots containing several Porsches and a pick-up camper (Figure 8.21).[41] Damage caused by this sinkhole exceeded $2 million.

Sinkholes form nearly every year in central Florida when the groundwater level is lowest. Although exact positions cannot be predicted, more sinkholes form during droughts. The Winter Park sinkhole and several smaller sinkholes developed during the 1981 drought when groundwater levels were at a record low.

More recently, a sinkhole 6 m (~20 ft.) deep by 5 m (~16 ft.) wide quickly collapsed above a home in Seffner, Florida, in February 2013 (Figure 8.22). A 36-year-old man in his bedroom was swallowed as the room disappeared underground, and his body was never recovered. His brother heard him scream and jumped in the hole, but the sinkhole continued to collapse and the brother had to be rescued by emergency crews. The rest of the home had to be demolished and the hole filled with gravel. By the time the sinkhole was filled, it had grown to 20 m (~66 ft.) deep. Such deaths are rare but have occurred several times in the past. A spectacular sinkhole, about 15 m (~49 ft) in diameter, opened in August 12, 2013, near Walt Disney World (Orlando, Florida, area). A three-story building of condominiums partially collapsed into the developing sinkhole. Quick evacuation of tourists was successful, and there were no deaths.

In urban and some rural areas, the evidence of a subsidence hazard is sometimes masked by human activities. For example, a sinkhole in the Lehigh Valley in eastern Pennsylvania was first identified in the 1940s and was then completely filled with dead stumps, blocks of asphalt, and other trash and covered with a cornfield. Although no longer recognizable as a sinkhole, the filled depression collected urban runoff from nearby pavement and buildings. The runoff loosened

< FIGURE 8.21 Sinkhole That Swallowed Part of a Town
This sinkhole in Winter Park, Florida, grew rapidly for three days in 1981, swallowing part of a community swimming pool as well as several businesses, houses, and vehicles. The sides of the sinkhole have since been stabilized and landscaped, and it is now a park with a small lake. *(AP Images)*

∧ FIGURE 8.22 Collapsing Sinkhole
This small sinkhole collapsed under a home in Seffner, Florida, in February 2013. A man in his bedroom was swallowed and killed (his body was never recovered). The sinkhole collapsed several more times as emergency crews responded. Incidences of people killed by collapsing sinkholes, while rare, have occurred several other times. *(AP Photo/The Tampa Bay Times, Dirk Shadd)*

the plug of soil and trash and continued to dissolve the underlying limestone. An increased demand for groundwater in the area also contributed to lowering of the groundwater table. These factors combined to cause a catastrophic collapse of the land surface in 1986. Within only a few minutes, the collapse produced a pit that was at least 30 m (~100 ft.) in diameter and 14 m (~46 ft.) deep. Although the damage was confined to a street, parking lots, sidewalks, and utility lines, subsequent stabilization and repairs cost nearly $500,000.

CHANGES IN GROUNDWATER CONDITIONS

Karst development creates a geologic environment in which groundwater is easily accessed and intensively used by humans and wildlife and is also easily polluted. Sinkholes, caves, and related karst features can form direct connections between surface water and groundwater (Figure 8.23). Such connections make the groundwater vulnerable to pollution and contribute to fluctuations in the groundwater table during droughts. These vulnerabilities are a major concern for both public and private sources of drinking water.

∧ FIGURE 8.23 Water from a Subterranean Stream
In karst topography, groundwater and surface water may be directly connected. This waterfall develops from a subterranean groundwater flowing from a cave. *(Kenneth Murray/Science Source)*

One common source of pollution comes from sinkholes that have been used for waste disposal, especially where the bottom of the depression is near the groundwater table. Groundwater can also become contaminated where polluted water from surface streams flows into caves and fractures. This water can reach the groundwater table without the natural filtration provided by soil or sand.

Groundwater-table fluctuations in karst areas affect humans, plants, and wildlife. For example, groundwater from karst limestone is heavily used along the edge of the Edwards Plateau in central Texas. In this area, frequent drought and heavy groundwater use can rapidly lower the groundwater table and cause springs to reduce or stop their flow. Diminished groundwater flow impacts communities that rely on it as a source of water, such as the city of San Antonio, as well as threatens a number of unique plants and animals that are found only in the springs.

DAMAGE CAUSED BY MELTING PERMAFROST

Early settlers in the Arctic built their homes directly on permafrost, only to find that heat radiated from the buildings thawed the soil. By the middle of the twentieth century, most Arctic structures were built on posts or pillars sunk into the permafrost. This technique elevates the floor of a building above the land surface and keeps its heat from melting the underlying soil. However, this use of pilings assumed that the permafrost would remain frozen. For the past several decades, that assumption has not always been the case.

Thawing of permafrost has caused roads to cave in, airport runways to fracture, railroad tracks to buckle, and buildings to crack, tilt, or collapse (Figure 8.24).[22] In two Siberian cities alone, an estimated 300 apartment buildings have been damaged. The state of Alaska now spends around 4 percent of its annual budget repairing permafrost damage.[42]

COASTAL FLOODING AND LOSS OF WETLANDS

Flooding of coastal areas and destruction of wetlands are two major effects of subsidence in marine deltas and bays. Subsidence of the Mississippi Delta during the past century has contributed to wetland loss and the sinking of New Orleans. The city of New Orleans and adjacent suburbs are subsiding relative to sea level at an average rate of 8 mm (~0.3 in.) per year (Figure 8.25). As a result, most of New Orleans is below sea level, and parts of east New Orleans have subsided to 3 to 5 m (~10 to 16 ft.) below sea level.[43] Rings of levees and flood walls surrounding New Orleans and many of its suburbs were established to keep urban areas from flooding by the Gulf of Mexico, Lake Pontchartrain, and the Mississippi River.

▲ **FIGURE 8.24 Melting Permafrost Destroys Building**
The foundation of this apartment building in Cherskii in eastern Siberia was undercut by thawing permafrost. Structural damage from melting permafrost is becoming common in Russia, Alaska, and Canada. *(Professor V. E. Romanovsky/University of Alaska Fairbanks Press)*

Land surface around 1956

10 to 16 inches of subsidence

Land surface today

▲ **FIGURE 8.25 Subsidence in New Orleans**
Land surrounding many New Orleans buildings has subsided since they were built. The land surface around this house in the Lakeview area appears to have subsided 25 to 50 cm (~10 to 20 in.) since 1956. *(Professor David J. Rogers)*

Outside the levees and flood walls are wetlands that help protect the city and its suburbs from ocean waves and storm surges. Without intervention, this marsh and swampland will subside or erode by 2090, and New Orleans will be directly on the sea.[33]

DAMAGE CAUSED BY SOIL VOLUME CHANGE

Expansive and frost-susceptible soils cause significant environmental problems. As one of our most costly natural hazards, soil volume change is responsible for many billions of dollars in damages annually to highways, buildings, and other structures. Annual costs from soil expansion and contraction exceed $15 billion in the United States and are substantial in Canada and Mexico as well.[44] Frost action on roads and streets in the United States alone costs several billion dollars each year for rebuilding thousands of miles of pavement.[45] In many years, damage caused by soil volume change exceeds the cost for all other natural hazards combined.

Every year, more than 250,000 new houses are constructed on expansive soils. Of these, about 60 percent will experience some minor damage, such as cracks in the foundation, walls, driveway, or walkway, and 10 percent will be seriously damaged, some beyond repair (see Figures 8.17, 8.18, and 8.26a).[46]

Underground water lines in expansive soils may rupture when there is a significant change in soil moisture. This rupture can result in a loss of water pressure, which makes the line vulnerable to contamination and requires customers to boil their water before using it.

In frost-susceptible soils, frost heaving has some of the same effects as swelling in expansive soils (Figure 8.26b). Cracking occurs in foundations and asphalt and concrete pavement, and pipelines break. Frost heaving also moves larger particles in the soil upward at a faster rate than smaller particles. This upward movement, as much as 5 cm (~2 in.) per year, can simply be an annoyance, such as the continual appearance of large stones in a garden or farmer's field each spring, or it can be destructive, such as the lifting of fence posts and utility poles out of the ground. Even coffins have been known to rise to the surface by frost heaving![47]

8.4 CHECK your understanding

1. List the five main effects of subsidence and soil volume changes.

2. What are some of the damages resulting from soil volume changes?

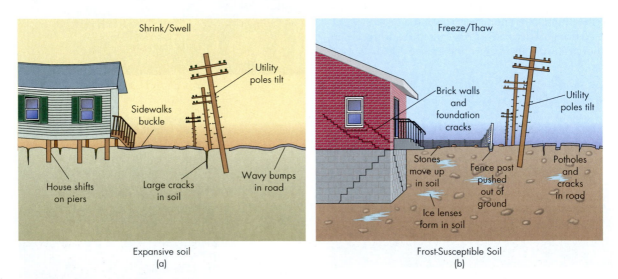

∧ FIGURE 8.26 Soil Movement Effects
Shrinking and swelling of expansive soils (a) and freezing and thawing of frost-susceptible soils (b) have similar effects. Both types of soil movement crack and shift foundations and pavement and tilt utility poles, road signs, and fence posts. Shrinking of the soil can produce deep cracks in the ground, and swelling of the soil can create small mounds on the land surface. Freeze and thaw cycles move buried objects, such as stones, upward in the soil to the surface. These cycles also contribute to the formation of potholes in asphalt pavement.

8.5 Linkages between Subsidence, Soil Volume Change, and Other Natural Hazards

As mentioned previously, subsidence can be a side effect of earthquakes, volcanoes, severe weather, and climate change. However, subsidence may also cause other natural hazards to occur. As described earlier for the Mississippi Delta, the link between subsidence and flooding is a common one. Soil expansion and contraction are also linked to increased rates of mass wasting on slopes.

In areas undergoing rapid subsidence, especially those that are also subsiding because of the overpumping of groundwater, flooding can be a major problem. As the land is lowered relative to surrounding bodies of water, the incidence of flooding increases (see the chapter opener and Applying the Fundamental Concepts). In many growing cities, there is a high demand for clean drinking water. Unfortunately, this demand leads to the mining of groundwater, often depleting the resource faster than it can be replenished. Because of groundwater mining and the resulting subsidence, coastal and river floods have become more common and severe in low-lying coastal cities such as Bangkok, Thailand, which experienced disastrous floods in 2011, and urban flooding has become more common in Mexico City.[48]

On sloping land surfaces, frost heaving and the shrinking and swelling of clay interact with mass wasting (see Chapter 7). This interaction can increase the rate of gravity-driven creep and produce an uneven land surface.

Subsidence also has direct linkages to climate change. This linkage is apparent in hot deserts, cold Arctic regions, and coastal areas. In hot arid areas, drought conditions commonly lower the groundwater table. The withdrawal of groundwater can further compact and shrink unconsolidated earth materials. In the deserts of the southwestern United States, the drying of earth materials is contributing to regional subsidence and to the formation of large, polygonal *desiccation cracks* (Figure 8.27). These cracks are similar in shape to the cracks that you see in mud after it has dried out. Except for their polygonal form, the large desiccation cracks are the same size and depth as large linear cracks, called *earth fissures*, which are produced by the overpumping of groundwater.[49,50] In cold Arctic areas, global warming is the primary cause of the melting of permafrost. In turn, permafrost melting releases the greenhouse gases carbon dioxide and methane into the atmosphere, which could accelerate climate change (see Chapter 12). Global warming is also the primary cause for the increased rate of sea-level rise. In coastal communities,

∧ **FIGURE 8.27 Large Desiccation Crack**
This crack in the road is part of a polygonal network of large desiccation (drying) cracks in Graham County, southeastern Arizona. Unlike similar earth fissures that result from overpumping of groundwater, these cracks appear to be the result of the natural lowering of the groundwater table in a drought. *(Arizona Geological Survey)*

such as Venice, New Orleans, Galveston Bay near Houston, and the Nile Delta, subsidence increases the local rate of sea-level rise and the resulting loss of land.

8.5 CHECK your understanding

1. In what important ways do subsidence and soil volume changes link with other natural hazards?

8.6 Natural Service Functions of Subsidence and Soil Volume Change

Although subsidence and soil volume change cause many environmental and economic problems, there are also benefits from these processes. Expansion and contraction of the soil form natural coherent blocks of soil material called peds that are similar to clods of soil produced by plowing or digging. The creation of peds contributes to soil productivity, especially in clay-rich soils, by facilitating root growth and soil drainage.

The same karst processes that cause subsidence also contribute water as well as aesthetic, scientific, and ecological resources. Karst terrains are some of the world's most productive sources of drinking water. Beautiful karst formations such as cavern systems and tower karst are important aesthetic and scientific resources (see Figures 8.12 and 8.13). Finally, the caves of karst areas are home to rare, specially adapted creatures, some of which are found nowhere else. For example, caverns and other karst features in the Edwards Limestone in Texas are home to more than 40 unique species, eight of which are legally designated as endangered.

WATER SUPPLY

Karst regions contain the world's most abundant groundwater supply, thus providing a critical world resource. About 25 percent of the world's population either lives on or gets its drinking water from karst formations. In the United States, an estimated 40 percent of the groundwater used for drinking comes from karst terrains.[51] For example, the Edwards Formation, discussed previously, provides drinking water for more than 2 million people in Texas.

AESTHETIC AND SCIENTIFIC RESOURCES

Rolling hills, extensive cave systems, and beautiful formations of tower karst are among the aesthetic resources found in karst areas. Unique landscapes, such as the tower karst regions of China, are areas of unparalleled beauty that offer stunning vistas. Caves, too, have proved to be a popular destination for both spelunkers and tourists. Mammoth Cave National Park, Kentucky, contains the world's longest cave system and attracts droves of visitors each year. Aesthetics aside, karst regions also provide scientists with a natural laboratory in which to study the record of climate change contained in cave formations. Caves also provide an ideal environment for preserving animal remains, making them important resources for paleontologists and archaeologists.

UNIQUE ECOSYSTEMS

Caves are home to rare creatures specially adapted to live in the karst environment. Karst-dependent species

known as *troglobites* have evolved to live in the total darkness of caves. Such species include eyeless fish, shrimp, salamanders, flatworms, and beetles. Other animals, such as bats, rely on caves to provide shelter. Karst areas generally support a diverse cross-section of species; preserving these areas and, hence, these organisms for the future should be a societal objective.

8.6 CHECK your understanding

1. Describe the main natural service functions of subsidence and soil volume change.

8.7 Human Interaction with Subsidence and Soil Volume Change

As discussed in the previous sections, subsidence can both cause problems and provide benefits. Commonly, when human beings live in areas underlain by karst formations, compacting sediment and soil, permafrost, and expansive or frost-susceptible soil, previously existing problems are exacerbated and new problems arise. Human beings contribute to problems associated with subsidence and soil volume change by withdrawing subsurface fluids, excavating underground mines, thawing frozen ground, restricting deltaic sedimentation, altering surface drainage, and using poor landscaping practices.

FLUID WITHDRAWAL

The withdrawal of subsurface fluids, such as oil with associated natural gas and water, groundwater, and mixtures of steam and water for geothermal power, have all caused subsidence.[52] In each case, the high fluid pressure within the sediment or rock helped support the earth material above. A large rock at the bottom of a swimming pool seems lighter for the same reason: Buoyancy produced by the liquid tends to lift the rock. Support or buoyancy by means of fluid pressure can be especially important in shallow or rapidly deposited earth material. In these instances, pumping out the fluid reduces support and causes surface subsidence.

Groundwater mining—extracting groundwater through wells and springs in volumes greater than that being replenished through the percolation of rain and surface water—is a leading cause of regional subsidence. A classic example can be found in the central valley of California where thousands of square kilometers have subsided from overpumping groundwater for irrigation and other uses (Figure 8.28a). One of the areas of greatest subsidence has been across the valley from Fresno, where most of the land surface has subsided more than 0.3 m (~1 ft.) and locally up to 9 m (~30 ft.) (Figure 8.28). As the groundwater was mined, water pressure was reduced, sedimentary grains compacted, and the surface subsided (Figure 8.29).[53,54] Similar examples of subsidence caused by overpumping have been documented near Phoenix, Arizona; Las Vegas, Nevada; the Houston–Galveston area in Texas; San Jose, California; and Mexico City, Mexico. Such subsidence has reactivated geologic faults and formed extremely long earth fissures in unconsolidated sediment (Figure 8.30).[54]

UNDERGROUND MINING

Serious subsidence events have been associated with the underground mining of coal and salt. Most subsidence in coal mining is caused by the failure of pillars of coal that have been left behind to support the mine roof. With time, these pillars weather, weaken, and collapse, causing the roof to cave in and the land surface above the mine to subside (see Case Study 8.4: Professional Profile). In the United States, more than 8000 km² (~3100 mi.²) of land, an area twice the size of Rhode Island, has subsided because of underground coal mining. This subsidence continues today, long after mining has ended. In 1995, a coal mine that was last operated in the 1930s collapsed beneath a 600 m (~1970 ft.) length of I-70 in Ohio; repairs of the highway took three months.[55] Although coal mine

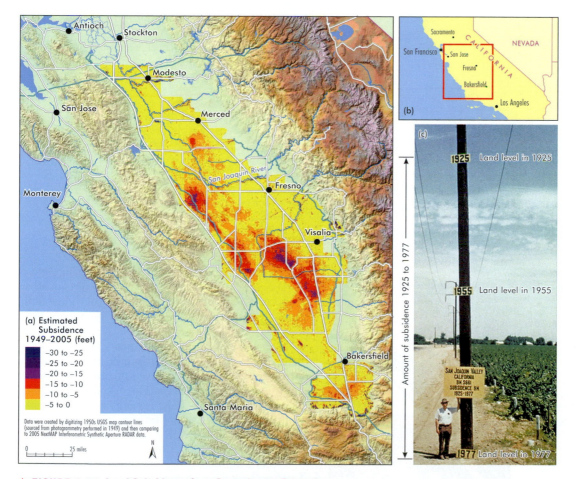

∧ FIGURE 8.28 Land Subsidence from Groundwater Extraction

(a) Principal areas of land subsidence in California resulting from groundwater withdrawal. *(Adapted from California Department of Water Resources)* (b) Location map showing the approximate location shown in (a). (c) Photograph illustrating the amount of subsidence in the San Joaquin Valley, California. The marks on the telephone pole are the positions of the ground surface in recent decades. The photo shows nearly 8 m (~26 ft.) of subsidence. *(Richard Ireland/USGS)*

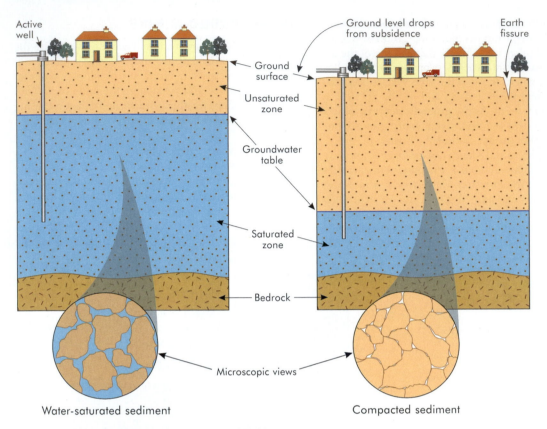

∧ FIGURE 8.29 Subsidence from Pumping of Fluids
Idealized diagram showing how surface subsidence results from pumping groundwater. The unsaturated zone is the area above the groundwater table where the holes (pores) between grains contain both air and water. In the saturated zone beneath the groundwater table, the pores are completely filled with water. When groundwater is pumped out, the pores collapse and the surrounding earth material compacts. This compaction can cause the land surface to subside.

> FIGURE 8.30 Earth Fissures Damage Road
A caution sign was erected after an earth fissure damaged this road in rural Pima County, Arizona, in 1981. Earth fissures have developed in the desert of the southwestern United States where overpumping has lowered the groundwater table. *(USGS)*

subsidence most often affects farmland and rangeland, it has also damaged buildings and other structures in urban areas such as Scranton, Wilkes-Barre, and Pittsburgh, Pennsylvania; Youngstown, Ohio; and Farmington, West Virginia.[55,56]

In the case of salt, subsidence has taken place over both solution and open-shaft mines. Solution mines, the source for most of our table salt, use wells to inject freshwater into salt deposits. The dissolved salt is then pumped out of the well, leaving a cavity behind. Collapse of this cavity and subsequent surface subsidence have taken place in solution mines in Kansas, Michigan, New York, and Texas.[39]

Open-shaft mining is used to extract rock salt from Earth and is taking place underground below Detroit, Michigan, and Cleveland, Ohio. In the past three decades, the catastrophic flooding of two underground salt mines has caused surface damage and subsidence. One, the Retsof Mine near Geneseo, New York, was the largest salt mine in the world. Collapse of its roof in

1994 allowed groundwater to flood the mine. Flooding produced two large sinkholes and subsidence damage to roads, utilities, and buildings.[57]

The second flooding event occurred on the Jefferson Island salt dome in southern Louisiana. In 1980, an oil rig drilling for natural gas accidentally penetrated an underground salt mine (Figure 8.31). The rig was mounted on a floating barge in a small lake above the salt dome. After drilling into the mine shaft, the 45 m tall (150 ft.) rig disappeared into the 3 m (10 ft.) deep lake, followed by eight large barges that were on the water that day. Fortunately, 50 miners and seven people on the drilling rig escaped injury. The mine was a total loss, and buildings and gardens were damaged. Within three hours the entire lake had drained, and a 90 m (~300 ft.) deep, 0.8 km (~0.5 mi.) wide subsidence crater formed above the flooded mine. The structural integrity of underground salt mines is of particular concern because the U.S. Strategic Petroleum Reserve is stored in four Gulf Coast salt mines. A fifth storage site below Weeks Island, Louisiana, was emptied of oil in 1999 because of groundwater seepage through a sinkhole.

MELTING PERMAFROST

Humans have contributed to the thawing of permafrost, with resulting changes in soil volume, through global warming and poor building practices. Shoddy construction practices, inadequate removal of heat from beneath buildings, and burial of warm utility lines have locally melted permafrost, broken pipelines, and damaged buildings.[22,58]

RESTRICTING DELTAIC SEDIMENTATION

Marine deltas require the continual addition of sediment to their surfaces to remain at or above sea level. This sediment comes from the distributary channels that carry river water, sand, and mud to an ocean. In many deltas, humans have stopped or reduced this sedimentation by constructing dams upstream, building levees on both sides of distributary channels, and diverting sediment-laden river water into canals. All these practices may contribute to subsidence of the delta plain.

ALTERING SURFACE DRAINAGE

People have altered surface drainage for agriculture and settlement for centuries. These alterations are designed to remove or add water to the soil. In organic soils, draining water causes or increases soil compaction and subsidence. In the United States, draining organic soils has caused or increased subsidence in the Mississippi Delta, the Florida Everglades, and the Sacramento–San Joaquin River Delta. Subsidence in the Sacramento–San Joaquin River Delta threatens the integrity of levees that protect cropland from flooding and keep saltwater from intruding into California's massive water supply system.[59] Adding water to the soil, such as through irrigation, can cause land subsidence on collapsible soils and, as described next, increase the swelling of expansive soils.

POOR LANDSCAPING PRACTICES

The shrinking and swelling of expansive soils are often amplified by poor landscaping practices. Planting trees and large shrubs close to foundations may cause damage from soil shrinkage during dry periods as plant roots pull moisture from the soil.[37] At the other extreme, planting a garden or grass that needs frequent watering close to foundations can cause damage from soil swelling. Rather than keeping the soil at a constant moisture level, irrigation systems commonly leave excess water in the soil. Excess water is considered the most significant cause of damage from swelling soil.[26]

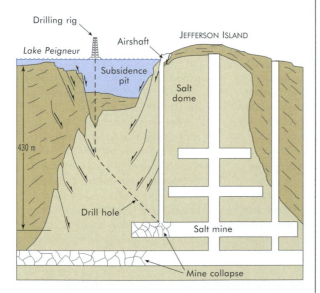

⋀ **FIGURE 8.31 Subsidence from Salt Dissolution**
Idealized diagram showing the Jefferson Island salt dome collapse that caused a large subsidence pit to form in the bottom of the lake. An estimated 15 million m³ (~530 million ft.³) of water flooded the salt mine after a natural gas well penetrated the mine shaft.

8.4 Professional Profile

Helen Delano, Environmental Geologist

For those looking to build a house in Pennsylvania, any number of natural hazards need to be considered. "We have something for everyone," says Helen Delano, from the Pennsylvania Bureau of Topographic and Geologic Survey. "As a geologist working for the state of Pennsylvania, one of my tasks is looking at where hazards may occur" (Figure 8.4.A).

Some of those dangers include deadly gases, such as methane and radon, which can migrate up from buried organic or radioactive material, respectively, but one of the more common problems is the potential for sinkholes and subsidence, particularly given the many abandoned coal mines in the area. According to Delano, the southwestern region of the state has

been mined for coal since the mid-1700s, and the shallower mines are only 15 to 30 m (~50 to 100 ft.) below the surface. Although the technology does exist to empirically test an area, Delano said a lot of information can be gained simply by consulting coal mining maps.

But when the maps do not provide a complete picture, relatively simple and inexpensive technology can help geologists determine whether there are any major cavities in the ground. Delano said geologists sometimes use what's called a "television borehole camera," literally drilling a hole and lowering a small camera into the ground to examine the underground contents. This technology, however, is available for only relatively shallow depths. Delano said geophysical methods for deeper depths are far more expensive.

In addition to old coal mines, shallow caves, which have the potential to form sinkholes, can be detected with ground-penetrating

radar—useful information for anyone considering a construction project in several parts of the state. "In eastern Pennsylvania, we get sinkholes from limestone subsidence," says Delano.

Delano began her career examining problems from natural hazards in the Pittsburgh area. "When I first started working, I would go out and look at people's backyard problems and help them understand what was going on and collect data for our use, trying to keep track of the scope of the problems for the state," she recalls.

Now she also deals with municipalities, advising them when natural hazards, including subsidence, pose problems for building projects. "I have a folder on my desk right now for a rural community that has a proposal for a housing project over eighteenth- and nineteenth-century iron mines," Delano said.

When a major construction project is slated to be built over a region where subsidence is a risk, there are several options, according to Delano. "Usually, the cheapest thing to do is build your house somewhere else," she said. But when the location is just too good to pass up, there are engineering

< FIGURE 8.4.A Fieldwork Required for Natural Hazard Assessments
Environmental geologist Helen Delano is the Pennsylvania Geological Survey contact for local governments in southeastern Pennsylvania. Here she examines an exposure of the Gettysburg Formation in her area of responsibility. *(James Shaulis/Pennsylvania Geological Survey)*

solutions, such as digging a deep foundation in firm bedrock.

The key, says Delano, is in knowing the risks ahead of time: "It's much, much cheaper to factor in those costs at the beginning." But for the most part, a close study of geological hazards, such as subsidence, is not yet routine in Pennsylvania when it comes to small construction projects. "Most residential construction doesn't take into account geological hazards," Delano said, adding that, as more and more valuable land is occupied, it will become increasingly necessary to consider such dangers.

— CHRIS WILSON

8.7 CHECK your understanding

1. Explain how land subsides from extracting fluids such as oil and groundwater.
2. Why can underground mining of coal and salt cause subsidence?
3. How may melting permafrost affect soils?
4. How can altering surface drainage cause subsidence?
5. What is the role of vegetation in causing shrinking or swelling of soil?

8.8 Minimizing Subsidence and Soil Volume Change

Minimizing the hazards related to subsidence and the expansion and contraction of soil requires an understanding of the landscape from a geologic perspective. Even with this understanding, it is difficult to prevent natural subsidence and changes in soil volume. However, there are steps that can be taken to minimize the damage associated with this hazard.[17]

Reducing Fluid Withdrawal We will always be plagued with subsidence problems in areas where bedrock is aggressively being dissolved or where groundwater levels are continuing to fall because of drought. Natural sinkholes will continue to open up. We can, however, prevent some human-caused subsidence associated with the mining of groundwater or pumping of oil and gas. Where groundwater, oil, or gas removal is causing the land surface to subside, it is possible to prevent or minimize further subsidence.

For example, from the early 1900s to the mid-1970s, groundwater mining in the Houston–Galveston area of Texas was the primary cause for up to 3 m (~10 ft.) of subsidence over an 8200 km² (~3200 mi.²) area.[60] This prompted the 1975 Texas legislature to create a regulatory district to issue well permits. Since creation of this district, subsidence has essentially stopped in areas where groundwater pumping has decreased. This has not been the case in parts of Florida, Arizona, and Nevada, where increasing use of groundwater continues to cause sinkholes or earth fissures.

Installation of *injection wells* is often suggested as a way to minimize or stop subsidence from fluid withdrawal. This technique was used with some success in the 1950s when water was injected at the same time that oil was being pumped from the Long Beach, California, oil field. However, this method does not work for most groundwater mining because irreplaceable water has been extracted from layers of fine earth material. Once fine earth material has been compacted, it is not possible to push the particles back apart.[38]

Regulating Mining The best way to prevent damage from subsidence caused by mining activities is to prevent mining in urban areas. Although such laws are currently in place in many countries, the threat from older mines still exists.

Preventing Damage from Thawing Permafrost Most existing engineering practices for building on permafrost have assumed that the permafrost will remain frozen if heat from a building or pipeline does not radiate into the ground. With the recent widespread thawing of permafrost, new and more costly practices are being developed, such as putting buildings on adjustable screw jacks or latticelike foundations to allow for the freezing and thawing of permafrost.[22]

Reducing Damage from Deltaic Subsidence Completely stopping further subsidence of human settlements on delta plains and restoring the deltas to natural conditions is unrealistic. Levees must continue to be elevated to protect urbanized areas, and adequate pumping systems must be maintained to remove excess surface water from the levee-protected enclosures. However, in undeveloped areas, levees could be breached to restore the sediment and freshwater supply necessary to rebuild marshes (see Case Study 8.3: Subsidence of the Mississippi Delta). The restored marshes would again help protect subsided urban areas from storms and rising sea levels.

Managing Drainage of Organic and Collapsible Soils
As with most groundwater-induced subsidence, restoration of drained organic soils or collapsed soils is not possible. Only proper water management of existing drainage to marshes and swamps will minimize future subsidence from organic soils. For collapsible soils, it is especially important to limit irrigation and modify the land surface so that water does not pond near buildings.

Preventing Damage from Expansive Soils
Proper design of subsurface drains, rain gutters, and reinforced foundations can minimize the damage from expansive soil. These techniques improve drainage and allow the foundation to accommodate some shrinking and swelling of the soil.[37] Another preventive method is to construct buildings on a layer of compacted fill that acts as a barrier between the structure above and the expansive soil below. This method helps control the moisture level in the soil and provides a stable base upon which to build. For larger buildings, roads, and airports, it may be cost-effective to excavate and redeposit the upper part of an expansive soil or to mix in calcium-based stabilizers, such as quicklime, to bind soil particles together.

8.8 CHECK your understanding

1. How might subsidence due to groundwater extraction be minimized?
2. Discuss ways to minimize subsidence from mining.
3. How can subsidence from thawing permafrost be prevented?
4. How might we minimize subsidence on deltas and collapsible soil?
5. List ways to minimize damage from expansive soil.

8.9 Perception of and Adjustment to Subsidence and Soil Volume Change

PERCEPTION OF SUBSIDENCE AND SOIL VOLUME CHANGE

Subsidence and expansion and contraction of the soil are natural hazards that get little media coverage. Few people living in the United States are concerned about subsidence or soil volume change hazards. However, people living in areas directly affected, such as those in areas of expansive soils, permafrost, or rapid groundwater withdrawal, are more likely to understand the hazard. Furthermore, people living in regions where sinkholes commonly develop are generally familiar with the hazard and perceive it to pose a real risk to property.

ADJUSTMENT TO SUBSIDENCE AND SOIL VOLUME CHANGE

The most appropriate adjustment to subsidence and soil volume change is to avoid building in areas prone to these hazards. Clearly, this is not always possible because a significant portion of the eastern United States is underlain by karst, large areas of the western United States are underlain by swelling soil, and areas of the northern United States and Canada have frost-susceptible or organic soils or permafrost. The best we can do is to identify high-risk areas in which construction should be prohibited or limited. Unfortunately, in areas of natural subsidence such as karst regions, subsidence may be difficult or impossible to predict. The groundwater system is constantly altering the subsurface rock, and sinkholes are common. There are, however, some methods that can help identify areas of potential subsidence and soil volume change.

Geologic and Soil Mapping
Detailed geologic and soil maps can be made to identify the hazards present as accurately as possible. An understanding of the geology and soil, coupled with the surface and groundwater systems in an area, will greatly aid in the prediction and avoidance of subsidence and soil expansion and contraction.

Observing Surface Features
In areas underlain by limestone, surface features such as cracking of the land surface should be noted. The appearance of cracks in the ground may indicate that sinkhole collapse is imminent, and appropriate steps should be taken to avoid damage and injury.

In the western United States, cracks in the ground may indicate expansive soils or, in some desert areas, a falling groundwater table due to over-pumping and aquifer compaction. To minimize subsidence and preserve precious aquifer pore space, rapid and detailed observations of ground surface elevation changes are needed in order to know where pumping should be limited and/or where ground water injection is necessary. Fortunately, new satellite-based radar technology are now capable of measuring millimeter-scale (0.4 in.) vertical movements over large areas of the Earth's surface over spans of days to years (Figure 8.32). It is clear that over the next few decades, such radar techniques and other technological advances in space-based Earth observing will provide a greater understanding of the causes and rates of subsidence around the world.

Subsurface Surveys When planners must make decisions about where to build structures in karst regions, knowledge of the subsurface environment is critical. Subsurface exploration with *ground penetrating radar* (GPR) and drilling boreholes to examine the subsurface geology are often desirable before construction begins. These techniques may help prevent structures from being built above shallow caves. Additional geologic surveys may be needed to evaluate high-risk areas encountered during construction. In areas of expansive and frost-susceptible soils or areas underlain by permafrost, geotechnical borings and soil testing may be needed to properly design foundations.

Some states, such as Colorado, require disclosure of the presence of expansive soils when houses are sold.

Disclosure requirements apply to new-home builders, homeowners, and real-estate brokers.[61] A recent lawsuit for construction defects on expansive soils, brought by owners of a 246-unit condominium complex in suburban Denver, Colorado, resulted in a settlement of more than $39 million. Homeowners who live in areas where subsidence hazards are present should check the hazard coverage in their insurance policies. For example, in many areas, neither sinkholes nor mine subsidence is covered in standard homeowners' policies, and they require additional coverage.

8.9 CHECK your understanding

1. What is the most appropriate adjustment to subsidence and soil volume change?

2. What three main methods can help identify areas where soils prone to subsidence exist?

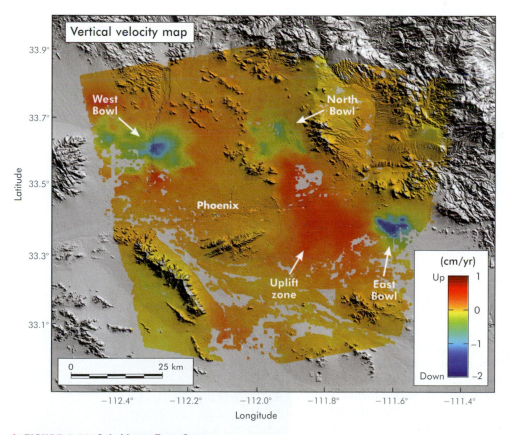

^ FIGURE 8.32 Subsidence From Space
InSAR (Interferometric synthetic-aperature RADAR) data of the greater Phoenix metropolitan area showing local annual rates of surface subsidence (blue) and uplift (red) across the region due to groundwater withdrawal between 2003 and 2010. *(Courtesey Megan Miller and Manoochehr Shirzael; adapted from J. Geophys. Res. Solid Earth, 120, 5822–5842)*

APPLYING the **5** Fundamental Concepts

Venice Is Sinking

1 **Science helps us predict hazards.**

2 **Knowing hazard risks can help people make decisions.**

3 **Linkages exist between natural hazards.**

4 **Humans can turn disastrous events into catastrophes.**

5 **Consequences of hazards can be minimized.**

The Italian people have studied the subsidence of Venice for decades and have kept records of flooding and subsidence for hundreds of years. The famed city, once a flourishing trading center and now an important tourist destination in Italy, is located on 118 small islands near the center of the 500 km² (~190 mi.²) Venetian Lagoon on the Adriatic Sea (Figure 8.33).[62] The city's structures are built on thousands of closely spaced wooden pilings driven through the marshy wetlands to rest on a dense clay layer on the lagoon floor and waterways between islands take the place of streets.

Flooding is an age-old problem in Venice. Rivers flowing into the lagoon were diverted to reduce the input of sediment that would decrease the space available for water flow. People were so concerned with flooding in the thirteenth century that the Venetian empire considered moving its city inland to Constantinople. Changes in water levels were recorded as early as the mid-1860s.

Tidal-induced seasonal flooding, along with storms from the ocean and inland, rising sea level due to global warming, and natural geologic subsidence and tilting, along with human-caused subsidence is increasing the incidents of flooding in Venice. Although only about one half that of the United States, the annual population growth rate of Italy during the early part of the twentieth century was more than 10 percent, doubling the country's population by 1960, many of whom settled near the coast. This growth lead to an increase in the demand for fresh water, so in the

1930s pumping of groundwater on the Italian mainland surrounding the Venetian Lagoon (Figure 8.33) commenced. For the next 30 years excess groundwater pumping caused compaction of the aquifer and regional subsidence. Of the 25 cm (~10 in.) of Venetian subsidence in historic times, it is estimated that about 15 cm (~6 in.) was caused by pumping. It wasn't until the great flood of 1966, which caused significant loss of life and property when lagoon water level rose more than 2 m (~7 ft.) that the Italian government was prompted to ban groundwater pumping to prevent further subsidence.[62]

Contrary to the belief a decade or so ago that subsidence had stabilized, the city has continued

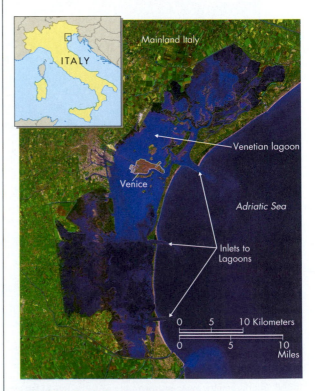

∧ FIGURE 8.33 The Venetian Coast
Satellite image showing the location of Venice and important geographic features of this part of Italy. *(NASA)*

∧ FIGURE 8.34 Gondola Pushes Away from Flooded Walkway in 2008
Built on a lagoon, Venice is known for its gondolas and water buses but accelerating subsidence is causing regular flooding of walkways and its famed architectural treasures. About 40 percent of the buildings in Venice were affected by the 2008 floods. *(Luigi Costantini/AP Images)*

to subside at a rate of about 1 to 2 mm (~0.04 to 0.08 in.) year over the past decade. The Venetian Lagoon is subsiding even faster, at about 2–4 mm (~0.08 to 0.16 in.) per year with an eastward tilt.

The most serious flooding in over 20 years occurred in 2008 (Figure 8.34), when days of heavy rain resulted in more than 95 percent of the historic city center being under water. Significant flooding also occurred in 2010 and 2012. Thus, plans to control the flooding must consider ongoing subsidence, as well as the effects of sea-level.[63]

To combat the flooding, the Italian government has started a $7 billion project that it calls the Moses Project (Moses, according to legend, parted the Red Sea). The public works project, which started in 2004 and may be finished as early as 2014, is the largest in the history of Italy.

The Moses Project will place 78 hinged, steel flood gates across the three tidal inlets to the Venetian Lagoon (Figure 8.33).[64] The floodgates are designed to swing upward using compressed air when a storm warning is received in order to prevent the wind-driven surge of seawater from the Adriatic Sea from entering the lagoon (Figure 8.35). The gates are

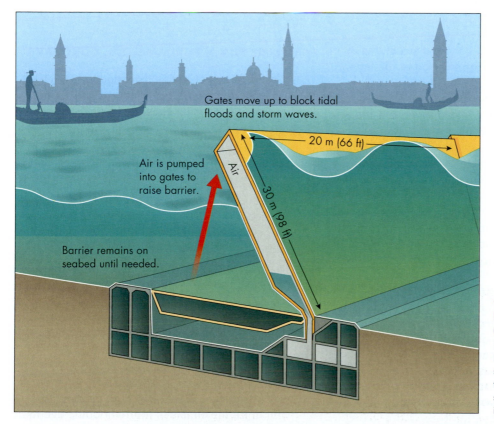

Gates move up to block tidal floods and storm waves.

20 m (66 ft)

Air is pumped into gates to raise barrier.

Air

30 m (98 ft)

Barrier remains on seabed until needed.

< FIGURE 8.35 Barriers to Protect the City of Venice
Idealized diagram shows how barriers in the seabed may be raised to block tidal floods and storm surges from flooding Venice.

∧ FIGURE 8.36 Flood Control Structures during Construction
Photograph of the Venetian lagoon beyond the construction site where retractable flood control structures are being built. Dashed white lines show the locations where the flood control structures will rise from the channel to restrict flow of sea water into the lagoon. *(Magistrato alle Acque di Venezia – Consorzio Venezia Nuova/Wikimedia)*

supported by steel and concrete structures 38 m (~125 ft.) long driven into the bed of the lagoon (Figure 8.36).

In addition to being expensive, the project has caused controversy. The lagoon, as is the case with many large wetlands, is a fragile environment. The waters of the lagoon are home to large numbers of migrating water birds. Any program to control flooding must also protect the greater lagoon environment. Moreover, these gates will neither slow the subsidence of the city nor completely stop flooding, and their long-term utility has been questioned.[65] As worldwide sea levels continue to rise, the future of Venice is uncertain.

The story of the subsidence of Venice involves linkages among soil, rock, compaction of sediment, tectonics, hydrology, and human use of the land. Although methods exist to minimize the subsidence hazard, application of these methods may address some but not all aspects of the problem when a complex set of factors is at work, as is the case in this historic city.

APPLYING THE **5** FUNDAMENTAL CONCEPTS

1. Describe how scientific investigations have contributed to understanding the processes involved in causing subsidence of Venice?

2. Risk is defined as the probability of an event occurring times the consequences. How can we evaluate the risk to Venice from flooding to help make decisions on future actions?

3. What other hazards are linked to the sinking of Venice? Will the effect of these hazards increase or decrease in the future? What do you need to know to make such a determination?

4. Name two or three human-induced factors that are linked to the flooding of Venice.

5. Human processes, such as including preparedness plans for an event, is part of the equation of determining the consequences of future flooding in Venice. Discuss the actions taken by the population and the Italian government to minimize the effects of subsidence? What else could be done to minimize the effects of future flooding events?

CONCEPTS in review

8.1   Soil and Hazards

LO:1 Describe what a soil is and the processes that form and maintain soils.

LO:2 Explain why soil erosion might threaten our society.

- Engineers define soil as earth material that may be removed without blasting; soil scientists define soil as solid earth material that can support rooted plant life. A basic understanding of soils and their properties is crucial to land-use planning, waste disposal, and evaluation of natural hazards such as flooding, landslides, and earthquakes.

- Soils result from interactions of the rock and hydrologic cycles and are affected by variables such as climate, topography, parent material, time, and organic activity. Soils tend to have distinctive layers, or horizons, defined by the processes that formed them and the type of materials present. Of particular importance are the processes of leaching, oxidation, and accumulation of materials in various soil horizons. Development of the argillic B horizon, for example, depends on the translocation of clay minerals from upper to lower horizons. Three important properties of soils are color, texture (particle size), and structure (aggregation of particles).

- An important concept in studying soils is relative profile development. Young soils are weakly developed. Soils older than 10,000 years tend to show moderate development, with more pronounced soil structure, redder soil color, and more translocated clay in the B horizon. Strongly developed soils are similar to those of moderate development but with better-developed B horizons. Such soils can range in age from several tens of thousands of years to several hundred thousand years or older. A soil chronosequence is a series of soils arranged from youngest to oldest in terms of relative soil profile development. A soil chronosequence in a region is useful in evaluating rates of processes and recurrence of hazardous events such as earthquakes and landslides.

- A soil may be considered as a complex ecosystem in which many types of living things convert soil nutrients into forms that plants can use. *Soil fertility* refers to the capacity of the soil to supply nutrients needed for plant growth.

- Water may flow vertically or laterally through the pores (spaces between grains) of a soil. The flow is either saturated (all pore space filled with water) or, more commonly, unsaturated (pore space partially filled with water). The study of soil moisture and how water moves through soils is an important topic in environmental geology.

- Soil erosion refers to the erosion process of grain by grain removal of soil or removal of soil at an unacceptable rate or where the removal rate exceeds that of soil formation. The natural rate of soil erosion is about 1.0 ton per hectare per year. Poor soil conservation in agricultural practices may increase the rate by 10 to 30 times.

KEY WORDS

soil (p. 300), **soil erosion** (p. 307), **soil horizons** (p. 300), **weathering** (p. 300)

8.2   Introduction to Subsidence and Soil Volume Change

LO:3   Summarize the causes and effects of subsidence and volume changes in the soil.

- Subsidence is a type of ground failure characterized by nearly vertical deformation, or the downward sinking, of earth materials. This failure may be caused by natural processes, human activities, or a combination of the two. Most subsidence is caused by the underground dissolution of soluble rocks. Other causes include lowering groundwater levels and fluid pressures in sediment, thawing permafrost, reduced sedimentation on delta plains, flooding of collapsible soils, and draining organic soils. Earthquakes, the deflation of magma chambers, and the drainage of lava tubes also may cause some subsidence.

- Underground dissolution of limestone by acidic groundwater creates a landscape of caves and sinkholes known as karst topography. Other karst features include disappearing streams, springs, and tower karst. Most sinkholes form by the slow dissolution of limestone. Other sinkholes form from the collapse of cave roofs. This collapse is often caused by a falling groundwater table during a drought or by an increase in groundwater pumping. Karst topography also develops where layers of highly soluble rock salt or rock gypsum are near the surface.

- During the past several decades, permafrost thawing has become a major hazard in Arctic and near-Arctic regions. Most of this melting is a direct result of climatic warming. Thawing permafrost causes subsidence and structural damage, as well as the formation of thermokarst, a terrain consisting of uneven ground with sinkholes, mounds, ponds, and caves.

- Loosely compacted fine sediment subsides where the groundwater table has fallen or fluid pressure has been reduced. Groundwater changes may be natural or result from human activities. This subsidence is often irreversible because of the drying of underground layers of very fine sediment. Surface features associated with this compaction include large earth fissures or desiccation cracks.

- Marine deltas are areas of natural compaction and subsidence. The continual deposition of sediment on the delta plain generally keeps up with the compaction of sediment underground. Reducing or stopping sedimentation on the delta plain by the construction of dams, levees, and canals causes the delta surface to subside below the sea. This subsidence destroys wetlands and can produce flooding in urban areas such as New Orleans.

KEY WORDS

cave systems (p. 310), **collapsible soil** (p. 317), **expansive soil** (p. 318), **karst topography** (p. 309), **permafrost** (p. 313), **sinkhole** (p. 309), **subsidence** (p. 309)

- Soil expansion and contraction can cause the land surface to either heave upward or sink downward. Both heaving and sinking occur in expansive and frost-susceptible soils, and sinking takes place in collapsible soils.

- Expansive soils are those that swell when wet and shrink when dry, commonly a result of changes in the amount of water clinging onto the surface of fine smectite clay. These volume changes can cause extensive structural damage. Factors that affect the moisture content of an expansive soil include climate, vegetation, topography, and drainage.

- Frost-susceptible soils are those that are likely to accumulate ice in pockets or lenses between silty earth material. Growth of these ice accumulations displaces the surrounding soil and produces frost heaving. The upward heaving and later thawing of the soil cause structural and pavement damage in areas underlain by permafrost and in soils that are only intermittently frozen.

- Collapsible soils are normally dry, with particles loosely packed or weakly cemented together. These soils are susceptible to subsidence when water ponds on the surface.

8.3 Geographic Regions at Risk for Subsidence and Soil Volume Change

LO:4 Locate on a map the geographic regions at risk for subsidence and volume changes in the soil.

LO:5 Discuss the hazards associated with karst regions.

- Subsidence is a hazard in more than 45 U.S. states and most Canadian provinces. Karst hazards include sinkhole collapse, groundwater pollution, and unreliable water supplies. Large areas of karst topography are found in a region extending through Tennessee, Virginia, Maryland, and Pennsylvania; parts of Indiana and Kentucky; the Salem-Springfield plateaus of Missouri and Arkansas; central Texas; central Florida; and Puerto Rico.
- Soil volume change is also a common hazard in many areas. Permafrost and frost-susceptible and organic soils are common in Alaska, Canada, and Russia. Organic soils are also abundant in the Upper Midwest, the Pacific Northwest, and the Gulf and Atlantic Coasts in the United States. Frost-susceptible soils are found in the northern contiguous United States and at high altitudes in mountainous areas. Collapsible soils occur in arid and semiarid regions, such as the southwestern United States. Expansive soils are a problem primarily in the western United States and Canada.

8.4 Effects of Subsidence and Soil Volume Change

LO:6 Summarize the effects of subsidence and soil volume changes.

- Sinkholes may cause damage to highways, homes, and other structures and affect water quality if they are used as disposal sites.
- Melting permafrost causes soil volume changes and soil instability that can damage structures such as roads and buildings.
- Subsidence on deltas causes flooding and loss of wetlands.
- Soil volume changes, especially from expansive soils, cause billions of dollars in damages to structures, mostly to highways, buildings, and pipelines.

8.5 and 8.6 Linkages between Subsidence, Soil Volume Change, and Other Natural Hazards
Natural Service Functions of Subsidence and Soil Volume Change

LO:7 Identify linkages between subsidence, soil expansion and contraction, and other hazards, as well as natural service functions of karst.

- Subsidence may be linked to volcanic, earthquake, flood, and landslide hazards.
- About 25 percent of the world's population gets its drinking water from karst formations.
- Karst regions offer important aesthetic and scientific resources. They are home to rare creatures that are specially adapted to living underground.

8.7, 8.8, and 8.9   Human Interaction with Subsidence and Soil Volume Change
Minimizing Subsidence and Soil Volume Change
Perception of and Adjustment to Subsidence and Soil Volume Change

KEY WORD
groundwater mining (p. 326)

LO:8   Explain how humans interact with subsidence and soil hazards.

LO:9   Evaluate what can be done to minimize the hazard from subsidence and volume changes in the soil.

- Human beings exacerbate subsidence by removing subsurface fluids, underground mining, melting permafrost, reducing sediment accumulation on deltas, and draining organic soils.
- Effects of soil volume change can be intensified by poor landscaping and drainage practices on expansive and collapsible soils.
- Natural subsidence and changes in the volume of the soil are difficult to prevent, but human-induced subsidence may be avoided or minimized. Methods for limiting human-induced subsidence include injecting water during crude oil production and regulating groundwater pumping and underground mining. Problems with soil volume change may be minimized with sound construction and landscaping techniques.
- An understanding of the local geologic and hydrologic systems can help prevent water pollution in karst areas.
- Adjustments to the subsidence and soil volume change hazard include identifying problem areas through geologic, soil, and subsurface mapping. Homeowners can protect themselves with insurance that covers the subsidence and soil volume change hazards in their area.

CRITICAL thinking questions

1. You are considering building a home in rural Kentucky. You know the area is underlain by limestone, and you are concerned about possible karst hazards. What are some of your concerns? What might you do to determine where to build your home?
2. You work in the planning department in one of the parishes (counties) close to New Orleans. What would you advocate in the long term and in the short term to protect your community from subsidence and flooding? Consider both regional and local solutions to the problem.
3. You have inherited a ranch house built on a concrete slab on clay soil in a suburb east of Denver, Colorado. How would you determine if the house is

built on expansive soil? If you found that the soil was expansive, how would you minimize damage from the shrinking and swelling of the soil?
4. You would like to build a home in a desert community in Arizona or New Mexico. What would you look for to determine if subsidence or soil volume change is a potential problem? What could you do to protect your investment?
5. As a town council representative in a small village in New England or Ontario, Canada, you have been asked to approve a building permit on property that is partly underlain by silty glacial deposits and partly by a marsh. What questions should you ask the permit applicant regarding planned construction on the silty

soil and the applicant's proposal to drain and build on the wetland?

6. Defend or criticize the contention that soil erosion is threatening our civilization, as it has done to previous societies that have disappeared or literally eroded away.

7. What might be the reactive policies following sudden formation of a large collapse sinkhole measuring 200 m (~650 ft.) in diameter and 30 m (~100 ft.) deep in a housing area? How might these policies compare to proactive steps that could be taken before the event? Which plan of action is better? Why?

References

1. **Bock, Y., Wdowinski, S., Ferretti, A., Novali, F.,** and **Fumagalli, A.** 2012. *Recent subsidence of the Venice Lagoon from continuous GPS and interferometric synthetic aperature radar.* Geochemistry, Geophysics, Geosystems, an American Geophysical Union publication.

2. **Waltham, T.** 2002. Sinking cities. *Geology Today* 18(3): 95–100.

3. **Birkland, P. W.** 1984. *Soils and geomorphology.* New York: Oxford University Press.

4. **Brady, N. C.,** and **Weil, R. R.** 1996. *The nature and properties of soils,* 11th ed. Upper Saddle River, NJ: Prentice Hall.

5. **Keller, E. A., Bonkowski, M. S., Korsch, R. J.,** and **Shlemon, R. J.** 1982. Tectonic geomorphology of the San Andreas fault zone in the southern Indio hills, Coachella Valley, California. *Geological Society of America Bulletin* 93: 46–56.

6. **Van der Woerd, J., Klinger, Y., Sieh, K., Tapponnier, P., Ryerson, F. J.,** and **Meriaux, A. S.** 2006. Long-term slip rate of the southern San Andreas fault from 10Be-26Al surface exposure dating of an offset alluvial fan. *J. Geophys. Res.*, 11 B04407: 10.1029/2004 JB00359.

7. **Krynine, D. P.,** and **Judd, W. R.** 1957. *Principles of engineering geology and geotechnics.* New York: McGraw-Hill.

8. **Singer, M. J.,** and **Munns, D. N.** 1996. *Soils,* 3rd ed. Upper Saddle River, NJ: Prentice Hall.

9. **Montgomery, D.** 2007. Is agriculture eroding civilization's foundation? *GSA Today* 17(10): 4–10.

10. **Montgomery. D. R.** 2007. *Dirt: The erosion of civilizations.* Berkeley, CA: University of California Press.

11. **Diamond, J.** 2010. Intra-island and inter island comparisons. In *Natural experiments of history,* ed. J. Diamond, and J. A. Robinson, pp. 120–4. Cambridge, MA: Belknap Press of Harvard University Press.

12. **Mongabay.** 2010. Deforestation figures for selected countries. http://rainforest.mongabay.com/deforestation. Accessed 5/17/10.

13. **Author unknown.** Erosion: Erosion from inappropriate agricultural practices on crop lands. http://people.oregonstate. edu/~muirp/erosion.htm. Accessed 6/25/13.

14. **Pimentel, D.** and **10 others.** 1995. Environmental and economic costs of soil erosion and conservation benefits. *Science* 267(5021): 1117–23.

15. **Wischmeier, W. H.,** and **Meyer, L. D.** 1973. Soil erodibility on construction areas. In *Soil erosion: Causes, mechanisms, prevention and control,* pp. 20–29. Highway Research Board Special Report 135. Washington, DC: Highway Research Board.

16. **Dunne, T.,** and **Leopold, L. B.** 1978. *Water in environmental planning.* San Francisco: W. H. Freeman.

17. **Veni, G., DuChene, H., Crawford, N. C., Groves, C. G., Huppert, G. N., Kastning, E. H., Olson, R.,** and **Wheeler, B. J.** 2001. *Living with karst: A fragile foundation.* Alexandria, VA: American Geological Institute.

18. **Galloway, D., Jones, D. R.,** and **Ingebritsen, S. E.** 1999. Introduction. In *Land subsidence in the United States,* ed. D. Galloway, D. R. Jones, and S. E. Ingebritsen, pp. 1–6. U.S. Geological Survey Circular 1182.

19. **Waltham, T., Bell, F.,** and **Culshaw, M.** 2005. Sinkholes and subsidence: Karst and cavernous rocks in engineering and construction. New York: Springer-Verlag.

20. **Bloom, A. L.** 1991. *Geomorphology: A systematic analysis of late Cenozoic landforms,* 3rd ed. Upper Saddle River, NJ: Prentice Hall.

21. **Nelson, F. E., Anisimov, O. A.,** and **Shiklomanov, N. I.** 2001. Subsidence risk from thawing permafrost. *Nature* 410: 889–90.

22. **Goldman, E.** 2002. Even in the high Arctic nothing is permanent. *Science* 297: 1493–94.

23. **Stanley, J.-D., Goddio, F., Jorstad, T. F.,** and **Schnepp, G.** 2004. Submergence of ancient Greek cities off Egypt's Nile Delta—a cautionary tale. *GSA Today* 14(1): 4–10.

24. **Penvenne, L. J.** 1996. The disappearing delta. *Earth* 5(4): 16–17.

25. **Roberts, H. H.** 1997. Dynamic changes of the Holocene Mississippi River delta plain: The delta cycle. *Journal of Coastal Research* 13(3): 605–27.

26. **Morton, R. A., Bernier, J. C.,** and **Barras, J. A.** 2006. Evidence of regional subsidence and associated interior wetland loss induced by hydrocarbon production, Gulf Coast region, USA. *Environmental Geology* 50: 261–74.

27. **Dokka, R. K.** 2006. Modern-day tectonic subsidence in coastal Louisiana. *Geology* 34: 281–84.

28. **Torbjörn, T. E., Bick, S. J., van de Borg, K.,** and **De Jong, A. F. M.** 2006. How stable is the Mississippi Delta? *Geology* 34: 697–700.

29. **Louisiana Coastal Wetlands Conservation and Restoration Task Force.** 1998. *Coast 2050: Toward a sustainable coastal Louisiana.* Baton Rouge, LA: Louisiana Department of Natural Resources.

30. **Penland, S., Wayne, L. D., Britsch, L. D., Williams, S. J., Beall, A. D.,** and **Butterworth, V. C.** 2000. *Process classification of coastal land loss between 1932 and 1990 in the Mississippi River delta plain, southeastern Louisiana.* U.S. Geological Survey Open-File Report 00–0418.

31. **Scheffe, K. F.** 2005. Collapsible soils in the Rio Grande Valley of central New Mexico. *Geological Society of America Abstracts with Programs* 37(7): 327.

32. **Scheffe, K. F.,** and **Lacy, S. L.** 2004. Hydro-compactible soils. In *Understanding soil risks and hazards: Using soil survey to identify areas with risks and hazards to human life and property,* ed. G. B. Muckel, pp. 60–64. Lincoln, NE: U.S. Department of Agriculture, Natural Resources Conservation Service, National Soil Survey Center.

33. **Fischetti, M.** 2001. Drowning New Orleans. *Scientific American* 285(10): 76–85.

34. **Ingebritsen, S. E., McVoy, C., Glaz, B.,** and **Park, W.** 1999. Florida Everglades: Subsidence threatens agriculture and complicates ecosystem restoration. In *Land subsidence in the United States,* ed. D. Galloway, D. R. Jones, and S. E. Ingebritsen, pp. 95–106. U.S. Geological Survey Circular 1182.

35. **Grunwald, M.** 2006. The swamp: The Everglades, Florida, and the politics of paradise. New York: Simon & Schuster.

36. **Wilding, L. P.,** and **Tessier, D.** 1988. Genesis of vertisols: Shrink-swell phenomena. In *Vertisols: Their distribution, properties, classification and management,* ed. L. P. Wilding and R. Puentes, pp. 55–81. College Station, TX: Texas A&M University Soil Management Support Services Technical Monograph No. 18.

37. **Noe, D. C., Jochim, C. L.,** and **Rogers, W. P.** 1999. *A guide to swelling soils for Colorado homebuyers and homeowners.* Colorado Geological Survey Special Publication 43.

38. **McCarthy, D. F.** 2002. *Essentials of soil mechanics and foundations: Basic geotechnics,* 6th ed. Upper Saddle River, NJ: Prentice Hall.

39. **Johnson, K. S.** 2005. Subsidence hazards due to evaporite dissolution in the United States. *Environmental Geology* 48: 395–409.

40. **Olive, W. W., Chleborad, A. F., Frahme, C. W., Schlocker, J., Schneider, R. R.,** and **Schuster, R. L.** 1989. *Swelling clays map of the conterminous United States.* U.S. Geological Survey Miscellaneous Investigations Series Map I-1940.

41. **Schmidt, W.** 2001. Sinkholes in Florida. *Geotimes* 46(5): 18.

42. **Comiso, J. C.,** and **Parkinson, C. L.** 2004. Satellite-observed changes in the Arctic. *Physics Today* 57(8): 38–44.

43. **Dixon, T. H., Amelung, F., Ferretti, A., Novali, F., Rocca, F., Dokka, R., Sella, G., Kim, S.-W., Wdowinski, S.,** and **Whitman, D.** 2006. Subsidence and flooding in New Orleans: A subsidence map of the city offers insight into the failure of the levees during Hurricane Katrina. *Nature* 441: 587–88.

44. **Wray, W. K.,** and **Meyer, K. T.** 2004. Expansive clay soil … a widespread and costly geohazard. *Geo-Strata* 5(4): 24–25, 27–28.

45. **DiMillo, A.** 2005. *A quarter-century of geotechnical research.* U.S. Department of Transportation Federal Highway Administration. www.fhwa.dot. gov/engineering/geotech/pubs/century/. Accessed 6/25/06.

46. **Jones, D. E.,** and **Holtz, W. C.** 1973. Expansive soils—the hidden disaster. *Civil Engineering—ASCE* 43(6): 49–51.

47. **Ritter, D. F., Kochel, R. C.,** and **Miller, J. R.** 2002. *Process geomorphology.* New York: McGraw-Hill.

48. **National Research Council (U.S.) Joint Academies Committee on the Mexico City Water Supply.** 1995. *Mexico City's water supply: Improving the outlook for sustainability.* Washington, DC: National Academies Press.

49. **Harris, R. C.** 2004. Giant desiccation cracks in Arizona. *Arizona Geology* 34(2): 1–4.

50. **Slaff, S.** 1993. *Land subsidence and earth fissures in Arizona.* Arizona Geological Survey Down-to-Earth Series 3.

51. **Karst Waters Institute.** 2002. *What is karst (and why is it important)?* www.karstwaters.org/kwitour/ whatiskarst.html/. Accessed 7/23/06.

52. **Holzer, T. L.,** and **Galloway, D. L.** 2005. Impacts of land subsidence caused by withdrawal of underground fluids in the United States. In *Humans as geologic agents,* ed. J. Ehlen, W. C. Haneberg, and R. A. Larson, pp. 87–99. Geological Society of America, Reviews in Engineering Geology XVI.

53. **Bull, W. B.** 1973. Geologic factors affecting compaction of deposits in a land subsidence area. *Geological Society of America Bulletin* 84: 3783–802.

54. **Kenny, R.** 1992. Fissures: Legacy of a drought. *Earth* 1(3): 34–41.

55. **Craig, J. R., Vaughan, D. J.,** and **Skinner, B. J.** 1996. *Resources of the Earth: Origin, use, and environmental impact,* 2nd ed. Upper Saddle River, NJ: Prentice Hall.

56. **Rahn, P. H.** 1996. *Engineering geology: An environmental approach,* 2nd ed. Upper Saddle River, NJ: Prentice Hall.

57. **Kappel, W. M., Yager, R. M.,** and **Miller, T. S.** 1999. The Retsof Salt Mine collapse. In *Land subsidence in the United States,* ed. D. Galloway, D. R. Jones, and S. E. Ingebritsen, pp. 111–20. U.S. Geological Survey Circular 1182.

58. **Péwé, T. L.** 1982. *Geologic hazards of the Fairbanks area, Alaska.* Alaska Division of Geological & Geophysical Surveys Special Report 15.

59. **Ingebritsen, S. E., Ikehara, M. E., Galloway, D. L.,** and **Jones, D. R.** 2000. *Delta subsidence in California: The sinking heart of the state.* U.S. Geological Survey Fact Sheet FS-005–00.

60. **Coplin, L. S.,** and **Galloway, D.** 1999. Houston-Galveston, Texas: Managing coastal subsidence. In *Land subsidence in the United States*, ed. D. Galloway, D. R. Jones, and S. E. Ingebritsen, pp. 35–48. U.S. Geological Survey Circular 1182.

61. **Galloway, D., Jones, D. R.,** and **Ingebritsen, S. E.** 1999. Mining ground water—Introduction. In *Land subsidence in the United States*, ed. D. Galloway, D. R. Jones, and S. E. Ingebritsen, pp. 7–13. U.S. Geological Survey Circular 1182.

62. **Merali, Z.** 2002. Saving Venice. *Scientific American*, August 19. www.scientificamerican.com/article.cfm?id=saving-venice

63. **Bock, Y., Wdowinski, S., Ferretti, A., Novali, F.,** and **Fumagalli, A.** 2012. Recent subsidence of the Venice Lagoon from continuous GPS and interferometric synthetic aperture radar. *Geochemistry Geophysics Geosystems* 13:Q03023: AGU 10.1029/2011gc003976

64. **Fletcher, C.,** and **Da Mosto, J.** 2004. *The science of saving Venice.* New York: Umberto Allemandi & Co.

65. **Nosengo, N.** 2003. Save our city! *Nature* 424: 608–09.

Anything in the direct path of the **tornado** . . . was **completely obliterated**

9

1

Science helps us predict hazards.

As you will see when we return to Moore, Oklahoma, advances in weather satellites and ground-based monitoring equipment over the past decade have enhanced prediction of tornado formation and extended the lead time for tornado warnings.

2

Knowing hazard risks can help people make decisions.

The frequency of severe weather events, such as tornadoes, for a particular region is fairly well known and, therefore, most severe weather phenomena are amenable to risk analysis.

APPLYING the **5** fundamental concepts 1999 and 2013 Tornadoes of Moore, Oklahoma

Atmospheric Processes and Severe Weather

1999 and 2013 Tornadoes of Moore, Oklahoma

Although the majority of weather-related deaths are caused by blizzards and heat waves, it is the tornado that strikes fear in the hearts of many people living in the midcontinental United States. This fear is well founded—a **tornado**, simply defined as a violently rotating column of air associated with extreme horizontal winds, can cause tremendous property damage and loss of life. Having been struck by extremely powerful tornadoes in both 1999 and 2013, the city of Moore, Oklahoma, is an excellent example of the frightening power of tornadoes and the human interaction with and adjustment to severe weather.

The "1999 Oklahoma tornado outbreak" included 152 tornadoes spread throughout the U.S. Great Plains over a seven-day period,[1]

◄ 2013 MOORE, OKLAHOMA, TORNADO
An extremely powerful tornado devastates the City of Moore just south of metropolitan Oklahoma City, Monday, May 20, 2013. *(Ks0stm/Wikimedia Commons)*

Atmospheric processes and energy exchanges are driven by Earth's energy balance and linked to climate and weather. Hurricanes, thunderstorms, tornadoes, blizzards, ice storms, dust storms, and heat waves, as well as flash flooding resulting from intense precipitation, are all natural processes that are hazardous to people. These severe hazards affect considerable portions of North America and cause significant death and destruction each year.

After reading this chapter, you should be able to:

LO:1 Differentiate the three different ways in which heat is transferred in the atmosphere.

LO:2 Discuss how Earth's energy balance and energy exchanges produce climate and weather.

LO:3 Compare and contrast the different layers of Earth's atmosphere.

LO:4 Explain the different types of severe weather events and discuss the local weather conditions that cause these events.

LO:5 Explain how movement of air masses control the distribution of different severe weather hazards.

LO:6 Discuss how humans interact with severe weather.

LO:7 Summarize the effects of severe weather events, as well as their linkages to other natural hazards.

LO:8 Recognize some natural service functions of severe weather.

LO:9 Identify how we can minimize the effects of severe weather hazards.

3

Linkages exist between natural hazards.

Similar to the linkage between severe thunderstorms and tornadoes, other severe weather phenomena, such as blizzards, ice storms, mountain windstorms, heat waves, and dust storms share linkages.

4

Humans can turn disastrous events into catastrophes.

Tornadoes have long been a menace to agricultural lands of the midcontinental United States, but as metropolitan suburbs like Moore expand, the risk grows.

5

Consequences of hazards can be minimized.

About the time of the 1999 Moore tornado, new standards for above-ground shelters were released. At the end of this chapter you can evaluate how successful these standards were for saving lives during the 2013 Moore tornado.

⋀ FIGURE 9.1 Neighborhood Flattened by 1999 Moore, Oklahoma, Tornado
A tossed motor vehicle trailer lies among the wreckage of countless homes that were completely demolished by the high winds generated by the 1999 tornado. *(Staff Sgt. Caroline Hayworth)*

with the most noteworthy tornado striking Moore. On the morning of May 3, the storm prediction center in Norman, Oklahoma, issued a warning for a slight risk of severe thunderstorms from Kansas to Nebraska. By late afternoon, mobile weather tracking systems indicated thunderstorm conditions were favorable for tornado formation. With the highest-ever recorded wind speed of 484 km (~301 mi.) per hour, the 1999 Moore tornado touched down at 6:23 P.M. and ripped a path 61 km (~38 mi.) long and, in places, 1.6 km (~1 mi.) wide through the city. In less than 90 minutes, the 1999 Moore tornado killed 36 people, injured 295, and damaged or destroyed about 10,000 homes.[1] Anything in the direct path of the tornado that was not below ground was completely obliterated (Figure 9.1).

On average about 1,000 tornadoes strike the United States each year, and of those only about one per year is close to the power of the 1999 Moore tornado. Although the probability of the same town being hit by a similarly large tornado less than 15 years later might seem unlikely, that is exactly what happened. On May 20, 2013, Moore was devastated by a tornado of comparable destructiveness (see chapter-opener photograph) that followed a remarkably similar path.

We will revisit Moore at the end of this chapter and evaluate the 2013 tornado within the context of the five fundamental concepts.

9.1 Energy

The concept of *energy* is fundamental to understanding severe weather. Energy is an abstract concept because we cannot see or feel it. We can, however, experience the part of energy referred to as *force*. Many people experience force by either pushing or pulling an object. For example, when we pull a box along the floor or push our car when it has stalled, we are exerting a force in a specific direction. The strength or magnitude of this force can be measured by how much the force accelerates the motion of the box or car. In the metric system, force is measured in newtons. A *newton* (N) is defined as the force necessary to accelerate a 1 kg (~2.2 lb) mass 1 m (~3.3 ft.) per second each second that it is in motion.

Another important concept in understanding energy is *work*. Work is done when energy is expended. In physics, work is done when a force is applied to an object and that object moves a given distance in the direction of the applied force. Work is thus calculated by multiplying the force times the distance over which it is applied. In the metric system, work is measured in joules. A *joule* is defined as a force of 1 N applied over a distance of 1 m (~3.3 ft.). To relate this concept to weather, the amount of work that is taking place in a typical thunderstorm is approximately 10 trillion joules, whereas an average hurricane expends approximately 100,000 times that amount of work.

The rate at which work is done is its *power*. Another way of stating this is that power is energy divided by time. In the metric system, power is expressed as joules per second, or *watts* (W), and 1 joule per second is equal to 1 W. Many people associate the latter unit of measurement with the power used by appliances and lightbulbs.

When we look at atmospheric processes, we are often concerned with large amounts of energy. For example, when we discuss global energy consumption, the amounts are so large that they are often expressed in exajoules (EJ). One EJ is 1 quintillion (10^8) joules. U.S. Department of Energy statistics indicate that global energy consumption is more than 440 EJ per year, a little more than 1.2 EJ each day. For comparison, the internal heat energy of Earth that reaches the surface is three times human global energy consumption, whereas solar energy reaching the surface of Earth is nearly 10,000 times more energy than humans consume each year.[2] Thus, it is primarily solar energy that heats the surface of our planet, evaporates water, and produces the differential heating that causes air masses to move across our landscape.

TYPES OF ENERGY

We must expend energy to do work; another way of stating this concept is to define energy as the "ability to do work." Three main types of energy are potential energy, kinetic energy, and heat energy. *Potential energy* is stored energy. For example, the water held behind a dam contains gravitational potential energy that, if released, would flow downward. *Kinetic energy* is the energy of motion. From the previous example, the water flowing from the dam loses potential energy but gains kinetic energy as it accelerates downward. In the case of a hydroelectric dam, the kinetic energy of the flowing water is used to turn turbines (doing *work*) to generate electricity.

Heat energy is the energy of random motion of atoms and molecules. Heat itself may then be defined as the kinetic energy of atoms or molecules within a substance. Heat may also be thought of as the energy that is transferred from one object to another due to a difference in temperature.[3] Two types of heat that are important in atmospheric processes are sensible heat and latent heat. As the name suggests, *sensible heat* is heat that may be sensed or measured by a thermometer. It is sensible heat that we feel in the air on a warm day. **Latent heat** is a more difficult concept to comprehend; it is the amount of heat that is either absorbed or released when a substance changes phase (from solid to liquid, for example). In the atmosphere, latent heat is related to the three phases of water: ice, liquid water, and water vapor (Figure 9.2). For example, the evaporation of water involves a phase change from liquid to water vapor, a gas. The energy required for this transformation is known as the *latent heat of vaporization*. This energy is recoverable when water vapor changes back to a liquid through condensation in the atmosphere, when moist air cools and water droplets form.[3]

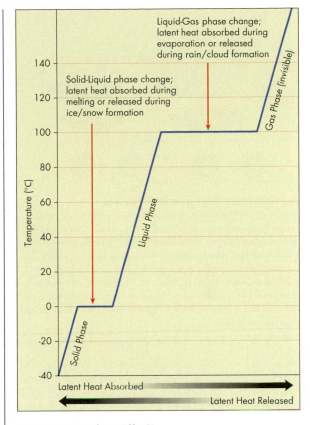

^ FIGURE 9.2 Latent Heat
Generalized graph showing how latent heat is either absorbed or released during phase changes of water. As you move from left to right—melting ice to water then vaporizing water to gas—heat is absorbed by the water. When the process is reversed, the previously absorbed heat is given off as water changes from gas, to liquid, to ice.

HEAT TRANSFER

To complete our discussion of energy, work, and power, we need to consider how heat energy is transferred in the atmosphere. Three major heat-transfer processes are conduction, convection, and radiation. A pot of boiling water on an electric range illustrates these three processes at work (Figure 9.3).

Conduction is the transfer of heat through a substance by means of atomic or molecular interactions. The process of conduction relies on temperature differences, causing heat to flow through a substance from an area of greater temperature to an area of lesser temperature. In our example, conduction of heat through the metal pot causes the handle to heat up. Although conduction also occurs in the atmosphere, on land, and in bodies of water such as the ocean, it is least important in the atmosphere. Air is actually a poor conductor of heat. This is why trapped

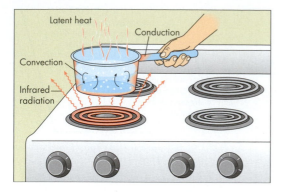

∧ FIGURE 9.3 Heat Transfer Processes
Idealized heat transfer for a pan of boiling water on an electric range. Conduction is taking place in the metal of the pan and its handle; circulating water is transferring heat by convection, and visible and invisible electromagnetic radiation is being transferred through the air from the glowing coil. Latent heat is being absorbed by water vapor as liquid water evaporates. *(Based on Christopherson, R. W.,* Geosystems: An Introduction to Physical Geography, *6th ed. Upper Saddle River, NJ: Pearson Prentice Hall, 2006)*

air is used to increase the insulation value of goose-down comforters, double-paned windows, and plastic foam. In the atmosphere, significant conduction is generally limited to very thin layers of air in contact with Earth's surface.

Convection is the transfer of heat by the mass movement of a fluid, such as water or air. In our example, water in the bottom of the pot is heated and rises upward to displace the cooler water at the surface. The cooler water then sinks downward to the bottom of the pot. This physically mixes the water by moving the heat energy and creates a circulation loop known as a *convection cell.* Convection is an important process for the transfer of atmospheric heat in thunderstorms and in the large-scale circulation of warm air away from the equator.

Finally, **radiation** refers to wavelike energy that is emitted by any substance that possesses heat. The transfer of energy by radiation occurs by oscillations in an electric field and a magnetic field, and thus the waves are generally called *electromagnetic waves.* In our example, the red glow from the heating element of the electric range transmits visible electromagnetic radiation as red light and nonvisible infrared radiation. Most electromagnetic waves are not visible to us.

Our example is summarized as follows: The heat transfer from the electric range includes electromagnetic radiation from the glowing heating coil on the stove, conduction through the solid metal pot, and finally convection bringing warm water bubbling upward from the bottom to the top of the pot.

9.1 CHECK your understanding

1. Describe the differences between force, work, and power.
2. What are the three types of energy? How do they differ from one another?
3. What is the difference between sensible heat and latent heat?
4. What are the three types of heat transfer? How do they differ from one another?

9.2 Earth's Energy Balance

The energy that Earth receives from the sun affects the atmosphere, oceans, land, and living things before being radiated back into space. This process creates *Earth's energy balance*—a general equilibrium between incoming and outgoing energy (Figure 9.4). In the shift from incoming to outgoing energy, some of the energy changes form, but as stated in the first law of thermodynamics, it is neither created nor destroyed.

Although Earth intercepts only a tiny fraction of the total energy emitted by the sun, the intercepted energy is adequate to sustain life. The sun's energy also drives the hydrologic cycle, ocean waves and currents, and global atmospheric circulation. Although Earth's energy balance or budget contains several important components, nearly all of the energy that is available at Earth's surface comes from the sun (Figure 9.4).

ELECTROMAGNETIC ENERGY

Much of the energy emitted from the sun is *electromagnetic energy.* This energy, a type of radiation, travels through the vacuum of space at the speed of light, a velocity close to 300,000 km (~186,000 mi.) per second. Electromagnetic radiation is commonly described as a wave, and the distance between the tops of two successive waves is referred to as the *wavelength.* The various types of electromagnetic radiation are distinguished by their wavelengths, and the collection of all possible wavelengths is known as the *electromagnetic spectrum* (Figure 9.5). Longer wavelengths, those greater than 1 m (~3.3 ft.), include radio waves and microwaves, whereas the shortest wavelengths are X-rays and gamma rays. Visible electromagnetic radiation, referred to as light, makes up only a very small fraction of the total electromagnetic spectrum. Other types of electromagnetic radiation with environmental significance include infrared (IR) and ultraviolet (UV) radiation. Infrared radiation is involved in global warming, and

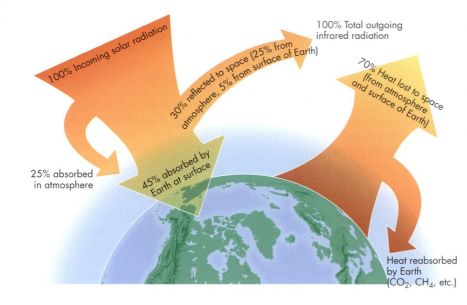

◄ FIGURE 9.4 Earth's Energy Balance
Most of the annual energy flow to Earth from the sun is either reflected or reradiated back into outer space. Only a small component of Earth's heat is actually coming from its interior. *(Based on Pruitt, N. L., Underwood, L. S., and Surver, W.,* Bioinquiry, Learning System 1.0, Making Connections in Biology. *New York: John Wiley & Sons, 1999)*

100% Incoming solar radiation

100% Total outgoing infrared radiation

30% reflected to space (25% from atmosphere, 5% from surface of Earth)

70% Heat lost to space (from atmosphere and surface of Earth)

25% absorbed in atmosphere

45% absorbed by Earth at surface

Heat reabsorbed by Earth (CO_2, CH_4, etc.)

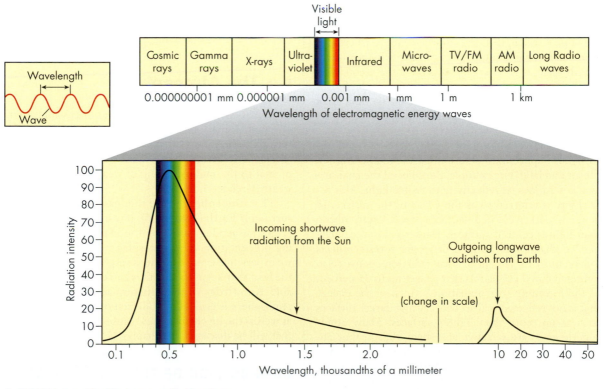

∧ FIGURE 9.5 The Electromagnetic Spectrum
Wavelength is the distance between one wave crest and the next. Values for wavelength in the electromagnetic spectrum have an enormous range from billionths of a meter for X-rays to thousands of meters for long radio waves. The incoming energy from the sun has shorter wavelengths than the radiation emitted from Earth. Only a small amount of the electromagnetic spectrum is visible light that can be seen by the human eye.

levels of UV radiation at Earth's surface are influenced by the depletion of ozone in the upper atmosphere (see Chapter 12).

ENERGY BEHAVIOR

Once electromagnetic energy from the sun reaches Earth, it is redirected, transmitted, or absorbed by the atmosphere, ocean, or land. In redirection, the energy is either reflected like a light bouncing off a mirror or scattered in different directions. Reflection—from the tops of clouds, the water, and the land—is one of two ways that solar energy returns to outer space (see Chapter 12). Scattering disperses the energy in many directions, with scattered energy generally weaker than reflected or transmitted energy. Transmission involves the energy passing through the atmosphere, like light passing directly through window glass. On a clear day, the atmosphere transmits most of the energy it receives from the sun. Absorption of electromagnetic energy either alters the structure of molecules or causes them to vibrate and emit energy. Alteration of molecular structure can take place with penetrating shortwave radiation, such as UV radiation. Emitted energy can take the form of heat or electromagnetic radiation. For many gases, the emitted electromagnetic energy is at the same wavelength as the absorbed energy, whereas in many solids, the emitted energy is at a longer wavelength, such as IR energy. Emission is the second way that solar energy returns to outer space; Earth absorbs shortwave energy from the sun, heats up, and then emits energy in longer wavelengths (see Figures 9.4 and 9.5).

The relative amount of energy reflected and absorbed is important in determining the temperature of the air and land. For example, the color of the land surface determines its reflectivity to visible light. This reflectivity is referred to as the *albedo* of the surface. In general, darker-colored surfaces have lower albedo than lighter-colored surfaces. Darker-colored coniferous woodland, such as pine or spruce forests, reflects 5 percent to 15 percent of solar radiation compared to lighter-colored grasslands that reflect up to 25 percent of the incoming solar radiation.[4] Much of the energy that is not reflected is absorbed, causing darker surfaces to heat up more than lighter surfaces. Many atmospheric gases are selective absorbers; that is, they absorb some wavelengths of energy and are transparent to others. For example, carbon dioxide and water vapor selectively absorb IR wavelengths and contribute to the warming of the lowest layer of the atmosphere (see Chapter 12).

Absorbed energy that is not emitted as heat is emitted as electromagnetic radiation. The temperature of

the object emitting the radiation also affects the wavelength of the electromagnetic radiation that it emits. Hotter objects radiate energy more rapidly and at shorter wavelengths. This fact explains why the sun emits mainly short-wave-length radiation, such as gamma rays and X-rays, and visible and UV light (Figure 9.5). In contrast, Earth's land surface, oceans, and clouds are so cool that they emit predominantly longer-wavelength IR radiation.

9.2 CHECK your understanding

1. Describe how Earth's energy balance works.

2. What is electromagnetic energy? How are the different types of electromagnetic energy distinguished?

3. List the following types of electromagnetic energy in order from shortest wavelength to longest wavelength: radio waves, ultraviolet radiation, gamma radiation, visible light, infrared radiation, X-rays, and microwaves.

4. Explain why incoming solar radiation does not overheat the Earth, using the energy balance diagram (Figure 9.4).

5. How is color related to energy absorption?

9.3 The Atmosphere

Now that we have finished our brief discussion of energy and Earth's energy balance, we will discuss the various components of Earth's atmosphere, along with atmospheric circulation. These concepts are fundamental to the understanding of weather processes and weather-related hazards.

The **atmosphere** is the thin gaseous envelope that surrounds Earth (Figure 9.6). It is made up of gas molecules, suspended particles of solid and liquid, and falling precipitation. The atmosphere causes the weather we experience every day and is responsible for trapping the heat that keeps the Earth warm enough to be habitable. Knowledge of the structure and dynamics of the atmosphere is critical to understanding severe weather, as well as the mechanisms and causes of global warming that will be discussed in Chapter 12.

COMPOSITION OF THE ATMOSPHERE

The atmosphere is mainly composed of nitrogen and oxygen; it contains smaller amounts of argon, water vapor, and carbon dioxide. Other trace elements and compounds exist in still lesser amounts. We will discuss

◄ **FIGURE 9.6 Earth's Thin Atmosphere**
Viewed from space, the atmosphere appears as
a thin layer surrounding Earth. *(NASA)*

these gases in greater detail in Chapter 12. The behavior and content of water vapor in the atmosphere is an important part of cloud formation and atmospheric circulation. We use the term *humidity* to describe the amount of water vapor. Humidity is largely a function of temperature; warm air has the capacity to hold more water vapor than does cold air. The amount of moisture in the air is commonly given as the **relative humidity**, the ratio of the water vapor present in the atmosphere to the maximum amount of water vapor that could be there for a given temperature. Relative humidity is expressed as a percentage and varies from a few percent to 100 percent.

Natural changes in relative humidity occur each day without significant changes in the amount of water vapor in the atmosphere. Relative humidity increases at night because of the cooler temperature and decreases with daytime heating. Changes in the actual water vapor content of the atmosphere take place where water evaporates from Earth's surface or where air masses mix, either vertically or horizontally.

Virtually all the water vapor in the air is derived from evaporation of water from Earth's surface. Water is constantly being exchanged between the atmosphere and the various parts of the Earth. Sleet, snow, hail, and rain remove water from the atmosphere and deposit it on Earth where it may enter groundwater, rivers, lakes, oceans, and glaciers. Eventually this water will evaporate and return to the atmosphere to begin the cycle again. This constant cycling of water between the atmosphere and Earth's surface is a major part of the hydrologic cycle (see Chapter 1).

STRUCTURE OF THE ATMOSPHERE

The water vapor content and temperature of the atmosphere vary from Earth's surface to its upper limits.

Images from orbiting spacecraft show that our atmosphere is very thin when compared to the size of Earth and that it is not easy to identify its upper limits (see Figure 9.7). Although the atmosphere has no well-defined upper boundary, most of the gas molecules are concentrated below a height of 100 km (~62 mi.).

Earth's atmosphere has a structure consisting of four major layers or spheres (Figure 9.7). The lowest layer, the **troposphere**, extends about 8 to 16 km (~5 to 10 mi.) above the surface of Earth. With the exception of some jet airplane travel, we spend our entire lives within the troposphere. Not even the highest mountains breach the upper boundary of the troposphere, known as the *tropopause*. The defining characteristic of the troposphere is a rapid upward decrease in temperature that results from decreasing air pressure with increasing altitude. However, the most visible characteristic of the troposphere is abundant condensed water vapor in the form of clouds. Most of the clouds and weather that directly affect us are found in the troposphere.

The formation and development of clouds are particularly important. Clouds develop when very small water droplets or ice crystals condense from the atmosphere. Without clouds there would be no rain, snow, thunder, lightning, or rainbows. You are probably familiar with two of the most common types of clouds— puffy, fair-weather *cumulus* clouds that may look like pieces of floating cotton, and the towering *cumulonimbus* thunderstorm clouds (Figure 9.8). Keep in mind that the formation of these large cumulonimbus clouds is associated with tremendous release of latent heat energy due to condensation of water vapor (see Figure 9.2).

Most of the atmospheric water vapor condenses in the troposphere, leaving little water in the upper layers of the atmosphere. In addition to water vapor, the

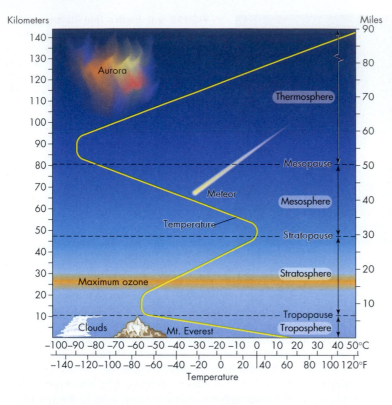

< FIGURE 9.7 Atmospheric Structure
Earth's atmosphere has a structure based on
changes in air temperature from the surface
upward. The curving yellow line shows the
average change in air temperature with height.
Weather develops in the lowest layer, the troposphere, where temperatures fall with height.
Temperatures rise in the stratosphere due to
the presence of ozone molecules that absorb
solar radiation. The rising temperatures act as
a "lid," keeping air from rising and condensing
past the tropopause. *(Based on Lutgens, F. K.,
and Tarbuck, E. J.,* The Atmosphere: An Introduction to Meteorology, *10th ed. Upper Saddle
River, NJ: Pearson Prentice Hall, 2007)*

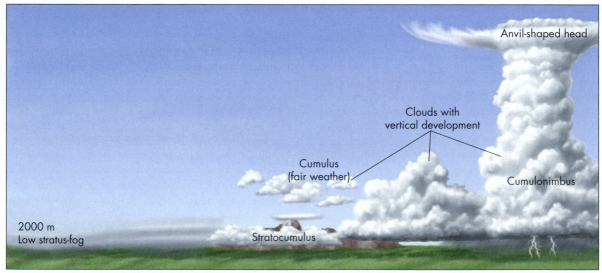

∧ FIGURE 9.8 Cloud Types Associated with Severe Weather
Clouds consist of small water droplets and are classified on the basis of altitude (low, middle, and high) and
form (cirriform, cumuliform, and stratiform). Cumulonimbus clouds are associated with severe thunderstorms
and tornadoes. Stratus clouds in contact with the ground form fog. *(Based on Christopherson, R. W.,* Geosystems: An Introduction to Physical Geography, *6th ed. Upper Saddle River, NJ: Pearson Prentice Hall, 2006)*

troposphere contains most of the atmospheric carbon
dioxide and methane, two gases that are important in
global warming (Chapter 12). Only ozone, a gas

composed of oxygen atoms (O_3), is significantly less
abundant in the troposphere when compared with the
upper atmosphere (see Figure 9.7).

9.3 CHECK your understanding

1. Describe the characteristics of the troposphere. How do meteorologists identify the top of the troposphere?

2. What is the tropopause? How high is it above Earth's surface?

3. How do clouds form?

9.4 Weather Processes

A complete discussion of the atmospheric conditions and processes associated with severe weather is beyond the scope of this book. Students who find this topic especially interesting should pursue coursework in meteorology, the scientific study of weather. Instead, we will focus on four aspects of the atmosphere that are directly related to severe weather: atmospheric pressure and circulation patterns, the vertical stability of the atmosphere, the Coriolis effect, and the interaction of different air masses.

ATMOSPHERIC PRESSURE AND CIRCULATION

Weather forecasters often talk about areas of low or high pressure. The pressure that they are referring to is *atmospheric pressure*. Also called *barometric pressure*, atmospheric pressure is the weight of the column of air that is above any given point (Figure 9.9a). This point

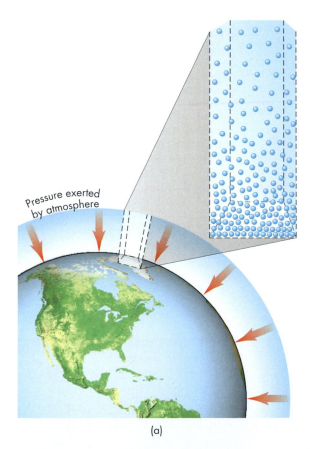

(a)

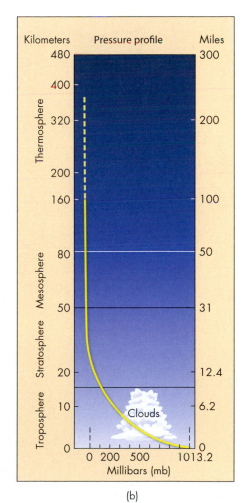

(b)

∧ FIGURE 9.9 Atmospheric Air Pressure Changes
Idealized diagram showing Earth and its atmosphere. Both the (a) density and (b) pressure of Earth's atmosphere decrease with increasing altitude. Air pressure at points on the Earth's surface also varies because the mass of overlying air varies from place to place. Variations in the mass of overlying air are caused by differences in the temperature and density of air masses. Air pressure can be expressed as atmospheric force per unit area. At sea level, average air pressure is about 1 kg per square centimeter (~14.7 pounds per square inch). *(Based on Lutgens, F. K., and Tarbuck, E. J., The Atmosphere: An Introduction to Meteorology, 10th ed. Upper Saddle River, NJ: Pearson Prentice Hall, 2007)*

may be on Earth's surface or above it, such as in an airplane. Atmospheric pressure can also be thought of as the force exerted by the gas molecules on a surface. As you might expect, atmospheric pressure is greater at sea level than at the top of a high mountain where there is less air (fewer gas molecules) above the surface (Figure 9.9b). Nearly all of the weight of the atmosphere and, thus, nearly all the pressure are in the lower atmosphere below an elevation of 50 km (~31 mi.). Atmospheric density and pressure decrease rapidly as one goes to higher elevations. You may have noticed this pressure change if your ears have ever "popped" while driving up or down a mountain or flying in an airplane when changing altitude.

Changes in air temperature and air movement are responsible for most of the horizontal variation in atmospheric pressure. Temperature change influences pressure because cold air is denser than warm air and exerts a higher pressure on the underlying surface. The density of cold air is higher because its gas molecules have lower kinetic energy and stay more closely packed together. Because air pressure and temperature are related, air pressure varies geographically, and this variation has a strong effect on the weather and global atmospheric circulation patterns.

Solar heating and evaporation of seawater at low latitudes near the equator creates warm air masses that have high relative humidity. This warm, low-density air near Earth's surface creates low-pressure zones at equatorial latitudes that convect upward in the atmosphere (Figure 9.10). The air masses cool as they rise; their relative humidity increases and water vapor within the

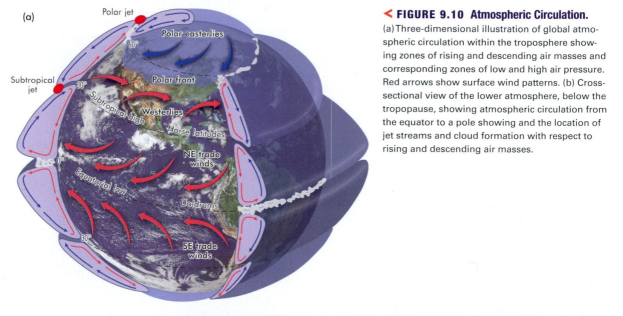

◀ FIGURE 9.10 Atmospheric Circulation.
(a) Three-dimensional illustration of global atmospheric circulation within the troposphere showing zones of rising and descending air masses and corresponding zones of low and high air pressure. Red arrows show surface wind patterns. (b) Cross-sectional view of the lower atmosphere, below the tropopause, showing atmospheric circulation from the equator to a pole showing and the location of jet streams and cloud formation with respect to rising and descending air masses.

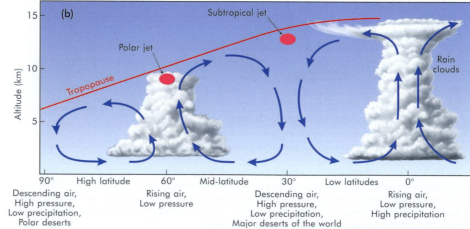

air masses condenses to form clouds and the abundant precipitation characteristic of the low latitude tropics (Figure 9.10b). Having dropped the rain, the now dry, cold, and dense air in the upper troposphere moves toward the poles and sinks at 30° north and south latitude. The subtropical midlatitude deserts of the world, such as the Sahara, Arabian, and Mojave, are, in fact, the result of these dry descending air masses. Similar vertical circulation cells are observed at middle and high latitudes, where low-pressure zones at 60° north and south latitude are characterized by rising humid air that spreads out at the tropopause and descends at the poles and 30° latitude (Figure 9.10b). Where midlatitude air masses of different temperatures collide near the tropopause, narrow fast-flowing jets of air occur. **Jet streams**, as they are called, are westerly winds (flowing from west to east) that encircle the globe and play an important role in creating severe weather and controlling the path of storms (Figure 9.11). Their westerly flow is due to the Coriolis effect (see Case Study 9.1: A Closer Look), and the greater the temperature difference between the colliding air masses, the faster the jet stream will flow.

Each hemisphere has two jet streams: one at an average altitude of 10 km (32,800 ft.) called the *polar jet stream*, and the other at an average altitude of 13 km (~42,660 ft.) referred to as the *subtropical jet stream* (Figure 9.11).[5] These jet streams vary from less than 100 km (~60 mi.) to more than 500 km (~310 mi.) in width and are typically several kilometers thick.[6] The Northern Hemisphere polar jet streams cross North America from west to east at an average speed of 180 km (~110 mi.) per hour in the winter and 90 km (~50 mi.) per hour in the summer, with peak flow within the jet streams being twice the average wind speed.[7] Because the jet streams flow at an altitude similar to that of commercial air traffic, you have likely interacted directly with these fast-moving jets of air. Next time you take a round-trip flight between the east and west coasts, notice that the length of the flight is almost always longer when flying westward. This is due to strong jet stream headwinds while traveling westward and tailwinds while flying eastward.

The polar jet stream is the stronger of the two jets because it occurs along the boundary between cold arctic polar air masses to the north and warm subtropical and tropical air masses to the south (Figure 9.11). The polar jet stream shifts from a path crossing the conterminous United States in the winter to one crossing southern Canada in the summer (Figure 9.12). This migration in its path causes a northward shift in the location of severe thunderstorm and tornado activity in the United States during the late winter, spring, and early summer. The subtropical jet stream that normally

∧ FIGURE 9.11 Polar and Subtropical Jet Streams
Strong winds near the boundary between the troposphere and stratosphere are concentrated in jet streams. The strongest winds are in the polar jet stream at around 10 km (~32,815 ft.), and generally less intense winds form the subtropical jet stream at about 13 km (~42,660 ft.). *(Based on Lutgens, F. K., and Tarbuck, E. J., The Atmosphere: An Introduction to Meteorology, 10th ed. Upper Saddle River, NJ: Pearson Prentice Hall, 2007)*

∧ FIGURE 9.12 Polar and Subtropical Jet Streams
The position of the polar jet stream shifts from about 30° latitude in the winter months to about 70° latitude in the summer months. Because the temperature difference between midlatitude and polar air masses is lower during the summer months, the velocity of the polar jet is also decreased. *(Based on Lutgens, F. K., and Tarbuck, E. J., The Atmosphere: An Introduction to Meteorology, 12th ed. Upper Saddle River, NJ: Pearson Prentice Hall, 2010)*

crosses Mexico and Florida is weak during summer months and strongest in the winter when the temperature gradient between low-latitude and midlatitude air masses is greatest.

Air movement can also cause changes in atmospheric pressure. If there is an overall flow of air into a region, *convergence* of the air occurs, and the air piles up to increase atmospheric pressure. In contrast, if there is an overall flow of air out of a region, then *divergence* occurs, and the loss of air lowers atmospheric pressure (Figure 9.13a).

The combined effects of temperature and air movement produce the *low-pressure centers* (L) and *high-pressure centers* (H) that you see on weather maps. At the surface and aloft, air flows from areas of high pressure to areas of low pressure. Where elongate regions of high and low pressure develop aloft—known as *ridges* and *troughs*, respectively—the jet stream is forced to meander northward and southward along its westerly path (Figure 9.13b). As the jet stream travels along the ridges and troughs, it causes local flow convergence and divergence aloft leading to sinking and rising air, which can affect near-surface atmospheric pressure conditions and weather patterns (Figure 9.13b).

UNSTABLE AIR

As stated in the previous section, air movement can also be caused by vertical differences in temperature and pressure in the atmosphere. We can understand how

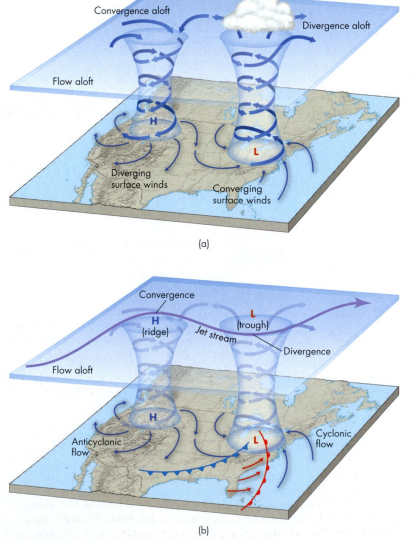

(a)

(b)

◄ FIGURE 9.13 Low- and High-Pressure Centers
This diagram illustrates airflow near low- and high-pressure centers in the Northern Hemisphere. (a) Horizontal changes in atmospheric pressure and temperature at Earth's surface causes areas of low (L) and high (H) pressure. In areas of low pressure, cyclonic surface winds converge causing air to rise and clouds to develop. Whereas diverging anticyclonic surface winds in high-pressure areas cause air to descend, resulting in clear skies. (b) Similar pressure conditions and wind patterns also result from convergence and divergence of air flowing in the upper troposphere due to high- and low-pressure ridges aloft. *(Based on Lutgens, F. K., and Tarbuck, E. J.,* The Atmosphere: An Introduction to Meteorology, *12th ed. Upper Saddle River, NJ: Pearson Prentice Hall, 2010)*

Coriolis Effect

The unequal distribution of solar energy that reaches the surface of Earth inevitably leads to temperature and pressure gradients that drive atmospheric circulation. You might expect air moving from a high-pressure area to a low-pressure area to follow a straight path; yet, when wind patterns are traced across Earth's surface, they appear to curve (Figure 9.1.A). This is because Earth, our frame of reference, rotates beneath the flowing air masses, causing a deflection of the wind to the right or to the left. This apparent change in motion or deflection is known as the **Coriolis effect**.

In order to understand the Coriolis effect, we must first step off our planet and get some perspective on our rotating home. As Figure 9.1.B shows, Earth rotates from west to east and, when viewed from the Northern Hemisphere, rotates in a counterclockwise manner. Conversely and perhaps surprisingly, when viewed from above the South Pole, the west to east rotation of Earth results in a clockwise rotation. Because of the different rotation directions, the Coriolis effect behaves differently in the Northern Hemisphere than in the Southern Hemisphere. Specifically, in the Northern Hemisphere the deflection will be to the right, and in the Southern Hemisphere the deflection will be to the left (Figure 9.1.A).

Let us now examine what causes the Coriolis deflection. Notice in Figure 9.1.B that the rotation speed of Earth varies with latitude, increasing from zero at the poles to 1675 km (~1041 mi.) per hour at

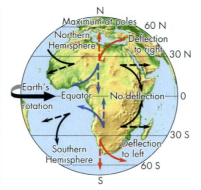

∧ FIGURE 9.1.A Surface Wind Patterns Resulting from the Coriolis Effect
Air masses moving across the surface of Earth (dashed lines) are deflected to the right in the Northern Hemisphere and to the left in the Southern Hemisphere (solid lines) due to the Coriolis effect. *(Based on Christopherson, R. W., Geosystems, 7th ed. Upper Saddle River, NJ: Pearson Prentice Hall, 2009)*

∧ FIGURE 9.1.B Earth's Rotation
Apparent deflection of an airplane traveling along a straight path between the North Pole and Quito, Ecuador, which is on the equator. *(Based on Christopherson, R. W., Geosystems, 7th ed. Upper Saddle River, NJ: Pearson Prentice Hall, 2009)*

the equator. It is this variability in the rotation speed that causes the apparent deflection of anything that flows or flies north or south above the rotating surface of Earth, including airplanes, oceans, and air masses. The Coriolis force acts at a right angle to the direction of motion, causing an object moving above the surface to have a curved path (Figure 9.1.C). A good way to visualize the Coriolis effect is to visualize an airplane flight to explain how Earth's rotation could cause air masses to be deflected across the surface of the Earth. The pilot plots a course due south along a straight path from the North Pole to Quito, Ecuador (Figure 9.1.B). Departing from the North Pole (the axis of

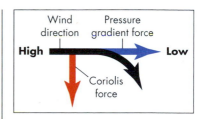

∧ FIGURE 9.1.C Forces Involved in Wind Patterns
The path of an air mass is controlled by the pressure gradient and the Coriolis forces. On a nonrotating Earth, air masses would flow along the pressure gradient (blue arrow). However, on a rotating Earth, the Coriolis force exerts a force to the right in the Northern Hemisphere (red arrow) perpendicular to the horizontal wind direction, which causes the path of an air mass to curve (black arrow).

Earth's rotation), the plane has zero eastward rotation. Although the plane flies along a straight path toward Quito, the greater rotation speed of Earth near the equator carries the city eastward. Upon arriving at the latitude of Quito, the pilot would be at a location over the Pacific Ocean. To an observer in Quito watching the airplane, the land surface would feel stationary, yet the airplane would appear to turn toward the west and head out over the sea, even though its path was straight. On the return flight from Quito, the greater eastward rotation at the equator compared to more northern latitudes would cause the airplane to be deflected toward the east. In both cases, when viewed from above, the plane appears to have been deflected to the right.

Let's now return to Figure 9.1.A to see how the Coriolis effect can be used to explain the predominant wind directions on Earth. Wind traveling southward from the Northern Hemisphere's subtropical high-pressure zone (30°N latitude) will be deflected toward the west, producing easterly winds, whereas winds moving toward the north will produce westerly winds. In contrast, air moving southward from the Southern Hemisphere's subtropical high-pressure zone (30°S latitude) will be deflected toward the west, producing easterly winds, whereas wind moving toward the north will produce westerly winds.

this occurs by examining the behavior of a small volume or *parcel* of air. You can visualize a parcel of air as an imaginary balloon similar to a small blimp circling over an outdoor stadium during a football game.

The tendency of a parcel of air to remain in place or change its vertical position is referred to as *atmospheric stability*. An air mass is stable if parcels of air within it resist vertical movement or return to their original position after they have moved. Alternatively, an air mass is considered unstable if parcels within it are rising until they reach air of similar temperature and density.[8] The atmosphere commonly becomes unstable when lighter warm or moist air is overlain by denser cold or dry air. Under these conditions, the instability causes some parcels of air to sink and others to rise like hot-air balloons. Severe weather, such as thunderstorms and tornadoes, is associated with unstable atmospheric conditions. Upper atmosphere winds can cause air masses to move over colder or warmer surfaces, increasing or decreasing atmospheric stability.

FRONTS

Weather forecasters refer to the boundary between a cooler and warmer air mass as a **front**. This boundary is called a cold front when cold air is moving into a mass of warm air and a warm front when the opposite occurs (Figure 9.14). Regardless of which air mass is advancing on the other, the warmer air will always be lifted by the colder, denser air mass. In addition to having different temperatures, the air masses may also have different humidity levels, densities, wind patterns, and stability. As a result of friction with the ground, high-density cold fronts are slowed at lower levels, compared to higher up in the air mass, causing the front to become steeper (Figure 9.14a). As the cold front advances, warm air is driven rapidly up the steep cold front causing localized cloud formation, strong updrafts, and, often, heavy precipitation. In contrast, warm fronts tend to be gentler as they override the receding cold air mass, and, therefore, updrafts and cloud formation are more scattered (Figure 9.14b). Fronts may remain stationary for a few hours or even days. A boundary between cooler and warmer air that shows little movement is called a *stationary front*. A fourth type of front, called an *occluded front*, develops when rapidly moving cooler air overtakes another cooler air mass and warm air is wedged above the frontal boundary. Each of these four types of fronts can cause inclement weather. The positions of fronts, as well as areas of high and low pressure, are shown on standard weather maps of surface atmospheric conditions (Figure 9.15).

9.4 CHECK your understanding

1. Why does atmospheric pressure decrease with increasing altitude?

2. What is the difference between stable and unstable air?

3. Explain the Coriolis effect. How does it influence weather?

4. Why is the polar jet stream stronger than the subtropical jet stream?

9.5 Hazardous Weather and Geographic Regions at Risk

The basic principles of atmospheric physics described earlier can help us understand severe weather and its associated hazards. Severe weather refers to events such as thunderstorms, tornadoes, hurricanes (see

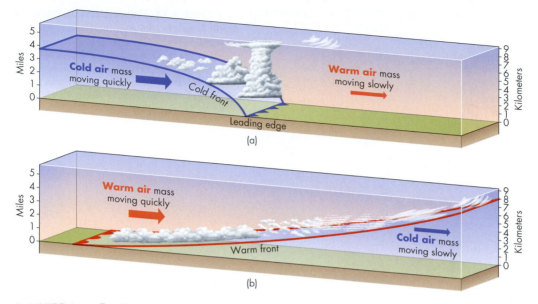

∧ FIGURE 9.14 Fronts

Weather fronts mark the boundary of air masses that have different densities, generally as the result of differences in temperature. (a) Advancing cold front forces warm air upward. The rising warm air can create clouds and heavy precipitation. (b) Advancing warm front forces warm air to rise over cooler air. Clouds and precipitation may develop. *(Based on McKnight, T. L., and Hess, D.,* Physical Geography, *8th ed. Upper Saddle River, NJ: Pearson Prentice Hall, 2004)*

Chapter 10), blizzards, ice storms, mountain windstorms, heat waves, and dust storms. These events are considered hazardous because of the energy they release and damage they are capable of causing.

THUNDERSTORMS

At any one time, thousands of thunderstorms are in progress on Earth. Most occur in the equatorial regions. For example, the city of Kampala, Uganda, near the equator in East Africa, holds the world record for thunderstorm frequency; it has thunderstorms nearly seven out of every ten days. In North America, the regions with the highest annual number of days with thunderstorms are in an area along the Front Range of the Rocky Mountains in Colorado and New Mexico and in a belt that encompasses all of Florida and the southern parts of Georgia, Alabama, Mississippi, and Louisiana (Figure 9.16).

Most readers are likely to have experienced at least one thunderstorm, because they occur in virtually every part of the United States and Canada. Although they can form at any time, thunderstorms are most common during afternoon and evening hours, especially in the spring and summer. Their abundance at this time of

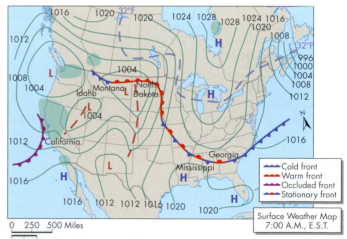

∧ FIGURE 9.15 Weather Map

The weather showing surface atmospheric conditions for the United States and southern Canada on April 13, 2003. Map shows positions of surface high- (H) and low- (L) pressure centers, two low-pressure troughs (reddish brown dashed lines), a cold front extending east from Georgia into the Atlantic Ocean (blue line with triangles), a warm front in North Dakota (red line with half circles), a stationary front in the lower Mississippi Valley, an occluded front off the California coast, the line marking freezing temperature (curving dashed blue line), and areas of precipitation (solid green). The green contour lines show atmospheric pressure in millibars (e.g., 1016). *(From the National Weather Service, NOAA)*

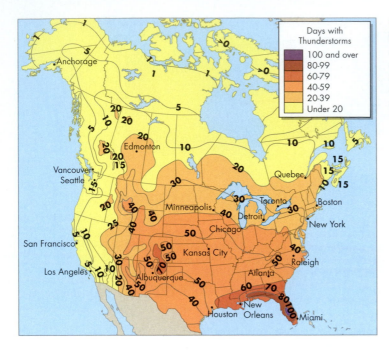

< FIGURE 9.16 Thunderstorm Occurrence in North America
Average number of days per year with thunderstorms in the United States and Canada. South-central Florida has the greatest number of days with thunderstorms, and Baffin Island in the Canadian Arctic has the least. *(From the National Weather Service, NOAA)*

maximum daytime heating is related to three basic atmospheric conditions that are required to produce a thunderstorm:

1. Warm humid air must be available in the lower atmosphere to feed clouds and precipitation and provide energy to the storm as it develops.

2. A steep vertical temperature gradient must exist in the environment such that the rising air is warmer than the air through which it is moving. This gradient places colder air over warmer, moist air.

3. An updraft must force moist air up to colder levels of the atmosphere.

These three conditions are common along cold fronts where the warm humid air is forced upward along the relatively steep front of the cold air mass (see Figure 9.14a).

Thunderstorm formation starts as moist air is forced upward and cools, and water vapor condenses to form a puffy cumulus cloud. Initially the cloud will form and then evaporate with little increase in height. If the moisture supply and updraft continue, the relative humidity increases in the air surrounding the cloud and it grows in size instead of evaporating. The upward growth in size of a cumulus cloud begins the *cumulus stage* (Figure 9.17) of thunderstorm development. In this stage, the cumulus cloud becomes a cumulonimbus cloud with the growth of domes and towers that look like a head of cauliflower. This growth requires a continuous release of latent heat from water vapor

condensation to warm the surrounding air and cause the air to rise farther.

As the domes and towers grow upward, precipitation starts by one of two mechanisms. First, growth of the cloud into colder air causes water droplets to freeze into ice crystals and snowflakes. The larger snowflakes fall until they enter air that is above freezing and melt to form raindrops. Second, in warm air in the lower part of the cloud, large cloud droplets collide with smaller droplets and coalesce to become raindrops. Once raindrops are too large to be supported by updrafts in the cloud, they begin to fall, creating a downdraft.

The *mature stage* of thunderstorm development begins when the downdraft and falling precipitation leave the base of the cloud (Figure 9.17). At this stage, the storm has both updrafts and downdrafts, and it continues to grow until it reaches the top of the unstable atmosphere. Commonly this upper limit of growth is the tropopause. At this point, the updrafts may continue to build the cloud outward to form a characteristic anvil shape (Figure 9.17). During the mature stage, the storm produces heavy rain, lightning and thunder, and occasionally large chunks of ice, known as *hail*.

The final or *dissipating* stage begins when the upward supply of moist air is blocked by downdrafts at the lower levels of the cloud (Figure 9.17). Downdrafts incorporate cool, dry air surrounding the cloud and cause some of the falling precipitation to evaporate. This evaporation further cools the downdraft and limits the updraft of warm humid air. Deprived of moisture, the

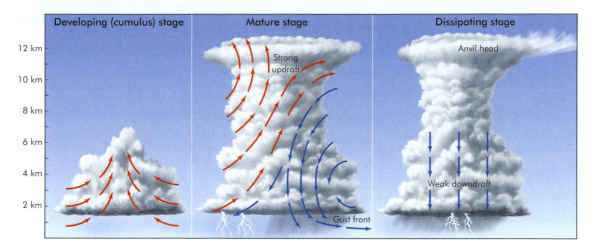

∧ FIGURE 9.17 Life Cycle of a Thunderstorm
Idealized diagram showing stages of development and dissipation (disappearance) of a thunderstorm. Red arrows show updrafts of warm air, and thin blue arrows show downdrafts of cold air, which are responsible for powerful gusts of wind (gust front) often felt in advance of an approaching storm. *(Based on Aguado, E., and Burt, J. E., Understanding Weather and Climate, 4th ed. Upper Saddle River, NJ: Pearson Prentice Hall, 2007)*

thunderstorm weakens, precipitation decreases, and the cloud dissipates. Most individual thunderstorms, sometimes called *air-mass thunderstorms*, last less than an hour and do little damage.

Severe Thunderstorms The scenario just described is typical of most thunderstorms, which never reach severe levels. However, under the right conditions, these storms can become severe. In the United States, the National Weather Service classifies a thunderstorm as *severe* if it has wind speeds exceeding 93 km (~58 mi.) per hour, or hailstones larger than 1.9 cm (~0.75 in.), or generates a tornado.[7] Severe thunderstorms, which rely on favorable atmospheric conditions over a large area, are able to self-perpetuate. They often appear in groups and can last from several hours to several days.

Conditions necessary for the formation of a severe thunderstorm include large changes in wind shear, high water vapor content in the lower troposphere, updraft of air, and the existence of a dry air mass above a moist air mass.[7] Of these four conditions, vertical wind shear is especially important. **Vertical wind shear** is produced by an increase in wind velocity with altitude, or change in the horizontal direction of wind with altitude, which causes a rolling motion—horizontal axis rotation—of the air in the lower troposphere (Figure 9.18). In general, the greater the vertical wind shear, the more severe a thunderstorm will become.[9]

Three types of severe thunderstorms have been identified on the basis of their organization, shape, and size. They include roughly circular clusters of storm cells called mesoscale convective systems (MCS), linear belts of thunderstorms called squall lines, and large cells with single updrafts called supercells.

The MCS is the most common of the three types. These systems are large clusters of self-propagating storms in which the downdraft of one cell leads to the formation of a new cell nearby. Unlike many single-cell, air-mass thunderstorms that last for less than an hour, these complexes of storms can continue to grow and move for periods of 12 hours or more.

Squall lines, which average 500 km (~310 mi.) in length, are long lines of individual storm cells. These lines commonly develop parallel to cold fronts at a distance of 300 to 500 km (~180 to 310 mi.) ahead of the front.[9] Updrafts in the advancing line of storms typically form anvil-shaped clouds whose tops extend ahead of the line. Downdrafts originating on the back side of the storms often surge forward as a *gust front* of cold air that arrives ahead of precipitation (see Figure 9.17). Squall lines can also develop along *drylines*, an air-mass boundary similar to a front, but in which the air masses differ in moisture content rather than air temperature. Drylines develop in the southwestern United States during the spring and summer, sometimes producing daily squall lines.

The most damaging of all severe thunderstorms is the **supercell storm**. A supercell thunderstorm is defined by the presence of an upward spiraling column of air—vertical axis rotation—known as a *mesocyclone*. This rotation is detectable with Doppler radar, which

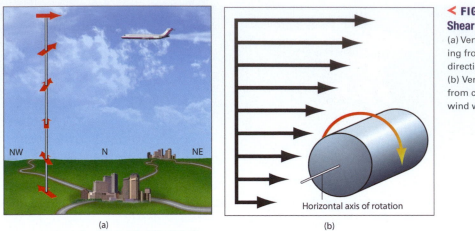

< **FIGURE 9.18 Vertical Wind Shear**
(a) Vertical wind shear resulting from changes in the wind direction with increasing altitude. (b) Vertical wind shear resulting from changes in the velocity of wind with increasing altitude.

is a primary tool used for identifying and predicting severe weather (see Section 9.9 and Case Study 9.2: Professional Profile). In the initial formation stage, vertical wind shear causes rotation to develop within the thunderstorm (Figure 9.19a). Strong updrafts develop ahead of the advancing cold front as the warm moist air is lifted. These updrafts, which are strengthened due to additional heating within the rising air caused by the latent heat of condensation, tilt the horizontally rotating air vertically (Figure 9.19b). At

this point in the storm's development, vertical wind shear causes the rotating mesocyclone to tilt and the rain and forward downdraft to shift downwind, which stabilizes the updraft (Figure 9.19c). Although smaller than an MCS or a squall line, supercell storms are extremely violent and are the breeding ground for most large tornadoes. They usually range from 20 to 50 km (~12 to 30 mi.) in diameter, last from 2 to 4 hours, and are capable of doing significant damage in that time.

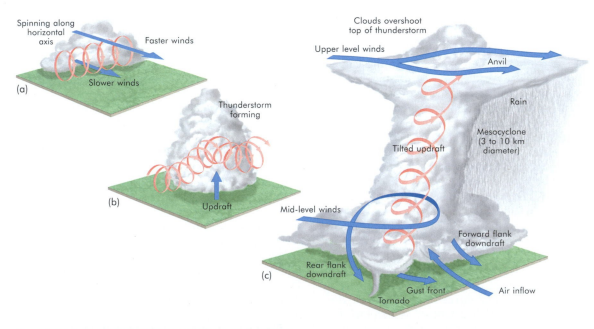

∧ FIGURE 9.19 Supercell Storm and Tornado Formation
(a) Vertical wind shear creates a horizontal axis rotation or rolling of air. (b) Updrafts produced by rising warm air lift the horizontally rolling air toward the vertical. (c) A vertical rotating mesocyclone develops. If conditions are right, a tornado will develop from the slowly rotating wall cloud toward the rear of the storm. *(Based on McKnight, T. L., and Hess, D., Physical Geography, 10th ed. Upper Saddle River, NJ: Pearson Prentice Hall, 2004)*

Sarah Tessendorf, Severe Storm Meteorologist

Sarah Tessendorf (Figure 9.2.A) has been a storm watcher for long as she can remember. "My favorite thing as a child was being woken up in the middle of the night by a really bright flash of lightning. I was scared to death of it—but it got my blood running every time," she said.

Tessendorf's fascination with extreme weather has never faltered. While still an undergraduate, she began studying tornadoes at SOARS (Significant Opportunities in Atmospheric Research and Science), a summer internship program at the National Center for Atmospheric Research (NCAR) in Boulder, Colorado.

Working with Jeffrey Trapp of Purdue University, Tessendorf studied two types of storms known to trigger tornadoes. Supercells—severe thunderstorms containing rotating updrafts—were considered responsible for the most severe tornadoes. Squall lines—storms that travel in one long line—were thought to spawn only weak twisters. Using radar data, Tessendorf and Trapp discovered that squall lines birth up to 20 percent of tornadoes, some of which can be violent. Their results sent atmospheric scientists scrambling to learn how to spot squall line tornadoes in time to warn the public.

"It was exciting to have my first project lay the groundwork that there is a threat from this type of storm," Tessendorf said.

Building on this expertise, Tessendorf went on to study lightning for her Ph.D. The high plains of Colorado are struck by unusual, positive cloud-to-ground lightning. To find out why, dozens of scientists from the federal government, universities, and private industry combined forces in 2000 to observe high plains storms with three radars, instrumented balloons, and even a chase plane. "We wanted to figure out what made these storms unique," Tessendorf said.

Tessendorf and her colleagues confirmed that these odd storms have what is known as inverted charge structures—a midlevel positively charged layer sandwiched in between negatively charged layers. She then hypothesized how such a charge structure could develop. Scientists believe that storms become electrified after small ice crystals collide and transfer charge with small hail-like particles. Updrafts separate the charge by carrying the small crystals skyward, leaving the graupel (ice pellets) at lower elevations. Tessendorf found that unusually strong and large updrafts were common features of Colorado storms with inverted charge structures and could aid in altering the charge transfer process.

Now a postdoctoral researcher at the University of Colorado at Boulder, Tessendorf uses computer models to study precipitation processes in storms. In her spare time, she enjoys storm chasing. "Out in front of the storm I can feel warm humid air being sucked into the clouds. Then it suddenly gets really still, and cool wind flows outward from the storm. That's the key to get moving again," Tessendorf said. "To feel and experience that calm is both eerie and amazing at the same time."

Tessendorf hopes to continue working in meteorology while sharing her love of storms with others. "The weather is something we all experience in common. I'd like to use it to help convey the excitement of science to the public," she said.

— KATHLEEN WONG

◀ FIGURE 9.2.A Dr. Sarah Tessendorf, Severe Storm Meteorologist
Dr. Tessendorf's smile comes from a successful day of storm observations on the High Plains in eastern Colorado. *(Sarah Tessendorf)*

Downbursts of air from severe thunderstorms, especially MCSs, can also generate strong, straight-line windstorms. These windstorms vary in size, with the largest ones, called *derechos*, producing severe, tornado-strength wind gusts along a line that is at least 400 km (~250 mi.) in length.[10] In general, derecho winds exceed 90 km (~55mi.) per hour and can cause numerous trees to fall, widespread power outages, serious injuries, and multiple fatalities. Strong derechos generate winds approaching 210 km (~130 mi.) per hour and produce the same damage as a medium-sized tornado. More than a dozen derechos strike North America each year, with most occurring in the eastern two-thirds of the contiguous United States and southern Canada.[10] Smaller thunderstorm downbursts, called *microbursts*, are more common than derechos and are a serious hazard to aviation.

Lightning A common occurrence during thunderstorms, **lightning** consists of flashes of light produced by the discharge of millions of joules of electricity (Figure 9.20). These discharges can heat the surrounding air to as high as 30,000°C (~54,000°F)—five times hotter than the surface of the sun.[7] This extreme heat causes the air to rapidly expand, producing the familiar accompanying sound of *thunder*.

Most lightning bolts are cloud-to-cloud; that is, they start and end within the thunderstorm or end in nearby clouds. Although cloud-to-ground lightning is less common, an estimated 25 million lightning bolts strike in the United States each year.[11] Cloud-to-ground lightning is more complex than it appears to an observer.

Most lightning comes from cumulonimbus clouds that have grown high enough for ice crystals and pellets to form. Similar to how dragging your feet on the carpet can cause you to get an electric shock when you touch something metal, interactions between rising ice crystals and falling ice pellets within the updraft are thought to create an electrical field strong enough to produce lightning.[12] Electrons moving between the two forms of ice build up positive electrical charge in the upper part of the cloud and negative charge in the lower part (Figure 9.21a). Since like electrical charges repel, the increased negative charge in the base of the cloud drives away the negative charge on Earth's surface below. This leaves the ground beneath the thunderstorm with a net positive charge. A strong difference in electrical charge between the cloud and the ground is required for lightning to overcome the natural insulating property of air.

The majority of lightning strikes begin as a narrow column of high-speed electrons moving downward from the base of a cloud toward a pocket of positive charge on Earth. Within milliseconds this column, called a *stepped leader*, branches downward until it is close to the ground while positively charged *streamers* move upward from the ground, emanating from a tall object like a tree or building (Figure 9.21b). As the stepped leader approaches, a spark jumps from the streamer and attaches to the leader (Figure 9.21c). The spark completes a conductive channel of ionized air molecules for an upward surge of positive electrical charge. This surge heats the air to create a brilliant *return stroke* of lightning to the cloud (Figure 9.21d). Within milliseconds, additional leaders and return strokes move along the same path and often cause the lightning flash to appear to flicker to the human eye.

Cloud-to-ground lightning

Cloud-to-cloud lightning

◀ FIGURE 9.20 Types of Lightning Nighttime photograph of cloud-to-ground lightning strokes and cloud-to-cloud lightning strokes. Although most lightning that we see goes from cloud-to-ground or cloud-to-cloud, more than half of all lightning occurs within thunderstorm clouds. *(NOAA)*

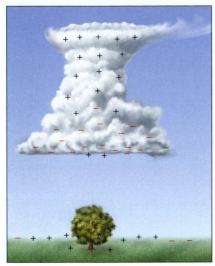

(a)

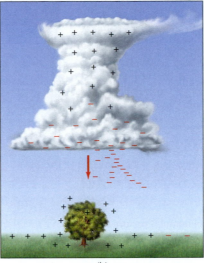

(b)

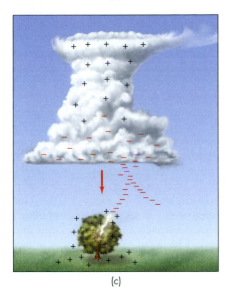

(c)

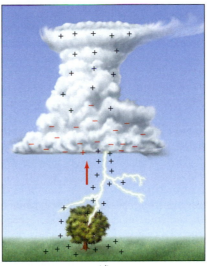

(d)

< FIGURE 9.21 Development of Cloud-to-Ground Lightning (a) Electrical charge separation occurs in a cumulonimbus cloud. The buildup of a negative electrical charge at the base of the cloud locally induces a positive electrical charge below the thunderstorm. (b) A nearly invisible, stepped leader forms a column of negatively charged air that branches downward from the cloud, while positively charged streamers move upward from the ground. (c) When the stepped leader and the streamer meet, a narrow channel causing a rapid flow of electrons results. (d) Flowing electrons heat the channel to extremely high temperature and create a return stroke back to the cloud.

Lightning strikes constitute a serious natural hazard even though the number of annual lightning deaths in the United States has decreased as more people leave rural areas to live in cities. The chances of being struck by lightning in the United States are estimated to be 1 in 240,000 each year. For an 80-year life span, this becomes 1 in 3,000.[13] Your risk, however, depends on where you live and work. Nevertheless, lightning kills an estimated 100 people and injures more than 300 per year in the United States and kills seven people and injures 60 to 70 each year in Canada.[12,14] Of those who survive a lightning strike, 70 percent suffer serious long-term health effects (see Case Study 9.3: Survivor Story).[6,13]

Hail Although many large thunderstorms can produce hard, rounded, or irregular pieces of ice called *hail-stones*, it is large hail from severe thunderstorms that is the greatest hazard (Figure 9.22a). Evidence for the origin of hailstones can be found by cutting a stone in half to reveal a bull's-eye pattern of concentric rings of ice (Figure 9.22b). These rings form as a hailstone is moved up and down in a thunderstorm. Starting with a small ice pellet as a nucleus, a hailstone gets a coating of liquid water in the lower part of the storm, and that coating freezes when a strong updraft carries the stone upward into cold air. This process is repeated many times to form a large piece of hail. The largest authenticated hailstone in North America fell from a severe

(a)

(b)

∧ FIGURE 9.22 Hail Damage and Internal Structure
(a) Woman repairs her roof damaged by hail. *(DANIL SEMYONOV/AFP/Getty Images)* (b) Internal structure of hailstone showing concentric rings, which develop due to accretion of successive ice layers around a central core within the thunderstorm. *(deepspacedave/Shutterstock)*

thunderstorm in Aurora, Nebraska, in June 2003. Nearly as large as a volleyball, the hailstone measured 18 cm (~7 in.) in diameter, weighted 0.8 kg (~1.8 lb), and is estimated to have hit the ground at a velocity of more than 160 km (~100 mi.) per hour![15]

Hailstorms generally cause more property damage than casualties. The damage caused by hail in the United States alone averages $1 billion per year. In North America, damaging hailstorms are most common in the Great Plains, particularly in northeastern Colorado and southeastern Wyoming, and in the Calgary area of Alberta, Canada. North-central India, Bangladesh, Kenya, and Australia also frequently experience damaging hailstorms. Although rare in North America, deaths from hailstones are not uncommon in Bangladesh and India in heavily populated areas with poorly constructed dwellings.[15]

TORNADOES

Usually spawned by severe thunderstorms, a tornado is one of nature's most violent natural processes. From 1992 to 2012, tornadoes in the United States killed an average of 62 people per year.[16] These spinning columns of wind can take on a variety of shapes, including a rope, funnel, cylinder, and wedge (Figure 9.23). Other names given to tornadoes include "twisters" and "cyclones," although the term *cyclone* also is used to refer to large spinning columns of wind within clouds, hurricanes, and very large atmospheric low-pressure systems. To be called a tornado, a spinning column of wind or a *vortex* must extend downward from a cloud and touch the ground. In small or newly developed tornadoes, the vortex can be almost invisible until water vapor begins to condense or the tornado picks up dust and debris. In the first stage of tornado development, funnel-shaped

vortices called *funnel clouds* that do not reach to the ground form below the thunderstorm (Figure 9.23a). Undoubtedly the number of funnels that develop each year far exceeds the number of tornadoes.

Tornadoes form where there are large differences in atmospheric pressure over short distances, as often results during a major storm such as a supercell thunderstorm. Although meteorologists do not completely understand precisely how tornadoes form, they have recognized that most tornadoes go through similar stages in development. Major updrafts, often in the rear or southwestern part of the storm, lower a portion of the cumulonimbus cloud to form a *wall cloud* (Figure 9.23b). This wall cloud may begin to slowly rotate, and a short funnel cloud may descend. A tornado has formed if dust and debris on the ground begin to swirl below the funnel (Figure 9.23d). Not all wall clouds and mesocyclones produce tornados, and tornados can develop without the formation of a wall cloud or mesocyclone.

In the second, *mature* stage, a visible condensation funnel extends from the thunderstorm cloud to the ground as moist air is drawn upward (Figure 9.23c). In stronger tornadoes, smaller intense whirls, called *suction vortices*, may form within the larger tornado. The suction vortices orbit the center of the large tornado vortex and appear to be responsible for its greatest damage.[17]

When the supply of warm moist air is reduced, the tornado enters the *shrinking stage*. In this stage, the funnel thins and begins to tilt. As the width of the funnel decreases, the winds can increase, making the tornado more dangerous.

In the final decaying stage, or *rope stage* (Figure 9.23d), the upward-spiraling air comes in contact with downdrafts and the tornado begins to move erratically. Although this is the beginning of the end for the tornado, it can still

∧ FIGURE 9.23 Stages in Tornado Development
(a) Airborne funnel clouds, like this one over the state capitol in Austin, Texas, are not considered tornadoes until they touch the ground. *(Dave Chapman/Alamy Stock Photo)* (b) This wall cloud near Miami, Texas, is the lighter gray cloud to the left of the lightning bolt. Wall clouds hang downward from a severe thunderstorm and are often where tornadoes form. *(NOAA)* (c) A tornado in the mature stage of development near Seymour, Texas. *(NOAA)* (d) A tornado in the decaying rope stage of development near Cordell, Oklahoma. A tornado is still dangerous at this stage. *(NOAA)*

be extremely dangerous at this point. Tornadoes may go through all the stages just described, or they may skip stages. Like other types of clouds, new tornadoes can form nearby as older tornadoes disappear.

As tornadoes move along the surface of Earth, they pick up dirt and debris. This debris gives the tornado cloud its characteristic dark color (Figure 9.23d). Tornadoes typically have diameters measured in tens of meters and wind speeds of 65 km (~40 mi.) to more than 450 km (~280 mi.) per hour.[7] Once they touch down, tornadoes usually travel 6 to 8 km (~4 to 5 mi.) and last only a few minutes before weakening and disappearing. However, the largest, most damaging tornadoes may move at speeds of close to 100 km (~60 mi.) per hour along a path several hundreds of kilometers long.

Classification of Tornadoes Tornadoes are classified by the most intense damage that they have produced along their path. Each tornado can be assigned a value on the *Enhanced Fujita* or **EF Scale** (Table 9.1) based on a post-storm damage survey. This survey determines the levels of damage experienced by 26 types of buildings, towers, and poles and by hardwood and softwood trees. Values on the EF Scale estimate the maximum 3-second wind gust in the tornado and range from 105 to 137 km (~65 to 85 mi.) per hour for an EF0 tornado, to more than 322 km (~200 mi.) per hour for an EF5 tornado.[18] The EF Scale replaces the F-scale developed by T. Theodore Fujita in 1971 and is similar to the Modified Mercalli Scale for assessing earthquake intensity.

Struck by Lightning

like most of us, Michael Utley didn't worry much about being struck by lightning. The odds, after all, are second to none. So at a charity golf tournament to benefit a local YMCA near his Cape Cod home, he was not overly apprehensive about the looming threat of a thunderstorm. "I didn't pay attention to it," he said.

Four holes into the game, the warning horn blasted, urging golfers to seek shelter. Utley replaced the flag in the hole and was several yards behind his three companions when he was struck by lightning.

Hearing a thunderous crack behind them, Utley's friends turned around. Utley says they reported smoke coming from his body, his shoes torn from his feet, and his zipper blown open.

Luckily, Utley remembers none of this. In fact, he remembers nothing from the next 38 days, which he spent in an intensive care unit. Fortunately for Utley, one of his companions had recently been retrained in CPR, which may very well have saved Utley's life while the EMS vehicles were on their way.

As for the various images we commonly associate with those struck by lightning, Utley said some are accurate and some are not.

∧ **FIGURE 9.3.A Lightning Safety**
This National Weather Service poster of professional golfer Vijay Singh is a reminder that more people are killed and injured by lightning strikes in outdoor recreation than in any other activity. *(NOAA)*

"The hair standing up, that's real. When that happens, you're pretty close to being dead," he said, "but I didn't see a white light at the end of the tunnel. I don't burn clocks off the wall when I touch them."

Utley's reaction, like that of many victims of lightning, was filled with a variety of ups and downs. His wife, Tamara, recalls that he did not slip into a coma-like state for several days. "It was very, very strange," she said. "When he first got to the ICU, he was able to move and he was very lucid." But within a few days, he began to lose consciousness for extended periods of time. Utley remembers none of this either.

Since the incident, Utley has worked to educate the community about the perils of lightning, including serving as a NOAA spokesman for lightning safety and working with the PGA on the danger electrical storms pose to golfers (Figure 9.3.A). He sums up his message to golfers in one easy slogan: "If you see it, flee it; if you hear it, clear it."

— CHRIS WILSON

Tornadoes that form over water are generally called *waterspouts* rather than tornadoes (Figure 9.24). Most waterspouts develop beneath fair-weather cumulus clouds and appear to be associated with wind shear along the boundary of contrasting air masses.[17] Wind shear means that the winds along a boundary are blowing in different directions and velocities. You can produce a similar kind of shear-induced rotation by holding a pencil vertically between your hands and moving your hands in opposite directions. Small, wind-shear–produced tornadoes can develop on land and, like waterspouts, are typically weak EF1 or EF0 tornadoes.[17] Although not as powerful as most tornadoes in supercell storms, waterspouts and related weak tornadoes on land can cause considerable damage and should be respected.

Occurrence of Tornadoes Although tornadoes are found throughout the world, they are much more common in the United States than in any other location on Earth. The United States has just the right combination of weather, topography, and geographic location to

▼ **TABLE 9.1 Enhanced Fujita Scale of Tornado Intensity**

Scale	Wind Estimate[a]	Typical Damage[b]
EF0	105–137 km/hr (~65–85 mph)	*Light damage.* Some damage to chimneys; branches broken off trees; shallow-rooted trees pushed over; sign boards damaged.
EF1	138–177 km/hr (~86–110 mph)	*Moderate damage.* Surfaces of roofs peeled off; mobile homes pushed off foundations or over-turned; moving autos blown off roads.
EF2	178–218 km/hr (~111–135 mph)	*Considerable damage.* Roofs torn off frame houses; mobile homes demolished; boxcars over-turned; large trees snapped or uprooted; light-object missiles generated; cars lifted off ground.
EF3	219–266 km/hr (~136–165 mph)	*Severe damage.* Roofs and some walls torn off well-constructed houses; trains overturned; most trees in forest uprooted; heavy cars lifted off the ground and thrown.
EF4	267–322 km/hr (~166–200 mph)	*Devastating damage.* Well-constructed houses leveled; structures with weak foundations blown away some distance; cars thrown and large missiles generated.
EF5	More than 322 km/hr (More than 200 mph)	*Incredible damage.* Strong frame houses leveled off foundations and swept away; automobile-sized missiles fly through the air in excess of 100 m (~110 yds.); trees debarked; incredible phenomena will occur.

[a] Wind speeds are the maximum estimated for a 3-second gust.
[b] Accurate placement on this scale involves expert assessment of the degree of damage to 28 indicators including homes, buildings, towers, poles, and trees.

Source: Information from Wind Science Engineering Center, "A Recommendation for an Enhanced Fugita Scale (EF-Scale)." Lubbock, TX: Texas Tech University, 2004. www.spc.noaa.gov/faq/tornado/ef-ttu.pdf. Accessed 6/7/13.

∧ FIGURE 9.24 Waterspouts
Two water spouts can be seen about a mile off shore south of downtown Honolulu Monday, May 2, 2011. Unstable weather conditions over the island of Oahu lead to the formation of of thunderstorms over Oahu. *(Eugene Tanner/AP Images)*

make it the perfect spawning ground for tornadoes.[17] Most U.S. tornadoes occur in the midwestern states, between the Rocky Mountains and the Appalachians. As with severe thunderstorms, spring and summer are the most common time for tornadoes, and most develop in the late afternoon and evening. The highest risk for tornadoes occurs along what is called "Tornado Alley," a belt that stretches from north to south through the Great Plains states (Figure 9.25). This is where all the ingredients for the formation of a tornado collide. Areas of high annual tornado occurrence include a huge region from Florida to Texas and north to the Dakotas, Indiana, and Ohio. Although Canada experiences far fewer tornadoes than the United States, it has several tornado-prone regions, such as Alberta, southern Ontario, and southeastern Quebec, as well as an area running from southern Saskatchewan and Manitoba to Thunder Bay, Ontario.[19]

Possibly with the exception of Bangladesh, violent tornadoes (EF4 or EF5) are rare or nonexistent outside of the United States and Canada.[17] Waterspouts are both less hazardous and more common than tornadoes on land. Most take place in tropical and subtropical waters, but they have been reported off the New England and California coasts as well as on the Great Lakes. They are especially common along the Gulf Coast, Caribbean Sea, Bay of Bengal in the Indian Ocean, and the South Atlantic. A study conducted in the Florida Keys counted 390 waterspouts within 80 km (~50 mi.) of Key West during a five-month period.[17]

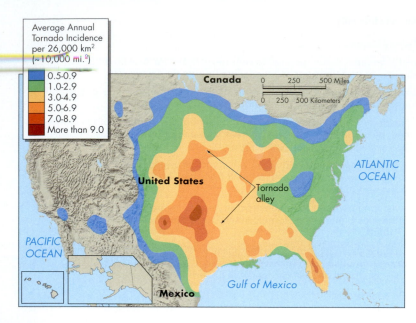

Average Annual
Tornado Incidence
per 26,000 km²
(~10,000 mi.²)

- 0.5-0.9
- 1.0-2.9
- 3.0-4.9
- 5.0-6.9
- 7.0-8.9
- More than 9.0

◄ **FIGURE 9.25 Tornado Occurrence in the United States**

Average occurrence of tornadoes per 26,000 km² (~10,000 mi.²) in the United States and southernmost Canada from 1950 to 2000. *(Based on Christopherson, R. W., Geosystems: An Introduction to Physical Geography, 6th ed. Upper Saddle River, NJ: Pearson Prentice Hall, 2006. Data from National Severe Storm Forecast Center, National Weather Service.)*

9.4 CASE study

Tri-State Tornado

In just 3½ hours on March 18, 1925, the Tri-State Tornado killed more people and destroyed more property than any other tornado in historic time. The tornado was unique in several ways. First, it was in contact with the ground for 349 km (~215 mi.), a distance of 183 km (~115 mi.) longer than any other tornado to date. Second, the average width of the tornado was 1.2 km (~0.75 mi.) and at times it reached 1.6 km (~1 mi.) in width. Only 73 of the more than 35,000 reported tornadoes have been as wide as or wider than the Tri-State Tornado.[17] This width produced extreme damage over at least 425 km² (~164 mi.²) (Figure 9.26). Third, the tornado's track was a straight line over much of its path through the states of Missouri, Illinois, and Indiana. This track contrasts with the curved track of most tornadoes. Finally, the tornado's average forward speed of 100 km (~62 mi.) per hour was one of the fastest ever reported. Most tornadoes travel at an average speed of 50 km (30 mi.) per hour.[7] People seeing the storm approach assumed it was a thunderstorm because it was so broad, and in fact, there were associated thunderstorms. The tornado's violent winds inflicted unprecedented damage, killing at least 695 people and injuring more than 2,000. Damages, expressed in year 2000 dollars, totaled about $170 million.

As the Tri-State Tornado moved across the landscape, it traversed hills as high as 425 m (1400 ft.) and crossed valleys and lowlands; the topography essentially had no effect on the tornado. Although much of the death and damage resulted from the collapse of buildings, flying debris was also responsible for significant destruction (Figure 9.27). As the storm moved northeast, the damaging rotational winds had velocities of up to 290 km (~180 mi.) per hour.

The extensive death and damage from the Tri-State Tornado resulted from several factors: (1) There were no tornado forecasts or warnings because the technology for them did not exist in 1925, (2) the storm destroyed telephone lines that could have been used to warn people in the projected path of the tornado, (3) the tornado was exceptionally large and strong, (4) massive amounts of flying debris and dust masked any funnel shape and made it hard to recognize the storm as a tornado, and (5) many of the homes and farms were poorly constructed and unable to withstand the strong winds.[20] Should such a tornado occur in the future, successful forecasting and warning, along with better construction techniques, would likely save countless lives.

BLIZZARDS AND ICE STORMS

Blizzards are severe winter storms in which large amounts of falling or blowing snow are driven by high winds to create low visibilities for an extended period of time. Extremely low visibility, commonly referred to as a *whiteout*, along with cold temperatures and high

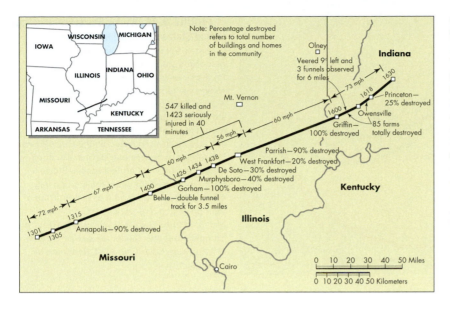

< FIGURE 9.26 Tornadoes Traveling Long Distances
In a span of 3½ hours, the Tri-State Tornado of March 18, 1925, traveled 350 km (~217 mi.) from Reynolds County, Missouri, across Illinois to near Princeton, Indiana. Traveling at an average speed of 100 km (~62 mi.) per hour, this tornado varied from 1 to 1.5 km (~0.6 to 0.9 mi. to 1 mi.) in width. The numbers from 1301 to 1630 along the tornado track refer to the time of day. *(Modified after Wilson, J. W., and Changnon Jr., S. A. 1971. Illinois tornadoes. Illinois State Water Survey Circular 103. Urbana, IL.)*

< FIGURE 9.27 Most Destructive Tornado in History
Ruins of the Longfellow School in Murphysboro, Illinois, where 17 children were killed by the Tri-State Tornado. The tornado struck the school around 2:34 P.M. on March 18, 1925. Most of the children were killed by the collapse of the unreinforced brick walls. Trees surrounding the school were also destroyed. A total of 234 people were killed in Murphysboro, the largest number of tornado deaths in a single town in U.S. history. *(NOAA)*

winds make a blizzard dangerous. Official thresholds for blizzard conditions differ in the United States and Canada. In the United States, winds must exceed 56 km (~35 mi.) per hour with visibilities of less than 0.4 km (~0.25 mi.) for at least 3 hours, whereas in Canada winds must exceed 40 km (~25 mi.) per hour with visibilities of less than 1 km (~1.6 mi.) for at least 4 hours.[21,22] In either case, a blizzard may involve no additional snowfall. High winds picking up previously fallen snow create a *ground blizzard* in which visibility can drop to a few meters or less for many hours. Ground blizzards develop numerous times each winter in Antarctica, Alaska and parts of Canada, and in the Great Plains states.

In North America, blizzards associated with heavy snowfall are most common in the Great Plains, the Great Lakes area, the U.S. Northeast, the Canadian Atlantic provinces, and the eastern Arctic. Storms producing heavy snowfall and blizzards form from an interaction between upper-level winds associated with a low-pressure trough and a surface low-pressure system.[23] In North America, blizzard-producing storms generally form in eastern Colorado; central Alberta, Canada; or along the coast of North Carolina or the northern Gulf of Mexico (Figure 9.28). Storms in Colorado and along the coast that produce heavy snowfall are derived from moist air originating over the Atlantic Ocean. In contrast, fast-moving storms

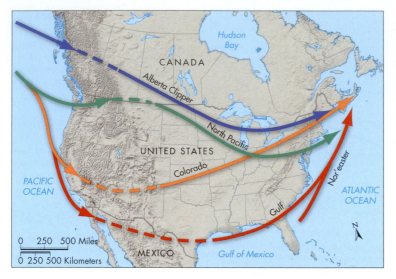

◀ **FIGURE 9.28 Winter Storm Track in the United States and Southern Canada**
In winter months, low-pressure centers and associated storms develop over the northern Pacific Ocean and come onshore along the west coast of the United States, Canada, and Mexico. The upper-level remnants of these storms (dashed lines) cross the western mountains, and low-pressure centers and associated storms redevelop on the Great Plains. Low-pressure centers and storms also develop along the Gulf and Carolina coasts. Storm tracks are named for geographic origin (e.g., Alberta Clippers, Colorado) or the direction of their prevailing high winds (e.g., nor'easter). (Adapted from Keen, R. A., Skywatch West: The Complete Weather Guide, revised ed. Golden, CO: Fulcrum Publishing, 2004. Reprinted by permission)

forming east of the Canadian Rockies, called *Alberta Clippers*, are generally drier with less snow and extremely cold temperatures.

Blizzards can produce lengthy periods of heavy snowfall, wind damage, and large snowdrifts. For example, the Saskatchewan blizzard of 1947 lasted for 10 days and buried an entire train in a 1 km (0.6 mi.) long snowdrift that was 8 m (~26 ft.) high.[24] Another famous storm, the "Blizzard of 1888," killed more than 400 people and paralyzed the northeastern United States for three days with snow drifts that reportedly were as high as the second floor of buildings.

Storms that move along the east coast of the United States and Canada, called *nor'easters*, can produce blizzards (see Chapter 10). Nor'easters wreak havoc with hurricane-force winds, heavy snows, intense precipitation, and high waves that damage coastal areas. These storms are most common between September and April and often create blizzard conditions in large cities such as New York and Boston. In March 1993, a severe nor'easter paralyzed the East Coast, causing snow, tornadoes, and flooding from Alabama to Maine. Damages from this storm, known as the "Blizzard of '93," totaled more than a billion dollars, and more than 240 people were killed.

Less than three years later, in January 1996, a strong winter storm brought another massive blizzard to the East Coast. The storm crippled the eastern United States for several days, produced record-breaking snowfall in Philadelphia and parts of New Jersey, and dropped 51 cm (~20 in.) of snow in New York City's Central Park (Figure 9.29). In all, the blizzard killed at least 100 people and caused an estimated $2 billion in damages.

One reason that blizzards are generally more dangerous than other snowstorms is the **wind chill** effect. Moving air rapidly cools exposed skin by evaporating

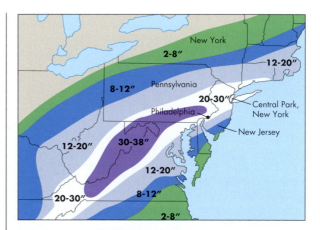

▲ FIGURE 9.29 Snowfall from the Blizzard of 1996
The greatest amount of snow in this storm fell along the Appalachian Mountains. Blizzard-force winds created snowdrifts much deeper than the snow depths indicated on this map.

moisture and removing warm air from next to the body. This chilling reduces the time it takes for *frostbite* to form. Thus, in blizzards the lower wind chill temperature is more important than the air temperature.

Ice storms, prolonged periods of freezing rain, can be more damaging than blizzards and just as dangerous. Ice accumulates on all cold surfaces and is especially harmful to utility lines and trees and to surface travel (see Case Study 9.5: The Great Northeastern Ice Storm of 1998). These storms typically develop during the winter in a belt on the north side of a stationary or warm front. In this setting, a combination of three conditions leads to freezing rain: (1) an ample source of moisture in the warm air mass south of the front, (2) warm air uplifted over a shallow layer of cold air,

∧ FIGURE 9.30 Thickness of Cold Air Determines Type of Precipitation
In winter storms, the thickness of cold air at the surface determines the type of precipitation. (a) Most precipitation begins falling as snowflakes and remains as snow if it does not pass through warm air before reaching the ground. (b) Falling snowflakes melt if they encounter warm air aloft, and the resulting rain refreezes to form ice pellets (sleet) if there is sufficient time and cold air before they reach the ground. (c) If there is only a thin layer of cold air at the surface, raindrops become supercooled and then freeze immediately upon contact with a cold surface such as a tree limb, house, or power line. (d) Snowflakes will melt and rain will fall if a thick warm air layer is at the surface. *(Based on Linda Scott and Austin American-Statesman)*

and (3) objects on the land surface at or very close to freezing. Under these conditions, snow begins to fall from the cooled top of the warm air mass. The snow melts as it passes through the warm air, and the resulting raindrops become supercooled when they hit the cold air at the surface (Figure 9.30). Upon contact with cold objects, such as roads, trees, and utility lines, the rain immediately freezes to form a coating of ice. Coatings of 15 to 20 cm (~6 to 8 in.) of ice have been produced by prolonged ice storms in Idaho, Texas, and New York. The regions most prone to ice storms include the Columbia River Valley in the Pacific Northwest, the south-central Great Plains, the Ohio River Valley, the Mid-Atlantic and New England states, and the Atlantic provinces and St. Lawrence River Valley in eastern Canada.[15]

9.5 CASE study

The Great Northeastern Ice Storm of 1998

One of the most destructive and long-lasting ice storms in North American history struck U.S./Canadian border states and provinces in January 1998. In most areas, the storm started late on January 4 and continued for six days. At least 45 people died in the United States and Canada as electric power was cut off to more than 5.2 million people.[25] The power loss contributed to most deaths, primarily from the loss of heat for an extended time. Causes of death ranged from carbon monoxide poisoning produced by improperly vented heat sources to hypothermia from cold temperatures.

Although workers from 14 power companies and military personnel worked long hours to restore power, more than 700,000 people were still without power in the third week after the onset of the storm.[26] The provinces of Quebec, Ontario, and New Brunswick and the states of New York and Maine were the worst hit, although damage was also considerable in Vermont, New Hampshire, and Nova Scotia (Figure 9.31). In addition to substantial damage to the power and telecommunications system, forestry, dairy farming, maple syrup production, and homeowners were severely affected by the storm. Overall losses exceeded $6.2 billion, with more than 80 percent of the losses occurring in Canada.[27] It was Canada's costliest disaster in history.

In many ways, the meteorological conditions that caused the massive storm were typical for ice storms of the Northeastern United States. A persistent flow of warm moist air from the Gulf of Mexico rose up over a thin wedge of cold air to the north of a stationary front (Figure 9.32). Cold air was driven south by northeasterly winds circulating around an Arctic high-pressure center over Hudson Bay, where it settled in river and mountain valleys, including the St. Lawrence and Ottawa River Valleys and the Lake Champlain Valley. Warm air was driven northward by a high-pressure center over Bermuda and a low-pressure trough over the lower Mississippi Valley.[28]

Weather conditions in the region were essentially unchanged for almost a week. The flow of moist air from the Gulf of Mexico acted like a conveyor belt to constantly deliver water vapor for precipitation both north and south of the stationary front. South of the front, heavy rains caused severe flooding in the mountains of North Carolina and Tennessee, killing nine people.[29] North of the front, most of the precipitation took the form of

∧ FIGURE 9.31 Ice Glazes Power Lines and Trees
Freezing rain turns to clear ice once it falls on a surface that
is at or below freezing. Many tree limbs and utility lines, like
these in the January 1998 ice storm in Watertown, New York,
bend and break with the weight of accumulated ice. *(Syracuse
Newspapers/Dick Blume/The Image Works)*

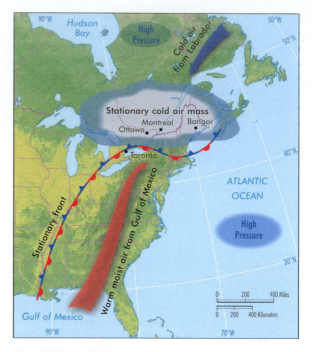

**∧ FIGURE 9.32 Weather Conditions That Caused the Ice
Storm of 1998**
A conveyor-belt–like flow of warm, moist air from the Gulf of
Mexico overrode a stationary cold air mass in Quebec, eastern
Ontario, Nova Scotia, Prince Edward Island, and the northeast-
ern United States to cause the ice storm of 1998. This system
remained in place for nearly a week, causing the worst disaster
in Canadian history. *(Based on Environment Canada)*

freezing rain with total liquid-equivalent accumulations
of 20 to 100 mm (~0.8 to 4 in.). Ice accumulations of more
than 250 mm (~9.8 in.) were reported from the freezing
rain at stations in Canada and the United States.[26,30]

Why was this storm so catastrophic? As with most
other modern catastrophes, both natural processes and humans contributed to the losses. For many Canadian communities, including Ottawa and Montreal, and some U.S.
towns, such as Burlington, Vermont, and Massena, New
York, the ice storm of 1998 was the most prolonged period
of freezing precipitation on record.[26,28] Historically most
episodes of freezing rain last only a few hours at a time;
however, the 1998 ice storm brought more than 80 hours
of freezing rain and drizzle over a six-day period, twice the
annual average.

On the human side, the ice storm demonstrated
how dependent our society has become on electricity,
especially the electrical grid. The storm destroyed
more than 1,000 high-voltage towers, 35,000 telephone poles, and more than 120,000 km (~75,000 mi.)
of power lines (Figure 9.33).[25] Montreal lost its water
supply after treatment plants and pumping stations

experienced a power failure on January 9. In many
places emergency backup electrical generators were
either absent, failed, or eventually ran out of fuel.

Also apparent was an increasing reliance on "just-
in-time" delivery for many items. It was difficult in
some cases to repair the electrical grid when needed
supplies, such as telephone poles, had to be shipped
long distances when transportation systems were
shut down. This was repeated more recently in
Austin, Texas, when an ice storm depleted deicing
fluid for several airlines at the city's airport, resulting
in the cancellation of more than 100 flights.[31] The
airline's reliance on ground delivery, rather than adequate on-site storage, caused the transportation
disruption.

Sometimes the lessons learned from a disaster or catastrophe are not conclusions drawn from scientific observations, technological malfunctions, or policy
failures. They are simple observations about human
interactions. Responses from Maine schoolchildren to
the question "What did you learn from the Ice Storm of
'98?" provide valuable lessons:[32]

△ FIGURE 9.33 Transmission Towers Collapse Under Weight of Ice
A series of Hydro-Quebec high-voltage power transmission towers collapsed under the weight of ice accumulations during the January 1998 ice storm near Bruno, Quebec, south of Montreal. More than 5.2 million people lost power in the ice storm and some power outages lasted more than three weeks. *(AP Photo/ Jacques Boissinot)*

"I learned that we take things like power, heat, and water for granted." (student at Readfield Elementary School)

"People must have been together as families a lot more before there was electricity." (13-year-old student at South Meadow School)

"I learned that if you are in a big storm it helps if you help others." (12-year-old student at Woodside Elementary School)

"I learned that ice can cripple a whole state." (12-year-old student at Woodside Elementary School)

Source: Federal Emergency Management Agency. 1998. Disaster Connections Kids to Kids, "Kids from Maine Talk About Surviving the Ice Storm of '98." www.fema.gov/ kids/me98_04.htm. Accessed 1/20/07.

FOG

Like many other natural phenomena, there is nothing innately hazardous about **fog**; it is simply a cloud in contact with the ground. Fog can become hazardous when it obscures visibility for travel and other human activities and when air pollutants are added to form smog or vog (see Chapter 5).

Clouds and, thus, fog form either by air cooling until condensation occurs or by evaporation—adding water vapor to already cool air. Cooling can produce fog as heat radiates from the land at night, as warm moist air blows over cold coastal water, and as humid air rises up a mountainside. Fog can also develop by evaporation when cold air flows over river, lake, or coastal water that is warmer than the air and when warm rain falls through cool air along a frontal boundary.

In North America, the foggiest areas are the Pacific, New England, and Atlantic Canada coasts and the valleys and hills of the Appalachians. Fog contributed to the worst aircraft accident ever when 583 people died in the 1977 collision of two Boeing 747 airliners on a runway in the Canary Islands.[33] Dense fog was also responsible for the worst maritime accident in Canadian history when 1,014 people died in the 1914 collision of two ships on the St. Lawrence River.[34] Although technological advancements make collisions in the fog less likely, dense fog still contributes to numerous accidents, injuries, and deaths each year.

DROUGHT

Drought is defined as an extended period of unusually low precipitation that produces a temporary shortage of water for people, other animals, and plants. More than 1 billion people live in semiarid regions where droughts are common, and more than 100 million people are threatened with malnutrition or death if drought causes their crops to fail. Droughts commonly contribute to regional food shortages, but today the worldwide food distribution system is usually able to prevent drought from causing widespread famine in an area.

In the United States, drought affects more people than any other natural hazard causing serious water and hydroelectric power shortages, as well as agricultural problems. Since the late 1990s, moderate to extreme drought conditions have persisted across a significant area of the United States (Figure 9.34). Particularly hard hit by drought conditions are regions of the western United States where precipitation amounts are relatively low during normal conditions. During the summer of 2016, Lake Mead—the largest reservoir in the United States that provides drinking water and electicity to more than 25 million people in the states of Arizona, California, and Nevada—recorded its lowest water level since Hoover Dam was completed in 1936. The reservoir is currently at about 37 percent capacity,

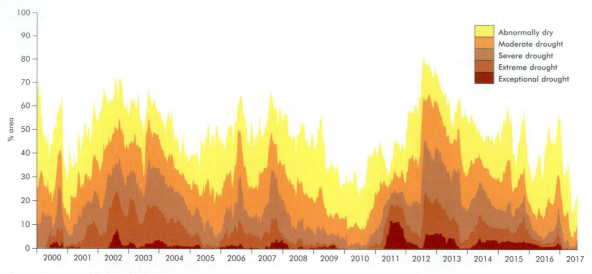

∧ FIGURE 9.34 United States Drought Data

United States time-series drought severity data since the beginning of the twenty-first century. Drought severity catagories are defined by probability of occurence over the next 100 years: Exceptional (1–2 percent), Extreme (2–5 percent), Severe (5–10 percent), Moderate (10–20 percent), Abnormally dry (20–30 percent). *(From the National Drought Mitigation Center)*

and was last full in 1983. Since 1999 the elevation of the water surface has plummeted more than 41 m (135 ft.) which can be clearly seen by the prominent "bathtub ring" preserved along its margins (Figure 9.35). This low water has caused about a about a 30 percent decrease in hydroelectric power output of Hoover Dam,

∧ FIGURE 9.35 Drought Depletion of Water Reservoirs

Behind Hoover Dam, a small fraction of the Lake Mead Reservoir is visible in the upper part of the photo. Sustained drought conditions have led to a significant reduction in water reserves and hydroelectric power generation over the past two decades. Rising 45 m (~150 ft) above the current lake level, the high-water mark is clearly visible as a white band rimming the reservoir.

which is the cheapest source of energy in the country. On average drought in the United States has caused an average revenue lose of approximately $7.8 billion per year, since 1980.[35]

MOUNTAIN WINDSTORMS

Mountain windstorms that develop seasonally on the downwind or lee side of mountain ranges and downslope of large glacial ice fields can be extremely destructive. Mountains act as barriers to prevailing winds and can, under specific temperature and pressure conditions, cause winds to roar down their slopes at speeds that can exceed 185 km (~115 mi.) per hour. Referred to as Chinooks east of the Rocky Mountains and Santa Ana winds west of the Transverse Ranges in southern California, these windstorms also occur downwind of the Sierra Nevada, Cascades, and mountain ranges on other continents such as the Alps of Europe. Mountain windstorms sometimes persist for a day or more, cause widespread roof and tree damage, can blow vehicles off highways, and contribute to large wildfires (see Chapter 13).

DUST STORMS AND SANDSTORMS

Dust storms are strong windstorms in which suspended dust that is carried by the wind reduces visibility for a significant period of time. Wind velocities in these storms exceed 48 km (~30 mi.) per hour, and visibility is reduced to less than 0.8 km (~0.5 mi.).[3] A typical dust storm is several hundred kilometers in diameter and may carry more than 100 million tons of dust. Most natural dust particles are pieces of minerals less than 0.05 mm (~0.002 in.) in diameter. Natural dust also contains minor amounts of biological particles, such as spores and pollen. In addition to being a safety hazard for travel, airborne dust particles can affect climate and human health. Once suspended in the air, fine dust particles of less than 0.01 mm (~0.0004 in.) can travel long distances in the upper atmosphere. Satellite images show dust storms from West Africa crossing the Atlantic Ocean to Florida. It is not uncommon to hear television weather reporters mistakenly call these events "sandstorms."

Unlike dust storms, *sandstorms* are almost exclusively a desert phenomenon in which sand is transported in a cloud of bouncing grains that rarely extends more than 2 m (~6.6 ft.) above the land surface.[36] Blowing sand is abrasive. One of your authors learned this firsthand when his windshield was frosted by blowing sand while driving in Colorado.

Dust storms and sandstorms occur mostly in midlatitude, semiarid, and arid regions. In the United States, huge dust storms in the southern High Plains during the 1930s produced conditions known as the "Dust Bowl." A combination of drought and poor agricultural practices during the Great Depression caused severe soil erosion in parts of five states centered on the Oklahoma panhandle. Frequent, sometimes daily, dust storms originated in this area and destroyed crops and pastureland (Figure 9.36).

HEAT WAVES

All of North America, and much of the world, is vulnerable to the effects of heat waves. In most areas, **heat waves** are considered to be prolonged periods of extreme heat that are both longer and hotter than normal. Since 2000, heat waves in the United States have killed

(a)

(b)

◀ **FIGURE 9.36 Dust Storms**
(a) Dust storm approaching Elkhart, Kansas, on May 21, 1937. Sometimes called "black blizzards" because of their color, these storms eroded topsoil from cropland in the southern High Plains during the Unites States "Dust Bowl." *(MPI/ Getty Images)* (b) Dust storm approaching Phoenix, Arizona in 2011. *(REUTERS/ Joshua Lott)*

◄ FIGURE 9.37 Heat Index Chart
This National Weather Service chart is used to calculate the heat index by combining information about air temperature and relative humidity. A similar chart, the Humidex Chart, is used by the Meteorological Service of Canada. (*From Excessive Heat Awareness and Safety, National Weather Service, NOAA, www. erh.noaa.gov/ rah/heat/*)

an average of 200 people per year, which is about equivalent to deaths from flooding, lightning, tornadoes, and hurricanes combined during the same 10-year period. Heat waves can be especially deadly in urban areas in cooler climates where residents are not acclimated to periods of prolonged heat and humidity. This was apparent in a heat wave in 2006, when more than 200 heat-related deaths occurred in the United States and Canada, and in August 2003, when there were an estimated 14,800 heat-related deaths in France (see Case Study 9.6: Europe's Hottest Summer in More Than 500 Years).[37]

Heat waves in the eastern United States and Canada are commonly associated with elongated areas of high pressure, called *ridges*. Wet conditions are generally found to the west of the ridge, whereas sunny, dry conditions prevail to the east. If such a ridge stays in place for several days, air temperatures below the ridge will rise to above-normal levels and cause a heat wave.

Heat waves can accompany either severe humidity or extreme dryness. In either case, it is important to monitor the **heat index** (Figure 9.37). This index measures the body's perception of air temperature, which is greatly influenced by humidity. For example, a temperature of 35°C (~95°F) will feel significantly hotter in parts of Florida when the relative humidity is 75 percent. In this example, the combination of high temperature and high humidity produces a heat index of 128, which is extremely dangerous. In addition to humidity, the length of time that someone is exposed to direct sunlight, the wind speed, and an individual's health will affect heat stress.

9.5 CHECK your understanding

1. What are the conditions necessary for both thunderstorms and severe thunderstorms to form? Describe the three stages of thunderstorm development.

2. What are supercells, mesoscale convective systems? And squall lines? How do they differ? Why are they significant natural hazards?

3. Characterize a tornado in terms of wind speed, size, typical speed of movement, duration, and length of travel.

4. Describe the five stages of tornado development.

5. What is a blizzard? How does a blizzard develop?

6. Describe the weather conditions that cause an ice storm.

7. How are the heat index and wind chill temperature alike? How are they different?

9.6 Human Interaction with Weather

Many natural hazards are clearly and significantly altered by human activities. For example, we have discussed how changes in land use affect flooding and landslides and how deep-well disposal and the filling of large water reservoirs may contribute to earthquakes. Land-use practices may also increase the effects of weather events. For example, the farming practice of plowing cropland after fall harvest and leaving the top-soil exposed to wind erosion during the winter significantly increased the size of dust storms in the Dust Bowl of the 1930s. Also, locating mobile homes in areas subject to frequent high winds and tornadoes greatly increases damages and loss of life from this type of severe weather. Finally, land-use practices in cities can intensify the effects of heat waves. Large areas of pavement, "big box" one-story buildings, and sparse park land contribute to the **urban heat island effect**, a local climatic condition in which a metropolitan area may become as much as 12°C (~22°F) warmer than the surrounding countryside (see Case Study 9.6: Europe's Hottest Summer).[7]

On a larger scale, human interaction with severe weather is taking place through global warming. On the basis of computer models, atmospheric scientists conclude that global warming is likely to increase the heat index and number of heat waves over land and the intensity of precipitation events in most areas.[38] Computer models also indicate that global warming is likely to increase the risk of drought in midlatitude continental interiors and increase wind and precipitation **intensities** in hurricanes, typhoons, and other tropical cyclones. The effect of global warming on small-scale events such as tornadoes, thunderstorms, hail, and lightning is still being studied. Overall, global warming will likely increase the incidence of severe weather.[39]

9.6 CASE study

Europe's Hottest Summer in More Than 500 Years

Heat waves are one of the deadliest of natural hazards but often get little publicity and are not long remembered. Many Americans have heard of the hurricane that destroyed Galveston, Texas, in 1900 and killed more than 6,000 people, but few people know that a year later, a midwestern July heat wave killed more than 9,500 people.[12] Even today, with rapid global communication, heat waves do not get as much attention as other less deadly natural hazards. This was even true for the 2003 European heat wave that killed more than 30,000 people—more than any other heat wave in recorded history.

Like most heat waves, the severe 2003 heat wave in Europe started gradually, many weeks before it peaked. Late spring and summer drought conditions set the stage for a long, hot summer. Record-setting temperatures began in Switzerland, breaking a 250-year record

for the hottest June ever reported.[40] Extreme daily high temperatures of 35° to 40°C (~95° to 104°F) continued through July in central and southern Europe (Figure 9.38). The heat wave peaked in early August, with August 10 the hottest day ever recorded in the United Kingdom—a high temperature of 38.1°C (~100.6°F) was measured in Gravesend, Kent, breaking records dating back to 1875.[41] That same day, Belgium had a high of 40°C (~104°F). Overall, climatologists have concluded that the summer of 2003 was probably the hottest in Europe in more than 500 years.[42]

In human terms, the 2003 heat wave was a disaster of major proportions. France was hit the hardest with more than 14,800 deaths, followed by 7,000 deaths in Germany, 4,200 in Spain, and 4,000 in Italy. Drought conditions severely affected agriculture, and Europe experienced more than 25,000 fires burning an estimated 647,069 hectares (1,599,000 acres) of forests. The global impact of the European drought and forest fires has been estimated at more than $17 billion.[40] Drought also had a major impact on power production: France, Europe's major energy producer, had to cut its power exports in half because of a lack of cooling water for its power plants.

European ecosystems were also significantly altered by the 2003 heat wave. Alpine glaciers lost up to 10 percent of their volume, the greatest loss of ice in any year on record. Ice also melted in the soil, as Alpine permafrost thawed to depths of up to 2 m (~6 ft.).[40] With permafrost ice no longer present to bind weathered rock, there were numerous rockfalls in the Alps. A rockfall on the Matterhorn, Switzerland's highest peak, forced the evacuation of 90 climbers.

As with many other heat waves, the 2003 event in Europe was enhanced by a stagnant pattern of atmospheric circulation. Stagnant air circulation centered around an anticyclone, a ridge of high atmospheric pressure, that remained nearly stationary over Western Europe for more than 20 days. The anticyclone produced nearly cloudless skies and a downward flow of dry air and also blocked rainfall from storms in the North Atlantic. Exceptionally dry soil moisture and clear atmospheric conditions lasted from May through August and contributed significantly to the heat wave.[43] With clear skies, solar radiation heated the ground day after day and dried out the soil. As soil moisture decreased, there was less water to evaporate and cool the air. Heat built up in the soil and then radiated back into the atmosphere each night to significantly warm nighttime temperatures. Hot days followed by hot nights greatly increased heat stress for many people.

For several reasons, the greatest number of deaths occurred in Paris and surrounding French communities. First, heat waves are generally worse in cities

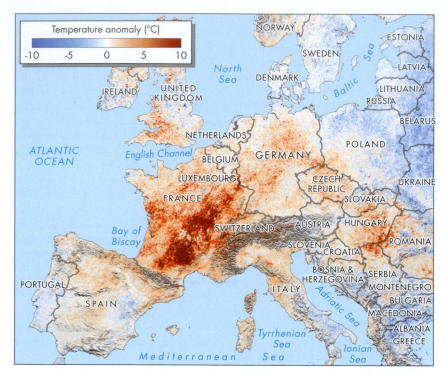

< FIGURE 9.38 Land Surface Heats Up in 2003 European Heat Wave
Difference between the daytime temperature of the land surface in July 2003 and the cooler, more normal temperatures of July 2001. Temperatures were measured by the Moderate Resolution Imaging Spectroradiometer (MODIS) on NASA's Terra satellite. The dark red areas were at least 10°C (18°F) hotter in 2003 than in 2001. *(NASA Earth Observing System)*

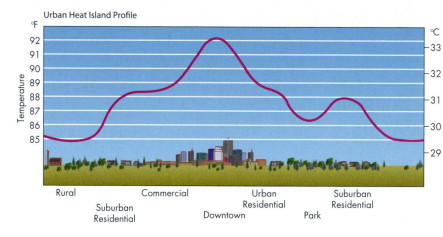

< FIGURE 9.39 Urban Heat Island Effect
Late afternoon temperature profile across urban, suburban, and rural areas on a summer day. The urban and suburban areas are significantly warmer than nearby rural areas, a phenomenon known as the urban heat island effect. This effect is greatest on calm, clear evenings when a city is often 5.6°C (~10°F) warmer than the surrounding countryside. *(Based on the U.S. Environmental Protection Agency, "Heat Island Effect." www.epa.gov/heatisland/about, 2006)*

because of the *urban heat island* effect. Environmental conditions associated with cities often make them 5.6°C (~10°F) warmer than surrounding natural areas (Figure 9.39).[44] The urban heat island effect is caused primarily by waste heat from vehicles, factories, and air conditioners; reduced airflow between buildings that traps heat; and more heat-absorbing concrete and other materials with less vegetation in urban areas. In contrast, evaporation from plant leaves and the soil in natural areas cools the air and mitigates processes that heat up urban areas. For Paris, the urban heat island effect during the first six days of August contributed to raising the air temperatures 7°C (~13°F) above normal (Figure 9.40).

A second reason that Paris and its suburbs experienced numerous casualties is that air pollution levels generally rise in cities during a heat wave. Stagnant air circulation associated with most heat waves allows air pollutants to accumulate and form photochemical smog. The smog contains ozone and other gases that cause or amplify respiratory problems. Finally, for Paris and other European cities, August is customarily a holiday month.

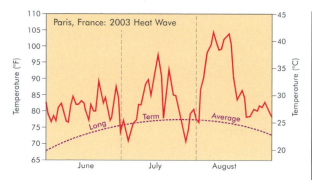

^ FIGURE 9.40 A Hot Time in Paris
The 2003 (orange line) and long-term average (purple line) daily high temperatures in Paris, France. Although the June and July daily high temperatures are not exceptional for many U.S. cities, they were a tremendous increase above the long-term average for Parisians. *(Based on the U.S. Environmental Protection Agency,* Excessive Heat Events Guidebook. *Publication EPA 430-B-06–005), 2006.)*

With many people on vacation, health services were short staffed and unable to handle the thousands of heat-related illnesses. Even morgues were unable to keep pace with the increased number of deaths.

Many Europeans wanted to know if the heat wave was caused by global warming. Meteorologists cannot tell if a single weather event is the result of changes that humans have caused in the atmosphere. They can, however, estimate the probability that human influence has increased the risk of a heat wave exceeding a given magnitude. For the 2003 European heat wave, computer climate models indicate that the greenhouse gases that humans have added to the troposphere (see Chapter 12) have doubled the probability of an event of this magnitude.[45]

LESSONS LEARNED

Warning and mitigation programs for heat waves have been slow to develop. Philadelphia, Pennsylvania, and Toronto, Ontario, have established model programs, and Lisbon, Portugal, and Rome, Italy, had systems in place before the 2003 European heat wave.

Warning Systems: At the time of the 2003 heat wave, most areas of Europe did not have heat health warning systems.[46] A similar situation existed in North America prior to the deadly 1995 heat wave in Chicago, Illinois. Today, a number of European cities and countries and more than 15 major metropolitan areas in the United States have heat health warning systems. These systems establish different criteria for warnings based on local meteorological conditions and the adaptation of

residents to high heat conditions. The conditions required to trigger excessive heat warnings in Phoenix, Arizona, which regularly experiences temperatures higher than 38°C (~100°F), differ from those in normally cooler Toronto.

Vulnerable Populations: The elderly, infants and young children, people with mental illness or mobility impairments, the homeless and poor, people engaged in vigorous outdoor exercise or work, and those who are socially isolated or physically ill are at greatest risk from periods of prolonged heat. Most of those who died in France in the 2003 heat wave were elderly, although an unusually high number of heat-related deaths occurred among younger people.[47] Vulnerable individuals need to be monitored regularly by caregivers or other concerned citizens for signs of heat exhaustion or heat stroke.

Cooling Systems: Heat waves, including the 2003 European event, are often accompanied by power failures. Loss of air conditioning and power for fans and refrigeration is especially problematic for people who live in urban areas, especially in high-rise buildings (Figure 9.41). Spending a few hours every day in air conditioning can lower heat stress. Mobile cooling centers, such as city buses, can be routinely dispatched to publicized locations to provide heat relief to urban residents.

Public Education: Government officials and the mass media must provide the public with timely and consistent information about heat conditions and mitigation

^ FIGURE 9.41 Parisians Cool Off in Record Heat
Children and adults cool off in the Trocadero Fountains near the Eiffel Tower in Paris on August 12, 2003. Temperatures in Paris reached 40°C (104°F). *(AP Photo/Franck Prevel)*

▼ TABLE 9.2 Personal Adjustments for Heat Waves

Do
• Minimize direct exposure to the sun
• Use air conditioners or spend time in air-conditioned places
• Use portable fans to exhaust hot air from or draw cooler air into rooms
• Take a cool bath or shower
• Stay hydrated by regularly drinking water or other nonalcoholic beverages
• Eat light, cool, and easy-to-digest foods such as fruit or salads
• Wear loose-fitting, light-colored clothes
• Check on older, sick, or frail people who may need help adjusting to the heat
• Know the symptoms of heat exhaustion and heat stroke and the appropriate responses

Don't
• Direct air from fans toward yourself when room temperature is hotter than 32°C (~90°F)
• Leave children and pets alone in cars for any amount of time
• Drink alcoholic beverages to try to stay cool
• Eat heavy, hot, or hard-to-digest foods
• Wear heavy or dark clothing

Source: Based on U.S. Environmental Protection Agency, *Excessive Heat Events Guidebook*. Publication EPA 430-B-06–005, 2006.

actions. Use of a heat index, which combines temperature and humidity (see Figure 9.37), is critical to communicating conditions that could result in heat-related illness. Individuals can take a number of actions (Table 9.2) to reduce their heat stress and prevent medical conditions caused by excessive heat exposure, such as heat cramps, heat exhaustion, and heat stroke (sunstroke).

PROSPECTS FOR THE FUTURE

Climate studies indicate that additional extreme heat events similar to the European heat wave of 2003 are likely to occur in the twenty-first century. Atmospheric scientists have calculated that, because of continued global warming, it is 100 times more likely that an event similar to the 2003 heat wave will take place in the next 40 years.[45] Environmental and public health agencies in the United States, Canada, and elsewhere have established programs to prepare for extreme heat events. In particular, these agencies advocate long-term programs to reduce the urban heat island effect that intensifies heat waves. Methods to reduce the effect include increasing the reflectivity of urban surfaces, such as by installing light-colored rooftops; increasing the amount of pervious pavement to provide the air access to moisture in the soil; and adding vegetation to urban areas, even to the flat roofs of buildings.

9.6 CHECK your understanding

1. How is global warming expected to affect severe weather?

2. Explain how drought, soil moisture, and a heat wave can be interrelated.

3. What causes the urban heat island effect? How can this effect be mitigated?

9.7 Linkages with Other Hazards

Severe weather is directly linked to short-term events such as flooding, mass movements, and wildfires, and to long-term changes in global climate. Linkages related to large tropical and extratropical cyclones are discussed in Chapter 10.

Flooding, often flash flooding, can be produced by one or more intense, slow-moving thunderstorms that produce a large amount of rain in a relatively short time (see Chapter 6). Thunderstorms move slowly if prevailing winds are relatively stagnant. Light prevailing winds can cause essentially the same storm, such as a mesoscale convective system, to remain fixed over a geographic area, or it can cause stagnation of a frontal boundary, such as a stationary front. Since thunderstorms commonly move parallel to a stationary front, the stagnation of a front can cause storms to track over the same area, dropping large amounts of rainfall. In 2016, for example, a week-long period of sustained thunderstorm activity along a stationary front near the capital city of Lousiana, Baton Rouge, caused local rainfall totals in excess of 61 cm (~24 in.), and about one-third of the state being declared a federal disaster area due to flooding. Stagnation of thunderstorms can also occur over mountain foothills and cause deadly flash floods, like those in Rapid City, South Dakota, in 1972; Big Thompson Canyon west of Longmont, Colorado, in 1976; and Fort Collins and Boulder, Colorado, in 1997 and 2013.

Thunderstorms also produce lightning, the primary, natural ignition source for wildfires (see Chapter 13). However, not all lightning strikes in undeveloped areas cause wildfires. Only an estimated one out of every four cloud-to-ground lightning strikes has the continuous current necessary to start a wildfire, and then the fire may spread only if moisture conditions are low.[48]

Drought, dust/sandstorms, and heat waves are all hazardous weather events that may become more prevalent in some regions with global warming (see Chapter 12). Heat waves in the Arctic have already contributed to the melting of permafrost, which in turn has caused subsidence (see Chapter 7) and the release of additional greenhouse gases (see Chapter 12).

9.8 Natural Service Functions of Severe Weather

For many of us, it may be difficult to envision any benefits from severe weather; this is probably because most of the natural service functions are long term and not obvious. For example, lightning is the primary ignition source for natural wildfires, and wildfires are a vital process in prairie, tundra, and forest ecosystems (see Chapter 13). Windstorms also help maintain the health of forests. These storms topple dead and diseased trees, which then are recycled in the soil. Fallen trees also create clearings that become new habitats for diverse plants and animals. Even ice storms, which appear to be so destructive to trees, are part of a natural ecological cycle that increases plant and animal diversity in the forest.

In the hydrologic cycle, blizzards and other snowstorms, thunderstorms, and tropical storms are important, and in some areas these storms constitute the primary water source for an area. A continual supply of water from snowmelt and seasonal rainfall reduces a region's vulnerability to drought.

Finally, there are some uniquely human benefits to severe weather. Snowfall, cloud formations, and lightning displays have long had an aesthetic value. Severe weather also excites many people. Thanks to movies and television, tornado chasing has become a popular avocation and a new form of tourism. Guided expeditions in specially equipped vehicles regularly drive into "Tornado Alley" to chase and photograph tornadoes. Tornado chasing, however, can be extremely dangerous—serious injury, and even death, can occur if a vehicle actually encounters a tornado.

9.8 CHECK your understanding

1. Describe some of the natural service functions of severe weather.

9.9 Minimizing Severe Weather Hazards

Thunderstorms, tornadoes, mountain windstorms, ice storms, blizzards, and heat waves will continue to threaten human lives and property. As long as people live in the path of such hazards, we must take steps to minimize the damage and loss of life associated with them. We must be able to predict these events accurately in order to reduce their hazard.

FORECASTING AND PREDICTING WEATHER HAZARDS

Timely and accurate prediction of severe weather events is extremely important if human lives are to be spared. Even with improvements in satellite sensors and computer modeling, severe weather events are still difficult to forecast, and their behavior is unpredictable. A network of *Doppler radar* stations across North America has significantly improved our ability to predict the path of severe storms. Doppler radar antennas send out electromagnetic radiation that has a wavelength a little longer than microwaves (see Figure 9.5). Clouds, raindrops, ice particles, and other objects in the sky reflect these electromagnetic waves back to a receiving antenna. The wavelength of the reflected waves changes depending on whether the objects are moving toward or away from the antenna. This change in wavelength, called the *Doppler effect*, is similar to the difference in the pitch of sound waves as an ambulance siren approaches you and then goes away from you. The changes in radar wavelength are analyzed and can be used to make short-term predictions about the weather, on the scale of hours. For example, Doppler radar can detect a mesocyclone within a thunderstorm and allow meteorologists to issue some tornado warnings up to 30 minutes in advance (see Applying the 5 Fundamental Concepts).

Watches and Warnings You may have heard on the news that a tornado **watch** has been issued for a given area. A tornado watch warns the public of the possibility of a tornado, or tornadoes, developing in the near future. A typical tornado watch might include an area of 52,000 to 104,000 km^2 (~20,000 to 40,154 mi.^{2}) and last from 4 to 6 hours.[49] A watch does not guarantee that the event will occur; rather, it alerts residents to the possibility of a severe weather event and suggests that they monitor local weather conditions and listen to radio or television stations for more information.

When a tornado has actually been sighted or detected by weather radar, the watch is upgraded to a **warning**. A warning indicates that the area affected is in danger, and people should take immediate action to protect themselves and others. Watches may be upgraded to warnings, or warnings may be issued for an area not previously under a watch. Both watches and warnings may be issued for any type of severe weather—severe thunderstorms, tornadoes, tropical storms, hurricanes, heat waves, blizzards, and others, with some variation in the area covered and duration of the watch or warning.

People's perception of the risk of severe weather hazards differs according to their experience. Someone who has survived a tornado is more likely to perceive the hazard as real than someone who has lived in a region at risk for tornadoes yet has never experienced one. Incorrect predictions of where or when a hazard will strike may also lower risk perception. For example, if people are repeatedly warned of severe thunderstorms that never arrive, they may become complacent and ignore future warnings. As with any other hazard, accurate risk perception by planners and the public alike is key to reducing the threats associated with severe weather events.

Finally, our current theoretical understanding and monitoring technology do not always allow accurate prediction of the intensity and extent of many severe weather events. For example, even though less than 1 percent of tornadoes, the severe EF4 and EF5 storms, are responsible for half of tornado-caused deaths, meteorologists are not yet able to include predictions of intensity in official tornado watches and warnings.[50] Likewise, predicting the amount of icing and depths of snow accumulation in winter storms is often difficult. However, the National Weather Service and private-sector meteorologists are developing computer-based techniques for making real-time predictions of the intensity of some severe weather events.[51] Referred to as *nowcasting*, these techniques use information from weather radar, satellites, and automated weather stations to predict the path and development of severe storms once they have formed.

ADJUSTMENT TO THE SEVERE WEATHER HAZARD

Although we cannot control Earth's atmospheric system to prevent severe weather, we can take a number of steps to reduce the associated death and damage. These actions include both long-term changes to community infrastructure and plans or procedures to be implemented for severe weather. Long-term actions to prevent or minimize death, injuries, and damage are considered *mitigation*. Mitigation activities include the safety-conscious engineering and building of structures, the installation of warning systems, and the establishment of hazard insurance. Establishing community and individual plans and procedures to deal with an impending natural hazard is considered *preparedness*.[52]

Mitigation Although mitigation techniques differ for each weather hazard, some general statements can be made. Building new structures and modifying existing buildings can save lives and protect property from weather hazards. For example, windproofing buildings may significantly reduce damage from severe storms such as tornadoes, derechos, mountain windstorms, and hurricanes. In the United States, the Federal Emergency Management Agency (FEMA) offers grants and architectural plans to establish community shelters and safe rooms in buildings for tornado protection.[53]

Ensuring that electric, gas, water, wastewater, and communications systems continue to function following storms or severe winter weather is a continuing challenge. For example, electrical transmission lines are generally designed to survive ice storms that have a 50-year recurrence interval in the area that they serve.[30] However, in wooded areas, winds can combine with ice storms to cause power outages for weeks because ice-loaded branches fall on low-voltage power lines. For many water and wastewater systems, the use of pump stations makes them vulnerable to power interruptions caused by lightning, wind, and ice storms. Without backup sources that can provide power for days until a power grid is restored, these systems remain vulnerable to weather-related interruption. Finally, communication systems have the greatest potential to survive severe weather events if they have redundancy. A combination of landline telephone, cell phone, voice-over-Internet, multifrequency radio, and satellite communication links has the best potential for surviving various combinations of wind, water, ice, and lightning damage.

Other mitigation techniques include developing and installing warning systems and ensuring that universal hazard insurance is available. The goal of warning systems is to give the public the earliest possible notification of impending severe weather. Announcements can be made by commercial radio, television, and the

Internet; by U.S. and Canadian government weather radio broadcasts; and by local warning sirens. Some local U.S. and Canadian governments have recently installed community notification systems that call telephone subscribers, including registered mobile/Internet phone users, and warn them of emergencies.

Finally, insurance policies should be available to property holders living in regions at risk for weather disasters. Basic policies cover damage from water and wind but often exclude specific types of hazards such as flooding. Residents of risk areas should determine whether extra coverage is required for a severe storm, tornado, hurricane, blizzard, or other natural hazards.

Preparedness and Personal Adjustments

Individuals can take several steps to prepare for severe weather. Many of these steps can and should be carried out before a watch or warning is issued, whereas others are more appropriate when the danger is imminent. In areas prone to severe weather, people should be aware of the times of year that are most hazardous and make adequate preparations for themselves and their home. Information about how to prepare for various weather-related disasters is available from the U.S. National Oceanic and Atmospheric Administration (NOAA) and its subsidiary, the National Weather Service; from FEMA; and from Environment Canada and its subsidiary, the Meteorological Service of Canada. For example, knowing when and where to take shelter during a storm and what supplies should be kept in emergencies would be useful for more than one severe weather event.

Wearing proper clothing and modifying travel plans are prudent adjustments to many types of severe weather. Protection from direct sunlight in heat waves and the need for insulated clothing in snowstorms and blizzards are intuitive for most adults. Where people often are not prepared is in cool (5° to 15°C [41° to 59°F]), wet, and windy weather during which the human body can lose heat rapidly by evaporation from the skin and lungs and by contact with cold air and water. Rapid loss of core body heat causes intense shivering, loss of muscle coordination, mental sluggishness, and confusion, a medical condition known as *hypothermia*. In cool, wet, and windy weather, an improperly dressed person can show symptoms of hypothermia within several hours. People can become hypothermic immersed in 25°C (~77°F) water, or if improperly dressed and wet, in 20°C (~68°F) air temperature.[54,55] Unless body heat is restored, the continued loss of heat from a hypothermic person will cause unconsciousness and eventually death.

Low-visibility conditions can develop rapidly in dust storms, blizzards, and fog and are especially hazardous for travel on freeways and other roads where vehicles normally travel at high speed. For example, on average in the United States 40,000 highway accidents each year occur in fog conditions, resulting in nearly 640 deaths and more than 16,000 injuries.[56] When traveling by automobile, "onward through the fog" is not the best course of action.

9.9 CHECK your understanding

1. What is the difference between a severe weather watch and warning?
2. How do preparedness planning and mitigation differ?
3. List ways to personally prepare for a severe weather event.

APPLYING the 5 Fundamental Concepts

Moore, Oklahoma, Tornadoes

1 **Science helps us predict hazards.**

2 **Knowing hazard risks can help people make decisions.**

3 **Linkages exist between natural hazards.**

4 **Humans can turn disastrous events into catastrophes.**

5 **Consequences of hazards can be minimized.**

The United States is the tornado capital of the world, experiencing about 1,000 tornadoes per year, with Canada in second place with about 100 per year.[57] About 40 percent of those tornadoes occur in states within "Tornado Alley," such as Oklahoma (see Figure 9.25). That said, the probability of your town being struck by a tornado is only a few percent on any particular day during the spring months, even in Tornado Alley—and the chances of being hit by a powerful tornado is even less. The webpage of the city of Moore, Oklahoma, discusses the powerful 1999 tornado (see the chapter opener) and states, "May 3 was an extremely unique event weather wise," and "There's an extremely small chance of Moore experiencing another 'May 3' type event."[58] Statistically speaking this is a true statement, but a 1 percent chance per day that parts of your town could be completely leveled still represents a significant risk to life and property. This point was made 14 years later when a another EF5 tornado destroyed about 1,200 homes and killed 23 of Moore's residents.[59]

The tornado that struck Moore on May 20, 2013, was only one of 22 tornadoes that touched down that day but was the most destructive of the bunch. The tornadoes were spun out of a severe thunderstorm system that generated a total of 60 tornadoes and caused flooding in numerous states across the Great Plains (Figure 9.42). The tornado that struck Moore had the destructive power of an EF5. It was on the ground for 39 minutes and cut a 27 km (~17 mi.) long path that was as wide as 2 km (~1.3 mi.) through parts of the city.[60] Fortunately, wind velocities along much of its path were below 322 km (~200 mi.) per hour and only reached EF5 strength for a short distance (Figure 9.43).

For several days the National Weather Service (NWS) had been forecasting a moderate risk of severe thunderstorms capable of producing baseball-sized hail with damaging winds and the potential for a few tornadoes. On the morning of May 20, the NWS issued a severe thunderstorm warning through 3:00 P.M. and a tornado watch through 10:00 P.M. for 30 counties in central Oklahoma. At 2:40 P.M. a tornado warning was issued for Moore and Newcastle, and by 2:52 P.M. Doppler radar had detected a distinct hook echo—the telltale sign of vertical axis rotation of a supercell with tornadic winds reaching for the ground (Figure 9.44).[61]

Moore is an NWS certified "Storm Ready City," which means the city (1) has a 24-hour warning and emergency operation center with multiple methods for receiving severe weather warnings and forecasts and for alerting the public; (2) has its own local weather monitoring system; (3) promotes public readiness through community seminars; and (4) has developed a formal hazardous weather plan, which includes trained weather spotters and regular emergency exercises. As soon as the tornado warning was issued, Moore's system of 36 sirens went off, automated mass calls to phone numbers, text messages, tweets, and emails went out, in addition to radio and TV public announcements.[62]

The tornado touched down 12 minutes after the warning went out. Nationally, the average

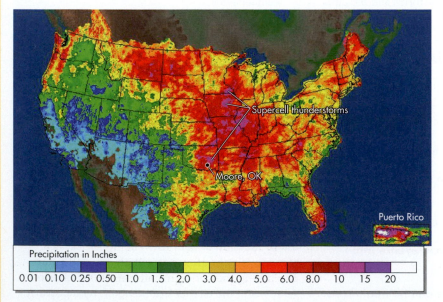

◄ FIGURE 9.42 Heavy Rain and Flooding
Doppler radar image showing heavy precipitation across the United States and supercell thunderstorm outbreaks in the mid-continental region on May 20, 2013. *(NOAA)*

Precipitation in Inches
0.01 0.10 0.25 0.50 1.0 1.5 2.0 3.0 4.0 5.0 6.0 8.0 10 15 20

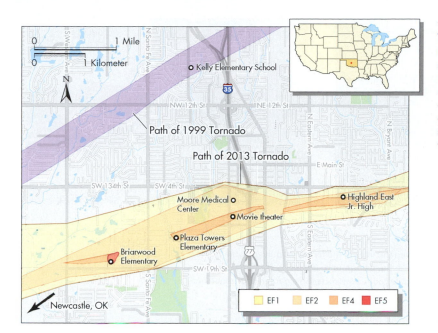

> **◀ FIGURE 9.43 Moore Tornado Paths**
> Map of Moore Oklahoma showing the paths of the 1999 and 2013 EF5 tornadoes and the locations of some of the key landmarks that experienced disastrous consequences due to high wind velocity.

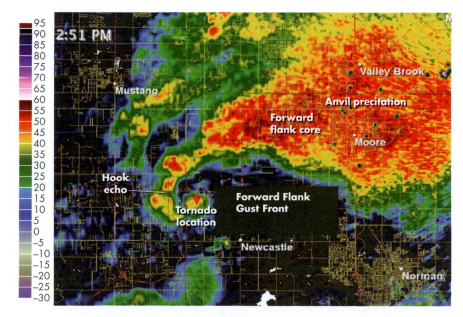

> **◀ FIGURE 9.44 Hook Echo Indicates Tornado**
> Doppler radar image of super cell thunderstorm passing over southeastern Oklahoma, and the city of Moore. Meteorologists have come to recognize that a characteristic hook shape develops in the southwest corner of supercell thunderstorm when tornadic wind conditions ideal for tornado formation are present. *(NOAA)*

early warning lead time is 13 minutes; when Moore was hit in 1999, lead time was about 10 minutes.[63] Many Moore residents had more than 30 minutes of lead time before the tornado struck densely populated regions of the city—in theory, long enough for everyone to get out of the tornado's path, assuming they had some place safe to go.

Although Moore has more experience with EF5 tornadoes than most cities, less than one-tenth of its homes have basements or below-ground shelters.[58] This lack of underground shelters in Oklahoma is largely the result of the underlying geology of the region. Specifically, the soil that underlies much of central Oklahoma is clay rich, which is problematic for building

(a)

(b)

⋀ FIGURE 9.45 Tornado Shelters
(a) Open hatch to below ground shelter that saved the lives of the family who lived in the home next to the shelter, the remnants of can be seen in the background. *(Photo by Paul Hellstern, The Oklahoman, copyright 2013)* (b) Above-ground safe room constructed in a garage of a house that was paritally destroyed by the 2013 Moore tornado. *(Julie Dermansky/Corbis)*

below-ground structures because of the way clay expands and contracts due to wetting and drying (see Chapter 8). Underground shelters can be built (Figure 9.45a) but may cost $15,000 to $20,000 to be correctly engineered for this type of soil, making them prohibitively expensive for most families. Although below ground is the safest place to be in a tornado, above-ground shelters are the less expensive route ($4,000–5,000), and all 16 above-ground shelters in the tornado's path survived (Figure 9.45b).[64] Unfortunately, these structures are small, usually holding less than six people, and are not appropriate for large groups of people gathered at a school or office building.

There are no community shelters in Moore, and most large buildings are not equipped with shelters big enough to hold all occupants. People at the elementary schools and the medical center (Figure 9.43), which laid along the path of the tornado, had no safe place to go other than into the interior of the buildings, as far away from outside walls and windows as possible. Briarwood Elementary and a few homes nearby took a direct hit from EF5 winds and were completely obliterated (Figure 9.46a), but fortunately the school and most of the homes had been evacuated. Unfortunately, Plaza Towers Elementary

(Figure 9.43), hit just a few minutes later, had not been completely evacuated and seven children were killed. The 1999 tornado similarly destroyed Kelley Elementary (Figure 9.43), but it has since been rebuilt with "safe hallways" with overhead doors that can be shut to enclose the halls, making them safe while the tornado passes overhead. The school district superintendent has proposed a similar plan for the schools destroyed in 2013 when they are reconstructed.[65]

Compared to the 1999 tornado, the loss of life due to the 2013 tornado was relatively light, killing one-third less people. This is in part due to a greater warning lead time as well as differences in the location and length of the tornado path. Some might say that Moore got lucky in 2013, while others might say the opposite. What we do know is that since 1999 the population of Oklahoma has grown by about 10 percent and the population of Moore has grown by about 30 percent.[66] Not surprisingly, the Moore school district, which includes Briarwood and Plaza Towers Elementary schools, is the fastest-growing district in Oklahoma.[65] As you will learn in Chapter 12, climate change will likely bring more tornadoes to this region in the future, and it seems inevitable that the citizens of Moore will be confronted with yet another destructive tornado.

(b)

< FIGURE 9.46 2013 Moore Tornado Devastation
(a) Arial view of the remains of Briarwood elementary after taking a direct hit by EF5 velocity winds. *(Tony Gutierrez/AP Images)*
(b) Cars piled up outside the heavily damaged Moore Medical Center. See Figure 9.43 for location of these landmarks. *(Brett Deering/ Getty Images)*

APPLYING THE 5 FUNDAMENTAL CONCEPTS

1. What precursor events and scientific tools were used to forecast the 2013 tornado? Discuss how successful or unsuccessful the forecasting and warnings were for Moore in 2013.

2. Explain the tornado risk you would face if you bought a home and moved your family, including young children, to Moore. How might you limit your personal tornado risk?

3. What other natural hazards are linked to tornadoes in Moore, Oklahoma?

4. Discuss the impact of population growth in Oklahoma and Moore with regard to tornado hazard. List ways that Moore could limit future loss of life due to tornadoes even with continued population growth.

5. Describe Moore's warning system and the city's preparedness to minimize the tornado hazard. How successful were these techniques and tools for saving lives and preventing property damage? If money was not an obstacle, what actions would you recommend to prevent future loss of life?

CONCEPTS in review

9.1 Energy

LO:1 Differentiate the three ways in which heat is transferred in the atmosphere.

- Energy can simply be thought of as the ability to do work. To move heavy boxes from one place to another, you must expend energy in order to do the work—energy is the force you must apply to the boxes to pick them up and move them.
- Three primary forms of energy are potential, kinetic, and heat energy. Potential energy is stored energy; kinetic energy is energy of motion; and heat energy is the energy produced by the random motion of molecules.
- In the atmosphere, heat energy occurs as sensible heat that can be measured and latent heat that is stored. Latent heat is either absorbed or released in phase changes. Evaporation absorbs latent heat and cools the air, whereas condensation, which forms clouds, releases latent heat and warms the air. Energy transfer occurs by convection, conduction, and radiation. Of these, convection is the most significant in producing clouds and severe weather.

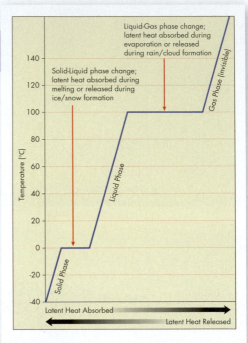

KEY WORDS

conduction (p. 347), **convection** (p. 348), **latent heat** (p. 347), **radiation** (p. 348)

9.2 Earth's Energy Balance

LO:2 Discuss how Earth's energy balance and energy exchanges produce climate and weather.

- Earth receives energy from the sun, and this energy affects the atmosphere, oceans, land, and all living things before being radiated back into space.
- Earth receives primarily short-wavelength, electromagnetic energy from the sun. This energy is reflected, scattered, transmitted, or absorbed on Earth. Dark-colored surfaces reflect less and absorb more solar energy and thus generally heat up. Most absorbed solar energy is radiated back into the atmosphere as long-wavelength, infrared radiation.

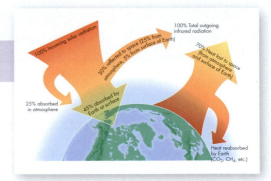

9.3   The Atmosphere

LO:3   Compare and contrast the different layers of Earth's atmosphere.

- Most weather occurs in the troposphere, the lowest of the five major layers of the atmosphere.
- Clouds are formed primarily in the troposphere and are composed of water droplets and ice crystals.

KEY WORDS

atmosphere (p. 350), **relative humidity** (p. 351), **troposphere** (p. 351)

9.4   Weather Processess

LO:4   Explain the different types of severe weather events and discuss the local weather conditions that cause these events.

- Changes in atmospheric pressure and temperature are responsible for air movement. Air flows from areas of high pressure to low pressure. Convergence of air produces low atmospheric pressure and divergence produces high pressure.
- Atmospheric stability is the tendency of a parcel of air to remain in place or change its vertical postion. The atmosphere is unstable if air parcels rise until they reach air of similar temperature and density. Severe weather is associated with unstable air.
- Winds blowing over long distances curve because Earth rotates beneath the atmosphere. Called the Coriolis effect, this curvature is to the right in the Northern Hemisphere. Low-level wind patterns and the jet streams are controlled by horizontal changes in atmospheric pressure; the Coriolis effect; and, for surface winds, friction.
- Boundaries between cooler and warmer air masses, called fronts, are described as cold, warm, stationary, or occluded. Many thunderstorms, tornadoes, snowstorms, ice storms, and dust storms are associated with fronts.

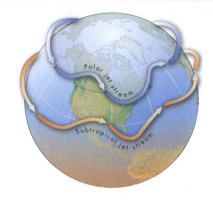

KEY WORDS

Coriolis effect (p. 357), **fronts** (p. 358), **jet streams** (p. 355)

9.5   Hazardous Weather and Geographic Regions at Risk

LO:5   Explain how movement of air masses control the distribution of different severe weather hazards.

- Rainstorms with lightning and thunder, called thunderstorms, form where there is moist air in the lower troposphere, rapid cooling of rising air, and updrafts to create cumulonimbus clouds. Thunderstorm development proceeds through cumulus, mature, and dissipative stages. Most thunderstorms form during maximum daytime heating, either as individual air-mass storms or as lines or clusters of storms associated with fronts.

KEY WORDS

blizzards (p. 370), **drought** (p. 375), **dust storm** (p. 376), **EF Scale** (p. 367), **fog** (p. 375), **hail** (p. 365), **heat index** (p. 378), **heat waves** (p. 377), **ice storms** (p. 372), **lightning** (p. 364), **supercell storm** (p. 361), **vertical wind shear** (p. 361), **wind chill** (p. 372)

- Severe thunderstorms have winds exceeding 93 km (~58 mi.) per hour, hail diameters greater than 1.9 cm (~0.75 in.), or a tornado. Severe thunderstorms form where there is strong vertical wind shear, uplift of air, and dry air above moist air. The three major types of severe thunderstorms are mesoscale convective systems (MCS), linear squall lines, and large individual storms called supercells.
- Lightning is an underestimated safety hazard associated primarily with thunderstorms. Large differences in electrical charge develop within a thunderstorm cloud or between the cloud and the ground. Flow of electrical charges within the cloud or between the ground and the cloud produces lightning.
- Tornadoes are generally funnel-like columns of rotating winds of 65 to more than 450 km (~40 to more than 280 mi.) per hour that extend downward from a cloud to the ground. With diameters in tens of meters, these storms usually travel at 50 km (31 mi.) per hour from southwest to northeast. Damage by tornadoes is rated on the Enhanced Fujita (EF) Scale, with EF5 as the most severe.

- Blizzards are severe winter storms in which large amounts of falling or blowing snow reduce visibilities for extended periods of time. Safety during cool and cold weather is based on the wind chill temperature, a combination of air temperature and wind speed.
- Ice storms occur with prolonged freezing rain along a stationary or warm front, below-freezing surface temperature, and a shallow layer of cold air at the surface. Sleet and snow occur where there is a progressively greater thickness of cold air at the surface.
- Weather conditions such as a dust storm or fog can greatly reduce visibility, resulting in deadly accidents. Fog is simply a cloud in contact with the ground. Other hazardous weather conditions include sandstorms, mountain windstorms, droughts, and heat waves.
- Drought and heat waves are often linked to high-pressure centers that stagnate over regions for extended periods of time. Heat waves are intensified in cities because of the urban heat island effect. Safety in hot weather requires knowing the heat index, which is based on a combination of air temperature and relative humidity.

9.6   Human Interaction with Weather

LO:6   Discuss how humans interact with severe weather.

- Potential human interactions with weather and its hazards are varied. At the local level, land use such as type of housing, landscaping, and agricultural practices may increase the effect of severe weather. On the global scale, atmospheric warming in response to burning of fossil fuels is changing our planet's weather systems. This warming of both the atmosphere and the oceans may feed more energy into storms, potentially increasing the incidence of severe weather events.

KEY WORDS
intensities (p. 379), **urban heat island effect** (p. 379)

9.7 and 9.8   Linkages with Other Hazards Natural Service Functions of Severe Weather

LO:7   Summarize the effects of severe weather events, as well as their linkages to other natural hazards.

LO:8   Recognize some natural service functions of severe weather.

- Severe weather is directly linked to short-term events such as tropical and extratropical cyclones, flooding, mass movements, and wildfires, as well as to long-term changes in global climate.
- Benefits of severe weather for the natural environment are primarily related to the transfer of water in the form of rain or snow. Fires from lightning strikes benefit landscapes in the long term.

9.9   Minimizing Severe Weather Hazards

LO:9   Identify how we can minimize the effects of severe weather hazards.

KEY WORDS
warning (p. 384), **watch** (p. 384)

- Minimizing hazards associated with severe weather such as thunderstorms, tornadoes, heat waves, droughts, blizzards, and ice storms requires a multifaceted approach. This approach should include (1) more accurate prediction that leads to better forecasting and warnings; (2) mitigation techniques designed to prevent or minimize death and loss of property, such as constructing buildings to better withstand severe weather; (3) hazard preparedness, such as short-term activities that individuals and communities can take once they have been warned of severe weather; and (4) education and insurance programs to reduce risk.

CRITICAL thinking questions

1. What severe weather events are potential hazards in the area where you live? What are some steps you might take to protect yourself from such hazards? Which of these hazards is your community the least prepared for?
2. Tornadoes can often be spotted on weather radar, whereas many other clouds cannot. What makes tornadoes visible?
3. Study the diagrams of cold fronts and warm fronts, and read the description about the development of ice storms. Explain why sleet (small pellets of ice) is more likely to accompany cold fronts than is freezing rain.
4. Why does hail form in thunderstorms and not in other rainstorms or snowstorms?
5. Has your community ever experienced a heat wave? If so, when did it occur and how were people affected? Does your community have a heat health warning system? If so, what actions do local officials and emergency personnel take when the system is activated? If not, what type of system would you recommend? What actions could you take to mitigate the effects of a heat wave on your living conditions?

References

1. **National Oceanic and Atmospheric Administration.** The Great Plains Tornado Outbreak of May 3–4, 1999: Storm A Information. National Weather Service Weather Forecast Office, Norman, Oklahoma. www.srh.noaa.gov/oun/?n=events-19990503-storma. Accessed 6/20/13.
2. **Smith, G. A.,** and **Pun, A.** 2006. *How does Earth work? Physical geology and the process of science.* Upper Saddle River, NJ: Pearson Prentice Hall.
3. **Smith, J.,** ed. 2001. *The facts on file dictionary of weather and climate.* New York: Facts on File, Inc.
4. **Lutgens, F. K.,** and **Tarbuck, E. J.** 2007. *The atmosphere: An introduction to meteorology,* 10th ed. Upper Saddle River, NJ: Pearson Prentice Hall.
5. **Ahrens, C. D.** 2005. *Essentials of meteorology: An invitation to the atmosphere.* Belmont, CA: Thomson Brooks/Cole.
6. **Burt, Christopher C.** 2004. *Extreme weather: A guide & record book.* New York: W. W. Norton.
7. **Aguado, E.,** and **Burt, J. E.** 2007. *Understanding weather and climate,* 4th ed. Upper Saddle River, NJ: Pearson Prentice Hall.
8. **Christopherson, R. W.** 2006. *Geosystems: An introduction to physical geography,* 6th ed. Upper Saddle River, NJ: Pearson Prentice Hall.
9. **Ackerman, S. A.,** and **Knox, J. A.** 2003. *Meteorology: Understanding the atmosphere.* Pacific Grove, CA: Thomson Learning.
10. **Ashley, W. S.,** and **Mote, T. L.** 2005. Derecho hazards in the United States. *Bulletin of the American Meteorological Society* 86: 1577–92.
11. **American Meteorological Society.** 2002. Updated recommendations for lightning safety— 2002. www.ametsoc.org/policy/Lightning_Safety_Article.pdf.
12. **Rauber, R. M., Walsh, J. E.,** and **Charlevoix, D. J.** 2005. *Severe & hazardous weather: An introduction to high impact meteorology,* 2nd ed. Dubuque, IA: Kendall/Hunt.

13. **National Weather Service.** 2004. Lightning—the underrated killer. www.lightningsafety.noaa.gov/overview.htm. Accessed 7/26/06.

14. **National Weather Service.** 2004. Lightning risk reduction outdoors. www.lightningsafety.noaa.gov/outdoors.htm. Accessed 7/26/06.

15. **National Weather Service.** 2008. Natural hazard statistics. www.nws.noaa.gov/om/hazstats.shtml. Accessed 1/11/09.

16. **Blumer, Brad.** 2012. A short history of violent tornadoes in the United States. *Washington Post*, May 21.

17. **Grazulis, T. P.** 2001. *The tornado: Nature's ultimate windstorm.* Norman: University of Oklahoma Press.

18. **National Weather Service.** 2006. *The Enhanced Fujita Scale (EF Scale).* Norman, OK: NOAA/National Weather Service Storm Prediction Center. www.spc.noaa.gov/efscale/. Accessed 9/17/06.

19. **Environmental Canada.** 2004. Tornados. www.pnrrpn.ec.gc.ca/air/summerservere/ae00s02.en.html. Accessed 7/26/06.

20. **Wilson, J. W.,** and **Changnon, S. A. Jr.** 1971. *Illinois tornadoes.* Urbana: Illinois State Water Survey Circular 103.

21. **National Weather Service.** 2001. Winter storms: The deceptive killers. www.nws.noaa.gov/om/brochures/winterstorm.pdf. Accessed 7/27/06.

22. **Environmental Canada.** 2002. Blizzards. www.pnrrpn.ec.gc.ca/air/winterservere/blizzards.en.html. Accessed 7/27/06.

23. **National Snow** and **Ice Data Center.** 2004. *The blizzards of 1996.* Boulder: University of Colorado. http://nsidc.org/snow/blizzard/plains.html. Accessed 7/27/06.

24. **Environment Canada.** 2004. Blizzards and winter weather hazards. www.mscsmc.ec.gc.ca/cd/brochures/blizzard_e.cfm. Accessed 7/27/06.

25. **Lecomte, E. L., Pang, A. W.,** and **Russell, J. W.** 1998. *Ice storm '98.* Toronto: Institute for Catastrophic Loss Reduction Research Paper Series No. 1.

26. **Environment Canada.** 2002. Ice Storm 1998: The worst ice storm in Canadian history? www.msc-smc.ec.gc.ca/media/icestorm98/icestorm98_the_worst_e.cfm. Accessed 1/17/07.

27. **Changnon, S. A.,** and **Changnon, J. M.** 2002. Major ice storms in the United States, 1949–2000. *Environmental Hazards* 4: 105–11.

28. **DeGaetano, A. T.** 2000. Climatic perspective and impacts of the 1998 northern New York and New England ice storm. *Bulletin of the American Meteorological Society* 81: 237–54.

29. **National Climate Data Center.** 2006. *Eastern U.S. flooding and ice storm.* National Oceanic and Atmospheric Administration. www.ncdc.noaa.gov/oa/reports/janstorm/janstorm.html. Accessed 1/20/07.

30. **Jones, K. F.,** and **Mulherin, N. D.** 1998. *An evaluation of the severity of the January 1998 ice storm in northern New England; Report for FEMA Region 1.* Hanover, NH: U.S. Army Corps of Engineers Cold Regions Research and Engineering Laboratory. www.crrel.usace.army.mil/techpub/CRREL_Reports/reports/IceStorm98.pdf. Accessed 10/9/06.

31. **Coppola, S.,** and **Wear, B.** 2007. Lack of de-icer backs up airport. *Austin American-Statesman* 136(174): A4.

32. **Federal Emergency Management Agency.** 1998. Disaster connections kids to kids: Kids from Maine talk about surviving the Ice Storm of '98. www.fema.gov/kids/me98_04.htm. Accessed 1/20/07.

33. **Whittow, J.** 1980. *Disasters: The anatomy of environmental hazards.* London: Penguin Books.

34. **Jones, R. L.** 2005. Canadian disasters—A historical survey. www.ott.igs.net/jonesb/DisasterPaper/disasterpaper.html. Accessed 10/7/06.

35. **NOAA National Centers for Environmental Information (NCEI)** U.S. Billion-Dollar Weather and Climate Disasters (2017). www.ncdc.noaa.gov/billions/. Accessed 10/7/06.

36. **Bagnold, R. A.** 1941. *The physics of blown sand and desert dunes.* New York: Chapman Hall.

37. **Saunders, G.** 2004. The silent killer. *Weatherwise* 57(2): 27.

38. **Houghton, J.** 2004. *Global warming: The complete briefing,* 3rd ed. Cambridge: Cambridge University Press.

39. **Trenberth, K. E., Jones, P. D., Ambenje, P., Bojariu, R., Easterling, D., Klein Tank, A., Parker, D., Rahimzadeh, F., Renwick, J.A., Rusticucci, M., Soden, B.,** and **Zhai, P.** 2007. Observations: surface and atmospheric climate change. In *Climate Change 2007: The physical science basis; Contribution of Working Group I to the Fourth Assessment Report of the Intergovernmental Panel on Climate Change*, eds. S. Solomon, D. Qin, M. Manning, Z. Chen, M. Marquis, K. B. Averyt, M. Tignor, and H. L. Miller. New York: Cambridge University Press.

40. **De Bono, A., Peduzzi, P., Giuliani, G.,** and **Kluser, S.** 2004. *Impacts of summer 2003 heat wave in Europe.* United Nations Environment Programme, Division of Early Warning and Assessment—Europe, Early Warning on Emerging Environmental Threats. www.grid.unep.ch/product/publication/EABs.php. Accessed 12/23/06.

41. **British Broadcasting Corporation News.** 2003. Sizzling temperatures break UK record. http://news.bbc.co.uk/1/hi/uk/3138865.stm. Accessed 12/24/06.

42. **Luterbacher, J., Dietrich, D., Xoplaki, E., Grosjean, M.,** and **Wanner, H.** 2004. European seasonal and annual temperature variability, trends, and extremes since 1500. *Science* 303: 1499–1503.

43. **Black, E., Blackburn, M., Harrison, G., Hoskins, B.,** and **Methven, J.** 2004. Factors contributing to the summer 2003 European heat wave. *Weather* 59: 217–23.

44. **U.S. Environmental Protection Agency.** 2006. Heat island effect. www.epa.gov/heatisland/about. Accessed 9/17/06.

45. **Stott, P. A., Stone, D. A.,** and **Allen, M. R.** 2004. Human contribution to the European heat-wave of 2003. *Nature* 432: 610–14.

46. **World Health Organization.** 2003. The health impacts of 2003 summer heat-waves; Briefing note for the delegations of the fifty-third session of the WHO Regional Committee for Europe. www.euro.who.int/document/Gch/HEAT-WAVES%20RC3.pdf. Accessed 1/7/07.

47. **Bouchama, A.** 2004. The 2003 European heat wave. *Intensive Care Medicine* 30: 1–3.

48. **Pyne, S. J., Andrews, P. L.,** and **Laven, R. D.** 1996. *Introduction to wildland fire.* New York: John Wiley & Sons.

49. **National Oceanic and Atmospheric Administration National Weather Service Storm Prediction Center.** 2002. Storm Prediction Center. www.spc.noaa.gov/misc/aboutus.html. Accessed 7/27/06.

50. **Golden, J. H.,** and **Adams, C. R.** 2000. The tornado problem: Forecast, warning, and response. *Natural Hazards Review* 1(2): 107–18.

51. **Roberts, R. D., Burgess, D.,** and **Meister, M.** 2006. Developing tools for nowcasting storm severity. *Weather and Forecasting* 21: 540–58.

52. **Godschalk, D. R., Brower, D. J.,** and **Beatly, T.** 1989. *Catastrophic coastal storms: Hazard mitigation and development management.* Durham, NC: Duke University Press.

53. **Federal Emergency Management Agency.** 2004. Saferooms: Know your risk and have a safe place to go … with time to get there. www.fema.gov/plan/prevent/saferoom/. Accessed 1/10/07.

54. **Auerbach, P. S.** 2003. *Medicine for the outdoors.* Guilford, CT: Lyons Press.

55. **Gill, P. G. Jr.** 2002. *Wilderness first aid: A pocket guide.* Camden, ME: Ragged Mountain Press.

56. **Federal Highway Administration Road Weather Management Program.** 2006. Low visibility. www.ops.fhwa.dot.gov/weather/weather_events/low_visibility.htm. Accessed 7/29/06.

57. **National Climatic Data Center.** National Ocean and Atmospheric Administration website. www.ncdc.noaa.gov/oa/climate/severeweather/tornadoes.html#overview. Accessed 7/3/13.

58. **City of Moore, Oklahoma.** Webpage. www.cityofmoore.com/storm-shelters. Accessed 7/3/13.

59. **Thompson, A.** 2013. Satellite picture reveals the scars left behind by Moore Tornado. June 5. NBC News Science. www.nbcnews.com/science/satellite-picture-reveals-scar-left-behind-moore-tornado-6C1021 9235?franchiseSlug=sciencemain. Accessed 7/3/13.

60. **The Tornado Outbreak of May 20, 2013.** National Weather Service Weather Forecast Office, Norman, Oklahoma. National Oceanic and Atmospheric Administration, www.srh.noaa.gov/oun/?n=events-20130520. Accessed 7/4/13.

61. **National Weather Service Weather Forecast Office.** A 20-slide presentation of the Event Timeline for the Newcastle-Moore tornado including NWS forecasts and warnings, as well as social media info from Twitter and Facebook. Norman, Oklahoma. National Oceanic and Atmospheric Administration.

62. **Emergency Operations Plan, Moore Oklahoma (PDF).** www.cityofmoore.com/sites/default/files/main-site/2012%20Emergency%20Operations%20Plan.pdf. Accessed 7/6/13.

63. **Freedman, A.** 2013. Oklahoma tornado shows progress in weather warnings, climate central. www.climatecentral.org/news/moore-tornado-showcases-advancements-in-weather-warnings-16026. Accessed 7/6/13.

64. **Watkins, T.** 2013. Basements are scarce in tornado-prone Oklahoma: Here's why. May 22. CNN U.S., edition. www.cnn.com/2013/05/22/us/oklahoma-tornado-basements. Accessed 7/6/13.

65. **Talley, T.** 2013. Moore, Oklahoma schools destroyed by tornado will be rebuilt. May 30. *Huffington Post.* www.huffingtonpost.com/2013/05/30/moore-oklahoma-schools-tornado_n_3362123.html. Accessed 7/6/13.

66. **Barker, S.** 2012. Demographic State of the State Report, Oklahoma State and County Population Projections Through 2075, Oklahoma Department of Commerce (PDF).

The storm had swelled to a diameter of 1800 km, making Hurricane Sandy the largest Atlantic hurricane on record

10

APPLYING the 5 fundamental concepts Hurricane Sandy

1

Science helps us predict hazards.

Hurricanes are among the most costly natural hazards. Fortunately, our ability to predict the path of these storms and their effects is getting better as hurricane scientists tracking Sandy showed in 2012.

2

Knowing hazard risks can help people make decisions.

Evaluation of the effects of rising sea level, with regards to the hurricane hazard, is just one example of how those living near to sea level along the U.S. eastern seaboard must assess their risk from the negative effects of hurricanes going forward.

Hurricanes and Extratropical Cyclones

Hurricane Sandy

It would take seven days from Sandy's formation in the Caribbean Sea for the hurricane to work its way up the U.S. Eastern Seaboard before finally making landfall just south of New York City, the center of the country's most densely populated metropolitan region. By that time, the storm had swelled to a diameter of 1800 km (~1100 mi.), making Hurricane Sandy the largest Atlantic hurricane on record.[1,2] Due to Sandy's great size, atypical path, and the fact that the storm merged with an arctic cold front before coming onshore, the unofficial name "Superstorm Sandy" was applied by the media.[1] Although the storm no longer had hurricane-force winds upon landfall, at an estimated cost of $71 billion Sandy was the second most expensive storm to ever strike the United States—second only to Hurricane Katrina in 2005.

In just one week, Hurricane Sandy damaged or destroyed more than 200,000 buildings across four countries.[3] In the United States, Sandy triggered intense snowstorms in North Carolina, Maryland,

◀ **Hurricane Sandy** Landsat image of Hurricane Sandy on October 28, 2012, headed toward landfall near New York City. By this time Sandy had merged with an arctic cold front causing the storm to swell in size, making Sandy the largest Atlantic hurricane on record. *(NASA)*

LEARNING Objectives

After reading this chapter, you should be able to:

LO:1 Describe the weather conditions that create, maintain, and dissipate cyclones.

LO:2 Explain why it can be difficult to forecast cyclone behavior.

LO:3 Locate several geographic regions at risk for hurricanes and extratropical cyclones.

LO:4 Give examples of the effects of cyclones in coastal and inland areas.

LO:5 Recognize linkages between cyclones and other natural hazards.

LO:6 List the benefits derived from cyclones.

LO:7 Describe adjustments that can minimize damage and personal injury from coastal cyclones.

LO:8 Propose prudent actions to take for hurricane or extratropical cyclone watches and warnings.

3

Linkages exist between natural hazards.

Flooding and coastal erosion along the eastern seaboard represent just a couple of the natural hazards linked to Hurricane Sandy.

4

Humans can turn disastrous events into catastrophes.

Making landfall just south of New York City along the New Jersey coastline, the most densely populated metropolitan region in the United States, clearly led to Hurricane Sandy's great financial impact.

5

Consequences of hazards can be minimized.

When we return to Hurricane Sandy at the end of this chapter it will become clear how both state and federal agencies are involved in minimizing the effects of future hurricane events such as Sandy, through reevaluation of flooding and evacuation maps and other more extreme measures.

(a)

(b)

∧ FIGURE 10.1 The Effects of Sandy on New York City and New Jersey
(a) FDR Drive in Manhattan's Lower East Side, flooded by Hurricane Sandy. *(Beth Carey, Wiki Commons)*
(b) Eroded New Jersey coastline and flooded neighborhood in the town of Tuckerton. *(2012 U.S. Coast Guard/ Getty Images)*

and the Virginias, which caused extensive power outages along the eastern seaboard. New York City and the New Jersey coast were hit particularly hard by the large waves and heavy wind and rain, which caused extensive flooding and coastal erosion (Figure 10.1). By the time the storm dissipated over Pennsylvania on October 31, 2012, Hurricane Sandy had killed more than 130 people in the United States, half of whom lived in New York.[3] For nearly a decade, scientists had warned that New York City was particularly susceptible to rising sea levels, changing tides, extreme weather patterns, and more frequent flooding due to climate change.[4] A warning came in 2011 when Hurricane Irene passed by New York City, flooding low-lying areas and inundating several sewage treatment plants that resulted in raw sewage spills into local waters. Hurricanes Irene and Sandy are responsible for two of the top three worst flooding events to hit New York City since 1900.[4] Rising sea level is likely to cause more extensive and frequent flooding later in the twenty-first century, raising the risk and insurance costs for those who live in low-lying areas adjacent to the coast or along river valleys.

10.1 Introduction to Cyclones

In meteorological terms, Hurricane Sandy was a **cyclone**, an area or center of low atmospheric pressure characterized by rotating winds. Because of the Coriolis effect (see Case Study 9.1: A Closer Look), winds in the Northern Hemisphere are deflected to the right and blow in a counterclockwise rotation around the low-pressure center (Figure 10.2), whereas in the Southern Hemisphere, the Coriolis effect causes winds to rotate clockwise. Cyclones rarely develop within 5° of the equator, where

the Coriolis effect is weakest. Cyclones are classified as tropical or extratropical based on their place of origin and the temperature of their center or core region. The modifier *extratropical* means "outside of the tropics." **Tropical cyclones** form over warm tropical or subtropical ocean water, typically between 5° and 20° latitude. Unlike extratropical cyclones, they are not associated with fronts (boundaries between warm and cold air masses) and have warm central cores. In contrast, **extratropical cyclones** develop over land or water in temperate regions, typically between 30° and 70° latitude. These midlatitude cyclones are generally associated with fronts and have cool central cores. Both types of storms are characterized by their **cyclone intensity**, which is indicated by their sustained wind speeds and lowest atmospheric pressure.

Cyclones are associated with the most severe weather in North America. Tropical cyclones include tropical depressions, tropical storms, and hurricanes that can produce high winds, heavy rain, and surges of rising seawater along the U.S. Eastern and Gulf Coasts. Extratropical cyclones can cause strong windstorms, heavy rains, and surges of rising seawater on both the West and East Coasts during warmer months and snowstorms and blizzards during cooler months. Extratratropical cyclones are typically less severe than tropical cylones, however, they can produce outbreaks of tornadoes and severe thunderstorms, especially east of the Rocky Mountains in the United States and Canada.

Although the destructive effects of tropical and extratropical cyclones can be similar, the two types of storms differ in their source of energy and their structure. Tropical cyclones derive energy from warm ocean water and the latent heat that is released as rising air condenses to form clouds (latent heat of condensation).

Extratropical cyclones obtain their energy from the horizontal temperature contrast between the air masses on either side of a front (see Chapter 9). In hurricanes, the most intense of the tropical cyclones, warm air rises to form a spiraling pattern of clouds (Figure 10.2). The rising and warming air surrounding the center of a hurricane heats the entire core of the storm. In contrast, most extratropical cyclones are fed by cold air at the surface and another flow of cool, dry air aloft. The resulting storm has a cool core from bottom to top. Tropical cyclones that move over land or cooler water lose their original source of heat—warm ocean water. These storms either dissipate or become extratropical cyclones moving along a front.

The system of classification and naming of cyclones is often debated because it is based on a combination of science, custom, and politics. Scientific classification and description of cyclones have their roots in regional names given to these storms. Naming of individual cyclones is generally limited to the more intense tropical cyclones and is a practice that began in the 1940s.[5]

CLASSIFYING CYCLONES

Although *cyclone* is the general meteorological term applied to a large mass of air circulating around an area of low-pressure, various terms are used to describe these systems in different parts of the world. For example, both forecasters and residents use the term *nor'easter* to describe an extratropical cyclone that tracks northward along the eastern coast of the United States and Canada. Onshore winds from these storms blow from the northeast and can sometimes reach hurricane strength.

The terminology for strong tropical cyclones is especially varied. In the Atlantic and eastern Pacific Oceans, these storms are called **hurricanes** after a Caribbean word for an evil god of winds and destruction.[6] In the Pacific Ocean west of the International Dateline (180° longitude) and north of the equator, the storms are called **typhoons** after a Chinese word for "scary wind" or "wind from four directions." Hurricanes in the Pacific Ocean south of the equator and Indian Ocean are referred to using some variation of *cyclone*, a term coined from the Greek word meaning "coil of a snake."[6] For simplicity, in this book we refer to all of these strong tropical cyclones as hurricanes.

Hurricanes are classified by their wind speed on a damage-potential scale developed by Herbert Saffir, a consulting engineer, and Robert Simpson, a National Weather Service (NWS) meteorologist, in the 1970s. The Saffir–Simpson Hurricane Scale is divided into five categories based on the highest 1-minute average wind speed in the storm and was first used for NWS public advisories in 1975 (Table 10.1).[5] A hurricane's category typically changes as it intensifies or weakens, and all but the weakest hurricanes will have more than one category assigned to them during their lifetime. Meteorologists describe Category 3 through 5 hurricanes as major hurricanes. As a hurricane's wind speed and, thus, its category increase, the atmospheric pressure in the storm's center drops. Category 5 hurricanes generally have a central atmospheric pressure of less than 920 millibars (~27.17 in. of mercury measured in a barometer). The record low pressure for an Atlantic hurricane, 882 millibars (~26.05 in.), was set by Hurricane Wilma in 2005.[7]

NAMING CYCLONES

Only a small percentage of all cyclones are given names, either to identify where they form or to track their movement. Extratropical cyclones, especially those that become

∧ FIGURE 10.2 Hurricane Rotation
High-pressure air masses outside the storm flow along the pressure gradient (blue arrows) from high pressure toward the central low-pressure zone (L) within the storm. The wind created by the moving air is deflected to the right (black arrows) in the Northern Hemisphere by the Coriolis effect (red arrows). These effects set up a counter counterclockwise rotation around the low pressure region. Because the Coriolis effect acts at right angles to the wind direction, Coriolis deflects winds circling the eye of the storm away from the central low-pressure zone, thereby opposing the pressure gradient and sustaining the storm. *(NASA)*

In the figure legend:
Pressure gradient
Coriolis deflection
Wind direction
L   Low pressure area

▼ **TABLE 10.1 The Saffir–Simpson Hurricane Scale**

The Saffir–Simpson Hurricane Scale is a rating of 1 to 5 based on the hurricane's present intensity. This rating is used to give an estimate of the potential property damage and flooding expected along the coast from a hurricane landfall. The highest current 1-minute average wind speed is the determining factor in the scale because storm surge values are highly dependent on the slope of the continental shelf where the hurricane makes landfall.

Category	Sustained Winds	Types of Damage Due Hurricane Winds
1	119 to 153 km (~74 to 95 mi.) per hour; storm surge generally 1.2 to 1.5 m (~4 to 5 ft.) above normal	No real damage to building structures. Damage primarily to unanchored mobile homes, shrubbery, and trees. Some damage to poorly constructed signs. Also some coastal road flooding and minor pier damage. Hurricanes Irene of 1999 and Katrina of 2005 were Category 1 hurricanes when they made landfall in southern Florida.
2	154 to 177 km (~96 to 110 mi.) per hour; storm surge generally 1.8 to 2.4 m (~6 to 8 ft.) above normal	Some roofing material, door, and window damage of buildings. Considerable damage to shrubbery and trees, with some trees blown down. Considerable damage to mobile homes, poorly constructed signs, and piers. Coastal and low-lying escape routes flood 2 to 4 hours before arrival of the hurricane center. Small craft in unprotected anchorages break moorings. Hurricanes Isabel of 2003 and Frances of 2004 were Category 2 hurricanes when they made landfall in North Carolina and Florida, respectively.
3	178 to 209 km (~111 to 129 mi.) per hour; storm surge generally 2.7 to 3.7 m (~9 to 12 ft.) above normal	Some structural damage to small residences and utility buildings with a minor amount of wall failures. Damage to shrubbery and trees with foliage blown off trees and large trees blown down. Mobile homes and poorly constructed signs are destroyed. Low-lying escape routes are cut off by rising water 3 to 5 hours before arrival of the hurricane center. Flooding near the coast destroys smaller structures, with larger structures damaged by battering of floating debris. Terrain continuously lower than 1.5 m (~5 ft.) above mean sea level may be flooded inland 13 km (~8 mi.) or more. Evacuation of low-lying residences within several blocks of the shoreline may be required. In 2004, Hurricanes Ivan and Jeanne were Category 3 hurricanes when they made landfall along the Alabama and Florida coasts, respectively.
4	210 to 249 km (~130 to 156 mi.) per hour; storm surge generally 4 to 5.5 m (~13 to 18 ft.) above normal	More extensive wall failures with some complete roof structure failures on small residences. Shrubs, trees, and all signs are blown down. Complete destruction of mobile homes. Extensive damage to doors and windows. Low-lying escape routes may be cut off by rising water 3 to 5 hours before arrival of the hurricane center. Major damage to lower floors of structures near the shore. Terrain lower than 3.1 m (~10 ft.) above sea level may be flooded, requiring massive evacuation of residential areas as far inland as 10 km (~6 mi.). Hurricanes Charley of 2004 and Dennis of 2005 were Category 4 hurricanes when they struck the coasts of Florida and Cuba, respectively.
5	greater than 249 km (~157 mi.) per hour; storm surge generally greater than 5.5 m (~18 ft.) above normal	Complete roof failure on many residences and industrial buildings. Some complete building failures with small utility buildings blown over or away. All shrubs, trees, and signs blown down. Complete destruction of mobile homes. Severe and extensive window and door damage. Low-lying escape routes are cut off by rising water 3 to 5 hours before arrival of the hurricane center. Major damage to lower floors of all structures located less than 4.6 m (~15 ft.) above sea level and within 458 m (~500 yd.) of the shoreline. Massive evacuation of residential areas on low ground within 8 to 16 km (~5 to 10 mi.) of the shoreline may be required. Three Category 5 hurricanes have made landfall in the United States since records began: the Labor Day Hurricane of 1935, Hurricane Camille of 1969, and Hurricane Andrew of 1992. In 2005, Hurricane Wilma was a Category 5 hurricane at peak intensity in the Gulf of Mexico and is the strongest Atlantic tropical cyclone of record.

Source: Based on Spindler, T., and Beven, J., *Saffir-Simpson Hurricane Scale*, NOAA, 1999. www.nhc.noaa.gov/aboutsshs.shtml. Accessed 12/12/18.

snowstorms, are sometimes named for the geographic area where they form (e.g., Alberta Clipper) (see Figure 9.28). In contrast, all tropical depressions that develop into tropical storms and hurricanes are given individual names by government forecasting centers. These names are established by international agreement through the World Meteorological Organization (WMO). The standardized naming system adopted by the WMO replaced a patchwork of national practices that once included naming hurricanes in honor of the saint on whose holy day the storm came ashore, the military phonetic alphabet, and the names of weather forecasters' girlfriends.

Today, an official name is assigned once the maximum sustained winds of a tropical depression exceed 63 km (~39 mi.) per hour and it becomes a tropical storm. Names are assigned sequentially each year from a previously agreed-upon list for the region in which the storm forms. For example, in the Atlantic Ocean the first three names for 2017 are Arlene, Bret, and Cindy; for 2018 they are Alberto, Beryl, and Chris, and for 2019 Andrea, Barry, and Chantal. Naming tropical storms and hurricanes helps forecasters keep track of multiple storms moving across the ocean at the same time.

In the Atlantic Ocean, the names come from one of six alphabetical lists of alternating men's and women's names. Each list has 21 names derived from English, Spanish, or French. The six lists are used in rotation; therefore, the 2017 list will be used again in 2023 and 2029. Names of especially deadly or damaging hurricanes, such as Sandy, Katrina, Rita, Ivan, and Charley, are retired from the rotating lists, similar to the retirement of the jersey number of a star baseball player.

In 2005, the number of tropical storms and hurricanes in the Atlantic Ocean exceeded the list for that year. This was the first time since lists were established by the NWS National Hurricane Center in 1953 that more than 21 tropical storms and hurricanes formed in the Atlantic Ocean in a single year. Having run through the entire list of names for 2005, NWS forecasters went to their backup plan of assigning letters of the Greek alphabet to the additional storms (Tropical Storm Alpha, Hurricane Beta, etc.).

10.1 CHECK your understanding

1. Describe the basic characteristics of a cyclone. What distinguishes a tropical cyclone from an extratropical cyclone?

2. What is the primary way that hurricanes are classified? Give an example.

3. What is the basic requirement that a storm must meet to be given a name?

4. How many named tropical storms preceded Hurricane Sandy in 2012?

10.2 Cyclone Development and Movement

Tropical and extratropical cyclones differ not only in their development, movement, and characteristics but also in their evolution. Most tropical and extratropical cyclones form, mature, and dissipate independently. Some tropical cyclones, however, transform into extratropical cyclones if they encounter an upper-level low-pressure trough as they weaken over land or over cooler seawater at higher latitudes.

TROPICAL CYCLONES

A *tropical cyclone* is a general term for large thunderstorm complex with winds that circulate inward toward an area of atmospheric low pressure that has formed over warm tropical ocean water. These complexes go by a variety of names depending on their intensity and location. Low-intensity tropical cyclones are called tropical depressions and tropical storms. High-intensity tropical cyclones are known as hurricanes. To be classified as a hurricane, a tropical cyclone must have sustained winds of at least 119 km (~74 mi.) per hour somewhere in the storm (see Table 10.1). Hurricanes require tremendous amounts of heat to develop and generally form only where the sea surface temperature is at least 26°C (~80°F) (Figure 10.3).

Most hurricanes start out as a **tropical disturbance**, a large area of unsettled weather that is typically 200 to 600 km (~120 to 370 mi.) in diameter and has an organized mass of thunderstorms that persists for more than 24 hours. Tropical disturbances form in a variety of ways, including lines of convection similar to squall lines, remnants of cold fronts, or troughs—elongated regions of low pressure (see Chapter 9). Air masses within a tropical disturbance exhibit weak partial rotation caused by the Coriolis effect. Most tropical disturbances in the Atlantic Ocean are related to **easterly waves**. Easterly waves are north–south elongatated troughs that form over west Africa and migrate westward across the Atlantic Ocean with the trade winds at a rate of 15–30 km (~10–20 mi.) per hour. Deflection of air masses around the troughs causes "kinks" in the trade winds and local convergence of air east of the trough axis, which encourages air to rise and clouds to form (Figure 10.4a). West of the trough axis, clear skies dominate as air masses diverge. Consequently, these tropical disturbances can be seen in satellite images as a series of sequential cloud patches moving from east to west across the Atlantic Ocean from Africa (Figure 10.4b).

A tropical disturbance may become a **tropical depression** if winds increase and spiral around the area of disturbed weather to form a low-pressure center. Warm moist air that is drawn into the depression behaves like spinning ice skaters who draw their arms in toward the body, thereby increasing their rate of spin. Once maximum sustained wind speeds increase to 63 km (~39 mi.) per hour, the depression is upgraded to a **tropical storm** and receives a name. Although the winds are not as strong in tropical storms as in hurricanes, their

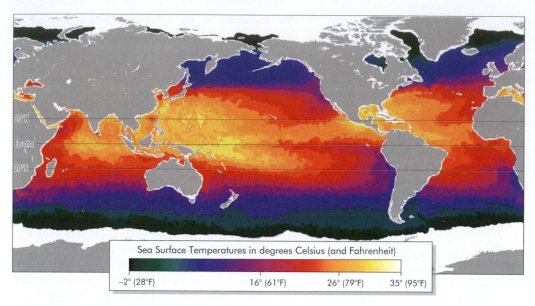

∧ FIGURE 10.3 Ocean Sea Surface Temperatures
Tropical cyclones typically form above oceanic region where sea surface temperatures are at least 26°C (~80°F). These temperature conditions are mostly found within the area of the tropics or about 20° north and south latitude. *(From J. Allen, "Global Sea Surface Temperature," NASA, http://earthobservatory.nasa.gov/IOTD/view. php?id=3513)*

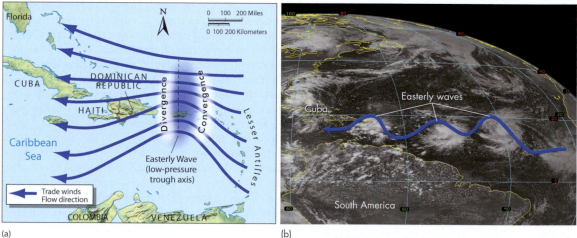

(a)   (b)

∧ FIGURE 10.4 Easterly Waves and Formation of Tropical Disturbances
(a) Convergence of trade wind flow paths east of the low pressure trough axis causes air masses to "pile-up" and rise leading to cloud formation and persistent thunderstorms, whereas divergence west of the axis leads to clear skies. *(From NASA)* (b) Doppler radar image of the equatorial Atlantic Ocean showing several tropical disturbances developed within a series of easterly waves between west Africa and the Caribbean Sea. *(NOAA)*

rainfall amounts can be as intense. If the winds continue to increase in speed, a tropical storm may become a hurricane.

Just as not all tropical depressions develop into tropical storms, even fewer tropical storms develop into hurricanes. Several favorable environmental conditions must be present to allow a hurricane to form. First, warm ocean waters of at least 26°C (~79°F) must extend to a depth of 46 m (~150 ft.) or more.[8] Warm surface temperatures alone are not enough for hurricane formation; there has to be an ample depth of warm water to provide energy for the storm. It is the latent heat stored within the evaporated water, which will be released as it condenses, that will fuel the storm. Second, the atmosphere must cool fast enough from the surface of the ocean upward to allow moist air to continue to be unstable and convect (uplift), causing condensation. Layers of warm air aloft stop or "cap" hurricane development. Third, there must be little vertical wind shear, that is, a change in wind speed between the ocean surface and the top of the troposphere (see Figure 9.18). Strong winds aloft will prevent hurricane development. Finally, the disturbance must be far enough away from the equator so that the Coriolis effect is strong enough to cause rotation around the region of low pressure.

Once developed, a mature hurricane averages around 500 km (~310 mi.) in diameter and consists of counterclockwise spiraling clouds that swirl toward the storm's center in the northern hemisphere and clockwise in the southern hemisphere.[9] This rotation gives hurricanes their characteristic circular appearance when viewed from satellites (see Figure 10.2). Hurricane-strength winds of 119 km (~74 mi.) per hour or greater are limited to the interior 160 km (~100 mi.) of an average hurricane. The outer area of the storm has gale-force winds, which are greater than 50 km (~30 mi.) per hour. Thus, much of a hurricane actually has winds that are less than hurricane strength.

The clouds that spiral around a hurricane are referred to as **rain bands** and contain numerous thunderstorms (Figure 10.5). Both these thunderstorms and the surface winds generally increase in intensity toward the center of the hurricane. The most intense rainfall and winds occur in the innermost band of clouds, known as the **eyewall**. It is not uncommon for rainfall rates directly beneath the eyewall to reach 250 mm (~10 in.) per hour.[9] A hurricane's eyewall is constantly changing as the storm progresses, and some intense hurricanes develop double

eyewalls. In these storms the inner eyewall may gradually dissipate and be replaced with a strengthening outer eyewall. This replacement of one eyewall with another often precedes intensification of the hurricane. Trying to predict changes in hurricane intensity is of great interest to meteorologists responsible for issuing hurricane warnings. Recently, NASA scientists discovered that tall clouds up to 12 km (~7 mi.) high, called *hot towers*, can develop within an eyewall 6 hours before a storm intensifies (Figure 10.6).[10]

A hurricane's eyewall surrounds a circular area of nearly calm winds and broken clouds known as the **eye**. Eye diameters range from 5 to more than 60 km (3 to more than 37 mi.).[11] Most hurricane eyes are smaller at the surface and widen upward to form a large, amphitheater-like area of nearly cloud-free blue sky surrounded by the white clouds of the eyewall (Figure 10.7). When a hurricane's eye passes directly overhead, there is a "calm before the storm," because the fierce winds of the other side of the eyewall will soon blow from the opposite direction. Birds may become trapped in the eye of a hurricane, and thousands of exhausted birds are sometimes reported roosting as an eye passes overhead.

In a typical hurricane, rising warm and moist air spirals upward around the eyewall. As the air rises, it condenses as clouds and releases large amounts of latent heat. The latent heat of condensation warms the surrounding area, which strengthens convective updrafts and encourages additional warm moist air to rise. This upward spiraling rotation draws air from the eye and causes some of the now dry air aloft to sink back downward in the center of the storm (see Figure 10.5). The upward moving air also flows out of the top of the storm and builds a large cloud platform. Outward flowing air at the top of the storm is concentrated in one or more "exhaust jets."[6] This exhaust system is critical for the continued survival of the

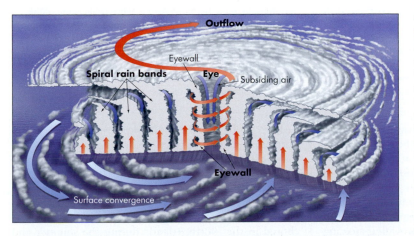

< **FIGURE 10.5 Cross-section of a Hurricane**

A vertically exaggerated cross-section of clouds and wind patterns in a hurricane. Warm, moist air at the surface spirals toward the center of the storm and rises to form rain bands. Sinking dry air creates spaces between the bands that have fewer clouds. The innermost rain band is the eyewall cloud that surrounds the eye of the storm. Rising warm air from the eyewall either leaves the storm in one of several outflow jets aloft or loses moisture and sinks back into the eye of the storm. Subsiding dry air in the eye warms by compression, giving the storm its characteristic "warm core," with clearer skies and calmer winds in the eye. (*Based on NOAA*)

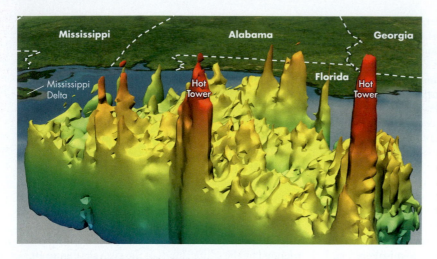

◄ FIGURE 10.6 "Hot Towers" in Hurricane Katrina
NASA's Tropical Rainfall Measuring Mission (TRMM) satellite spotted a pair of extremely tall thunderstorms in the eyewall and outer rain band of Hurricane Katrina shortly before it intensified to Category 5. Referred to as "hot towers," these clouds were more than 15,000 m (~50,000 ft.) high. Like cylinders in an engine, these towers release huge amounts of heat and contribute to the rising winds in the storm. Detection of hot towers may help predict changes in hurricane intensity. *(NASA/JAXA)*

▲ FIGURE 10.7 Stadium-Like Clouds Form a Hurricane's Eyewall
Looking into the eye of Hurricane Katrina, the area of broken clouds in the lower part of the photo mark the eye of the storm where dry air is descending from the top of the troposphere. High winds and rotating clouds in the upper part of the photo compose the eyewall hurricane. Photo taken from a National Oceanic and Atmospheric Administration (NOAA) hurricane hunter aircraft flying through the eye the day before the storm struck the Gulf Coast. Aircraft measurements of a hurricane's wind speed and atmospheric pressure help predict its behavior. *(NOAA)*

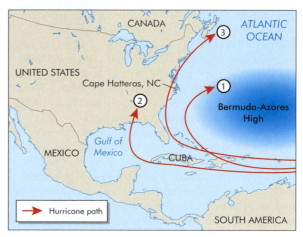

▲ FIGURE 10.8 Hurricane Paths
Three common hurricanes paths, or tracks. Hurricane tracks in the North Atlantic are controlled by the Coriolis effect, direction and strength of steering atmospheric currents, and the location and strength of the Bermuda-Azores High. While south of the Bermuda-Azores High, hurricanes track westward with the trade winds across the Atlantic before being deflected northward by Coriolis west of the high pressure region. All three paths threaten Caribbean islands; then path 1 can affect Bermuda, path 2 the United States Gulf coast, and path 3 the East Coast of the United States and Atlantic Canada.

hurricane because it allows additional warm, moist air to converge inward in the lower levels of the storm, providing more energy. Outflow of warm air at the top of hurricanes is also a mechanism for the transfer of heat from Earth's tropics to the polar regions.

Movement of a hurricane is controlled by the Coriolis effect, which deflects the storm to the right in the Northern Hemisphere, and by providing steering winds that flow 8 to 11 km (~5 to 7 mi.) above the surface. In the Northern Hemisphere, this means that hurricanes commonly track westward in the trade winds that cross the Atlantic Ocean, before curving northward due to the Coriolis effect (Figure 10.8). Where this northward deflection occurs is strongly influenced by the location and strength of the *Bermuda-Azores High*, a persistent high-pressure anticyclone that remains anchored in the North Atlantic during the summer and early fall (see Case Study 10.1: A Closer Look). Many

North Atlantic Oscillation

As you have learned from this chapter and Chapter 9, the westerly winds and polar jet stream play a significant role in the development and movement of severe weather including extratropical cyclones (see Applying the Fundamental Concepts). Recall that winds flow from regions of high pressure to regions of low pressure, and due to the Coriolis effect, these winds blow from west to east across the United States and southern Canada toward Europe (Figure 10.1.A). Variations in the location and magnitude of high and low atmospheric pressure regions, therefore, play a significant

role in the distribution and severity of weather around the globe. The **North Atlantic Oscillation** (NAO) is a fluctuating atmospheric pressure phenomenon in the North Atlantic Ocean that affects the flow of westerly winds in the midlatitudes primarily during winter months. The NAO index describes atmospheric pressure anomalies between the subtropical Bermuda-Azores High and the subarctic low-pressure center near Iceland (Figure 10.1.A). The NAO index can be either positive (+) or negative (–) depending on deviations from long-term averages.[1]

During positive NAO conditions, higher than average atmospheric

pressure in the subtropics and lower than average pressure in the subarctic low persist (Figure 10.1.Aa). The enhanced pressure gradient strengthens the midlatitude westerly winds, including the polar jet stream, which carry winter storms from the North Atlantic Ocean over northern Europe. This leads to a dry and mild winter in southern Europe and a wet and mild winter in northern Europe (Figure 10.1.Aa).

In contrast, negative NAO conditions are characterized by a relatively weak subtropical high-pressure center and a weak subpolar low-pressure center. This

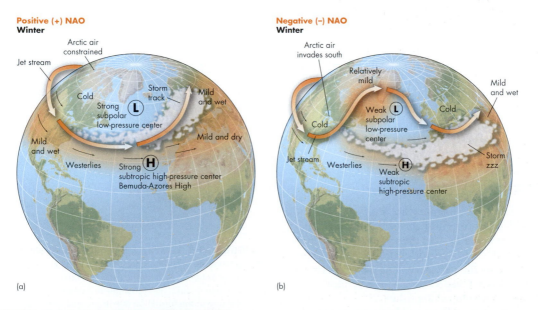

Positive (+) NAO
Winter

Negative (–) NAO
Winter

(a)   (b)

⋀ FIGURE 10.1.A North Atlantic Oscillation (NAO)
Generalized (a) positive NAO and (b) negative NAO winter weather patterns. *(Based on "Linking Climate and Weather," Scientific America 307(6): 54–55, December 2002)*

atmospheric situation weakens the flow of winds toward the east, causing the paths of the jet stream and the westerly winds to diverge (Figure 10.1.Ab). After leaving eastern North America, the jet stream turns sharply northward flowing over Greenland before heading toward southern Europe, while the westerlies carry the winter storm track directly from the North Atlantic toward the Mediterranean (Figure 10.1.Ab). This brings a wet and mild winter to southern Europe and leads to cold dry conditions to the north. Although the effects of fluctuation in the NAO are less pronounced in North America than in Europe, during negative NAO conditions the northeast and eastern seaboard of North America often experience more frequent snowstorms and cold outbreaks including sub-freezing temperatures as far south as Florida (Figure 10.1Ab).

The period of oscillation of the NAO is not constant, but positive conditions appear to be broadly decadal. Although somewhat controversial, some scientists have suggested that long periods of stability of either positive or negative NAO could have a significant effect on global climate and may explain past climate phenomena such as the Medieval Warm Period (A.D. 800–1300) and the Little Ice Age (A.D. 1350–1850) (see Chapter 12).[2]

hurricane paths tend to curve around the west side of the Bermuda-Azores High before turning northeastward upon encountering Northern Hemisphere westerly winds. However, deviations from these paths are common when steering atmospheric currents are weak (see Applying Fundamental Concepts: Hurricane Sandy). Hurricanes have even been known to reverse course and make an entire loop in their path.

A mature hurricane will typically have a forward speed of 19 to 27 km (~12 to 17 mi.) per hour.[12] In most cases, this slow forward speed will mean that a hurricane is a two-day event for communities directly in the path of the storm. Toward the end of a hurricane's life, its forward speed may suddenly increase to 74 to 93 km (~46 to 58 mi.) per hour.[12] This increase in speed happens to many hurricanes that move up the East Coast past Cape Hatteras, toward coastal Canada (see Figure 10.8). A rapid increase in forward speed can be especially dangerous if it results in an early landfall for the hurricane. Once over land or having moved into the cooler waters of the North Atlantic, hurricanes rapidly lose strength and usually dissipate within a few days.

EXTRATROPICAL CYCLONES

Two key ingredients contribute to the formation and movement of an extratropical cyclone—a strong temperature gradient in the air near the surface and wind patterns within the jet stream. Surface temperature gradients are generally strongest along a cold, warm, or stationary front (see Chapter 9). Therefore, most extratropical cyclones develop along fronts.

Large high-pressure ridges and low-pressure troughs in the upper troposphere cause jet streams to bend north or south of their normal path, producing long meanders or waves in their flow. A jet stream may also split in two around isolated high-pressure centers and reunite down flow (Figure 10.9). Extratropical cyclones often develop in an area where jet stream winds curve cyclonically or diverge, such as on the east side of a trough.

Bending or splitting can cause the polar jet stream to dip southward and the subtropical jet stream to flow northeastward into the conterminous United States. The southern branch of a split polar jet stream in the Pacific Ocean brings warm moist air out of the tropics and can be

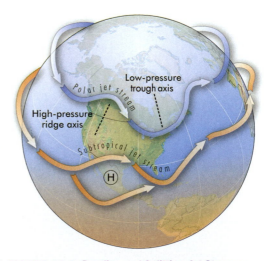

∧ FIGURE 10.9 Bending and Splitting Jet Streams
When jetstreams encounter localized or elongate areas of high or low atmospheric pressure the eastward flow of air tends to bend, dip, and/or split. Such changes in jetsream flow play a fundamental role in the formation of extratropical cylones and the location and path the storm follows. *(Based on E. J. Tarbuck and F. K. Lutgens,* The Atmosphere: An Introduction to Meteorology, *10th ed. © 2007 Upper Saddle River, NJ: Pearson Prentice Hall)*

> **FIGURE 10.10 The Pineapple Express Feeding Moisture to West Coast Extratropical Cyclone**
In this image, the Pineapple Express, a stream of warm moist air carried along by a southern branch of the polar jet stream, feeds an extratropical cyclone in the Pacific Ocean. Storms following this stream of tropical air hit the West Coast, causing coastal erosion, flooding, and land-slides. This color-enhanced infrared image was taken by the NOAA GOES-9 satellite. *(From the National Oceanic and Atmospheric Administration/National Climatic Data Center)*

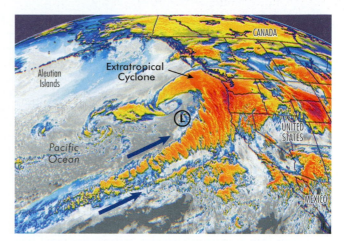

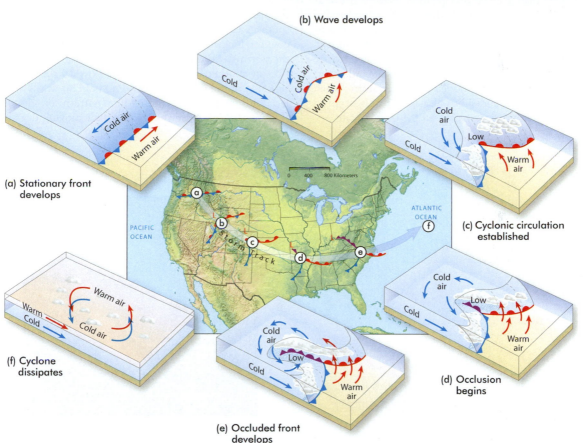

∧ **FIGURE 10.11 Development of an Extratropical Cyclone**
Stages in the development of an extratropical cyclone: (a) Air masses, moving in opposite directions along a stationary front providing the temperature contrast for cyclone development; (b) a wave forms along the front at a place where upper-level winds diverge, such as a bend in the jet stream; (c) a surface low-pressure center develops as cold air pushes south on the west side and warm air pushes north on the east side of the cyclone; (d) the cold front advances faster than the warm front, catches up with the warm front, and displaces the warm air upward; (e) an occluded front develops with warm air held aloft by cold and cool air masses at the surface; (f) the temperature contrast at the surface disappears, fric-tion slows winds, and atmospheric pressure rises as the cyclone dissipates. *(Based on E. J. Tarbuck and F. K. Lutgens,* The Atmosphere: An Introduction to Meteorology, *10th ed. © 2007 Upper Saddle River, NJ: Pearson Prentice Hall)*

recognized on infrared satellite images as a band of clouds extending northeastward from the equatorial Pacific Ocean (Figure 10.10). These narrow bands of moist air sometimes contain more water than Earth's largest rivers, which is why meteorologists refer to them as *atmospheric rivers*, or as the *"Pineapple Express"* by West Coast forecasters because of the origin water being near Hawaii. On average, 10 to 15 extratropical cyclones may develop in the eastern Pacific Ocean and are responsible for 20–50 percent of the precipitation delivered to western North American. Between October 2016 and February 2017, "Pineapple Express" fueled storms delivered to California of more than 76 cm (~30 in.) average precipitation (more than twice the annual average) making it the wettest six month period in 122 years of record keeping. Although the higher than average precipitation ended a five-year drought, heavy storm runoff threatened to undermine the tallest dam in the United States, leading to the evacuation of almost 200,000 people living downstream, and damage to California infrastructure estimated to be greater than $1 billion. On the Atlantic Seaboard, extratropical cyclones called nor'easters often form when bends of the polar and subtropical jet streams begin to merge off the southeastern coast of the United States.

Extratropical cyclones can intensify when they cross areas with strong low-level temperature gradients. For example, extratropical cyclones that hit the West Coast often weaken as they cross the Rocky Mountains and then deepen and strengthen as they leave the mountains and enter the Great Plains.

Because extratropical cyclones develop in the midlatitudes within the band of westerly winds, unlike tropical cyclones that form within the equatorial easterly winds, these storms can be seen developing in the west and tracking eastward across the United States. Most extratropical cyclones start as a low-pressure center along a frontal boundary, with a cold front developing on the southwest side of the cyclone and a warm front on the east (Figure 10.11). In the case of a polar front, cold polar easterly winds are to the north of the front and warm moist westerly winds are to the south. A conveyor belt-like flow of cold air circulates counterclockwise around the cyclone, wedging beneath the warm air to the southeast. Lighter, warm moist air rises in another conveyor belt-like flow on the southeast side of the cyclone, overriding cold air at the surface and creating a comma-like pattern of clouds (Figure 10.12a). A third conveyor belt moving dry air

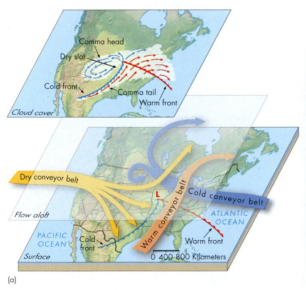

∧ FIGURE 10.12 Structure of an Extratropical Cyclone
(a) Once an extratropical cyclone develops, it is fed by three "conveyor belts" of air: Warm, generally moist air feeds the storm from the south and rises up over a stream of cold air coming from the east, and dry air aloft feeds the storm from the west, sinking behind the advancing cold front at the surface. *(Based on E. J. Tarbuck and F. K. Lutgens, The Atmosphere: An Introduction to Meteorology, 10th ed. © 2007 Upper Saddle River, NJ: Pearson Prentice Hall)* (b) This NASA Geostationary Observational Environmental Satellite (GOES) image shows the comma-shaped cloud pattern of a mature extratropical cyclone moving across the eastern United States. Clouds in the comma head are formed by counterclockwise circulation of cold air around the cyclone. Sinking dry air behind a cold front forms the dry slot. The comma tail is created by thunderstorms and other clouds ahead of the cold front. *(NOAA)*

aloft often feeds the cyclone and the dry air sinks behind the cold front, forming a *dry slot* that is sometimes visible on satellite images (Figure 10.12b). As the cyclone matures, the cold front wraps around the cyclone and merges with the warm front to become an *occluded front* with warm air trapped aloft (see Chapter 9 and Figure 10.11d). At first, the storm intensifies; then, as the cold air completely displaces the warm air on all sides of the cyclone, the pressure gradient weakens and the storm dissipates within a day or two.

Predicting the birth, development, direction of movement, and death of extratropical cyclones is a challenge for forecasters. If conditions are right, the development and strengthening of an extratropical cyclone can be rapid, occurring within 12 to 24 hours. Once formed, an extratropical cyclone's movement is typically steered by winds in the middle of the troposphere at around the 5600 m (~3.5 mi.) level.[9] The forward movement of an extratropical cyclone is generally at half the speed of the steering winds.[11]

10.2 **CHECK** your understanding

1. What are easterly waves, and how are they related to tropical cyclones?

2. What are the four requirements for a tropical disturbance to become a hurricane?

3. What factors control the path a hurricane takes in the Atlantic Ocean?

4. Explain the atmospheric factors needed for the formation of an extratropical cyclone.

10.2

CASE study

Hurricane Katrina: "The most anticipated natural disaster in American history"

Unlike the massive earthquake that struck Japan in 2011, which took many scientists and the world by surprise, the flooding of New Orleans by a major hurricane was widely predicted for decades in scientific publications, including the first edition of this book, and in magazine and newspaper articles. A Pulitzer Prize-winning series about the future catastrophe was published in New Orleans's major newspaper, the *Times-Picayune*, in 2002. Overall, as David Brooks of the *New York Times* later wrote, "Katrina was the most anticipated natural disaster in American history."[13]

Why was this disaster so widely anticipated? First, the southeastern U.S. coast has a high incidence of hurricanes; in the past 50 years, hurricanes have flooded New Orleans and devastated coastal Mississippi, Alabama, and Louisiana on more than one occasion. Second, 95 percent of New Orleans has subsided an average of 1.5 m (~5 ft.) below sea level and the city relies on a patchwork of levees, flood walls, and pumps for protection.[14] Parts of the flood protection and drainage system are more than 100 years old. Finally, in the past century, coastal Louisiana has lost wetlands and barrier islands that help mitigate the effects of hurricane storm surges.

Contrary to initial reports indicating that Katrina was a Category 4 storm when it hit New Orleans, it was actually a Category 3 hurricane when it made its first landfall on the Mississippi Delta about 80 km (~50 mi.) southeast of New Orleans.[15] The storm's intensity then dropped, and most of the city experienced the effects of a Category 1 or weak Category 2 hurricane. Both Katrina's drop in intensity and its path were fortunate for New Orleans. Less than 24 hours before landfall, Hurricane Katrina had been a Category 5 storm, one of the most powerful ever observed in the Gulf of Mexico.[15]

From a hazard-warning standpoint, the National Weather Service did an excellent job of forecasting Katrina's landfall and alerting government officials and the public. NWS forecasters monitored Katrina from its inception almost a week before it came ashore in Louisiana. Air Force and National Oceanic and Atmospheric Administration (NOAA) reconnaissance aircraft, satellites, and radar tracked the progress of the storm as it intensified and then doubled in size over the Gulf of Mexico. A day and a half before Katrina made landfall, Dr. Max Mayfield, then director of the NWS National Hurricane Center, called the governors of Louisiana and Mississippi and the mayor of New Orleans urging evacuation of particularly vulnerable areas.[14] This was only the second time in Mayfield's 36-year career that he believed it necessary to call a governor to warn of a hurricane.[16] The next day he briefed the president and other top federal officials indicating "very, very grave concern" that the New Orleans levees would be overtopped.[17] Special NWS advisories for Katrina accurately warned of storm surge heights of up to 8.5 m (~28 ft.) and floodwaters overtopping levees, resulting in a presidential emergency declaration before the storm hit land and the first-ever mandatory evacuation of New Orleans. Katrina's final landfall was within 35 km (~22 mi.) of the forecast location.[15]

Hurricane Katrina breached the New Orleans flood protection system, the single most costly catastrophic failure of an engineered system in history.[18] Flood walls and levees were breached in at least eight locations by water levels lower than design specifications (Figure 10.13).[17,18] On the east side of New Orleans, levees and flood walls were overtopped and breached by the storm surge because navigation canals acted like funnels to focus the storm surge.[14] Levee and flood wall failures caused 85 percent of the greater New Orleans metropolitan area to be inundated with 0.3 to 6.1 m (~1 to 20 ft.) of water (Figure 10.14).[18] Much of the city remained flooded for two to three weeks before levees could be temporarily repaired and floodwaters pumped out. Other engineering failures included collapse of the I-10 causeway east of New Orleans due to the storm surge; wind damage to the Super-dome roof; loss of cell phone communications and power for residences, businesses, and hospitals; and collapse of several offshore oil and gas production platforms and refineries responsible for 17 percent of U.S. capacity.

If Katrina was not particularly strong or unexpected, and there was ample advance warning, why did more than 1,800 people die and damage exceed $125 billion, making it the most costly natural disaster in U.S. history? Answers to this question are complex and multifaceted, including failure in the design, construction, and maintenance of levees and flood walls; overreliance on technology to protect life and property; social and psychological denial of the hazard; the poverty and limited education of many residents in the affected area; diversion of military and government resources overseas; and

< FIGURE 10.13 Aerial View of Temporary Repair of Breached Flood Wall and Levee on New Orleans's 17th Street Canal
Army helicopters lowered white sand bags to temporarily fill this breach in the flood wall and levee along the 17th Street drainage canal. The failure, caused by the Hurricane Katrina storm surge, occurred at a water level lower than the design limit for the structure. Both faulty design and incomplete geotechnical information may have been responsible for the failure. *(Bob McMillan/UPI/Newscom)*

< FIGURE 10.14 New Orleans Inundated after Flood Protection System Fails
Levees and flood walls designed to protect New Orleans from a Category 3 storm were breached in at least six places by Hurricane Katrina. As a result, 85 percent of New Orleans was flooded. This photo, taken the day after landfall, shows cars abandoned on freeway overpasses. *(AP Photo/David J. Phillip)*

failures in political leadership, communication, and public policy at all levels. The policy failures include a system of flood insurance and postdisaster aid that encourages people to live in coastal hazard areas and rewards politicians, developers, businesses, and individuals for rebuilding in previously flooded areas.[19]

From a social standpoint, Hurricane Katrina was especially devastating to the Gulf Coast. An estimated 1.2 million people were evacuated, and more than 400,000 people ended up in public shelters in 18 states.[16] Although evacuation is a difficult, costly, and uncertain process, especially for the poor and elderly, it is the most powerful tool emergency managers have to prevent loss of life in a hurricane.[19,20] For those who had no way to evacuate, or chose not to, there were few places to go. In the first 10 days of the disaster, Coast Guard boats, ships, and aircraft rescued more than 23,000 people stranded along the Gulf Coast, and other rescuers saved tens of thousands of lives (see Case Study 10.3: Survivor Story).[21] Katrina also damaged or destroyed more than 800,000 homes in four states.[16]

More than a decade after Katrina struck the New Orleans, the social and economic impact of the storm can still be seen, particularly in poor communities. With about 70 percent of New Orleans occupied housing (~135,000 units) damaged or destroyed, more than half of the city's population (~250,000 residence) were forced to temporarily relocate. Although much of the population and city has since rebounded, many of the low-income communities have not. For example, the bustling neighborhood of the Lower 9th Ward, where more than 15,000 people lived prior to Katrina, had a population of just 3,000 twelve years later.[22] The residences who returned now live in sparse neighborhoods filled with empty lots and abandoned house, with very few stores and no schools (Figure 10.15a). Although the reasons for the lack of recovery in such communites are many, the most clear reason why most of the former residence did not return was they could not afford to return. For those who owned a home, many did not have flood insurance or were under insured and consequently could not afford home repair costs (Figure 10.15a). Nor was returning and renting anywhere in New Orleans an option, due to sky-rocketed rental prices driven by a historic low vacancy rate due to so many homes having been destroyed by Katrina.

Although certainly not agreed upon by all New Orleanians, many Americans feared that the loss of such communites like the 9th Ward—home to many historically famous musicians—would kill the heart and soul of the city responsible for the creation of jazz. Consequently, a number of philanthropic organizations in the decade that followed Katrina stepped in to bring back the original residence of such communities. One

such example is the *Make It Right Foundation* founded by the actor Brad Pitt. He and his foundation have pledged to build 150 sustainable houses for former home owners of the Lower 9th Ward who could not afford to return after the storm. As of 2017, more than 350 former residence are back home living in the 107 houses that have been constructed thus far (Figure 10.15b).

(a)

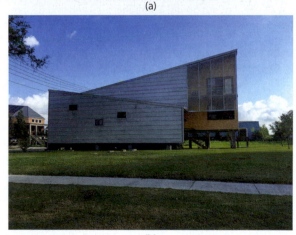

(b)

∧ FIGURE 10.15 A Slow Painful Recovery
In 2017, more than a decade after near complete destruction by Hurricane Katrina, the Lower 9th Ward struggles still to recover. (a) This newly constructed home lies between an empty lot on the right where a Katrina damaged home was demolished, and on its left by an uninhabitable house that suffered extensive wind and water damage in 2005. (b) Sustainable home built by the *Make it Right Foundation*, which hopes to revive this struggling community by helping former owners rebuild. Note that both newly constructed homes in this figure are elevated on 1-m (~3 ft) high pylons to protect them from future flood events. (*Duane E. DeVecchio*)

Hurricane Katrina

Abdulrahman Zeitoun's Family Evacuated— He Stayed to "Mind the Damage"

After living in New Orleans for more than 30 years, contractor Abdulrahman Zeitoun has seen more than his share of tropical storms. So when forecasters began advising Gulf Coast residents to evacuate for Hurricane Katrina in late August 2005, Zeitoun declined to leave. "I say, 'Look, I'm not going. Someone has to stay behind and mind the damage.'"

On Saturday, August 27, Zeitoun waved goodbye to his wife and four children as they drove to relatives in Baton Rouge to wait out the storm. The next day, the wind and rain grew stronger and stronger, the power died, and the roof began to leak. But by late Monday, the weather had calmed and the danger seemed over. "I see water on the street a foot and a half deep, a few trees down. I call my wife, 'I say, '"Everything's done, if you want to come back."'"

In fact, the worst disaster in the history of New Orleans was only just beginning. Nearby, ocean waters surging inland started to overtop city levees, bursting through in several places to inundate entire neighborhoods.

Zeitoun heard the floodwaters before he saw them. "There were noises like when you sit by a river. I stick my head out the window and see water in the backyard rising very fast." He moved family photographs, the first-aid kit, flashlights, and small items upstairs, and then he stacked the first-floor furniture in tall piles. "I sacrificed one thing to save another one; anything I could do to minimize the damage."

The water rose for another 24 hours. Zeitoun called his wife and told her not to return, as the city would not be habitable for a long time. Then he climbed in his canoe to check on his tenants elsewhere in town.

What he saw as he paddled the silent streets was beyond his worst imaginings. Water on his block had reached the top of the stop sign, more than 6 feet high. A neighbor asked for a lift to check on his truck. They found the vehicle in water up to its roof. An older couple waving white flags asked to be rescued. Unable to fit them in the canoe, Zeitoun promised to find help. Further along, they heard a voice from a one-story house. Jumping in the water, Zeitoun managed to open the door. An elderly woman, skirts ballooning around her like a Civil War belle, stood in water up to her shoulders. To rescue her, Zeitoun and his neighbor flagged down one military boat after another, to no avail. Finally, a few civilians with a hunting boat stopped. Zeitoun positioned a ladder beneath her legs, and together the men levered the woman into the vessel. They went on to rescue the older couple and delivered all three seniors to a local hospital.

At his rental, Zeitoun found his tenant healthy and, by some miracle, in possession of a working phone. Neighborhood residents began converging to call friends and relatives.

Zeitoun helped several more people the next day, obtaining a boat to help rescue an older man in a wheelchair. In the evening, Zeitoun, his tenant, a friend, and a man using the phone were gathered at the rental when a military boat floated by. "They jump into the house with machine guns and say, 'What are you doing here?' I say, 'I own the place, it's my house,'" Zeitoun said. But the soldiers arrested all four and took them to a bus station being used as a temporary jail. They were thrown into a makeshift cage of chicken wire, interrogated for three days, and moved to a nearby prison. After spending a month behind bars, Zeitoun was released on bail. All charges—for suspected looting—have since been dropped.

Despite his ordeal, Zeitoun plans to stay in New Orleans. "My business is here. I have good relations here with my associates, friends, customers. I feel like I have a family in this city," he said.

— KATHLEEN WONG

10.3 Geographic Regions at Risk for Cyclones

In general, tropical and extratropical cyclones pose threats to different parts of the United States and Canada. Tropical cyclones have the greatest impact on coastal areas with warm offshore waters, such as the Gulf of Mexico and the Gulf Stream along the East Coast. Hurricanes pose the most serious threat to the eastern contiguous United States, Puerto Rico, the Virgin Islands, and U.S. territories in the Pacific Ocean. They are a lesser but real threat to Hawaii and Atlantic Canada. On the Pacific Coast, hurricanes frequently strike Baja California and the west coast of the Mexican mainland. Moist, unstable air from the remnants of Pacific tropical storms and hurricanes that moves inland over Mexico often contributes to torrential rains in the southwestern United States as far east as Texas.

The East and Gulf Coasts of the United States have the highest risk for tropical storms and hurricanes in North America and experience, on average, five hurricanes each year. Although hurricanes and tropical storms sometimes form in the Gulf of Mexico and Caribbean Sea, most hurricanes that threaten the East and Gulf Coasts form off the western coast of Africa and take one of three tracks (see Figure 10.8):

1. Westward toward the eastern coast of Florida, sometimes passing over Caribbean islands such as Puerto Rico and the Virgin Islands; these storms then move out to the Atlantic Ocean to the northeast without striking the North American continent.

2. Westward over Cuba and into the Gulf of Mexico to strike the Gulf Coast.

3. Westward to the Caribbean and then northeastward, skirting the East Coast; these storms may strike the continent from central Florida to New York. A few storms continue north as hurricanes to strike coastal New England or Atlantic Canada.

The geographic regions at risk for hurricanes and tropical storms include the entire Gulf Coast from southern Texas to southern Florida and the Atlantic Coast from southern Florida northward through the coastal provinces of Canada (Figure 10.16). This risk can be assessed with a map showing the probability that a hurricane will strike a particular 80 km (~50 mi.) coastal segment in a given year (Figure 10.17). Hurricane-strike probabilities are particularly high along the coasts of southern Florida, Alabama, Mississippi, Louisiana, eastern Texas, and Cape Hatteras.

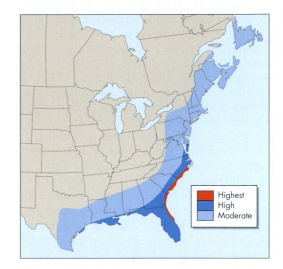

⋀ **FIGURE 10.16 North American Regions at Risk for Hurricanes**
Areas shown are based on the likelihood that a hurricane will pass within 160 km (~100 mi.). Highest-risk area (red area) is likely to experience 60 hurricanes, the high-risk area 40 to 60 hurricanes, and the moderate-risk area fewer than 40 hurricanes in the next 100 years. Map based on observations from 1888 to 2005. *(Based on the U.S. Geological Survey and NOAA Atlantic Oceanographic and Meteorological Laboratory)*

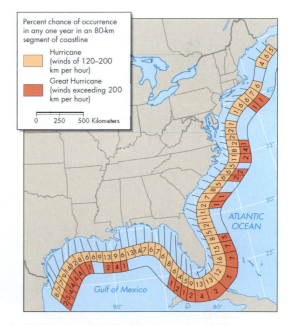
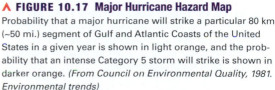

⋀ **FIGURE 10.17 Major Hurricane Hazard Map**
Probability that a major hurricane will strike a particular 80 km (~50 mi.) segment of Gulf and Atlantic Coasts of the United States in a given year is shown in light orange, and the probability that an intense Category 5 storm will strike is shown in darker orange. *(From Council on Environmental Quality, 1981. Environmental trends)*

With the greatest hurricane threat in the southeastern United States, the threat to New England is often underestimated. In 1815, a major hurricane made landfall on Long Island, New York, and then crossed into Massachusetts and New Hampshire. The Great Hurricane of 1938, one of the fastest-moving hurricanes on record, had an intensity and path nearly identical to the 1815 hurricane. The hurricane struck Long Island without warning and killed an estimated 680 people in New York, Connecticut, and Massachusetts. An estimated 80 percent of New England lost electrical power from the storm. Winds were clocked at 130 km (~80 mi.) per hour in New York City and 160 km (~100 mi.) per hour or more from Providence, Rhode Island, to the Boston area.[6] Then in 2012, Hurricane Sandy, the largest hurricane on record, made landfall in New Jersey just 100 km (~60 mi.) south of New York Harbor (see Applying the Fundamental Concepts: Hurricane Sandy).

Our emphasis on North America might imply that the North Atlantic Ocean is the most hazardous area for hurricanes and tropical storms. In fact, that is far from the case; the northwest Pacific generally has three times the number of hurricanes as the North Atlantic, and the Pacific and Indian Oceans overall have more hurricanes (Figure 10.18).[5] Two areas, the South Atlantic and southeast Pacific, rarely have hurricanes because of their cold surface waters. Hurricanes do not form close to the equator because of the absence of the Coriolis effect and, once formed, do not cross the equator.

Hurricanes affecting North America generally develop in the summer and early fall with the length of the hurricane season varying from year to year. The official Atlantic hurricane season starts on June 1 and ends November 30. Most Atlantic hurricanes occur in August, September, and October when the sea surface is the warmest. In contrast, the season for tropical cyclones in the Southern Hemisphere is from January to April.

In North America, the geographic region at risk from extratropical cyclones is far larger than from tropical cyclones. Severe weather from extratropical cyclones is greatest in the interior of the continent but may also occur in coastal areas. The greatest threat for this severe weather varies with location and time of year. In winter months, extratropical cyclones create strong windstorms along the Pacific Coast and nor'easters on the Atlantic Coast, as well as produce heavy snowstorms and blizzards, especially in the Sierra Nevada Mountains and within and east of the Rocky Mountains. During spring and summer months, extratropical cyclones are responsible for the severe thunderstorm and tornado hazard in much of the United States and Canada east of the Rocky Mountains (see Chapter 9).

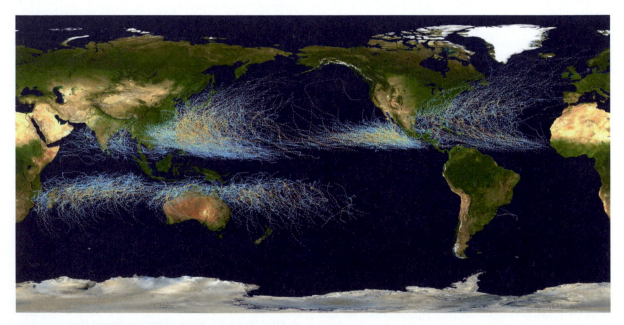

∧ FIGURE 10.18 Typical Tropical Cyclone Paths and Regions of Formation
Most tropical cyclones develop between 5° and 20° latitude. Colored lines indicate cyclone paths, with colors correlating to storm intensity. Tropical storms are shown in blue, and cyclones with intensities from 1 to 5 on the Saffir–Simpson hurricane scale are shown in shades that range from yellow (Category 1) to red (Category 5). *(From NASA and Nilfanion)*

10.3 CHECK your understanding

1. Where are tropical cyclones most common, and why?

2. Which areas of the United States and Canada have the highest risk for hurricanes?

3. How do the chances of a Category 1 hurricane making landfall in the U.S. South (Florida–Georgia) compare with that in the Northeast (New York–Massachusetts)?

10.4 Effects of Cyclones

Tropical and extratropical cyclones claim many lives and cause enormous amounts of property damage every year. Both types of cyclones produce flooding, thunderstorms, and tornadoes, and extratropical cyclones can create snowstorms and blizzards (see Chapter 6 and 9). Three additional effects of cyclones are especially damaging: storm surge, high winds, and heavy rains. In the case of hurricanes, storm surge by far causes the greatest damage and contributes to 90 percent of all hurricane-related fatalities.[12]

STORM SURGE

Storm surge is the local rise of sea level that results primarily from water that is pushed toward the shore by the winds that swirl around a storm. Major hurricanes and intense extratropical cyclones approaching a coastline often generate storm surges of more than 3 m (~10 ft.). Storm surges of 12 m (~40 ft.) or greater have occurred with hurricanes in Bangladesh and Australia.[23] Larger and more intense cyclones build an elevated dome of water beneath the storm that often creates a higher storm surge as the storm makes landfall. Higher storm surges also develop when the coastal water depth gradually grows shallower toward shore. On these coasts, wind-driven water piles up until the wind shifts direction or decreases in speed to allow the surge to flow back to the sea. For most hurricanes in the Northern Hemisphere, the storm surge is greatest in the right forward quadrant of the storm as it makes landfall (Figure 10.19). The height of the surge is generally greatest near the time of maximum wind speed and is also greater if landfall takes place at high tide (see Applying the Fundamental Concepts: Hurricane Sandy).

Two mechanisms in an intense cyclone cause the storm surge. The first and by far the more significant is stress exerted by wind on the water surface. As winds grow stronger, the water level rises quickly, with the

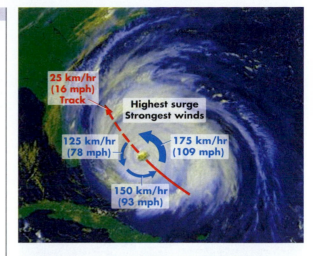

∧ FIGURE 10.19 Hazard Greatest in Right Forward Quadrant of Atlantic Hurricanes

For hurricanes in the Northern Hemisphere, such as Hurricane Floyd shown here, the right forward quadrant of the cyclone typically has the highest storm surge, the strongest winds, most tornadoes, and heaviest rainfall. Image processed from NOAA GOES satellite. *(Goddard Space Flight Center/NASA)*

height of the water proportional to the square of the wind speed. The larger the area over which the wind blows, referred to as the *fetch*, the higher the water will rise. Thus, large storms such as Hurricane Sandy in 2012 generally produce greater storm surges. The second but considerably less important mechanism is the low atmospheric pressure in the storm that pulls up the sea surface. For every millibar that atmospheric pressure drops, the sea surface rises 1 centimeter (~0.5 in.). In most intense hurricanes, the atmospheric pressure drops around 100 millibars, increasing the storm surge height by around 1 m (~3 ft.).[6]

Storm surge is also affected by the shape of a coastline. In a narrow bay, lagoon, or lake, the height of the storm surge may increase as water sloshes back and forth in the enclosed or partially enclosed body of water. This produces *resonance*, a phenomenon that occurs when a wave reflecting from one shore is superimposed on a wave moving in another direction. The net effect is an amplification of the surge, similar to the amplified shaking from reflected seismic waves in a sedimentary basin (see Chapter 3).

Contrary to depictions in Hollywood movies, a storm surge is not an advancing wall of water; instead, it is a continual increase in sea level as the storm approaches landfall. Much of the storm surge damage actually comes from the large storm waves that are superimposed upon the surge. These storm waves combined with ocean currents erode beaches, islands, and roads.

< FIGURE 10.20 Hurricane Overwash Deposits
Overwash sand from Hurricane Katrina cover Dauphin Island when the storm surge overtopped the narrow strip of land. The storm surge destroyed more than half the houses on the island and damaged and displaced the oil platform in the foreground more than 106 km (~66 mi.). Although Dauphin Island was flooded by storm surges from Hurricanes Lili (2002), Ivan (2004), Dennis (2005), and Katrina (2005), some home owners want to rebuild. *(Adrien Lamarre/US Army Corps of Engineers)*

channels are naturally repaired by deposition of sand by coastal currents and wave action in the months following the storm. When Hurricane Allen struck South Padre Island on the Texas coast in 1980, a 4 m (~13 ft.) storm surge flooded part of the island and cut numerous washover channels. Geologists visiting the area shortly after the hurricane found that only the larger sand dunes had remained above the storm surge. When they climbed up on the dunes, they were faced with a new natural hazard—the coyotes and rattlesnakes that had taken refuge there during the storm.

Sand eroded from the beach and coastal sand dunes is carried landward by the storm surge and forms deposits known as **overwash** (Figure 10.20). Many overwash deposits form as broad fans or deltas at the end of channels as the storm surge cuts through the beach and dunes perpendicular to the shoreline. Called *washover channels*, these features can be problematic when they cut entirely through islands or peninsulas and isolate one area from another (Figure 10.21). Most washover

HIGH WINDS

Although the storm surge from a hurricane, and sometimes an extratropical cyclone, is generally more deadly, the damage caused by high winds is more obvious, in part because the winds affect a much larger area. The Saffir–Simpson Hurricane Scale provides a good summary of the various types of wind damage that commonly occur in a hurricane (see Table 10.1). As

< FIGURE 10.21 Hatteras Island Breached by Hurricane Isabel Storm Surge
Three channels, totaling 500 m (~1649 ft.) in width, were cut through Hatteras Island, North Carolina, by Category 2 Hurricane Isabel in September 2003. The breach, located close to where the eyewall crossed the island, is just west of Cape Hatteras. These washover channels were created by a 2.3 m (~7.6 ft.) storm surge in the right forward quadrant of the storm. *(USGS)*

mentioned previously, hurricane winds will be the strongest in the eyewall and the northwest quadrant of the storm. Because of this, forecasters try to predict where the eye of the hurricane will make landfall.

The highest sustained wind speed that has been measured in a hurricane is close to 310 km (~190 mi.) per hour. This speed was recorded for three Category 5 hurricanes: Camille in the Gulf of Mexico in 1969, Tip in the northwest Pacific in 1979, and Allen in the Caribbean Sea in 1980. Even higher gusts of 320 km (~200 mi.) per hour have been estimated for three hurricanes making landfall in the United States: the Labor Day Hurricane in the Florida Keys in 1935; Camille in Mississippi in 1969; and Andrew near Homestead, Florida, in 1992 (Figure 10.22).[24] The destruction wrought by winds of 320 km (~200 mi.) per hour is nearly total. In the 1935 Labor Day Hurricane, a train carrying World War I veterans working in the Florida Keys was blown entirely off the tracks (Figure 10.23).[25] Some of the 423 people who died in the Labor Day Hurricane appeared to have been sandblasted to death by sand blown from the beach. The Labor Day Hurricane was the most intense storm to hit the United States in recorded history and served as the inspiration for the classic Humphrey Bogart movie *Key Largo*.[23]

Wind speeds in most hurricanes diminish exponentially once the storm makes landfall; wind speed is generally reduced by one half within about 7 hours after the storm crosses the coastline.[12] However, a few hurricanes do not lose all their strength when they move over land or cold water and instead transition to become extratropical cyclones. During the transition, some of these storms can maintain or even increase

▲ FIGURE 10.23 Hurricane Blows Train off Tracks in Florida Keys
Wind gusts of more than 320 km (~200 mi.) per hour from the Labor Day Hurricane of 1935 blew the cars of this railroad train off its tracks in the Florida Keys. Hundreds of World War I veterans working in the Florida Keys were killed on the train as they were being evacuated from the storm. *(AP Photo)*

their wind speed. In 1954, Hurricane Hazel made landfall in North Carolina, yet it maintained 160 km (~100 mi.) per hour winds and caused extensive damage as far north as Toronto, Canada. Transition to an extratropical storm may occur if a dying hurricane merges with an upper-level extratropical cyclone or cold front or if it rapidly moves from warm to cold water.[6]

To many people's surprise, the strongest recorded wind in the United States has been from an extratropical cyclone, not a hurricane. This record was set by a 372 km (~231 mi.) per hour gust at a weather station on Mount Washington in New Hampshire's White Mountains in April 1934. On average, Mount Washington experiences

◄ FIGURE 10.22 Aerial View of Hurricane Wind Damage
This mobile home community near Homestead Air Force Base in southern Florida was devastated by wind gusts of up to 320 km (~200 mi.) per hour during Hurricane Andrew in 1992. Injuries were numerous, and 250,000 people in Florida were left homeless from the hurricane. *(AP Photo/Ray Fairall)*

hurricane-strength winds 104 days each year because of the exposed location and mountainous topography.[23]

Strong winds and heavy rains from extratropical cyclones cause most of the severe weather on the West Coast from San Diego to Vancouver. One of the most destructive of these storms was the Big Blow of 1962, which produced wind gusts of 260 km (~160 mi.) per hour along the Washington coast and 290 km (~179 mi.) per hour on the Oregon coast.[6]

In the central United States and Canada, extratropical cyclones are responsible for the high winds in blizzards and tornadoes (see Chapter 9) and for deadly windstorms on the Great Lakes. One 1913 storm sank or grounded 17,300-foot ore ships on Lakes Huron and Erie.[23] Another extratropical cyclone in 1975 sank one of the largest ore ships ever to sail on the Great Lakes, the *S.S. Edmund Fitzgerald*, killing the entire crew. Close to the time the *Fitzgerald* sank in Lake Superior, the captain of a nearby ship estimated wind gusts of more than 160 km (~100 mi.) per hour and waves up to 11 m (~36 ft.).[11]

HEAVY RAINS

An average hurricane produces about a trillion gallons of rainwater each day, nearly three times the amount of freshwater consumed in the United States each year.[6] The heaviest rainfalls in history have occurred where hurricanes have passed over, or close to, mountainous islands. La Reunion Island in the Indian Ocean has five rainfall records from tropical cyclones including 1 m (~45 in.) in 12 hours, 1.8 m (~70 in.) in 24 hours, and 2.5 m (~100 in.) in 48 hours.[6] Unlike the storm surge and very high winds that are generally limited to hurricanes and intense extratropical cyclones, heavy rains occur with cyclones of lower intensity. Tropical storms and even weaker tropical depressions can cause extensive flooding. In 2001, Tropical Storm Allison, which never became a hurricane, dumped up to 940 mm (~40 in.) of rain on Houston, Texas, and more than 690 mm (~30 in.) in southern Louisiana.[24] More than 45,000 homes and businesses were flooded in Houston, the fourth largest city in the United States. The storm killed 41 people, caused more than $5 billion in damages, and was the costliest tropical storm in U.S. history.[24] Rainfall from tropical storms and remnants of hurricanes is often most intense when the storms encounter topographic barriers, such as the Appalachian Mountains, and during nighttime hours when the rotational circulation of the storms contracts.

Intense rainfall from cyclones can cause widespread inland flooding (see Chapter 6). Four factors affect the extent of inland flooding: the storm's speed; changes in land elevation over which the storm moves; interaction with other weather systems; and the amount of water in the soil, streams, and lakes before the storm arrives.[26] In 2004, inland flooding from Hurricane Ivan extended from Georgia north through Ohio and Pennsylvania to southern New York State (see Figure 6.19). This flooding was caused, in part, by the slow movement of the storm, the hilly and mountainous terrain of the Appalachians, and saturation of the soil by heavy rains from the remnants of Hurricane Frances less than 10 days earlier. For some hurricanes, the damage and death toll associated with inland flooding exceed the storm damage to the coastal zone. Inland flooding along the coast from a hurricane often lasts for many days as floodwaters draining the land surface flow slowly into coastal bays and lagoons.

10.4 CHECK your understanding

1. What are three major causes of hurricane damage? Which is typically the most deadly?

2. Why are wind speeds the highest in the right forward quadrant of a hurricane?

3. Explain the causes and effects of storm surges. What will cause a storm surge to increase?

10.5 Linkages between Cyclones and Other Natural Hazards

Although cyclones themselves present numerous direct hazards to human life and property, their effects are magnified by linkages to other natural hazards. Cyclones are closely linked to coastal erosion, flooding, mass wasting, and other types of severe weather, such as tornadoes, severe thunderstorms, snowstorms, and blizzards. In coastal areas, and perhaps other regions, the cyclone hazard is directly linked to climate change.

Some of the fastest rates of coastal erosion occur during the landfall of cyclones. This observation applies to both hurricanes on the Gulf and East Coasts and winter extra-tropical cyclones on the West Coast. Wind-driven waves superimposed on a storm surge actively erode beaches and coastal dunes, predominantly on the Atlantic Coast, and sea cliffs, primarily on the Pacific Coast (see Chapter 11). Some of the eroded sand in beaches and dunes is replaced during fair-weather conditions, whereas other sand is removed entirely from the coastal zone by waves and currents during storms.

Most cyclones making landfall cause both saltwater flooding from a storm surge and freshwater flooding from heavy rains. Areas where the coast has subsided are more vulnerable to both types of flooding (see Chapter 8). In other cases on land, cyclones cause destructive flooding because weak steering winds aloft allow the storms to stagnate over the same area for many days. Slow-moving cyclones often cause flash flooding because heavy rains continue to fall on already saturated soil.

In mountainous areas, heavy rains associated with cyclones can cause devastating landslides and debris flows. For example, in 1998, heavy rains from Hurricane Mitch in Guatemala, Honduras, El Salvador, and Nicaragua triggered widespread flooding, landslides, and debris flows, which killed at least 11,000 people.[27] Likewise, many landslides and debris flows in California, the Pacific Northwest, and the Appalachian Mountains are associated with heavy rains produced by extratropical cyclones.

Wind damage from hurricanes is not limited to the winds circulating around the eye of the storm. Hurricanes may generate *downbursts* of precipitation-driven winds with speeds of more than 160 km (~100 mi.) per hour and numerous tornadoes.[5] Approximately half the hurricanes that come ashore produce tornadoes. Most of these tornadoes occur within 24 hours after the hurricane makes landfall. Tornado formation is most likely from north to east-southeast of the center of the storm.[25]

Finally, as global sea-level rises because of global warming (see Chapter 12), the negative effects of cyclones will become more severe along coastlines. This will be especially true in areas of the U.S. Gulf and East Coasts that have relatively flat coastal plains. For example, along much of the Texas coast, a 0.3 m (~1 ft.) rise in sea level will result in a landward shift of the shoreline by approximately 300 m (~1000 ft.). In the coming decades, storm surges from cyclones will be able to penetrate farther inland than ever before in history.

10.5 CHECK your understanding

1. Explain how cyclones are linked with other natural hazards.

10.6 Natural Service Functions of Cyclones

Cyclones, and the weather fronts associated with them, are the primary source of precipitation in most areas of the United States and Canada. In the eastern and southern United States, hurricanes and tropical storms often provide much-needed precipitation to moisture-starved areas. This moisture may come from tropical cyclones that formed in the Atlantic Ocean and Gulf of Mexico, or it may be carried by upper-level winds from hurricanes along the Pacific Coast of Mexico. Along the West Coast, extratropical cyclones are the major source of precipitation during the winter rainy season. Extratropical cyclones are also responsible for many heavy snowstorms during winter months; many regions depend on spring snowmelt to recharge streams.

As mentioned previously, tropical cyclones also serve an important function in distributing the Sun's energy, equalizing the temperatures of our planet. Cyclones elevate warm air from the tropics and distribute it toward polar regions.

Cyclones have important natural service functions in ecosystems. Winds from these storms carry plants, animals, and microorganisms long distances, helping populate volcanic islands with flora and fauna once the islands rise above sea level. Cyclone-generated waves can also stir up deeper, nutrient-rich waters, resulting in plankton blooms in the open ocean and in estuaries. Cyclones rejuvenate ecosystems ranging from old-growth forests to tropical reefs. High winds topple weak and diseased trees in forests, and strong waves break apart some types of coral. Overall, these storms contribute to species diversity in many ecosystems.

In summary, cyclones are important sources of precipitation and contribute to the ecological health of many plant and animal communities. When viewed from space, cyclones are awe-inspiring sources of beauty and reminders of the power of natural processes.

10.6 CHECK your understanding

1. What is the primary natural service function cyclones provide the western United States?

2. Briefly summarize the contribution of cyclones to the health and diversity of Earth.

10.7 Human Interaction with Cyclones

Human interaction with cyclones, especially hurricanes, has increased markedly in the past four decades. Prior to greater population concentrations in coastal areas, and certainly before the advent of global satellite imagery, cyclones simply fulfilled their natural functions and some hurricanes went undetected for all or part of their existence.

In the past 50 years, population growth in the United States has been greatest in coastal areas that are especially vulnerable to the effects of cyclones. Today, approximately 53 percent of the people in the United States live in coastal counties.[28] Urban areas continue to expand along coastlines that have been devastated by coastal cyclones in historic times. This includes extensive development in Galveston, Texas; the greater Miami–Ft. Lauderdale and Tampa–St. Petersburg areas of Florida; and the Hamptons on eastern Long Island, to name a few. The growth in coastal population has also increased property values in coastal zones, which result in continual increases in the cost of damages from hurricanes and coastal extratropical cyclones.

As more people build in coastal counties, the nature of such construction can potentially make the cyclone hazard more dangerous and costly to humans. Destruction of coastal dunes for building sites increases the vulnerability of a shore to storm surge. Seawalls and bulkheads that are built to protect property reflect storm waves and contribute to beach erosion (see Chapter 11), further increasing the risks from storms. Finally, improperly attached building materials, such as roofing, often become projectiles in hurricane-strength winds. These projectiles damage other buildings and injure and sometimes kill people.

Many scientists have recently become concerned about the effects of our changing climate on tropical cyclones. Global warming, caused by gases that humans are putting into the atmosphere, is raising the temperature of the sea surface and contributing to rising sea level (see Chapter 12). Warm sea-surface temperature is a major factor in tropical cyclone development, and it is possible that warmer ocean water will increase hurricane intensity.[29] Rising sea level will also increase the reach of a storm surge and extend the effects of the large waves that ride the surge.

10.7 CHECK your understanding

1. Explain how population trends and global warming will interact with cyclones in coastal areas.

10.8 Minimizing the Effects of Cyclones

Since we cannot prevent the cyclone hazard, the primary way to reduce property damage and avoid loss of life is to accurately forecast these storms and issue advisories to warn people in their path. Other ways to

minimize the effects, such as enforcing building codes and evacuation procedures, are adjustments to living with hazard and are discussed in the next section. Although the following discussion focuses on hurricanes, many of the challenges faced in forecasting their behavior, such as monitoring changes in storm path and intensity, also apply to intense extratropical cyclones.

FORECASTS AND WARNINGS

Minimizing property damage and casualties from hurricanes requires accurate forecasts that predict the behavior of these great storms as they approach landfall. The public must be warned in time to prepare or evacuate. Hurricanes can be difficult to forecast because they encompass many different weather processes, and they generally develop far from shore where there are few, if any, observers or weather stations. Once a hurricane has formed, meteorologists must predict if it will reach land, where and when it will strike, how strong the winds will be, how wide an area will be affected, and how much rainfall and storm surge will accompany the storm. Because of the complexity of hurricane prediction, hurricane forecast centers must use a variety of tools.

Hurricane forecast centers are located all across the globe and include the U.S. National Hurricane Center (NHC) on the campus of Florida International University in Miami, Florida, and the Canadian Hurricane Centre (CHC) in Dartmouth, Nova Scotia. The NHC watches the Atlantic Ocean, Caribbean Sea, Gulf of Mexico, and eastern Pacific Ocean between May 15 and November 30 each year, and the CHC monitors tropical and post-tropical cyclones that are likely to affect Atlantic Canada.[30,31] Both centers issue watches and warnings to the public and support hurricane research. In the United States, a *hurricane watch* is issued when a hurricane is likely to strike within the next 48 hours and a *hurricane warning* is given when the storm is likely to make landfall within the next 36 hours or less.[32] The CHC also issues watches and warnings at appropriate times.[31] Both centers utilize information from weather satellites, hurricane-hunter aircraft, Doppler radar, weather buoys, reports from ships, and computer models to detect, forecast, and predict hurricanes.

Hurricane Forecasting Tools Weather satellites are probably the most valuable tools for hurricane detection and tracking, since the great storms form over the open ocean. Satellites can detect early warning signs of hurricanes long before the storm actually begins (see Figure 10.4b) and alert meteorologists to areas that should be watched closely. Although satellites are excellent for monitoring the growth and track of a storms over the open ocean, they cannot give accurate detailed information about wind speed and other conditions within the storms. When within range, other tools are used to develop and refine predictions about the storm's behavior.

Aircraft are an invaluable tool, and special planes are flown directly into a hurricane to gather data, especially regarding intensity. U.S. Air Force hurricane-hunter aircraft perform most of the reconnaissance, although NOAA research aircraft also fly missions through hurricanes. These aircraft can begin their data collection when Atlantic tropical cyclones cross 55° W longitude, a north–south line approximately 2600 km (~1600 mi.) east of Miami, Florida (see Case Study 10.4: Professional Profile). During winter months, Air Force and NOAA aircraft also collect meteorological data on extratropical cyclones in the Pacific Ocean and along the East Coast.

Doppler radar systems provide another means of collecting data about hurricanes that come within about 320 km (~200 mi.) of the U.S. mainland and some islands, such as Puerto Rico and Key West. Radar can provide estimates of rainfall amount, wind velocity, and reveal hurricane rain bands and the eye of the storm (Figure 10.24).

Weather buoys floating along the Atlantic and Gulf Coasts are also used. These buoys are automated weather stations that continuously record weather conditions at their location and transmit information to the U.S. National Weather Service or the Meteorological Service. Some information about offshore weather conditions is also obtained from ships that are in the vicinity of a hurricane.

Meteorologists today use computer models that take into account all available data to make predictions about

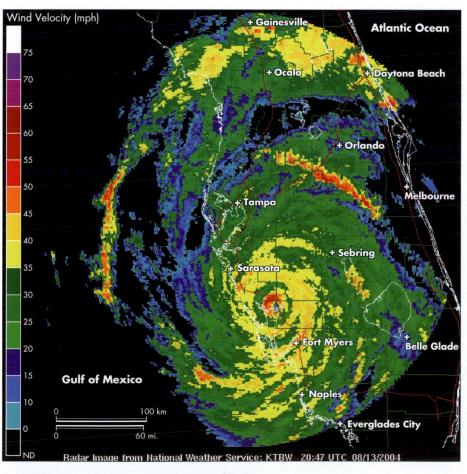

∧ FIGURE 10.24 Radar Image of Hurricane Charley
Image from the Tampa Bay weather radar after Hurricane Charley, a strong Category 4 storm, had made landfall between Fort Myers and Sarasota on the west coast of Florida in 2004. Red and orange colors in the spiraling rain bands have the highest winds. *(National Weather Service)*

hurricane tracks (Figure 10.25). Computer models are constantly being updated with millions of weather observations that are collected each day by the aforementioned tools. Processing such large amounts of data requires a team of scientists and large supercomputers. The most advanced mathematical model used by the NHC that includes weather data from around the globe is the Global Forecast System (GFS) model, which runs four times a day and can make weather predictions up to 16 days in advance. However, forecasts made by the GFS more than seven days out are not very accurate and are not widely accepted. Although such models have vastly improved our ability to predict where and when a storm will strike, they are still not completely accurate in predicting the intensity of the storm; therefore, reconnaissance flights remain an integral part of hurricane prediction.[5] Although such flights are dangerous and extremely expensive, their expense pales in comparison to the cost of evacuating additional miles of coastline with less accurate forecasts. Currently only the U.S.

government uses such flights, and for hurricanes their use is limited to the western Atlantic Ocean, Caribbean Sea, Gulf of Mexico, and an area around Hawaii.[5,33]

Storm Surge Predictions Forecasts of the track and the intensity of cyclones making landfall on coastlines are used to help predict the height and extent of the storm surge. As mentioned previously, forecasters can use wind speed, fetch, and average water depth to get a general idea of the elevation of the storm surge. Forecasters must also predict the time of day that the surge will arrive, because the additional water piled up by the storm is added to the level established by the astronomical tides. Along the Gulf of Mexico, with its tidal range of only a few feet, this is less important than in parts of New England and Atlantic Canada, where the tidal range is 3 m (~10 ft.) or more. Along these parts of the Atlantic Coast, a storm surge from a tropical or extratropical cyclone at high tide would be especially devastating.

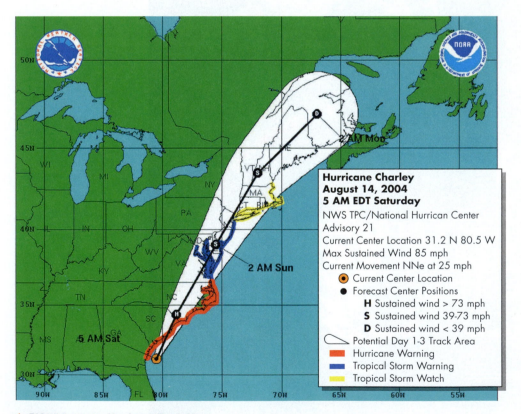

∧ **FIGURE 10.25 Hurricane Track Map**
This National Hurricane Center map shows the predicted path of Hurricane Charley for the following three days. The width of the white area indicates the potential error in the predicted track. Severe weather watches and warnings are shown with red indicating a hurricane warning, blue a tropical storm warning, and yellow a tropical storm watch. *(From NOAA)*

The Hurricane Hunters

Even though amateur tornado chasers occupy an established, if eccentric, niche in the world of weather enthusiasts, hurricane hunting is still a job for the pros.

In fact, Major Chad "Hoot" Gibson boasted that his crew is the only "weather reconnaissance" squadron in the world. The aptly named "Hurricane Hunters" are a crew of Air Force members who fly planes directly into the center of hurricanes while the storms are still lingering over the ocean (Figure 10.4.A). During their wild ride, they gather important data about the strength of the storm, which the National

Weather Service uses in its prediction of where the storm will strike. The crew begins 170 km (~110 mi.) out from where they estimate the eye of the hurricane will be, and they do their best to fly directly for the center in a straight line. In the course of doing so, Gibson said, they've encountered just about every form of severe weather out there: hail, updrafts and downdrafts, turbulence, and even tornadoes. "We avoid those," he said.

At the center of the hurricane is the furious eyewall, whose winds commonly blow in the neighborhood of 225 km (~140 mi.) per hour. And breaching the wall is no

smooth ride. "There are times when the plane vibrates so badly that you can't read your computer screen," Gibson said.

But when they've made it through, the view from the eye, Gibson said, is awe inspiring. "Amazing. Fantastic," he said. "It's like sitting at the 50-yard line of a football stadium." In a well-formed eye, the sky is blue overhead, although the water below—the plane flies at an elevation of around 3050 m (~10,000 ft.)—still "looks like a washing machine."

The Hurricane Hunters don't dally too long in the eye. One reason is that the winds at the center are so

∧ FIGURE 10.4.A Hurricane Hunter Aircraft
This U.S. Air Force Reserve Lockheed Martin WC-130 aircraft, based at Kessler Air Force Base in Mississippi, is specially equipped for taking weather measurements in hurricanes. WC-130 aircraft similar to this one investigate hurricanes at the direction of the National Hurricane Center. *(U.S. Air Force photo/Tech. Sgt. Ryan Labaden)*

calm, at around 4 to 7 km (~2 to 4 mi.) per hour, that the plane itself can throw off the data if it moves too much. Instead, they make a quick pass through the eye as the plane drops a dropsonde, a "weather station in a can," as Gibson described it, to take necessary measurements. Then it's back through the opposite wall and another bumpy ride.

To frame the sharp contrast in wind speed, Gibson recalled a mission through the center of Hurricane Floyd in September 1999 in which the crew entered the first eyewall, experienced 220 km (~137 mi.) per hour winds, then reached the eye 4 minutes later where the winds were blowing at 8 km (~5 mi.) per hour, and then 5 minutes later were again confronted with 225 km (~140 mi.) per hour gusts at the other side.

Not all eyes are perfectly formed, and some storms will have several smaller centers, so the hurricane hunters use the telltale 180-degree shift in the winds—the trademark of the center of a vortex—as the official core of the storm.

The squadron now flies the WC130J, a specially equipped version of the propeller-driven, Lockheed Martin Hercules military transport plane. Air Force pilots, and their predecessors in the Army Air Corps, have been flying scientific missions through hurricanes since 1944.

— CHRIS WILSON

Predicting the geographic extent of the storm surge requires detailed information about land elevation along a coastline. An airborne laser surveying technique, known as LIDAR (Light Detection and Ranging), is used to prepare detailed 3D digital elevation models of coastline elevation that are accurate to approximately 30 cm (~12 in.). This detailed survey data can be entered into sophisticated computer programs to predict both the height and the extent of the storm surge (Figure 10.26). Computer programs such as SLOSH (Sea, Land, and Overland Surges from Hurricanes) and ADCIRC (ADvanced CIRCulation) take into account the forecasts for central atmospheric pressure of the storm, its size, forward speed, track, wind speed, and the seafloor topography (i.e., bathymetry). ADCIRC, which can also take into account the wind and wave stress, tides, and additional water discharged from rivers, was used successfully to predict the storm surge for Hurricane Katrina (see Case Study 10.2: Hurricane Katrina).[14] Overall these models are accurate to within 20 percent of forecast values. This means that if the computer model calculates a 3.0 m (~10 ft.) storm surge, you can expect the observed storm surge to be between 2.5 and 3.7 m (~8 and 12 ft.).

∧ FIGURE 10.26 Hurricane Storm Surge Flooding of Galveston, Texas
An estimated 65,500 residents live on Galveston Island, a Gulf of Mexico barrier island south of Houston, Texas. These before (a) and after (b) images of part of the island show flooding that will occur when a 5.8 m (~19 ft.) storm surge hits the island from the next Category 5 hurricane. In 1900, catastrophic storm-surge flooding by a Category 4 hurricane drowned more than 6,000 people in the most deadly natural disaster in U.S. history. (Center for Space Research and Bureau of Economic Geology, The University of Texas at Austin)

Hurricane Prediction and the Future Until Hurricane Katrina in 2005, hurricane deaths in the United States had dropped dramatically, largely because of better forecasting, improved evacuation, and greater public awareness. Unfortunately, coastal populations are skyrocketing, and coastal development continues in hazard zones. This trend may result in more hurricane-related deaths if increased traffic slows evacuations, if inexperienced residents are not adequately prepared, or if new residents ignore evacuation procedures.[5]

Whereas increased coastal populations may lead to more hurricane-related deaths, property damage costs have already increased dramatically (Table 10.2). As more and more people build homes and businesses in the coastal zone, we can expect to see these costs continue to rise.

10.8 CHECK your understanding

1. What are the five main tools used to predict hurricanes and extratropical cyclones?
2. How do hurricane watches and warnings differ?

▼ **TABLE 10.2** **Property Damage Costs Associated with Hurricanes—U.S. Atlantic and Gulf Coasts**

Decade	Property Damage[a]
1900–1909	$2.4 billion
1910–1919	$5.7 billion
1920–1929	$3.6 billion
1930–1939	$9.8 billion
1940–1949	$9.0 billion
1950–1959	$21.8. billion
1960–1969	$40.5 billion
1970–1979	$34.8 billion
1980–1989	$35.1 billion
1990–1999	$92.2 billion
2000–2010	$279.5 billion

[a] In 2017 dollars.

Source: Blake, E. S., Landsea, C. W., Gibney, E. J., and I.M. Systems Group, NCDC Asheville, 2011. The deadliest, costliest, and most intense United States tropical cyclones from 1851 to 2010 (and other frequently requested hurricane facts). NOAA Technical Memorandum NWS TPC-4, Miami, FL: NOAA/National Weather Service, Tropical Prediction Center, National Hurricane Center. www.nhc.noaa.gov/pdf/nwsnhc-6.pdf. Accessed 10/30/17.

CASE study 10.5

2017 Atlantic Hurricane Season: Record Breaking

The 2017 hurricane season was catastrophic and the costliest U.S. hurricane season in history, both financially (~$300 Billion) and in loss of life (~3,000 killed). Estimates from Hurricane Harvey alone are between $180 and $200 billion (USD).[34,35] But Hurricane Harvey was just one of several hurricanes that ravaged the United States and Caribbean Islands during the record-breaking hurricane activity of September 2017 (Figure 10.27). With 17 named storms that year, 2017 was the fifth-most active season since records began, more than 160 years ago. It was also the only time that the continental United States was struck by more than one Category 4 hurricane in a single hurricane season.

Hurricane Harvey was the first Atlantic storm with hurricane strength to make landfall in the United States. Crossing the Texas shoreline 217 km (135 mi.) southwest of Houston as a Category 4 Hurricane. Persisting for almost five days after making landfall, doubling the longevity of any prior recorded hurricane on land, which consequently led to record-breaking rainfall and flooding in Texas (see Chapter 6 Applying the Fundamental Concepts). High winds and a 2–3 m (7–12 ft.) storm surge also resulted in extensive damage along the Gulf coast. By September 1 Hurricane Harvey had moved inland and was dissipating over Tennessee, and has ultimately been credited with 90 deaths in the United States—most of which were from drowning.[35]

Within one week of Harvey's dissipation, Hurricanes Irma, Jose, and Katia were fully formed and working their way westward (Figure 10.27). September 8, 2017, in meteorological records, represents the calendar day with the greatest recorded accumulated cyclone energy in the Atlantic Basin.[35] Although Jose and Katia resulted in thousands of resident evacuations of the island of Bermuda and coastal Mexico, respectively, neither storm resulted in any recorded deaths and neither hurricane made landfall in the United States. Hurricane Irma, however, is unofficially responsible for at least 90 people killed in the continental United States and more that 45 killed in the Caribbean Islands. On September 9, Irma made landfall as the first recorded Category 5 hurricane to strike the island of Cuba since 1924, and was the strongest-ever hurricane to impact the Caribbean Island (Figure 10.27).[35] Three days later, Irma made landfall in Florida as a Category 4, making it the first time that two hurricanes of this magnitude

< FIGURE 10.27 Highest Recorded Accumulated Cylcone Energy
With Hurricanes Katia (left), Irma (center), and Jose (right), September 8, 2017 holds the record for the greatest accumulated cyclone energy in the Atlantic Basin since records began. *(VIIRS image captured by the National Oceanic and Atmospheric Administration's Suomi NPP satellite)*

ever struck the United States in a single year. Unofficial estimates suggest that Hurricane Irma resulted in more that $66 billion in damage across the Caribbean and the U.S. mainland, but the Atlantic Basin had yet another final punch—Hurricane Maria.

After crossing the Caribbean Islands, headed toward Puerto Rico between September 16 and 18, tropical storm Maria doubled its wind strength to 175 km (109 mi.) per hour after making it the strongest tropical storm thus far of the 2017 Atlantic hurricane season.[35] Making landfall on September 20 as the strongest hurricane to strike Puerto Rico since 1928, Hurricane Maria caused widespread destruction across the U.S. Commonwealth. Still recovering from Hurricane Irma, which struck just 12 days earlier, more than 70,000 residence of the Puerto Rico were still without power and were about to be hit with 30 hours of high wind and rain. Subsequently, Maria completely destroyed Puerto Rico's electrical grid, and more than

three months after the hurricane, 45 percent of the population (~3.5 million residence) were still without electricity and the crisis is ongoing at the time of this writing.[36] As of late 2017, the Puerto Rican death toll estimates from hurricane Maria are still unknown, and range from a few tens to several thousand. Logistical and political challenges have hindered recovery to date, but aid from around the globe is slowly flowing into the U.S. Commonwealth of Puerto Rico as of November 2017.

Fueled by anomalously high ocean sea-surface temperatures in the Atlantic Basin and in the Gulf of California, the 2017 Atlantic Hurricane Season was extraordinarily active.[35] With record levels of hurricane rainfall and total cyclone energy in the Northern Hemisphere, some scientists speculate that such weather events are a sign of what is to come, as global climate change results in seasonally warmer tropical atmospheric conditions (see Chapter 12).

10.9 Perception of and Adjustment to Cyclones

PERCEPTION OF CYCLONES

Residents of regions at risk for hurricanes or coastal extra-tropical cyclones often have significant experience with them, yet surprisingly, they do not always perceive the danger. During every hurricane, a number of people choose not to evacuate. People's perception of severe storm risks varies according to their experience. Many

people now living in coastal areas are relatively new residents. Although they may have experienced a few hurricanes, they do not comprehend the threat of a major hurricane, and they may underestimate the hazard. Incorrect predictions of where or when a hurricane will strike may also lower risk perception. For example, if people are repeatedly warned of storms that never arrive, they may become complacent and ignore future warnings. As with any other hazard, accurate risk perception by planners and the public alike is key to reducing the threats associated with coastal hazards to those who live there.

ADJUSTMENT TO HURRICANES AND EXTRATROPICAL CYCLONES

Warning systems, evacuation plans and shelters, insurance, and building design are key adjustments to hurricanes. Emergency warning systems are designed to give the public the maximum possible advance notice that a potential hurricane is headed their way. Warning methods include media broadcasts of watches and warnings and, in cases of immediate danger, the local use of sirens. Efficient evacuation plans must be developed and distributed prior to hurricane season to ensure the most well-organized escape possible. At the time of evacuation, public transportation must be provided, shelters opened, and the number of outbound traffic lanes increased on evacuation routes. As was apparent with Hurricanes Katrina and Rita in 2005, it can take days to evacuate heavily populated areas, especially offshore barrier islands that require the use of ferries or have few bridges.

Homes and other buildings can be constructed to withstand hurricane-force winds and elevated to allow passage of a storm surge (Figure 10.28). Hurricane and other disaster-resistant building recommendations are available from the Partnership for Advancing Technology in Housing (PATH), a public–private partnership between homebuilding, product manufacturing, insurance, and financial industries and representatives of U.S. government agencies.

Personal adjustments for shoreline residents of the East and Gulf Coasts include being aware of the hurricane season and understanding the risks of owning and insuring shoreline property. Before the season starts, people should prepare their homes and property by trimming dead or dying branches from trees, obtaining flood insurance, and installing heavy-duty shutters that can be closed to protect windows. They should learn the evacuation routes and discuss emergency plans with their loved ones. Disaster preparedness for a hurricane is similar to that for an earthquake and includes having flashlights, spare batteries, a radio, first-aid kit, emergency food and water, can opener, cash and credit cards, essential medicines, and sturdy shoes.[37] As with a tornado watch, once a hurricane watch has been issued, it is imperative to stay tuned to local radio and/or TV stations. If an evacuation order is given, gather emergency supplies and follow the assigned evacuation route.

10.9 CHECK your understanding

1. Describe the adjustments that people need to make when living in hurricane-prone areas.

∧ FIGURE 10.28 Hurricane-Resistant House
Home in the Florida Keys that has been constructed with strong blocks and space below the living area to allow a hurricane storm surge to pass through the building. *(Edward A. Keller)*

APPLYING the **5** Fundamental Concepts

Hurricane Sandy Case Study

1 Science helps us predict hazards.

2 Knowing hazard risks can help people make decisions.

3 Linkages exist between natural hazards.

4 Humans can turn disastrous events into catastrophes.

5 Preparedness can mitigate hazard effects.

Unlike Katrina, which struck early in the 2005 Atlantic hurricane season and ravaged the southern states, Hurricane Sandy developed in late October 2012 and impacted much of the eastern seaboard of the United States and Canada. On October 22, Sandy began as a tropical wave in the southern Caribbean and was quickly upgraded to a tropical storm several hours later (Figure 10.29). Strengthening to a Category 1 hurricane, Sandy's eye passed west of Haiti and the Dominican Republic but still dropped more than

20 inches of rain on the islands. With more than 400,000 Haitians still living in poorly constructed shantytowns due to a slow recovery following the 2010 Haiti earthquake (see Chapter 3), more than 50 people on the island were killed by flooding and mudslides (Figure 10.30).[38] With sustained winds of 110 mph, 1 mph less than a major Category 3 hurricane, Sandy struck Cuba on October 26. Approximately 55,000 Cubans were evacuated from the projected path of the storm and low-lying coastal areas, but on landfall waves up to 9 m (~29 ft.) pounded Cuba's southern coast and a 2 m (~7 ft.) storm surge caused extensive flooding and coastal erosion, completely destroying more than 15,000 homes and killing 11 people.

As Sandy moved northward through the Caribbean, scientists at the NHC and other weather agencies applied computer modeling to weather data to predict Sandy's eventual path. About the time Sandy was ravaging Cuba, just three days before U.S. landfall, most models were not in agreement on what path Sandy might

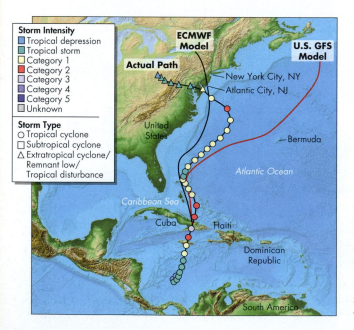

◄ FIGURE 10.29 Modeled and Actual Path of Hurricane Sandy
Computer models are the best way to predict the track of a developing storm, but just three days before Hurricane Sandy was to make landfall in New Jersey, the computer models did not agree. Although the European model (ECMWF) accurately predicted where Sandy would come ashore, the U.S. NHC model (GFS) suggested the storm would head northeast after leaving the Carribean Sea.

◄ FIGURE 10.30 Flooding in Haiti
Residents of Leogane, Haiti find higher ground as the water level continues to rise Friday, October 26, 2012. Residents of Leogane have had five consecutive days of rain in the aftermath of Hurricane Sandy. *(AP Photo/The Miami Herald, Carl Juste)*

take. The most trusted NHC global model (GFS) projected that Sandy would head east into the Atlantic Ocean to dissipate (see Figure 10.29). Consequently, much of the local weather media downplayed the potential impact that Sandy would have on the United States. However, the European Centre for Medium-Range Weather Forecasts (ECMWF) projected a very different outcome: Its model showed Sandy turning to the northwest in the Bahamas, followed by another turn that would take the hurricane northeast, parallel to the East Coast and making landfall near New York City.[39] Even more disconcerting, the ECMWF high-resolution forecast predicted that Sandy would join with an arctic front that was moving southward out of Canada. The rare occurrence of an arctic cold front and a hurricane joining forces was cause for great concern all over the Atlantic northeast.

On October 27 Sandy tracked northwest and then northeast, as the ECMWF model had predicted. Although the storm's eyewall was offshore, strong winds, rain, and waves caused widespread loss of power and closure of several highways due to coastal erosion and flooding in Georgia, the Carolinas, and Virginia. In Bermuda, winds from Sandy caused an F1 tornado, which touched down and damaged homes and businesses.[40]

As Sandy moved up the mid-Atlantic Coast as a Category 1 hurricane, meteorological conditions changed. Atlantic hurricanes at this latitude

typically begin to lose energy as they move into cooler waters and are forced eastward into the North Atlantic by the prevailing westerly winds (see Figure 10.8). However, an anomalous intense high-pressure ridge over Greenland, due to negative North Atlantic Oscillation conditions, was blocking the westerly flow of air and deflecting Hurricane Sandy westward (Figure 10.31).[41] The high pressure region also caused the arctic cold front to remain stationary over the northeastern United States. Not only was Sandy now heading toward the Northeastern Seaboard, but by early October 28 it was clear that the hurricane would merge with the cold front, forming a hybrid "superstorm" just as the ECMWF model had predicted (Figure 10.29).

Preparations for heavy widespread snow in the Appalachian Mountains and Midwest and massive flooding in the northeastern states was not being anticipated. A power outage model built by researchers at Johns Hopkins University predicted that 10 million people from northern Virginia to Pennsylvania would lose power.[42] Government officials and power companies called on independent contractors from the southeastern states to New England to be ready to help repair downed power lines and restore electricity to the eastern seaboard in the wake of the hurricane.

Twenty-four hours before Sandy was projected to make landfall, the ECMWF model predicted that the storm would come ashore

429

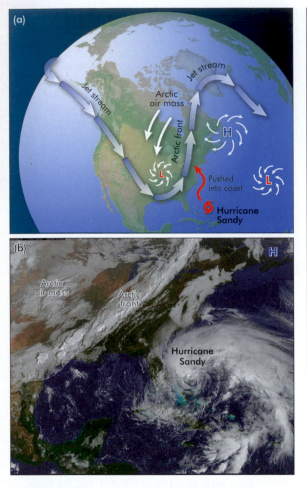

∧ FIGURE 10.31 Hurricane Sandy's Unusual Path
(a) Hurricanes moving north along the eastern seaboard typically head back into the Atlantic away from the United States and Canadian coasts due to the prevailing westerly winds north of 35° latitude. However, in October 2012 an atypical intense high-pressure ridge over Greenland blocked Sandy's path toward the northeast and deflected the storm westward. These atmospheric conditions drove Hurricane Sandy into the U.S. Eastern Seaboard where the storm combined with an arctic cold front coming out of Canada to form a hybrid "superstorm." *(From Climate Central. Reprinted by permission.)* (b) Satellite image of Sandy in the Carribbean Sea one day after being upgraded from a tropical depression and one day before becoming a Category 1 hurricane. *(NASA)*

within 25 miles of Atlantic City, New Jersey (see Figure 10.29). This path would place New York City in the right forward quadrant of the hurricane, where winds would be at a maximum (see Figure 10.19). Before making landfall, however, Sandy would run across 480 km

(~300 mi.) of open water, which when combined with the storm's great diameter meant a huge storm surge could be expected. Furthermore, Sandy's arrival was timed with the full moon and high tide, which would add an additional 0.6 m (~2 ft.) to the anticipated 3.6 m (~12 ft.) storm surge. Of particular concern was flooding in New York City, the largest city in the United States with a population density of more than 27,000 people per square mile!

On the afternoon of October 28, New York City Mayor Bloomberg ordered the evacuation of 375,000 citizens living in coastal and low-lying areas along rivers in Evacuation Zone A (Figure 10.32).[43] This region is susceptible to flooding from any hurricane and broadly correlates to FEMA's Flood Zone A.

As Sandy's eye came ashore in New Jersey at 8:00 P.M. on October 29, the storm no longer had sustained hurricane-strength wind velocities and was downgraded to a post-tropical cyclone.[44] However, the storm's huge size meant high winds, rain, and storm surge battered the coasts of New York and New Jersey through three 12-hour cycles of high and low tide. Sandy's storm surge in New York Harbor was the highest ever recorded at 4.3 m (~14 ft.); the previous record was 4 m (~13 ft.) in 1821.[45] The surge topped seawalls in lower Manhattan, causing widespread power outages and flooding of parts of the city's subway system and the Hugh Carey Tunnel connecting Manhattan with Brooklyn. In the community of Breezy Point on the Rockaway Peninsula, flooding triggered a six-alarm fire that destroyed 111 homes and damaged 20 more (Figure 10.32).

Of the approximately 130 U.S. citizens in eight states killed by Hurricane Sandy, 43 were in New York City, half of whom were on Staten Island (see Figure 10.32). Staten Island was particularly vulnerable to the storm surge because the Long Island and New Jersey coasts meet at a 120° angle, which focused the surge directly toward Staten Island.[46] The storm surge also ravaged much of the New Jersey coast, destroying several coastal boardwalks, numerous homes and businesses, and killing 37 people (see Figure 10.1b).
By the morning of October 30, Sandy's center had moved inland and would ultimately dissipate

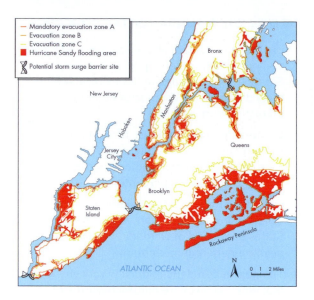

◁ **FIGURE 10.32 New York City Flood Evacuation Plan and Sandy Inundation**
This map shows the areas flooded by Hurricane Sandy with respect to New York City's three flood evacuation zones. Zones are differentiated by their likelihood of flooding as a result of hurricane intensity: Zone A—potential for flooding from any hurricane; Zone B—potential for flooding from a Category 2+ hurricane; and Zone C potential for flooding from a Category 3–4 hurricane hitting just south of New York City. Evacuation of Zone A was made mandatory 24 hours before Sandy made landfall. *(Based on FEMA ABFE maps of New York City)*

over western Pennsylvania on October 31. In the wake of the hurricane, about 380,000 homes were destroyed or damaged, more than 8.5 million residences were without power over 16 states, and heavy snow blanketed five states with parts of Virginia, West Virginia, and North Carolina receiving 0.6–0.9 m (~2–3 ft.). Although the number of hurricanes in the 2012 season fell short of predictions, the season tied for the third most active season on record and Hurricane Sandy became the second most expensive hurricane to hit the United States.

As with Katrina, flooding caused the greatest monetary damage and loss of life. Building codes based on FEMA's Advisory Base Flood Elevation (ABFE) maps for New York City had been in place since 1983.[47] Structures built after 1983 that were flooded by Hurricane Sandy fared better than those built earlier, thereby saving hundreds of millions of dollars paid out by insurance companies to those living in FEMA Zone A. However, flooding was more extensive than was predicted for a storm with Sandy's wind velocities, and areas beyond Zone A were inundated (see Figure 10.32). In the aftermath of Sandy, FEMA released new ABFE maps for all of New York City and for 11 New York and New Jersey counties; this expands the area covered by Zone A in order to mitigate future flood damage.[47] For those living in the newly redistricted areas the cost of flood insurance

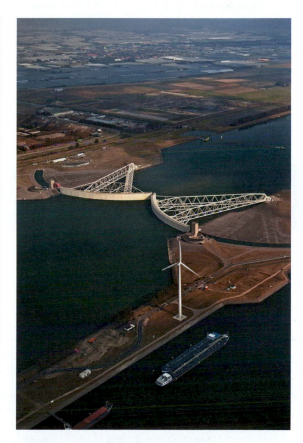

⋀ FIGURE 10.33 Storm Surge Barriers
Aerial view of storm surge barriers in the Netherlands, similar to what could be installed in critical locations to protect parts of New York and New Jersey (see Figure 10.32 for locations). Barriers are closed in this image. *(Frans Lemmens/Corbis)*

will likely rise accordingly if they choose to rebuild. Alternatively, New York Governor Cuomo proposed a more ambitious long-term plan to mitigate future damage: home buyouts at full prestorm market value for the more than 100,000 homeowners affected by Sandy, as well as for homeowners unaffected by Sandy but who live in FEMA Zone A. The land would never be built on again, with some properties turned into public parks or wetlands, dunes, and other natural buffers that would help protect coastal communities from future storms.

In addition to Governor Cuomo's proposal to buy out homeowners and businesses located in the ever-expanding flood-prone areas, more drastic mitigation procedures include constructing movable storm surge barriers (Figure 10.32).[48] These barriers would be built at three key locations (see Figure 10.32) that would stop the flooding of parts of New York and New Jersey and could protect as much as 50 percent of the areas in the flood-prone zone.[18] But with an estimated price tag of $10 billion, is the cost of these barriers too great? Or, in the aftermath of Hurricane Sandy, is the risk of inaction even greater?

APPLYING THE 5 FUNDAMENTAL CONCEPTS

1. Explain how scientists predicted the size, path, and potential problems posed by Hurricane Sandy from its origins in the tropical Atlantic to its dissipation. What factors complicated prediction and warnings?

2. Describe the preparedness of areas affected by Hurricane Sandy in terms of risk—to what extent did the affected regions perceive the risks and act on this perception?

3. Link the various weather-related events that occurred in the Caribbean and eastern United States at the end of October 2012 to Sandy's development and movement.

4. The population density of New York city means that when any severe natural hazard strikes the region the consequences are likely to be high. Discuss a few examples of how urban development of the region has added to the risk posed by large hurricanes.

5. Provide examples of how the adverse consequences of Sandy were minimized and what new strategies have been proposed to minimize the effects of future hurricanes.

CONCEPTS in review

10.1 Introduction to Cyclones

LO:1 Describe the weather conditions that create, maintain, and dissipate cyclones.

- Cyclones, large areas of low atmospheric pressure with winds converging toward the center, are associated with most severe weather. Cyclones are described as either tropical or extratropical based on their characteristics and origin.
- Tropical cyclones have warm cores, are not associated with weather fronts, and form over tropical and subtropical oceans between 5° and 20° latitude. Called typhoons, severe tropical cyclones, or cyclonic storms in most of the Pacific and Indian Oceans, these storms are called hurricanes in the Atlantic and northeast Pacific Oceans and in this book.
- Extratropical cyclones have cool cores, develop along weather fronts, and form between 30° and 70° latitude over either the land or ocean. These midlatitude cyclones produce windstorms,

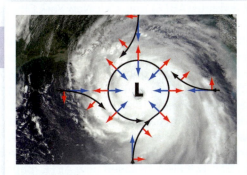

KEY WORDS

cyclone (p. 398), **cyclone intensity** (p. 398), **extratropical cyclones** (p. 398), **hurricanes** (p. 399), **tropical cyclones** (p. 398), **typhoons** (p. 399)

snowstorms, blizzards, and severe thunderstorm outbreaks. Both types of cyclones can have high-velocity straight-line winds, tornadoes, heavy precipitation, and coastal storm surges.

- Tropical cyclones are classified as tropical depressions, tropical storms, and hurricanes with increasing wind speed. Tropical storms must have sustained winds of at least 119 km (~74 mi.) per hour to become a hurricane. A hurricane is assigned a category based on measured or inferred sustained winds. Hurricanes reaching Category 3 or above are considered major hurricanes.
- Tropical storms and hurricanes are given names from lists for the region where they form, whereas extratropical cyclones are sometimes named for their geographic area of origin or prevailing wind direction.

10.2 Cyclone Development and Movement

LO:2 Explain why it can be difficult to forecast cyclone behavior.

- Most tropical cyclones start out as a tropical disturbance, a large thunderstorm complex associated with a low-pressure trough. In the Atlantic Ocean, many of these disturbances form off the west coast of Africa as easterly waves in the trade winds. A tropical disturbance that develops a circular wind pattern becomes a tropical depression and may become a tropical storm if winds exceed 63 km (~39 mi.) per hour. Only a few tropical storms become hurricanes.
- Hurricanes develop spiraling rain bands of clouds around a nearly calm central eye. The eye is surrounded by an eyewall cloud where the strongest winds and most intense rainfall occur. Warm, moist air condensing in rain bands and the eyewall releases latent heat and provides continual energy for the hurricane. Excess heat is vented out of the storm's top.
- Most hurricanes are steered by winds in the middle and upper troposphere and these storms generally lose intensity over land or cooler water, whereas extratropical cyclones typically form along fronts where there are strong, diverging winds in the upper troposphere.

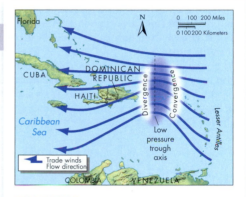

KEY WORDS

easterly waves (p. 401), **eye** (p. 403), **eyewall** (p. 403), **North Atlantic Oscillation** (p. 405), **rain bands** (p. 403), **tropical depression** (p. 401), **tropical disturbance** (p. 401), **tropical storm** (p. 401)

10.3 Geographic Regions at Risk for Cyclones

LO:3 Locate several geographic regions at risk for hurricanes and extratropical cyclones.

- In North America, the Gulf and Atlantic Coasts are at highest risk for tropical cyclones, with hurricane-strike probabilities being the greatest in southern Florida, the northern Gulf Coast, and Cape Hatteras.
- Extatropical cyclones are the primary severe weather hazard on the Pacific Coast and cause storms on the Great Lakes. For the eastern United States and Atlantic Canada, these cyclones are known as nor'easters because of their strong northeasterly winds that cause severe weather.

10.4   Effects of Cyclones

LO:4   Give examples of the effects of cyclones in coastal and inland areas.

- Cyclones produce coastal storm surges, high winds, and heavy rains with major hurricanes and intense extratropical cyclones generating storm surges of more than 3 m (~10 ft.). Higher surges are produced by larger or more intense storms and on shallow coastlines or narrow bays.
- In the Northern Hemisphere, the greatest storm surge and highest winds are typically in the right forward quadrant of tropical cyclones. Much of the structural damage from coastal storms comes from wind-driven waves on top of the storm surge.
- Strong currents from storm surges cut channels through islands and peninsulas and deposit sand as overwash.
- Wind damage from hurricanes is more widespread but less deadly than the storm surge.
- Although hurricanes hold most rainfall records, tropical storms also produce intense rains and flooding. Inland flooding occurs from hurricane remnants if storms move slowly, encounter hills or mountains, interact with other weather systems, or track over already saturated ground.

KEY WORDS
overwash (p. 416), **storm surge** (p. 415)

10.5, 10.6, and 10.7   Linkages between Cyclones and Other Natural Hazards
Natural Service Functions of Cyclones
Human Interaction with Cyclones

LO:5   Recognize linkages between cyclones and other natural hazards.

LO:6   List the benefits derived from cyclones.

LO:7   Describe adjustments that can minimize damage and personal injury from coastal cyclones.

- Cyclones are closely linked with other severe weather, flooding, landslide, and debris flow hazards.
- The hazard posed by storm surges and erosion from coastal cyclones will increase as sea-level rises worldwide from global warming. Hurricane intensity is also projected to increase with warmer global temperatures.
- Both hurricanes and coastal extratropical cyclones will become more destructive to our society as coastal populations and per capita wealth grow. On the positive side, cyclones are major sources of precipitation, help maintain global heat flow, and contribute to long-term ecosystem health.

10.8 and 10.9   Minimizing the Effects of Cyclones
Perception of and Adjustment to Cyclones

LO:8   Propose prudent actions to take for hurricane or extratropical cyclone watches and warnings.

- Accurate forecasts, effective storm warnings, strict building codes, and well-planned evacuations can minimize the effects of coastal cyclones.
- Hurricane forecasting relies on weather satellites, aircraft flights, Doppler radar, and automated weather buoys. Computer models predict hurricane tracks more accurately than their intensity. Storm surge predictions are based on wind speed, fetch, average water depth, and timing of landfall in relation to astronomical tides.
- Perception of the coastal cyclone hazard depends on individual experience and proximity to the hazard.
- Community adjustments to hurricanes in developed countries involve building protective structures or modifying people's behavior through land-use zoning, evacuation procedures, and warning systems.
- Individual adjustments to hurricanes include having emergency supplies on hand; preparing once a storm prediction is made; and if required, evacuating before the storm hits. Homes can also be constructed to withstand hurricane-force winds and elevated to allow passage of storm surges.

CRITICAL thinking questions

1. If you had to evacuate your home and go to a nearby public shelter, where would it be? If you had to evacuate your home and travel at least 160 km (~100 mi.) from where you live, where would you go? What would you take with you in either case? What problems might you or your community face in an evacuation (e.g., people in poor health or with disabilities, people without a means of transportation, visitors, pets, domesticated animals)? What would be your concerns?

2. The southern tip of Florida has a very large area of marsh and swampland—the Everglades—and a large lake, Lake Okeechobee. Based on your understanding of how a hurricane works, what effect would these wetlands have on a hurricane that was tracking across southern Florida? What effect would the hurricane have on Lake Okeechobee?

3. What conditions required for hurricane development could explain why hurricanes generally do not form between 5° N and 5° S of the equator, or in the southeastern Pacific and South Atlantic Oceans?

4. Obtain a topographic map (see Appendix C) for an ocean beach you visit or for an urban area along the Atlantic shore or Gulf Coast. Shade or color the area that would be affected by a 5 m (~20 ft.) storm surge. Examine your completed map and assess the damage that would occur, the routes people would take for evacuation, and land-use restrictions you would recommend for future development.

References

1. **Why Is Sandy Unusual?** 2013. www.cnn.com. Accessed 3/17/13.
2. Sandy Brings Hurricane-Force Gusts After New Jersey Landfall. 2013. *Washington Post*. Accessed 3/31/13.
3. **Marin, D.** 2012. Hurricane Sandy's Storm Surge Mapped ... Before It Hit. *Huffington Post*. November 11, 2012. Accessed 3/25/13.
4. **Chen, D.,** and **Navarro, M.** 2012. For Years, Warnings That It Could Happen Here.

New York Times. October 30, 2012. Accessed 3/10/13.

5. **Fitzpatrick, P. J.** 1999. *Natural disasters: Hurricanes: A reference handbook.* Santa Barbara, CA: ABC-CLIO.

6. **Emanuel, K.** 2005. *Divine wind: The history and science of hurricanes.* New York: Oxford University Press.

7. **Pasch, R. J., Blake, E. S., Cobb, H. D. III,** and **Roberts, D. P.** 2006. *Tropical cyclone report: Hurricane Wilma 15–25 October 2005.* National Hurricane Center, National Weather Service, National Oceanic and Atmospheric Administration. www.nhc.noaa. gov/2005atlan.shtml/. Accessed 3/26/07.

8. **Vescio, M., Cooper, S.,** and **Cain, D.** 2006. *Tropical cyclones: Introduction.* National Weather Service Jetstream; An online school for weather. www.srh. noaa.gov/jetstream/tropics/tc.htm. Accessed 4/9/07.

9. **Ahrens, C. D.** 2005. *Essentials of meteorology: An invitation to the atmosphere.* Belmont, CA: Thomson Brooks/Cole.

10. **Aguardo, E.,** and **Burt, J. E.** 2007. *Understanding weather and climate,* 4th ed. Upper Saddle River, NJ: Pearson Prentice Hall.

11. **Ackerman, S. A.,** and **Knox, J. A.** 2003. *Meteorology: Understanding the atmosphere.* Pacific Grove, CA: Thomson Brooks/Cole.

12. **Flanagan, R.** 1993. Beaches on the brink. *Earth* 2(6): 24–33.

13. **Brooks, D.** 2005. The best-laid plan: Too bad it flopped. *The New York Times,* September 11, 2005.

14. **van Heerden, I.,** and **Bryan, M.** 2006. The storm: What went wrong and why during Hurricane Katrina—the inside story from one Louisiana scientist. New York: Viking Penguin.

15. **Knabb, R. D., Rhome, J. R.,** and **Brown, D. P.** 2006. *Tropical cyclone report: Hurricane Katrina 23–30 August 2005.* National Hurricane Center, National Weather Service, National Oceanic and Atmospheric Administration. www.nhc.noaa.gov/2005atlan.shtml. Accessed 4/9/07.

16. **The White House.** 2006. *The federal response to Hurricane Katrina: Lessons learned.* Washington, DC: The White House. www.whitehouse. gov/reports/ katrina-lessons-learned/. Accessed 3/18/07.

17. **Committee on Homeland Security and Governmental Affairs.** 2006. *Hurricane Katrina: A nation still unprepared.* Washington, DC: United States Senate.

18. **Seed, R. B., Bea, R. G., Abdelmalak, R. I., Athanasopoulos, A. G., Boutwell, G. P., Bray, J. D., Briaud, J.-L., Cheung, C., Cobos-Roa, D., Cohen-Waeber, J., Collins, B. D., Ehrensing, L., Farber, D., Hanemann, M., Harder, L. F., Inkabi, K. S., Kammerer, A. M., Karadeniz, D., Kayen, R. E., Moss, R. E. S., Nicks, J., Nimmala, S., Pestana, J. M., Porter, J., Rhee, K., Riemer, M. F., Roberts, K., Rogers, J. D., Storesund, R., Govindasamy, A. V., Vera-Grunauer, X., Wartman, J. E.,**

Watkins, C. M., Wenk, E. Jr., and **Yim, S. C.** 2006. *Investigation of the performance of the New Orleans flood protection systems in Hurricane Katrina on August 29, 2005,* Volume I: *Main text and executive summary.* Berkeley, CA: University of California at Berkeley, Department of Civil Engineering, Independent Levee Investigation Team Final Report.

19. **Waugh, W. L. Jr.,** ed. 2006. Shelter from the storm: Repairing the national emergency management system after Hurricane Katrina. *The Annals of the American Academy of Political and Social Science* 806: 288–332.

20. **Wolshon, B., Urbina, E., Wilmot, C.,** and **Levitan, M.** 2005. Review of policies and practices for hurricane evacuation. 1: Transportation planning, preparedness, and response. *Natural Hazards Review* 6: 129–42.

21. **Johns, C.,** ed. 2005. Katrina; Why it became a man-made disaster; Where it could happen next. National Geographic Special Edition.

22. **Stewart, S. R.** 2005. *Tropical cyclone report: Hurricane Ivan 2–24 September 2004.* National Hurricane Center, National Weather Service, National Oceanic and Atmospheric Administration. www.nhc.noaa. gov/2004ivan.shtml? Accessed 3/18/07.

23. **Burt, C. C.** 2004. *Extreme weather: A guide & record book.* New York: W. W. Norton.

24. **Stewart, S. R.** 2002. *Tropical cyclone report: Tropical storm Allison 5–17 June 2001.* National Hurricane Center, National Weather Service, National Oceanic and Atmospheric Administration. www.nhc. noaa. gov/2001allison.html. Accessed 4/10/07.

25. **Edwards, R.** 1998. Tornado production by exiting tropical cyclones. Preprints, *23rd Conference on Hurricanes and Tropical Meteorology,* pp. 485–88. Dallas, TX: American Meteorological Society.

26. **National Weather Service Southern Regional Headquarters.** 2002. *Tropical cyclones & inland flooding for the southern states.* Forth Worth, TX: National Oceanic and Atmospheric Administration.

27. **Negri, A. J., Burkardt, N., Golden, J. H., Halverson, J. B., Huffman, G. J., Larsen, M. C., McGinley, J. A., Updike, R. G., Verdin, J. P.,** and **Wieczorek, G. F.** 2005. The hurricane–flood–landslide continuum. *Bulletin American Meteorological Society* 86: 1241–47.

28. **Crossett, K. M., Culliton, T. J., Wiley, P. C.,** and **Goodspeed, T. R.** 2004. *Population trends along the coastal United States: 1980–2008.* National Ocean Service, National Oceanic and Atmospheric Administration.

29. **Intergovernmental Panel on Climate Change, Working Group II.** 2007. *Climate change 2007: Climate change impacts, adaptation and vulnerability; Summary for policy makers.* World Meteorological Organization. www.ipcc.ch/. Accessed 4/8/07.

30. **National Hurricane Center.** 2007. *About the National Hurricane Center.* National Weather Service,

National Oceanic and Atmospheric Administration. www.nhc.noaa.gov/aboutnhc.shtml/. Accessed 4/8/07.

31. **Environment Canada.** 2004. *Canadian Hurricane Centre.* Meteorological Service of Canada. www.ns.ec.gc.ca/weather/hurricane/chc1.html. Accessed 4/8/07.

32. **National Hurricane Center.** 2006. *Hurricane preparedness; Hurricane basics.* National Weather Service, National Oceanic and Atmospheric Administration. www.nhc.noaa.gov/HAW2/english/basics.shtml/. Accessed 4/8/07.

33. **U.S. Air Force 53rd Weather Reconnaissance Squadron.** 2004. Tropical cyclone mission. www.hurricanehunters.com/cyclone.htm/. Accessed 4/8/07.

34. **National Oceanographic and Atmospheric Administration** website www.ncdc.noaa.gov/billions/events/US/1980–2017. Accessed 11//20/17.

35. **Hicks, B.,** and **Burton, M.** 2017. Hurricane Harvey: Preliminary estimates of commercial and public sector damages on houston metropolitan area. Center for Business and Economic Research Miller College of Business.

36. **Greenemeier, L.,** 2017, *Repair or Renovate? Puerto Rico Faces Stark Power Grid Options.* Scientific American. www.scientificamerican.com/article/repair-or-renovate-puerto-rico-faces-stark-power-grid-options/. Accessed 11/20/17.

37. **National Hurricane Center.** 2004. *Hurricane preparedness; Disaster supply kit.* National Weather Service, National Oceanic and Atmospheric Administration. www.nhc.noaa.gov/HAW2/english/prepare/supply_kit.shtml/. Accessed 4/8/07.

38. **Watts, Jonathan.** 2013. Hurricane Sandy: Haiti in emergency aid plea as disaster piles on disaster.

The Guardian. www.theguardian.com/world/2012/oct/30/hurricane-sandy-haiti-emergency-aid Accessed 03/02/13.

39. Is U.S. Global Weather Prediction Falling Behind? www.weatherchannel.com. Accessed 3/21/13.

40. **Hurricane Sandy Spawns a Tornado in Bermuda.** http://bermudaweather.wordpress.com. Accessed 3/21/13.

41. **Revkin, A.** 2012. The #Frankenstorm in Climate Context. *New York Times.* October 28, 2012. Accessed 4/1/13.

42. **Johns Hopkins University Media advisor:** Hurricane Sandy—10 million could lose power. October 27, 2012. Accessed 3/19/13.

43. **Barron, J.** 2012. Sharp Warnings as Hurricane Churns In. *New York Times.* October 23, 2012. https://cn.nytimes.com/world/20121029/c29storm/en-us/ Accessed 3/21/13.

44. **Blake, E. S., Kimberlain, T. B., Berg, R. J., Cangialosi, J. P., Beven II, J. L.,** and **National Hurricane Center**. 2013. *Tropical cyclone report.* February 12, 2013.

45. **Emanuel, K.** 2005. *Divine wind: The history and science of hurricanes.* New York: Oxford University Press.

46. **Chen, D.,** and **Navarro, M.** 2012. For Years, Warnings That It Could Happen Here. *New York Times.* October 30, 2012. Accessed 3/10/13.

47. **Kaplan, T.** 2013. Cuomo Seeking Home Buyouts in Flood Zone. *New York Times.* February 2, 2013. Accessed 3/10/13.

48. **Van Lenten, C.** 2005. *Bracing for Super Floyd: How storm surge barriers could protect New York City.* Academye Briefngs (PDF). Posted 5/31/05. Accessed 3/21/13.

The beaches along the Eastern Seaboard are facing a serious **erosion** problem. **Some beaches are shrinking landward by 0.3 to 1 m per year**

11

APPLYING the 5 fundamental concepts Folly Island

1

Science helps us predict hazards.

Erosion of Folly Island and other coastal communities is a serious and costly natural hazard. By understanding the science behind ocean currents and sand migration along the world's coasts, we are better able to interpret the causes of coastal erosion and defend against it.

2

Knowing hazard risks can help people make decisions.

The risk of damage to land, property, and the livelihood of the 3,000 or so residents of Folly Island is being evaluated on a personal, state, and federal level.

Coastal Hazards

Folly Island

In many ways, Folly Island, 11 km (~7 mi.) south of Charleston, South Carolina, is a typical Atlantic or Gulf Coast barrier island. The island is essentially an accumulation of sand that acts as a barrier to ocean waves that would otherwise strike the mainland (Figure 11.1). A little more than 10 km (~6 mi.) long and generally less than 1 km (~0.6 mi.) wide, most of the island has an elevation of only 1.5–3 m (~5–10 ft.) above sea level. Folly Island formed 4,000 to 5,000 years ago from sand dune ridges that accumulated as the rise in sea level from the melting of Pleistocene glaciers slowed.[1] A pine forest once covered the island and was the source of its Old English name *folly*, meaning "a tree-crested dune ridge." Today Folly Island is home to about 3,000 South Carolinians.

One characteristic of Folly Island that makes it typical of Atlantic barrier islands is that it is eroding at a high rate, with some beaches shrinking landward by 0.3 to 1 m (~1 to 3 ft.) each year. The island's east-facing ocean shoreline has been retreating for at least 160 years and, in the early twentieth century, was receding at an average rate of 2 m (~7 ft.) per year.[1] In addition to natural processes that cause coastal erosion, such as hurricanes, Folly is eroding at an accelerated rate due to a pair of rocky barriers, known as jetties, that were constructed northeast of island in 1896 by the U.S. Army Corps of Engineers (Figure 11.1). The jetties were built to prevent the entrance of Charleston Harbor from being blocked by sand, which is carried southward along the coast by ocean currents. The jetties were successful

◀ EROSION AT FOLLY ISLAND SOUTH CAROLINA

Coastal erosion may occur slowly over tens to hundreds of years or may happen rapidly during brief periods when combined with hurricane-force winds and storm surges. Folly Island, shown here in 2012, experienced extensive erosion following Hurricanes Irene and Sandy in 2011. *(Alamy Photos)*

The coast, where the sea meets the land, is one of the most dynamic environments on Earth. Beaches, composed of sand or pebbles, and rocky coastlines attract tourists and new residents like few other areas, yet most of us have little understanding of how ocean waves form and change coastlines. In this chapter, we remove the mystery of the processes at work in coastal areas while retaining the wonder. We also explain the hazards resulting from waves, currents, and rising sea level and discuss how we can learn to live in the ever-changing coastal environment while sustaining its beauty.

After reading this chapter, you should be able to:

LO:1 Discuss the role of plate tectonics influencing coastal zone morphology and process.

LO:2 Explain coastal processes, such as waves, beach forms and processes, and rising sea level.

LO:3 Summarize the effects of sea-level rise on coastal processes.

LO:4 Explain why coastal erosion rates vary along different U.S. coastlines.

LO:5 Synthesize the coastal erosion hazard.

LO:6 Summarize the potential link between coastal processes and other natural hazards.

LO:7 Evaluate how human use of the coastal zone affects coastal processes.

LO:8 Summarize what we can do to minimize coastal hazards.

LO:9 List options available for coastal management.

3

Linkages exist between natural hazards.

Coastal erosion of Folly Island is linked to several other natural hazards, such as hurricanes which were responsible for extensive erosion of the island in both 2011 and 2012.

4

Humans can turn disastrous events into catastrophes.

The world's coasts are attractive places to live. Once developed, great expense is taken to preserve these narrow strips of land. Yet coastlines are some of the most dynamic landscapes on Earth, and human attempts to modify these environments often lead to unexpected negative consequences.

5

Consequences of hazards can be minimized.

When we return to Folly Island at the end of this chapter, you will learn about several strategies attempted to minimize the effects of coastal erosion on the island. But like the unexpected consequences of the Charleston jetties, some attempts only exacerbated the coastal erosion problem.

at disrupting the natural southward flow of sand down the Atlantic coast and keeping the harbor entrance open. Unfortunately, this southward flow of sand from the north regularly replenished the beaches to the south. As a consequence, severe shoreline erosion occurred on Folly and Morris Islands, which lie south of the entrance to Charleston Harbor and the jetties (Figure 11.1).

Folly Island, like coastline communities worldwide, is facing tough decisions about how to adapt to and manage coastal erosion in the future. Attempts to minimize the effects of coastal erosion of Folly Island over the past century have been diverse and have cost many tens of millions of dollars, with limited success. These investments are being made on a personal level to protect private property, at the state level to protect the island's economy, and at the federal level to reverse the damage caused by the Charleston Harbor jetties.

11.1 Introduction to Coastal Hazards

Coastal areas are dynamic environments that vary in their topography, climate, and organisms. In these areas, continental and oceanic processes converge to produce landscapes that are capable of rapid change.

Overall, coastal topography is greatly influenced by plate tectonics. The east coasts of the United States and Canada, as well as the Canadian Arctic coast, are described as tectonically passive because they are not close to a convergent plate boundary (see Chapter 2). Tectonically passive coasts typically have wide continental shelves with barrier islands and sandy beaches. Less typical are rocky shorelines, which are mostly restricted to New England, Atlantic Canada, and Arctic islands where ancient mountain ranges, such as the Appalachians, meet the Atlantic and Arctic Oceans. Along the Great Lakes, most shorelines are sandy, with rocky shores generally limited to northern areas that have experienced greater uplift following retreat of the Pleistocene glaciers.

In contrast, the west coasts of the United States and Canada are tectonically active because they are close to transform boundaries between the North American and Pacific plates (e.g., the San Andreas fault) and to convergent boundaries between the North American and Juan de Fuca or Pacific plates (e.g., the Cascadia subduction zone). Hawaii, Puerto Rico, the Virgin Islands, and southern Alaska are also tectonically active. In tectonically active regions, mountain building produces coasts with sea cliffs and rocky shorelines. Although long sandy beaches are present in these regions, they are not as abundant as along the eastern and Arctic coasts or the southern Great Lakes.

The topography of some American and Canadian coasts is significantly influenced by climate and organisms. For example, shorelines in parts of Alaska, Canada, and the Great Lakes are affected by seasonal ice or by the movement of present-day glaciers. Coastlines in temperate regions are influenced by marsh vegetation, and subtropical and tropical shores are affected by the growth of mangroves and offshore coral reefs.

The impact of hazardous coastal processes is considerable because many populated areas are located near the coast. In the United States, it is expected that most of the population will eventually be concentrated along

< FIGURE 11.1 Eroding Barrier Islands along the South Carolina Coast
Folly Island is a shrinking barrier island located south of the inlet to Charleston Harbor. One potential cause of the diminishing size of Folly Island is the jetties built to keep the harbor entrance free of sand, which consequently block sand being transported southward along the coast that would normally replenish the beaches of Folly Island. *(M-Sat Ltd./Science Source)*

the nation's 150,000 km (~93,000 mi.) of shoreline, including the Great Lakes. Today, the nation's largest cities lie in the coastal zone, and approximately 40 percent of the population lives in coastal counties, even though those counties account for only 10 percent of the land area (not including Alaska). In fact, population density for coastal counties is six times greater than that of inland counties. The population in coastal counties becomes even larger during peak vacation periods. For example, Ocean City, Maryland, receives an estimated 4 million visitors between Memorial Day and Labor Day.[2] Coastal problems will increase as more people live in coastal areas where the hazards occur. Once again, our activities continue to conflict with natural processes. Hazards along the coasts, such as coastal erosion, may become compounded by global warming, which is contributing to a worldwide rise in sea level (see Chapter 12).

The most serious coastal hazards include

> Strong coastal currents, including rip currents generated in the surf zone and tidal currents in narrow bays and channels

> Coastal erosion, which continues to produce considerable property damage that requires human adjustment

> Storm surge from tropical and extratropical cyclones, which claims many lives and causes enormous amounts of property damage every year (see Chapter 10)

> Tsunamis, which are particularly hazardous to coastal areas of the Pacific Ocean (see Chapter 4).

11.1 CHECK your understanding

1. How does plate tectonics influence coastlines?
2. What are the most serious coastal hazards?

11.2 Coastal Processes

WAVES

Waves that batter the coast are generated by offshore winds, sometimes thousands of kilometers from where the waves reach the shoreline. Wind blowing over the water produces friction along the water surface. Because the air is moving much faster than the water, the moving air transfers some of its energy to the water and produces waves. The waves, in turn, eventually expend their energy at the shoreline.

Waves vary in both their size and their shape. The size of waves in the ocean or on a lake depends on a combination of the following:

> The velocity or speed of the wind; the stronger the wind speed, the larger the waves.

> The duration of the wind. Winds that last longer, such as during storms, have more time to impart energy to the water, thereby producing larger waves.

> The distance that the wind blows across the water surface. This distance is referred to as the fetch. A longer fetch allows larger waves to form. This relationship is one reason that waves are generally larger in the ocean than in a lake.

Waves develop in a variety of sizes and shapes within an area that is experiencing a storm. As waves move away from their place of origin, they become sorted out into groups or sets of waves that have similar sizes and shapes. These sets may travel for long distances across the ocean and arrive at distant shores with little energy loss. If you stand on a beach watching the surf, you may be able to recognize these "sets" of similar-sized waves. Interactions between different sets from different sources produce a regular pattern or "surf beat." This pattern tends to repeat itself over time, allowing surfers to wait and take advantage of the sets of larger waves that they know will eventually arrive. Occasionally, however, a wave will appear that is much larger than the rest. Known as a *rogue wave*, such a huge wave is created by a number of factors.

Most rogue waves appear to form by *constructive interference*. In this process, multiple, similarly sized waves intersect to create a much larger wave. If they intersect just right, with both their crests and their troughs matching, the new wave may be as high as the sum of the intersecting waves. Undersea irregularities, as well as currents, also influence the formation of a rogue wave. Such waves can be extremely dangerous to the unsuspecting beachgoer.

When rogue waves strike the shore, lives may be lost. Such waves may appear out of nowhere, crashing down over a pier or cliff. Sadly, each summer, unsuspecting beach visitors are swept into the ocean by rogue waves and killed. Coastal zones are not the only dangerous areas. Rogue waves can also appear out of nowhere in the open ocean—in the middle of hundreds of kilometers of calm water (Figure 11.2). Such waves may be large enough to break in open water and threaten ships. In stormy seas, where almost all waves are large, rogue waves may reach 30 m (~100 ft.) in height. An initial study of three weeks of global satellite radar data found 10 separate rogue waves in the world's oceans that were taller than 25 m (~80 ft.). This is critical information for the shipping industry, since most ships

Crest of rogue wave

< FIGURE 11.2 Rogue Wave
This huge wave is approaching the bow of the *JOIDES Resolution*, the scientific drilling ship of the Ocean Drilling Program. Known as rogue waves, these large waves are responsible for sinking supertankers and large cargo-container ships. In April 2005, a 20 m (~65 ft.) rogue wave struck the cruise ship *Norwegian Dawn* about 400 km (~250 mi.) off the coast of Georgia. *(Personnel of NOAA Ship DELAWARE II)*

are designed to withstand 15 m (~50 ft.) waves.[3] Massive rogue waves were apparently common during the October 1991 "Perfect Storm," the intense extratropical cyclone depicted in both a book and a movie, that was responsible for sinking the fishing boat *Andrea Gail*, killing its entire six-person crew.

Waves moving across deep water have a similar basic shape or wave form (Figure 11.3a). Three parameters describe the size and movement of a wave: *wave height* (*H*), which is the difference in height between the trough and crest of a wave; *wavelength* (*L*), the distance between successive wave crests; and *wave period* (*T*), the time in seconds for successive waves to pass a reference point. Recall that we also used wavelength earlier in this book to describe the electromagnetic radiation and wave height of seismic waves as one indicator of earthquake magnitude. The reference point used to determine wave period could be a pier or similar object, which is anchored in the seafloor or lake bottom.

To understand how waves transmit energy through the water, it is useful to study the motion of an object on the water surface and one below the surface (Figure 11.3b). For example, if you were floating with a life preserver in deep water and could record your motion as waves moved through your area, you would find that you bob up, forward, down, and back in a circular orbit, always returning to the same place. If you were below the water surface with a breathing apparatus, you would still move in circles, but the circles would be smaller. That is, you would move up, forward, down, and back in a circular orbit that would remain in the same place while the waves traveled through.

The shape of the orbital motions changes as waves enter shallow water. At a depth of less than about one-half their wavelength, the waves begin to "feel bottom." This lack of room causes the circular orbits to become ellipses. Motion at the bottom may eventually be a narrow ellipse with a back-and-forth movement that is essentially horizontal (Figure 11.3c). You may have experienced this if you have stood or swum in relatively shallow water on a beach and felt the water repeatedly push you toward the shore and then away from the shore.

Far out at sea, the wave sets generated by storms are called *swell*. As swell enters shallower and shallower water, transformations take place, and the waves eventually become unstable and break on the shore. In deep-water conditions, mathematical equations can be used to predict wave height, period, and velocity, based on the fetch, wind speed, and duration of time that the wind blows over the water. This information has important environmental consequences: By predicting the velocity and height of the waves generated by a distant storm, we can estimate when storm-generated waves with a particular erosive capability will strike the shoreline and when to expect dangerous coastal currents.

When waves enter the coastal zone and shallow water, they impinge on the bottom and become steeper. Wave steepness is the ratio of wave height to wavelength. Waves are unstable when the ratio exceeds 1/7 (~0.14). That is, waves are unstable when wave height is about 14 percent (or more) of wavelength.

As waves move into shallow water, the wave period remains constant, but wavelength and velocity decrease, and wave height increases. The waves change shape from the rounded crests and troughs in deep water to peaked crests with relatively flat troughs in shallow water close to shore. Perhaps the most dramatic feature of waves entering shallow water is their rapid increase in height. The height of waves in shallow water, where they break, may be as much as twice their deep-water height (Figure 11.3c). Waves near the shoreline, just outside the surf zone, reach a wave steepness that is unstable. The instability causes the waves to break and expend their energy on the shoreline.[4]

We have said that waves expend their energy when they reach the coastline, but just how much energy are

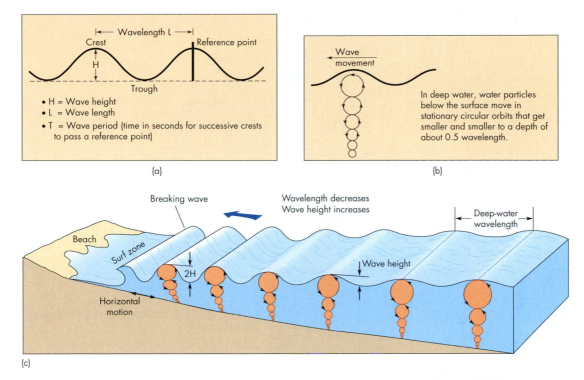

∧ FIGURE 11.3 Waves and Beaches
(a) Deep-water wave forms (water depth is greater than 0.5 L, where L is wavelength). The curving black line is the water surface, and the thick vertical black line is a pier or some other fixed object that can act as a reference point for determining the wave period (T). A black dashed line connects the bottom of the troughs and is the reference line for calculating wave height (H). (b) Motion of water particles associated with wave movement in deep water. The water particles follow the path of the arrows in the black circles. Wave movement is from right to left. (c) Motion of water particles as waves approach the shore. The water particles follow the path of the arrows in the circles. The waves are approaching the shore from right to left.

we talking about? The amount is surprisingly large. For example, the energy expended on a 400 km (~250 mi.) length of open coastline by waves with height of about 1 m (~3 ft.) over a given period of time is approximately equivalent to the energy produced by an average nuclear power plant over the same time period.[5]

Wave energy is approximately proportional to the square of the wave height. Thus, if wave height increases from 1 m (~3 ft.) to 2 m (~7 ft.), the wave energy increases by a factor of 4. If waves continue to grow to the 5 m (~16 ft.) height that is typical for large storms, then the energy expended, or wave power, will increase by 25 times as powerful as waves with a height of 1 m (~3 ft.).

Variations along a Coastline Although the heights of waves offshore from a coastline are relatively constant, the wave height along a coast may increase or decrease as waves approach the shore. These variations are caused by irregularities in the offshore topography of the seafloor and by changes in the shape of the coastline. One way to understand the variations in wave height is to

examine the behavior of the long, continuous crest of a single wave, or *wave front*, as it approaches an irregularly shaped coastline (Figure 11.4a).

Irregular coastlines have small rocky peninsulas known as *headlands*. The shoreline between headlands may be relatively straight or somewhat curved. Underwater, the offshore topography surrounding the headland is similar to that of the rest of the coast; that is, the water gets progressively shallower close to shore. This shallowing means that as a wave approaches the coast, it will first slow down in the shallow water off the headland. The slowdown will cause a long wave front to bend when it reaches a headland. This bending, referred to as *refraction*, causes wave fronts to become more nearly parallel to the shoreline.

Effects of Wave Refraction To visualize the effects of wave refraction, draw a series of imaginary lines, called *wave normals*, perpendicular to the wave fronts, and add arrows pointing toward the shoreline. The resulting diagram (see Figure 11.4a) shows that wave

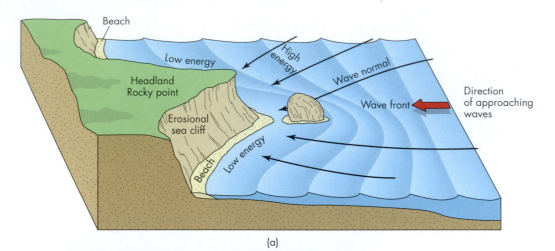

(a)

(b)

∧ FIGURE 11.4 Convergence and Divergence of Wave Energy

(a) Idealized diagram of the process of wave refraction and concentration of wave energy at rocky points called *headlands*. Refraction, or bending of wave fronts, causes the convergence of wave normals on the headland and their divergence in low-energy areas along the coast away from the headland. Wave normals are the imaginary long, curving black arrows. The red arrow indicates that the waves are approaching the shore from right to left. (b) Large waves striking a rocky headland along the Pacific coast at Pebble Beach in San Mateo County, California. *(Robert H. Blodgett)*

refraction causes a *convergence* of the wave normals at the headland and a *divergence* of the wave normals along the shoreline away from the headland. Where wave normals converge, both wave height and the energy expended by the waves increase. Thus, the largest waves along a shoreline are generally found at the end of a rocky headland (Figure 11.4b). The long-term effect of greater energy expenditure on protruding areas, such as headlands, is that wave energy is used to erode the coast, and the resulting erosion tends to straighten the shoreline.

Breaking Waves Waves also vary in how they break along a coastline. Breaking waves may peak quickly and plunge or surge, or they may gently spill, depending on local conditions. Waves that plunge are called *plunging breakers* (Figure 11.5a). They typically form on steep beaches and tend to be more erosive. Waves that spill are referred to as *spilling breakers* (Figure 11.5b). They commonly develop on wide, nearly flat beaches and are

more likely to deposit sand. The type of breaker that occurs along a coastline can change seasonally and with changes in underwater slope and topography. Overall, large plunging breakers that form as the result of storms cause much of the coastal erosion.

BEACH FORM AND PROCESSES

A **beach** is a landform consisting of loose material, such as sand or gravel, which has accumulated by wave action at the shoreline. Beaches may be composed of a variety of loose material. For example, the color and composition of beach sand are directly tied to the source of the sand: Many white sand Pacific island beaches are made of broken bits of shell and coral, Hawaii's black sand beaches are composed of fragments of volcanic rock, and brown Carolina beaches contain grains of the minerals quartz and feldspar.

(a)

(b)

ᐱ FIGURE 11.5 Types of Breakers
Idealized diagrams and photographs showing (a) plunging breakers on a steep beach and (b) spilling breakers on a gently sloping beach. *([a] Peter Cade/Stone/Getty Images [b] StockPhotosArt—Nature/Alamy Stock Photo)*

The Beach Onshore Beaches have a number of features, both onshore and offshore (Figure 11.6). Onshore, the landward extent of the beach is generally a cliff, called a **sea cliff** along the seashore and a *bluff* along a lakeshore, a line of sand dunes, or a line of permanent vegetation. Sea cliffs and lakeside bluffs develop by the erosion of rock or unconsolidated sediment. In contrast, coastal sand dunes are created by the deposition of wind-blown beach sand.

The onshore portion of most beaches can be divided into two areas: one that slopes very gently landward, called a berm, and another that slopes toward the water, called the beach face. *Berms* are nearly flat backshore areas formed by deposition of sediment as waves rush up and expend the last of their energy. They are often places where you'll find people sunbathing. Beaches may have more than one berm, or there may be no berm at all. A *beach face* begins where the beach slope changes direction and steepens toward the water. The part of the beach face that experiences the up-rush and backwash of waves is called the *swash zone*. This zone shifts in size and location with changes in water level resulting from storms or, on seacoasts, from tides.

The Beach Offshore Directly offshore from the swash zone are two distinctive zones in the water—the surf zone and breaker zone (see Figure 11.6). The *surf zone* is that portion of the nearshore environment where turbulent water moves toward the shore after the incoming waves break. Beyond the surf zone is the *breaker zone*, the area where incoming waves become unstable, peak, and break. These conditions on the water surface are reflected in the underwater topography. A sandbar, called a *longshore bar*, forms beneath each line of breakers in the breaker zone. Landward from a longshore bar is a *longshore trough*, which is formed by wave and current action. Both the bar and trough are usually elongate, parallel to the breaking waves. Wide and gently sloping beaches may have several lines of breakers and longshore bars.[4]

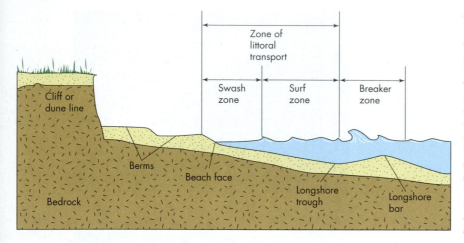

< FIGURE 11.6 Beach Terms
Basic terminology for landforms and wave action in the beach and near-shore environment. A sea cliff or line of coastal sand dunes on the left marks the landward extent of the beach. Two berms are shown in the beach sand, each gently sloping toward the cliff or dunes. The beach face, where the land meets the water, is more steeply sloping toward the water. One longshore sandbar and longshore trough are shown underwater; the sandbar forms below the breaker zone. The zone of littoral transport includes both beach drift in the swash zone and longshore drift in the surf zone.

Sand Transport The sand on beaches is not static; wave action constantly keeps the sand moving in the surf and swash zones. Storms erode sand from the beach and redeposit the sand either offshore or landward from the shoreline. Most of the sand moved offshore during a storm returns to the beach during fair-weather conditions.

A great deal of sand movement occurs parallel to the shoreline by **littoral transport** (*littoral* means "shore") in the swash and surf zones (Figure 11.7). This movement includes two processes, beach drift and longshore drift. In *beach drift*, the up-and-back movement of beach material in the swash zone causes sediment to move along a beach in a zigzag path, whereas *longshore drift* refers to the transport of sediment by ocean currents that flow essentially parallel to the shoreline. These currents, called *longshore currents*, are the primary mechanism for littoral transport. Both longshore drift and beach drift occur when waves strike the coast at an angle. The terms *up-drift* and *downdrift* are often used to indicate the direction in which sediment is moving or accumulating along the shore. For example, up-drift in Figure 11.7 is shown in the lower-left part of the diagram.

The direction of littoral transport along both the east and the west coasts of the United States is most often to the south, although it can be quite variable. Along U.S. seacoasts, most rates of longshore drift are between 150,000 and 300,000 m^3 (~200,000 and 400,000 yd.^{3}) of sediment per year. Although rates of longshore drift in the Great Lakes are much less, on the order of 6000 to 69,000 m^3 (~8000 to 90,000 yd.^{3}) per year, these rates are still substantial when you consider

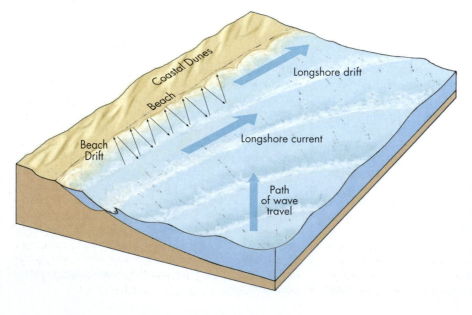

< FIGURE 11.7 Transport of Sediment along a Coast
Block diagram illustrating the processes of beach drift and longshore drift. Direction of beach drift is shown by the many arrows in the swash zone. The longshore drift direction is shown by the straight, thick (dark blue) arrow in the surf zone. Collectively, these two types of drift move sand along the coast in a process known as littoral transport.

that a typical dump truck carries a meager 8 m^3 (~10 yd.^{3}) of sand. Transported sediment is eventually deposited on shore in sand dunes, deposited on sandbars or barrier islands, or carried further out to sea.

11.2 CHECK your understanding

1. What three factors determine the size of waves?
2. What are rogue waves?
3. Define wave height, wavelength, wave period, and wave energy.
4. What is wave refraction?
5. Define plunging and spilling breakers.
6. What defines a beach?
7. What is littoral transport?
8. Differentiate between beach drift and longshore drift.

11.3 Sea-Level Change

Like the beach, the level of the sea at the shore is constantly changing. This change in sea level is caused by a number of processes, some of which operate locally and others that affect all the world's oceans. The position of the sea at the shore, referred to as **relative sea level**, is influenced by both the movement of the land and the movement of the water. These movements can be local, regional, or global in extent. Global sea level, also called **eustatic sea level**, is controlled by processes that affect the overall volume of water in the ocean and the shape of the ocean basins. Changes in eustatic sea level are just one of many factors that cause a change in relative sea level.

Eustatic Sea Level Global sea-level rises or falls when the amount of water in the world's oceans increases or decreases or when there is a change in the overall shape of ocean basins. Climate, primarily the average air temperature, exerts the greatest control on the amount of water in the ocean today. Air temperature influences both the average temperature of the ocean and the amount of water that is stored in ice on land. As the average temperature of the ocean increases, the volume of water expands; as it cools, the volume of water contracts. Referred to as *thermal expansion* or *contraction*, global warming or cooling of the atmosphere is responsible for this phenomenon. For example, present eustatic rise in sea level is about 3 mm or 0.12 in. per year, and about half of the rise in eustatic sea level in the world's oceans (currently about 1.6 mm or 0.06 in. per year)

results from thermal expansion caused by the warming of the atmosphere.[6]

Changes in air temperature also cause ice on land to melt or snowfall to increase. Thus, the volume of water frozen in glacial ice, ice caps, ice sheets, and permafrost is closely related to the average air temperature over years or decades. Most recently, global warming of our atmosphere has contributed to melting glaciers, ice caps, and the ice sheets in Greenland and Antarctica. This melting has increased the amount of water in the world's oceans and has contributed to about half of the eustatic rise in sea level (~1.5 mm per year).[6]

Over longer geologic time spans, ocean basins change shape as the result of plate tectonic processes. These large-scale processes, such as the rate of sea-floor spreading at mid-ocean ridges (see Chapter 2), influence eustatic sea level over long periods of time and are unlikely to contribute significantly to changes in relative sea level observed over decades or centuries.

Relative Sea Level Superimposed on eustatic sea level are local or regional processes that influence the movement of the land and water. The land can rise or fall slowly, as Earth's crust responds to the weight of the now-melted Pleistocene glaciers (see Chapter 12), or rapidly, in response to tectonic movements during an earthquake. For example, on Good Friday, 1964, the shoreline of Montague Island in Prince William Sound southeast of Anchorage, Alaska, rose close to 10 m (~33 ft.) in just a few minutes during a **M** 9.2 earthquake.[7] Many coastlines that are tectonically active and that experience frequent uplift in earthquakes may be less strongly influenced by eustatic sea-level rise caused by global warming.

Movement of the shoreline is also influenced by the rates of deposition, erosion, or subsidence along the coast. Input and output factors in the beach budget described in Section 11.5 determine whether a coastline is experiencing net erosion or deposition. Rates of coastal subsidence have a major influence on relative sea level on some shorelines, such as the Louisiana coastline described in Chapter 8.

On a daily basis, astronomical tides and weather conditions primarily control relative sea level. Astronomical tides are produced by the gravitational pull of the moon and, to a lesser extent, the sun. These tides cause relative sea level to fluctuate daily and seasonally as the position of a coast relative to the moon and sun constantly shifts. Although tidal fluctuations are entirely predictable, they can create hazardous currents and affect the height of storm surges. As mentioned previously, storm surge that occurs at high tide is generally more damaging than one that occurs at low tide (see Chapter 10).

Weather conditions also change relative to sea level over a period of hours or days. Changes in wind speed and atmospheric pressure influence the level of the sea. Wind speed has a greater effect than atmospheric pressure on relative sea level. In the open ocean, high winds pile up water and increase wave height, producing the swell described earlier. This swell increases both water level and wave heights when it reaches the shore. Storm surge from tropical storms, hurricanes, and extratropical cyclones crossing the coast are extreme cases of the swell moving landward. As mentioned in Chapter 10, the significant drop in atmospheric pressure in major hurricanes can add a meter or more to the height of the storm surge. For weaker tropical and extratropical cyclones, there is a smaller drop in atmospheric pressure and, thus, less of an effect on the storm surge. Finally, over short time periods (days to months), sea level may change up to about 30 cm (~12 in.) with El Niño conditions (an atmospheric and ocean event characterized by warm water, and along the U.S. West Coast, stormy conditions; see Chapter 12).

In summary, rapid changes in relative sea level contribute to coastal flooding and hazardous nearshore currents, whereas rising eustatic sea level over decades increases hazards from storm surge and coastal erosion. Over the long term, rising sea level threatens coastal cities and the very existence of many islands (see Chapter 12). We may need to rethink how to best reduce coastal erosion in a world of rising sea level.

11.3 CHECK your understanding

1. Define relative and eustatic sea-level rise, and explain the corresponding coastal hazards.
2. How does sea level change temporarily for very short periods of hours or days?

11.4 Geographic Regions at Risk from Coastal Hazards

Coastal hazards are present on both seacoasts and lakeshores. In particular, large lakes, such as the Great Lakes, Great Bear Lake, Great Slave Lake, and Lake Winnipeg, develop coastal conditions similar to those on the ocean. In North America, coastal processes may have hazardous consequences along the Atlantic, Pacific, Gulf, and Arctic coasts, as well as along the shores of the Great Lakes. Strong nearshore currents, sea-level rise, storm surge from cyclones, and tsunamis are not hazards on all coastlines. For example, tsunamis and hurricanes are generally absent in lakes.

In North America, only Lake Okeechobee in Florida and Lake Pontchartrain in Louisiana have experienced significant storm surges from hurricanes in historic times.

Coastal erosion is, however, a more universal hazard. As discussed later, there are various causes for this erosion, but the net effect is the horizontal retreat of the shoreline, sometimes at the rate of meters per year. All 30 U.S. states and the Canadian provinces bordering the ocean or the Great Lakes have problems with coastal erosion. Average erosion rates along some barrier islands in the southeastern United States may reach as high as 8 m (~25 ft.) per year but average closer to 0.3 to 1.0 m (~1 to 3 ft.) per year. Florida and the Gulf Coast states have highly variable erosion rates of 0.3 to 2 m (~1 to 7 ft. per year). Shorelines of the Great Lakes have experienced erosion rates up to 15 m (~50 ft.) per year[8] but average about 0.3 to 1 m (~1 to 3 ft.) per year. Along the U.S. West Coast erosion rates are also highly variable, from 0.1 to 0.3 m (~0.3 to 1 ft.) per year. Whatever the average rate of erosion at a particular coastline, local conditions (rock type and amount of sand on a beach, for example) at a particular time (when storm waves are present) and specific location (close to human structures such as a seawall) make average rates difficult to interpret.

Strong nearshore currents, especially the rip currents described later, are common on coastlines that have regular, strong surf conditions, such as those in California and Hawaii. These currents can develop on any coast with breaking waves, including the Great Lakes. Strong ocean currents are also a hazard on U.S. Gulf and East Coast beaches, where many people are rescued from rip currents each year. Strong coastal currents are especially dangerous around artificial structures, such as groins, jetties, and piers.

Astronomical tides can also produce strong currents in narrow bays and channels. On coastlines with a large range between high and low tides, the incoming flood tide can rush in rapidly. For example, the Bay of Fundy, between Nova Scotia and New Brunswick, Canada, has a tidal range that is 15 m (~50 ft.), and the incoming tide can rise 1 m (~3 ft.) in 23 minutes. Tidal currents rushing through the Golden Gate beneath the Golden Gate Bridge in San Francisco can reach 3 m per second (~7 mi. per hour). Although most tidal currents are predictable, the location and intensity of strong coastal currents can take even the strongest swimmers by surprise.[9]

The geographic areas at greatest risk for rising sea level are coastal areas and islands whose elevation is closest to present-day sea level or whose coastlines are subsiding. Coastal areas in North America that are close to sea level include most of the Gulf Coast, much

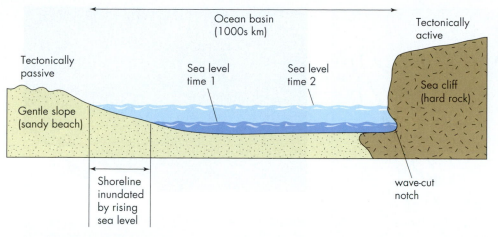

Rising Sea Level Has the Greatest Effect on Low Coastlines
Low, tectonically passive coastlines are most affected by rising sea level. As sea-level rises, the shoreline shifts the farthest distance inland on shallow coastlines.

of the East Coast from Long Island south to Florida, and parts of Arctic Canada and the North Slope of Alaska. In these areas, a 0.3 m (~1 ft.) rise in sea level can result in a landward shift of the shoreline anywhere from 30 to 300 m (~100 to 1000 ft.) (Figure 11.8). Areas that are rapidly subsiding, such as parts of the Louisiana coast, may experience greater inundation from rising sea level (see Chapter 8). Many low islands in the Pacific, called *atolls*, are in danger of becoming inundated with rising sea level. Some examples associated with the United States include the Marshall Islands, the Federated States of Micronesia, and the Republic of Palau in the western Pacific Ocean (see Chapter 12).

11.4 CHECK your understanding

1. List factors that help determine the rate of coastal erosion at a particular beach.

2. Why are average rates of erosion at the regional scale useful but difficult to interpret?

11.5 Effects of Coastal Processes

Coastal processes create hazards for both individuals and communities. Individuals face safety hazards from strong coastal currents when they are swimming or wading and hazards to both safety and property during storm surges and tsunamis. Communities face long-term hazards related to coastal erosion and rising sea level. Strong coastal currents can be produced by the backflow of water in the surf zone and by the rising and falling of the tide. Coastal erosion affects natural features, such as beaches, dunes, cliffs, and bluffs, and threatens beachfront houses and businesses, lighthouses and docks, as well as roads, streets, and utility lines running close to the shore.

On *barrier islands* and sandy peninsulas called *spits*, coastal processes constantly shift the location of the shoreline, eroding material in one location and depositing material in another. This can involve movement of an entire island, the inlets between islands, or the separation of a spit from the mainland. The southern end of Hog Island along the Delmarva Peninsula of Virginia has shifted nearly 2.5 km (~1.6 mi.) in historic times (Figure 11.9).[10] In Texas, Aransas Pass, the tidal inlet that provides ships access to Corpus Christi Bay, shifted more than 3 km (~2 mi.) between 1859 and 1911 before its location was stabilized with the construction of jetties.[11] Bayocean spit near Tillamook, Oregon, and the Nauset Beach spit near Chatham, Massachusetts, were both breached by storms in the twentieth century and separated from the mainland. The continual movement of barrier islands and spits is one of several characteristics that make them poor places for coastal development.

Our discussion of the effects of coastal hazards focuses on the effects of strong nearshore currents, called rip currents, and the effects of beach and cliff erosion. Discussion of the effects of tsunamis (Chapter 4), subsiding coastlines (Chapter 8), storm surge (Chapter 10), and rising sea level (Chapter 12) appears elsewhere.

RIP CURRENTS

The longshore currents described previously are not the only currents that develop along a beach. Under specific conditions along a seacoast or lakeshore,

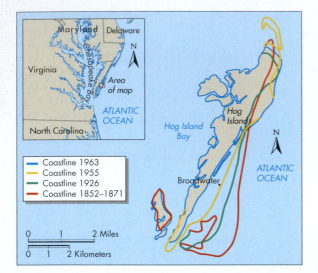

∧ **FIGURE 11.9 Barrier Islands Continually Change Shape and Location**
Hog Island in Northampton County, Virginia, shifted location and changed form from the mid-nineteenth to mid-twentieth centuries. The island was once covered with pine forest and was the site of Broadwater, a town of 300 people, 50 houses, a school, church, and lighthouse. A hurricane inundated the entire island in 1933, and by the early 1940s, the inhabitants had left. The town site is now under several meters of water. *(Based on Williams, S. J., Dodd, K., and Gohn, K. K. 1991. "Coasts in Crisis," U.S. Geological Survey Circular 1075.)*

(a)

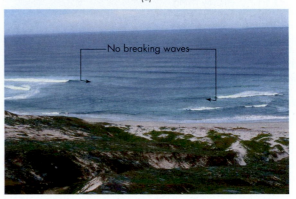

(b)

∧ **FIGURE 11.10 Rip Currents**
(a) The rip current is indicated by the blue milky water spreading outward from this rocky point. The blue milky water has sediment suspended in it. Dark stringers in the water are areas of kelp. *(Edward A. Keller)* (b) The existence of a rip current on this sandy beach is indicated by an area of smooth water where no waves are breaking. *(Edward A. Keller)*

powerful currents form that carry large amounts of water away from the shore. These currents, called **rip currents**, develop when a series of large waves piles up water between the longshore bar and the swash zone. The water does not return offshore the way it came in but, instead, is concentrated in narrow zones to form rip currents (Figure 11.10). Beachgoers and lifeguards often incorrectly call this flow "riptide" or "undertow." These currents are not tides and don't pull people under the water, but they can pull people away from shore (see Case Study 11.1: Survivor Story).

In the United States, more than 100 people are killed by rip currents annually, and rip currents are the cause of more than 80 percent of the 60,000 surf rescues each year. Therefore, rip currents constitute a serious coastal hazard, on average killing more people in the United States on an annual basis than do hurricanes or earthquakes; deaths from rip currents are equivalent to those from river flooding. People drown in rip currents because they panic and fight them by trying to swim directly back to shore. This feat is nearly impossible because the current can exceed 6 km

(~4 mi.) per hour, which even strong swimmers cannot maintain for long. A swimmer trying to fight a rip current soon becomes exhausted and may not have the energy to keep swimming.

Rip currents vary in width from about 15 m (50 ft.) to about 100 m (300 ft.) wide. They usually form in the surf zone at breaks in the longshore bar, but also around human structures, such as piers. They flow out perpendicular to the shoreline for a distance of tens to hundreds of meters offshore. Rip currents are fed by incoming breakers and longshore currents and may make their own channel as they pass through the longshore bar (Figure 11.11). Fortunately, they widen and dissipate once they have passed the line of breaking waves.

Rip Current: Two Experienced Swimmers Rescued on Florida Beach

I t was a hot summer's day in Pensacola, Florida, and Jennifer Kleinbaum and her fiancé Ernie decided to go to the beach. The waters of the Emerald Coast were still as glass, but a red flag warned visitors that it was too dangerous to swim. To be on the safe side, the couple decided to go for a stroll instead (Figure 11.1.A).

But the sand scorched their feet, and they waded ever farther into the water to stay cool. Before they knew it, they were treading water.

Even so, neither was worried. At 26, both were strong swimmers and certified SCUBA divers. Kleinbaum was certified as a master SCUBA diver, whereas Ernie had grown up in Pensacola, swimming at this very beach.

On the horizon, they spotted a triangular fin. A marine biologist, Kleinbaum dismissed it as a harmless dogfin shark. But Ernie, unnerved, suggested returning to shore.

It took Kleinbaum some time to notice they weren't making any headway. "They say if you're swimming and not getting any closer, don't fight it; swim parallel or float out of the path of the rip current. But I was fighting without realizing it," Kleinbaum said.

They drifted into an area where the waves crashed directly over their heads and found themselves unable to swim away. "I was treading water and trying to keep myself up. I can do this for hours in a pool. But I was struggling trying to avoid the waves, and that made me much more tired," Kleinbaum said.

Fatigued and frightened, Kleinbaum began to panic. She knew she couldn't hang on to Ernie, for fear of drowning him. "But I wanted to stay close to him, because there was nothing else I could see—he was my mental life raft." Gasping, she told him she couldn't keep her head above water much longer.

Ernie began waving his arm back and forth, yelling for help. Kleinbaum joined him. "Even though I was so tired, I somehow had no problem finding energy for this; it was something I could focus on," Kleinbaum said.

◀ FIGURE 11.1.A A Happier Day on the Beach
Jennifer Kleinbaum plays with her nephew Jake on a Florida beach. An earlier trip to a Pensacola, Florida, beach with her fiancé Ernie nearly ended in tragedy when they were both caught in a rip current while swimming near shore. *(Aaron Kleinbaum)*

On the beach, they saw people jump up and run away. The next thing Kleinbaum remembers, a man was pushing a red, sausage-shaped lifebuoy toward them, then helping tow the pair toward shore. An off-duty lifeguard, he sat them down on the sand to rest and called an ambulance.

Only then did Kleinbaum realize they had been caught in a rip current. "After he got us out of the water, our overriding emotion wasn't relief or thank you—it was oh, my god, I am so embarrassed. We kept looking at each other saying, how did this happen to us?"

She had thought it couldn't happen to a strong swimmer. Ernie believed it happened only to tourists. "There couldn't have been two people who knew more about the ocean and rip tides, what they were and how to get out of them. Yet we still made all the wrong choices," she said.

— KATHLEEN WONG

(a)

(b)

To safely escape a rip current, a swimmer must first recognize the current and then swim parallel to the shore until he or she is outside the current. Only then should the swimmer attempt to swim back to shore. If you cannot reach shore, wave an arm and yell for help. The key to survival is not to panic. When you swim in the ocean, watch the waves for a few minutes before entering the water and note the "surf beat," the regularly arriving sets of small and larger waves. Rip currents can form quickly after the arrival of a set of large waves. They can be recognized as a relatively quiet area in the surf zone where fewer incoming waves break (see Figure 11.10b). You may see the current as a mass of water and debris moving out through the surf zone. The water in the current may also be a different color because it carries suspended sediment. Remember, if you do get caught in a rip current, don't panic—swim parallel to shore before heading back to the beach, and you should be able to safely escape.

COASTAL EROSION

As a result of the continuing global rise in sea level and extensive development in the coastal zone, coastal erosion is becoming recognized as a serious national and worldwide problem. Coastal erosion is generally a more

◄ **FIGURE 11.11 Flow in a Rip Current**
(a) Bird's-eye view of the surf zone showing a rip current, which is the return flow of water that forms as a result of incoming waves. Return flow starts in the surf zone and goes through a low area in the longshore bar; flow expands outward once the current gets beyond the breaker zone. (b) Large rip cell on a popular beach; notice the center of the photo where the line of breakers have a gap – that is the rip current. (Prof. Hyeong-Dong Park, Seoul National University, Korea)

continuous, predictable process than other natural hazards, such as earthquakes, tropical cyclones, and floods. Large sums of money are spent in attempts to control coastal erosion, but many of the fixes are only temporary. If extensive development of coastal areas for vacation and recreational living continues, coastal erosion will certainly become a more serious problem.

Beach Erosion   An easy way to visualize erosion at a particular beach is to take a beach budget approach. A **beach budget** is similar to a bank account. In a bank account, deposits made to the account are input; the account balance reflects the storage of money in the account. Withdrawals by cash, check, or debit card are output from the account. Similarly, we can analyze a beach in terms of input, storage, and output of the sand or gravel that makes up the beach. Most input of sediment to a beach is caused by the coastal processes that move the sediment along the shore (see Figure 11.7) or by local wave erosion. Coastal processes, particularly beach drift and longshore drift, bring sediment from sources updrift of a beach. Local wave erosion of sand dunes or sea cliffs landward of the beach also results in inputs of sediment to the beach. Erosion of sand dunes and cliffs is generally most intense during storms when water floods the entire beach and large waves can reach inland. Sediment moving along a shoreline comes from one or more updrift sources, such as a river delta or erosion of another beach. The sediment that is in storage on a beach is what you see when you visit the site. Output of sediment occurs when coastal processes move sand or gravel away from the beach. These output processes can include littoral drift, a return bottom flow from storm waves that carries sediment into deeper water, and onshore winds that erode sand from the beach and blow it inland.

If input of sand to a beach exceeds output, the beach will grow as more sediment is stored and the beach widens. When input and output are about the same, the beach will stay fairly constant in width. If output of sediment exceeds input, the beach will erode and there will be fewer grains of sand or pieces of gravel on the beach. Thus, we see that the budget represents a balance of sediment on the beach over a period of years. Short-term changes in sediment supply from the attack of storm waves will cause seasonal or storm-related changes to the balance of sediment on a beach. Long-term changes in the beach budget caused by climate change or human impact result in either deposition or erosion of a beach.

A long-term change on some Gulf and East Coast beaches in the United States has been a reduction in sand supply from rivers. Construction of dams on rivers reaching the coast, such as the Savannah River in Georgia and the Rio Grande in Texas, has decreased the sand supply to beaches by trapping sediment behind the dam, thereby amplifying the coastal erosion problem. We now will consider the concept of the beach budget in greater detail with an idealized example.

Imagine a simple coast with rivers supplying sand, a sea cliff, a beach, and a submarine canyon (Figure 11.12). This coast defines a *littoral cell*, which is a segment of coast (a system) that includes sources and transport of sand to and along the beach (in this simple case, rivers deliver sand, as does sea cliff erosion). Sand is transported along the coast, and some is transported from the nearshore environment down a *submarine canyon* (an offshore canyon that may head in the surf zone and remove sand from the beach transport system). Now imagine that a dam was built in the area 15 years ago, which affects the amount of sediment supplied to the cell. Here, we determine the budget before and after the dam was constructed, confirming the erosion observed on beaches south of the submarine canyon where homes are threatened. Sources of sand for the beach are as follows:

(Sl) Littoral transport (+):
South (200,000 m^3/yr) – North (50,000 m^3/yr) = Net (150,000 m^3/yr) to the south

(Scf) Sea cliff erosion (+):
(0.55 m/yr)(6 m)(3000 m)(0.5) = 4500 m^3/yr
(Erosion rate is 0.5 m/yr; average height of sea cliff is 6 m; total length 3000 m. Assume 50 percent of material eroded remains on beach.)

(Scy) Down submarine canyon (–):
Estimated sediment removal from offshore observations: 220,000 m^3/yr

(Sr) River source (+):
Original drainage area of 800 km^2

To estimate the sediment delivered to the coast from the river, we can use equations (numerical models) or regional graphs. In this case, the area of the basin contributing sediment before the dam was constructed was about 800 km^2, and we estimate that the sediment delivered to the beach from the river was about 300 m^{3}km^2/yr. Following dam construction, the sediment-contributing area is reduced to 500 km^2 with a yield of 400 m^{3}km^2/yr. The increase in sediment per unit area results because smaller tributaries below the dam have a higher sediment yield.

Assuming that 30 percent of sediment delivered from the river will remain on the beach, and that it is sand sized and larger, we can calculate the sediment yield from the river before and after dam construction:

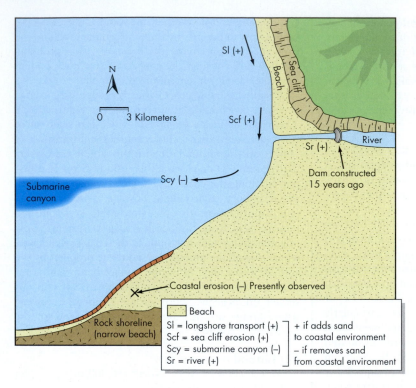

Beach erosion is observed at Point X. See text for calculation of the beach budget before and after construction of the dam.

Before dam:

$Sr(+) = (300 \ m^3km^2/yr)(800 \ km^2)(0.3) = 72,000 \ m^3/yr$

After dam:

$Sr(+) = (400 \ m^3km^2/yr)(500 \ km^2)(0.3) = 60,000 \ m^3/yr$

Now that we know the amount of sediment added to the system from the river before and after the dam was constructed, we can calculate the budget:

Budget before dam:

Longshore drift (Sl):	$+150,000 \ m^3/yr$
Cliff erosion (Scf):	$+4,500 \ m^3/yr$
River (Sr):	$+72,000 \ m^3/yr$
Submarine canyon (Scy):	$-220,000 \ m^3/yr$

Budget is $+6500 \ m^3/yr$ because more sand was arriving at point X on Figure 11.12 than is leaving; the beach was growing before the dam was built.

Budget after dam:

Longshore drift (Sl):	$+150,000 \ m^3/yr$
Cliff erosion (Scf):	$+4,500 \ m^3/yr$
River (Sr):	$+60,000 \ m^3/yr$
Submarine canyon (Scy):	$-220,000 \ m^3/yr$

Budget is $-5500 \ m^3/yr$; therefore, more sand is leaving point X than is arriving, and erosion is observed. The beach will get smaller over time.

Sea Cliff Erosion   A sea cliff or lakeshore bluff along a coastline may be subject to additional erosion problems. Both features are exposed to a combination of wave action and land erosion by running water and landslides that erode the cliff at a greater rate than either process could do alone. The problem is further compounded when people alter a cliff through poor development.

Erosion by plunging breakers during storms causes many sea cliffs and lakeshore bluffs to retreat. This susceptibility to erosion is not always apparent during low-water periods when a beach is present at the base of the cliff (Figure 11.13). The beach sediment helps protect the cliff from wave erosion during periods between storms. On the U.S. West Coast, most of the sea cliff erosion takes place during winter storms, and in the northeastern United States and Atlantic Canada, considerable cliff erosion occurs during nor'easters (see Chapter 10). Storm waves also help erode lakeshore bluffs on the Great Lakes.

A variety of human activities can increase the erosion of sea cliffs and lakeshore bluffs. These activities create effects that include uncontrolled surface runoff, increased groundwater discharge, and the addition of weight to the top of a cliff. Increased surface runoff, as often occurs with urbanization, can enhance the erosion of coastal cliffs and bluffs unless the drainage is controlled and diverted from the cliff face. Erosion can also intensify if there is an increase in the amount

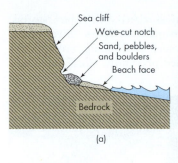

(a)

⋀ FIGURE 11.13 Sea Cliff and Beach

(a) Generalized cross-section and (b) photograph of sea cliff, beach, and bedrock exposed at low tide at Santa Barbara, California. *(Donald W. Weaver)*

(b)

of groundwater discharging from the base of the cliff. Watering lawns and gardens on top of a sea cliff or lakeside bluff can increase the amount of water in the earth material. When this water emerges as seeps or springs from the cliff, it reduces its stability and facilitates both erosion and landslides.[12] Structures such as walls, buildings, swimming pools, and patios may also decrease the stability of a cliff or bluff by increasing the driving forces for mass wasting (Figure 11.14). Strict regulation of development in many areas of the coastal zone now forbids most risky construction, but we continue to live with some of our past mistakes.

› FIGURE 11.14 Sea Cliff Erosion

Apartment buildings on the edge of a sea cliff in the university community of Isla Vista, California. The sign states that the outdoor deck is now open. Unfortunately, the deck is not particularly safe, as it is overhanging the cliff by at least 1 m (~3 ft.). Notice the exposed cement pillars in the sea cliff, originally in place to help support the buildings. These decks and apartment buildings were in imminent danger of collapsing into the sea and were condemned in 2004. *(Edward A. Keller)*

The rate of sea cliff and lakeshore bluff erosion is variable, and few measurements have been available. More are on the way because of the increasing use of remote sensing devices, such as an aircraft-mounted laser system called "light detection and ranging" (LIDAR). This system can record several thousand elevation measurements each second with a vertical resolution of 15 cm (~6 in.). Once a baseline set of elevations is recorded, subsequent LIDAR flights can detect changes in the coastal zone, such as the shape of the beach, dunes, bluff, or cliff. For example, LIDAR flights along the coast near Pacifica, California, about 10 years apart showed beach erosion of about 1 to 2 m (~3 to 6 ft.) and sea cliff erosion of about 1 to 3 m (~3 to 10 ft.) over the 10-year period.

Moderate rates of erosion of 0.1 to 0.3 m (~4 to 12 in.) per year occur along other parts of the California coast near Santa Barbara, in comparison to 2 m (~7 ft.) per year along parts of the Norfolk coast in the United Kingdom and up to 4.6 m (~15 ft.) per year along the eastern side of Cape Cod, Massachusetts. The rate of erosion depends primarily on the resistance of earth materials and the energy of the waves,[12] as well as sand supply.

11.5 CHECK your understanding

1. How are rip currents produced, and how can you escape from one?

2. Why is the beach budget such a valuable tool in risk analysis for coastal erosion?

3. Assume there are several beach homes near point X in our beach budget example. What advice would you give them based on the calculations? What would be your response if there were high-rise coastal resorts near point X?

11.6 Linkages between Coastal Processes and Other Natural Hazards

Coastal processes such as waves and currents present numerous hazards to human life and property. These hazards are also often linked to other natural hazards, such as earthquakes, volcanic eruptions, tsunamis, cyclones, flooding, landslides, subsidence, and climate change. Earthquakes such as the **M** 9.1 Sumatran earthquake of 2004, coastal volcanic eruptions, and tsunamis can radically change the shape of the shoreline. Storm waves, storm surge, and coastal flooding from cyclones are closely linked to coastal erosion (Figure 11.15).

Intense precipitation associated with tropical storms, hurricanes, and extratropical cyclones drives many other coastal hazards, such as flooding, erosion, and landslides. As mentioned previously, storm surge and

⋀ FIGURE 11.15 Storms Contribute to Coastal Erosion Two hurricanes, 21 days apart, eroded sand from this coastline in Vero Beach on Florida's east coast. The top picture, taken in August 1997 before the hurricanes, shows a sand dune providing some protection for the homes. In the middle picture, taken on September 8, 2004, after Hurricane Frances, the dune has migrated next to the houses. The third picture, taken on September 29, 2004, after Hurricane Jeanne, shows a house that has been undermined and damaged by storm waves. A red arrow points to the same place in all three photographs. *(USGS)*

heavy rainfall inland combine to cause widespread coastal flooding. Areas where the coast has subsided are more vulnerable to both freshwater flooding and storm surge.

Another common hazard caused by coastal processes is landslides. Wave erosion at the base of sea cliffs and lakeside bluffs undercuts the slope and frequently produces landslides. These cliffs may be more susceptible to sliding if structures, such as homes, add weight at the top (Figure 11.16).

Coastal erosion is also linked to climatic conditions that change over decades. For example, an interaction between surface winds and sea surface temperatures in the Pacific Ocean produces a climatic condition known as El Niño (see Chapter 12). Strong El Niño conditions can increase the intensity and frequency of extratropical cyclones making landfall on the U.S. West Coast. Large waves and storm surge from storms produced by a strong El Niño in 1982 to 1983 significantly increased coastal erosion in California and Oregon.[13]

Coastal erosion has significant linkages with global warming which is warming the ocean, raising sea level and increasing the energy and, thus, the intensity of high-energy storms. As sea-level rises and more intense storms occur, coastal erosion is a growing concern (see Chapter 12).

Overall, coastal processes are closely linked to other natural hazards that affect the coast: tsunamis; freshwater flooding of coastal plains, bays, and lagoons; hurricanes; and landslides of sea cliffs along the ocean and lakes.

11.6 CHECK your understanding

1. How are coastal processes linked to other hazards, such as earthquakes, tsunamis, landslides, volcanic eruptions, hurricanes, and flooding?

11.7 Natural Service Functions of Coastal Processes

As is true of many of the other hazards we have discussed, it is difficult to imagine the benefits of rip currents, coastal erosion, and sea-level rise; however, some coastal processes provide significant benefits.

∧ FIGURE 11.16 Hazardous Room with a View
These homes perched on a sea cliff along Puget Sound west of Port Townsend, Washington, were spared in this 1997 landslide. Future landslides will cause further retreat of the cliff face. *(Geology and Earth Resources Washington)*

Although erosion is a problem for property owners in coastal areas, the attraction of the coastal zone results in part from wave action and erosion. Many tourists drive the Pacific Coast highway from Washington to California each year to experience the beauty of the coastal zone. The stunning cliffs, rocky headlands, and sea arches found along this scenic road are the direct result of erosion and are important aesthetic resources (Figure 11.17).

For many beaches, coastal erosion is the only significant input of sand. Without erosion of dunes, cliffs, and bluffs inland from the beach, or the erosion of updrift beaches, there would be little sand to form a beach. Such is the case for most beaches on the eastern shore of Cape Cod in Massachusetts.

Sandy beaches have a number of natural service functions, including:

> Transporting and storage of sand in the coastal zone

> Supplying sediment for ecosystems, such as coastal lagoons, estuaries, and sand dunes

> Dissipating waves as a buffer to coastal erosion

(a thick mantle of sand helps protect the sea cliff from erosion)

> Breaking down pollutants and organic materials; cycling nutrients

> Providing nesting and roosting sites for shorebirds and seabirds

> Filtering and discharge of groundwater through beach sand and gravel

> Maintaining biodiversity in the coastal zone, including California grunion (a small fish that rides the high-tide waves to spawn in upper beach sand on summer nights)

> Providing food sources (sand crabs, beach hoppers, bloodworms, etc.) for migrating and resident shorebirds and fish, such as corvine and surf perch

> Providing a valuable cultural and aesthetic resource for people, as well as recreational opportunities.

The beach processes described earlier in the chapter maintain sandy beaches on all coasts, including lakeshores. Disturbance of the coastal zone and coral reefs by storm waves renews these ecosystems, keeping them

Sea arch

∧ FIGURE 11.17 Sea Arch Eroded from Rocky Headland
Sea arches, like this one along the Pacific coast in Mendocino, California, are formed by the erosion of a rocky headland. Located in a public park, this coastal erosion feature, at least temporarily, adds to the beauty of the coastline. (*Robert H. Blodgett*)

healthy and maintaining their diversity. Finally, coastal processes provide aesthetic value and much-loved recreational opportunities, including swimming, surfing, sailing, fishing, and sunbathing.

11.7 CHECK your understanding

1. List some of the natural service functions of beaches.

11.8 Human Interaction with Coastal Processes

Human interference with natural shore processes has caused considerable coastal erosion. Most problems arise in areas that are highly populated and developed. Efforts to stop coastal erosion often involve engineering structures that impede littoral transport. These artificial barriers interrupt the movement of sand, causing beaches to grow in some areas and erode in others, thus damaging valuable beachfront property. This type of human interaction with coastal processes is especially prevalent along the more densely settled parts of the Atlantic, Gulf, and Pacific coasts of the United States, the Great Lakes, and in some parts of Canada.

THE ATLANTIC COAST

The Atlantic coast from northern Florida to New York is characterized by barrier islands, long narrow islands of sand that are separated from the mainland by a lagoon or bay (Figure 11.18). Many barrier islands have been altered to a lesser or greater extent by human use.

∧ FIGURE 11.18 The Outer Banks of North Carolina
View looking north from the *Apollo-9* mission. The barrier islands appear as thin white ribbons of sand separating the Atlantic Ocean on the right from Pamlico Sound and the mainland on the left. The yellowish brown color in the water is caused by suspended sediment, such as clay, moving within the coastal system. Lobe-shaped fans of suspended sediment are entering the Atlantic Ocean at Ocracoke Inlet and the inlet to the northeast. Kitty Hawk, the site of the Wright Brothers' first powered aircraft flight, is near the top of the image. The distance from Cape Lookout to Cape Hatteras is approximately 100 km (~62 mi.). *(NASA)*

Maryland Barrier Islands

The history of barrier islands along the Maryland coast illustrates the interaction of human activity with coastal processes. Demand for the 50 km (~30 mi.) of oceanfront beach in Maryland is **very** high. This limited resource is used seasonally by residents of the Washington and Baltimore metropolitan areas (Figure 11.19). Probably the greatest repercussions from this use have taken place in Ocean City on Fenwick Island and on Assateague Island to the south of Fenwick Island. Since the early 1970s, Ocean City has promoted high-rise condominium and hotel development on its waterfront. To make way for new construction, the coastal dunes of this narrow island were removed in many locations. This activity and serious beach erosion have increased

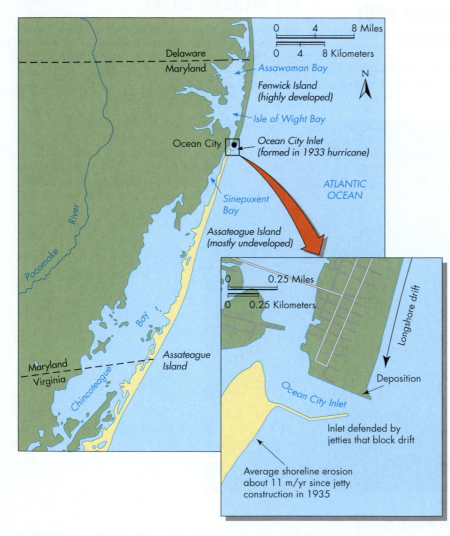

▲ **FIGURE 11.19 Urban Development with Jetty Increases Beach Erosion**
Fenwick Island, a barrier island on the Maryland coast, is experiencing rapid urban development, and potential hurricane damage is a concern. What if a new inlet forms during a hurricane in Ocean City to the north of the Ocean City inlet? Inset shows details of the Ocean City inlet and effects of jetty construction. The northern end of Assateague Island, shown in yellow, is relatively undeveloped and has experienced rapid shoreline erosion since a jetty was constructed along the inlet.

the vulnerability of the island to hurricanes. As mentioned previously, coastal dunes act as natural barriers to storm waves and partially protect structures built behind them. Without coastal dunes, a shoreline has no protection from storm waves or from the formation of washover channels by storm surge.

Ocean City experienced the effects of hurricane storm surge in 1933 when the Ocean City inlet formed to the south of the city. Despite attempts to stabilize the inlet by coastal engineering, there is no guarantee that a new inlet will not destroy part of the city in the future.[14]

Assateague Island is a barrier island located to the south, across the Ocean City inlet. It encompasses two-thirds of the Maryland coastline and contrasts with the highly urbanized Fenwick Island. Assateague is in a much more natural state and is used for passive recreation, such as sunbathing, swimming, walking, and wildlife observation. However, both islands are in the same littoral cell, meaning they share the same sand supply. At least that was the case until 1935, when jetties were constructed to stabilize the Ocean City inlet. Since construction of the jetties, beaches on the north end of Assateague Island have lost about 11 m (~36 ft.) per year, nearly 20 times the long-term rate of shoreline retreat for the Maryland coastline. During the same time, beaches immediately north of the inlet became considerably wider, requiring the lengthening of a recreational pier.[15]

Observed changes on Maryland's Atlantic coast are clearly related to human interference with the longshore drift of sand. Construction of the Ocean City inlet jetties interfered with the natural southward movement of sand. The jetties diverted sand offshore rather than allowing it to continue southward to nourish the beaches on Assateague Island. Starved of sand, the northern portions of the island have experienced serious shoreline erosion during the past 50 years. This example of beach erosion associated with engineering structures has been cited as the most severe in the United States.[15]

THE GULF COAST

Coastal erosion is also a serious problem along the Gulf of Mexico, which, like the Atlantic coast, has numerous barrier islands. One study in the Texas coastal zone suggests that in the past 100 years, human modification of the coastal zone has accelerated coastal erosion by 30 to 40 percent as compared with prehistoric rates.[16] Much of the accelerated erosion appears to be caused by coastal engineering structures, subsidence as a result of groundwater and petroleum withdrawal, and damming of rivers that supply sand to the beaches.

THE PACIFIC COAST

Human modification of the Pacific coast has not had as long a history of development as the Atlantic coast. Nevertheless, in coastal development, such as highways and harbors from San Diego to Seattle and in numerous small communities in between, the trend to armor the coast (usually by building seawalls) is very evident. In the past few decades, this trend for more and accelerated coastal defenses using rock and concrete has stalled in many areas. However, with the present rise in sea level and increased human use of beaches and interest in beach development, the debate about the future of the Pacific coast beaches and what action to take continues.

THE GREAT LAKES

Erosion is a periodic problem along the coasts of the Great Lakes and has been particularly troublesome on the Lake Michigan shoreline. Damage is most severe during high lake levels caused by above-normal precipitation. Measurements by the U.S. Army Corps of Engineers since 1860 show that the level of Lake Michigan has fluctuated about 2 m (~7 ft.). During periods of high water, the considerable wave erosion has destroyed many buildings, roads, retaining walls, and other structures (Figure 11.20).[17] For example, in 1985, high lake levels from fall storms caused an estimated $15 to $20 million in damages.

During periods of below-average lake level, wide beaches develop that dissipate energy from storm waves and protect the shore. However, as lake levels rise, beaches become narrower, and storm waves exert considerable energy against coastal areas. Even a small rise in water level on a gently sloping shore will inundate a surprisingly wide section of beach.[17]

Erosion of lakeshore bluffs has also been a problem. Many bluffs along Lake Michigan have eroded back at an average rate of 0.4 m (~1.3 ft.) per year.[18] Severity of erosion at a particular site depends on several factors:

> Coastal dunes—dune-protected bluffs erode at a slower rate

> Coastline orientation—sites exposed to high-energy storm winds erode faster

> Groundwater seepage—seepage at the base of a bluff causes slope instability and increased erosion

> Protective structures—engineering structures are locally beneficial but often accelerate coastal erosion in adjacent areas.[17,18]

In recent years, sand has been artificially added to some Great Lakes beaches. For some projects, the added sands have been deliberately chosen to be coarser and heavier than the natural sands, theoretically reducing erosion potential.

CANADIAN SEACOASTS

Canada has the longest ocean coastline in the world, and like the United States, it faces a serious erosion problem. As is true elsewhere, Canadian sandy and muddy beaches are eroding more rapidly than rocky beaches. Although most erosion rates are less than 1 m (~3 ft.) per year, some areas of Atlantic Canada are eroding at rates of up to 10 m (~30 ft.) per year! Even at close to the average rate of erosion, entire islands off Nova Scotia have disappeared into the sea. In the Canadian Arctic along the Beaufort Sea, erosion rates are a bit higher, at 1 to 2 m (~3 to 7 ft.) per year. Each year, the Canadian government spends about $1 billion on projects related to the coastline, including erosion control.[19]

11.8 CHECK your understanding

1. Describe the main causes of erosion at Assateague Island, located just south of Ocean City on Fenwick Island.

2. What factors affect the severity of erosion on Great Lake shorelines? How applicable are these factors to general coastal erosion?

11.9 Minimizing the Effects of Coastal Hazards

On the surface, it may appear that coastal hazards would be much easier to control than other natural hazards that release tremendous amounts of energy, such as earthquakes, volcanoes, and hurricanes. In practice, interactions among coastal processes are complex, and efforts to combat coastal erosion and rising sea level have often met with failure or unintended consequences. The following discussion focuses on minimizing beach erosion, one of the most widespread coastal hazards.

Most efforts to minimize beach erosion focus on stabilizing a beach at its present location. These efforts use two approaches:

> Hard stabilization: engineering structures to protect a shoreline from direct wave erosion

> Soft stabilization: adding sand to replace sand that has been eroded from the beach, a process known as beach nourishment.

HARD STABILIZATION

Engineering structures in the coastal environment, such as seawalls, groins, breakwaters, and jetties, are generally constructed to improve navigation or retard erosion. However, because these structures interfere with the littoral transport of sediment along the beach, they all too often cause undesirable deposition and erosion in their vicinity.

Seawalls Seawalls (or sea walls) are structures built on land parallel to the coastline to help retard erosion and protect buildings from damage. They may be constructed with concrete, large stone blocks called *riprap*, pilings of wood or steel, cemented sandbags, or other materials. Such walls are also called rock revetments. The construction of a seawall at the base of a sea cliff or lakeside bluff may not be particularly effective because the sea cliff is eroded from both the land and the water. Seawalls can fail for several reasons. For example, rock seawalls (riprap), unless carefully engineered, may fail as the rock blocks are washed away by large waves. Also, they may be overtopped (if not properly keyed in at the top) and erode behind the wall, leaving the rock intended to protect the coast stranded on the beach (Figure 11.20).

⋀ FIGURE 11.20 Rock seawall (rock revetment) at Goleta Beach in Southern California
This seawall was overtopped by waves in 2017 and a ~2m (6 ft) sea cliff constructed on decades old fill is eroding back, stranding the rock blocks (riprap) on the beach. The failure of the seawall reflects a poor design and engineered structure. There is controversy as to whether or not the seawall should ever have been constructed. (*Edward A. Keller, 2017*)

The use of seawalls has been criticized because their vertical design reflects incoming storm waves and redirects their energy to the shore. Therefore, seawalls, over a period of decades, promote beach erosion and produce a narrower beach with less sand (Figure 11.21). Unless carefully designed to complement existing land uses, seawalls generally cause environmental and aesthetic degradation. The design and construction of seawalls must be carefully tailored to specific sites. Because of these difficulties, many geologists believe that seawalls cause more problems than they solve and should be used rarely, if ever, where beach preservation is a goal.

Groins Groins are linear structures placed perpendicular to the shore, usually in groups called *groin fields*.

Each groin is designed to trap a portion of the sand from the longshore drift (Figure 11.22). A small amount of sand accumulates updrift of each groin, thus building an irregular but wider beach.

However, there is an inherent problem with groins—although trapped sand is deposited updrift of a groin, erosion occurs downdrift of these structures because the longshore drift has been depleted of sand. Thus, a groin or groin field results in a wider, more protected beach in a desired area but causes erosion of the adjacent shoreline. Once a groin has trapped all the sediment it can hold, sand in the groin is transported around its offshore end to continue its journey along the beach. Therefore, erosion may be minimized by artificially filling each groin. Known as **beach nourishment**, this process requires extracting sand from

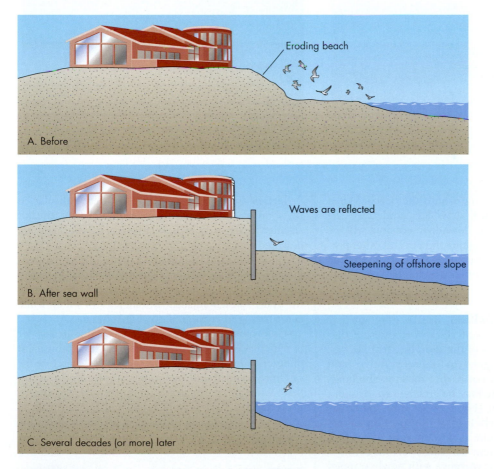

A. Before

Eroding beach

B. After sea wall

Waves are reflected

Steepening of offshore slope

C. Several decades (or more) later

∧ FIGURE 11.21
Seawalls Increase Beach Erosion
Construction of seawalls to mitigate coastal erosion and protect buildings from rising sea level results in beach erosion. A seawall acts as a reflector for storm waves and can result in the permanent loss of a beach.

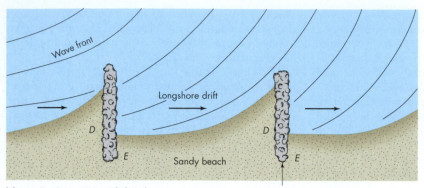

(b)

(c)

(a) Diagram of two beach groins built to trap sand; updrift depositional areas (D) are on the left, and the downdrift erosional areas (E) are on the right of the groins. Waves approach from the upper left, and longshore drift is from left to right. (b) Close-up of the eroded area downdrift of a groin. *(Edward A. Keller)* (c) A groin extends seaward in the center of the photograph; contrast the width and height of the beach on the updrift side to the right and the downdrift side to the left. *(Edward A. Keller)*

(a)　　D = Deposition, wide beach
　　　　E = Erosion, narrow beach

Beach groin, barrier to longshore drift, constructed of large rock blocks or other materials

the ocean floor and placing it onto the beach. When nourished, the groins will draw less sand from the longshore drift, and the downdrift erosion will be reduced.[4,13] Despite beach nourishment and other precautions, groins often cause undesirable erosion in the downdrift area; therefore, their use should be carefully evaluated.

Breakwaters and Jetties

Breakwaters and jetties are linear structures of riprap or concrete that protect limited stretches of the shoreline from waves. A **breakwater** is designed to intercept waves and provide a protected area or harbor for mooring boats or ships. This structure may be attached to, or separated from, the beach (Figure 11.23a,b). In either case, a breakwater blocks littoral transport and alters the shape of the coastline as new areas of deposition and erosion develop. This can cause serious erosion, or it can block a harbor entrance with newly trapped sand. The trapped sand may have to be moved by *artificial bypass*, that is, the dredging and pumping out of sand from around the breakwater to redeposit it downdrift.

Jetties are usually built in pairs, perpendicular to the shore at the mouth of a river or at the entrance of an inlet to a bay or lagoon (Figure 11.23c). They are designed to keep a ship or boat channel open and in a fixed location, with minimal dredging to remove sediment. Jetties also shelter a channel from large waves. Like groins, jetties block littoral transport, thus causing the updrift beach adjacent to one jetty to widen while contributing to erosion of downdrift beaches.

Unfortunately, it is impossible to build a breakwater or jetty that will not interfere with longshore movement of beach sediment. These structures must, therefore, be carefully planned and incorporate protective measures that eliminate or at least minimize adverse effects. Protective measures may include a dredging and artificial sediment-bypass system, a beach-nourishment program, seawalls, or some combination of these measures.[4,13]

SOFT STABILIZATION

Beach Nourishment

The use of beach nourishment to reduce rates of shoreline retreat is not limited to adding sand to engineering structures. In fact, in

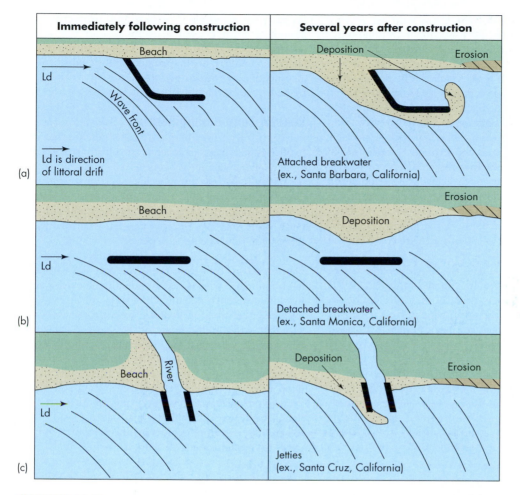

◄ FIGURE 11.23
Engineering Structures Built in the Surf Zone Cause Change
Diagrams illustrating the effects of breakwaters (the thick black lines in (a) and (b)) and jetties (the short black lines in (c)) on local patterns of deposition (tan) and erosion (brown). The left column of the diagrams shows the structures immediately following their construction, and the right column shows deposition and erosion after the structures were built. Curved, thin black lines show approaching waves, and the arrows indicate the direction of littoral transport (Ld).

most cases, it is used as an alternative to engineering structures. In its purest form, beach nourishment consists of artificially placing sand on beaches in the hope of constructing a positive beach budget, where more sand remains in storage than leaves in output. Beach nourishment is sometimes referred to as the "soft solution" to coastal erosion, as contrasted with "hard solutions" such as groins or seawalls. Ideally, a nourished beach will protect coastal property from the attack of waves. Beach nourishment is aesthetically preferable to many engineering structures, and it provides a recreational beach as well as some protection from shoreline erosion. Unfortunately, beach nourishment is expensive

and must be repeated at regular intervals to stabilize a beach. Considerable care must be taken in selecting the appropriate range of sand sizes for a beach replenishment project, and locating a nearby source for this sand is sometimes difficult.

In the mid-1970s, the city of Miami Beach, Florida and the U.S. Army Corps of Engineers began an ambitious beach nourishment program to reverse serious beach erosion that had plagued the area since the 1950s. The program was designed to produce a positive beach budget that would widen the beach and protect coastal resort areas from storm damage.[20] At a cost of approximately $62 million over 10 years, the project added about 160,000 m³ (~210,000 yd.³) of sand per year to replenish erosion losses. By 1980, about 18 million m³ (~24 million yd.³) of sand had been dredged and pumped from an offshore site onto the beach, producing a 200 m (~660 ft.) wide beach.[21] The change that took place over a decade is dramatic (Figure 11.24).

Part of the Miami project included building vegetated coastal dunes to function as a buffer to wave erosion and storm surge (Figure 11.25). Special wooden walkways allow public access through the dunes, but other areas of the dunes are protected. The successful Miami Beach nourishment project has functioned for more than 20 years and survived major hurricanes in 1979 and 1992; it is certainly preferable to the fragmented erosion control methods that preceded it.[4]

More than 600 km (~370 mi.) of U.S. coastline has received some sort of beach nourishment. Not all this nourishment has had the positive effects reported for Miami Beach. For example, in 1982, Ocean City, New Jersey, nourished a stretch of beach at a cost of just over $5 million. A series of storms that subsequently struck the beach eroded the sand in just 2½ months. The beach sands of Miami may yet be eroded and require replacement at a greater cost. Beach nourishment remains controversial, and some consider it nothing more than "sacrificial sand" that will eventually be washed away by coastal erosion.[21] Most beaches require frequent nourishments to remain intact. Nevertheless, beach nourishment has become a preferred method of restoring or even creating recreational beaches and protecting the shoreline from coastal erosion around the world. Additional case histories are needed to document the success or failure of the projects. Public education is also needed to inform people about what can be expected from beach nourishment.[4]

Coastal Zone Bioengineering *Coastal zone bioengineering* (CZB) is an emerging coastal defense that uses living or dead plants to assist in erosion control. CZB is a relatively new concept being tested. Sand dunes constructed along a coast to provide a barrier to ocean intrusion have been used many times, often as part of beach nourishment. The dunes may be stabilized with native dune grassland or other vegetation, insofar as the plants help hold the dune in place.

The Sargasso Sea, located in the North Atlantic Ocean east of the Carolinas, is the only sea without a land boundary. The sea is bounded by clockwise rotation of ocean currents, including the Gulf Stream, and is named after the free-floating Sargassum seaweed (algae) that grows there. Some of the seaweed is eventually carried westward to Galveston, Texas, where it periodically piles up on beaches. The seaweed provides valuable nutrients to the beach, but it is a nuisance to people visiting the beach, and crews often collect the decaying seaweed. Galveston is experimenting with utilizing the seaweed to help retard coastal erosion. The seaweed is collected with machinery similar to that used to gather and roll hay in fields. It is then

(a)   (b)

∧ Figure 11.24 Beach Nourishment
Miami Beach (a) before and (b) after beach nourishment. *(New Orleans District, U.S. Army Corps of Engineers)*

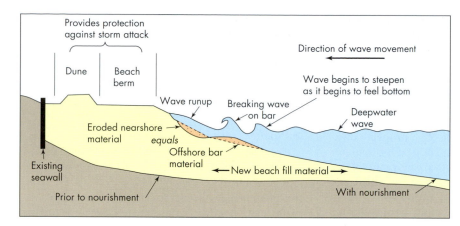

◀ **Figure 11.25 Miami Beach Shape after Nourishment**
Diagrammatic cross-section of the Miami Beach nourishment project. The dune and beach berm system provides protection against storm attack. An existing seawall is indicated by the thick black line on the left. New beach fill material is in yellow. Orange areas are beach sand that may be eroded and deposited during a storm. A dashed line shows a pre-storm profile of the beach. *(From the U.S. Army Corps of Engineers)*

compressed and buried with sand to form a kind of dune. The dune provides a first, although not a very strong, defense from storm waves and high tides.

Another example of CZB can be found at Refugio State Beach near Santa Barbara, California, where palm trees are located near the high-tide line. We do not know if the palms were planted with erosion control in mind, but the palm trees have a dense, strong root system that retards erosion. The tough, interwoven roots provide a first defense against beach erosion from high tides and storms.

11.9 CHECK your understanding

1. How can a seawall increase beach erosion?

2. Why are beach groins used? What problems can groins cause?

3. Why do breakwaters and jetties cause beach erosion problems?

4. What are the advantages and disadvantages of beach nourishment?

5. What is coastal zone bioengineering, and what are its advantages and possible shortcomings?

11.10 Perception of and Adjustment to Coastal Hazards

PERCEPTION OF COASTAL HAZARDS

People generally perceive land as being stable, and, in many cultures, it is viewed as a permanent asset that can be bought and sold.[10] This perception of stability is appropriate for most land where geologic changes are relatively slow and alterations to the land can take

centuries or millennia. Coastlines, however, are dynamic, and shorelines erode, accumulate, and move in time frames of months and years (see Figure 11.15).

Even if individuals understand that erosion can be rapid, their past experience, proximity to the coastline, and the probability of suffering property damage influence their perception of coastal erosion as a natural hazard. One study of sea cliff erosion near Bolinas, California, 24 km (~15 mi.) north of the entrance to San Francisco Bay, found that coastal residents in an area that was likely to experience damage in the near future were generally well-informed and saw the erosion as a direct and serious threat.[22] Other people living a few hundred meters from a possible hazard, although aware of the hazard, knew little about its frequency, severity, and predictability. Still further inland, people were aware that coastal erosion exists but had little perception of it as a hazard.

Many people do not perceive waves and currents as hazards. Often, their first visits to beaches are as children, where their parents closely limit their activities. In fact, waves that have large heights and wavelengths and plunging waves can be dangerous for wading and swimming. These waves are considerably more powerful than the spilling waves that generally provide safe conditions for waders and swimmers. Rip currents, which are commonly produced by plunging waves with large wavelengths, are responsible for about 80 percent of all lifeguard rescues on the seashore. Many beachgoers do not recognize the danger and try to enter the water where there are few breaking waves (see Figure 11.10b), only to be caught in a rip current.[23]

ADJUSTMENT AND MANAGEMENT OF COASTAL HAZARDS

Coastal erosion is usually a complex problem involving natural coastal processes of sand supply and people interfering with natural sand flow by building jetties, seawalls, rock revetments, small boat harbors, and other structures. Social, economic, and political forces often are present, so decisions about what to do about an erosion problem can be controversial. As a result, coastal erosion is often viewed as a wicked problem (see Chapter 1) for which a simple solution is elusive and difficult. A problem is that the various interests and values regarding a particular beach often are in conflict. Some people (e.g., property owners) want to protect the homes and other developments such as a beach park, while others are more interested in preserving a natural beach for recreational and ecologic reasons. Two sorts of mistakes are often made: implementation of a simple hard solution, such as sea walls (that may damage the beach), and doing little or nothing while declaring that the problem is too complex. The proper response is often to take an incremental approach with monitoring and periodic review. This may not be acceptable to homeowners or parks that want to preserve development at all costs.

Adjustments As with most other natural hazards, there is a variety of ways that we can adjust to shorelines with strong currents, coastal erosion, and rising sea level. Since strong currents cannot be prevented—they exist naturally, as well as around engineering structures—the best adjustments are improving education about and awareness of the conditions that create this hazard. For many ocean beaches and some beaches on large lakes, it may be best to restrict swimming to supervised locations when lifeguards are on duty. On these beaches, warnings about rip currents and tidal currents are often posted. In the United States, general beach and surf forecasts are issued by the National Weather Service. Information about coastal wave and weather conditions is also available through NOAA's now-COAST website (nowcoast. noaa.gov).

For the coastal erosion hazard, adjustments generally fall into one of three categories:[24]

1. Beach nourishment that tends to imitate natural processes—a "soft solution."

2. Land-use change that attempts to avoid the problem by not building in hazardous areas or by relocating threatened buildings—the "managed retreat solution" (also a soft solution).

3. Shoreline stabilization through structures such as groins and seawalls—a "hard solution." For example, a seawall may be constructed along an entire beach, or a coastal point of particular interest may be defended by coastal engineering.

Management Management of the coastal zone includes *adaptive management* and *integrated coastal zone management*. **Adaptive management** is a structured management concept with the following features:[25]

> Includes the use of science, statistical analysis, and modeling with the goal of reducing management uncertainty

> Promotes sustainable coasts and beaches—sustainability is a long-term intergenerational process to help ensure that future generations inherit a quality environment

> Communicates alternatives for discussion and negotiation—that is, the management is an open process

> Recognizes that future change is inevitable and that management can change as conditions change.

Integrated coastal zone management is similar to adaptive management and has the following goals:[26]

> To maintain the ecological function and integrity of the coastal zone

> To encourage sustainable solutions to coastal zone management

> To reduce or minimize potential and present conflicts resulting from development in the coastal zone, use of coastal resources, and recreational use of the coastal zone.

One of the first tasks in managing coastal erosion is determining the rates of erosion. Estimates of future erosion rates are based on historic shoreline change or on statistical analysis of the oceanographic environment, such as the waves, wind, and sediment supply that affect coastal erosion. Once local erosion rates have been determined, the Committee on Coastal Erosion Management, a special committee of the National Research Council (NRC), recommends that maps be made showing erosion lines and zones, referred to as *E-lines and E-zones*[24] (Figure 11.26). An *E-line* is the location of expected erosion in a given number of years; for example, the E-10 line is the location in 10 years. *E-zones* are similar to hazard zones in floodplains; that is, the E-10 zone would be the area between sea level and the E-10 line. Rising sea level is considered to be an imminent hazard in the E-10 zone, and no new habitable structures should be allowed in such areas.

NRC recommendations are then made concerning setbacks, which are considered to be minimum standards for state or local coastal erosion management programs. A *setback* is the distance from the shoreline beyond which development, such as the construction of homes, is allowed. The system of E-lines and E-zones

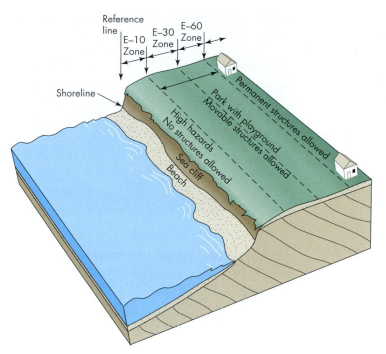

Reference line
E-10 Zone
E-30 Zone
E-60 Zone
Shoreline
Permanent structures allowed
Park with playground Movable structures allowed
High hazards No structures allowed
Sea cliff
Beach

◀ **FIGURE 11.26 E-Lines and E-Zones**
Conceptual diagram of E-lines and E-zones, based on the rate of coastal erosion from a reference point, such as the sea cliff or dune line. Width of the zones depends on the rate of erosion and defines setback distances. After 60 years of erosion, the structures will be much closer to the shoreline and vulnerable to erosion. It is a form of planned obsolescence. *(Based on National Research Council,* Managing Coastal Erosion. *Washington, DC: National Academy Press, 1990)*

can be used to establish setback distances. For example, if the erosion rate is 1 m (~3.3 ft.) per year, the E-10 setback is 10 m (~30 ft.). Movable structures are allowed in the intermediate and long-term hazard zones (E-10 to E-60), while permanent large structures are allowed at setbacks greater than the E-60 line. With the exception of those on high bluffs or sea cliffs, all new structures built seaward of the E-60 line should be constructed on pilings. Their design should withstand erosion associated with a high-magnitude storm with a recurrence interval of 100 years.

Only a few states, such as California, Florida, New Jersey, New York, North Carolina, and South Carolina, use a setback distance for buildings based upon the rate of erosion.[27] Nevertheless, the concept of E-lines and E-zones has real merit in coastal zone management and is at the heart of land-use planning that minimizes damage from coastal erosion.

We are at a crossroads today with respect to adjustment to coastal erosion. One road leads to ever-increasing coastal defenses in an attempt to control erosion. The second path involves learning to live with coastal erosion through flexible environmental planning and wise land use in the coastal zone.[24,28] The first path follows history in our attempt to control erosion by building engineering structures, such as seawalls. In the second path, most structures in the coastal zone are considered temporary and expendable; only a few critical facilities may be considered permanent. Development in the coastal zone must be in the best

interests of the general public rather than for a few who profit from it. This concept is at odds with the attitude of developers who consider the coastal zone "too valuable not to develop." In fact, development in the coastal zone is not the problem; rather, the problem lies in building in hazardous areas and areas better suited for other uses. In other words, beaches belong to all people to enjoy, not only those fortunate enough to purchase beach-front property. The states of Hawaii and Texas have taken this philosophy to heart. In Hawaii, all beaches are public property, and local property owners cannot deny access to others. Likewise, in Texas, virtually all of the ocean beaches are public property, and coastal zoning now requires avenues for public access.

Accepting the philosophy that, with minor exceptions, coastal zone development is temporary and expendable and that consideration should first be given to the general public requires an appreciation of the following five principles:[28]

1. Coastal erosion is a natural process rather than a natural hazard.

2. Any shoreline construction causes change. Such change interferes with natural processes and produces a variety of adverse secondary and tertiary changes.

3. Stabilization of the coastal zone through engineering structures protects property, not the beach itself. Most protected property belongs to relatively few people, at a large expense to the general public.

4. Engineering structures designed to protect a beach may eventually destroy it.

5. Once constructed, shoreline engineering structures produce a costly trend in coastal development that is difficult, if not impossible, to reverse.

If you consider purchasing land in the coastal zone, remember these guidelines: (1) allow for a good setback from the beach, sea cliff, or lakeshore bluff; (2) be high enough above the water level to avoid flooding and take into account rising sea level in making this determination; (3) construct buildings to withstand adverse weather, especially high winds; and (4) if hurricanes are a possibility, be sure there are adequate evacuation routes.[28] Remember that it is always risky to buy property where land meets water.

We conclude our discussion of coastal processes and hazards with two case histories: Cape Hatteras Lighthouse, where the decision was to move the structure, and Pointe du Hoc, France, where the decision was very different.

(a)

(b)

⋀ FIGURE 11.27 Lighthouse Moved
The historic Cape Hatteras Lighthouse was moved during summer 1999 to keep it from being destroyed by coastal erosion. (a) Cape Hatteras Lighthouse before it was moved. Only around 100 m (~330 ft.) from the water, the lighthouse could have been destroyed in a major storm. *(Don Smetzer/ PhotoEdit)* (b) Cape Hatteras lighthouse being moved; at its new location, the lighthouse should be safe for the next 50 to 100 years. *(AP Photo/Bob Jordan)*

11.3

The Cape Hatteras Lighthouse

Managed retreat is one method for dealing with coastal erosion and rising sea level. That was the option chosen by the National Park Service for protecting the historic Cape Hatteras Lighthouse. When the North Carolina lighthouse was built in the late nineteenth century—the tallest brick lighthouse in the world—it was approximately 500 m (~1650 ft.) from the Atlantic Ocean. By the early 1990s, it was only 100 m (~330 ft.) from the sea and in danger of being destroyed by a major storm. As you may recall from Chapter 10, Cape Hatteras is frequently a target for both hurricanes and nor'easters.

One probably would not expect the task of moving a 4400-ton lighthouse to be a matter of simple hydraulics. But this is how Rick Lohr, president of the International Chimney Corporation (ICC), described the task of moving the Cape Hatteras Lighthouse more than 800 m (~2700 ft.) when coastal erosion threatened to claim the historic structure (Figure 11.27).

The ICC was contracted to move the lighthouse to a safe location away from the shoreline in 1998, and it immediately set about making the necessary calculations to ensure that the project would go smoothly.

Among the preparatory tasks was testing the 880 m (~2900 ft.) path to the structure's new location for any hidden regions, such as subsurface holes that could collapse during the move. In order to find out about possible subsidence, the ICC conducted what is known as "proof rolling," in which a similar amount of weight (in this case the weight of the lighthouse) was hauled across the move route to test for any such holes. This did not require an extraordinary amount of weight. The total surface area of the structure (lighthouse) at ground level, including the hydraulic jacks and rollers the ICC installed, was about 455 m² (~4900 ft.²), making the average pressure at any given point roughly 10 tons per m² (~1 ton per ft.²).

As for the move itself, the company used a system known as the "Unified Jack System," invented in the 1950s by ICC employee Pete Freisen, who, in his mid-80s, served as a technical director for the Hatteras project. The system used 120 individual jacks, which allowed for every point to rise equally, no matter what the pressure.

After cutting the lighthouse from its foundation, the 120 jacks were placed under the structure and divided into three groups of 40 jacks each so that, during transport, a downward tip in any of the three groups would not stress the structure. The same method was used to move an entire terminal at Newark International Airport in 2000.

Even though computers now aid companies such as ICC in their initial calculations, such as determining the center of gravity of a structure, the science of moving large objects has existed since the turn of the century. Thanks to Rick Lohr and his engineers, the lighthouse is expected to be safe in its new location until the middle or end of the twenty-first century.

—CHRIS WILSON

11.4 Pointe du Hoc, France

CASE study

Pointe du Hoc is a promontory on the English Channel, several hundred kilometers northwest of Paris on the Normandy coast. The point is a historical site located on a coastal plain of low relief. The ocean side is a sea cliff, approximately 30 m (~98 ft.) high (Figure 11.28). Each year, many thousands of people come to see the location where the invasion to free Europe in World War II began on June 6, 1944. However, the point has experienced as much as 10 m (~33 ft.) of erosion in the past 60 years. As a result, there is concern that the observation post (OP) might actually fall into the sea. Some predictions indicated that the OP could not last more than a few decades at most.

The site is considered by many to be so important that expending a large amount of money to preserve the most vulnerable parts of the point is worthwhile. Of course, this is a value judgment. Those who wanted to preserve the last bit of rock on the point, rather than move some of the structures, obtained funding to evaluate whether the point could be preserved for another few decades by reinforcing it. This is an interesting approach to coastal erosion because it is a site-specific solution to a particular problem that involves coastal stabilization, rather than adjustment to the hazard, such as managed retreat. The case is also interesting because it involves detailed geologic evaluation and study of processes of sea cliff retreat, linked to a specific strategy to reduce the rate of coastal erosion, which has been estimated at 0.1–0.2 m per year.[29] Scientists and engineers studying the sea cliff realized that any attempt to absolutely stop sea cliff erosion was futile, but their objective was a temporary solution to buy as much as 50 years. Total erosion in the past 60 years has been approximately 10 m (~33 ft.), but in future decades, that rate might be significantly reduced.

The sea cliff was carefully evaluated and its geology studied in detail. The idealized profile of the sea cliff is shown in Figure 11.29, illustrating three geologic layers:[29]

> The top layer, which is up to about 9.0 m (~30 ft.) thick, is a silty, gravelly, sandy, clay soil. This layer has relatively low strength and is prone to small landslides and grain-by-grain erosion resulting from precipitation.

> The middle layer is a fractured limestone about 15 m (~50 ft.) thick that, with depth, becomes sandy and enriched with calcium carbonate muds. The limestone beds are nearly horizontal but have numerous nearly vertical fractures that are filled, to a lesser or greater extent, with weathered products of the limestone that are rich in clay. Precipitation from the surface can remove some of the clay, resulting in the fractures being more open. The fractured limestone is vulnerable to larger slope failures, particularly if the rock is underlain by areas of coastal erosion that have produced wave-cut notches and sea caves. In fact, investigators studying the site have observed that this is the dominant failure mechanism that results in the most rock being lost from the sea cliff. When wave erosion undermines the rocks at the base, they may fail along the fractures, falling onto the beach or into the sea.

> The lowest layer present is a black marl, which is a calcium carbonate–rich mudstone. Both the

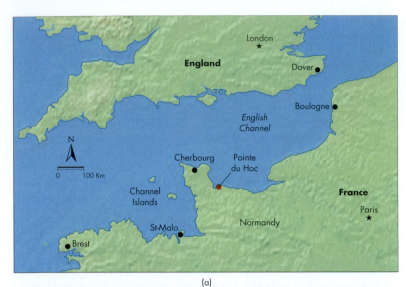

(a)

◀ FIGURE 11.28 Pointe du Hoc
(a) Location of Pointe du Hoc. (b) The historical site of Pointe du Hoc is on a coastal plain underlain by cretaceous limestone. Pits are bomb craters from World War II. *(Forget Patrick/Sagaphoto. com/Alamy)*

(b)

fractured limestone and the black marl have relatively high rock strength, compared to that of the overlying soil.

The sea cliff study suggested that the attack of waves at the rock face near the beach dislodges blocks of rock and creates caverns or sea caves. These are clearly visible along the base of the sea cliff and can be quite large, as illustrated in Figure 11.29. The caves grow larger and deeper with time and, eventually, leave the overlying rock unsupported, and a failure may occur along the fractures. The study concluded that the primary reason for collapse of the sea cliff was the presence of the caverns at the base of the cliff. As a result, the primary recommendation for reducing the rate of

erosion was to fill the caverns and stabilize the base of the sea cliff.

Work on this project began in 2010 with the goal of filling the caverns with several hundred cubic meters of concrete. Rock bolts were also used to secure the concrete and stabilize the rocks. After filling the caverns, an outermost layer of cement, which was specially designed to look like native rock, would be added. This was obviously a temporary solution because new sea caves could form, and the cement itself might be undermined and eroded. As a result, the study recommended a maintenance program based on inspections, recognizing that future work would undoubtedly be necessary. It is also important to recognize that the development of sea caves and large landslides was not the only process

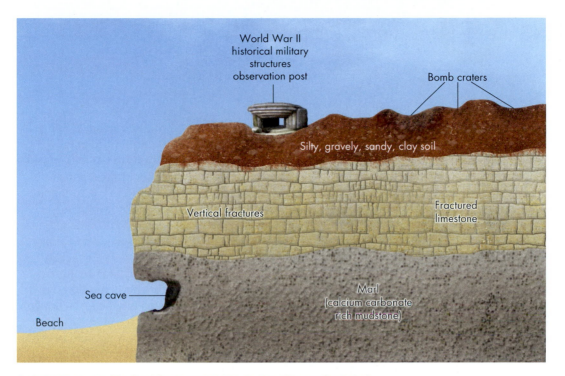

∧ FIGURE 11.29 Idealized Geology at Pointe du Hoc (Observation Point)
(Data from Briand, J.-L., Nouri, H. R., and Darby, C., Pointe du Hoc Stabilization Study. Texas A&M University, 2008.)

affecting the sea cliff. The top unit of the sea cliff is weak soil, and numerous small landslides were occurring on a regular basis. In fact, the top of the sea cliff has a wavy form, resulting from the many small landslides. Therefore, stabilizing the base of the sea cliff would not completely eliminate sea cliff retreat but would probably reduce it by about one-half. Thus, the work on the lower part of the sea cliff, which was completed by September 2010, was a temporary solution at best.

The decision to use a "hard solution" at Pointe du Hoc is an interesting one, based on the value judgment that the historical significance of the physical site is so great that measures needed to be taken to preserve it. The cost of the work is modest to date, at about $6 million; however, future maintenance costs will be incurred, and structures that are built in the sea cliff environment are often fortified even more in subsequent years, so the total cost in the future is unknown. The solution to coastal erosion and the decision to minimize it at Pointe du Hoc are obviously special circumstances for stabilizing a small length of sea cliff. Should such an approach be applied to a long section of sea cliff, the cost would be far too great to maintain. For now, the point is better protected, and the erosion rate will probably decrease, but, in the long run, the

environmentally preferable solution will be to move historically important structures back from the edge of the cliff. This is essentially the same solution that was implemented for the Cape Hatteras Lighthouse.

11.10 CHECK your understanding

1. What are the main adjustments to managing coastal erosion?

2. What are the attributes of adaptive management?

3. What are the goals of integrated coastal zone management?

4. What are E-lines and E-zones?

5. Why was moving the Cape Hatteras Lighthouse successful?

6. Could erosion control measures for Pointe du Hoc in France be widely applied to other eroding sea cliffs? Why or why not?

APPLYING the **5** Fundamental Concepts

Folly Island

1 Science helps us predict hazards.

2 Knowing hazard risks can help people make decisions.

3 Linkages exist between natural hazards.

4 Humans can turn disastrous events into catastrophes.

5 Consequences of hazards can be minimized.

Coastal erosion along much of the world's coastlines is a serious and costly problem, and nowhere is this truer than on Folly Island, South Carolina, where some beaches are receding at a rate of 2 m (~7 ft.) per year. The high rates of shoreline retreat on Folly Island is the result of both natural and human-caused changes to the environment that began in the late nineteenth century with the construction of the Charleston jetties, as discussed at the beginning of this

chapter. Just 7 km (~4.3 mi.) northeast of Folly Island, the rock jetties are perpendicular to the coastline and extend for 5 km (~3 mi.) into the Atlantic Ocean from the entrance to the Charleston Harbor (see Figure 11.1). By impeding the south-flowing longshore drift, which for thousands of years replenished Folly, the jetties upset the natural balance between sand supply and beach erosion on the island. The effects of this seemingly minor imbalance due to human attempts to minimize nature's will has combined with sea-level rise to create Folly's coastal erosion problem.

Folly Island has seen not only conflict between civilization and nature, but also conflicting approaches to long-term solutions to coastal erosion. By the middle of the twentieth century, it was clear that Folly's beaches were washing away, so construction of walls both perpendicular (groins) and parallel (seawalls) to the shoreline began following World War II. Between 1949 and 1970, the state highway department built 40

◀ FIGURE 11.30 Groins on Folly Island
Pictured are eight of the more than 40 groins on Folly Island, South Carolina, that trap sand moving by coastal currents. Groins create an irregular pattern of shoreline erosion and deposition and, in the end, contribute to coastal erosion downdrift from Folly Island. *(Bob Krist/Corbis)*

Pilings to elevate
house above
storm surge

Riprap

Seawall

◀ FIGURE 11.31 Attempts to Mitigate Coastal Flooding and Erosion on Folly Island
Property owners use a variety of methods to mitigate the effects of storms and coastal erosion on Folly Island, South Carolina, including this seawall constructed of large boulders, called riprap, and wood and the elevation of this house on wood pilings. Although seawalls may help protect buildings, they generally contribute to beach erosion. *(Robert H. Blodgett)*

groins along Folly's coast, which were composed of wood pilings sunk deep into the sandy island coastline (Figure 11.30).[1] Although the groins resulted in updrift beach deposition, they caused accelerated erosion downdrift of their locations. Property owners have also built walls of concrete, boulders, or wood parallel to the beach or have added sand on the beach in front of their homes to protect the houses from erosion and flooding (Figure 11.31). Although the seawalls provided temporary protection for buildings, over the long term these structures similarly led to excess erosion and damage elsewhere by locally amplifying the effects of storm waves.[1] For example, in 1985, a North Carolina land developer was permitted to construct a nine-story hotel on the Folly coast if a seawall was constructed to protect it. However, the hotel's seawall ultimately caused excess beach erosion and failed to protect the building from the 3.7 m (~12 ft.) storm surge of Hurricane Hugo in 1989.[1] State regulations have since been put in place to limit the construction of such coastal erosion structures.

Folly Island homeowners sued the U.S. Army Corp of Engineers in 1979 for constructing the Charleston jetties, and in 1987, the federal government agreed to pay for more than 70 percent of the cost of stabilizing Folly's beaches through the year 2037. To mitigate this erosion,

engineers, business interests, and some politicians have favored beach nourishment, the periodic addition of sand to the beach. Sand was added to the beach three times, in 1979, 1993, and 2005, by contractors for the Corps (Figure 11.32). In fact, in 1996, Congress made promotion of beach nourishment part of the Corp's mission.[30] The Corps estimates that, on average, sand will have to be added to Folly Beach every eight years. Although this frequency may seem high, it is not as high as estimated for Jupiter Beach, Florida, and Ocean City, New Jersey, where beach nourishment is projected for every three years![31] A study of sand added to Folly Beach in 1993 indicates that significant quantities of the sand placed on the beach are being carried out onto the continental shelf and completely removed from the beach system.[32]

Beach nourishment was again deemed necessary in 2012 after sustaining severe erosion on both ends of the island, due to Hurricanes Irene and Sandy in 2011 and 2012 (see opening photograph). However, federal budget battles have hindered procurement of the funds to pay for the replenishment. Beach nourishment on Folly Island has cost more than $24 million to date, up from the $15 million originally estimated. The Corps announced that $2 million in federal funds would be available

∧ FIGURE 11.32 Nourishment of Folly Beach
(a) A floating hydraulic cutter-head suction dredge in the distance sucked sand from the continental shelf and pumped a sand-and-water slurry to the shore in 2005. The slurry flowed along the beach in the steel pipe that appears in the foreground. *(Robert H. Blodgett)* (b) Sand and water discharged from the pipe on the beach, and the sand was moved in position by the bulldozer shown in (a). The U.S. Army Corps of Engineers spent more than $12 million to add approximately 1.7 million m³ (~2.3 million yd³) of sand to the beach in 2005. An estimated 1.9 million m³ (~2.5 million yd³) was added at a cost of $12.5 million in 1993. *(Robert H. Blodgett)*

for the project, but as of late 2013, the funds had not yet been provided.[31] With so much beach erosion and no certain plan for funding beach nourishment, residents began to fight regulations that limited construction of seawalls by private landowners. Ultimately the South Carolina Department of Health and Environmental Control overruled the coastal regulations in the spring of 2013, making it possible for a small number of residents to build mitigation structures to protect their homes. By that summer, the City of Folly Beach also decided it could no longer wait for federal assistance and announced that it would move forward with its own $3 million nourishment plan. Beach nourishment finally occurred in 2014 when about 765,000 cubic meters (1 million cubic yards) of sand was added to the beaches of the island. Unfortunately, Hurricane Joaquin hit Folly Island in October of 2015 with high waves and flooding. The storm removed about one-third of the sand just added in 2014. Beach nourishment is a gamble, insofar as the sand is considered sacrificial. Sometimes it lasts for a few years, but one large storm can remove it at any time.

Geologists study shoreline behavior over the long term and take into account the natural movement of barrier islands and rising sea level. They have observed that without hard structures, such as seawalls and groins, beaches generally do not disappear; they just shift location. Although shifting location may sound better than severe coastal erosion, shifting is just as serious once humans populate and develop these islands. Sea level in Charleston, just north of Folly Island, has been rising on average 3 mm (~0.12 in.) per year since 1920,[33] which is causing sand deposition to shift westward toward mainland South Carolina. Therefore, even in the absence of man-made structures such as the Charleston jetties or the groins on the island, natural processes are adding to Folly's coastal erosion problem. Moreover, the rate of sea-level rise is likely to increase as the effects of humans increase the rate of global warming.[6]

Although most ocean beaches in the United States are eroding like Folly Beach, few states and coastal communities are considering sea-level rise in their planning.[34] South Carolina's Beachfront Management Act, implemented in 1990, recognizes the coastal erosion hazard and establishes setback distances for new construction from the shoreline. However, because it has been developed since the

⌃ FIGURE 11.33 New Home Construction on Folly Island

After Hurricane Hugo's 3.7 m (~12 ft.) storm surge in 1989, more recent construction on Folly Island has been on higher piers. The base flood elevation on the 2004 FEMA Flood Insurance Rate Maps is 4.9 to 7 m (~16 to 23 ft.) above sea level for most of the island. With coastal erosion and rising sea level, this elevation may be too low to protect this house from storm surge for the remainder of the century. *(Robert H. Blodgett)*

1920s, Folly Beach is at least temporarily exempted from the setback provision.[1] What has yet to be addressed is the effect of rising sea level. New houses built on Folly Island are elevated on tall piers (Figure 11.33) at a height that is determined from the base flood elevation shown on FEMA flood insurance rate maps. This base flood elevation is based on present-day sea level, not the rise in sea level that is likely to occur during the lifetime of the structure.

Erosion on Folly Island varies from place to place, and there are erosion hot spots (particularly at the east and west ends). Although beach nourishment is a big part of management at Folly Island and additional hard structures, such as seawalls and beach groins, are also being considered, some homeowners are angry that regulations do not allow or encourage them to protect private property. Individual homes are now essentially on the high-tide beach and vulnerable to erosion and flooding.

The challenge for Folly Island, and much of the Atlantic and Gulf coasts, will be adopting coastal management policies and regulations that are forward-looking and take into consideration coastal erosion, local subsidence, and rising sea level. In some areas, difficult choices will have to be made between preserving beaches and coastal wetlands and protecting buildings and infrastructure.

APPLYING THE ❺ FUNDAMENTAL CONCEPTS

1. How can we better predict the amount of coastal erosion Folly Island will experience by 2050?

2. Assuming we can predict the amount of erosion that will occur on Folly Island in the next decade, how can this prediction be used by the city of Folly Beach and property owners to define the risk?

3. How is coastal erosion of Folly Island linked to hurricanes, flooding, and rising sea level from climate change?

4. In what ways have human processes impacted coastal erosion at Folly Island?

5. Given the history and information about Folly Island presented here, what would you suggest to minimize the erosion hazard?

CONCEPTS in review

11.1 Introduction to Coastal Hazards

LO:1 Discuss the role of plate tectonics influencing coastal zone morphology and process.

- The coastal environment is incredibly dynamic and capable of rapid change.
- Coastal morphology around the globe is greatly influenced by plate tectonics, and there are major differences between a coast on the leading edge of a continent and near a plate boundary (such as the U.S. West Coast) and a more passive coastline far from a major plate boundary (such as the U.S. East Coast).
- Impacts of coastal hazards are increasing for a variety of reasons, including the fact that about 50 percent of the population of the United States today lives in large cities in the coastal zone, or at least in coastal counties.
- The most serious coastal hazards include strong coastal currents, coastal erosion of beaches and sea cliffs, storm surge from hurricanes and cyclones (see Chapter 10), and tsunamis (see Chapter 4).

11.2 Coastal Processes

LO:2 Explain coastal processes, such as waves, beach forms and processes, and rising sea level.

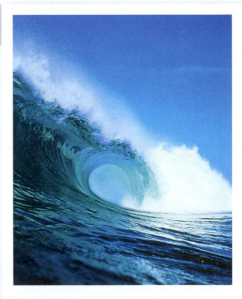

- Waves that arrive at coastlines are generated by offshore winds, and the size of the waves is related to the velocity or speed of the waves, the duration that the wind blows in the storm area, and the distance that the wind blows across the water surface (fetch).
- Waves develop in a variety of sizes and shapes. Rogue waves are unusually large waves formed by currents and undersea irregularities.
- As waves enter shallow coastal water, remarkable transformations take place. Wave period remains constant, but velocity decreases and height increases. When a wave reaches a critical steepness, it is no longer stable at the surface, and the wave breaks.
- Waves seldom arrive at the coasts with their wave fronts perfectly parallel to the coast. Because they arrive at an angle, a component of the wave power is directed along the shore, producing the littoral transport system, which includes beach drift along the beach face, as well as longshore drift in the surf zone.
- The two main types of breakers are spilling and plunging breakers. Spilling breakers build beaches, while plunging breakers are more destructive and may erode a beach.
- Basic terms used to describe a beach are *berms*, *beach face*, *longshore trough*, *longshore bar*, *swash zone*, *surf zone*, and *breaker zone*. At the back of a beach, there may be either sand dunes or, sometimes, a sea cliff.

KEY WORDS

beach (p. 444), **littoral transport** (p. 446), **sea cliff** (p. 445)

11.3 Sea-Level Change

| **LO:3** | Summarize the effects of sea-level rise on coastal processes. |

- Sea level is always changing, and it can change over time scales of hours and days to many thousands of years.
- Relative sea level refers to sea level influenced by tectonic movements of the land, as well as movements of the water and climatic factors. Relative sea level may change very rapidly in response to storm surge from hurricanes or to storm events, such as El Niño events, which can move a mass of warmer water toward the western coast of the United States.
- Eustatic sea-level curves refer to worldwide changes that are linked to climatic factors and global change affecting the overall volume of water in the ocean.

11.4 Geographic Regions at Risk from Coastal Hazards

| **LO:4** | Explain why coastal erosion rates vary along different U.S. coastlines. |

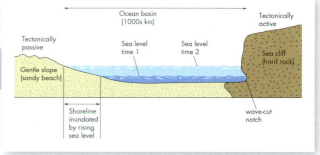

- Coastal erosion is essentially a worldwide hazard. This results because most coastlines are experiencing a rise in sea level, and that may drive coastal erosion.
- Rates of coastal erosion related to beaches and sea cliffs are highly variable. Coastal erosion along the east and west coasts of the United States vary from about 0.3 m (~1 ft.) per year to as much as 2 m (~7 ft.) or more per year. The lower rates, however, tend to dominate in the United States, with notable exceptions that are usually related to human processes that affect erosion.

11.5 Effects of Coastal Processes

| **LO:5** | Synthesize the coastal erosion hazard. |

- Individual hazards for people swimming in the ocean are related to strong currents. The most familiar of these is the rip current that develops as incoming breakers pile up water at the coast on the beach. This water moves out in discrete flow paths known as rip currents.
- Coastal erosion is a serious natural hazard. It is a generally continuous predictable process. The changes may be dramatic during storms, but, over a period of years, it is a relatively slow process.
- One way to visualize erosion at a particular beach is to compute the beach budget, which accounts for sources and sinks of sand at a particular beach. If input of sand exceeds output, then the beach will grow, and if output exceeds input, then erosion occurs.

- Sea cliff erosion occurs along many coastlines in the United States and around the world, as well as along the U.S. Great Lakes. Human processes can increase coastal erosion and sea cliff erosion through uncontrolled surface runoff, increased groundwater discharge, and added weight to the top of a sea cliff.

11.6 and 11.7 Linkages between Coastal Processes and Other Natural Hazards
Natural Service Functions of Coastal Processes

LO:6 Summarize the potential links between coastal processes and other natural hazards

- Coastal erosion is typically linked to other natural hazards such as earthquakes, volcanic eruptions, tsunamis, cyclones, flooding, landslides, and subsidence, as well as climate change.
- The coastal zone provides a variety of service functions for the natural and human-built environment. For example, coastlines provide aesthetic resources for people, but they do much more in terms of maintaining ecosystems, eroding and transporting sands, and building coastal sand dunes.

11.8 Human Interaction with Coastal Processes

LO:7 Evaluate how human use of the coastal zone affects coastal processes.

- Human use of and interest in the coastal zone is a major driving force that affects coastlines around the world. Human processes have caused considerable coastal erosion, particularly near highly populated and developed coastlines.
- Humans often attempt to control coastal erosion by building structures that are designed to minimize erosion. Often, these end up causing damage and further erosion to beaches.
- Coastal processes include both physical and human-induced factors. The physical factors include the existence of coastal dunes or sea cliffs, as well as the orientation of the coastline (exposure of the coast to waves); human factors include constructing protective structures (engineering structures) that might be locally beneficial but might accelerate erosion at adjacent areas.

11.9 Minimizing the Effects of Coastal Hazards

LO:8 Summarize what we can do to minimize coastal hazards.

- The two major approaches to coastal erosion and attempts by people to minimize it are the so-called hard solutions and the soft solutions.
- Examples of hard stabilization include construction of seawalls, groins, breakwaters, and jetties. Examples of soft stabilization include beach nourishment, land-use planning, and bioengineering approaches.

KEY WORDS
adaptive management (p. 468), **integrated coastal zone management** (p. 468)

11.10 Perception of and Adjustment to Coastal Hazards

LO:9 List options available for coastal management.

- People have attempted to adjust to coastal erosion in a variety of ways, and the predominant ones fall into three categories: beach nourishment, shoreline stabilization through engineering structures, and land-use change.
- Management of coastlines and beaches encourages the use of adaptive management, which utilizes science as an integral part, and integrated coastal zone management, which may include a mixture of hard and soft solutions.
- A number of general principles related to coastal erosion and processes must be considered: Coastal erosion is a natural process; any shoreline construction causes change; stabilizing coastal zones through engineering may protect property but not the beach; engineering structures designed to protect the beach may eventually damage or destroy it, and, once constructed, shoreline engineering structures produce a trend that is difficult to reverse.

KEY WORDS
beach nourishment (p. 464), **breakwater** (p. 464), **jetties** (p. 465)

CRITICAL thinking questions

1. Do you think that human activity has increased coastal erosion? Outline a research program that could test this question.
2. Do you agree or disagree with the following statements: (1) All structures in the coastal zone, with the exception of critical facilities, should be considered temporary and expendable. (2) Any development in the coastal zone must be in the best interest of the general public rather than the few who developed the oceanfront. Explain your position on both statements.
3. You have been asked by a coastal community to evaluate the feasibility of a beach nourishment project. Describe the types of information that you would require for your evaluation and how you would determine how often nourishment will be needed in the future.
4. Compare and contrast the shoreline features of the tectonically active and passive coasts of the United States and describe how coastal hazards differ in the two coasts that you have evaluated.

5. Assume you are living on a barrier island in the Gulf of Mexico that is experiencing rapid urbanization. Nearly everyone wants to live near the edge of the ocean, in other words, as close to the water as possible. It is expected that, in 10 years, the number of homes and people will increase dramatically from a few hundred now to a few thousand. You are hired by the community to help evaluate land use and potential effects of coastal erosion. What advice would you give to the communities on the barrier island, and what sorts of data would you need to gather to make conclusions concerning future land use? In addition, how would you handle the people problems? That is, how would you talk to people about erosion and what the consequences are likely to be and how the process might be minimized?

References

1. **Lemmon, G., Neal, W. J., Bush, D. M., Pilkey, O. H., Stutz, M.,** and **Bullock, J.** 1996. *Living with the South Carolina coast.* Durham, NC: Duke University Press.

2. **National Coastal Population Report.** Population Trends from 1970 to 2020. March 2013. NOAA State of the Coast Report Series, National Oceanic and Atmospheric Administration. http://stateofthecoast. noaa. gov/features/coastal-population-report.pdf. Accessed 7/30/13.

3. **Minkel, J. R.** 2004. Surf's up—way up. *Scientific American* 291(4): 38.

4. **Komar, P. D.** 1998. *Beach processes and sedimentation,* 2nd ed. Upper Saddle River, NJ: Prentice Hall.

5. **Davis, R. E.,** and **Dolan, R.** 1993. Nor'easters. *American Scientist* 81: 428–39.

6. **Intergovernmental Panel on Climate Change, Working Group I.** 2013. Contribution of Working Group I to the Fourth Assessment Report of the Intergovernmental Panel on Climate Change; Summary for policy makers. World Meteorological Organization. www.ipcc.ch/. Accessed 4/25/07.

7. **Bolt, B. A.** 2006. *Earthquakes,* 5th ed. New York: W. H. Freeman and Company.

8. **National Oceanic and Atmospheric Administration (NOAA).** 1998. *Population at risk from natural hazards* by S. Ward and C. Main. NOAA's State of the Coast Report. Silver Spring, MD: NOAA. http:// oceanservice.noaa.gov/websites/retiredsites/sotc_pdf/ PAR.PDF. Accessed 4/20/07.

9. **Garrison, T.** 2005. *Oceanography; An invitation to marine science.* Belmont, CA: Thomson Brooks/Cole.

10. **Williams, S. J., Dodd, K.,** and **Gohn, K. K.** 1991. *Coasts in crisis.* U.S. Geological Survey Circular 1075.

11. **Morton, R. A., Pilkey, O. H. Jr., Pilkey, O. H. Sr.,** and **Neal, W. J.** 1983. *Living with the Texas shore.* Durham, NC: Duke University Press.

12. **Norris, R. M.** 1977. Erosion of sea cliffs. In *Geologic hazards in San Diego,* ed. P. L. Abbott and J. K. Victoris. San Diego, CA: San Diego Society of Natural History.

13. **Komar, P. D.** 1997. The Pacific Northwest Coast: Living with the shores of Oregon and Washington. Durham, NC: Duke University Press.

14. **U.S. Department of Commerce.** 1978. *State of Maryland coastal management program and final environmental impact statement.* Washington, DC: U.S. Department of Commerce.

15. **Leatherman, S. P.** 1984. Shoreline evolution of North Assateague Island, Maryland. *Shore and Beach,* July: 3–10.

16. **Wilkinson, B. H.,** and **McGowen, J. H.** 1977. Geologic approaches to the determination of long-term coastal recession rates, Matagorda Peninsula, Texas. *Environmental Geology* 1: 359–65.

17. **Larsen, J. I.** 1973. *Geology for planning in Lake County, Illinois.* Illinois State Geological Survey Circular 481.

18. **Buckler, W. R.,** and **Winters, H. A.** 1983. Lake Michigan bluff recession. *Annals of the Association of American Geographers* 73(1): 89–110.

19. **Geological Survey of Canada.** 2006. *CoastWeb: Facts about Canada's coastline.* http://gsc.nrcan.gc.ca/coast/ facts_e.php. Accessed 4/20/07.

20. **Carter, R. W. G.,** and **Oxford, J. D.** 1982. When hurricanes sweep Miami Beach. *Geographical Magazine* 54(8): 442–48.

21. **Flanagan, R.** 1993. Beaches on the brink. *Earth* 2(6): 24–33.

22. **Rowntree, R. A.** 1974. Coastal erosion: The meaning of a natural hazard in the cultural and ecological context. In *Natural hazards: Local, national, global,* ed. G. F. White, pp. 70–79. New York: Oxford University Press.

23. **Leatherman, S. P.** 2003. Dr. Beach's survival guide: What you need to know about sharks, rip currents, and more before going in the water. New Haven, CT: Yale University Press.

24. **Committee on Coastal Erosion Zone Management.** 1990. *Managing coastal erosion.* National Research Council. Washington, DC: National Academy Press.

25. **Resilience Alliance.** *Adaptive management.* www. resalliance.org/index.php/adaptive_management. Accessed 6/30/13.

26. **European Commission.** 2013. *Integrated coastal zone management*. ec.europa.eu/environment/iczm. Accessed 6/30/13.

27. **Neal, W. J., Blakeney, W. C. Jr., Pilkey, O. H. Jr.,** and **Pilkey, O. H. Sr.** 1984. *Living with the South Carolina shore*. Durham, NC: Duke University Press.

28. **Pilkey, O. H.,** and **Dixon, K. L.** 1996. *The Corps and the shore*. Washington, DC: Island Press.

29. **Briand, J-L, Nouri, H. R.,** and **Darby, C.** 2008. *Pointe du Hoc stabilization study*. Texas A&M University.

30. **Dean, C.** 1999. *Against the tide: The battle for America's beaches*. New York: Columbia University Press.

31. **Petersen, B.** 2013. Folly Beach gets $20 M in federal beach renourishment money. *Post and Courier*, June

25. www.postandcourier.com/article/20130625/PC16/130629585. Accessed 7/31/13.

32. **Thieler, E. R., Gayes, P. T., Schwab, W. C.,** and **Harris, M. S.** 1999. Tracing sediment dispersal on nourished beaches: Two case studies. In *Coastal Sediments '99*, ed. N. C. Kraus and W. G. McDougal, pp. 2118–36. Reston, VA: American Society of Civil Engineers.

33. **Morton, R. A.,** and **Miller, T. L.** 2005. National assessment of shoreline change, Part 2: Historical shoreline changes and associated coastal land loss along the U.S. Southeast Atlantic Coast. U.S. Geological Survey Open-File Report 2005–1401.

34. **Dean, C.** 2006. Next victim of warming: The beaches. *New York Times*, June 20, 2006.

Sunny Isles Beach near Miami **has more than $10 billion in resort** and other property that is **becoming vulnerable** to climate change with rise **in sea level that causes** coastal flooding events.

12

APPLYING the 5 fundamental concepts Sea-level rise: Miami, Florida

①

Science helps us predict hazards.

Studying sea level change in the Miami area and predicting future changes are key to understanding the hazards linked to sea-level rise.

②

Knowing hazard risks can help people make decisions.

Understanding potential risk of damage to land and property from sea-level rise is critical in making decisions to manage the hazards of coastal flooding in Miami.

Climate Change and Natural Hazards

Poster Child for Climate Change in a Time of Sea-level rise and Denial

Miami Beach and adjacent areas in southern Florida, from Miami north to Fort Lauderdale and Palm Beach (including the famous Mar-a-Lago Estate), is at the forefront of climate change and sea-level rise that are causing ever higher and more frequent coastal flooding. Population is growing rapidly in South Florida, and property worth many billions of dollars are at risk from coastal flooding, especially during the highest tides of the year known as King Tides (Figure 12.1).

As a result of accelerating sea-level rise in South Florida, the number of coastal (tidal) flooding events has increased about 400 percent since about 2006.[1] The famous beach resorts of Sunny Isles

< Sunny Isles Beach, South Florida

With its high-rise hotels, Sunny Isles Beach has more than $10 billion in development. The beach is on a narrow barrier island between the Atlantic Ocean and Coastal Waterway. Accelerating rise in sea level as a result of human-induced climate change is threatening much of the South Florida coastline. *(Alamy)*

Many natural hazards, such as drought, heat waves, floods, hurricanes, blizzards, and wildfire, are related to climate and climate change. A basic understanding of climate science is needed to comprehend the mechanisms of these hazards. After reading this chapter, you should be able to:

LO:1   Explain the difference between climate and weather and how their variability is related to natural hazards.

LO:2   Describe the basic concepts of atmospheric science, such as the structure, composition, and dynamics of the atmosphere.

LO:3   Synthesize how we have studied past climate change.

LO:4   Summarize how climate has changed during the past million years through glacial and interglacial conditions and how human activity is altering our current climate.

LO:5   Name and explain the potential causes of climate change and link these causes to natural hazards.

LO:6   Speculate on how we might better predict future climate change.

LO:7   Propose ways to mitigate climate change and associated hazards.

3

Linkages exist between natural hazards.

Coastal erosion and flooding from sea-level rise are linked to rising sea level due to global warming. Sea-level rise also increases the cyclones and tsunami hazard for coastal areas.

4

Humans can turn disastrous events into catastrophes.

Worldwide population growth has compounded the effects of climate change, such as sea-level rise due to warming, in addition to increasing the global risk from rising seas due to dense urbanization of low-lying coastal regions.

5

Consequences of hazards can be minimized.

Consequences of sea-level rise can be temporarily minimized at specific locations by engineering projects, but global solutions will require international cooperation to reduce carbon emissions that are driving much of recent sea-level rise.

∧ FIGURE 12.1
Tidal flooding in South Florida State. Coastal highway is flooded in 2013 during a high tide. Such flooding has increased dramatically in recent years. *(Alamy)*

and Hollywood Beach (see opening photograph), with property value in excess of $10 billion, that are located on a narrow barrier island between the Atlantic Ocean and the inland waterway are particularly vulnerable to future sea-level rise. All the red flags are up, but the real crunch and impact from rising sea level in Florida is most likely to be 50 to 100 years in the future. Thus, South Florida has some time, as does the world, to attempt to reduce the impact from climate change, but such preparations could take decades to enact.

Ironically, top Florida State Officials in 2015 all but outlawed and censored the use of the terms "global warming" and "sea-level rise," as if not saying the words would make the problem go away. This situation made national news before being partly backed away from. Denying climate change and sea-level rise is a doomed strategy. Officials finally said something must be done other than talk and state their personal beliefs. The science is clear: sea level is rising, largely from human-induced global warming. What officials say or believe has no merit when the data speaks for itself. It is estimated that by the year 2050, there will be many more coastal flood events than today.[2] We will return to the South Florida coastal flooding issue at the end of the chapter and apply the fundamental concepts.

12.1 Global Change and Earth System Science: An Overview

Preston Cloud, a famous Earth scientist who studied the history of life on Earth, human impact on the environment, and the use of resources, proposed two central goals for the Earth sciences:[3]

1. Understand how Earth works and how it has evolved from a landscape of barren rock to the complex landscape dominated by the life we see today.

2. Apply that understanding to better manage our environment.

Until recently, it was thought that human activity caused only local or, at most, regional environmental change. It is now generally recognized that the effects of human activity on Earth are so extensive that we are actually involved in an unplanned planetary experiment. To recognize and perhaps modify the changes we have initiated, we need to understand how the entire Earth works as a system. The discipline, called **Earth System Science**, seeks to further this understanding by learning how the various components of the system—the atmosphere, oceans, land, and biosphere—are linked on a global scale and interact to affect life on Earth.[4] A major goal of Earth System Science is to predict future changes that are relevant to society and people today.

12.1 CHECK your understanding

1. Define Earth System Science and describe its goals.

12.2 Climate and Weather

Many people who hear about global warming form opinions based on day-to-day or week-to-week changes in the weather. They do not appreciate the important distinction between climate and weather. **Climate** refers to the characteristic atmospheric conditions of a given region over long periods of time, such as years or decades. **Weather** refers to the atmospheric conditions of a given region for short periods of time, such as days or weeks. For example, when we think of the coast of the Pacific Northwest, we think of mild temperatures, high humidity, and lots of rain. Thus, when planning a trip to Vancouver, British Columbia, or Seattle, Washington, we might use our knowledge of the climate to help us remember to bring an umbrella! It is possible, however, that during a weeklong stay in Vancouver or Seattle, we may enjoy only bright, sunny, dry weather because day-to-day weather conditions can vary greatly.

Although we generally think of temperature and precipitation as climatic factors, climate is more than average high and low temperatures and/or the amount of rainfall at a given location. In fact, it is possible for two locales to have the same average annual temperature but be in different climate zones. For example, San Diego, California, has an average annual temperature of around 18°C (~64°F) and small temperature changes from month to month. In contrast, El Paso, Texas, has close to the same average annual temperature but experiences large changes in temperature, ranging from a January average of 7°C (~45°F) to a July average of 28°C (~82°F). Although the two areas have the same average annual temperature, it is clear that they have different climates.

CLIMATE ZONES

Climate may be classified in many different ways. The simplest classification is by temperature and precipitation, but climate at a particular place or region may be complex.

For example, it may depend on infrequent or extreme seasonal patterns, such as rain in the monsoon season in much of India. The major climatic zones on Earth are shown in Figure 12.2. Selected processes and changes that produce and maintain the climate system are shown in Figure 12.3. For example, ocean currents can affect the temperature and precipitation of areas far removed from the coast; warm currents traveling northward along the eastern Atlantic bring mild temperatures to northern Europe. Mountain ranges and plateaus can significantly affect precipitation patterns and seasons, which can influence temperature as well; for example, Asia's Tibetan Plateau is partly responsible for the Asian monsoon, and the Rocky Mountains are important factors in the southwestern U.S. monsoon.

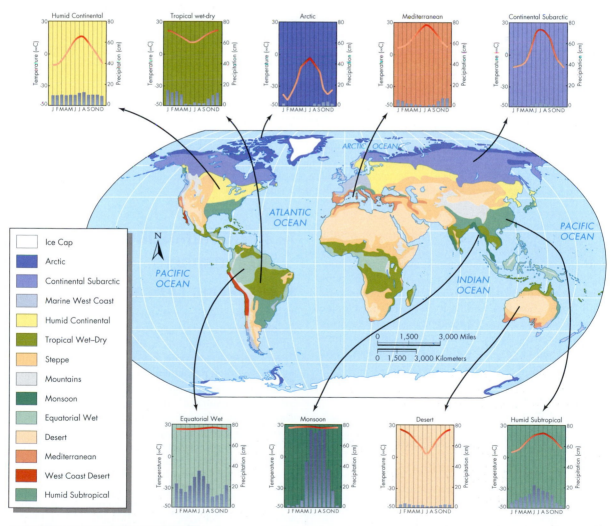

∧ FIGURE 12.2 Climates of the World

Characteristic temperature and precipitation conditions. Temperature is represented by the red line, and precipitation is shown as bars. *(Based on Marsh, W. M., and Dozier, J. Landscapes: An Introduction to Physical Geography. Copyright © 1981. John Wiley & Sons.)*

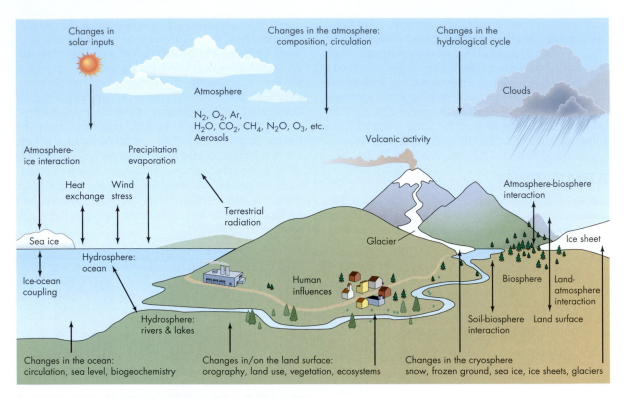

∧ FIGURE 12.3 Complex components of the climate system
Idealized diagram showing the complex, linked components and changes of the climate system that produce, maintain, and change the climates of the world. *(IPCC. 2007. The Physical Science Basis: Working Group I. Contribution to the Fourth Assessment Report. IPCC. New York: Cambridge University Press.)*

Climate zones exert a major influence on natural processes and, thus, on natural hazards. Flooding is, in part, dependent on rainfall amount and intensity, landslides may be more common in areas with rainy climates, and wildfires are more likely to occur in areas with a dry season. A familiarity with Earth's climate zones is a first step toward recognizing the threat from natural hazards. Applying climate classification provides not only a map of world climates but also information about the relationship between climate and vegetation.[5] Not all dry areas, for example, are prone to wildfires, but some dry areas are at significantly more risk than others.

12.2 CHECK your understanding

1. Distinguish weather from climate.
2. What are the two main variables used to define climate zones? What other factors influence climate in a region?
3. How is climate classification useful to understanding natural hazards?

12.3 The Atmosphere and the Cryosphere

ATMOSPHERIC COMPOSITION

As described in Chapter 9, our atmosphere is composed mainly of nitrogen and oxygen with smaller amounts of other gases. Compounds, such as nitrogen (N_2), that form a constant proportion of the mass of the atmosphere are called permanent gases, whereas gases whose proportions vary in time and space, such as carbon dioxide, are considered variable gases. The atmosphere also contains microscopic particles called aerosols, whose proportions also vary in time and space.

The major permanent gases, which are those that remain essentially constant by percentage of total volume, include nitrogen (~78 percent), oxygen (~21 percent), and argon (~1 percent). Together, these gases comprise about 99 percent by volume of all atmospheric gases. Nitrogen generally occurs as molecules of paired nitrogen atoms (N_2) and composes about 78 percent of the volume of all permanent gases. Although N_2 molecules make up the greatest volume of the atmosphere, they are relatively unimportant in atmospheric dynamics unless nitrogen forms compounds with other gases, such as oxygen.

The second largest component is oxygen, composing 21 percent of the atmosphere by volume. Like nitrogen, oxygen molecules consist mainly of paired atoms of oxygen (O_2). As we humans well know, oxygen is critical to the existence of most life forms on Earth. Like nitrogen, O_2 is relatively unimportant to atmosphere dynamics. However, as discussed later, other oxygen compounds, such as ozone (O_3), play an extremely important role in the atmospheric-climate system.

Argon makes up most of the remaining 1 percent of the permanent gases, along with lesser amounts of neon, helium, krypton, xenon, and hydrogen. With the exception of hydrogen, these permanent gases are not chemically reactive and have little or no effect on climate.

Although the variable gasses, with the exception of water vapor (0.2–4 percent), account for only a small percentage of the total mass of the atmosphere, some of them play important roles in atmospheric dynamics. These gases include carbon dioxide, ozone, methane, and nitrous oxide. The atmosphere also contains a variable amount of aerosols. An aerosol is not a gas but a microscopic liquid or solid particle. In the atmosphere, these particles act as nuclei around which water droplets condense to form clouds. Aerosols have a wide variety of natural and anthropogenic sources. Natural sources include desert dust, wildfires, sea spray, and major periodic contributions from volcanoes.[6] Anthropogenic sources include the burning of forests in land clearing and sulphate particles and soot from the burning of fossil fuels.

HYDROSPHERE AND CRYOSPHERE

A study of climate change is, to a great extent, the study of changes in the atmosphere and linkages between the atmosphere and the lithosphere, hydrosphere, cryosphere, and biosphere (see Chapter 1). The hydrosphere is the water part of the global Earth system and includes the oceans, lakes, and rivers, as well as the water in the armosphere. Of particular significance are the processes that transport the water from the ocean to the land and back to the ocean known as the hydrologic cycle (see Figure 1.8 and Table 1.3).

The cryosphere is the part of the hydrosphere in which most of the water stays frozen year-round. Major components of the cryosphere include permanently frozen ground (permafrost), sea ice, and glacial ice. The ice in permafrost and sea ice is either fixed in place or floating. In contrast, glacial ice is an accumulation of ice, snow, rock, sediment, and liquid water that flows from high areas to low areas due to gravity acting on the weight of this accumulated ice. Thus,

glaciers are landbound, moving masses of glacial ice. In this chapter, we will discuss the giant ice sheets in polar areas, as well as valley and mountain glaciers. Like beaches, glaciers have budgets with inputs and outputs (see Chapter 11). Inputs consist of new snow recrystallizing to form ice at higher elevations, and outputs occur at lower elevations where ice is lost by melting, evaporation, and blocks breaking off the front of the glacier or ice sheet. Glaciers and ice sheets advance when their budget is positive and retreat when they have a net loss of ice.

Over the past 2.5 million years, Earth's climate system has fluctuated greatly and alternated between periods of major continental glaciation, referred to as glacial intervals, and times of warmer climate with significantly less glaciation, referred to as interglacial intervals. During glacial intervals, large ice sheets covered northern North America and Europe, and glaciers expanded in mountainous regions. In interglacial intervals, individual glaciers and large ice sheets retreated. Today, we live in interglacial conditions, with warm temperatures not experienced since the last interglacial interval 125,000 years ago. We will discuss why Earth's climate changes from glacial to interglacial later in this chapter with a more specific discussion of why climate changes.

Major continental glacial events (ice ages) have been relatively rare during the 4.6 billion years of Earth history. However, during the past 1 billion years, several major glacial events have occurred. We are now living during one of these events—one that began approximately 2.5 million years ago when glaciers began to repeatedly advance and retreat across much of the landscape. The last series of glacial and interglacial intervals took place during the Pleistocene Epoch, which included multiple ice ages when continental glaciers advanced. During the Pleistocene Epoch, glaciers covered as much as 30 percent of the land area of Earth, including the present sites of major cities such as New York, Toronto, and Chicago (Figure 12.4). Pleistocene continental glaciers achieved their maximum extent about 21,000 years ago. Because large quantities of the world's freshwater were stored in the ice of these glaciers, global sea level was more than 100 m (~330 ft.) lower than its current level. Today, glacial ice covers only about 10 percent of Earth's land. Nearly all this ice is contained in the Antarctic ice sheet; lesser amounts are found in the Greenland ice sheet and the mountain glaciers of Alaska, British Columbia, southern Norway, the Alps, New Zealand, and elsewhere. We are presently in an interglacial interval, so the abundance of ice is relatively low. Nevertheless, we are probably still living in a glacial event (ice age), and ice sheets may again advance in the future.

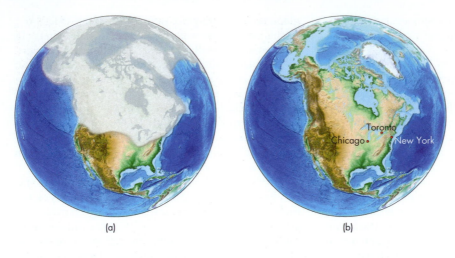

(a)   (b)

FIGURE 12.4 Extent of North American Ice Sheets (a) Ice sheets about 21,000 years ago. (b) Ice sheets today. *(From NOAA. www.ncdc.noaa.gov. Accessed 9/25/10.)*

12.3 CHECK your understanding

1. What are the major permanent and variable gases in the atmosphere?

2. Define glacier and how the extent of glaciers has changed in the past 21,000 years.

3. Differentiate between the hydrosphere and the cryosphere

12.4 How We Study Past Climate Change and Make Predictions

The data gathered to document and better understand climate change are from a variety of time scales and variable regions, from continents, oceans, hemispheres, and the entire planet. Data are available for three main time periods:[7]

> *The instrumental record.* Starting about 1880, temperature measurements have been made at various locations on land and in the oceans. Earliest records are from the late seventeenth and early eighteenth centuries, and the network of stations has significantly increased over time. About 1,000 records exist from the late nineteenth century. Today, temperature is measured at about 7,000 stations around the world. The concentration of carbon dioxide in the atmosphere has been measured since about 1960. Accurate measurement of the production of solar energy has been taken over the past several decades.

> *The historical record.* A variety of historical records go back several hundred years. Included are people's written recollections (books, newspapers, journal articles, personal journals, etc.) of the Medieval Warm Period and the Little Ice Age, as well as ships' logs, travelers' diaries, and farmers' crop records. These are not generally quantitative data, but they contain qualitative information about the climate of the past.

> *The paleo-proxy record.* The instrumental record is short, and most of the historical information is not quantitative. As a result, there has been a need to extend the record back further. Paleoclimatology (the study of past climates) is part of Earth Science. It is clear that the paleo-record of Earth's climate has provided some of the strongest data to support and test recent climate change. The term **proxy data** refers to data that are not strictly climatic but can be correlated with climate, such as temperature of the land or sea.[7] Some of the information gathered as proxy data includes natural records of climate variability, as indicated by tree rings, ocean sediments, ice cores, fossil pollen, corals, and carbon-14 (^{14}C).

The disadvantages of paleoclimate proxy data are obvious; the data are not a direct measurement of temperature, and, therefore, temperature must be inferred from the data. In spite of this, the paleoclimate proxy data preserved in the geologic record provide the best evidence of change that predates the historical and instrumental records. A discussion of the various data sources for paleo-proxy records follows.

Tree Rings The growth of trees is influenced by climatic conditions, such as the amount of rainfall and the variability in temperature. Most trees put on one growth ring per year, and the width, density, and isotopic composition of annual rings provide information about past climate—referred to as dendroclimatology.

Thus, tree rings contain proxy climate information, such as relative rainfall or periods of drought. By counting rings within living and older dead trees that overlap with the living record, scientists have developed a dendroclimatology proxy record that extends about several thousand years.

Sediments   The oceans of the world are a repository for the sediments from the land delivered by rivers, wind, and volcanic eruptions, as well as sediment from the ocean itself (such as shells of dead organisms). Lakes, bogs, and ponds accumulate sediment over time that can be sampled for a climate signal. When sediment deposits are sampled and dated, these provide paleo-proxy data sources for climate change. Sediments are recovered by drilling in the bottom of an ocean basin or lake (Figure 12.5). From the cores, samples may be taken of small fossils and chemicals that are contained within the sediments that may be interpreted to better understand past climate change. Some of the strongest evidence for past climate change comes from these proxy records.

Ice Cores   Glaciers contain an accumulation record of the snow that has been transformed into glacial ice

∧ **FIGURE 12.6 Ice Core**
Preparing a 2005 ice core from the Greenland Ice Cap for examination of the climate record it contains. (*Simon Fraser/Science Source*)

∧ **FIGURE 12.5 Examining Cores**
Marine sediment core from seafloor under the Antarctic Ice Shelf is being examined. (*Peter West, Office of Polar Programs, National Science Foundation*)

over hundreds of thousands of years. Ice cores are obtained by drilling into the ice and obtaining a core, which can be studied in detail in order to learn about past conditions (Figure 12.6). Ice cores from glaciers often contain small bubbles of air deposited at the time of the snow. The composition and ratio of past atmospheric gases preserved in the ice may be studied and used to infer a number of paleoclimatic variables. Ice cores also contain a variety of chemicals and other materials, such as volcanic ash and dust, which can provide proxy data to assist in evaluating climate change. The ice itself may be studied to determine the paleo-isotopic composition of the water, which provides information about the volume of ice on the land, as well as processes occurring in the paleo-oceans (oceans of the past as determined from sediments).

Pollen   Pollen (from flowers, trees, and other plants), along with other sediment, accumulates in a variety of environments, including oceans, bogs, and lakes. Scientists study the abundance and types of pollen in order to investigate past climate. For example, if the climate cools, there will be a change in the types and quantities of pollen found in sediments that reflect the change in climate. Pollen, if found in sufficient quantity, may be dated. Since the grains are preserved in sedimentary layers that also might be dated, a chronology can be developed. Based on the types of plants found at different times, the climatic history can be reconstructed.

Corals Coral reefs consist of corals (and other organisms) that have hard skeletons composed of calcium carbonate ($CaCO_3$) extracted by the corals from the seawater. The calcium carbonate contains isotopes of oxygen, as well as a variety of trace metals, that can be used to estimate the temperature of the water in which the coral grew. Thus, corals are a source of paleo-proxy data that can help us interpret climate change. Corals may be dated by several dating techniques, and a chronology of change over time may be constructed.

Carbon-14 The production of carbon-14 (^{14}C) produced in the upper atmosphere is caused by collisions between neutrons and nitrogen-14 (^{14}N). The nitrogen is part of what are called cosmic rays that come from outer space and are a product of the energy from the sun. Solar activity can be observed from the frequency of sunspots, which are dark areas on the sun surrounded by a lighter area. As sunspot activity increases, the energy from the sun that reaches Earth also increases. The frequency of sunspots has been accurately measured for decades and observed by people for about 1,000 years. When sunspot activity is high, an associated solar wind, which produces ionized particles consisting predominately of protons and electrons, emanates from the sun. The solar wind reflects cosmic rays (including N-14, ^{14}N). As a result, the amount of ^{14}C is reduced. The record of ^{14}C in the atmosphere is correlated to tree ring chronology, known as dendrochronology. Each ring of wood of known age contains carbon, and the amount of ^{14}C can be measured. If the climatic record for a period of time is known, it may be correlated with ^{14}C. Thus, we can examine solar energy output, extending back thousands of years, by studying tree rings and the carbon-14 they contain.

Based on the ^{14}C record and its link to solar energy, it appears that the production of solar energy was slightly higher around A.D. 1000, during the Medieval Warm Period, and was slightly lower during the Little Ice Age from about A.D. 1400 to 1800. The effect of solar radiation on recent climate change (during the instrumental record) can account for a small percentage of observed changes in climate. However, the variability of solar energy is not sufficient to explain the warming since about 1960.

Carbon Dioxide The concentration of global carbon dioxide in the atmosphere is arguably the most important proxy for global temperature change. There are carbon dioxide measurements from the instrumental record, and these have been extended back through the measurements from trapped air in glacial ice (Figure 12.7). Figure 12.8 illustrates strong positive correlation between CO_2 in the atmosphere collected

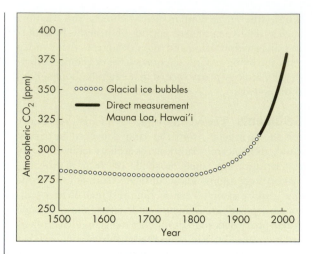

⋀ FIGURE 12.7 Carbon Dioxide in the Atmosphere
Atmospheric concentration of carbon dioxide measured at Mauna Loa, Hawaii since 1958, as well as average concentration of atmospheric carbon dioxide from 1500 to 2000, based on measurements of air bubbles trapped in glacial ice. Source: NOAA Mauna Loa Observatory in Hawaii. *(Data in part from Post, W. M. et al.,* American Scientist, *78(4): 210–26, 1990)*

from ice cores to the paleo-temperature record. The time between carbon dioxide emission and most of the warming from the emission is about one decade and will persist for about a century.[8] This short period from emission to warming suggests that if we act to reduce emissions of carbon dioxide (from burning fossil fuels), we may reap the benefits of avoiding climate change within our lifetime.

GLOBAL CLIMATE MODELS

Scientists develop mathematical models to represent real-world phenomena. These models describe numerically the linkages and interactions between natural processes. Mathematical models have been developed to predict the flow of surface water and groundwater, erosion and deposition of stream sediment, and the global circulation of water in the ocean and air in the atmosphere.

The mathematical models used by scientists to study climate and climate change have their origin in the first attempts to use computers to prepare weather forecasts in the 1950s.[9] Computers in the 1950s were in their infancy—large and extremely slow machines with vacuum tubes that could take 24 hours of processing time to produce a 24-hour weather forecast. Early mathematical models run on computers were, by necessity, regional in extent and used primarily to describe the general circulation of the atmosphere. Called general circulation models, these mathematical models were the foundation for the first truly global

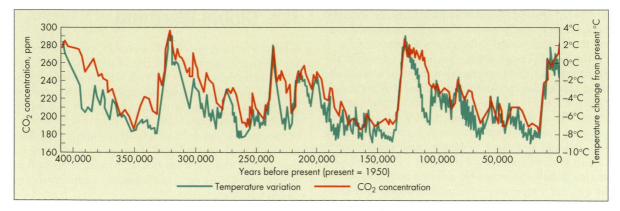

∧ FIGURE 12.8   Air Temperature Changes Correspond Closely to Atmospheric CO_2
Measurement of the carbon dioxide content in air bubbles in glacial ice from cores taken at Vostok, Antarctica, show that atmospheric CO_2 levels have corresponded closely to air temperature for more than 410,000 years. Records from other ice cores show a similar pattern extending back more than 800,000 years. *(Based on Petit, J. R., Jouzel, J., et al., "Climate and Atmospheric History of the Past 420,000 Years from the Vostok Ice Core in Antarctica,"* Nature *399: 429–36, © 1990 AAAS)*

climate models. From the 1980s to the present, atmospheric general circulation models have been coupled with mathematical models of other Earth subsystems, such as the land surface, ocean and sea ice, aerosols, and the carbon cycle.[10]

The framework for the general circulation model is that of a large stack of boxes (Figure 12.9a). Each box is a three-dimensional cell that is several degrees of latitude and longitude on each side. A typical cell is generally rectangular in shape and has an area about the size of the state of Oregon. Each cell varies in height, depending on how many cells are used to subdivide the lower atmosphere. Most models use 6 to 20 layers of cells to represent the atmosphere up to an

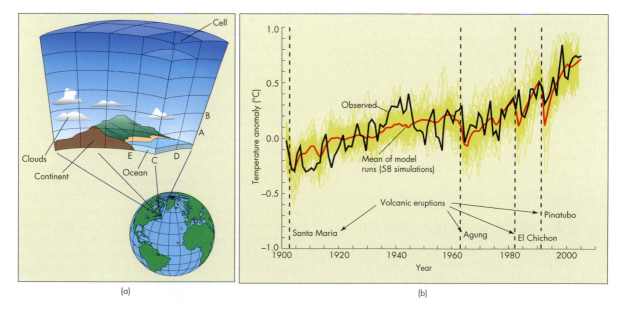

∧ FIGURE 12.9 Modeling Climate
(a) Idealized diagram illustrating the cells used in climate models. (b) Observed (black) and predicted temperature changes 1900–2005. The predicted mean (red line) is the average of 58 different simulations, using 14 different climate models. The yellow is the variability in simulations. *(IPCC. 2007. The Physical Science Basis: Working Group I.* Contribution to the Fourth Assessment Report. IPCC. *New York: Cambridge University Press.)*

altitude of around 30 km (~19 mi.). Data are arranged into each cell, and boundary conditions are established for the overall model. Mathematical equations based on the principles of physics are used to describe the major atmospheric processes that interact between the cells. Models are run backward to see if they accurately describe historic, and even prehistoric, changes in climate. Like general circulation models for forecasting weather, global climate models are constantly being adjusted and evaluated to better make predictions.[11] Global climate models are primitive in terms of predicting details of climate but are reasonably consistent with global temperature change from 1900 to the present (Figure 12.9b).

Global climate models are valuable in solving the complex problems of climate change. These models provide information for evaluating Earth as a system. The models also identify additional data that are needed to produce better and more detailed models in the future. Today, global climate models provide useful guidance on regions that are likely to be relatively wetter or drier and hotter or colder in the future. Current models are sophisticated, and predictions from modeling are generally consistent. However, one of the greatest difficulties in using global climate models to predict future climate change is our inability to anticipate human behavior and the paths that our civilization will take in dealing with climate change.

Modeling the climate generates hypotheses that scientists can test with data. It is important to acknowledge that these models do not produce data. Models use data-linked mathematical calculations with the purpose of better understanding the global climate systems and making predictions. The most important evidence (and only data) for global warming is from the measurements in the past 150 years derived from proxy evidence (such as ocean sediments and tree rings over the past few thousand to few hundred thousand years).

12.4 CHECK your understanding

1. For climate data, define and differentiate between the instrumental record, the historical record, and the paleo-proxy record.

2. Describe several sources of paleo-proxy climate data.

3. Why is the study of carbon dioxide in the atmosphere so important?

4. What are global climate models, and why are they important?

12.5 Global Warming

Global warming is defined as the observed increase in the average temperature of the near-surface land and ocean environments of Earth during the past 60 years. A growing volume of evidence suggests that we are now in a period of global warming, resulting from burning vast amounts of fossil fuels. Does this mean we are experiencing human-induced global warming? Yes! The data collected over several decades strongly suggest that climate change from human processes, as well as natural processes, are contributing significantly to global warming.[12–14]

THE GREENHOUSE EFFECT

For the most part, the temperature of Earth is determined by three factors: the amount of sunlight Earth receives, the amount of sunlight Earth reflects (and, therefore, does not absorb), and the amount of reradiated heat retained by the atmosphere[13] (see Chapter 9, which discusses the basics of Earth's energy balance between incoming and outgoing energy).

Earth's energy balance today is slightly out of equilibrium, with about 1 watt per m^2 (W/m^2) more energy coming from the sun than is lost to space. This measure of energy, W/m^2, is energy per unit time (joules/sec) per unit area m^2. When we speak of solar energy, we use these units to represent solar power per unit area. This measure is widely used in global warming and climate change research.[12–14] Earth receives energy from the sun in the form of electromagnetic radiation (including visible light, infrarad radiation, radio waves, and X rays). Radiation from the sun is relatively short in wavelength and mostly visible, whereas Earth radiates relatively longwave infrared radiation. The hotter an object, whether it is the sun, the Earth, a rock, or a lake, the more electromagnetic energy it emits. The sun, with a surface temperature of 5800°C (~10,472°F), radiates much more energy per unit area than does Earth, which has an average surface temperature of 15°C (~59°F).

Absorbed solar energy warms Earth's atmosphere and surface, which then reradiate the energy back into space as infrared radiation.[12] Water vapor and several other atmospheric gases, including carbon dioxide (CO_2), methane (CH_4), and chlorofluorocarbons (CFCs, human-made chemicals used in air conditioners and refrigerators), tend to trap heat. That is, they absorb some of the energy radiating from Earth's surface and are thereby warmed. As a result, the lower atmosphere of Earth is much warmer than it would be if all of the infrared radiation escaped into space without this intermediate absorption and warming. This effect is somewhat analogous to the trapping of heat by the windows of a

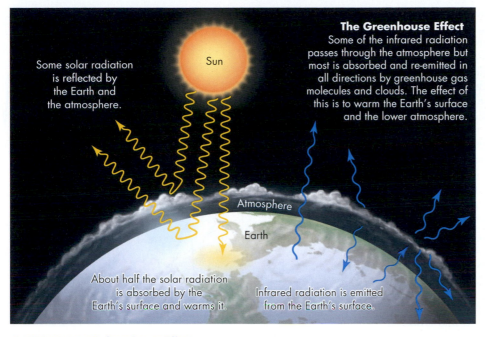

Some solar radiation is reflected by the Earth and the atmosphere.

Sun

The Greenhouse Effect
Some of the infrared radiation passes through the atmosphere but most is absorbed and re-emitted in all directions by greenhouse gas molecules and clouds. The effect of this is to warm the Earth's surface and the lower atmosphere.

Atmosphere

Earth

About half the solar radiation is absorbed by the Earth's surface and warms it.

Infrared radiation is emitted from the Earth's surface.

∧ FIGURE 12.10 Greenhouse Effect
Idealized diagram showing the greenhouse effect. Incoming visible solar radiation (sunshine) is absorbed by Earth's surface, warming it. Infrared radiation is then emitted at the surface of Earth as earthshine to the atmosphere and outer space. Most of the infrared radiation emitted from Earth is absorbed by the atmosphere, heating it and maintaining the greenhouse effect. *(Based on Climate Change Connection, www.climatechangeconnection.org/science/Greenhouseeffect_diagram.htm. Accessed 11/17/13.)*

greenhouse, which allows the sun's warmth in and keeps it from reradiating back out, and is therefore referred to as the **greenhouse effect** (Figure 12.10).

It is important to understand that the greenhouse effect is a natural phenomenon that has been occurring for millions of years on Earth, as well as on other planets in our solar system. Without heat trapped in the atmosphere, Earth would be much colder than it is now, and all surface water would be frozen. Most of the natural "greenhouse warming" is due to water vapor and small droplets of water in the atmosphere. However, potential global warming resulting from human activity is related to carbon dioxide, methane, nitrogen oxides, and chlorofluorocarbons. In recent years, the atmospheric concentrations of these gases and others have been increasing because of human activities. These gases tend to absorb infrared radiation from Earth, and it has been hypothesized that Earth is warming because of the increases in the amounts of these so-called greenhouse gases. Table 12.1 shows the rate of increase of these atmospheric gases resulting from human-induced emissions and their relative contribution to the anthropogenic, or human-caused, component of the greenhouse effect. Notice that carbon dioxide accounts for 60 percent of the relative contribution.

▼ TABLE 12.1 Rate of Increase and Relative Contribution of Several Gases to the Anthropogenic Greenhouse Effect

	Rate of Increase (% per year)	Relative Contribution (%)
CO_2	0.5	60
CH_4	<1	15
N_2O	0.2	5
O_3[a]	0.5	8
CFC-11[b]	4	4
CFC-12[b]	4	8

[a] In the troposphere.

[b] CFC—chlorofluorocarbon (human-made chemical).

Source: Data from Rodhe, H. "A Comparison of the Contribution of Various Gases to the Greenhouse Effect," *Science* 248: 1218, table 2. Copyright 1990 by the AAAS.

Measurements of carbon dioxide trapped in air bubbles of the Antarctic ice sheet suggest that, during most of the past 160,000 years, the atmospheric concentration of carbon dioxide has varied from a little less than 200 parts per million (ppm) to about 300 ppm.[13]

The highest recorded levels are during major interglacial periods that occurred approximately 125,000 years ago and today. Major interglacials occurred about four times during the past 400,000 years—about every 100,000 years. During each of these periods, the concentration of CO_2 in the atmosphere was similar to that of the most recent interglacial event about 125,000 years ago (see Figure 12.8).[15]

At the beginning of the Industrial Revolution, the atmospheric concentration of carbon dioxide was approximately 280 ppm. Since 1860, fossil fuel burning has contributed to the exponential growth of the concentration of carbon dioxide in the atmosphere. The change, based on measurements made from air bubbles trapped in glacial ice from approximately 1500 to 1958, is shown in Figure 12.7. The concentration of carbon dioxide in the atmosphere today is approximately 400 ppm, and it is predicted to reach at least 450 ppm— more than 1.5 times the preindustrial level—by 2050. Changes in atmospheric concentration of carbon dioxide at Mauna Loa, Hawaii, from the mid-twentieth century to today, are also shown in Figure 12.7.

GLOBAL TEMPERATURE CHANGE

The Pleistocene ice ages began approximately 2 million years ago; since then, there have been numerous changes in Earth's mean annual temperature. Figure 12.11 shows the changes of approximately the past million years on several time scales. The first scale shows the entire million years (Figure 12.11a), during which major climatic changes involved swings of several degrees Celsius in mean temperature. Low temperatures have coincided with major glacial events that have greatly altered the landscape; high temperatures are associated with interglacial conditions. Interglacial and glacial events become increasingly prominent in the scales, showing changes over 150,000 and 18,000 years. The last major interglacial warm period, even warmer than today, was the Eemian (Figure 12.11b). During the Eemian, sea level was a few meters higher than it is today. The cold period that occurred about 11,500 years ago is known as the Younger Dryas; it was followed by rapid warming to the

> **FIGURE 12.11 Changes in Global Temperature**
Changes in temperature over different periods of time during the past million years. Graphs (a) to (d) are over time periods of 100,000 to 1,000 years. The rapid rise from about 1900 to 2008 of nearly 0.9°C is shown in (e). Note the very rapid rise, since about 1970, of about 0.12°C per decade. See text for further explanation. *(Based on University Corporation for Atmospheric Research, Office for Interdisciplinary Studies, Science Capsule, "Changes in Time in the Temperature of the Earth," EarthQuest 5(1), 1997; and the UK Meteorological Office, "Climate Change and its Impacts: A Global Perspective," 1997)*

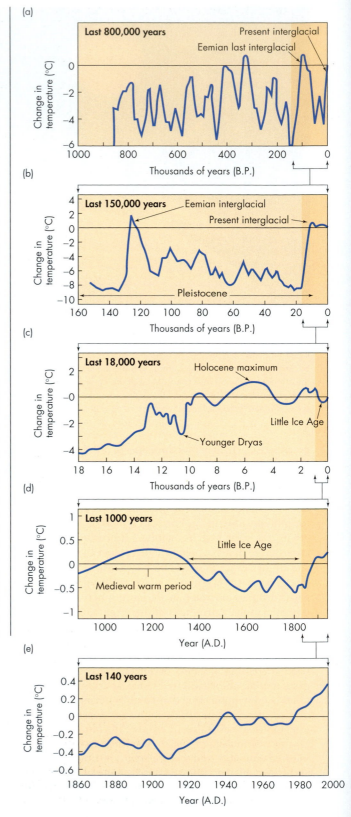

▼ **TABLE 12.2** Evidence Supporting the Late Twentieth and Early Twenty-First-Century Rise in Global Temperature: Global Temperature Data from the United States (NOAA) and Europe (WMO)

Earth's surface temperature in 2016 was the warmest year on record (since 1880) and the third year in a row to set a new record global average. Average global surface temperature in 2016 was about 0.99 degrees Celsius warmer than the mid-twentieth-century mean.

The Earth's surface temperature has risen about 1.1 degrees Celsius since the late nineteenth century.

Most of the warming occurred in the past 35 years, with 16 of the 17 warmest years occurring since 2001.

Note: A few years of high temperatures with drought, heat waves, and wildfires are not by themselves an indication of longer term global warming. Likewise, a halt in warming for a decade is not sufficient to justify saying that global warming over the next few decades will not continue.

Holocene maximum, which preceded the Little Ice Age (Figure 12.11c).

A scale of 1,000 years shows several warming and cooling trends that have affected people (Figure 12.11d). For example, a major warming trend from approximately A.D. 1100–1300 allowed the Vikings to colonize Iceland, Greenland, and northern North America. When glaciers made a minor advance around 1400 during a cold period known as the Little Ice Age, the Viking settlements in North America and parts of Greenland were abandoned.

In approximately 1750, an apparent warming trend began that lasted until approximately the 1940s, when temperatures cooled slightly. Over 1860–2000 more changes are apparent, and the event of the 1940s is clearer (Figure 12.11e). It is evident from the record that from 1860–2000, global mean annual temperature increased by approximately 0.8°C (~1.4°F). Most of the increase has been since the 1970s, with the 1990s and the first two decades of the twenty-first century having the warmest temperatures since global temperatures have been monitored. Table 12.2 lists evidence for the current warming trend.

The persistent trend of increasing temperatures over several decades (1970–2017) is compelling evidence that global warming is real and is happening. Figure 12.12a shows observed globally averaged combined land and ocean surface temperture change from 1850 to the present. Figure 12.12b shows decadal averaged data. Notice the sharp increase in the past four decades. This is the late twentieth- to early twenty-first-century rise in temperature that continues today.

WHY DOES CLIMATE CHANGE?

The question that begs to be answered is this: Why does climate change? Examination of Figure 12.11 suggests that there are cycles of change lasting 100,000 years, separated by shorter cycles of 20,000 to 40,000 years in duration. These cycles were first identified by

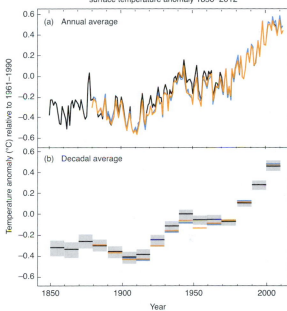

▲ **FIGURE 12.12** Changes in Global Temperature 1850–2015 (a) Annual data where different colors are different data sets. (b) Decadal average values where shaded areas are uncertainty. Notice the dramatic increase in the last four decades. *(IPCC. 2015. Climate Change 2014: Synthesis Report. Intergovernmental Panel on Climate Change. Geneva, Switzerland.)*

Milutin Milankovitch in the 1920s as a hypothesis to explain climate change. Milankovitch realized that the spinning Earth is like a wobbling top, unable to keep a constant position in relationship to the sun; this instability partially determines the amount of sunlight reaching and warming Earth. He discovered that variability in Earth's orbit around the sun follows a 100,000-year cycle that is correlated with the major glacial and interglacial periods shown in Figure 12.11a. Earth's orbit varies from a nearly circular ellipse to a more elongated ellipse. Over the 100,000-year cycle, when Earth's orbit is most elliptical, solar radiation reaching Earth is greater than during a more circular orbit. Cycles of approximately 40,000 and 20,000 years are the result of changes in the tilt and wobble of Earth's axis.

Milankovitch cycles reproduce most of the long-term cycles observed in the climate, and they do have a significant effect on climate. However, the cycles are not sufficient to produce all of the observed large-scale global climatic changes. Therefore, these cycles, along with other processes, must be invoked to explain global climatic change. Thus, the Milankovitch cycles that force (push) the climate in one direction or another can be looked at as natural processes (forcing) that, when linked to other processes (forcing), produce climatic change.[16,17] We now will consider this concept of climate forcing in more detail.

Climate forcing is defined as an imposed change of Earth's energy balance. The units for the forcing are W/m^2, and they can be positive if a particular forcing increases global mean temperature or negative if temperature is decreased. For example, if the energy from the sun increases, then Earth will warm (this is positive climate forcing). If CO_2 were to decrease, causing Earth to cool, that would be an example of negative climate forcing.[18] Climate sensitivity refers to the response of climate to a specific climate forcing after a new equilibrium has been established, and the time required for the response to a forcing to occur is the climate response time.[19] A significant implication of climate forcing is that if you maintain small climate forcing for a long enough time, large-scale climate change can occur.[12] Figure 12.13 shows the negative climate forcing that produced the last ice age 22,000 years ago (last glacial maximum).[12,19] Notice that a

negative energy input change of 1 W/m^2 produces a temperature change of about 0.75°C (~1.3°F). Climate forcing in the industrial age is shown in Figure 12.14. Total positive forcing is about 1.6 W/m^2, most of which is due to greenhouse gas inputs (CO_2, CH_4, and N_2O). Figure 12.15 clearly shows that the atmospheric concentrations of these greenhouse gases have dramatically increased in the past 100 years.[13]

We now believe that our climate system may be inherently unstable and capable of changing quickly from one state to another in as short a time as a few decades.[20] However, very short or **abrupt climate change** is unlikely. Part of what may drive the climate system and its potential to change is the ocean conveyor belt, a global-scale circulation of ocean waters, characterized by strong northward movement of 12° to 13°C (~54° to 55°F) near-surface waters in the Atlantic Ocean that are cooled to 2° to 4°C (~35° to 39°F) when they arrive near Greenland (Figure 12.16).[20] As the water cools, it becomes saltier due to sea ice formation at the ocean surface. The salinity increases the water's density and causes it to sink to the bottom. The current then flows southward around Africa, adjoining the global pattern of ocean currents. The flow in this conveyor belt current is huge, equal to about 100 Amazon Rivers. The amount of warm water and heat released to the atmosphere, along with the stronger effect of relatively warm winter air moving east and northeast across the Atlantic Ocean, is sufficient to keep northern Europe 5° to 10°C (~8.5° to 17°F) warmer than it would otherwise be. If the conveyor belt were to shut down, it would have an effect on the climate of Europe. However, the effect would not be catastrophic to England and France in terms of producing extreme cold and icebound conditions.[21]

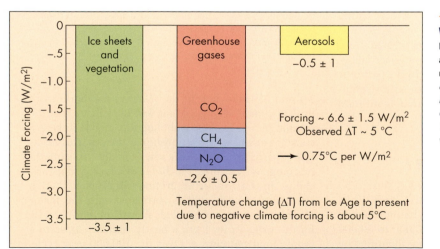

◄ **FIGURE 12.13 Climate Forcing – W/m^2 During Last Ice Age**
Negative climate forcings 22,000 years ago, when glacial ice was last at a maximum. *(Based on Hansen, J. 2003. Can we defuse the global warming time bomb? Edited version of presentation to the Council on Environmental Quality. June 12. Washington DC; also Natural Science. www.naturalscience.com.)*

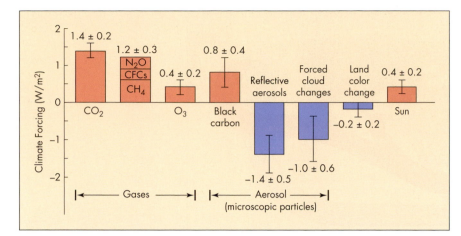

< FIGURE 12.14 Climate Forcings in the Industrial Age Starting in 1750
Positive forcings warm, and negative forcings cool. Human-caused forcings in recent years dominate over natural forcings. Total forcing is about 1.6 ± 0.1 W/m², consistent with observed rise in air surface temperature over the past few decades. *(Based on Hansen, J. NASA Goddard Institute for Space Studies and Columbia University Earth Institute, 2003)*

- Increases of greenhouse gases (except O_3) are known from observations and bubbles of air trapped in ice sheets. The increase of CO_2 from 285 parts per million (ppm) in 1850 to 368 ppm in 2000 is accurate to about 5 ppm. The conversion of this gas change to a climate forcing (1.4 W/m²), from calculation of the infrared opacity, adds about 10 percent to the uncertainty.

- Increase of CH_4 since 1850, including its effect on stratospheric H_2O and tropospheric O_3, causes a climate forcing about half as large as that by CO_2. Main sources of CH_4 include landfills, coal mining, leaky natural gas lines, increasing ruminant (cow) population, rice cultivation, and waste management. Growth rate of CH_4 has slowed in recent years.

- Tropospheric O_3 is increasing. The United States and Europe have reduced O_3 precursor emissions (hydrocarbons) in recent years, but increased emissions are occurring in the developing world.

- Black carbon ("soot"), a product of incomplete combustion, is visible in the exhaust of diesel-fueled trucks. It is also produced by biofuels and outdoor biomass burning. Black carbon aerosols are not well measured, and their climate forcing is estimated from measurements of total aerosol absorption. The forcing includes the effect of soot in reducing the reflectance of snow and ice.

- Human-made reflective aerosols include sulfates, nitrates, organic carbon, and soil dust. Sources include burning fossil fuel and agricultural activities. Uncertainty in the forcing by reflective aerosols is at least 35 percent.

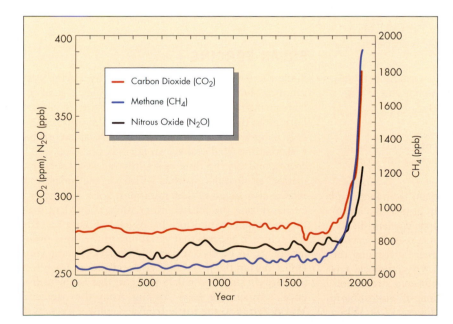

< FIGURE 12.15
Concentrations of Greenhouse Gases During the Past 2000 Years
Atmospheric concentrations of carbon dioxide, methane, and nitrous oxide have increased rapidly in recent decades. *(Based on IPCC, The Physical Science Basis: Working Group I Contribution to the Fourth Assessment Report, IPCC. New York: Cambridge University Press. p. 989, 2007.)*

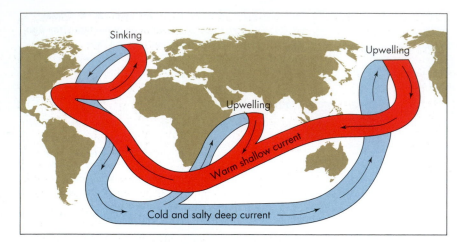

∧ FIGURE 12.16 Ocean Conveyor Belt
Idealized diagram of the ocean conveyor belt. Although the actual system is more complex
than the diagram, in general, warm surface water (red) is transported westward and northward
(increase in salinity, owing to sea ice formation, is very important in the high salinity of sinking water) to near Greenland, where it cools from contact with cold Canadian air. As the water
increases in density, it sinks to the bottom and flows south, then east to the Pacific, where upwelling occurs. The masses of sinking and upwelling waters balance, and the total flow rate is about
20 million m^3 (706 million $ft.^3$) per second. The heat released to the atmosphere from the warm
water helps keep northern Europe warmer than it would be if the oceanic conveyor belt were not
present. *(Based on Broecker, W., "Will Our Ride Into the Greenhouse Future be a Smooth One?"
Geology Today 7(5): 2–6, 1997)*

Although some scientific uncertainties still exist,
sufficient evidence exists to state that:[13,14]

> There is a significant, discernible human influence
on global climate

> Warming is now occurring and probably
accelerating

> The mean surface temperature of Earth will likely
increase by at least 2° C during the twenty-first
century

> The Arctic region will continue to warm more rapidly
than at midlatitude areas (as for example, the USA)

> The mean warming over the ocean will be less than
over most land areas

> At daily and seasonal timescales over most land ar-
eas, there will be fewer cold and more hot temperature extremes.

Human-induced global warming results from increased
emissions of gases that tend to trap heat in the atmosphere. There is good reason and lots of scientific evidence to confidently state that increases in carbon
dioxide and other greenhouse gases are increasing the
mean global temperature of Earth. Over the past few
hundred thousand years, there has been a strong correlation between the concentration of atmospheric
CO_2 and global temperature (see Figure 12.8). When

CO_2 has been high, temperature has also been high
and, conversely, low concentrations of CO_2 have been
correlated with a low global temperature. However, in
order to better understand global warming, we need to
consider major forcing variables that influence global
warming, including solar emission, volcanic eruption,
and anthropogenic input.

SOLAR FORCING

Since the sun is responsible for heating Earth, solar
variation should be evaluated as a possible cause of climate change. When we examine the history of climate
during the past 1,000 years, the variability of solar energy plays a role. Examination of the solar record reveals that the Medieval Warm Period (A.D. 1000–1300)
corresponds with a time of slightly increased solar radiation, comparable to that which we see today. The
record also suggests that minimum solar activity occurred during the fourteenth century, coincident with
the beginning of the Little Ice Age (see Figure 12.11d).
Therefore, it appears that variability of the input of solar energy to Earth can partially explain climatic variability during the past 1,000 years. However, the effect
is relatively small, only 0.25 percent; that is, the difference between the solar forcing from the Medieval
Warm Period to the Little Ice Age is only a fraction of

1 percent.[22,23] Brightening of the sun is unlikely to have had a significant effect on global warming since the beginning of the Industrial Revolution.[24]

VOLCANIC FORCING

Upon eruption, volcanoes can hurl vast amounts of particulate matter, known as aerosols, as high as 15–25 km (~9–16 mi.) into the atmosphere. The aerosol particles are transported by strong winds around Earth. They reflect a significant amount of sunlight and produce a net cooling that may offset much of the global warming expected from the anthropogenic greenhouse effect.[13,25] For example, increased atmospheric aerosols over the United States (from air pollution) are probably responsible for mean temperatures being roughly 1°C (~1.7°F) cooler than they would be otherwise. Aerosol particle cooling may, thus, help explain the disparity between model simulations of global warming and actual recorded temperatures that are lower than those predicted by models.[26]

⋀ FIGURE 12.17 Volcanic Eruption Temporarily Cools Climate
This eruption of Mount Pinatubo in the Philippines in 1991 ejected vast amounts of volcanic ash and sulfur dioxide aerosols up to about 30 km (~19 mi.) into the stratosphere. (*USGS*)

Volcanic eruptions add uncertainty in predicting global temperatures. For instance, consider the cooling effect of the 1991 Mt. Pinatubo eruption in the Philippines (Figure 12.17). Tremendous explosions sent volcanic ash to elevations of 30 km (~19 mi.) into the stratosphere, and as with similar past events, the aerosol cloud of ash and sulfur dioxide remained in the atmosphere, circling Earth, for several years. The particles of ash and sulfur dioxide scattered incoming solar radiation, resulting in a climatic forcing of about 3 W/m^2, cooling Earth about 2.3°C (~4°F) in 1991 and 1992. Calculations suggest that aerosol additions to the atmosphere from the Mt. Pinatubo eruption counterbalanced the warming effects of greenhouse gas additions through 1992. However, by 1994, most aerosols from the eruption had fallen out of the atmosphere, and global temperatures returned to previous higher levels.[27] Volcanic forcing from pulses of volcanic eruptions is believed to have significantly contributed to the cooling of the Little Ice Age (see Figure 12.11d).[23]

ANTHROPOGENIC FORCING

Anthropogenic climate forcing refers to human-related forcing. Most important are emissions of greenhouse gases (especially carbon dioxide and emissions of aerosol particles). Evidence of anthropogenic forcing resulting in a warmer world is based, in part, on the following:

> Recent warming over the past few decades of 0.12°C (~0.24°F) per decade cannot be explained by natural variability of the climate over recent geologic history.

> Industrial age forcing of 1.6 W/m^2 (see Figure 12.14) is mostly due to emissions of carbon dioxide that, with other greenhouse gases, have greatly increased in concentration in the past few decades (see Figure 12.15).

> Climate models suggest that natural forcings in the past 100 years cannot be responsible for what we know to be a nearly 1°C (~1.8°F) rise in global land temperature (see Figure 12.14). When natural and anthropogenic forcings are combined, the observed changes can be explained (Figure 12.14).

Human processes are also causing a slight cooling. Reflection from air pollution particles (aerosols) has reduced incoming solar energy by as much as 10 percent. This is termed global dimming. Negative forcing from aerosols in the industrial age is −1.4 W/m^2 (Figure 12.14) and may be offsetting up to 50 percent of the expected warming resulting from greenhouse gases.

In summary, abundant evidence supports the hypothesis that recent (1970–2015) warming greatly exceeds the natural variability (climate forcing) and closely agrees with the response predicted from models of human induced greenhouse gas forcing.[13]

12.5 CHECK your understanding

1. What is meant by global warming?

2. What are the major greenhouse gases? Explain how the greenhouse effect works.

3. What are Milankovitch cycles, and why are they important?

4. Define climate forcing. What can you conclude from studying climate forcings in the last ice age with climate forcings in the industrial age?

5. Explain the ocean conveyor belt.

6. What is the impact of solar forcing since the Industrial Revolution?

7. Define volcanic forcing.

8. Give examples of anthropogenic forcing.

9. What are the three lines of evidence that support the hypothesis that anthropogenic forcing rather than natural forcing is responsible for most of the warming in the past few decades?

12.6 Effects of Climate Change

All climate models are consistent in predicting that warming will continue as a result of the greenhouse gases now in the atmosphere and will possibly accelerate in the coming decades. That holds true even if greenhouse gas emissions by people are almost completely eliminated—warming of about 0.5 to 1.0°C (~0.9 to 1.8°F) would still occur in coming decades. Therefore, we need to carefully examine the potential effects of such warming.

If carbon dioxide in the atmosphere doubles from pre-industrial levels, as expected, it is estimated that the average global temperature will rise about 2.0° to 4.5°C (~3.6° to 8.1°F), with significantly greater warming at the polar regions.[13] Specific effects of this temperature rise in a specific region are difficult to predict, but they include: changes in glaciers and sea ice; change in climate pattern; rise in sea level; change in the biosphere; and change in occurrences of natural hazard events such as floods, droughts, hurricane, landslides, glacial movements, wildfires, earthquakes, tsunamis, and volcanic eruptions. The above changes and hazards are not mutually exclusive and there are linkages between them. For example, melting of glacial ice contributes to sea-level rise that is linked to coastal erosion, coastal flooding, landsliding, and even volcanic eruptions and tsunamis. We will discuss the influence of human-induced global change on natural hazards individually and illustrate important linkages between them. Remember from Chapter 1 the principle of environmental unity: that our actions (in this case, burning huge amounts of fossil fuels) is setting in motion in our dynamic Earth a chain of likely and possible events that constitutes perhaps the largest challenge to our future well being and safety that humanity has faced as a result of our own activity. The famous human ecologist Garrett Harden emphasized that everything affects everything else, and "we can never merely do one thing."

GLACIERS AND SEA ICE

Global warming is resulting in accelerated melting of glacial ice of the Greenland ice sheet (Figure 12.18 and Figure 12.19) and mountain glaciers. The latter are of particular importance in Europe and South America as

Λ FIGURE 12.18 Shrinking of Greenland Ice Sheet
This summer meltwater stream on the Greenland ice sheet illustrates that surface melting is one of several ways that ice is being lost from the large Greenland ice sheet. The ice sheet is also losing large amounts of ice through outflow glaciers shedding icebergs into the ocean. *(Joe Raedle/Getty Images)*

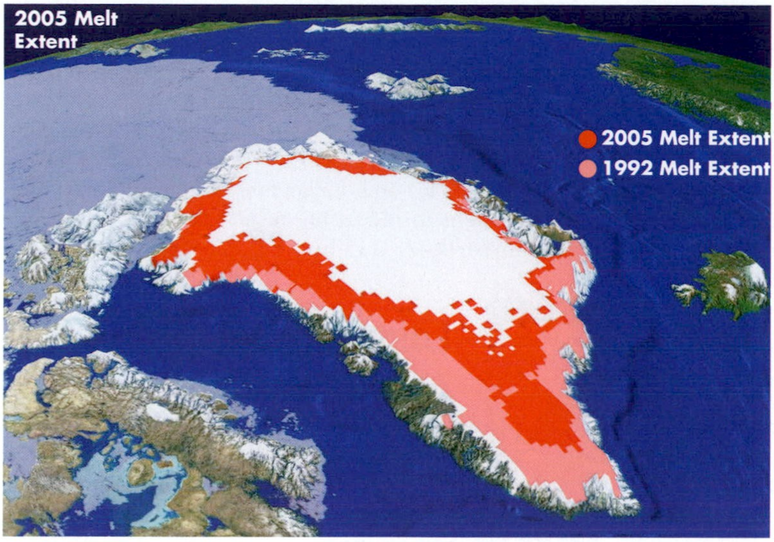
2005 Melt Extent
2005 Melt Extent
1992 Melt Extent

FIGURE 12.19 Increased Melting of Greenland Ice Sheet
Between 1992 (salmon colored) and 2005 (salmon and red color combined), the portion of the Greenland ice sheet that experienced surface melting during the summer increased considerably. The area of greatest increase was in the southern third of Greenland where the entire ice sheet experienced surface melting during summer months. An extreme melt event occurred in July 2012, due to a ridge of warm air over Greenland. *(Courtesy Russell Huff and Konrad Steffan/University of Colorado CIRES)*

water resources for people and ecosystems farther down the mountain. In Bolivia, a 70-year-old ski resort on a glacier is all but shut down because the small glacier on which the ski runs are built has just about disappeared.

Many glaciers in the Pacific Northwest of the United States are currently retreating (Figure 12.20); those in Europe, China, and Chile are retreating or decreasing more, in terms of thickness of ice (typically 10–30 percent of ice thickness since 1977), rather than advancing or increasing in thickness.[28,29] The increase in the number of retreating glaciers is accelerating in the Cascades of the Pacific Northwest, in Switzerland, and in Italy. Evidently, this acceleration is in response to a mean global temperature that has averaged 0.4°C (~0.68°F) above the long-term mean temperature from 1977 to 1994. On Mt. Baker in the Northern Cascades, for example, all eight glaciers were advancing in 1976. By 1990, all eight were retreating. In addition, four of 47 alpine glaciers observed in the Northern Cascades have disappeared since 1984. Most of the glaciers in Glacier National Park may be gone by 2030 and those of the European Alps by the end of the century.[28,30]

Annual melting of the Greenland ice sheet from 2002 to 2011 was about six times as great as from 1992 to

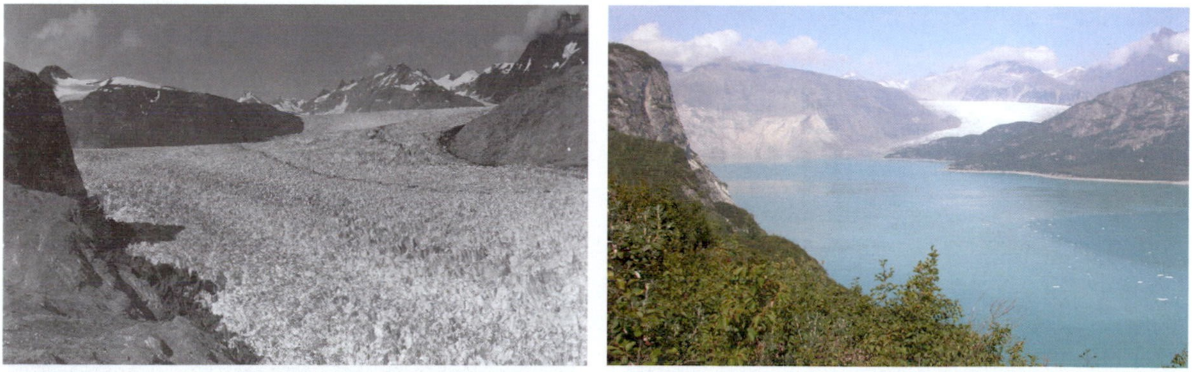

(a) (b)

FIGURE 12.20 Most Glaciers Are Retreating
Global warming is the primary cause for retreat of most of the world's glaciers, like Muir Glacier in Glacier Bay National Park, Alaska, shown here. (a) Photograph taken on August 13, 1941. *(USGS)* (b) Photograph taken from the same vantage on August 31, 2004. Between 1941 and 2004, this glacier retreated more than 12 km (~7 mi.) and thinned by more than 800 m (~875 yd.). Vegetation in the foreground of (b) is now growing where there was only bare rock in 1941. *(USGS)*

2001. Melting produces surface water that flows through openings and fractures (crevasses) and down to the base of the glacier, where it lubricates the bottom of the ice, resulting in an acceleration of glacial movement. Most glaciers in the southern half of the ice sheet are accelerating and losing ice more rapidly than previously. In 2005 alone, about 200 km³ (~48 mi.³) of glacial ice was lost.[28]

When glacial ice melts and bare ground is exposed, there is a positive feedback because white ice reflects sunlight (absorbing less energy), whereas darker rock reflects less sunlight while absorbing more. The more ice that melts, the faster the warming and increased melting occur. This is a classic positive feedback cycle—warming leads to more warming. This explains, in part, why warming at higher latitudes and elevations may exceed that in other areas.[13]

Sea ice in the Arctic Ocean is at a minimum each September. Based on satellite images, sea ice in September has declined an average of about 10 percent per decade (Figure 12.21). If these trends continue, the Arctic Ocean might be seasonally ice free on or before 2030.[31] Sea ice covered about 3.4 million km² (~1.3 million mi.²) in September 2012, a record low for the Arctic Ocean (Figure 12.21b). The total lost ice would cover an area five times the size of Texas. However, the impact of sea ice loss is more than just total area covered by the ice. For example, the depth and age of the ice are also important. Newer ice is thinner and, as a result, stores a lesser amount of water.

Methods of studying changes of glacial ice in Antarctica and Greenland have significantly improved in recent years. Scientists now use repeated measurements from satellites to estimate the mass of ice sheets. Change in ice mass can also be measured from satellites by observing polar gravity, which changes with the mass of ice.[31]

The loss of glacial ice on Greenland between 1992 and 2011 was approximately 2700 +/– 930 billion metric tons (Gt) of ice. During the same period, loss of glacial ice on Antarctica was about 1350 +/– 1010 Gt of ice. Thus, during the period of measurement, Greenland lost about twice as much ice as did Antarctica. The loss of the glacial ice on both land masses during this period is equivalent to increasing mean sea level by 11.2 +/– 3.8 mm (0.44 +/– 0.15 in.).[31]

CASE study 12.1 Loss of Glacial Ice in Antarctica

Antarctica is also losing sea ice and glacial ice. The loss of glacial ice is greatest at the west end of Antarctica, which has been losing ice since at least 2000. The Antarctic Peninsula (the relatively narrow land that points toward South America) is known to be one of the most rapidly warming regions on Earth. New studies suggest that the warming extends well beyond the peninsula to include most of West Antarctica. Although the trend is partly offset by East Antarctic's cooling in the autumn, the trend for Antarctica is positive climate forcing, resulting in warming. The causes of warming are complex, but they require changes in atmospheric circulation with temperature changes (warmer sea surface water and sea ice). The most likely direct cause is anthropogenic greenhouse gas forcing, resulting from increasing concentrations of greenhouse gases in the atmosphere.[31,32] On the other hand, the East Antarctic ice sheet has been growing in volume, albeit at a relatively low rate compared to the rate of loss of the West Antarctic ice sheet. The increase in the mass of

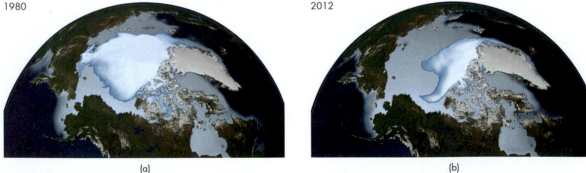

1980 (a)   2012 (b)

⋀ **FIGURE 12.21 Decreasing Arctic Sea Ice**
Arctic sea ice is seasonal, growing during the northern hemisphere winter and shrinking to a minimum in late summer (September). Since arctic sea ice data collection began in 1980 (a), annual minimum sea ice has decreased by about 10 percent per decade, reaching an all time low in September of 2012 (b). *(From NOAA)*

glacial ice in the East Antarctic ice sheet is the result of increased snowfall in that part of Antarctica. Increased snowfall results because of increased moisture in the atmosphere as temperatures warm (keep in mind that "warm" here is a relative term, as temperatures in Antarctica are always very cold).

Melting of glacial ice has increased on the West Antarctic ice sheet, due in part to acceleration of the collapse of the ice shelf in the sea. Large icebergs break off from the ice sheets at the coast, falling into the sea. There was a massive 6200 square kilometer (2400 square mile) collapse of part of the Larson ice shelf on the Antarctic Peninsula in July of 2017. The area of ice that broke off was about the size of the state of Delaware, and the weight of that ice was about 1 trillion metric tons. The collapse was predicted for years but occurred suddenly, much to the surprise of the general public. More large collapses are expected as the Antarctic continues to warm. When broad ice shelves are present, they tend to buttress the glaciers and slow down the movement of the ice toward the ocean. With the loss of ice sheets, glacier movement increases, sometimes surging (like the flow of maple syrup when the cap is removed), and more glacial ice enters the sea.

Melting sea ice does not raise sea level, but melting glacial ice does.

As Earth warms, more snowfall on Antarctica is predicted. Satellite measurement from 1992 to 2003 suggested that the East Antarctic ice cap on Antarctica increased in mass during the measurement period by about 50 billion tons per year[33] (a very small amount, given the enormous size of Antarctica—less than 0.5 percent of the land mass, covered to a depth of 1 m (~3 ft.)). Another study, using computer simulation and ice core records, reported that there has been no statistically significant increase in snowfall over Antarctica since the 1950s, suggesting that there is neither a reduction of sea-level rise by increased Antarctic snowfall nor an increase in the mass of glacial ice that would store water.[34] What is clear is that Antarctica is a complex place where more basic data research is needed to better understand the consequences of global warming.

In summary, although there is considerable variability in the rate of loss of glacial ice in Greenland and West Antarctica and, in some years, there may be an increase in glacial ice, measurements from 1992 to 2012 clearly suggest a continuous loss of glacial ice at both polar regions.[31]

Glacial Hazards Although at first thought, glaciers may not seem to present a particularly threatening hazard to people, that is not the case. Glaciers are huge, actively flowing masses of ice and rock debris whose movement and melting have been responsible for property damage, injuries, and deaths. Their irregular, cracked surfaces have always presented hazards to people crossing them for scientific exploration or mountaineering, and surface snow may hide deep cracks in the ice, known as crevasses, into which a traveler may fall. On some steep slopes, glacial ice falling from above is an additional hazard.

Changes in the flow of glaciers present hazards that have been well-documented, particularly in the Alps of southern Switzerland.[35] Glaciers can expand or surge and overrun villages, fields, roads, or other structures. They may also advance and cross a stream valley, producing an ice dam and temporary lake. Such a dam can lead to major downstream flooding if lake water undercuts or spills over the ice dam or if the dam is otherwise destroyed. When glaciers move from the land into the sea, large blocks of ice often break off from the front of the glacier and drop into the water in a process known as calving. This process is spectacular to observe—at a distance, that is—because the blocks are often as large as a house or a small building. Calving produces blocks

of floating ice, known as icebergs, which can float into shipping lanes and create a hazard to navigation. This danger was dramatically illustrated on April 14, 1912, when the *Titanic*, a luxury ocean liner on its maiden voyage from England to the United States, struck an iceberg and sank in the North Atlantic. That tragic accident claimed 1,517 lives. It is interesting that large icebergs may also be beneficial to us. They are being looked at as potential sources of fresh water if they could be towed economically to areas where water is scarce.

CLIMATE PATTERNS AND METEOROLOGICAL HAZARDS

The nature and extent of future impacts from climate change with more extreme meteorological events such as floods and hurricanes result from risk. The risk depends not only on the extreme events that occur, but also on exposure to people and property at risk, as well as their vulnerability (see Chapter 1).

Wet regions will get wetter and dry regions drier in the future. Global warming is changing the frequency and increasing the intensity of violent storms as warming oceans feed more energy into the atmosphere.[13] For example, hurricanes are apparently becoming more intense, as are the energy and precipitation in atmospheric rivers (see Chapter 6) that may deliver extreme rain events with a warmer world. More or larger coastal storms will increase the hazard of living in low-lying coastal areas, many of which are experiencing rapid growth of human populations. Global and regional climatic changes also have an effect on the incidence of hazardous natural events, such as storm damage, landslides, drought, and fires. These hazards are illustrated by El Niño, which is a natural climatic event that occurs on an average of once every few years. **El Niño** is both an oceanic and an atmospheric phenomenon, involving unusually high surface temperatures in the eastern equatorial Pacific Ocean and droughts and high-intensity rainstorms in various places on Earth.[36]

El Niño events probably start from random, slight reductions in the trade winds that, in turn, cause warm water in the western equatorial Pacific Ocean to flow eastward (Figure 12.22). This change further reduces the trade winds, causing more warm water to move eastward, until an El Niño event is established.[36]

El Niños are thought to bring about an increase in some natural hazards on a nearly global scale by putting a greater amount of heat energy into the atmosphere. The heat energy in the atmosphere increases as more water evaporates from the warm ocean to the atmosphere. The increase of heat and water in the atmosphere produces more violent storms, such as hurricanes. Figure 12.23 shows the extent of natural disasters that have been attributed, in part, to the El Niño event of 1997–98, when worldwide hurricanes, floods, landslides, droughts, and fires killed people and caused billions of dollars in damages to crops, ecosystems, and human structures. Australia, Indonesia, the Americas, and Africa were particularly hard hit. There is some disagreement about how much damage

and loss of life are directly attributable to El Niño, but few disagree on its significance.[37,38] We do not yet completely understand the cause of El Niño events. They occur every few years, to a lesser or greater extent, including the large El Niño events of 1982, 1997, and 2015.

El Niño events can cause havoc by increasing the occurrence of hazardous natural events, and there is concern that human-induced climate change may, through global warming, produce more and stronger El Niño events in the future. This effect would result, in part, because burning more fossil fuels and emitting more greenhouse gases into the atmosphere will further warm the oceans, and differential warming in various parts of the ocean may increase the frequency and intensity of El Niño events in the Pacific Ocean.

The opposite of El Niño is La Niña, in which eastern Pacific waters are cool, and droughts rather than floods may result in Southern California. The alternation from El Niño to La Niña is a natural Earth cycle that has only recently been recognized.[39]

Decadal Oscillations Storm tracks and, thus, climate and weather in North America and Western Europe are influenced by the natural climate phenomena known as decadal oscillations. Examples include the Pacific Decadal Oscillation (PDO), the Arctic Oscillation (AO), and the North Atlantic Oscillation (NAO).[40]

The PDO is recognized in the North Pacific Ocean by a relatively warm surface water (positive) phase or relatively cool surface water (negative) phase. The PDO changes phases on average every 20 to 30 years, alternating relatively wetter and warmer (positive phase) or drier and cooler (negative phase) cycles. The PDO has important climatic and ecologic implications

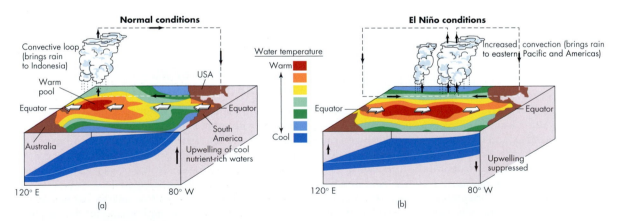

∧ FIGURE 12.22 Effects of El Niño
Idealized diagrams contrasting (a) normal conditions and processes with (b) those of El Niño. *(Based on National Oceanic and Atmospheric Administration. (www.elnino.noaa.gov)*

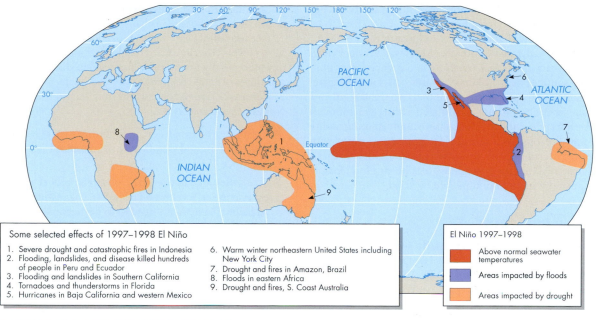

Some selected effects of 1997–1998 El Niño

1. Severe drought and catastrophic fires in Indonesia
2. Flooding, landslides, and disease killed hundreds of people in Peru and Ecuador
3. Flooding and landslides in Southern California
4. Tornadoes and thunderstorms in Florida
5. Hurricanes in Baja California and western Mexico
6. Warm winter northeastern United States including New York City
7. Drought and fires in Amazon, Brazil
8. Floods in eastern Africa
9. Drought and fires, S. Coast Australia

El Niño 1997–1998

- Above normal seawater temperatures
- Areas impacted by floods
- Areas impacted by drought

⋀ **FIGURE 12.23 The 1997–1998 El Niño Event**
Map showing the general extent of the El Niño effects and the regions damaged by floods, fires, or drought.
(Data from National Oceanic and Atmospheric Administration, 1998)

for people (rainfall and drought) and marine resources (especially salmon).

The AO and NAO work in the same direction. The positive states of the AO and NAO are characterized by the existence of a strong polar vortex that helps to constrain cold arctic air further north. On the other hand, a negative AO and NAO is characterized by weaker atmospheric pressure in both the subtropics and subarctic. This produces a weakened polar vortex, which allows breakouts of cold air from the Arctic to move further south. This brings cold temperatures to the East Coast of the United States. There is not a general movement to the south of cold air masses, but there are periodic breakouts of cold that extend southward as the southern jet stream dips toward lower latitudes. This allows cold air to move south, both in the United States and in parts of Europe.[40]

In summary, we now know that natural oscillations of the ocean, linked to the atmosphere, can produce warmer or cooler periods for a few years or a decade or more. The effects of the oscillations can be as much as 10 times as strong in a given year as the long-term warming that we have observed over the last century. The changes may be considerably larger over a period of a few decades than our human-induced climate change. By comparison, the annual increase in warming, thought to be due to human activity (1970–2000), was about 0.02°C (~0.04°F) per year. Some scientists

have speculated that the cold winter of 2009–10 was a result of natural ocean-atmospheric oscillations. Scientists have learned by studying the phenomena of periodic cycles in the Pacific, Atlantic, and Arctic areas that our climate system is incredibly complex. As a result, interactions with global warming can also be complex. Putting it all together, we now have a better explanation for changes in winter storms from year to year in terms of interactions between both natural cycles and global warming.

Global rise in temperature is changing rainfall patterns, soil moisture relationships, and other climatic factors that are important to agriculture. Scientists have predicted that some northern areas such as Canada and Eastern Europe may become more productive, whereas others, including the southwestern United States, will become more arid. Stable or expanding global agricultural activities are crucial to people throughout the world who depend on the food grown in the major grain belts. Hydrologic changes associated with climatic change resulting from global warming might seriously affect food supplies worldwide. Less water may be available in California and the San Joaquin Valley, where much of the fruits and vegetables for the United States are raised. Much of the irrigation water comes from runoff from spring snowmelt in the Sierra Nevada to the east of the valley. With global warming, winter rainfall will likely increase, but the snowpack will be less. As a result, runoff will be more rapid, filling reservoirs before it is needed to irrigate crops. Water may have to be released from filled reservoirs and lost to the sea.

A lesson learned from the Medieval Warm Period 1,000 years ago is that warming and reduced rainfall in the southwestern United States, Mexico, Central America, and parts of Africa, China, India, and Southeast Asia may result in prolonged (intergenerational) drought that will threaten water supply and the ability to grow crops. As a result, if long droughts occur with global warming, the most serious potential threat is that we will not have an adequate supply of food for the expected 9 to 10 billion people by 2050.

SEA-LEVEL RISE

A rise of sea level is a potentially serious problem related to global warming. Global ocean temperature to a depth of at least 3 km (~1.9 miles) has increased since about 1960. The oceans of the world, in that time, have absorbed about four-fifths of the heat that has been added to Earth's climate system.

Warming of the oceans causes the water to expand (thermal expansion), producing a rise in sea level. The rate of sea-level rise resulting from thermal expansion from 2005 to 2015 is estimated to be about 1.0 mm (0.04 in.) per year (Figure 12.24).[41]

Added water (predominantly melting of glacial ice) is the largest contributor to sea-level rise, at about 2.0 mm (0.08 in.) per year from 2005 to 2015 (approximately two-thirds of observed rise of about 3 mm (0.12 in.) per year from 2005 to 2015 (Figure 12.24).[41] Keep in mind that sea-level rise in a specific location such as Southern California, Hawaii, or South Florida may be more or less than the global average of about 3 mm per year. This may result from subsidence, erosion, or ocean currents. For example, at Santa Barbara,

California, the rate is less than 1 mm per year, and at South Florida, the rate probably exceeds 5 mm per year.

Melting of sea ice (floating ice) does not cause a rise in sea level (you can verify this by placing ice in a glass of water and observing the water level in the glass while waiting for the ice to melt).

Several tentative conclusions may be put forth concerning rise in sea level:

> Both thermal expansion and melting glacial ice have contributed significantly to observed sea-level rise since 1993.

> The difference between the observed global and local rise in sea level is considerable, suggesting that additional research is needed to better understand sea-level rise.

> The rates of both thermal expansion and melting glacial ice are accelerating.

> The contribution of the Greenland Ice Sheet to global sea-level rise has increased several times in recent decades, consistent with surface observations of accelerated melting of glacial ice.

Estimates of the rise expected in the next century vary widely. Global mean sea level, with high confidence from scientists, will rise at least 20 cm (~8 in.), but no more than 2.0 m (6.6 ft.), by 2100.[42] Precise estimates are not possible at this time. However, a 20 cm (~8 in.) rise in sea level would have significant environmental impacts. Such a rise could cause increased coastal erosion on open beaches of up to 40 m (~130 ft.), rendering buildings and other structures more vulnerable to waves generated from high magnitude storms. In areas with coastal estuaries, a 20 cm (~8 in.) sea-level rise would cause a landward migration of existing estuaries, again putting pressure on human-built structures in the coastal zone. Communities would have to choose

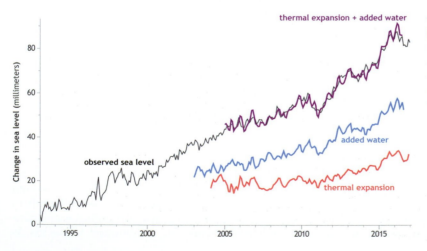

< **FIGURE 12.24 Relative Contribution to Sea-Level Rise** Melting glacial ice and thermal expansion are the two main components of sea-level rise. *(Lindsey, R. 2016. Climate change: global sea level. www. climate. gov)*

between making substantial investments in controlling coastal erosion or allowing beaches and estuaries to migrate landward over wide areas.[13,42,43] A sea-level rise of about 40 cm (~16 in.) would flood the San Francisco International Airport and parts of San Francisco Bay (Figure 12.25).

The coastline of northwestern and northern Alaska is experiencing rapid erosion from rising sea level, melting permafrost soil, and loss of sea ice that can protect a coastline from erosion.[44] Some small islands may disappear, and some villages have already been moved inland.

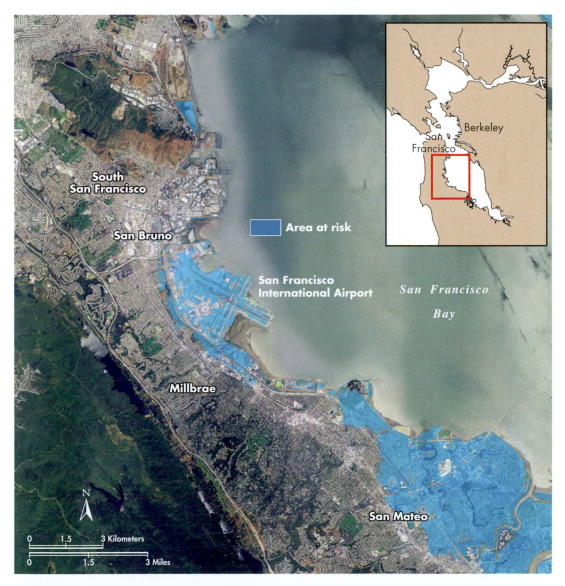

∧ FIGURE 12.25 San Francisco Airport at Risk
A sea-level rise of 40 cm (~16 in.) would inundate densely developed low-land regions along the San Francisco Bay, including the San Francisco International Airport. *(Inundation data used in these maps do not account for existing shoreline protection or wave activity. These maps are for informational purposes only. Users agree to hold harmless and blameless the State of California and its representatives and its agents for any liability associated with the use of the maps. The maps and data shall not be used to assess actual coastal hazards, insurance requirements, or property values or be used in lieu of Flood Insurance Rate Maps issued by the Federal Emergency Management Agency (FEMA).)*

Disappearing Islands: Tuvalu, South Pacific

Rising sea level is already threatening some small islands in the tropical Pacific Ocean. Let's think a little deeper. The island nation of Tuvalu consists of about six atolls and three islands (Figure 12.26). People first inhabited the islands of Funafuti and other atolls of Tuvalu about 2,000 years ago when stable islets formed.[45–47] The population now totals nearly 11,000, with well over half living on the main island of Funafuti. Because total land area is small, only 26 km^2 (~10 mi.^{2}), Tuvalu's population density is a very high 423 persons per square kilometer. Global rise in sea level during most of the twentieth century was about 1.8 mm (~0.07 in.) per year as a result of global warming. The rate increased to about 2.5 mm (~0.1 in.) per year during the last decade of the twentieth century and about 4 mm (~0.16 in.) per year during the first decade of the twenty-first century. In 50 years' time, sea level will be about 200 mm (~8 in.) higher. Although scientists use models to predict the amount of global sea-level rise, current sea-level rise at Tuvalu for a particular year cannot be recognized, as the change is small relative to short-term seasonal effects of changing trade winds (200 mm) and natural climate cycles that span a few years, such as El Niño

and La Niña events (500 mm [~20 in.]). However, effects of rising sea level over several decades will become apparent as increasing erosion begins to threaten the stability of the islets of Tuvalu by eroding the more resistant underpinnings of these islets.[45–47]

The prediction of future changes of Tuvalu is complex and related to the geologic history of the atolls, linked to past changes in sea level that are responsible for the atolls we observe today. Recent sea-level rise is now threatening Funafuti Atoll, especially during high tides that enhance wave attack and flooding. During high tide flood events, people report water bubbling up from the ground, contributing to the flooding of low-lying areas (Figure 12.27). With continued sea-level rise, the future ability of Funafuti and other atolls to support people is uncertain. The 11,000 people on Funafuti could become the first to be displaced by rising sea level.

The islands of Tuvalu do not have streams or rivers to supply fresh water. As with most islands, there is a lens (of variable thickness) of fresh groundwater that rides on top of salt water.[45,46] Therefore, Tuvalu has little potable (drinking) water, and most of the land will not support agriculture. Study of groundwater at the Fongafale Islet, a small island 16 km long and 400 m wide (~10 mi. long and 0.25 mi. wide) on the northeastern side of Funafuti Atoll, suggests that the area is vulnerable to rising sea level and tidal changes (tidal forcing). As much as half the islet is reclaimed swampland, filled with porous coral blocks. Shallow,

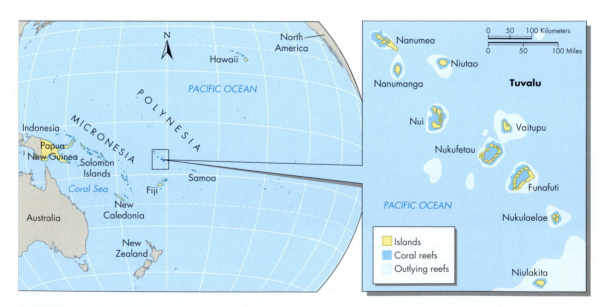

∧ FIGURE 12.26 Islands of Tuvalu, South Pacific Ocean
Many South Pacific islands are at risk of losing land due to sea-level rise that will result from global warming. Tuvalu, a nation of low-elevation Polynesian islands, is especially vulnerable to hazards linked to climate change.

thin sheets of freshwater and brackish water (less salty than seawater but saltier than freshwater) are present in the islet, but there are no thick freshwater lenses that are often present in other atoll islands of similar size. Thin freshwater lenses are found at elevations above general high tide or below taro swamps (taro is a cultivated tuber plant, a food staple that can be grown in swampy ground), but at lower elevations, periodic salinization (increase in salinity of the soil) occurs in phase with tidal cycles. The thin nature of the brackish and freshwater lenses suggests that taro fields and fresh groundwater resources are very vulnerable to increasing salinization as climate changes and sea-level rises.[46] Assessing the vulnerability of the water resources will require continued study of the linkages between topography, geology (uplift, subsidence, erosion, and deposition), and potential climate change with sea-level rise.

On February 28, 2006, maximum sea level on Funafuti was about 3.4 m (~11 ft.) on an island where the average elevation is about 2 m (~7 ft.). Widespread flooding occurred (Figure 12.27). Local people on the island call this event the "King Tide," which now occurs once or twice per year. These floods damage plants by bringing salt water into contact with the roots, and they degrade the freshwater resources as the water table rises. Tide gauge data, as well as satellite altimetry data, suggests that in the past 18 years, sea level has risen above 3.2 m (~10 ft.) 23 times. Compilations of sea-level trends from 1977 to 2011, utilizing tide gauge records, suggest that sea-level rise is about 3.7 ± 2.9 mm (~0.15 ± 0.11 in.) per year, which is very imprecise. What the record does show is that, if this rate is correct, sea level will rise approximately 0.4 m (~1.3 ft.) per century. However, the tide gauge records show large fluctuations above 0.6 m (~2 ft.) every few years, which is about twice as high as the projected rise of sea level over the next 100 years. These transient (short-lived) rises in sea level are likely due to El Niño and La Niña events, which push the water first east and then west across the Pacific Ocean.[47]

Recent coastal erosion of the Funafuti Atoll shorelines is modest because the emergent mid-Holocene reef rubble that is cemented together protects modern unconsolidated calcareous sand from direct wave attack (Figure 12.28). The main shoreline modification over recent years has been accretion (building up) rather than erosion of large rubble ramparts as, for example, those installed during Cyclone Bebe in 1972.[47]

Atoll formation in Tuvalu is intimately related to reef growth and erosion and to changing sea level. During the past 20,000 years, sea level has been anything but stable. Atolls such as Funafuti in Tuvalu have formed by the following general processes:[47]

> During the Last Glacial Maximum (LGM), sea level was about 120 m (~400 ft.) lower than today, with erosion of the carbonate (limestone) platforms present in the area.

> Following the LGM, sea level rose, and carbonate platforms grew with the sea-level rise. The modern Holocene reef began building about 8,000 years ago, and sea level reached a peak known as the mid-Holocene highstand (the time during which sea level is at its highest) about 4,000 years ago. The reefs grew above present sea level during this highstand and are the backbone of the stable atoll islets that rim the atoll today.

> Since the mid-Holocene highstand, sea level has dropped about 2.3 m (~7.5 ft.) on Tuvalu. This has allowed the emergence of the reef islets, which were discovered and colonized by seafaring people about 2,000 years ago.

∧ FIGURE 12.27 "King Tide" on Funafuti, 2006
Regular seasonal flooding on Funafuti affects everyone on this densely populated island. *(Ashley Cooper/Corbis)*

Reef rubble

∧ FIGURE 12.28
White material is sand on top of older mid-Holocene coral reef, which provides a buffer to erosion. *(William Dickenson)*

Today, the atolls of Tuvalu are marginal environments for people. The surfaces (reef islets) are generally less than 3 m (~10 ft.) above the sea. As sea level continues to rise, sometime between 2070 and 2140, the islets will likely be overtopped. It should be pointed out that there are major uncertainties involved in predicting future sea level, and these dates are more of an indication of when that will likely happen than a conclusion based upon analysis. However, before overtopping takes place, the reef islets will become unstable and increasingly vulnerable to wave attack. In fact, they might well be seriously eroded before overtopping occurs. When this happens, the islets will likely be unfit for human occupation. If the indicated rise in sea level occurs, then the people of these atoll islets have between 60 and 130 years to prepare for and adapt to changing conditions.

Adaption to changing conditions on Tuvalu islands may eventually become unfeasible, and people will need to migrate to places higher above sea level. Some residents have already left the country, migrating to Fiji or New Zealand. Potential adjustments in coming decades to sea-level rise and erosion, along with flooding, on Tuvalu include the following:

> Defending the coast with seawalls
> Planting mangrove trees along the coast to buffer islets from wave attack
> Planting salt-resistant crops for human consumption, so that agricultural activities may continue
> Improving storage of rainwater to augment freshwater supply or importing water.

WILDFIRES

There is a complex relationship between climate change and fires, and the two phenomena interact on a variety of levels. Global warming is predicted to lead to an increase in droughts, which will set the stage for wildfires. For example, scientists in Spain studied a strip of Mediterranean coastline between 1968 and 1994, comparing the number of forest fires to the temperature and aridity of the area. They found that as the coastal area warmed and became more arid, both the number of fires and the size of the area burned increased.[48] Another investigation used four different global circulation models (GCMs) for the atmosphere. Historical weather data were input into the models to project forest fire danger levels in Canada and Russia under a warmer climate. These models suggested that there will be more frequent and severe forest fires, and that the number of years between successive fires in a given location will decrease.[48,49]

LANDSLIDE HAZARD

Worldwide, landslides kill several tens of thousands of people per year. Future global change that involves increase in temperature, as well as amount, frequency, and intensity of precipitation events in landslide-prone areas, will likely increase the occurrence of many types of landslides and from shallow slides, such as soil slips, debris flows, and rockfalls (see Chapter 7) associated with intense precipitation. Deeper, larger slides might involve much larger areas and cubic kilometers of slide material. For example, if the flank of a volcano on the coast or island were to collapse (slide) suddenly into the sea, the slide could cause a tsunami.[50,51] Several landslide-linked tsunami in the past few thousand years occurred (before the human-induced warming), causing widespread damages and deaths (catastrophes). Global warming linked to melting ice, rock weathering, raising ground water tables, and more intense rainfall reduces slope stability on high, steep slopes of volcanoes or other coastal rock slopes. These changes in rock and water will increase the probability of large, damaging landslides. Finally, if the leading edge of a glacier retreats upslope, blocks of ice may fall off in ice avalanches into the valley below, destroying property and perhaps taking human lives. For example, in 1962, a combined ice and debris avalanche, known as the Huascaran Avalanche, killed 4,000 people and destroyed six villages in Peru. The avalanche moved approximately 16 km (~10 mi.) from its source as it descended almost 4000 m (~13,000 ft.) of vertical elevation. A similar avalanche in Peru, triggered by a great earthquake in 1970, killed an estimated 20,000 people while burying several villages in debris (Figure 12.29).[35] More recently, in September 2002, a huge mountain glacier broke loose in southern Russia, sending a torrent of ice, mud, and rocks downslope. The debris buried villages and killed at least 100 people.

Rapidly moving landslides that cause many deaths in the tropics and subtropics are associated with high steep slopes, shallow soil, and abundant vegetation, as, for example, in Central and South America. A common landslide trigger is intense rainfall on saturated soil that will likely increase with global warming. Landslide-prone areas that have rapid population growth and people pushed by poverty to live on steep slopes (as, for example, in Brazil) will be faced with increasing exposure and vulnerability to the landslide hazard. As a result, the total risk increases (see Chapter 1).[50]

High mountain areas in the Alps, Himalayas, and Alaska are also warming. As ice that previously stabilized broken rock on slopes melts, the slope stability will make large to small landslides more likely.[51]

Debris flows in southern California and other fire-prone areas will likely become more common as wildfires increase in frequency and intensity. More and more

(a)

(b)

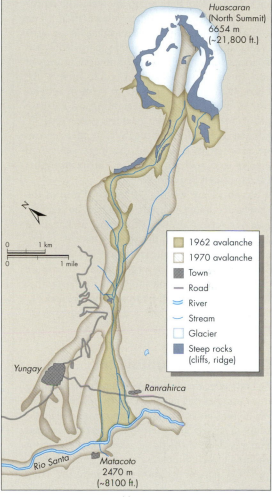

(c)

∧ FIGURE 12.29 Glacial Ice Contributes to Debris Avalanches

(a) Yungay, Peru, prior to the earthquake and debris avalanche. Snow-covered Mount Huascaran is on the skyline. *(Lloyd S. Cluff/PEER Center, University of California at Berkeley)* (b) The town site after the 1970 earthquake and avalanche. *(Lloyd S. Cluff/PEER Center, University of California at Berkeley)* (c) Tracks of the 1962 (yellow) and 1970 (area shaded with diagonal lines) avalanches down Mount Huascaran. *(From Welsh, W., and Kinzl, H., "Der Gletschersturz vom Huascaran [Peru] am 31. Mai 1970, Die grosste Gletscherkatastrophe der Geschichte." Zeitschrift für Gletscherkunde und Glazialgeologie 6: 181–92, 1970. Reprinted by permission of the author.)*

people are exposed to the debris flow hazard as new development continues in and adjacent to brush lands and forests of the American Southwest and similar areas around the world. We will return to wildfire and debris flows in Southern California in Chapter 13.

VOLCANIC, EARTHQUAKE, AND TSUNAMI HAZARDS

The crust of Earth often seems solid and steady. In fact, it is always flexing, uplifting, subsiding, and changing because of a variety of geologic processes that are loading the crust or releasing a load. For example, glacial ice in polar ice sheets have depressed (pushed down) the crust by many meters, and when the glacial ice thins and melts from a warming world, the crust rebounds (uplifts), often hundreds to a thousand kilometers from the ice sheet. The unloading may reactivate active faults, producing moderate to large earthquakes. The unloading (removal of pressure) may also trigger subglacial volcanic eruption. It is more difficult for magma that is formed and stored in subglacial volcanoes to reach the surface when covered by thick glaciers. Thus, with global warming and melting of glaciers, pressure is lessened and rock can melt quicker. More magma can lead to more eruptions, and dormant volcanoes may come to life. Iceland is being split apart along the Mid-Atlantic

Ridge (see Chapter 2). There are many Icelandic volcanoes that are linked to the glaciers above them. The Vatnajökull glacier (ice cap) in Iceland that has an area of about 7800 square kilometers (3000 square miles) has many volcanoes beneath the ice. Melting (thinning) of the glacier causes about 14 million cubic meters (18 million cubic yards) of new magma to form in subglacial volcanoes. Thus, with global warming, an increase in volcanic eruptions below the glacier is predicted.[50,52] The Eyjafjallajokull volcano beneath a smaller ice cap in Iceland erupted in 2010, and the huge ash cloud caused airlines to cancel thousands of European flights for several days (see Chapter 5).

Many volcanoes are within 250 km (150 mi.) of the sea, and there are many volcanic islands. The loading from rising sea level may bend the crust and lead to magma being squeezed and, therefore, easier to erupt. This may lead to more explosive volcanism and even lateral blasts that may produce large, sudden submarine or sea cliff landslides (a volcano flank collapse that could produce tsunamis). While this scenario is speculative, there are examples in the recent geologic past,[50] and future events cannot be ruled out.

CHANGES IN THE BIOSPHERE

A growing body of evidence indicates that global warming is initiating a number of changes in the biosphere, threatening both ecological systems and people. An ecosystem is a community of species and their nonliving environment in which energy flows through the system and chemicals cycle. Sustained life is a function of ecosystems, not individual species. Changes to ecosystems resulting from global warming include risk of regional extinction of species in the system as land and hydrology change (e.g., forested land changes to grassland). Other changes include shifts in the range of plants and animals, with a variety of potential consequences. For example, mosquitoes carrying diseases, including malaria and dengue fever in Africa, South America, Central America, and Mexico, are migrating to higher elevations; butterfly species are moving northward in Europe; some bird species are moving northward in the United Kingdom; subalpine forests in the Cascade Mountains in Washington State are migrating to higher meadows; alpine plants in Austria are shifting to higher elevations; sea ice melting in the Arctic is placing stress on seabirds, walruses, and polar bears; and warming and increasing acidity of shallow water in the Florida Keys, Bermuda, Australia's Great Barrier Reef, and many other tropical ocean areas are believed to be contributing to the bleaching of coral reefs.[13,49,53]

Ocean Acidification Approximately one-third of the carbon dioxide emitted into the atmosphere by human activity becomes dissolved into the world's ocean, thereby minimizing the input of greenhouse gas into the atmosphere. Although this process minimizes the positive forcing of global warming, it does so at the expense of the oceans. Dissolved carbon dioxide and water form carbonic acid $CO_2 + H_2O \rightarrow H_2CO_3$. The acid dissociates to bicarbonate ion HCO_3^- and hydrogen ion H^+. As H^+ increases, ocean waters become more acidic (Figure 12.30) and their pH lowers; over the past 15 years, ocean acidity has increased about 30 percent. The increased acidity of the oceans and ecosystems is fundamentally changing the structure and function of ocean chemistry by stressing marine organisms, ranging from plankton to coral and shellfish, as they attempt to adapt

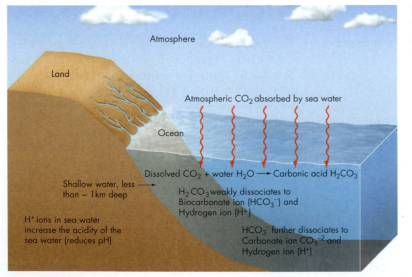

◄ FIGURE 12.30 Changes in Seawater
Idealized diagram showing processes that change the acidity of seawater.

to increasing acidity. The stress for some species may be manifested by reduced growth rates and reproduction. These changes could occur if organisms need to expend more energy to maintain their internal pH, resulting in less energy for growth and reproduction. How individual species and ecosystems will respond to further acidification in the future is a complex problem that is the subject of increasing research.[54,55] Increased acidity of seawater may make it more difficult for some marine organisms that produce carbonate shells, such as oysters, coral, and some plankton, to draw carbonate from the seawater to make their shells.

In response to the warming atmosphere over the past 100 years, heat has been transferred to the upper water layers of the oceans. As a result, the average temperature of the upper 700 m (~2300 ft.) of the sea has increased by about 0.6°C (~1°F) over the past century. Temperature increase has a fundamental effect on biological processes of marine organisms through increased metabolic rates that impact ecosystem processes of energy expenditure and chemical cycling. Marine organisms have the ability to adapt to temperature change within a particular range of temperatures. If the range is exceeded, the ability to adapt is reduced, and populations may decline or be driven to local extinction.[56] Mobile species such as fish can swim to other areas or deeper water where the temperature is closer to what they prefer. Those species tied to the bottom, such as shellfish, may slowly migrate if change is sufficiently slow. The change of temperature and acidity of seawater appears to be accelerating, which may make adaptation more difficult for some species. On the other hand, some species of coral are apparently adapting to more acidic seawater.[57]

Adaptation of Species to Global Warming As a result of global warming some species will experience stress. Most vulnerable will be those species that are not mobile, such as some vegetation on land and shellfish in the ocean. Many species will migrate toward higher altitudes in an attempt to adapt to warming. With global warming, temperature and precipitation change in different regions, which stresses ecosystems. During the past 25 years or so, plants and animals have shifted their ranges by about 6 km (~3.7 mi.) per decade toward the polar areas. In addition, spring is arriving earlier, and plants, as a consequence, are blooming earlier, frogs are breeding earlier, and migrating birds are arriving earlier in various places. The rate of change has been about 2.3 days per decade. In addition, it has been reported that tropical pathogens have moved up in latitude and elevation, affecting species that may not be adapted to them. Climate change may have caused the first extinction of a mammal, a species of opossum that has

apparently disappeared from Queensland, Australia. It was reported that the opossum lived only at an elevation of 1000 m (~3280 ft.) and was susceptible to being exposed to higher temperatures, even for a short period of time.[58]

A problem with past evaluations of the potential threats to species or their possible regional extinction was that they were based on simple models that primarily considered temperature and precipitation, perhaps along with soil type and hydrology, that determine, in part, where a particular species lives. This information would then be put into a standard climate model to predict where a particular species might migrate in order to avoid adverse effects of climate change. We now know that these simple envelope models that use a few climate and topographic variables are not sufficient, and newer models that incorporate additional biological elements, including competition among species and genetics, are being used to evaluate evolutionary responses to climate change.[59]

One controversial suggestion is that we might assist migration of some species that are unable to migrate with climate change. To some, this is a drastic step that may be unacceptable because of the risk of, essentially, creating an invasive species. Assisted migration worries ecologists and conservation biologists because they have spent a lot of time and effort working against invasive species, some of which cause ecosystem problems and the extinction of other species. Before seriously considering assisted migration, much more research is needed to better understand the habitats of threatened and endangered species, including where they can do well, what might threaten them, and what they might threaten.[58]

12.6 **CHECK** your understanding

1. How are glaciers and sea ice changing in the Arctic and Antarctic?

2. How might global warming change the intensity and frequency of violent storms? Describe the process.

3. What are the two main physical processes in the warming of seawater that cause sea-level rise?

4. How might global warming impact landslides, volcanic eruptions, earthquakes, and tsunamis?

5. How is global warming affecting the occurrence of wildfires?

6. How does ocean acidification occur, and why is it a potentially significant problem?

7. How is global warming influencing the biosphere, and how is life adapting?

12.7 Predicting the Future Climate

WHAT DOES OUR RECENT HISTORY TELL US ABOUT POTENTIAL CONSEQUENCES OF FUTURE GLOBAL WARMING?

The famous philosopher George Santayana stated in 1905 that "those that cannot remember the past are condemned to repeat it." It is a matter of debate whether cycles in human history actually repeat themselves, but the repetitive nature of many natural processes and hazards, even if they result from different reasons, is indisputable. So, what can we learn by examining the past with reference to global climate change and, in particular, global warming? It turns out that over an approximate 300-year period from A.D. 950 to 1250, Earth was considerably warmer than normal (normal meaning the average surface temperature during the past century or some shorter interval, such as between 1961 and 1990). This period is known as the Medieval Warm Period (MWP), and we can learn some lessons from that event. Although it occurred only a few hundred years ago, we don't know much about the MWP except that it was, in fact, warm and perhaps nearly as warm as it is today. Most scientists who study indicators of past climate state that it probably wasn't quite as warm, but certain parts of the world, Western Europe and the North Atlantic in particular, may have been warmer some of the time than they were in the last decade of the twentieth century. During the MWP, there were winners and losers. In Western Europe, there was a flourishing of culture and activity, as well as population growth, as harvests were plentiful and people generally prospered. Many of Europe's grand cathedrals were constructed during this period, and the first global trade routes opened from Europe to China, helped by favorable climate and connected by camel caravans.[60,61]

During the MWP, sea temperatures in the North Atlantic evidently were warmer, and there was less sea ice. The famous Viking explorer Erik the Red embarked on a voyage of exploration near the end of the tenth century. When he arrived at Greenland with his ships and people, they set up settlements that flourished for several hundred years, and they were able to grow a variety of crops, including corn, that had never before been cultivated in Greenland. They were also able to raise their animals and, in general, enjoy a prosperous life. During the same warm period, Polynesian people in the Pacific, taking advantage of winds flowing throughout the ocean, were able to sail to and colonize islands over vast areas of the Pacific, including Hawaii.[61]

Although some prospered in Western Europe and the Pacific during the MWP, other cultures were not so fortunate. Associated with the warming period were long, persistent droughts (lasting perhaps a human generation) that appear to have been partially responsible for the collapse of sophisticated cultures in North and Central America. The collapses were not sudden but occurred over a period of many decades; in some cases, the people just moved away. These included the people living near Mono Lake on the eastern side of the Sierra Nevada in California, the Chacoan people in what is today Chaco Canyon in New Mexico, as well as the Maya civilization in the Yucatan of southern Mexico and Central America.

Today, we are worried about the present warming trends. Should they continue and accelerate or increase, we may face a prolonged drought in the southwestern United States again, as well as in other parts of the world, which may wreak havoc on the ability to produce the food that the 9 billion people on Earth will soon require.[61]

When the MWP ended, it was followed by the Little Ice Age (LIA) that lasted several hundred years from the mid-1400s to 1700.[60] The cooling of the LIA made it more difficult for people in Southeast Asia and Western Europe. Angkor Watt, with its palace and temple in Cambodia, was constructed in 1181. The region flourished for several hundred years as canals and reservoirs irrigated vast rice fields. The water supply from summer monsoon rains was reliable, and people prospered. With climate change and the transition to the LIA, monsoon rains weakened, and water became scarce. The Ankorian civilization was in decline by the fifteenth century and collapsed by the end of the sixteenth. The collapse was gradual as people left the city, moving to smaller communities. Why the collapse occurred is not fully understood, but it was in part due to the onset of the Little Ice Age and a more volatile climate with reduced monsoon rains and drought. In short, the food supply diminished as water resources became scarce.[61]

During the Little Ice Age, there were storms, wet periods, extremes of heat and cold, and climate change that caused a number of problems in Europe. Some winters were very cold, and during one such winter in 1683–84, the River Thames in London was frozen solid for two months and sea ice extended for kilometers into the North Sea, closing harbors and significantly restricting shipping. Crop failures occurred in Western Europe, and the population was devastated by the Black Plague that reached out into the Atlantic and Iceland by about 1400.

With the cooling times, increasing sea ice, and more environmental stress, trade was restricted with Greenland. Eventually, the colonies in North America and Greenland were mostly abandoned. Part of the reason for the abandonment in North America and, particularly, in Newfoundland was because the Vikings may not have been able to adapt to the changing conditions, as did the Inuit peoples living there. As times became tough, the two cultures collided, and the Vikings, despite their fierce reputation, were not as able to adapt to the changing, cooler times as were the Inuit.

We do not know what caused the Medieval Warm Period, and the details about it are somewhat obscured by the lack of sufficient climate data during that period to estimate temperatures. We do know that it was warm in Western Europe and the North Atlantic, and we can't blame that on the burning of fossil fuels. What this may suggest is that more than one factor can cause warming; the present warming, for the most part, is clearly the result of the emission of carbon dioxide into the atmosphere from burning fossil fuels and land-use changes. Perhaps a lesson for us is that changes in the most recent past warming, the MWP, resulted in winners and losers. Today, the world population is greater than 7 billion, and half of us live in cities. The world population during the MWP was about 500 million, and agriculture was booming. However, persistent droughts during the MWP apparently caused pain and suffering over much of the semiarid areas of the world, from coastal California to the southwestern United States, as well as in Central and South America. This is a red flag! The potential effects of prolonged, persistent drought and loss of food production for the world, should present warming continue, is a serious threat.[61]

PREDICTING FUTURE CLIMATE IS PROBLEMATIC

Predicting the future climate is problematic because predicting the future has always been difficult. People who make predictions have often been wrong. For climate and its effects on living things, we can attempt to apply the geological concept of uniformitarianism, which assumes that processes that occurred in the past occur today and will occur in the future. This approach has led to investigations of atmospheric concentrations of greenhouse gases in the geologic record, which we have explored in this chapter. However, the time of particular interest to us is the one that is the most difficult from which to collect climate data. That time is the past few centuries to 1,000 years ago, during the scientific and technological revolutions.

The problem we face is that temperature records have been kept, at best, for only about 150 years, and the early records are from only a few places. Until the advent of satellite remote sensing, measurements of air temperature over the oceans were taken only where ships traveled, which did not provide the kind of sampling that satisfies statisticians. Many places on Earth have never had good, long-term, ground-based temperature measurements. So, when we want to know what the temperatures were like in the 19 centuries

before carbon dioxide concentrations began to rise from the burning of fossil fuels, experts have to find ways to extrapolate, interpolate, and estimate. The Hadley Meteorological Center in Great Britain is an outstanding example of a group of scientists that is attempting to reconstruct such temperature records from the mid-nineteenth century to the present.

The situation has improved greatly in recent years, with the establishment of ocean platforms with automatic weather-monitoring equipment, coordinated by the World Meteorological Organization. Thus, we have records that are particularly good since about 1960. What is emerging from the climate data is that the warming we have experienced over the past few decades (1970–2010) exceeds that of anytime in the past 400+ years. This is supported by a large variety of reliable data. There is less confidence in temperature reconstructions from about A.D. 950 to 1250 because the data are sparse and not yet quantified. This period of sparse data includes the Medieval Warm Period. Limited data suggest that some specific locations during the MWP may have been as warm as or even a bit warmer than today. Data available for specific locations suggest that the latter part of the twentieth century was warmer than the MWP. Prior to A.D. 900, there is little confidence in global mean surface or water temperatures, because uncertainties associated with the proxy evidence for past temperatures are large over both space and time.[13]

12.7 CHECK your understanding

1. What have we learned from the Medieval Warm Period?

2. Why is it difficult to predict future climate change?

12.8 Strategies for Reducing the Impact of Global Warming

An important unsolved question about climate change being vigorously researched is the following: What is the origin of rapid climate change over decades to about 100 years that we believe has happened during geologic time (past 1 million years)? Two other important questions concerning the Earth climate system and people are (1) What changes have occurred? and (2) What changes could occur in the future? Answering these questions requires geologic evaluation of prehistoric changes, and the prediction, through modeling and simulation, of future change (Figure 12.31).

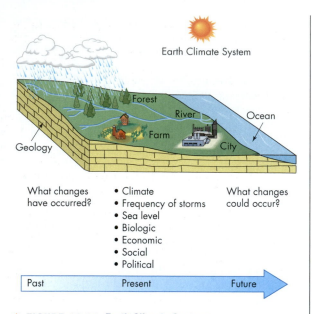

▲ FIGURE 12.31 Earth Climate System
Two big questions concerning Earth's climate system are linked to people and environment: What changes have occurred? What changes could occur? *(Based on International Panel on Climate Change 2001, at www.ipcc)*

Because we now know global warming is due, in part, to the increased concentration of greenhouse gases, reduction of these gases in the atmosphere is a primary management strategy. This was the subject of the 2016 Paris Climate Agreement. The objective of the United Nations Convention was to produce an international agreement to reduce emissions of greenhouse gases, especially carbon dioxide. The agreement was signed by 195 countries including the United States, but in 2017, the United States pulled out of the agreement, much to the disappointment of other nations, especially its European allies. As a result, the leadership in controlling global warming has shifted from the United States to the European Union.

International Agreements The United States is not committed to any international agreements to address climate change. A comprehensive energy plan that would have included climate change was killed in the Senate. Powerful lobbies from the fossil fuel industry, combined with some senators from states with abundant fossil fuels or those who fear changing energy policy or do not believe global warming from burning fossil fuels is occurring, used a filibuster to stop the legislation that would have initiated a more rapid change to alternative energy sources. In spite of this, the United States has significantly reduced carbon emissions to levels not thought likely, due to drilling technology that has allowed for greatly increasing the production of natural gas. Many coal-burning power plants are switching to natural gas systems that emit significantly less carbon and are a less expensive fuel.

Dilemma We are faced with a dilemma with respect to burning fossil fuels. On the one hand, fossil fuels are vital to our society and necessary for continued economic development and growth, as well as for human well-being. On the other hand, scientific evidence suggests that burning fossil fuels is contributing significantly to global warming. Burning fossil fuels is linked to environmental problems that include a rise in sea level, increased surface temperatures, air pollution, and increased frequency and intensity of storms such as hurricanes. Impacts associated with fossil fuels depend on how much the global temperature will actually rise. If the increase is close to 2°C, we can probably adapt with minimal disturbance. If the increase in temperature is higher, more adverse impacts are likely. Even if carbon emissions were reduced to zero, warming will continue during this century. There is 0.5° to 1.0°C warming in the Earth climate system now.

Assuming that the reduction of carbon emissions is necessary to reduce impacts of global warming, we need to take action. The reduction must occur at a time when human population and energy consumption are increasing. There are several ways that we might reduce emissions of carbon dioxide into the environment: improved engineering of fossil fuel-burning power plants (develop and use coal-burning power plants that sequester the carbon and place it in safe storage); use those fossil fuels that release less carbon into the atmosphere, such as natural gas, which, on burning, releases less carbon dioxide than does coal or oil (as is occurring in the United States); conserve energy to reduce our dependence on fossil fuels; use more alternative energy sources; and store carbon in Earth's systems, such as forests, soils, and rocks below the surface of Earth.

Sequestration of Carbon Sequestration is attractive because the residence time of carbon in the geologic environment is generally long, potentially thousands to hundreds of thousands of years.

The general principle of geologic sequestration of carbon is fairly straightforward. The idea is to capture carbon dioxide from power plants and industry and inject the carbon into the subsurface geologic environment. Two geologic environments have received considerable attention in this regard. The first is sedimentary rocks that contain salty water. These rocks, known as saline aquifers, are fairly widespread at numerous locations on Earth and have large reservoir capacity with the potential to sequester many years of human-produced carbon dioxide emissions. They could provide the necessary time to transition to an energy economy not dependent on fossil fuels.[62] The second (but related) geologic environment for sequestration of carbon is depleted oil and gas fields. The process of placing carbon dioxide in the geologic environment involves compressing the gas to a mixture of liquid and gases and then injecting it underground, using wells. Injecting carbon dioxide into depleted oil and gas fields has an added advantage, in that the carbon dioxide is not only stored but also serves as a way to enhance recovery of remaining oil and gas in the reservoir. The injected carbon dioxide helps move the oil toward production wells.

A demonstration project is now ongoing in Saskatchewan, Canada. The Weyburn oil field started production in the 1950s and was considered to be depleted. However, with enhanced recovery and storage of carbon dioxide, production is likely to last for several more decades. The source of carbon dioxide for the Weyburn field comes from a coal-burning plant in North Dakota by way of a pipeline that delivers several thousand tons of carbon per day.[63]

Another carbon sequestration project is located beneath the North Sea. Carbon dioxide is injected into a saltwater aquifer located below a natural gas field. The project, which started in 1996, injects about 1 million tons of carbon dioxide into the subsurface environment each year. It is estimated that the entire facility can hold about the amount of carbon dioxide that is projected to be produced from all of Europe's fossil fuel plants in the next few hundred years.[62] The cost of sequestering the carbon beneath the North Sea is high. However, it saves the company from having to pay taxes for emitting carbon dioxide into the atmosphere. Finally, pilot projects to demonstrate the potential and usefulness of storing carbon in the United States have been initiated in Texas beneath depleted oil fields. The good news is that immense salt aquifers are common beneath many areas of the United States, including the Gulf Coast, Texas, and Louisiana, and the potential to store carbon is immense.[64]

If we decide that we must stabilize the concentration of atmospheric carbon dioxide in the future, it will be necessary to go through a transition from fossil fuel energy sources to alternative sources that produce much less carbon dioxide. However, based on studies of glacial ice cores from Greenland, there are indications that significant climatic change may occur abruptly, perhaps in as short a time as a few years. A quick natural or human-induced warming or cooling is probably unlikely, but, if one does occur, the potential impacts could be fast and serious.

ABRUPT CLIMATE CHANGE

A large-scale change in the global climate system that takes place over a few decades or less is considered an abrupt climate change. Rapid changes are thought to have occurred in the past, but little is known about them or why they happened. Such change is anticipated to persist for at least a few decades and will cause substantial disruptions to both human and natural systems.[65] Several types of abrupt climate change that could cause a serious risk to humans and the natural environment in terms of our ability to adapt include the following:

> A rapid change of sea level as a result of changes in the glaciers and ice sheets.

> Droughts and floods, resulting from widespread rapid changes to the hydrologic cycle.

> Abrupt change in the pattern of circulation in the Atlantic Ocean that is characterized by northward flow of warm, salty water in the upper layers of the ocean.

> A rapid release of methane (a strong greenhouse gas) to the atmosphere from both melting permafrost and the ocean's sediment.

The U.S. Climate Change Science Program has addressed these four potential abrupt climate changes.[65] A major question is whether or not there will be an abrupt change in sea level. We know that even small changes in sea-level rise may have significant repercussions for society with serious economic impacts, including coastal erosion as well as an increase in coastal flooding and a loss of coastal wetlands. Current climate models do not adequately capture all aspects of sea level changes resulting from melting glacial ice. However, there is concern that projections for sea-level rise in the future, as presented by the Intergovernmental Panel on Climate Change,[13] probably underestimate the amount of sea-level rise during the twenty-first century.

The second question involves the potential for abrupt changes in the hydrologic cycle, especially those changes that affect water supply, especially through protracted droughts. Of particular significance is recognizing that droughts can develop faster than people and society are able to adapt to them. Thus, droughts that last from several years to a decade or more have serious consequences to society. It has been pointed out that long droughts have occurred in the past and are still likely to reoccur in the future, even in the absence of global warming, as a result of increased greenhouse gas forcing.[65]

A third question concerning abrupt climate change involves the Atlantic Ocean circulation system, which carries warm water to the North Atlantic, where it becomes saltier and sinks to become a cold water bottom current that moves south. The warm water is partly responsible for keeping Western Europe more comfortable and hospitable. However, it is primarily the large expanse of the ocean itself that accomplishes this. The strength of the current is expected to decrease approximately 25 to 30 percent as a result of global warming in the twenty-first century. However, it appears unlikely that the ocean current system in the Atlantic will undergo a collapse or an abrupt transition to a weakened state in the next 100 years.[65]

A final question is whether or not there will be a rapid change in atmospheric methane. This is a significant question because methane is a strong greenhouse gas that, if greatly increased in concentration in the atmosphere, would accelerate global warming. It is generally concluded that a very rapid change in the release of methane during the next 100 years or so is very unlikely. However, ongoing warming will increase emissions of methane from both ocean sediments and wetlands. Wetlands in the northern high latitudes are particularly susceptible to releasing additional methane because there is accelerated warming, along with enhanced precipitation, in permafrost areas that contain a lot of stored methane. Thus, it appears that, in future decades, methane levels will likely increase and cause additional warming.[65]

In conclusion, it appears that abrupt climate change during the next century is unlikely and, thus, we will probably have time to respond to potential adverse consequences of global warming. However, time is growing short and, because it takes time to initiate policy changes, a serious response to global warming from all countries is necessary in the near future.

A better understanding of past abrupt changes in climate, through the collection and analysis of geologic data, is necessary to fully understand what may cause future change. Geologic data provide the most direct evidence of past change. Geologic data from sediments and glacial ice, along with monitoring, help us understand the causes of long-term changes in ocean and glacial conditions and how these are linked to atmospheric response. With this knowledge, we may be better able to forecast both long- and short-term droughts that have serious consequences to humans and the natural environment.

Human civilization began and has developed into our present highly industrialized society in only about 7,000 years. That period has been characterized by a relatively stable, warm climate that is probably not characteristic of longer periods of Earth history. It is difficult to imagine the human suffering that might result late in the twenty-first century from a quick climate change to harsher conditions, when there are as many as 9 to 10 billion or more people on Earth to feed.

12.8 CHECK your understanding

1. What two big questions concerning Earth's climate system are linked to people and environment?

2. Define abrupt climate change.

3. If we control greenhouse gas emissions (especially carbon dioxide) in the next few decades, why might we not see real improvements for years, decades, or even 100 years after significant control?

APPLYING the 5 Fundamental Concepts

South Florida

1. **Science helps us predict hazards.**

2. **Knowing hazard risks can help people make decisions.**

3. **Linkages exist between natural hazards.**

4. **Humans can turn disastrous events into catastrophes.**

5. **Consequences of hazards can be minimized.**

Tidal floods in Miami Beach and South Florida are much more common today than they were a decade ago. South Florida has more property at risk from climate change than any other location in the United States. Figure 12.32 shows estimates of future rise in sea level at South Florida from three different agencies. Estimates reported by the Southeast Florida Regional Climate Change Compact (SFRCC) of sea-level rise by 2030 will be at about 15 to 25 cm (6 to 10 in.) above the 1992 level. By 2060, sea-level rise (from SFRCC) in Florida may be about 36 to 66 cm (14 to 26 in.) above the 1992 level. A recent study suggests that sea-level rise in Miami was about 3 +/–2 mm per year before 2006, but since 2006, is about 9 +/– 4 mm per year.[1] A rise of about 30 cm (1 ft.) over 35 years into the future may seem to be a small number, but if standing water from tidal flooding surrounds your beach home, condominium, or hotel, it is a big problem that damages property and disrupts the daily lives of people.

The Southeast Florida Regional Climate Change Compact is working with communities to reduce the coastal flood hazard. Miami Beach has a plan to minimize future flooding that involves seawalls, drains, and pumps. The construction will cost several $100 million over the next few years. Some pumps have already been installed to pump floodwaters into Biscayne Bay, and there is fear that the water quality will suffer as urban floodwater is pumped into the bay. As sea level continues to rise, controlling floodwater may become more and more difficult as the volume of water to pump increases. Decisions will also have to be made concerning what neighborhoods in South Florida will be protected by engineering structures and what areas will not. All this will occur as population continues to increase. People are becoming worried about their investments in beach property that is becoming exposed and vulnerable to flooding. One man I met last year said that his parents had a beach house in Miami Beach, and sea-level rise was on his mind. He was attempting to persuade his elderly parents, who probably love the beach retirement, to sell

their home and move before the flood threat drives property values down. He was worried about his inheritance.

APPLYING THE 5 FUNDAMENTAL CONCEPTS

1. Climate modeling techniques have predicted continuing sea-level rise in South Florida. Can the timing and extent of inundation of future floods be accurately predicted? Why or why not?

2. The people of South Florida are making decisions for their future, based on perceived and identified risks. Summarize their options and discuss which adaptions are likely to work best and why.

3. Link the known hazards of sea-level rise at South Florida to other natural hazards, such as water pollution.

4. Describe how South Florida's tidal flooding is a result of human activities, both at a local scale and at the global scale.

5. Propose ways to minimize the hazards facing South Florida.

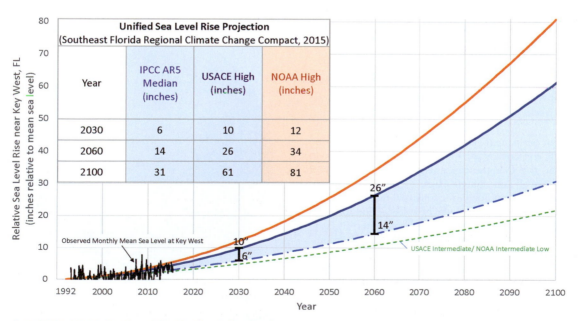

Unified Sea Level Rise Projection (Southeast Florida Regional Climate Change Compact, 2015)			
Year	IPCC AR5 Median (inches)	USACE High (inches)	NOAA High (inches)
2030	6	10	12
2060	14	26	34
2100	31	61	81

∧ FIGURE 12.32 Southeast Florida Rise in Sea Level
Projected rise in sea level from 1992 reference level to 2100. A rise of 15 cm (6 in.) to 25 cm (10 in.) is considered a likely scenario. *(Southeast Florida Regional Climate Change Compact)*

CONCEPTS in review

12.1 and 12.2 Global Change and Earth System Science: An Overview Climate and Weather

LO:1 Explain the difference between climate and weather and how their variability is related to natural hazards.

- The concept of Earth system science has the goal of understanding how Earth works and applying that knowledge to better manage the environment.
- Global changes on a time frame significant to people (such as temperature increase) and sea-level rise are apparently occurring. Because these are short-term predictions, Earth System Science is relevant to people everywhere.
- Climate refers to the characteristic atmospheric conditions of a given region over long periods of time, such as years or decades. Weather refers to the atmospheric conditions of a given region for short periods of time, such as days or weeks.
- Climate exerts a major influence on natural processes; a familiarity with Earth's climate zones is a first step toward recognizing the threat from natural hazards.

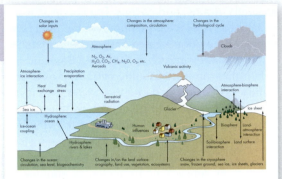

KEY WORDS
Earth System Science (p. 486), **climate** (p. 486), **weather** (p. 486)

12.3 The Atmosphere and the Cryosphere

LO:2 Describe the basic concepts of atmospheric science, such as the structure, composition, and dynamics of the atmosphere.

- The major permanent gases, which are those that remain essentially constant by percentage, include nitrogen, oxygen, and argon. Together, these gases comprise about 99 percent by volume of all atmospheric gases.
- Over the past 2.6 million years, Earth's climate system has fluctuated greatly and alternated between periods of major continental glaciation, referred to as glacial intervals, and times of warmer climate with significantly less glaciation, referred to as interglacial intervals.
- Glaciers are huge, actively flowing masses of ice and rock debris whose movement and melting have been responsible for property damage, injuries, and deaths.

KEY WORDS
glacier (p. 489)

12.4 How We Study Past Climate Change and Make Predictions

LO:3 Synthesize how we have studied past climate change.

- Methods of studying climate change include examination of the geologic record from lake sediments, glacial ice, and other Earth materials; gathering of real-time data from monitoring stations; and development of mathematical models to predict change.

KEY WORDS
proxy data (p. 490)

12.5 and 12.6 Global Warming Effects of Global Climate Change

LO:4 Summarize how climate has changed during the past million years through glacial and interglacial conditions and how human activity is altering our current climate.

LO:5 Name and explain the potential causes of climate change and link these causes to natural hazards.

- Global warming is defined as the observed increase in the average temperature of the near-surface land and ocean environments of Earth during the past 60 years. A growing volume of evidence suggests that we are now in a period of global warming, resulting from burning vast amounts of fossil fuels. Many scientists now believe that human processes, as well as natural ones, are contributing significantly to global warming.
- The trapping of heat by the atmosphere is generally referred to as the greenhouse effect. Water vapor and several other gases, including carbon dioxide, methane, and chlorofluorocarbons, tend to trap heat and warm Earth because they absorb some of the heat energy radiating from Earth.
- Climate forcing is defined as an imposed change of Earth's energy balance. The units for the forcing are W/m^2, and they can be positive if a particular forcing increases global mean temperature or negative if temperature is decreased.
- Natural climate forcing mechanisms that may cause climatic change include Milankovitch cycles, solar variability, and volcanic activity. Anthropogenic causes include air pollution and an increase in greenhouse gases, especially carbon dioxide.
- The potential effects of global warming are sea-level rise; increased wildfire; more intense storms; increase in drought; changes in landslide, volcanic, and tsunami hazards; and changes in the biosphere.

KEY WORDS
climate forcing (p. 498), **global warming** (p. 494), **greenhouse effect** (p. 495), **El Niño** (p. 506)

12.7   Predicting the Future Climate

LO:6   Speculate on how we might better predict future climate change.

- Over an approximately 300-year period from a.d. 950 to 1250, Earth was considerably warmer than normal (normal meaning the average surface temperature during the past century or some shorter interval, such as between 1961 and 1990). This period is known as the Medieval Warm Period (MWP).
- During the MWP, sea surface temperatures in the North Atlantic evidently were warmer, and there was less sea ice.
- Associated with the warm period were long, persistent droughts (lasting perhaps as long as a human generation) that appear to have been partially responsible

for the collapse of sophisticated cultures in North and Central America. The collapses were not sudden but occurred over a period of many decades, and, in some cases, the people just moved away.
- Perhaps a lesson for us from the MWP is that changes in the most recent past warming (MWP) resulted in winners and losers.
- The potential effects of prolonged, persistent drought and loss of food production for the world, should present warming continue, is a serious threat.
- Predicting the future climate is problematic because predicting the future has always been difficult. People who make predictions have often been wrong.

12.8   Strategies for Reducing the Impact of Global Warming

LO:7   Propose ways to mitigate climate change and associated hazards.

- We now understand that global climate can change rapidly over a time period of a few decades to a few hundred years.
- Two important questions are (1) What is the nature and extent of past climate change? and (2) What climate changes will occur in the future?
- The science of global warming is well-understood. Human-induced global warming is occurring. There is no reason for gloom and doom, but we need to take appropriate action soon to slow or stop global warming and associated environmental consequences.
- Adjustments to global warming will range from adapting to change to reducing emissions of carbon dioxide and sequestration of carbon. Several different and simultaneous adjustments are likely.

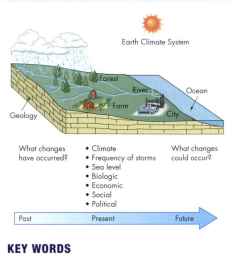

KEY WORDS
abrupt climate change (p. 498)

CRITICAL thinking questions

1. In this chapter, we discussed some possible effects of continued global warming. Which types of hazards would you expect to increase in your area as the climate changes? Would you expect to see any decrease in the types of hazards? Think about the ways that climate change might alter the lives and lifestyles of people living in your area. Do you see any evidence of climate change where you live?

2. Assessing the rate and cause of change is important in many disciplines. Have a discussion with a parent or someone of similar age and write down

the major changes that have occurred in his or her lifetime and in your lifetime. Characterize these changes as gradual, abrupt, surprising, chaotic, or another descriptive word of your choice. Analyze these changes and discuss which ones were most important to you personally. Which of these changes affected your environment at the local, regional, or global level?

3. How do you think climate change is likely to affect you in the future? What adjustments will you have to make? What can you do to mitigate the effects?

4. Some people, for cultural, political, religious, or other reasons, do not accept the conclusion that climate change is primarily caused by human activities. What do you see as the basis for their opinions or beliefs?

How do you think that they might be convinced otherwise? If you share their opinions or beliefs, indicate why you do and what it would take to convince you that climate change is the result of human activities.

References

1. **Wdowinski, S., Bray, R., Kirtman, B.P.,** and **Zhaohua, W.** 2016. Increasing flooding hazard in coastal communities due to rising sea level: Case study of Miami Beach Florida. *Ocean and Coastal Management* 126: 1–6.
2. **Union of Concerned Scientists.** 2015. The truth about Florida's attempt to censure climate change. *Got Science* April.
3. **Cloud, P.** 1990. Personal written communication.
4. **National Aeronautics and Space Administration (NASA).** 1990. *EOS: A mission to planet Earth.* Washington, DC: NASA.
5. **Aguado, E.,** and **Burt, J. E.** 2007. *Understanding weather & climate,* 4th ed. Upper Saddle River, NJ: Prentice Hall.
6. **Houghton, J.** 2004. *Global warming: The complete briefing,* 3rd ed. New York: Cambridge University Press.
7. **NOAA.** 2009. Paleo-proxy data. In *Introduction to paleoclimatology.* www.ncdc.noaa.gov. Accessed 3/24/10.
8. **Ricks, K. L.** and **Caldeire, K.** 2014. Maximum warming occurs about one decade after a carbon dioxide emission. *Environmental Research Letters* 9: 124002.
9. **Weart, S. R.** 2003. *The discovery of global warming.* Cambridge, MA: Harvard University Press.
10. **Oldfield, F.** 2005. *Environmental change: Key issues and alternative approaches.* New York: Cambridge University Press.
11. **Rahmstorf, S., Cazenave, A., Church, J. A., Hansen, J. E., Keeling, R. F., Parker, D. E.,** and **Somerville, R. C. J.** 2007. Recent climate observations compared to projections. *Science* 316: 709.
12. **Hansen, J.** 2004. Defusing the global warming time bomb. *Scientific American* 290(3): 68–77.
13. **IPCC.** 2015. Climate Change 2014: Synthesis Report. Intergovernmental Panel on Climate Change. Geneva, Switzerland.
14. **Ruddiman, W. F.** 2008. *Earth's climate past and future,* 2nd ed. New York: W. H. Freeman.
15. **Hansen, J.,** and **Sato, M.** 2004. Greenhouse gas growth rates. *Proceedings of the National Academy of Sciences* (PNAS) 101: 16109–14.
16. **Hansen, J., Sato, M., Ruedy, R., Lo, K., Lea, D.,** and **Medina-Elizade, M.** 2006. Global temperature change. *Proceedings of the National Academy of Sciences* (PNAS) 103: 14288–93.
17. **Kennett, J.** 1982. *Marine geology.* Englewood Cliffs, NJ: Prentice Hall.
18. **Hansen, J.,** and **44 others.** 2005. Efficacy of climate forcing. *Geophys Res.* 110: D18104.
19. **Hansen, J.** 2003. Can we defuse the global warming time bomb? Edited version of presentation to the Council on Environmental Quality. June 12. Washington DC. www.naturalscience.com.
20. **Broecker, W.** 1997. Will our ride into the greenhouse future be a smooth one? *GSA Today* 7(5): 1–7.
21. **Seager, R.** 2006. The source of Europe's mild climate. *American Scientist* 94: 334–41.
22. **Mann, M. E., Zhang, Z., Rutherford, S., Bradley, R. S., Hughes, M. K., Shindell, D., Ammann, C., Faluvegi, G.,** and **Ni, F.** 2009. Global signatures and dynamical origins of the Little Ice Age and Medieval Climate Anomaly. *Science* 326: 1256–60.
23. **Crowley, T. J.** 2000. Causes of climate change over the past 1000 years. *Science* 289: 270–77.
24. **Foukal, P., Frohlich, C., Spruit, H.,** and **Wigley, T.** 2006. Variations in solar luminosity and their effect on the earth's climate. *Nature* 443(14): 161–66.
25. **Charlson, R. J., Schwartz, S. E., Hales, J. M., Cess, R. D., Coakley, J. A. J., Hansen, J. E.,** and **Hofmann, D. J.** 1992. Climate forcing by anthropogenic aerosols. *Science* 255: 423–30.
26. **Kerr, R. A.** 1995. Study unveils climate cooling caused by pollutant haze. *Science* 268: 802.
27. **McCormick, P. P., Thomason, L. W.,** and **Trepte, C. R.** 1995. Atmospheric effects of the Mt. Pinatubo eruption. *Nature* 373: 399–436.
28. **Appenzeller, T.** 2007. The big thaw. *National Geographic* 211(6): 56–71.
29. **Pelto, M. S.** 1996. Recent changes in glacier and alpine runoff in the North Cascades, Washington. *Hydrological Processes* 10: 1173–80.
30. **U.S. Geological Survey. USGS repeat photography project documents retreating glaciers in Glacier National Park.** www.nrmsc.usgs.gov/repeatphoto/. Accessed 4/10/09.
31. **Shepherd, A.,** and **44 others.** 22012. A reconciled estimate of ice sheet mass balance. *Science* 338: 1183–89
32. **Steig, E. J.,** and **5 others.** 2006. Warming of the Antarctic ice-sheet surface since the 1957 International Geophysical Year. *Nature* 457: 459–62.

33. **Davis, C. H.**, and **4 others.** 2005. Snowfall-driven growth in East Antarctic ice sheet mitigates recent sea-level rise. *Science* 308(5739): 1898–901.

34. **Monagham, A. J.**, and **15 others.** 2006. Insignificant change in Antarctic snowfall since the International Geophysics year. *Science* 313(5788): 827–31.

35. **Tufnell, L.** 1984. *Glacier hazards*. New York: Longman.

36. **University Corporation for Atmospheric Research.** 1994. *El Niño and climate prediction*. Washington, DC: NOAA Office of Global Programs.

37. **Dennis, R. E.** 1984. A revised assessment of worldwide economic impacts: 1982–1984 El Niño/southern oscillation event. *EOS, Transactions of the American Geophysical Union* 65(45): 910.

38. **Canby, T. Y.** 1984. El Niño's ill winds. *National Geographic* 165: 144–81.

39. **Philander, S. G.** 1998. Who is El Niño? *EOS, Transactions of the American Geophysical Union* 79(13): 170.

40. **Greene, C. H.** 2012. The winters of our discontent. *Scientific American* 307(6): 50–55.

41. **Lindsey, R.** 2016. *Climate Change: Global Sea Level*. www. climate .gov.

42. **Titus, J. G., Leatherman, S. P., Everts, C. H., Moffatt and Nichol Engineers, Kriebel, D. L.,** and **Dean, R. G.** 1985. *Potential impacts of sea-level rise on the beach at Ocean City, Maryland*. Washington, DC: U.S. Environmental Protection Agency.

43. **Anderson, D. R.**, and **5 others.** 2009. *Coastal sensitivity to sea-level rise*. Washington DC: U.S. Climate Change Program.

44. **Kumar, M.** 2007. Alaska melting into the sea. *Geotimes* 52(9): 8–9.

45. **Frontline/World.** 2005. *Tuvalu: That sinking feeling*. www.pbs.org/frontlineworld.

46. **Nakada, S.**, and **3 others.** 2012. Groundwater dynamics of Fongafale Islet, Funafuti Atoll, Tuvalu. *Groundwater* 50: 639–44.

47. **Dickinson, W. R.** 2009. Pacific atoll living: How long already and until when. *GSA Today* 19(3):4–10.

48. **Pinol, J., Terradas, J.,** and **Lloret, F.** 1998. Climate warming, wildfire hazard, and wildfire occurrence in coastal eastern Spain. *Climatic Change* 38: 345–57.

49. **U.S. Climate Change Program.** 2008. *Climate change and ecosystems*. Washington DC: Author.

50. **McGuire, B.** 2012. *Waking the Giant*. Oxford: Oxford University Press.

51. **Gariano, S. L.**, and **Guzzetti, F.** 2016. Landslides in a changing climate. *Earth-Science Reviews* 162: 227–252.

52. **Handler, E.** 2015. Iceland's volcanic activity to increase from climate change. *Yale Scientific*. www.yalescientific.org

53. **Union of Concerned Scientists.** 2003. *Early warning signs: Coral reef bleaching*. www.ucsusa.org. Accessed 4/10/09.

54. **Doney, S.** C. 2010. The growing human imprint on coastal and open-ocean biogeochemistry. *Science* 238: 1512–16.

55. **Hoegh-Gulburg, O.**, and **Bruno, J. F.** 2010. The impact of climate change on the world's marine ecosystems. *Science* 238: 1523–28.

56. **Hardt, M. J.**, and **Safina, C.** 2010. Threatening ocean life from the inside out. *Scientific American* 303(2): 66–73.

57. **Brashis, D. J.** and **6 others.** 2013. Genomic basis for coral resilience to climate change. *Proceedings of the National Academy of Sciences* 110(4): 1387–92.

58. **Appell, D.** 2009. Can "assisted migration" save species from global warming? *Scientific American* 30(3): 378–30.

59. **Botkin, D. B.**, and **18 others.** 2007. Forecasting effects of global warming on biodiversity. *BioScience* 57(3): 227–36.

60. **Mann, M. E., Zhang, Z., Rutherford, S., Bradley, R. S., Hughes, M. K., Shindell, D., Ammann, C., Faluvegi, G.,** and **Ni, F.** 2009. Global signatures and dynamical origins of the Little Ice Age and Medieval Climate Anomaly. *Science* 326: 1256–60.

61. **Fagan, B. M.** 2008. *The great warming: Climate change and the rise and fall of civilizations*. New York. Bloomsbury Press.

62. **Friedman, S. J.** 2003. Storing carbon in Earth. *Geotimes* 48(3): 16–20.

63. **Nameroff, T.** 1997. The climate change debate is heating up. *GSA Today* 7(12): 11–13.

64. **Bartlett, K.** 2003. Demonstrating carbon sequestration. *Geotimes* 48(3): 22–23.

65. **McGeehin, J. P., Barron, J. A., Anderson, D. M.,** and **Verardo, D. J.** 2008. *Abrupt climate change. Final Report, Synthesis and Assessment Product 3.4.* Washington, DC: U.S. Climate Change Science Program.

Flames of the **2017 fire** that destroyed part of **Santa Rosa, California**

13

APPLYING the **5** fundamental concepts Santa Rosa Fire of 2017

1

Science helps us predict hazards.

Wildfire is a very serious and costly natural hazard. Fortunately, fire science is a mature subject, and our ability to predict and forecast seasons with high potential for wildfires has greatly improved.

2

Knowing hazard risks can help people make decisions.

The 2017 wildfire demonstrates the risk of damage to land and property from future wildfires. Hazard reevaluation allows people to view future wildfires from a risk management perspective. In many forest and brush locations, wildfire is the greatest risk many communities face.

Wildfires

Wildfire (Tubbs Fire) near Santa Rosa, CA: October, 2017

One of the most catastrophic wildfire in Californian history occurred over a few days in early October 2017. The wildfire grew in size and intensity very fast, surprising homeowners who believed they lived beyond the fire hazard area that had been defined. People were left running or driving for their lives. About 40 people were killed. There were several fires that grew together into a much larger, more intense fire. Insured losses were about $3 billion and total damages will probably top $5 billion. The fires destroyed 5,000 homes, damaging 10,000 structures on farms and the regions wineries. At Santa Rosa entire blocks of homes burnt to the ground melting steel, aluminum, and glass. Entire automobiles melted (over 3,000 personal vehicles were lost). An important lesson learned from the fires was that the regulations to minimize the impact of fire are not adequate when urban areas burn. Santa Rosa neighborhoods well outside fire zones designated as having a mild risk burned. The density of homes was high, and each home can be considered a large bundle of fuel for fire. The fire in the urban area probably did not result from wildfire sweeping down hills into houses, but from embers blowing into the neighborhood from fire as far as 0.6 km (1 mi.) away. When wind-driven hot embers enter

As Earth's population grows, more people are moving into and near brushlands, forest, and other wildlands where wildfires naturally occur. This trend increases the risk of property damage and loss of life from wildfires.

After reading this chapter, you should be able to:

LEARNING Objectives

LO:1 Summarize the role of wildfire during the past 11,500 years.

LO:2 Explain why wildfire is a natural process that would occur with or without people.

LO:3 Explain why the wildfire hazard varies from region to region.

LO:4 Summarize how wildfires are linked to other natural hazards.

LO:5 Make the link between global change (warming) and the number and intensity of wildfires.

LO:6 List the potential benefits provided by wildfires.

LO:7 Evaluate the methods employed to minimize the fire hazard.

LO:8 Summarize the potential adjustments to the wildfire hazard and the role of human processes.

< FIGURE 13.0
The Tubbs Fire (2017) burned several thousand homes including those shown here. *(Kent Porter/The Press Democrat/AP Images)*

3

Linkages exist between natural hazards.

Wildfire is directly linked to soil erosion, flooding, and landslides, including debris flows.

4

Humans can turn disastrous events into catastrophes.

Factors such as population growth and land-use decisions result in more people living in areas with a high fire risk. For a variety of reasons, wildfires are becoming larger and more intense.

5

Consequences of hazards can be minimized.

Consequences of wildfire can be minimized but never eliminated through understanding of the fire cycle and better land-use decisions. The Santa Rosa Fire emphasized that management policies must be reevaluated if we are to minimize wildfire destruction.

homes with small openings such as unscreened roof vents or land on roofs that are old and flammable, a fire starting is very likely, especially if crawl space is used to store flammable materials. Once ignited, homes can burn extremely hot and fire is spread from home to home. Sometimes trees, swimming pools, and even trashcans were not damaged, suggesting a house-to-house pattern of fire. The Santa Rosa fire has opened many questions concerning the fire hazard of densely populated neighborhoods near wildlands where fire occurs periodically.

13.1 Introduction to Wildfire

Wildfires are one of nature's oldest phenomena, dating back more than 350 million years when trees evolved and spread across the land. Wildfire behavior changed markedly about 20 million years ago when grasses evolved. Grasses provided a new type of fuel that grew quickly and could sustain more frequent fires. Then, about 11,500 years ago (the beginning of the Holocene Epoch), when the climate warmed after the last major glacial advance, wildfire behavior intensified again. The geologic record shows a significant increase in the amount of charcoal found in sediment from this time, indicating an increase in fires. Part of this increase probably reflects activity by early humans, who used fire to clear land and assist in hunting.[1,2]

Before humans evolved, fires that were ignited by lightning strikes or volcanic eruptions would burn until they ran out of fuel or were extinguished naturally. After a fire, plant regrowth begins from roots, spores, and seeds, and the cycle starts again. This cycle is so ancient that some plants evolved to rely on and use fire to their advantage. For example, oak and redwood trees have bark that resists fire damage, and some species of pine trees have seed cones that open only after a fire. In the Mediterranean climate of Southern California, chaparral (brush) species have adapted to fire and re-generate new foliage after being burned.

Natural fires started by lightning strikes and volca-nic eruptions allowed early humans to harness fire for heat, light, and cooking; these benefits of fire, in turn, helped human populations spread across continents.[1] The ability to use fire for cooking and for warmth al-lowed humans to expand their diet and settle in colder areas. Fires were also set to encourage new plant growth and attract game. Native Americans used fires as a tool in hunting, warfare, and agriculture. Early European settlers, including members of the famous Lewis and Clark expedition that explored North America from coast to coast, commented on the many fires set by native peoples for a variety of purposes. This practice continues today, sometimes with serious consequences.

13.1 Wildfires in the U.S. Southwest, Colorado, Arizona and California

CASE study

Two destructive and deadly wildfires occurred during the summer of 2013 (see chapter-opener photograph). The first, known as the Black Forest fire, occurred outside of Colorado Springs, Colorado. The Black Forest fire burned about 6,000 ha and destroyed more than 500 homes (Figure 13.1), making it the most de-structive fire in Colorado's history. Prior to the Black Forest fire, the most destructive fire had been the Waldo Canyon fire, which had burned through the Colorado Springs area just one year before, reflecting a disturbing trend in Colorado over the past few decades. Colorado wildfires in the 1970s and 1980s burned about 40,000 ha (100,000 acres) per decade, in the 1990s it was about 81,000 ha (200,000 acres), and in the 2000s the total burned area was more than 40,000 ha (100,000 acres). On a positive note, all the recent experience with wild-fires enabled authorities to act quickly during the Black Forest fire, and thousands of people were safely evacu-ated and only two people were killed.

The second wildfire occurred near Yarnell, Arizona. Like the Colorado wildfire, this fire was a swift-moving event that burned at least 30,000 ha (8,000 acres) very quickly. Although the fire destroyed about 200 homes, the even greater tragedy was the death of 19 firefight-ers, part of a professional group of firefighters known as the "hotshots" often called in to fight the most serious wildfires. The fire overrode (burned through) the posi-tion of the firefighters, and their protective equipment and attempts to dig and cover in the ground were futile. Autopsies after the fire showed that the firefighters

∧ FIGURE 13.1
Home destroyed by the 2013 wildfire near Colorado Springs, Colorado (*AP Photo/Ed Andrieski*)

died from a combination of burns, carbon monoxide poisoning (lack of oxygen), and smoke inhalation. Only one member of the 20-person crew escaped the fire, as he was moving the fire truck at the time.

Three large recent fires in California killed over one hundred people and destroyed more than 20,000 homes and other buildings. The first was the Thomas Fire in late December of 2017. The fire burned about 120,000 ha (300,000 acres) while destroying more than 1,000 homes and buildings in Southern California. Intense rain just weeks after the fire produced a catastrophic debris flow that killed 23 people in the town of Montecito. Damages from the fire and debris flow exceeded $2 billion.

Early November of 2018 brought the most damaging fire in California history. The wind-driven Camp Fire burned about 19,000 buildings (most in the first few hours of the fire). Nearly the entire town of Paradise in northern California was destroyed, killing more than 80 people. The fire burned about 62,000 ha (150,000 acres).

Estimated total losses are as high as $10 billion. The third fire, also in early November 2018, was in Southern California and burned about 39,000 ha (97,000 acres), including parts of Malibu. Named the Woosley Fire, flames destroyed about 1,600 homes and killed three people, while causing about $5 billion in damages. The three fires together caused about $15 billion in damages, prompting the evacuation of hundreds of thousand of people while taking an unprecedented number of lives. These fires are promoting fire scientists to re-think what we know about fires near urban areas and how best to limit future damages, and loss of life.

13.1 CHECK your understanding

1. How has the nature of wildfires and human interaction changed over geologic and historic time?

2. Why are recent wildfires in the United States so damaging?

13.2 Wildfire as a Process

Wildfire is a self-sustaining, rapid, high-temperature biochemical oxidation reaction that releases heat, light, and other products.[1,2] This reaction requires three things—fuel, oxygen, and heat—that are sometimes visualized as a fire triangle (Figure 13.2). For the chemical reaction to proceed, a fire must have appropriate fuel, adequate oxygen, and external heat that is high enough to start the reaction. If any one of the three requirements of this triangle is removed, the fire goes out.

Wildfires typically burn at the forest floor with temperatures of 200–1000°C (450–2000°F), but hot

∧ FIGURE 13.2 The Fire Triangle
If any part of this triangle is missing, a fire cannot start; likewise, if any part of the triangle is removed, the fire goes out. *(Based on Cottrell, W. H. Jr, The Book of Fire, 2nd ed. Missoula, MT: Mountain Press, 2004.)*

spots may exceed 1500°C (2700°F) which melts glass and aluminum and steel. Wood bursts into flames at about 300°C (570°F). At the temperatures of wildfire, plant tissue and other organic material (biomass) are rapidly oxidized and broken down by combustion, or burning. Grass, brush, and forestlands burn because, over long periods of time, these systems establish a balance between plant productivity and decomposition. Microbes alone do not decompose plants fast enough to balance plant growth, so wildfire helps in this decomposition. Therefore, in a sense, the primary cause of fire is vegetation, which has removed carbon dioxide from the atmosphere to produce organic matter by photosynthesis.[3,4]

Fire, like the rotting or decomposition of dead plants by bacteria, essentially reverses the process of photosynthesis. In this simplified view of wildfire, carbon dioxide, water vapor, and heat are released when plants burn. Actual combustion, however, is complex and releases numerous chemical compounds in solid, liquid, and gas form. Of the gases released, carbon dioxide and water vapor are the most abundant; other gases occur in trace quantities.[1] Common trace gases released are nitrogen oxides, carbonyl sulfide, carbon monoxide, methyl chloride, and hydrocarbons such as methane.[4] These gases, along with solid particles of ash and soot, comprise part of the smoke observed during a wildfire. Both ash and soot are powdery residues that accumulate after burning. Ash consists primarily of mineral compounds, and soot is made of unburned carbon.

The image of flames and smoke is accurate as a mental picture of wildfire, but a closer look reveals three phases of a wildfire:

1. Preignition.
2. Combustion.
3. Extinction.

In the first phase, *preignition*, fuel is brought to both a temperature and a water content that favors ignition. Preignition involves two processes: preheating and pyrolysis.

In preheating, the fuel loses a great deal of water and other volatile chemical compounds. A *volatile compound* is one that is easily vaporized to a gas. For example, because gasoline contains many volatile chemical compounds, spilled gasoline evaporates quickly to a vapor that you can smell.

The other important process of preignition is **pyrolysis**, which literally means "heat divided." Pyrolysis is actually a group of processes that chemically degrade the fuel. Degradation takes place as heat divides, or splits, large fuel molecules into smaller ones. The products of pyrolysis include volatile gases, mineral ash, tars, and carbonaceous char.[1,2] You have probably caused pyrolysis unintentionally. Pyrolysis takes place when you scorch a piece of toast and turn it black. The burnt toast is covered with char, and smoke coming out of the toaster contains small black droplets of tar. A similar result occurs when you scorch cotton fabric with a hot iron. The processes of preignition—preheating and pyrolysis—operate continuously in a wildfire. Heat, radiating from flames, causes both preheating and pyrolysis in advance of the fire. These processes produce the first fuel gases, which can ignite in the next phase of a fire.

The second phase, **combustion**, begins with ignition. Combustion marks the start of a set of processes completely different from those related to preignition. Preignition processes absorb energy. In the combustion phase, external reactions involving flaming or glowing liberate energy in the form of heat and light.[2] Although there are many sources of ignition, such as lightning, volcanic activity, and human action, ignition does not automatically lead to a wildfire. In fact, many more ignitions occur than do full-blown wildfires because a wildfire will not become established unless sufficient fuel is present. Wildfires develop when vegetation is mature and has accumulated in sufficient quantities to carry fire across the land.[3] Ignitions are common at a time scale that is relevant for mature grass, brush, or forestlands; in a period of 50 to 100 years, nearly every acre of land is struck by lightning. On this time scale, wildfires set by humans are not nearly as relevant. If a person starts a large wildfire, he or she is simply preceding a lightning strike that would soon cause the fire anyway. Once wildfire has crossed an area and most of the vegetation has been destroyed, the low fuel supply reduces future wildfires, even with an ignition source such as lightning. Fire will not threaten again until there is sufficient new fuel. This argument is contrary to twentieth-century fire management concepts, which commonly held that only people could prevent forest fires. On an ecological time scale, we have seen that fires will occur whether or not people start them, and humans play a relatively minor role in affecting large wildfires.

During a wildfire, ignition is not a simple or single process but rather a continuum of processes marked by an acceleration of the pyrolysis that began in the preignition phase (Figure 13.3). Most people think of a wildfire as having a single ignition that starts the fire. Actually, ignition repeats time and time again as a fire moves, like sparks and embers popping out of a fireplace, campfire, or barbeque grill. The dominant types of combustion are flaming combustion and glowing or smoldering combustion (Figure 13.3). These two types of combustion are distinct from each other in that they proceed by different chemical reactions and have different appearances. Flaming combustion is the rapid, high-temperature conversion of fuel to thermal energy and is characterized by a large amount of residual unburned material. With wildfire, flaming combustion dominates during the early stages of a fire as fine fuel and volatile gases produce rapid oxidation reactions. These reactions sustain flames as the fire advances across the landscape. As volatile gases are removed from the fuel, woody materials continue to decompose through pyrolysis, and carbon and ash begin to blanket new fuel. This blanket of noncombustible material can hinder flaming combustion and lead to glowing or smoldering combustion instead. This glowing or smoldering combustion takes place at lower temperatures and does not require rapid pyrolysis for its growth. The ultimate product of glowing or smoldering combustion can be charcoal.

As wildfire moves across the landscape, three processes control the transfer of heat, and as topography and wind direction change, the zone of flaming combustion can take on several forms (Figure 13.4). The three primary ways by which heat is transferred are *conduction*, which is the transmission of heat through molecular contact; *radiation*, heat transfer though electromagnetic waves; and *convection*, heat transfer by the movement of heated gases driven by temperature differences in a fluid, such as air. In Chapter 9, you

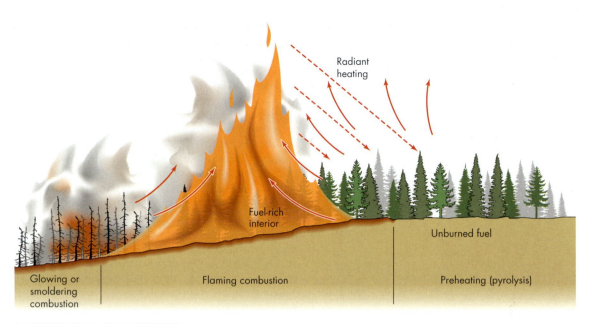

∧ FIGURE 13.3 Parts of a Wildfire

Idealized diagram of an advancing wildfire showing phases of a wildfire: (1) an area of pyrolysis where un-burned vegetation is preheating, (2) an area of flaming combustion with rising air currents, and (3) an area of glowing or smoldering combustion. Fire is advancing from left to right. Solid arrows indicate motion of air currents and dashed lines show radiant heating from the fire. *(Modified after Ward, D. 2001. Combustion chemistry and smoke. In* Forest fires: Behavior and ecological effects, *ed. E. A. Johnson and K. Miyanishi, pp. 55–77. San Diego, CA: Academic Press.)*

learned that convection is a dominant process in atmospheric transfer of heat. Wildfires also transfer heat primarily by convection, although radiation, which generates radiant heat, also plays a role (Figure 13.4). Transfer of both convective heat and radiant heat increases the surface temperature of the fuel. As heat is released, the air and other gases become less dense and rise. In wildfires, the oxidation chemical reaction in the fire releases heat, and this heat is transferred by convection of the air. The rising air and other hot gases remove both heat and combustion products from the zone of flaming. This process shapes the fire as it pulls in fresh air that is needed to sustain combustion.[2] Convection, thus, aids in continually providing the fire with oxygen.

Finally, *extinction* is the point at which combustion, including smoldering, ceases. When there is no longer sufficient heat and fuel to sustain combustion, the fire is considered extinct.

FIRE ENVIRONMENT

The behavior of a large wildfire can be explained by three factors in its environment: fuel, topography, and weather (Figure 13.4). With sufficient information concerning these three factors, we can better understand and predict wildfire behavior.[3,5,6]

Fuel Wildfire fuels are complex and differ in type, size, quantity, arrangement, and moisture content. Types of fuel include leaves, twigs, and decaying material on the forest floor; grass and shrubs of various size; small trees; large woody debris; and large, living trees. Where forests, swamps, bogs, or marshes are underlain by soils with high organic content, the fuel for a wildfire can also include underground accumulations of *peat*, an unconsolidated deposit of partially decayed wood, leaves, or moss.

The size of the fuel can influence ignition, as well as how rapidly a fire moves. Finer fuels, such as grass and pine needles, burn more readily than large woody material. For example, if disease or a storm downs a large number of trees, after 15 to 30 years or so, the larger pieces of wood may dry out and become partly decayed, allowing them to sustain combustion during a wildfire. Fuel arrangement can also be important. Landslides, tornadoes, and hurricanes arrange downed woody debris in clumps or accumulations in various parts of the landscape.[6] Finally, during droughts, even organic-rich swamp and marsh soils, which are usually saturated with water, dry out and become fuel for wildfires. These fires, commonly referred to as *peat fires*, burn slowly and can last for many months.

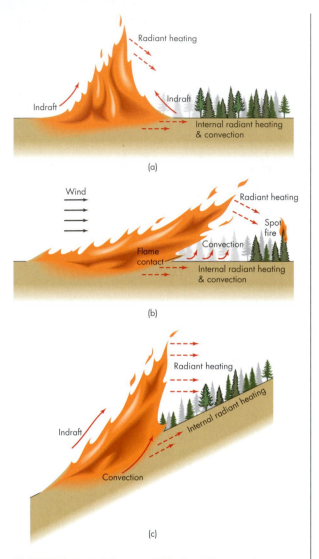

(a)

(b)

(c)

⋀ FIGURE 13.4 Influence of Wind and Topography on Wildfires
Idealized diagrams of flame shape and processes associated with a spreading fire under three conditions: (a) flat topography and no wind, (b) flat topography with wind, and (c) hillslope topography where slope drives fire uphill. Solid red arrows indicate motion of air currents, and dashed red lines show radiant heating from the fire. *(Based on Rothermel, R. C. 1972. "A Mathematical Model for Predicting Fire Spread in Wildland Fuels," U.S. Forest Service Research Paper INT-115)*

Topography Topography can have a profound effect on fire for several reasons. First, the moisture content of fuel is influenced by its location on the landscape. In the Northern Hemisphere, south-facing slopes are relatively warm and dry because they face the sun for a longer period of time. Thus, fuel on a south-facing slope is typically drier and burns more easily. Topography also influences air circulation. Slopes exposed to prevailing winds tend to have drier vegetation than do slopes sheltered from the wind and, thus, are more prone to combustion. Also, in mountainous areas, winds tend to circulate up canyons during the daytime, providing an easy path for wildfires. Finally, once a fire starts, topography can strongly influence its movement. Wildfires burning on steep slopes tend to preheat fuel upslope from the flames (Figure 13.4c). This makes it easier for a fire to move upslope and often causes it to spread more rapidly in that direction. Fires may also advance downslope, especially if they are driven by the wind.[6]

Weather Weather, especially temperature, precipitation, relative humidity, and wind, has a dominant influence on wildfire. Large wildfires are particularly common following droughts that reduce fuel moisture content. Hot, dry conditions associated with drought, as well as with summers in the southwestern United States, lead to the formation of "dry thunderstorms." During these storms, rain evaporates before it reaches the ground and is not available to extinguish fires started by lightning. Changes in relative humidity, even during a single day, influence the moisture content of fuel and how fast a fire burns. Wildfires burn more vigorously when relative humidity is at its lowest point, generally during the midafternoon.[2] Finally, winds greatly influence the spread, intensity, and form of a wildfire. Strong winds and changing wind conditions help a fire preheat unburned fuel in the surrounding area. Winds can also carry burning embers away from a wildfire to ignite *spot fires* well ahead of the flaming front (see Figure 13.4b).[6]

The influence of weather on fires was dramatically illustrated during the 1998–2002 drought in Florida, one of the worst in the state's history (Figure 13.5).[7] Hot, dry conditions during the drought reduced fuel moisture in swamps and wetlands. This resulted in wildfires in normally wet bald cypress swamps and grassy marshes, in addition to the more typical fires in upland pine/oak forests and scrublands. These wetland fires included the Mallory Swamp Fire in 2001, which burned more than 250 km² (~62,000 or 97 mi.²) of swamp and commercial woodland west of Tallahassee, and the Deceiving Fire in 1999, which charred 700 km² (~173,000 acres or 270 mi.²) of marsh in the Everglades and surrounding upland near Miami.[8] The name "Deceiving" was given to the fire because changing wind directions continually shifted its movement.[7] Smoke from the Deceiving Fire closed I-75 (Alligator Alley), the main east–west highway across southern Florida, for several weeks and drifted into Miami, where cars had to drive with headlights on in the middle of the day.[7] Windblown smoke was also a major problem in

∧ FIGURE 13.5　Drought Contributes to Florida Wildfires
During the 1998–2002 drought, over 25,000 wildfires burned more than 6100 km² (~1.5 million acres or 2300 mi.²) in Florida. Fires burned parts of the Everglades in south Florida and Mallory Swamp in north-central Florida. *(MIKE EWEN KRT/Newscom)*

2001, when it smothered central Florida theme parks, including Walt Disney World; closed I-4, the main east–west route across central Florida; and caused a 20-vehicle accident.[8,9]

TYPES OF FIRES

Wildfire scientists and firefighters classify fire behavior according to the layer of fuel that is allowing the fire to spread. Although there are three fire types—ground, surface, and crown—most fires are complex blends of all three. *Ground fires* creep along slowly just under the ground surface with little flaming and more smoldering combustion (Figure 13.6). In forests, these fires burn in *duff*, decaying organic matter in the soil, and in drained or temporarily dry swamps and marshes and in thicker peat deposits below the soil.

Surface fires, which move along the ground, may vary greatly in their intensity, that is, the amount of heat energy released by the fire (Figure 13.7). Low-intensity surface fires burn grass, shrubs, dead and downed limbs, leaf litter, and other biomass. They burn relatively slowly with glowing or smoldering combustion and limited flaming combustion. Some surface fires, however, such as those in the Mediterranean (dry summer subtropical) climate of Southern California that burn chaparral brushlands, can be extremely intense and release large amounts of heat energy as they move swiftly across a landscape.

Finally, *crown fires* are those in which flaming combustion is carried through the canopies of trees (Figure 13.8). Crown fires may begin when a surface

∧ FIGURE 13.7　Surface Fire
A surface fire, such as this one in a ponderosa pine forest, burns low vegetation rather than the upper part of tall trees. *(Kent Dannen/Science Source)*

∧ FIGURE 13.6　Ground Fire
A ground fire burns roots and some surface vegetation. Commonly, there is more smoldering combustion than flame. *(© Getty Images/Anadolu Agency/Contributor)*

∧ FIGURE 13.8　Crown Fire
Crown fires burn the upper part of tall trees. They sometimes spread ahead of the fire on the surface. In 1995, high winds turned this crown fire, the Green Mountain Fire on Mount Lemmon outside of Tucson, Arizona, into a firestorm. *(A.T. Willett/Alamy)*

fire moves up trunks into limbs of trees through various layers of fuel, or it may spread independently of surface fires. Large crown fires are generally driven by strong winds and are often aided by steep slopes. Such fires will grow and expand as long as conditions for combustion are favorable. Large wind-driven crown fires are nearly impossible to stop; humans and other animals need to evacuate from their path.

13.2 CHECK your understanding

1. Define wildfire.
2. How are wildfires related to plant photosynthesis and decomposition?
3. What are the major gases and solid particles produced by a wildfire?
4. What are the three requirements for fire to start and for combustion to continue? What happens when one of these requirements is removed?
5. Describe the three phases of a wildfire.
6. Explain how processes in the preignition phase prepare plant material for combustion.
7. What are sources for the initial ignition of wildfires? How often does ignition occur?
8. Explain how the two types of combustion differ. How do the processes of combustion differ from those of ignition?
9. What are the three processes of heat transfer in a wildfire?
10. Link the types of fuel to wildfire behavior.
11. How does topography influence a wildfire?
12. Describe the weather conditions that are most favorable for wildfires.
13. Describe the three types of fire.

13.3 Geographic Regions at Risk from Wildfires

Wildfires in the United States burn 2 to 5 million acres (~8,000 to 20,000 km^2) per year, comparable to the area of New Jersey. The vast majority of wildfires are in the western USA where there are large forests and brushlands. Another hot spot for wildfire is southern Florida where there are abundant forests and grasslands.[10]

Most areas of the United States and Canada that are within or in close proximity to grasslands, shrublands, woodlands, or tundra are at risk for wildfires during dry weather or drought conditions. Even the sparsely vegetated Sonoran Desert in Arizona, rainforests in the Pacific Northwest, Hawaii, and Puerto Rico, and normally wet marsh and swamplands in the southeastern United States, Alaska, and Canada can experience wildfire. Large wildfires, however, are most common in Alaska and the western contiguous United States (Figure 13.9), in the Canadian Rockies, and in a belt that extends from Canada's Yukon Territory southeast to Lake Superior and then east to Labrador.[11,12] The geographic region at greatest risk for wildfires shifts from year to year due to factors such as weather conditions and fuel availability.

13.3 CHECK your understanding

1. Which North American regions are most vulnerable to wildfire?
2. Why does the wildfire risk shift from year to year?

13.4 Effects of Wildfires and Linkages with Other Natural Hazards

Wildfires affect many aspects of the local environment: They burn vegetation, release smoke into the atmosphere, char soil, create favorable conditions for landslides, increase erosion and runoff, and harm wildlife. Although we may primarily be concerned with the effects a fire has on the local biota of a region, it is important to consider the changes that may also occur in the geologic and atmospheric environments.

EFFECTS ON THE GEOLOGIC ENVIRONMENT

Wildfires have differing effects on soils, depending on the type and moisture content of the soil and the duration and intensity of the fire. The amount and intensity of precipitation after a fire also play a role in determining how a wildfire affects the soil.[13]

Extremely hot fires that scorch dry, coarse soil in semiarid brushlands, such as southern California, may leave a nonwettable, or water-repellent, layer—called a **hydrophobic layer**—in the soil. Water repellency is caused by an accumulation of water-repellent chemicals in the lower layers of debris on the forest or brushland floor. These chemicals are derived from the burning vegetation and penetrate the soil as a gas that, when cool, forms a "waxy" substance that coats soil particles (Figure 13.10).[2] This water-repellent layer in the soil increases surface runoff and, thus, erosion because

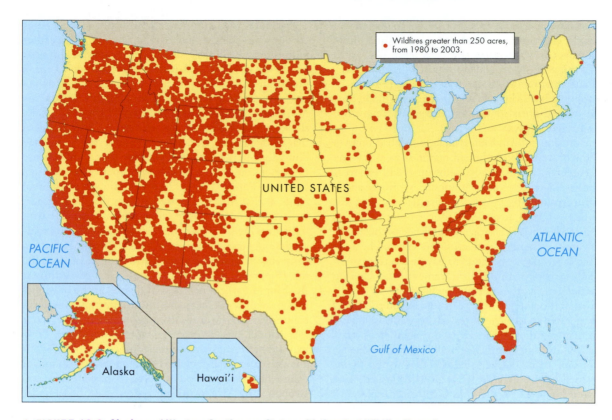

∧ FIGURE 13.9 Alaska and Western Contiguous States with Greatest Wildfire Hazard
Map depicts areas of the United States that experienced wildfires greater than 1 km² (~250 acres or 0.4 mi.²) from 1980 to 2003. There were no fires of this size in Puerto Rico during this time period. *(U.S. Geological Survey)*

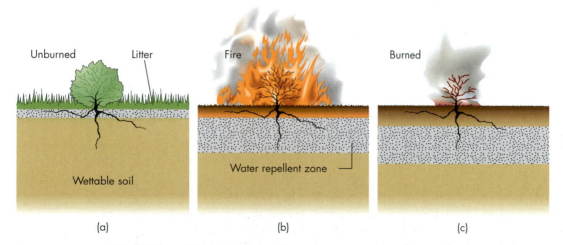

∧ FIGURE 13.10 Effect of Fire on Water-Repellent Soil
Idealized diagram showing the response of a water-repellent soil layer to fire. A water-repellent layer normally resides under the litter layer, atop wettable soil. During a fire, the vegetation and leaf litter are burned, and the water-repellent layer moves downward. After the fire has passed, the water-repellent layer remains at depth, and wettable soil lies above and below it. *(Based on DeBano, L. F., "Water Repellent Soils: A State-of-the-Art Review," U.S. Forest General Technical Report PSW-46, 1981 in Pyne, S. J., Andrews, P. L., and Laven, R. D.,* Intro-duction to Wildland Fire, *2nd ed. New York: John Wiley & Sons, Inc., © 1996)*

the burned surface now lacks vegetation to anchor the loose soil above the hydrophobic layer. Water flows rapidly over a water-repellent layer, just like water down a raincoat. Loose soil at the surface is easily saturated and then moved by the water.[14] The presence of hydrophobic layers in soils may increase the flood hazard in burned areas for several years following a fire until the layer decays.

Soil erosion and landslides, including debris flows (a fast flow of water and sediment confined to a channel; see Chapter 7), are common following wildfires. Areas naturally subject to erosion commonly experience an increase in erosion rates for a few years after a fire, whereas areas that normally experience little erosion show less increase after a fire.[2] Erosion and landslide problems are significantly increased on steep slopes charred by a severe burn. Steeper slopes tend to have a greater erosion problem to begin with, and the removal of anchoring vegetation only adds to the erosion potential.

Precipitation helps trigger landslides, and this effect is exaggerated in a burned area. Heavy rains following a fire may significantly increase the number of landslides. Wildfires in California in 1997 denuded vegetation from many slopes before the winter rains intensified by the El Niño effect. Of 25 burned areas that were mapped by the U.S. Geological Survey, 10 produced debris flows during the first winter storm.[15] On other slopes, sediment was flushed through streams and rivers because the rate of sediment erosion on burned slopes was greatly increased. Wildfire in Southern California is different from forest fires insofar as the dominant vegetation is often a type of shrubland known as chaparral rather than tall trees. Southern California is also home to millions of people who live on the edge of or within areas that have a significant wildfire hazard.

How likely are catastrophic floods and debris flows following wildfire in Southern California? We can't answer this question precisely because the historical record is short, and the time between wildfires, their *return interval*, is about 40 years. However, if catastrophic flooding and debris flows generally are responses to intense rainfall after a very large wildfire, then the return period at a particular site is probably several hundred to several thousand years. This estimate assumes that the probability of intense precipitation in the first year following a fire is low.

Although catastrophic flooding and large debris flows certainly do not follow every fire, flushing (scouring and downstream transport) of sediment from stream channels is common. Following the 1985 Wheeler Fire and the 1990 Painted Cave Fire (both in southern California), scientific studies documented the flushing of sediment from small burned watersheds.[16,17]

In Southern California and, probably, in most similar settings, sediment moves downslope in dry weather following a wildfire. This gravity-driven process, known as **dry ravel**, moves a large volume of sand, gravel, and organic material that was stored upslope of brush vegetation before the fire (Figure 13.11a, b). Much of this material accumulates at the base of the slope or in adjacent stream channels. Other material that remains on the slope is often easily washed downslope in the first rainfall event after a fire because the water-repellent layer is now deeper below the surface (see Figure 13.10). The voluminous amount of material that moves downslope from dry ravel or surface runoff subsequently clogs stream channels with sediment (Figure 13.12a, b). In a second moderate rainstorm and stream-flow event, this sediment is flushed from the channel, leaving it much as it was before the fire (Figure 13.12c). This process is called **sediment flushing**. Less sediment was available on the hillslopes in the small watershed, and the stream flow was sufficient to transport the available sediment and also scour the channel. Less often, large catastrophic debris flows may be produced when intense precipitation of sufficient duration falls on burned slopes with abundant coarse debris. Figure 3.11c shows a part of the catastrophic 2018 Montecito debris flow that filled and destroyed homes. The large boulders were easily transported by the debris flow because the density of the boulders and mud matrix of the viscous, fast-moving flow are close to the same. Observers reported the boulders bobbed along near the top of the flow. The Montecito debris flows mostly came down three channels and spread out on the flatter land below the mountain where the flows originated. The flows closed the main highway along the coast for several weeks, killed 23 people and damaged or destroyed about 200 homes. The flows occurred within minutes following a 5-minute burst of 14 mm (0.54 in.) of intense rain that seemed to observers like a waterfall. The flow happened so fast that some people were trapped. There was an evacuation notice, earlier, but many people had just returned from a two-week or more evacuation order for the Thomas Fire. They probably had evacuation fatigue, and did not know what a debris flow was. As a result, not all people evacuated.

Studies of wildfire in Southern California brushland have shown that fire is only the first half of an event that alters the environment. There is often a one–two punch. Following a fire (first punch), runoff and erosion often increase (the second punch), even in response to rather modest rainfall. The result is often sediment flushing of gravel-sized particles. Engineering solutions are often used to mitigate the high rates of sedimentation that can occur following a wildfire.

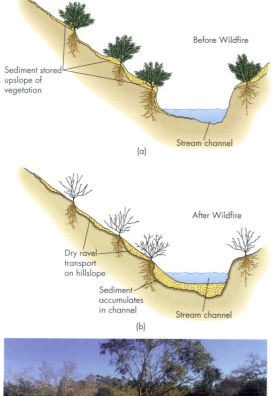

(a)

(b)

(a)

(b)

(c)

⋀ FIGURE 13.11 Debris Flows and Sedimentation after Wildfires

(a) Sediment stored on hillslopes between fires is trapped upslope behind chaparral vegetation stems. (b) After fire burns the stems, sediment moves downslope by the process of dry ravel and accumulates in irregularities on hillslopes or in channels. Moderate stream flows later flush the sediment from the channel. (c) Catastrophic debris flow in early January of 2018 in Montecito, CA. The flow killed 23 people and destroyed or damaged about 200 homes. Large boulders were easily carried near the top of the debris flow as the density of the boulders is only about 25 percent greater than that of the fast-moving, viscous mud-matrix that caries the flow. Thus, large boulders appear to float and bob in the flow, looking like corks on water. *(a and b: modified from Florsheim, J. L., Keller, E. A., and Best, D. W. 1991. Fluvial sediment transport in response to moderate storm flows following chaparral wildfire, Ventura County, southern California.* Geological Society of America Bulletin, *103: 504–11; c: Edward A. Keller.)*

(c)

⋀ FIGURE 13.12 Increased Sediment in Streams

(a) A small stream channel in Southern California shortly after a wildfire that burned vegetation in the drainage basin. Some trees near the channel survived the fire. *(Edward A. Keller)* (b) The scene after the first winter rainstorm. Note the voluminous amount of sediment deposited. *(Edward A. Keller)* (c) The same stream channel after a second winter rainstorm scoured the channel, removed much of the deposited sediment, and returned the channel to a shape that is similar to its shape before the fire. *(Edward A. Keller)*

Sediment control basins are constructed at the mouths of canyons to catch sediment flushed from stream channels, as well as debris flows. These basins, sometimes known as debris basins, have been constructed on many streams in Southern California (Figure 13.13a). For example, a debris basin was constructed following the Painted Cave Fire of 1990 near Santa Barbara, California (Figure 13.13b). Sediment was flushed from stream channels in the winter following the fire, and the debris basin trapped approximately 23,000 m^3 (~30,000 yd.^{3}) of sediment that otherwise would have been transported downstream, where it would have reduced the carrying capacity of the stream channel and increased the flood hazard.

Also, important to understanding wildfire in Southern California is reconstructing fire history for grassland, brush, and forest ecosystems. European settlers living in the Los Angeles area prior to the twentieth century were not, evidently, very concerned about wildfires that periodically burned for months in the mountains. This attitude changed as agriculture and urbanization transformed the land. As early as 1892, it was recognized that fire suppression was a potential management tool to preserve native vegetation in order to control erosion and flooding. This recognition led to the idea of setting small, controlled fires to reduce excessive accumulation of fuel and, thereby, reduce the risk of large, catastrophic wildfires.[18-20] However, large, catastrophic wildfires

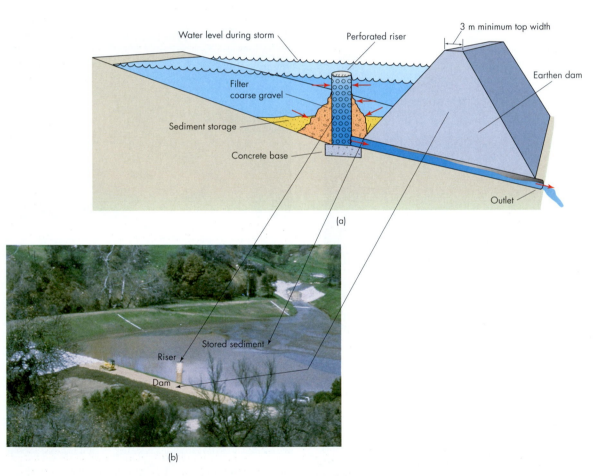

(a)

(b)

∧ FIGURE 13.13 Sediment Control Basin
A sediment control basin can mitigate the effects of increased erosion and sedimentation after a wildfire. (a) Cross-section of a sediment control basin. Sediment carried by storm water settles out in the basin, while the water filters through loose gravel into a riser, where it then drains through a pipe to an outlet. Accumulated sediment is periodically removed mechanically. Red arrows indicate direction of water flow. *(Based on DeBano, L. F., "Water Repellent Soils: A State-of-the-Art Review," U.S. Forest General Technical Report PSW-46, 1981 in Pyne, S. J., Andrews, P. L., and Laven, R. D., Introduction to Wildland Fire, 2nd ed. New York: John Wiley & Sons, Inc., © 1996)* (b) A sediment basin constructed to trap sediment that eroded after a wildfire in Southern California. *(Edward A. Keller)*

occur during periods of warm weather and high winds when suppression of fire is nearly impossible. Therefore, prescribed burns to reduce the fuel load of Southern California brushlands are not likely to be successful in reducing or stopping a catastrophic wildfire and resulting debris flows.[18]

EFFECTS ON THE ATMOSPHERIC ENVIRONMENT

In the short term, large wildfires create their own clouds (Figure 13.14) and release smoke, soot, and invisible gases that contribute to air pollution. Smoke and soot significantly increase the concentration of very fine particles, referred to as *particulates*, in the atmosphere. This increase can be observed thousands of miles downwind of large, long-lasting fires. Several times during the past decade, fires burning in Mexico, Guatemala, and other countries in Central America have increased particulates in Texas and other parts of the southern United States, causing some cities to violate federal Clean Air Act standards. Smoke plumes from large wildfires can be easily seen from space and are now regularly monitored from satellite images. The information provides real time data on the type of fire—fast-moving surface or crown (flaming) or slow-moving ground (smoldering); the size; the intensity; and the amount and types of particles and gasses being emitted. The analysis can be used to help direct firefighting activities, as well as to decide if a fire may be too intense and fast moving to fight and control.[21]

∧ FIGURE 13.14 Smoke from Forest Fire Creates Clouds Helicopter water tanker takes off with 3800 liters (1000 gallons) to fight the Hayman Fire in Colorado. Large pyrocumulus clouds formed over the fire and grew to an estimated height of 6400 m (~21,000 ft.). *(FEMA)*

Wildfires can also contribute to the formation of smog through the release of large quantities of carbon monoxide, volatile organic compounds, and nitrogen oxides. In the presence of sunlight, these gases form harmful ground-level ozone. The significance of wildfires adding to global air pollution was demonstrated in an international study of gases released in central Alaska and the Yukon Territory of Canada during wildfires in 2004. Wildfires in these areas burned more than 45,000 km^2 (~11 million acres or 17,000 mi.^{2}) of coniferous forest and organic soil, a land surface equivalent to New Hampshire and Massachusetts combined.[22] Using satellites and computer models, atmospheric scientists estimated that these fires increased ozone levels in parts of Canada and the northern United States by 25 percent and up to 10 percent in Europe.[22,23]

On a much larger scale, wildfires and fires set by humans in Indonesia and Malaysia in 1997 and 1998 created smog conditions that affected millions of people in Southeast Asia. The favorable conditions for these fires in Southeast Asia were created by an El Niño climate event in the Pacific Ocean. Thus, wildfires have both short-term interactions with weather and long-term interactions with climate.

LINKAGES WITH CLIMATE CHANGE

One projected effect of climate change in this century is an increase in the intensity and frequency of wildfires. This increase may be brought about by changes in temperature, precipitation, and the frequency and intensity of severe storms. It may also be linked to biological changes that have taken place in the type and quantity of fuel available for wildfires. Some of these changes may already be underway in the western United States and Canada.[24,25] For example the wildfire, known as "The Beast" and formally called the Fort McMurray Fire, began on May Day 2016. The cause of the fire is believed to be due to someone who visited a remote trail system and probably started the blaze. Fort McMurray is a boomtown that grew to a pre-fire urban population of about 88,000 people in the rush to develop the vast petroleum resources of the Alberta oil sands. The area is basically an enormous industrial complex built in a forest ecosystem. The advent of increased population resulted in more people recreating in the fire-prone forests near Fort McMurray, and the number of fires in that area increased significantly in the past 50 years. The fire when first noticed was very small, but it exploded in area quickly and eventually burned about 5900 km^2 (2300 mi.^{2}). Over 80,000 people were evacuated between May 3 and June 1, 2016 in Fort McMurray and surrounding areas. About 2,500 buildings were destroyed, as were over 600 work camps that were mostly associated with the oil sand operations. Total losses were about $10 billion. Global temperature records were broken in several years prior to the fire, including 2015 and 2016. Increasing temperature and drought, with a predicted decrease in return period (time between fires) and increase in intensity of wildfires in the future, led some to suggest that the Fort McMurray Fire is an example of wildfire linked to climate change. Global warming is a symptom of climate change. Other symptoms include change in precipitation patterns, volume, and intensity that are also linked to wildfire, sometimes in complex ways. The climate change link to the Fort McMurray Fire is debatable because it is a region with a natural fire regime as a result of the boreal forest (a mixture of pine, spruce, and fir trees) in which wildfire is the dominant disturbance. Increasing human population and more forest recreation has become a factor in more fires.[26] The Fort McMurray Fire is consistent with climate change, but no one fire can be attributed to climate change.

More of the wildfires today are catastrophic fires known as "runaway" fires that escape initial firefighting efforts and then spread to damage large areas. The Santa Rosa Fire of 2017 that opened this chapter is an example. In Southern California, autumn and winter are particularly hazardous seasons for catastrophic wildfires. Wildfires often accompany hot, dry Santa Ana winds, a large-scale pattern of atmospheric circulation that produces dangerous conditions over the entire Southern California region.[27,28]

Although the number of large wildfires in Southern California has not changed significantly in the past few hundred years, the size and intensity of fires and their effect on people certainly have. Human population has doubled in Southern California in the past 50 years, and fire suppression has led to the increase in availability of natural fuels; as a result, the severity and intensity of burns and their ecological consequences are increasing.[27] The association of wildfire and Santa Ana winds is well known, but the exact timing of fires is related to complex patterns of rainfall during winter months and drought conditions prior to the fires. The relationship of these winds to global warming is not particularly well understood. However, if the trend toward drier, windier conditions continues, wildfires will likely become larger and more intense.[28,29]

Climate change is also likely to increase the number and intensity of wildfires in Northern California. Regional climatic conditions are projected to become warmer, drier, and windier—three factors that promote wildfires. Warmer, drier climatic conditions will result in changes to plant communities, with grasslands and chaparral (scrub forest) replacing the larger coniferous trees. These changes in plant types will increase fire frequency and intensity because grasslands and brush burn more frequently than do forests. As a result of these changes, more property will be destroyed, more lives lost, more floods and debris flows will follow fires, and ash and soot will reduce air and water quality. In addition, the cost of firefighting and insurance against fire will increase as damages increase.[28]

Increases in the size or intensity of severe storms may also promote wildfires by increasing the incidence of lightning, one method of igniting wildfires. In addition, warmer and drier conditions will make trees more susceptible to beetle infestations, weakening them and rendering them more vulnerable to fire.

As mentioned previously, changes in vegetation resulting from global warming may also affect wildfire frequencies and intensities. Precipitation changes in parts of the western United States are also likely to change some landscape dominated by sagebrush into widespread grasslands. As a whole, grasslands tend to burn more frequently than the sagebrush that grows there today.[27,28]

These trends in fire frequency and intensity are not limited to California and the Southwest. Similar changes are taking place in Alaska and Canada.[25,26] For example, temperatures on Alaska's Kenai Peninsula have risen several degrees in the past 30 years. Warmer temperatures have weakened the coniferous forest and led to a large infestation of beetles. An estimated 40 million trees have been killed throughout an area the size of Connecticut. This insect infestation may well be the largest ever recorded, and as the trees die and become fuel, catastrophic wildfire will be much more likely.[28]

Finally, the linkage between wildfires and climate change will also affect human health. As mentioned previously, increases in airborne particulates from wildfires can lower air quality and harm human health. For example, wildfires in Florida in 1998, which were related to La Niña conditions in the Pacific Ocean causing drought in the southeastern United States, resulted in a marked increase in the number of people seeking hospital emergency room treatment for asthma, bronchitis, and chest pain.[28] Recently, atmospheric scientists and geologists have become concerned with increases in peatland fires in both the boreal forests of Canada and Russia and the tropical forests of Malaysia and Indonesia. Burning peat releases smog-producing gases and higher levels of particulates than the combustion of wood or grass, as well as significant quantities of toxic mercury into the environment (Figure 13.15).[30,31]

⋀ FIGURE 13.15 Burning Peatlands Contribute to Smog and Global Warming
Fires in organic soils, such as this one in a bog in the Poca das Antas Biological Reserve northeast of Rio de Janeiro, Brazil, release gases into the atmosphere that cause smog and contribute to global warming. Carbon, originally stored for many hundreds or thousands of years in organic soil or underlying peat deposits, is released into the atmosphere as carbon dioxide, a greenhouse gas contributing to global warming. *(AP Photo/ Douglas Engle)*

EFFECTS ON THE BIOLOGICAL ENVIRONMENT

Fires have many direct and indirect effects on the biological environment, which includes plants, animals, and human beings. The effects may vary from moderate to severe, depending on the type, size, location, duration, and intensity of the fire.

Vegetation Fire's effects on vegetation are numerous, varied, and complex. Some plants, such as scrub oak, juniper, and mesquite, are susceptible to fire, whereas others are not. For example, Douglas fir and ponderosa pine are adapted to frequent fires with thick bark that insulates them from fire damage.[32] Even if a plant does not die in a wildfire, a fire can make it vulnerable to later destruction by disease or drought. On the other hand, some plants, such as a number of native prairie wildflowers, require fires for the continuation of their species. Fire also helps to propagate many woody plants. For example, seeds of sumac and bearberry remain dormant in the soil until opened by the heat from a fire.[2] Overall, fire is an important long-term control on the types of plants that grow in many areas of North America. The type, intensity, and frequency of fires selectively favor some plant species and eliminate others.[2]

It has been observed that, in some cases, dense large oak trees provided a partial barrier to a fire's advance. In some cases, fire burns some of the oaks only halfway through, stopping with a knife-edge line (a vertical line between burned and unburned tree). Oaks can release visible water vapor from heating as the fire approaches. Large oaks live several hundred years and usually survive a number of fires. Even oaks that appear dead after a fire will usually sprout quickly and recover. Of course, extremely hot, intense, and lingering fire may kill oak trees and set homes on fire, but the homes with oak trees around them have a better chance of surviving. It is much more difficult for fire to advance through oak trees than through chaparral. The possible natural service function of oak trees to retard fire moving across the land needs to be studied in more detail.

The eucalyptus, a tree native to Australia but imported to the United States and elsewhere, is also well adapted to wildfire but in an entirely different way from oak trees. Eucalyptus trees have loose bark that hangs down, inviting fire to move up into the canopy. Sometimes the trees literally explode, sending burning embers through the air for hundreds of meters. With the high winds that often accompany fire, the embers can travel even greater distances, spreading spot fires. However, fire is important in enabling eucalyptus seedpods to open, and wind and flames may disperse the pods. On hot days, a faint blue flammable gas from the rich oil the trees produce has been reported. Burning eucalyptus trees can significantly increase the temperature of the fire, greatly increasing the hazard in the immediate area. During the 2008 Tea Fire in Santa Barbara, California, the Mount Calvary Monastery burned (Figure 13.16a). Just downslope of the structure is a group of eucalyptus trees (Figure 13.16b) that the fire moved through just before encountering the monastery. The trees are still there, having sprouted following the fire. There is ongoing discussion about removing eucalyptus trees from areas with a wildfire hazard. Some investigators have stated that eucalyptus trees are the most dangerous trees in a wildfire. Certainly, that is the experience in Australia.

Animals Although extremely large, intense wildfires may kill even the fastest-running animals, in general, fire has few direct effects on wildlife. Most animals are able to escape or hide from the danger posed by a fire. Mobile species, such as birds, deer, cougars, and bears, are generally able to move away from a fire as it advances. Rodents that nest beneath the ground also have a good chance of survival because soil is a good insulator. However, wildfires may have indirect effects on wildlife. In fire-denuded areas, loss of plant cover exposes the soil to solar heating, thereby increasing ground temperatures and altering habitats. In intensely burned areas, the pioneering plant species that initially colonize the land will determine both the type and the

(a)

(b)

∧ **FIGURE 13.16 Mount Calvary Monastery**
(a) Located above the city of Santa Barbara, California, this
monastery burned during a wildfire in 2008. *(Edward A. Keller)*
(b) A grove of eucalyptus trees just downslope contributed to
the intense fire that claimed the structure. *(Edward A. Keller)*

number of vertebrates that can thrive in the area. In
streams, fish and other aquatic species may suffer from
increased sedimentation resulting from wildfires, and
water temperatures may increase because plants along
the stream banks have been destroyed.[13]

Human Beings   When a fire removes much or all of
the vegetation in a watershed, subsequent rains will
have much greater erosive potential. Erosion produces
large quantities of sediment and plant debris that affect
the water quality of streams and lakes. For example, in
May 1996, a fire burned two watersheds that contribute
runoff into the drinking water reservoir for Denver and
Aurora, Colorado. Two months after the fire, a storm
produced floods in the watersheds and sent large
amounts of floating debris and high dissolved concentrations of manganese into the reservoir. Even two
years after the fire, the water quality was still not as
good as it had been before the fire.[15]

Smoke and haze produced by fires can harm human
health. Exposure to smoke and haze commonly produces
eye, respiratory, and skin problems. After prolonged exposure to smoke, people, especially firefighters, may experience respiratory problems that become permanent
or even fatal.

Wildfire, of course, has the potential to destroy personal property. As human population continues to increase, more and more people are living on the fringes of
wildlands, in an area called the **wildland–urban interface**. Homes are built in undeveloped canyons and on
wooded slopes susceptible to fire, and naturally occurring fires are suppressed to protect homes and businesses,
resulting in longer return intervals and accumulations of
fuel. When a wildfire occurs in brushland or forest areas
near a city, such as in the Santa Monica Mountains of Los
Angeles or the Front Range near Denver, Colorado,
hundreds to thousands of homes are at risk.

13.4 CHECK your understanding

1. How does burned vegetation and soil interact to
   repel water?

2. What is dry ravel?

3. Describe the process of sediment flushing
   following wildfire. How common is the flushing?

4. Why might high-magnitude debris flows occur
   following wildfire?

5. Describe how wildfire affects air and water
   quality.

6. How is wildfire linked to climate change?

13.5 Natural Service Functions of Wildfires

Although fires threaten human lives and property, they
are also beneficial to ecosystems. Fire is a part of the
natural cycle of the landscape and, as mentioned earlier,
has been so since terrestrial plants evolved.

BENEFITS TO SOIL

Fires tend to leave an accumulation of carbon on the
surface in the form of ash and increase the nutrient content of a soil. Under the right conditions, when erosion
does not remove the ash from the environment, a nutrient reservoir may form that is beneficial to local plants.
For plants that are well adapted to fire, such as those that
release seeds only after a fire, these nutrients provide a
valuable resource for new seedlings. Fires also tend to

reduce populations of soil microorganisms. Decline in microorganisms may benefit those plants with which they compete for nutrients. In some cases, the destruction of microorganisms may also be beneficial since some microorganisms are parasites or carry diseases.[2]

BENEFITS TO PLANTS AND ANIMALS

When a wildfire reduces the number of individuals of a species of plant in a given area, the result may be beneficial to the plant community as a whole. The removal of some species temporarily reduces competition for moisture, nutrients, and light, allowing the surviving species as well as new species to thrive.

In species that depend on fire for reproduction, a wildfire may trigger the release of seeds or stimulate flowering. Two species of trees, lodge pole pine and aspen, are strongly dependent on fire for reproduction and growth. In aspens, fire stimulates the release of a growth enzyme that can cause an aspen tree to produce as many as a million sprouts per acre.[33] These sprouts are important food for elk and moose. In lodgepole pines, seeds remain sealed in pine-cones for decades until fire melts a resin surface coating on the cones to release the seeds.[2] The seeds then fall to a nutrient-rich, ash-covered forest floor that is devoid of competition for the growth of pine seedlings. Species with these reproductive adaptations tend to exist mainly in environments where fire is common and frequent. If wildfires were eliminated from their environment, some species would eventually become extinct.

Fire is also important to ecosystems as a whole. The North American prairie is an example of an ecosystem that is well adapted to wildfires. Grasses, the dominant class of prairie plant, are especially suited for growth after a fire because they grow from their base rather than leaf tips, and 90 percent of their biomass is underground and generally undamaged by fire. Fire also removes surface litter and allows nutrients and moisture to reach the roots of grasses, allowing them to flourish. Native Americans long ago understood the importance of fire, wind, drought, and grazing animals for the prairie ecosystem.[34] Following a fire, colonization and replacement of plant species follow a regular pattern known as *secondary succession*.

Burning of plant material recycles nutrients in the system by quickly decomposing organic matter and allowing new plants to grow. Fire also reduces competition among species for sunlight, water, and nutrients, and crown fires allow more sunlight to reach the understory.

Many species of birds, insects, reptiles, and mammals may benefit from wildfire. For example, new grassland and forage for deer and bison emerge following removal of trees by fire. Also burned, decaying logs provide homes for insects and supply food for mice, coyotes, birds, and bears.

Because wildfires potentially benefit ecosystems, many scientists believe that natural fires in woodland areas should not be suppressed. In 1976, Yellowstone National Park, which covers 9000 km^2 (~2.2 million acres or 3500 mi.^{2}) of Wyoming and Montana, began a policy of allowing natural burns in all areas of the park managed as wilderness. This meant that any fire that started naturally was allowed to burn, provided it did not endanger human life, threaten visitor areas such as Old Faithful, or risk spreading to areas around the park. Any human-caused fire would be extinguished immediately. Before the summer of 1988, the worst fire in Yellowstone's history had burned about 100 km^2 (~25,000 acres or 29 mi.^{2}). By the end of the 1988 fire season, more than 3200 km^2 (~800,000 acres or 1250 mi.^{2}) had burned, causing a major controversy over the natural burn policy.

Lightning strikes during the summer of 1988 ignited 50 fires in the park. Of those, 28 were allowed to burn, according to the natural burn policy. Initially small, the fires quickly expanded in the hot, dry conditions, fueled further by high winds in mid-July. On July 17, bowing to political pressure, Yellowstone officials sent fire crews in to fight a natural fire. By July 21, multiple natural fires in and around Yellowstone were being fought; by July 22, however, it was clear that the fires were beyond the control of the crews. The fires did not slow until September 11, when rains fell throughout the area. The snows of November finally extinguished the fires.

The large area burned at Yellowstone National Park in 1988 generated much criticism of the national park's natural burn policy. Although most scientists agree that fire is good for the natural environment, it was difficult for some people to sit by and watch the park burn. Eventually, 9,500 firefighters were deployed at a cost of $120 million to taxpayers. Nonetheless, the fires continued until they were naturally extinguished. Critics stated that the fires would not have been so big if park officials had fought them from the beginning. Others argued that the fires would not have been so severe if years of fire suppression prior to the mid-1970s had not allowed fuel to accumulate in the area. Postfire studies have shown that the effects of the 1988 Yellowstone wildfires were beneficial to the environment, and the park still adheres to a natural burn policy. This policy is correct because Yellowstone's ecosystems have, through geologic time, adapted to and become dependent on wildfire. The fires of 1988 did not destroy the park; rather, they revitalized ecosystems through natural transformations that cycle energy and nutrients through soils, plants, and animals.

13.5 CHECK your understanding

1. Why is wildfire necessary for some plants and animals?
2. Was the impact from the 1988 wildfires in Yellowstone harmful or beneficial? Why?

13.6 Minimizing the Wildfire Hazard

Clearly, it is desirable to minimize the potential of wildfires to destroy human life and property. There are several components of fire management.

FIRE MANAGEMENT

The task of *fire management* is a difficult one because large wildfires cannot be prevented. Furthermore, not all fires should be suppressed; remember the benefits of fire to the natural environment. However, suppressing fires when they threaten human lives and property is desirable. The primary approaches to fire management include science, education, data collection, and use of prescribed burns.

Science Scientific understanding of wildfire and its role in the structure and function of ecosystems is critical to fire management; we cannot manage what we do not understand. Of particular importance in fire ecology is an understanding of the **fire regime** of an ecosystem. The fire regime is broadly defined to include (1) the types of fuel that are found in plant communities; (2) typical fire behavior as described by fire size, intensity, and amount of biomass removed; and (3) the overall fire history of the area, including fire frequency and recurrence interval. Reconstruction of fire histories is difficult in many areas where fires have been suppressed for nearly a century.[17,18] Nevertheless, fire management is more likely to be successful if fire regimes for specific ecosystems can be defined.

Education Public education is an integral part of fire management. People can be made aware of wildfire as a hazard and what they can do to minimize their personal risk.

Data Collection Remote sensing has become an important tool for fire management, especially the use of satellite imagery to map vegetation and determine fire potential. Since the early 1990s, the U.S. Geological Survey (USGS), in cooperation with the National Oceanographic and Atmospheric Administration (NOAA), has prepared weekly and biweekly maps for the contiguous United States and Alaska, illustrating plant growth and vigor, vegetation cover, and biomass production. These maps, combined with determinations of the moisture content of vegetation in a given area, are invaluable resources for fire managers. A management tool called the Fire Potential Index (FPI) was developed by the USGS and the U.S. Forest Service to characterize the relative fire potential of forests, rangelands, and grasslands. The FPI takes into account the total amount of burnable plant material or fuel load, the water content of dead vegetation, and the percentage of the fuel load that is living vegetation. Regional and local FPI maps are prepared daily in a geographic information system (GIS) to help land managers develop plans for minimizing the threat from fires. Satellite images are also used to monitor the intensity of wildfire growth in real time. This information can help direct fire control measures.

Prescribed Burns For the past century, fire management in the United States has been guided by a policy of wildfire suppression, mainly to protect human interests. In forests, this policy has led to a buildup of fuel and the potential for larger, high-intensity fires. One way to counter this dangerous buildup of fuel is to ignite **prescribed burns**. The use of prescribed or controlled burns for forest management is not new; such fires have been used for years as an alternative to absolute fire suppression (Figure 13.17).[18]

The main goal of prescribed burns is to reduce the amount of fuel and, thus, the likelihood of a catastrophic wildfire.[15] Fire ecologists have found that more frequent, smaller fires will lead to fewer large fires.

Each prescribed burn has a written plan that outlines the objectives, where and how it is to be carried out, under whose authority it will be carried out, and how the burn will be evaluated. Those in charge of a prescribed burn take on a great deal of responsibility—they have to predict the natural behavior of the fire and successfully keep it under control. Planners face the difficult task of predicting the acceptable fuel and weather conditions under which they can safely control the fire. Factors such as temperature, humidity, and wind must be taken into account.

Unexpected changes in wind direction during a prescribed burn can have catastrophic results. The unexpected happened in spring 2000 when fire managers lost control of a prescribed burn at Bandelier National Monument in New Mexico. A prescribed burn ignited on May 4 quickly raged out of control and sliced through 80 km^2 (~20,000 acres or 31 mi.^{2}). Within 24 hours, the burn was declared a wildfire as a result of sporadic, unexpected changes in wind direction. After two weeks, the fire had engulfed 190 km^2 (~46,000 acres or 73 mi.^{2}) and was still on the move. Nearby drought-parched pine forest and grassland supplied ample fuel

< FIGURE 13.17 Forest Management with Prescribed Burns
This old-growth longleaf pine forest near Thomasville, Georgia, has been managed with annual prescribed burns during lightning season by the Tall Timbers Research Station for more than 25 years. The result is a healthy, diverse ecosystem similar to what existed when Europeans first explored the region in the sixteenth century. *(©Getty Images/ Andrew Kornylak)*

for the fire that destroyed 280 homes and forced 25,000 people to evacuate the town of Los Alamos, just 110 km (~70 mi.) north of Albuquerque.

After the flames were extinguished and the damage assessed, the town and the nation were left only with questions. How could a controlled burn, ordered by the National Park Service itself, end up nearly destroying a town? A formal review following the fire called for changes in policy. The changes included a more careful analysis and review of burn plans, better coordination and cooperation among federal agencies in developing burn plans, and use of a standard checklist before setting a prescribed fire. These and other changes have been instituted. Furthermore, because of the near-disaster at Los Alamos, the existing 1995 Federal Fire Management Policy was reviewed and then updated in 2001. Specific changes include emphasizing the role of science in developing and implementing fire management programs.

In 2003, the U.S. Congress passed the Healthy Forests Restoration Act, a forest management plan to reduce large damaging wildfires by thinning (logging) trees on federal lands. The act reduces or limits the environmental analysis that is normally required to initiate logging projects. Those in support of the act state that the new management procedures will reduce the risk of catastrophic fires to towns in and around national forests, will save lives of forest residents and firefighters, and will protect wildlife, including threatened and endangered species. Those opposing the act argue that large-scale logging will be promoted far from communities at risk from wildfire and, overall, will damage our forests. They further argue that the best approach to minimizing risk from wildfire is through the selective removal of vegetation around communities and homes and through education and planning.

13.6 CHECK your understanding

1. Define fire regime.
2. What is the role of science in helping control wildfire once it has started?
3. Define prescribed burns and explain why they are part of fire management.

13.7 Perception of and Adjustment to the Wildfire Hazard

PERCEPTION OF THE WILDFIRE HAZARD

In general, people who live or work in the wildland–urban interface do not adequately perceive a risk from wildfires. California provides a good example of this apparent lack of concern. Residents of California, where fires burn almost every year, should be familiar with the risks associated with wildfires, yet development continues on brush-covered hillsides. The demand for hillside property has increased property prices in these areas, which means that the people whose property is most at risk from fire have paid a premium for that "privilege." Fire insurance and disaster assistance may actually worsen the situation; if people believe that the government will reimburse their fire losses, they may not see any reason not to live where they choose, regardless of the risk. In the past hundred years, dozens of fires have struck brushlands of California, while population and property values continue to increase. The results of not perceiving a real risk from fires was tragically illustrated in October 1991, when a wildfire destroyed almost 6.5 km^2 (~1600 acres or 2.5 mi.^{2}) and about 3,800 houses and apartments in the cities of Oakland and Berkeley (Figure 13.18). The fire killed 25 people and caused more than $1.68 billion in damages. When it was over, it was labeled one of the worst urban disasters in U.S. history.

The fire started on October 19 from cooking fires in a camp of homeless people who thought their fires were extinguished and left the site around midnight. The next day was hot with high winds, and embers left from the night before reignited the fire. Urbanization had added additional fuel to the area, which was previously composed of grass-covered slopes with a small number of oak and redwood trees. The additional fuel included numerous wood structures and imported trees, such as eucalyptus. For decades prior to the fire, the density of trees and structures had increased. Open land was reduced from 47 percent of the area in 1939 to 20 percent in 1988, and the number of plant species increased from a few to a few hundred. When the fire reignited, it quickly became uncontrollable. All that firefighters could do was evacuate residents. The fire moved through the urban landscape very quickly; during the first hour, it consumed a home every five seconds! Although the fire was started by people, other factors contributed: an ample fuel supply of buildings, brush, and trees and hot, windy weather. Furthermore, from a fire hazard perspective, land-use planning in the area was inadequate. Structures were not required to be fire resistant, the density of buildings and placement of utilities with regard to fire hazard were not restricted, and there were no legal requirements for removal of excess vegetation around buildings.[2]

Λ FIGURE 13.18 Killer Wildfire
A wildfire in the hills of Oakland, California, in 1991 devastated an entire neighborhood, killing 25 people and destroying about 3,800 homes. Damages were more than $1.68 billion. *(AP Photo/ Olga Shalygin)*

ADJUSTMENTS TO THE WILDFIRE HAZARD

Adjustment to the fire hazard may be accomplished through hazard mapping, fire danger alerts and warnings, education, codes and regulation, insurance, and evacuation. When wildfire breaks out, the mode shifts to firefighting, which is a coordinated, cooperative effort that, with large fires, involves firefighter units from across the region converging on the fire. Fire trucks may come from hundreds of miles away to assist in controlling a fire and protecting property. Firefighting is a dangerous profession.

Wildfire Hazard Maps Maps in California and other areas routinely indicate areas (zones) with moderate to very high wildfire hazard. These are often at the wildland–urban interface. Zones with very high hazard are subject to rules, regulations, and building standards. The Santa Rosa Fire suggests these zones may be too narrow. Hazard maps should also have to reflect dense urban development in areas adjacent to areas previously thought to have a mild to low hazard.

Fire Danger Alerts and Warnings Both U.S. and Canadian federal agencies have developed fire danger rating systems to alert land managers, residents, and visitors to daily changes in conditions affecting the development of wildfires. Fire danger ratings combine information about fuel conditions, topography, weather, and risk of ignition to assess the wildfire hazard. National Weather Service forecast offices also issue fire weather watches 12 to 72 hours in advance of extreme fire conditions and **red flag warnings** when extreme fire conditions either are occurring or will take place in less than 24 hours. These watches and warnings put citizens, public officials, and firefighters on alert.

Fire Education Community awareness programs and presentations in schools about fire safety may help reduce the fire hazard. Unfortunately, for many people who have never experienced a fire, the risk may not seem "real," even with the help of fire education programs.

Codes and Regulations One way to reduce risk in fire-prone areas is to enforce building codes that require structures be built with fire-resistant materials. For example, using stone or brick for building and roofing materials, rather than wooden shingles, helps to fireproof homes. Unfortunately, in California, stone roofing tiles are not always safe during earthquakes. Making appropriate fire-resistant materials mandatory in new structures could significantly reduce the amount of damage caused by fires. Many local governments and state/province and federal land management agencies also exercise their regulatory authority by issuing bans on outdoor burning and fireworks when conditions are favorable for grass, brush, and forest fires. These bans help reduce the number of fires ignited by people.

Fire Insurance Fire insurance is another adjustment to the threat of fires. Such insurance ensures that people whose property has been destroyed by a fire will be reimbursed. However, insurance may provide a false sense of security and prompt more people to live in fire-prone areas.

Evacuation Evacuation of people from danger zones is probably the most common adjustment to the fire hazard. Evacuation of people from homes helps ensure their safety, but it does not protect their homes or personal belongings. In some areas, reverse 911 calls warn people that a fire has started and update them with advised or required evacuations. The role of social media is becoming more important in people getting real-time fire conditions and warnings and evacuation options that are open. Evacuation of a large number of people is often a difficult process, and community-planning groups work on this issue, so that when evacuation is necessary, it can possibly be done in a more orderly manner. Many people may wait until the last moment, and that can be dangerous to property owners and firefighters entering a fire zone.

News reporters on the ground who are filming the fire and burning homes and those in the air who are covering the story may also be in personal danger and may impede firefighters. Risking the lives of reporters should not be a part of covering the news!

PERSONAL ADJUSTMENT TO THE FIRE HAZARD

There are many things you can do to protect yourself from fire (Table 13.1). If you live or work in the wildland–urban interface, you may want to obtain literature that describes how to prepare for the inevitable wildfire.[35]

▼ TABLE 13.1 Reducing Your Fire Hazard at Home

Maintain Home Heating Systems

- Have your chimney regularly inspected and cleaned.
- Remove branches hanging above and around the chimney.

Have a Fire Safety and Evacuation Plan

- Install smoke alarms on every level of your home.
- Test smoke alarms monthly and change the batteries at least once a year.
- Practice fire escape and evacuation plans.
- Mark the entrance to your property with signs that are clearly visible.
- Know which local emergency services are available and have those numbers posted.
- Provide emergency vehicle access through roads and driveways at least 3.7 m (~12 ft.) wide with adequate turnaround space.

Make Your Home Fire Resistant

- Use fire-resistant and protective roofing and materials such as stone, brick, and metal to protect your home. Avoid using wood materials, which offer the least fire protection.
- Keep roofs and eaves clear of debris.
- Keep windows and damper to the chimney tightly closed to keep smoke out.
- Cover all exterior vents, attics, and eaves with metal mesh screens having openings no larger than 6 mm (~0.2 in.)
- Install multiplane windows, tempered safety glass, or fireproof shutters to protect large windows from radiant heat.
- Use fire-resistant draperies for added window protection.
- Keep tools for fire protection nearby: 30 m (~100 ft.) of garden hose, shovel, rake, ladder, and buckets.
- Make sure that water sources, such as hydrants and ponds, are accessible to the fire department.

Let Your Landscape Defend Your Property

- Trim grass on a regular basis up to 30 m (~100 ft.) surrounding your home.
- Create defensible space by thinning trees and brush within 10 m (~30 ft.) around your home.
- Beyond 10 m (~30 ft.), remove dead wood, debris, and low tree branches.
- Landscape your property with fire-resistant grasses and shrubs to prevent fire from spreading quickly.
- Stack firewood at least 10 m (~30 ft.) away from your home and other structures.
- Store flammable materials, liquids, and solvents in metal containers outside the home, at least 10 m (~30 ft.) away from structures and wooden fences.

Source: Based on Federal Emergency Management Agency, "Wildfire: Are you Prepared?" FEMA Publication L-203, 1993. www.usfa.fema.gov/downloads/pdf/publications/wildfire.pdf.

13.7 CHECK your understanding

1. List ways in which communities can make adjustments to the wildfire hazard.
2. Suggest several ways that individuals can prepare for the wildfire hazard.

13.2 Survivor Story

Elderly Couple Survives Wildfire by Jumping into a Neighbor's Swimming Pool

The evening sky at 10:00 P.M. on Sunday October 8, 2017 was clear on the hills above Santa Rosa, California, and an elderly woman was on her deck, enjoying the view in a home she and her husband had built. After checking her tomatoes and enjoying the moonlit night, she took a shower, and when she got out, she could smell smoke. Her husband went outside, and the sky was red, but he thought it was probably just the sunset, and the couple got into bed. A phone call came from their daughter who advised them to evacuate because there was a wildfire in the area, but they didn't make a move at that time. The daughter called again at midnight, advising them that they had to get out. They looked outside, and the wind had kicked up and flames were approaching. They loaded their car and tried driving from the home, but there was a wall of flames and they had to return. They then had to make a decision of what would be a likely plan to survive. No one plans for such emergencies, but sometimes they happen. They called 911, but

help was not available and likely not to come until the flames had swept through the area. They were trapped, but they remembered that their neighbor had a swimming pool, and they wondered if they could possibly survive in the pool, still thinking that someone would come to get them soon. The 911 operator told them to get to a safe place—the pool was the only choice as flames raged through the area. At first, they hesitated getting in the pool because the water was cold and the fire wasn't quite there. But then the fire began licking on some wood near the pool. The heat was intense, so they jumped in. They were lucky that the pool was only 4 feet deep, so they didn't have to tread water, but it was cold as they clung to each other for warmth. Cold water that lowers body temperature can result in hypothermia which is a medical emergency that can kill people, especially the elderly. They wondered how long it was going to take for the house to burn and when they could get out of the water. The entire neighborhood (homes, cars, and trees) was on fire. They had to keep going

under the surface to avoid all the soot and ash and burning embers in the air and falling in the pool. They'd go underwater for a bit, then come up, take a breath, and go back down. Hours later, when the neighbor's home had burned but was still warm, they decided to get out and warm up because the flames had moved on. They looked for the phone that the wife had left in her shoe along the edge of the pool. It had melted! The couple then walked back to their own home, which was just a short distance away, and it was completely burned, along with their vehicles. In the meantime, their daughter was still trying to call but couldn't get through and prepared for the worst as she had heard that there had been numerous deaths from the wildfire. About 8:00 A.M. the next morning, officials were able to tell their daughter that they had survived the fire. They were extremely lucky.

(ABCARIAN, *R. THEY SURVIVED SIX HOURS IN A POOL AS A WILDFIRE BURNED THEIR NEIGHBORHOOD TO THE GROUND.* LOS ANGELES TIMES, *OCTOBER 12, 2017.*)

APPLYING the **5** Fundamental Concepts

Santa Rosa Fire of 2017

1 Science helps us predict hazards.

2 Knowing hazard risks can help people make decisions.

3 Linkages exist between natural hazards.

4 Humans can turn disastrous events into catastrophes.

5 Consequences of hazards can be minimized.

Wildfires in North America are apparently getting larger and more intense and are causing more damage to forest resources and homes inhabited by people living in the nation's wildlands. Wildfires that were considered to be unusually large in size and intensity 40 years ago are no longer thought to be surprising events. In addition, droughts are expected to be longer and more severe in coming decades. If vegetative cover is reduced by more frequent fires, and less frequent but more intense precipitation occurs, there may be an increase in flash flooding and increased erosion and movement of sediment to rivers and streams. Debris flows may become more common as sediment deposited in channels is mobilized by less frequent but more intense precipitation following wildfire.

Although the wildlands in the western United States are public lands and generally undeveloped, a larger percentage of the private land that is available for development is being turned into suburban housing estates. If the trend toward increased development continues, then changes in fire policy and management to further reduce the fire risk will be necessary.

Since 1960, the six most serious fire seasons have occurred since 2000, and the total amount of land burned, is about 1.5 times the size of the entire state of Oregon. During the next century,

the total U.S. land area burned per year may double.[36]

Larger, more intense wildfires result from several factors:[36,37]

> Biomass fuels (trees and brush) have increased in the past few decades as a result of land management practices that include overgrazing, which reduces ground cover (grass) while encouraging the growth of young trees.

> Timber harvesting (logging) of the largest pine trees has resulted in less fire-tolerant understory (smaller trees).

> Fire suppression has eliminated the smaller, less intense fires that previously reduced the amount of biomass (fire fuel) available.

> Higher temperatures and drought are widespread in recent years, perhaps related to global climate changes.

> Warmer summer and winters have occurred in North America. As a result, runoff is quicker and soil dries out earlier. When the heat of spring and summer arrives, vegetation is drier, which increases the fire hazard.

> While climate change can't be blamed for any one fire, the trend for larger, more intense fires is clearly expected with warming.

> In recent years as spring growth is extended, there has been a noted increase in insect and disease occurrences that can weaken trees.

> The fire season has grown in length by at least two months in recent decades.

> Continued human use and interest in the natural, rural forest environment of the North America has resulted in more homes and other structures being built at

∧ FIGURE 13.19 The 2017 Santa Rosa Fire
The fire rapidly moved from home to home. This urban development with high density of homes was well out of the identified
high fire hazard zone and, therefore, not subject to rules and
regulations. That fire hazard zones should be expanded based
on new criteria is being explored. *(California National Guard)*

the wildland–urban interface (previously
undeveloped land close to towns and cities).
Depending on the density of homes, these
homes do not necessarily change the fire
hazard (although some fires are started by
carelessness or arson); they do increase
the risk as defined in part by the product of
the probability of a fire occurring and the
consequences should a fire actually occur.
Building additional homes in areas with fire
hazards obviously also increases the risk
through greater exposure and vulnerability
that leads to greater financial losses from
wildfire. The Santa Rosa Fire in 2017 is
particularly worrisome because thousands of
homes (Figure 13.19) were burned in areas
previously thought to be outside the fire
danger area. The fire will require rethinking
how we define the wildland–urban interface.

As people responsible for fire safety and risk
reduction become more familiar with large,
damaging wildfires, better evacuation and

warning systems are being developed. When
wildfire breaks out, authorities are quicker
to call in emergency firefighting assistance,
including aircraft support that drops fire
suppressants.

During recent large wildfires, authorities
generally moved quickly, sending out telephone
messages and sending deputies door-to-door
to make sure that everyone had evacuated the
area. In Santa Rosa the fire moved so quickly
that little warning was possible. People in their
homes were completely surprised as they
believed they lived in a much more fire-safe
area. Social media will take on a larger roll
in real-time warning in future fires. A quick
response is absolutely necessary, as large forest
fires can double in size in just a few hours.
However, with fast-moving fires in hot and
windy conditions, there is always the potential
for loss of life. There is growing concern that
the largest and most intense wildfires are
so dangerous that we will just have to let
them burn rather than send firefighters into
dangerous, unpredictable situations.

APPLYING THE **5** FUNDAMENTAL CONCEPTS

1. Based on what you have learned in this chapter
   about wildfires, can the wildfire hazard be
   predicted?

2. How does the assumption that future
   wildfires will be larger and more intense
   in the future affect state fire management
   agencies and the decisions that developers
   make?

3. How are wildfires semiarid and other regions
   linked to soil erosion and landslides?

4. Explain how human processes and decisions
   are changing the risk and economic impact of
   wildfires in North America

5. List ways that loss of lives and property from
   the recent wildfires (United States and Canada)
   might have been further minimized.

CONCEPTS in review

13.1 Introduction to Wildfire

LO:1 Summarize the role of wildfire during the past 11,500 years.

- Wildfire, one of nature's oldest natural processes, is a self-sustaining, rapid, high-temperature, biochemical oxidation reaction that releases heat and light. Most wildfires in natural ecosystems maintain a rough balance between plant productivity and decomposition.

13.2 Wildfire as a Process

LO:2 Explain why wildfire is a natural process that would occur with or without people.

KEY WORDS
combustion (p. 532), **pyrolysis** (p. 532), **wildfire** (p. 531)

- Wildfire is a natural process that can be defined as a rapid, self-sustaining, high-temperature biochemical reaction that releases heat, light, gases, and other products. Wildfire, by burning organic matter, is part of the carbon cycle that returns carbon dioxide to the atmosphere.
- For a fire to burn, it must have fuel, oxygen, and heat. The two main processes that generate wildfires are preignition and combustion. Preignition involves heating and the pyrolysis of fuel to drive off moisture and break down large carbon molecules into smaller ones. The smaller molecules create a cloud of flammable gas directly above the fuel, which then ignites. Ignition often starts with a lightning strike and then continues with windblown embers from the existing fire. Combustion typically occurs first by flaming, followed later by glowing and smoldering.
- Wildfires can be classified based on what part of the landscape burns. Surface fires burn along the forest floor or across the surface of brush and grasslands. Ground fires burn beneath the forest floor by smoldering. In forests, swamps, and marshes with organic soils, ground fires can smolder in peat deposits for many months. Fast-moving crown fires begin when surface fires ignite treetops. Spot fires are ignited ahead of the main fire by embers carried on the winds.
- Wildfire behavior is influenced by fuel, weather, topography, and the fire itself. Fuel varies in size, shape, arrangement, and moisture content, ranging from fast-burning grasses and conifer needles to slow-burning logs and organic soil matter. Weather conditions favoring wildfires include high winds and high temperatures, low humidity, and dry thunderstorms. Longer-term drought conditions are especially favorable for wildfires. Fires spread rapidly up steep slopes and are driven by winds up canyons.

13.3   Geographic Regions at Risk from Wildfires

LO:3   Explain why the wildfire hazard varies from region to region.

- Most areas of the United States and Canada that are within or in close proximity to grasslands, shrublands, woodlands, or tundra are at risk for wildfires during dry weather or drought conditions. Even the sparsely vegetated deserts, rainforests, and normally wet marsh and swamplands can experience wildfire.

13.4   Effects of Wildfires and Linkages with Other Natural Hazards

LO:4   Summarize how wildfires are linked to other natural hazards.

LO:5   Make the link between global change (warming) and the number and intensity of wildfires.

- Fire can increase runoff, erosion, flooding, and landslides.
- Wildfire is linked to air pollution, reduction of water quality, and vegetation changes through damage and fire adaption.
- Wildfire occurrence and intensity are linked to global warming through drought, increasing surface temperature, and extended fire seasons.

KEY WORDS
dry ravel (p. 538), **hydrophobic layer** (p. 536), **sediment flushing** (p. 538), **wildland–urban interface** (p. 544)

13.5   Natural Service Functions of Wildfires

LO:6   List the potential benefits provided by wildfires.

- Natural service functions of fires include increasing the nutrient content of soils, initiating regeneration of plant communities, creating new habitat for animals by altering landscapes, and potentially reducing the risk of large fires in the near future.

13.6   Minimizing the Wildfire Hazard

LO:7   Evaluate the methods employed to minimize the fire hazard.

- Wildfire management is difficult because large wildfires cannot be prevented or, in some instances, even controlled.
- In recent decades, there has been a move to suppress wildfire. As a result, natural fires that occur may become larger and more intense. From that perspective, events that were previously disasters may now be catastrophes. The same change from disaster to catastrophe can occur because more and more people are living in the wildland–urban interface in areas with a high fire hazard and increasing the impacts of wildfire in the coming decades.

KEY WORDS
fire regime (p. 546), **prescribed burns** (p. 546)

13.7   Perception of and Adjustment to the Wildfire Hazard

LO:8   Summarize the potential adjustments to the wildfire hazard and the role of human processes.

- Adjustments to wildfire at the institutional level include establishing alerts and warnings, firefighting by fire crews, quick evacuation, and education.
- Most important in fire management is for homeowners to utilize fire-resistant construction, develop a defendable space around their homes, and be prepared to evacuate. Emergency planning at the community level is also important.

KEY WORD
red flag warnings (p. 548)

CRITICAL thinking questions

1. You live in an area with a significant wildfire hazard. What can you to do protect your home and belongings from fires? Make a list of actions you can take to protect yourself.
2. A large national park is reviewing its fire policy. As a wildfire expert, you have been asked for advice. The park's current policy is to suppress all fires as soon as they begin. It does not use prescribed burns and is considering switching to a policy of allowing natural burns. What would you suggest? List the pros and cons of each policy before making your decision.
3. Most discussion of the wildfire hazard focuses on the potential destruction, injury, or death that can take place from the flames. Discuss the hazards to humans and the environment that come from the smoke produced by wildfires.

References

1. **Rossotti, H.** 1993. *Fire*. Oxford, UK: Oxford University Press.
2. **Pyne, S. J., Andrews, P. L.,** and **Laven, R. D.** 1996. *Introduction to wildland fire*, 2nd ed. New York: John Wiley & Sons, Inc.
3. **Minnich, R. A.** 2002. Personal written correspondence.
4. **Ward, D.** 2001. Combustion chemistry and smoke. In *Forest fires: Behavior and ecological effects*, eds. E. A. Johnson and K. Miyanishi, pp. 57–77. San Diego, CA: Academic Press.
5. **Mennich, R. A.** and **Chou, Y. H.** 1997. Wildland fire patch dynamics in the chaparral of southern California and nothern Baja California. *Int J. of Wildland Fire* 7(3): 221–248.
6. **Arno, S. F.,** and **Allison-Bunnell, S.** 2002. *Flames in our forests*. Washington, DC: Island Press.
7. **Verdi, R. J., Tomlinson, S. A.,** and **Marella, R. L.** 2006. *The drought of 1998–2002: Impacts on Florida hydrology and landscapes*. U.S. Geological Survey Circular 1295.
8. **Florida Division of Forestry.** 2006. Wildland fire; Fire data reports. http://tlhforweb1.doacs.state.fl.us/PublicReports/. Accessed 1/28/07.
9. **New York Times.** 2001, May 31. Rain clears some smoke, but fires persist in Florida. Sec. A, p. 14.
10. **Kaufman, Yoram J.,** and **seven others**. 1998: Potential global fire monitoring from EOS-MODIS. *Journal of Geophysical Research*, *103*, pp. 32, 215–38.
11. **U.S. Geological Survey.** 2006. *Wildfire hazards—A national threat*. U.S. Geological Survey Fact Sheet 2006–3015.
12. **Canadian Forest Service.** 2006. Canadian Wildland Fire Information System: Historical analysis—Large fire database. http://cwfis.cfs.nrcan.gc.ca/en/historical/ha_lfdb_maps_e.php. Accessed 1/29/07.
13. **Chandler, C., Cheney, P., Thomas, P., Trabaud, L.,** and **Williams, D.** 1983. *Fire in forestry, Volume I: Forest fire behavior and effects*. New York: John Wiley & Sons.
14. **DeBano, L. F.** 1981. *Water repellent soils: A state-of-the-art*. U.S. Forest Service General Technical Report INT-79.
15. **U.S. Geological Survey.** 1998. *USGS wildland fire research*. U.S. Geological Survey Fact Sheet 125–98.
16. **Florsheim, J. L., Keller, E. A.,** and **Best, D. W.** 1991. Fluvial sediment transport in response to moderate storm flows following chaparral wild fire, Ventura County, southern California. *Geological Society of America Bulletin* 103: 504–11.
17. **Keller, E. A., Valentine, D. W.,** and **Gibbs, D. R.** 1997. Hydrologic response of small watersheds following the southern California Painted Cave Fire of June 1990. *Hydrological Processes* 11: 401–04.
18. **Keeley, J. E., Fotheringham, C. J.,** and **Morais, M.** 1999. Reexamining fire suppression impacts on brushland fire regimes. *Science* 284: 1829–32.
19. **Minnich, R. A.** 1983. Fire mosaics in southern California and northern Baja California. *Science* 219: 1287–94.
20. **Minnich, R. A., Garbour, M. G., Burk, J. H.,** and **Sosa-Ramiriz, J.** 2000. California mixed-conifer forests under unchanged fire regimes in Sierra San Pedro Martir, Baja California, Mexico. *Journal of Biogeography* 27: 105–29.
21. **Huff, A. K.,** and **Kondragunta, S.** 2017. Meteorologists track wildfires using satellitesmoke images. *Eos* 98. April.
22. **Beitler, J.** 2006. Tracking nature's contribution to pollution. NASA Earth Observatory. http://earthobservatory.nasa.gov/Study/ContributionPollution/. Accessed 1/29/07.
23. **Pfister, G., Hess, P. G., Emmons, L. K., Lamarque, J.-F., Wiedinmyer, C., Edwards, D. P., Pétron, G., Gille, J. C.,** and **Sachse, G. W.** 2005. Quantifying CO emissions from the 2004 Alaskan wildfires using MOPITT CO data. *Geophysical Research Letters* 32: L11809.
24. **Westerling, A. L., Hidalgo, H. G., Cayan, D. R.,** and **Swetnam, T. W.** 2006. Warming and earlier spring increase western U.S. forest wildfire activity. *Science* 313: 940–43.
25. **Kasischke, E. S.,** and **Turetsky, M. R.** 2006. Recent changes in the fire regime across the North American boreal region—Spatial and temporal patterns of burning across Canada and Alaska. *Geophysical Research Letters* 33: L09703.
26. **Stocks, B. J. and five others.** 2001. Boreal forest regimes and climate change. In *Remote sensing and climate modeling: synergies and limitations*, ed. M. Beniston, and M. M. Verstraete, pp. 233–46. Dordrecht, Netherlands: Kluwer Academic.
27. **Westerling, A. L., Cayan, D. R., Brown, T. J., Hall, B. L.,** and **Riddle, L. G.** 2004. Climate, Santa Ana winds and autumn wildfires in southern California. *EOS, Transactions, American Geophysical Union* 85(31): 294, 296.
28. **Tolmé, P.** 2004. Will global warming cause more wildfires? *National Wildlife* 42(5): 14–16.
29. **Fried, J. S., Torn, M. S.,** and **Mills, E.** 2004. The impact of climate change on wildfire severity: A regional forecast for northern California. *Climatic Change* 64: 169–91.
30. **Barber, C. V.,** and **Schweithelm, J.** 2000. *Trial by fire: Forest fires and forestry policy in Indonesia's era of crisis and reform*. Washington, DC: World Resources Institute.

31. **Turetsky, M. R., Harden, J. W., Friedli, H. R., Flannigan, M., Payne, N., Crock, J.,** and **Radke, L.** 2006. Wildfires threaten mercury stocks in northern soils. *Geophysical Research Letters* 33: L16403.

32 **Miller, M.** 2000. Fire autecology. In *Wildland fire in ecosystems: Effects of fire on flora*, ed. J. K. Brown, J. Smith, and J. Kapler. U.S. Forest Service General Technical Report RMRS-GTR-42-vol. 2.

33. **Fuller, M.** 1991. *Forest fires: An introduction to wildland fire behavior, management, firefighting and prevention.* New York: John Wiley.

34. **Jones, S. R.,** and **Cushman, R. C.** 2004. *A field guide to the North American prairie*. Boston: Houghton Mifflin.

35. **Arrowood, J. C.** 2003. *Living with wildfires: Prevention, preparation, and recovery.* Denver, CO: Bradford Publishing Company.

36. **U.S. Department of Agriculture, Forest Service.** 2012. Effects of climatic variability and change on forest ecosystems: A comprehensive science synthesis for the U.S. Forest Sector. General Technical Report PNW-GTR-870.

37. **Gorte, R.** 2013. The rising cost of wildfire protection. http://headwaterseconomics.org/wildfire/fire-costs-background/

No one knew the object was coming. Over 7,000 buildings in **six cities across the region were** damaged

14

APPLYING the **5** fundamental concepts Chelyabinsk Meteor, February 2013

① Science helps us predict hazards.

The potential hazard from extra-terrestrial impact is well known; NASA scientists and meteor astro-physicists from around the world currently track more than 10,000 objects in our solar system that could potentially impact Earth in the future, like the one that passed by the same day as the 2013 Chely-abinsk meteor.

② Knowing hazard risks can help people make decisions.

Earth is bombarded by thousands of tiny extraterrestrial objects every day, but the effect of these impacts is negligible, so the risk is low. In con-trast, a large extraterrestrial impact would be globally catastrophic, but the probability of such an impact is extremely low and so is the risk.

Impacts and Extinctions

Chelyabinsk Meteor, February 2013

Imagine driving to work early one morning, and observing through your car window a small streak of light moving across the sky that explodes in a flash more brilliant than the sun. On February 15, 2013, this scenario was recorded on countless car-mounted security cameras all over northwestern Russia, when a small asteroid traveling about 20 times faster than a speeding bullet entered Earth's atmosphere and exploded (chapter-opener photograph). People living in Chelyabinsk (Figure 14.1), the city closest to the impact, could feel the intense heat from the fireball, as well as the shock wave from the explosion. Because the object entered Earth's atmosphere at a low angle and at incredible velocity, the meteor did not impact the Earth directly but rather exploded in the stratosphere about 23 km (~15 mi.) above the surface, releasing energy equivalent to 20 to 30 World War II–era nuclear explosions.[1] Such impacts are known as airbursts. The Chelyabinsk meteor is the largest known asteroid to have entered Earth's atmosphere since a much larger airburst in 1908, which also occurred in Siberia near the town of Tunguska (Figure 14.1).

◄ **2013 CHELYABINSK METEOR, RUSSIA**
A car-mounted camera captures the Chelyabinsk meteor exploding into in a fireball in the morning sky over Russia on February 15, 2013. *(AP Images)*

Earth has been bombarded by objects from space since its birth. Such impacts have been linked to the extinction of many species, including the large dinosaurs. The risk of impact from asteroids, comets, and meteoroids continues today. The hazard from extraterrestrial impact is very different from other hazards we have discussed in this book and for which we have the ability to mitigate damages. We do have the technology to identify large, 1.5 km (1 mi.) or greater, extraterrestrial objects that could impact Earth several decades or longer in the future. However, with present technology, we cannot minimize this hazard.

After reading this chapter, you should be able to:

LO:1 Differentiate between asteroids, meteoroids, and comets.

LO:2 Describe the physical processes associated with airbursts and impact craters.

LO:3 Suggest the possible causes of mass extinction and provide examples.

LO:4 Evaluate the evidence for the impact hypothesis that produced the mass extinction at the end of the Cretaceous Period.

LO:5 List the physical, chemical, and biological consequences of an impact from a large asteroid or comet.

LO:6 Analyze the risk of impact or airburst of extraterrestrial objects and suggest how that risk might be minimized.

3

Linkages exist between natural hazards.

Extraterrestrial impact is directly linked to many other hazards, which may vary depending on the size of the impact and whether the impact occurs in the ocean or on land.

4

Humans can turn disastrous events into catastrophes.

Humans have no control over the occurrence of extraterrestrial impact, but a large airburst or direct impact of a Chelyabinsk-sized meteor in a heavily populated area would be catastrophic.

5

Consequences of hazards can be minimized.

Consequences of an extraterrestrial impact can be understood, but our ability to minimize the effects depends on the magnitude of the event and our ability to predict when and where a damaging impact might occur. The two extraterrestrial events of February 15, 2013, illustrate both the potential and the shortcomings of our prediction capabilities.

^ FIGURE 14.1 Meteor Impact Locations
Locations of the Chelyabinsk event of 2013 and the Tunguska event of 1908, both in Russia (light brown).

Before the atmospheric entry and explosion, no one knew that the object was coming. Coincidentally, meteor astrophysicists from around the world at that time were, literally, looking the other way because they were tracking a much larger and completely unrelated asteroid (2012 DA14) that was to make a close fly-by of Earth just sixteen hours later. The Chelyabinsk meteor damaged more than 7,000 buildings in six cities across the region and injured about 1,500 people seriously enough to require medical treatment, mostly due to flying glass from windows broken by the shock wave. Following the explosion, people were confused and frightened about what was happening, and mobile phone networks were soon overloaded with calls. Office buildings in Chelyabinsk were evacuated, and schools were closed with parents rushing to pick up their children. In all, about 100,000 homes sustained broken windows, requiring immediate attention along with maintaining the region's central heating systems at a time of year when subzero temperatures are the norm.

14.1 Earth's Place in Space

Preston Cloud, a famous geologist, wrote in 1978:

> Born from the wreckage of stars, compressed to a solid state by the force of its own gravity, mobilized by heat of gravity and radioactivity, clothed in its filmy garments of air and water by the hot breath of volcanoes, shaped and mineralized by 4.6 billion years of crustal evolution, warmed and peopled by the Sun, this resilient but finite globe is all our species has to sustain it forever.[2]

Cloud was referring to the original evolution of our planet that started about 4.6 billion years ago. One can go back even further and consider the start of the universe—some 14 billion years ago. The starting point is thought to have been an explosion, known as the "Big Bang." This explosion produced the atomic particles that later formed galaxies, stars, and planets (Figure 14.2). The first stars formed within the first 1 billion years after the Big Bang, and stars continue to form today. A star's life span depends on its mass; large stars have higher internal pressure and burn up more quickly than small stars. Objects with a mass equal to about that of our sun last around 10 billion years, and large stars with masses of 100 times that of our sun last only about 100,000 years. Thus, our sun is now a middle-aged star, about halfway through its life span.

When one of the massive, short-lived stars dies, it does so in a particularly spectacular way—as a supernova. A *supernova* occurs when the star is no longer capable of sustaining its mass and collapses inward, resulting in a high-energy explosion that scatters its mass into the void of space, creating a vast *nebula*. Ultimately, the force of gravity wins out, and the matter within the nebula begins to collapse back inward on itself, forming new stars in what is called a *solar nebula* (Figure 14.3). It is from such a nebula that our sun formed 5 billion years ago. The sun grew by accretion of matter from a flattened pancake-like rotating disk that was dominated by hydrogen and helium molecules, along with trace amounts of dust composed of the other elements of the periodic table. As the solar nebula condensed under gravitational forces, our sun, which contains approximately 99.8 percent of the mass of the solar system, formed at the center; the remaining mass became trapped in solar orbits as rings, similar to the rings around the planet Saturn today. Gravitational forces from the largest, densest particles attracted other particles in the rings until they collapsed into the planetary system we have today (Figure 14.4). During their early history, all the planets in our solar system, including Earth, were bombarded by extraterrestrial objects that ranged from dust-sized particles to objects many kilometers in diameter. However, the bombardment did not cease 4.6 billion years ago; it has continued, although at a lesser rate, to the present.

ASTEROIDS, METEOROIDS, AND COMETS

Trillions of particles remain in our solar system. Astronomers group these particles on the basis of diameter and composition (Table 14.1). They range in size from interplanetary dust a fraction of a millimeter in diameter to larger bodies such as **asteroids** that range in diameter from about 10 m (~30 ft.) to 1000 kilometers (~620 mi.). For the most part, asteroids are

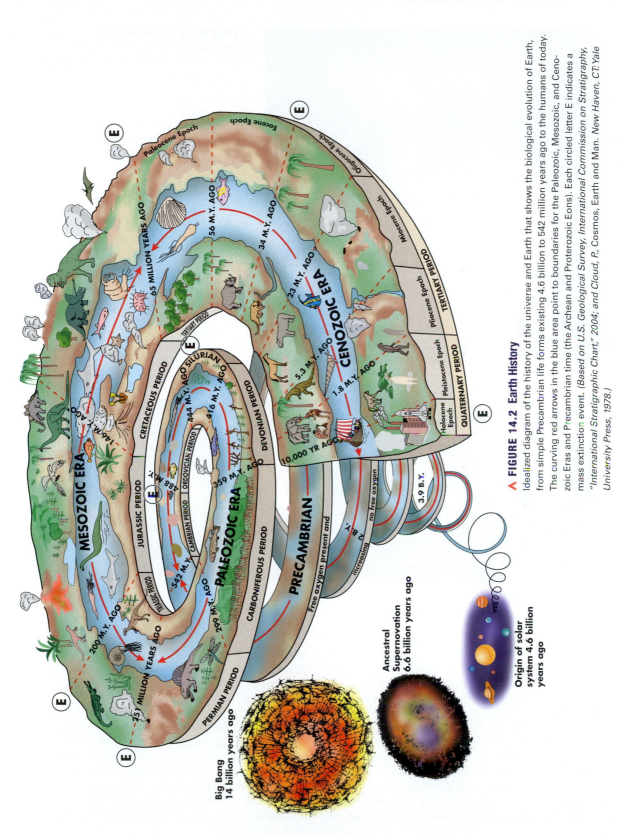

^ FIGURE 14.2 Earth History

Idealized diagram of the history of the universe and Earth that shows the biological evolution of Earth, from simple Precambrian life forms existing 4.6 billion to 542 million years ago to the humans of today. The curving red arrows in the blue area point to boundaries for the Paleozoic, Mesozoic, and Cenozoic Eras and Precambrian time (the Archean and Proterozoic Eons). Each circled letter E indicates a mass extinction event. *(Based on U.S. Geological Survey, International Commission on Stratigraphy, "International Stratigraphic Chart," 2004; and Cloud, P., Cosmos, Earth and Man. New Haven, CT: Yale University Press, 1978.)*

◄ FIGURE 14.3 Solar Nebula
Hubble Space Telescope image of the Trifid Nebula—a solar nebula within our own Milky Way Galaxy. The Trifid Nebula is named for the three bands of obscuring interstellar dust. Near the intersection of the dust bands, a group of recently formed bright stars is easily seen. The nebula is about 9,000 light years from Earth. *(Stocktrek Images, Inc./Alamy Stock Photo)*

ʌ FIGURE 14.4 Sizes in the Solar System
Artist's depiction of objects in our solar system, showing the relative sizes of the sun, planets, and Pluto (dwarf planet). *(NASA)*

found in the *asteroid belt*, which is a region between Mars and Jupiter (Figure 14.5). The asteroids, which are composed of rock material, metallic material, or rock-metal mixtures, would pose no threat to Earth if they remained in the asteroid belt. Unfortunately, they move around and collide with one another, and a number of them are now in orbits that intersect Earth's orbit. Asteroids sometimes break into smaller particles known as **meteoroids** (Table 14.1), which range in size from dust particles to objects a few meters in diameter. When a meteoroid enters Earth's atmosphere, it is known as a **meteor**. Frictional heating of meteors

as they streak through the atmosphere produces light, so meteors are sometimes called *shooting stars*. Meteors occurring in large numbers produce the familiar *meteor showers*.

The last type of particle is a **comet**. Comets are distinguished from meteoroids and asteroids by their glowing tail of gas and dust (Figure 14.6). They range in size from a few meters to a few hundred kilometers in diameter, and they are thought to be composed of a rocky core surrounded by ice and covered in carbon-rich dust. In addition to frozen water, the ice contains carbon dioxide (dry ice), carbon monoxide, and smaller

▼ TABLE 14.1 Meteorites and Related Objects

Type	Diameter	Composition	Comments
Asteroid	10 m to 1000 km	Stony or metallic	Strong and hard if made of metal or some rock types; hard types may hit Earth's surface, and weak types will explode in Earth's atmosphere at heights of several to hundreds of kilometers. Most originate in asteroid belt between Mars and Jupiter.
Comet	A few meters to a few hundred kilometers	Frozen water and/or carbon dioxide form an icy core that is surrounded by rock fragments and dust; like a "dirty snowball."	Weak, porous, will often explode in Earth's atmosphere at heights of kilometers to hundreds of kilometers. Most originate outside solar system in the Oort Cloud, 50,000 AU[a] from the sun, or within the solar system in the Kuiper Belt of comets; a tail, pointed away from the sun, forms as the icy core vaporizes and dust particles are shed from the object.
Meteoroid	Larger than dust size to under 10 m	Stony, metallic, or carbonaceous (containing carbon)	Most originate from collisions of asteroids or comets. May be made of strong or weak rock.
Meteor	Dust size to centimeters	Stony, metallic, carbonaceous, or icy	Destroyed in Earth's atmosphere as shooting stars; their light is produced by frictional heating in the atmosphere.
Meteorite	Centimeters to asteroid size	Stony, metallic, or carbonaceous	Object that has actually hit Earth's surface. Stony variety, called a *chondrite*, is most abundant.[b]

[a] AU is the distance from Earth to the sun, about 240,000,000 km (150,000,000 mi.).

[b] There are many types of chondrites. They contain chondrules, which are small (less than 1 mm) spheroidal inclusions that are glassy or crystalline. Planets are constructed from chondrite meteorites (asteroids).

Source: Data from Rubin, A. F. 2002. *Disturbing the solar system.* Princeton, NJ: Princeton University Press, 2002.

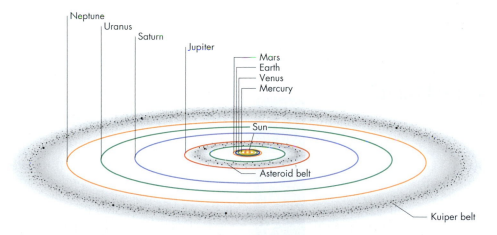

Λ FIGURE 14.5 Diagram of Our Planetary System Showing Asteroid Belt and Kuiper Belt
The Oort Cloud is too far away from the sun to be seen in this perspective. Orbits of the planets and belts are not to scale.

amounts of other compounds. As sunlight heats the comet, the "dirty ice" evaporates to form a mixture of gases. Comets are believed to have originated far out in the solar system, beyond the planet Neptune, and to have been thrown into an area called the Oort Cloud. The Oort Cloud lies beyond the Kuiper Belt and extends out as far as 50,000 times the distance from Earth to the sun.[3] Early in the history of the evolution of Earth, bombardment by asteroids and comets contributed the building blocks of our planet, which was built up from the collision of innumerable smaller bodies.

◀ FIGURE 14.6 Comet Hale-Bopp
Comet Hale-Bopp, pictured here in the night sky in 1997, remained visible to the naked eye for 18 months. Comets have a nucleus of ice and dust and a tail of evolved gases and dust grains. *(Aaron Horowitz/Corbis)*

14.1 CHECK your understanding

1. Explain how our solar system formed.
2. Describe the differences between an asteroid, meteor, comet, meteoroid, and meteorite.
3. What are meteorites and comets made of?
4. Where do comets and asteroids originate?

14.2 Airbursts and Impacts

When entering Earth's atmosphere, asteroids, comets, and meteoroids travel at a velocity of about 12 to 72 km per second (~7 to 45 mi. per second).[4] The composition of asteroids and meteoroids varies (see Table 14.1). Some contain carbonaceous material, whereas others are composed of native metals, such as iron and nickel. Others are stony, consisting of silicate minerals such as olivine and pyroxene—common minerals in igneous rocks. Stony meteoroids and asteroids are *differentiated*, meaning that they have undergone igneous, and sometimes metamorphic, processes as part of their geologic histories. As previously mentioned, meteoroids and asteroids are derived from the asteroid belt, between the orbits of Mars and Jupiter, whereas comets come from the Oort Cloud. Regardless of where they come from, when they intersect Earth's orbit and enter our atmosphere, meteoroids, asteroids, and comets undergo remarkable changes as they heat up due to atmospheric friction during descent and produce bright light. A meteoroid entering the atmosphere at about 85 km (~53 mi.) above Earth's surface will become a meteor and emit light (Figure 14.7). It will then either explode in an **airburst** in the atmosphere at an altitude between 12 and 50 km (~7 and 31 mi.), or collide with Earth.[5,6] Once a meteoroid actually strikes Earth, we speak of it as a **meteorite**. Many meteorites have been collected from around the world, particularly from Antarctica where they are concentrated because of particular meteorological conditions. An example of a modern era (1908) large airburst is the Tunguska event of 1908.

14.1 CASE study

The Tunguska Event

On June 30, 1908, shortly after 7 A.M., witnesses in Siberia (see Figure 14.1) reported observing a blue-white fireball with a glowing tail descending from the sky. The fireball exploded above the Tunguska River Valley in a heavily forested, sparsely populated area. Later, calculations would show that the explosion had the force of 10 megatons of TNT—equivalent to 10 hydrogen bombs. Although there were few witnesses close to the event, the sounds from the explosion were heard hundreds of kilometers away, and the blast wave was recorded at meteorological recording sites throughout Europe. A tremendous air blast caused more than 2000 km^2 (~770 mi.^{2}) of forest to be flattened and burned (Figure 14.8). The devastated area was more than twice the size of New York City.

A herdsman in the vicinity of the blast was one of the few people who witnessed the devastation on the

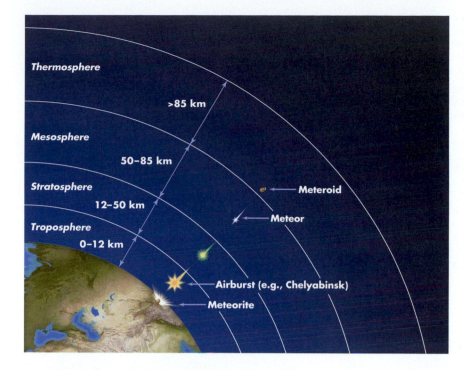

Idealized diagram showing a meteoroid's entry into the atmosphere, producing a light phenomenon called a meteor. Most meteors are extremely small dust- to sand-size meteoroids. A large meteoroid may break apart in an aerial burst or crash into Earth's surface. *(Modified after R. Baldini: www. th.bo.infn.it/tunguska/impact/ fig1_2jpg)*

∧ FIGURE 14.8 Tunguska Forest, Siberia, 1908
An aerial blast downed trees over an area of about 2000 km² (~770 mi.²). *(World History Archive/Newscom)*

ground. His hut was completely flattened by the blast, and its roof was blown away. Other witnesses a few tens of kilometers from the explosion reported that they were physically blown into the air and knocked unconscious; they awoke to find a transformed landscape of smoke and burning trees that had been blasted to the ground.

At the time of the explosion, Russia was in the midst of political upheaval. As a result, there was no rapid response or investigation of the Tunguska event. Finally, in 1924, geologists who were working in the region interviewed surviving witnesses and determined that the blast from the explosion was probably heard throughout an area of at least 1 million km² (~386,000 mi.²), an area the size of Texas and New Mexico combined! They also found that the fireball had been witnessed by hundreds of people. Russian scientists went into the area in 1927, expecting to find an impact crater produced by the asteroid that had apparently struck the area. Surprisingly, they found no crater, leading them to conclude that the devastation had been caused by an aerial explosion (airburst), probably at an elevation of about 7 km (~4.3 mi.). Later calculations estimated the size of the asteroid responsible for the explosion to be about 25 to 50 m (~80 to 160 ft.) in diameter. It was most likely composed of relatively friable (easy to crumble) stony material.[3,4]

The people on Earth were lucky that the Tunguska event occurred in a sparsely populated forested region. Given the rotation of Earth, if the asteroid explosion had occurred only a few hours later it would have threatened St. Petersburg and flattened it by direct impact and with a huge loss of human life. Worse yet, if the fireball had exploded over a large metropolis such as London, Paris, or Tokyo, millions of people would have died. Tunguska-type events are thought to occur on the order of once every 1,000 years.[5]

IMPACT CRATERS

The most direct and obvious evidence for impacts on the surface of Earth comes from studies of the **impact craters** they produce. More than 175 individual craters or small crater fields have been identified worldwide (Figure 14.9).[4,6,7] The 49,000-year-old Barringer Crater in Arizona is perhaps the most famous impact crater in the United States. Also known as "Meteor Crater," it is an extremely well-preserved, bowl-shaped depression with a pronounced, upraised rim (Figure 14.10a). The crater is approximately 1.2 km (~0.7 mi.) in diameter, with a depth of about 180 m (~590 ft.). The rim of the crater rises about 260 m (~850 ft.) above the surrounding Arizona desert. This rim is

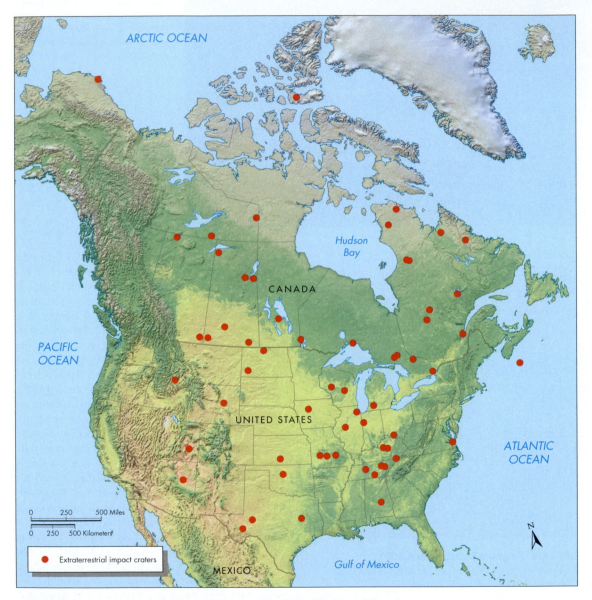

∧ FIGURE 14.9 Extraterrestrial Impact Craters of the United States and Canada
Map showing locations of individual craters that have been identified at the surface and in the subsurface. Not shown are more than a half million Carolina Bay craters that cover the coastal plain from Florida to New Jersey. Recently a young crater was discovered in northern Greenland. *(Based on Grieve, R. A. F., Impact Structures in Canada, St. Johns, Newfoundland and Labrador: Geological Association of Canada GEOtext 5, 2006; and Evans, K. R., Horton, Jr., J. W., Thompson, M. F., and Warme, J. E, "The Sedimentary Record of Meteorite Impacts: An SEPM Research Conference," The Sedimentary Record, March 2005, pp. 4–69)*

(a)

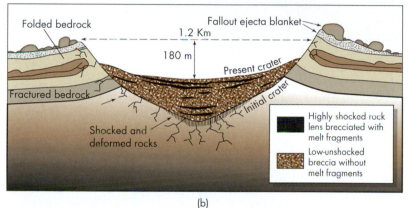

(b)

◀ **FIGURE 14.10 Simple Impact Crater in Arizona**

(a) Barringer Crater, Arizona (about 49,000 years old). The crater is about 1.2 km (~0.7 mi.) across and 180 m (~590 ft.) deep. *(Charles O'Rear/Corbis)* (b) Generalized cross-section of features associated with the Barringer Crater. Simple impact craters like this typically have raised rims and no central uplift or peak. *(Based on Grieve, R., and Cintala, M, "Planetary Impacts," in P. R. Weissman, L. McFadden, and T. V. Johnson, eds., Encyclopedia of the Solar System. San Diego, CA: Academic Press, 1999)*

Barringer Crater was later established through careful study and evaluation. This study concluded that the crater formed from the impact of a small asteroid, probably about 25 to 100 m (~80 to 330 ft.) in diameter.[8]

Features that form at the time of impact differentiate impact craters from craters that result from other processes, such as volcanic activity. Impacts involve extremely high velocity, energy, pressure, and temperatures, which are normally not experienced or produced by other geologic processes. Most of the energy of the impact is in the form of kinetic energy, or energy of movement. This energy is transferred to Earth's surface through a shock wave that propagates into Earth. The shock wave compresses, heats, melts, and excavates earth materials. It is this transfer of kinetic energy that produces the crater.[9] The shock can metamorphose some rocks in the impact area, while others are melted and mixed with the materials of the impacting object itself. Most of the metamorphism consists of high-pressure modifications of minerals, such as quartz. These modifications are characteristic of meteorite impact, so they are extremely helpful in confirming the origin of an impact crater.

Impact craters can be grouped into two types, simple and complex. *Simple craters* are typically small, a few kilometers in diameter. Barringer Crater has the features characteristic of a simple impact crater (Figure 14.10b). *Complex impact craters* experience the same processes of vaporization, melting, ejection of material, formation of ejecta rims, and later infilling typical of

overlain by a layer of debris, referred to as an *ejecta blanket*, which was blown out of the crater upon impact. Today, the ejecta blanket can be identified by the irregular terrain of mounds and depressions that surrounds the crater. The crater we see at the surface today is not nearly as deep as the initial impact crater (Figure 14.10b). Material fragmented by the impact has fallen back into the crater, and some of the walls have collapsed. Rocks that the asteroid hit were shattered and deformed, and the angular, broken pieces have been naturally cemented or fused together to form a rock type known as *breccia*.

When the existence of the Barringer Crater became widely known in the late nineteenth century, there was considerable debate concerning its origin. Ironically, G. K. Gilbert, the famous geologist who postulated that the majority of the moon's craters were formed by impacts, did not believe that Barringer Crater had been formed by impact. The impact origin for the

(a)

< **FIGURE 14.11 Chesapeake Bay Impact Crater**
(a) Map showing 35.5-million-year-old impact crater at the mouth of Chesapeake Bay, formed during the Eocene Epoch of the Paleogene Period. Approximate shoreline of the Atlantic Ocean during the Eocene is shown as a solid blue line between the Appalachian Mountains and the labeled U.S. cities. The blue dashed line offshore is the seaward edge of the continental shelf. (b) Note that there is about a 13× vertical exaggeration in the cross-section. The crater is approximately 85 km (~53 mi.) in diameter and 1.3 km ~0.8 mi. deep. Most of the crater is filled with angular rock fragments that have been cemented together to form a rock called breccia. This buried crater was discovered by the U.S. Geological Survey during a groundwater study. *(From Williams, S., Barnes, P., and Prager, E. J., U.S. Geological Survey Circular 1199, 2000)*

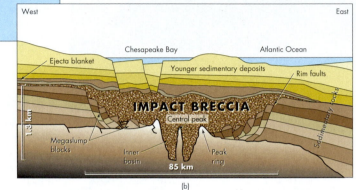

(b)

simple craters, but the shape of a larger, complex crater may be quite different (Figure 14.11). During a period of seconds to several minutes following impact, complex craters may grow to sizes of tens of kilometers to more than 100 km (~60 mi.) in diameter. In these craters, the rim collapses more completely, and the central crater uplifts following the impact. Typically, impact craters on Earth larger than about 6 km (~4 mi.) are complex, whereas smaller craters tend to be of the simple variety.

Geologically, ancient impact craters are difficult to identify because they are commonly either eroded or filled with sedimentary deposits that are younger than the impact. For example, subsurface imaging and drilling below the present Chesapeake Bay has identified a crater about 85 km (~53 mi.) in diameter, now buried by about 1 km (~0.6 mi.) of sedimentary deposits (Figure 14.11). The crater was produced by the impact of a comet or asteroid about 3 to 5 km (~2 to 3 mi.) in diameter about 35.5 million years ago.[10] Compaction and faulting of earth materials above the buried crater may be, in part, responsible for the location and shape of Chesapeake Bay.

A good example of an eroded impact crater, the Manicouagan Crater, can be found northeast of Quebec (Figure 14.12). At this crater, a ring-shaped reservoir about 65 km (~40 mi.) across fills a glacially eroded valley in the impact breccia. The lake outlines

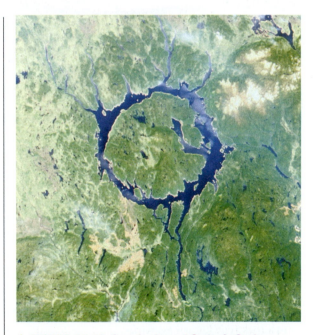

∧ **FIGURE 14.12 Complex Impact Crater in Quebec**
Satellite image of the Manicouagan impact structure northeast of Quebec City in Quebec, Canada. Complex craters typically have a faulted rim and central peak. This type of crater develops when the impact is so large that Earth's crust rebounds from the impact, forming the central peak. The ring-shaped lake is about 65 km (~40 mi.) in diameter. *(NASA)*

▼ **TABLE 14.2** Notable Impacts and Airbursts of Extraterrestrial Objects

Age[a]	Feature/Event	Location	Significance
4,450,000,000	Birth of the moon	Earth	Mars-size planetary body collides with Earth to form the moon.
2,023,000,000	Vredefort Dome	South Africa	Earth's oldest and largest terrestrial impact crater.
1,850,000,000	Sudbury Crater	Ontario, Canada	Earth's second largest terrestrial impact crater; rich in nickel and copper ores.
214,000,000	Manicouagan Crater	Quebec, Canada	Tied with Popigai Crater for fourth largest terrestrial impact crater.
64,980,000	Chicxulub Crater	Below Yucatán Peninsula and Gulf of Mexico	Contributed to extinction of large dinosaurs; produced enormous tsunamis; third largest terrestrial impact crater.
35,700,000	Popigai Crater	Siberia, Russia	Tied with Manicouagan Crater for fourth largest terrestrial impact crater; has produced numerous industrial-grade diamonds.
35,500,000	Chesapeake Bay Crater	Buried below Virginia, United States	Largest U.S. impact crater; affects regional groundwater flow.
49,000	Barringer Crater (aka Meteor Crater)	Arizona, United States	First impact crater identified on Earth; training site for Apollo astronauts.
12,800	Younger Dryas Airburst (hypothesis)	Southern Canada	Contributed to extinction of large Pleistocene mammals and the Clovis civilization.
1908	Tunguska Airburst	Siberia, Russia	Destroyed millions of trees
2013	Chelyabinsk Airburst	Russia	Injured about 1,500 people.

[a] Years before 1950, except for 1908 and 2013 events.

Source: Based on Grieve, R. A. F., *Impact Structures in Canada*. St. Johns, Newfoundland and Labrador: Geological Association of Canada, GEOtext 5, 2006.

most of the crater, which is estimated to have been originally about 100 km (~62 mi.) in diameter. One of the five largest terrestrial impact craters in the world, the Manicouagan Crater is close to 215 million years old (Table 14.2).[6]

Most detailed remote sensing studies of craters have been conducted on craters on the moon and Mars. Impact craters are much more common on the moon than on Earth for three reasons:

1. Most impact sites on Earth are in the ocean where craters are subsequently buried by marine sediment or destroyed by plate tectonic processes.

2. Impact craters on land are now generally subtle features that are eroded, buried, or covered with vegetation.

3. Smaller meteoroids and comets tend to burn up and disintegrate in Earth's atmosphere before striking the surface.

In 1993, Gene and Carolyn Shoemaker and David Levy discovered a comet, later known as Shoemaker-Levy 9, from photographs they had taken through a telescope on Palomar Mountain in Southern California. Less than a year and a half later, astronomers watched Shoemaker-Levy 9 explode in one of the most tremendous impacts ever witnessed. From the start, the comet was unusual in that it was composed of several discrete fragments with several bright tails, and it was reported as a misshapen comet. Shoemaker-Levy 9 was one of about 50 comets in the Jupiter family that circle between Jupiter and the sun. The comet's orbit was tied to that of Jupiter, and, after years of orbiting the planet, it separated into 21 fragments known as a "string of pearls."

From telescopes on Earth and the Hubble Space Telescope in Earth orbit, astronomers watched as fragments of the comet entered Jupiter's atmosphere at speeds of 60 km (~37 mi.) per second. The fragments then exploded, releasing between 10,000 and 100,000 megatons of energy, depending on their size. More energy was released than if all of Earth's nuclear weapons were detonated at the same time. Hot, compressed gases expanded violently upward from the lower part

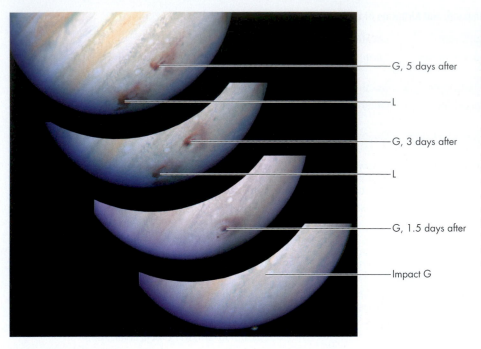

G, 5 days after

L

G, 3 days after

L

G, 1.5 days after

Impact G

< FIGURE 14.13

Spectacular Impact of the Comet Shoemaker-Levy 9G on the Planet Jupiter in 1994

The comet was composed of 21 fragments known as a "string of pearls." One fragment after another entered Jupiter's atmosphere and exploded. From bottom to top, these four images were taken several days apart. The reddish-brown dots and rings mark the impact sites of fragments G and L. The diameter of these rings exceeded the diameter of Earth. *(ESA)*

of Jupiter's atmosphere at speeds as high as 10 km (~6 mi.) per second. Gas plumes from the larger impacts reached elevations of more than 3000 km (~1900 mi.), a height that is around 340 times as high as Mt. Everest, the tallest mountain on Earth. Tremendously large rings developed in Jupiter's atmosphere around the impact sites (Figure 14.13). These rings exceeded the diameter of the Earth! It was truly a remarkable show for astronomers and a sobering event for those who consider that impacts such as this might one day occur on Earth.[3]

After the impact of the fragmented comet on Jupiter and investigations of the Barringer Crater in Arizona and several other craters on Earth, the idea that there could be a catastrophic impact on Earth was finally accepted. At the time of the Tunguska event, several bizarre ideas were suggested to explain it, such as nuclear explosions and the explosion of an alien spaceship!

UNIFORMITARIANISM, GRADUALISM, AND CATASTROPHISM

The idea that Earth could be hit by large objects from outer space was not accepted for many hundreds of years, despite numerous eyewitness accounts in various parts of the world. These accounts may have started with a lethal meteorite that landed in Israel in the year 1420 B.C.[4] Up until the fifteenth century, when Galileo invented the telescope and clearly demonstrated that Earth and the other planets revolve around the sun, the religious establishment did not accept Galileo's ideas; in fact, he was jailed for his beliefs. In 1654, Irish Archbishop Ussher proclaimed that Earth was created in 4004 B.C., making it approximately 6,000 years old. This was the dogma of the time and not to be disputed. However, people studying the formation of mountains, such as the Alps, large river valleys, and other features, had a hard time understanding how these physical features could be formed in only 6,000 years. Based on Archbishop Ussher's "young Earth" belief, early earth scientists were forced to conclude that most of the processes that formed our planet were catastrophic in nature (*catastrophism*). This perspective explained biblical events, such as Noah's Flood, but not much else.

Finally, in 1785, the Scottish doctor James Hutton wrote an influential book that introduced the concept of *gradualism*, or *uniformitarianism*. This concept states that present geological processes may be studied to learn the history of the past—"the present is the key to the past." Hutton also argued that Earth must be much older than 6,000 years to allow the gradual processes of erosion, deposition, and uplift to form mountain ranges and other features on Earth's surface. In 1830, Charles Lyell wrote another influential book on geology, popularizing the role of gradual processes. He proclaimed that Earth had a long history that could be understood by studying present-day processes and the rock record. After Lyell and other scientists overturned the young Earth idea, they could then explain Earth's history in terms of processes that could be carefully observed, such as uplift and erosion. Charles Darwin was impressed with Charles Lyell's book and applied the ideas of an ancient Earth and uniformitarianism to his concept of biological evolution. The concept of gradualism lasted into the twentieth century and culminated with the discovery of plate tectonics, perhaps the crowning glory of uniformitarianism. The theory of plate tectonics (see Chapter 2) explains the present position and origin of continents, based on the relatively slow processes of seafloor spreading and uplift.

But occasionally, scientists also discovered evidence of the role of catastrophic events. Some scientists pointed to large craters on Earth's surface, which they believed were the result of asteroid impacts. Also, it was well known that there was a number of relatively rapid extinctions during Earth's history in which a large percentage (commonly half or more) of the species of plants and animals perished relatively suddenly. Five of these extinctions are in the relatively distant geologic past; the most recent mass extinction is ongoing today as a result of human activity (see Figure 14.2). The mass extinction that occurred 65 million years ago at the K-Pg (Cretaceous-Paleogene) boundary appears to have been caused, to a great extent, by the impact of an asteroid. The asteroid that struck the Yucatán Peninsula in Mexico was approximately 10 km (~6 mi.) in diameter. Even faced with direct evidence for the impact, many scientists were skeptical that a mass extinction might be triggered by the impact of an asteroid. Also, it wasn't until 1947 that Barringer Crater (Meteor Crater) in Arizona was finally listed as a probable impact event. Up to that time, it was still often referred to as a *cryptovolcanic* event. This means that the volcanic activity was "hidden," yet it produced a crater or disturbed area at the surface. Eventually, Barringer Crater was shown to be the result of an asteroid impact, probably with airbursts; since then, more than 175 other large impact features have been identified on all the continents and on the ocean floor.

These observations have led to a new concept, *punctuated uniformitarianism*. This concept states that although uniformitarianism explains the long geologic record of gradual mountain building, canyon erosion, and landscape construction, periodic catastrophic events do occur and can cause mass extinctions.

14.2 CHECK your understanding

1. Define *airburst*.
2. Where is Tunguska? What happened there? Why is it important to our discussion of natural hazards? How often do events like this one occur?
3. Describe the general characteristics of an impact crater. How can it be distinguished from other types of craters?
4. Why does Earth apparently have so few impact craters?
5. Differentiate between simple and complex craters.
6. Why was Barringer Crater controversial, and why is it important?
7. Explain the significance of Comet Shoemaker-Levy 9.
8. Differentiate between uniformitarianism, gradualism, and catastrophism.
9. What is punctuated uniformitarianism?

14.3 Mass Extinctions

A **mass extinction** is characterized by the sudden loss of large numbers of plants and animals relative to the number of new species being added.[11] Because the geologic time scale was originally based on the appearance and disappearance of various fossil species, mass extinctions generally coincide with the boundaries of geologic periods or epochs of the time scale (Figure 14.14). Most hypotheses for mass extinctions involve relatively rapid climate change. This climate change can be triggered by plate tectonics, volcanic activity, or extraterrestrial impact. Plate tectonics is generally a relatively slow process that moves the position of continents and, thus, habitats to different locations. On occasion, plate movements create new patterns of ocean circulation, which have a major effect on climate. Extremely large volcanic eruptions can also cause significant climate change. Large basaltic eruptions producing huge volumes of *flood basalts* (see Chapter 5) can release large quantities of carbon dioxide, a greenhouse gas, into the atmosphere and cause global warming. In contrast, large explosive volcanic eruptions of more silica-rich lava can inject tremendous quantities of volcanic ash into the upper atmosphere and cause global cooling. Finally, climate change is one of several effects of extraterrestrial impacts or airbursts that can contribute to extinctions.

During the past 550 million years of Earth's history, at least six major mass extinction events have occurred. The earliest mass extinction occurred approximately 446 million years ago, near the end of the Ordovician Period (Figure 14.14). One of the largest mass extinctions in Earth's history, with around 100 families of animals becoming extinct, this event coincided with a major continental glaciation in the Southern Hemisphere. In fact, it appears that this event was actually two extinctions: one when the climate cooled, and the other when the climate warmed following the glacial interval. The largest recorded mass extinction occurred near the end of the Permian Period, about 250 million years ago, when 80 to 85 percent of all living species died out (Figure 14.14).[12] Although there is now evidence for an impact at the Permian-Triassic boundary, scientists also believe that this mass extinction may not have been caused by a single catastrophe but may have spanned a period of about 7 million years. The extinction also corresponds to eruption of the massive Siberian flood basalts, which likely emitted voluminous greenhouse gases to the atmosphere and may have dramatically changed the climate. In addition, sea-level fall and assembly of the supercontinent of Pangaea at this time would have dramatically reduced the area of shallow marine environment, where many of the affected species lived.

The third mass extinction at the Triassic–Jurassic boundary (Figure 14.14), 202 million years ago, also appears to have been related to volcanic activity and climate change. Large amounts of carbon dioxide were released from basaltic volcanic eruptions associated with the breaking apart of the supercontinent Pangaea. These eruptions may have produced the largest volume of basaltic lava ever released by volcanoes on land. Earth's temperatures, which were already 3°C (~5°F) warmer than today, increased another 3° to 4°C (~5° to 7°F) at the end of the Triassic Period. This rapid climate change resulted in the extinction of plants and animals on land and in the ocean. Half of all the genera of marine animals became extinct in this event (Figure 14.14).[12]

Another tremendous mass extinction event took place at the end of the Cretaceous Period (K) and beginning of the Paleogene Period (Pg). Known as the *K-Pg boundary*, this event was sudden and most likely caused by the impact of a giant asteroid. The K-Pg extinction brought an end to the large dinosaurs, which had been at the top of the food chain for 100 million years or more. Their demise allowed small mammals to diversify and evolve into the approximately 4,000 species, including humans, that are alive today. Species in the oceans of the world also evolved and diversified following the K-Pg extinction.

A fifth mass extinction took place near the end of the Eocene Epoch about 34 million years ago. Although there is some evidence for an asteroid or comet impact at that time, most scientists link this extinction to climate change brought on by plate tectonics. The movement of tectonic plates at this time allowed a cold ocean current to develop around Antarctica; the resulting cooling produced the Antarctic ice cap, which is still present today. This resulted in a global cooling and drying of the climate.[12]

The final mass extinction occurred near the end of the Pleistocene Epoch. This extinction of mammals, reptiles, amphibians, birds, fish, and plants continues today. The initial cause of this event was first thought to be some combination

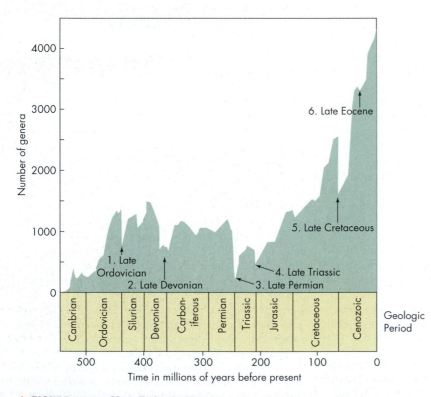

∧ FIGURE 14.14 Mass Extinction Events
Biodiversity diagram illustrating the increasing diversification of life through the Phanerozoic Period, punctuated by six mass extinction events which are named for the geologic period in which they occurred.

of overhunting by Stone Age man and climate change. However, new evidence indicates that it may have been initiated by the airburst of a comet, discussed below. In many ways, this mass extinction continues today due to human activities, including loss of habitat from land-use changes, widespread deforestation, the application of pesticides and other chemicals, and human-induced global warming.[11]

14.2 K-Pg Boundary Mass Extinction

CASE study

One of the great geologic detective stories of the past 50 years is the investigation of the K-Pg (Cretaceous-Paleogene) mass extinction. We now have evidence that 64,980,000 years ago, an asteroid with a diameter of about 10 km (~6 mi.) hit Earth along the northern shore of what is now the Yucatán Peninsula. That event altered Earth's history forever. Although much of the physical landscape of Earth remained unchanged after the impact, the planet's inhabitants were largely affected. Approximately 70 percent of all land and marine genera and their associated species died off (see Figure 14.14). Some reptiles, such as turtles, alligators, and crocodiles, and some birds, plants, and smaller mammals survived, either because of their habitat requirements or their location at the time of the impact. Not all areas experienced widespread wildfires following the impact, and some plants and animals were better adapted to subsequent changes in climate.[13] The impact contributed to the demise of the large dinosaurs on land and swimming reptiles in the oceans. Extinction of these animals set the stage for the evolution of mammals. This evolution eventually produced primates and humans. What would the world look like today if the K-Pg extinction had never happened? There's a good chance that humans never would have evolved!

Since this extinction has been so important for our evolution, we will look more closely at how scientists developed the hypothesis that the K-Pg mass extinction was caused by the impact of a large asteroid. The story is full of intrigue, suspense, rivalries, and cooperation—typical of many of the great scientific discoveries.[14] Scientists with backgrounds in geology, physics, chemistry, biology, geophysics, and astronomy worked together to develop and test the hypothesis that the K-Pg mass extinction was triggered by an impact. Walter Alvarez, a professor at the University of California, asked the question that started it all: What is the nature of the boundary between rocks of the Cretaceous and Paleogene Periods?

Alvarez was interested in Earth's history and, particularly, in reading that history as recorded by rocks. Early in his studies of the K-Pg boundary, Alvarez teamed up with his physicist father, Luis, and they decided to measure the concentration of a platinum-group metal called *iridium* in the thin clay layer that marks the K-Pg boundary in Italy. Walter Alvarez and colleagues had initially gone to the site in Italy to study the magnetic history of Earth. What they found at the K-Pg boundary was a very thin layer of clay (Figure 14.15), and it seemed that the extinction of many species occurred abruptly at the clay layer—fossils found in rocks below this clay boundary marker were simply not there in the rocks above the clay.[14]

Alvarez and his fellow scientists then asked the question: How much time was involved in the deposition of the clay layer? Was it a few years, a few thousand years, or millions of years? The approach they took was to measure the amount of iridium in the clay. They chose iridium because it is found in small concentrations in meteorites and because the global rate of accumulation of meteorite dust on Earth is constant. The deep ocean floor accumulates meteoritic dust like a tabletop or picture frame in your home—the longer ocean floor sediment sits undisturbed, the more iridium it will accumulate. Clay washing into the ocean from the land can cover previously deposited meteorite dust. The faster the rate of sedimentation, the more diluted the meteorite dust becomes. Slow rates of sedimentation allow time for more meteorite dust to accumulate in ocean deposits.

What Walter Alvarez and his colleagues found was entirely unexpected. The team had anticipated measuring approximately 0.1 part per billion of iridium in the

∧ FIGURE 14.15 Evidence for Impact
The Cretaceous/Paleogene (K-Pg) boundary in Italy lies in a thin clay layer. The scientist on the left is pointing to the layer below his left knee. This layer has an anomalously high concentration of the metal iridium. High concentrations of this metal are found in extraterrestrial objects such as meteorites. Its presence here is consistent with an asteroid impact. *(Walter Alvarez)*

clay layer, which they thought would represent slow accumulation through time. If the clay layer were deposited rapidly, then the amount of iridium would be even less. What they actually found was about 3 parts per billion—30 times as much as expected. As you recall from our discussion of CO_2 concentration in the atmosphere (see Chapter 12), 3 parts per million (ppm) is a very small quantity, and 3 parts per billion (ppb) is 1,000 times smaller than 3 ppm! Although 3 ppb is an extremely small amount of iridium, it was much more than could be explained by their previous hypothesis of slow deposition over time. They reevaluated the data, this time including samples that had been removed for treatment before measurement, and got a final value of about 9 ppb, which is nearly 100 times as great as expected. This discovery led them to a new hypothesis: The iridium might be the remains of a single asteroid impact. The team's iridium discovery, along with its hypothesis of an extraterrestrial cause for the extinction at the K-Pg boundary, was published in 1980.[15] In that paper, the team also reported elevated concentrations of iridium in deep sea sediments in Denmark and in New Zealand, all at the K-Pg boundary. Although the discovery of the iridium anomaly at several places around the world made the team more confident of its impact hypothesis, the scientists still had no crater. This absence prompted other scientists to search for potential craters that formed 65 million years ago.[14]

Would it be possible to find the crater? The K-Pg crater might have been completely filled with sediment and no longer recognizable, or the crater might have been on the ocean floor and subsequently destroyed by plate subduction. Fortunately, the site of the crater was identified in 1991.[16] Geologists in Mexico studying the structural geology of the Yucatán Peninsula discovered what they determined to be a buried impact crater with a diameter of at least 180 km (~112 mi.). The crater, named Chicxulub for a nearby village, is nearly circular and has a clear boundary between unfractured rocks within the crater and fractured rocks outside the crater. About half of the crater lies beneath the seafloor of the Gulf of Mexico and half beneath sedimentary rocks and vegetation on the northern end of the Yucatán Peninsula (Figure 14.16).

On land, researchers found a semicircular pattern of sinkholes, known to the Mayan people as *cenotes*, which corresponded directly to the edge of the proposed impact crater. The cenotes range from about 50 to 500 m (~160 to 1640 ft.) in diameter. They were presumably formed by the chemical weathering of fractured limestone on the outside of the crater boundary. Thus, both the distribution of fractured rock and the cenotes must be related to the circular structure because there are no other geologic features in the area that could explain the curved pattern. The crater, which is now filled with other rock, is believed to have been as deep as 30 to 40 km (~19

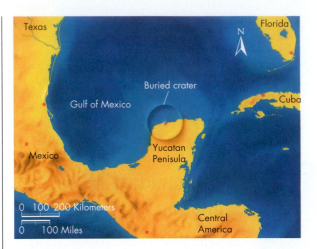

⋀ FIGURE 14.16 Large Impact Crater in Mexico
Map showing location of the Chicxulub impact crater on the north shore of the Yucatán Peninsula in Mexico. This crater is buried below sedimentary rocks on the peninsula and the seafloor of the Gulf of Mexico. Formed by an asteroid impact about 65 million years ago, the age of this crater coincides with the Cretaceous/Paleogene (K-Pg) mass extinction. *(Detlev van Ravenswaay/Science Source)*

to 25 mi.) at the time of impact. However, subsequent slumping and sliding of materials from the sides soon filled in much of the crater, and sedimentation over the past 65 million years completely buried the structure.

When drilling within the crater, geologists found glassy melt rock, interpreted to be impact breccia. The force of the impact excavated the crater, fractured the rock on the outside, and produced the breccia. The glassy nature of the breccia suggests that sufficient heat was present to melt rocks immediately following impact.[17] Another study of the crater found glass mixed in with and overlain by the breccia, as well as evidence of shock metamorphism commonly associated with impact features.[18] Most scientists now believe that the asteroid impact that struck the area nearly 65 million years ago did, in fact, contribute significantly to the K-Pg mass extinction.

After identifying the site of the crater and the evidence to support its existence, questions naturally arose regarding how such an event could cause a global mass extinction.[14] The asteroid that formed the Chicxulub crater was huge; its diameter is estimated at about 10 km (~6 mi.). By comparison, many jet aircraft fly at cruising altitudes of around 10 km above Earth's surface. The summit of Mt. Everest is not as high as 10 km, but it is close. Consider, too, that the asteroid struck the atmosphere of Earth at a speed of about 30 km (~19 mi.) per second, which is about 150 times faster than a jet airliner travels. The amount of energy that was released is estimated to have been about 100 million megatons, roughly 10,000 times as great as the entire nuclear arsenal of the world.

Walter Alvarez and numerous other geologists, physicists, paleontologists, and astronomers have reconstructed a likely sequence of events for the impact and its aftermath (Figure 14.17). The hole blasted in Earth's crust was nearly 200 km (~125 mi.) across and 40 km (~25 mi.) deep. At an altitude of 10 km (~6 mi.) in the upper atmosphere, and moving at 30 km (~19 mi.) per second, the asteroid would have taken less than half a second to reach Earth and almost instantaneously produce the large crater. When it

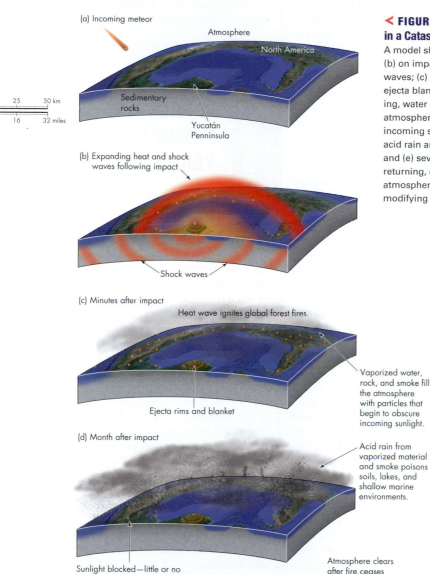

(a) Incoming meteor

Atmosphere

North America

Sedimentary rocks

Yucatán Penninsula

(b) Expanding heat and shock waves following impact

Shock waves

(c) Minutes after impact

Heat wave ignites global forest fires.

Vaporized water, rock, and smoke fill the atmosphere with particles that begin to obscure incoming sunlight.

Ejecta rims and blanket

(d) Month after impact

Acid rain from vaporized material and smoke poisons soils, lakes, and shallow marine environments.

Sunlight blocked—little or no photosynthesis "impact winter"

Atmosphere clears after fire ceases and particulate matter settles or is rained out.

(e) Months to years after impact

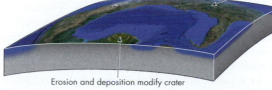

Erosion and deposition modify crater

Life begins to recover?

◄ FIGURE 14.17 Sequence of Events in a Catastrophic Impact

A model showing (a) incoming asteroid; (b) on impact, with a fireball and shock waves; (c) minutes after impact with an ejecta blanket forming, wildfires starting, water and nitrogen vaporizing in atmosphere, and a dust cloud blocking incoming sunlight; (d) a month later with acid rain and little or no photosynthesis; and (e) several months later with sunlight returning, dust and acid washed out of the atmosphere, and erosion and deposition modifying the crater.

contacted Earth, shock waves quickly crushed the rocks beneath, filling in all the cracks, partially melting the rocks, producing the breccia, and blasting bits and pieces high into the atmosphere. All this activity probably took about 2 seconds as the shock wave and heat vaporized rocks on the outer fringes of the impact. The tremendous amount of debris built a huge ejecta blanket around the crater. A gigantic cloud of vaporized rocks and gases would have produced an equally large fireball that rose up and formed a huge mushroom cloud. The explosion itself and the rising materials would likely have been sufficient to accelerate and eject material far beyond Earth's surface. Particles of rock were blasted into ballistic trajectories before they fell back to the ground. The fireball would have produced sufficient heat to set fires around the globe. Vaporization of gypsum salts in the bedrock and ocean water produced sulfuric acid in the atmosphere. Additional acids were added as a result of burning nitrogen in the atmosphere. Thus, following the impact, acid rain probably fell for a long period.

The dust ejected in the atmosphere circled Earth, and for months, essentially no sunlight reached the lower atmosphere. The lack of sunlight stopped photosynthesis on land and in the ocean. Acid rain is toxic to many living things, particularly terrestrial and shallow marine plants and animals. As a result, the food chain virtually stopped functioning because the base of the chain had been greatly damaged. Part of the impact occurred in the ocean and would have significantly disturbed the seafloor, generating tsunamis that could have reached heights of more than 300 m (~1000 ft.). These waves raced across the Gulf of Mexico and inundated parts of North America.[14,19] Wildfires ravaged southern North America, all of Central America, and parts of South America, Africa, Asia, and Australia.[13] The climate first cooled from lack of sunlight and then significantly warmed as aerosols and carbon dioxide from the pulverized limestone of the Yucatán Peninsula enhanced the greenhouse effect. Finally, large numbers of ferns restored plant cover on the burned landscape.[12]

The ultimate result of the asteroid impact was a global catastrophic killing, which we refer to as a mass extinction. Although evidence indicates that some species of dinosaurs were dying off before the impact, the event certainly seems to be responsible for wiping out the remaining large dinosaurs. So many other species of terrestrial and marine animals and plants died that there is little doubt that a massive impact was the likely cause of their extinction. Should such an event occur again, the loss of species would be immense. It might well mean the extinction of humans and many of the larger mammals and birds. These extinctions would lead the way to yet another period of rapid evolution.

What we have learned from the K-Pg extinction is sobering. We know that impacts from objects as large as 10 km (~6 mi.) in diameter are rare and occur only every 40 to 100 million years or so. However, large impacts are not the only hazard from comets, asteroids, and meteoroids. Smaller impacts are much more probable, and they can wreak havoc on a region, causing great damage and loss of life.

14.3 CASE study

IMPACT AND MASS EXTINCTION 12,800 YEARS AGO

Imagine a North American landscape with grasses, low shrubs, and clusters of coniferous spruce trees, along with mammoths, short-faced bears bigger than grizzly bears, camels larger than today's two-humped Bactrian camels, dire wolves, and large, lumbering ground sloths. Add to your image a small hunting party of Clovis people—men and women dressed in fur clothing with large spears, waiting near a watering hole for an unsuspecting mammoth (Figures 14.18 and 14.19). This would have been a typical day toward the end of the Ice Age, the Pleistocene Epoch, around 12,800 years ago. Along the horizon to the north, you would have seen the massive white and blue Laurentide ice sheet, a vast continental glacier about 100 m (~330 ft.) high at its edges. Returning to the same area just years later, you would find that everything has changed—the large animals and people are gone, and the ice sheet has advanced a considerable distance; a glacial interval has begun, known as the *Younger Dryas*—a geologically abrupt period of even colder climatic conditions that lasted about 1,300 years, named after tundra fossil pollen of the dryas plant.[20,21] What happened? This is a question that has perplexed geologists, paleontologists, and archeologists for more than two centuries.

Until the 1790s, most scientists and scholars didn't seriously entertain the idea that large Ice Age mammals, such as the mammoths and mastodons, were extinct. Even President Thomas Jefferson, who had written about the buried remains of a ground sloth found in West Virginia, thought that the large animals might be alive somewhere in the western frontier. It wasn't until 1796 that the French naturalist Georges Cuvier demonstrated that mammoths and mastodons, relatives of elephants, were indeed extinct.[22]

<FIGURE 14.18 Mastodons Die in Mass Extinction**

Large mammals, such as these mastodons in an Indiana bog, roamed North America at the end of the Pleistocene Epoch. They, and their elephant relatives, the mammoths, disappeared in a mass extinction at the end of the Pleistocene. A new hypothesis proposes that an airburst of a comet over the Laurentide ice sheet contributed to their extinction. *(Daniel Eskridge/Stocktrek Images/Getty Images)*

The first hypothesis has shortcomings. Human overkill seems unlikely in light of the large numbers of animals that became extinct, including many species that the paleo-americans evidently did not regularly hunt. Abrupt cooling definitely happened, but only at higher latitudes; what is being debated is the cause. Was the cooling due to regular climate shifts that have occurred frequently during Earth's history, or was it something very different?[20]

The conclusion is that the extinction of Pleistocene megafauna during the latest Pleistocene period was unique and was too abrupt and too widely distributed to have resulted from human overkill or climatic cooling.[20]

At many Clovis sites, archaeologists have identified what is known as a "black mat," a thin layer of carbonaceous, dark, organic-rich clays (Figure 14.20). The base of the black mat coincides with the Younger Dryas Boundary (YDB), the geologic boundary between earth material older and younger than 12,800 years ago, after which there is no evidence for "in-place" remains of the megafauna or artifacts from the Clovis culture. Murray Springs, Arizona, is a well-known and well-studied Clovis site, where the last (youngest) mammoth bones and Clovis tools are directly in contact with the base of the black layer. At that site it is apparent that the termination of Pleistocene megafauna and Clovis culture was sudden, coinciding with the beginning of deposition of the black mat.[20,23,24]

Another geologic mystery about the time of the Younger Dryas is the thousand-yearlong Younger Dryas cooling episode that brought the Pleistocene Epoch to a close. Several events occurred at the YDB that suggest that the Laurentide, European (Fennoscandian) and Greenland ice sheets that once covered parts of the Northern Hemisphere became especially unstable:

> Meltwater drainage from a huge North American glacial lake, Lake Agassiz, shifted abruptly from the Mississippi to other outlets via the St. Lawrence River to the Atlantic Ocean and the McKenzie River to the Arctic Ocean.

⋀ FIGURE 14.19 Clovis Projectile Points

The Clovis people fashioned distinctive fluted projectile points from chert and used them as spear tips and butchering knives to hunt and butcher mammoths, mastodons, and other now extinct animals. For several centuries, this technology was widespread in North America until it disappeared around 12,900 years ago at the start of the Younger Dryas glaciation. *(Robert McGouey/Alamy)*

The possible cause of the extinction of the megafauna and the termination of the Clovis culture has been a longstanding and controversial subject. Major competing hypotheses to explain the extinction of the megafauna are (1) overkill by humans; (2) abrupt cooling (i.e., rapid climate change); and (3) extraterrestrial impact.

< FIGURE 14.20 Black Layer at the Younger Dryas Boundary
Called the "black mat," this carbon-rich layer at the Murray Springs, Arizona, mammoth-kill site is draped over mammoth bones that were butchered by Clovis hunters. The base of the black mat (white arrow) contains a layer (the Younger Dryas boundary) with abundant materials of cosmic impact origin. A black layer of similar composition marks this Younger Dryas boundary in many places in North America and Europe. *(GeoScience Consulting)*

> All the margins of the ice sheet began to destabilize and melt rapidly with significant freshwater discharge into the surrounding oceans.

> A large armada of icebergs calved off the ice sheets and floated south into the North Atlantic.

> This polar and subpolar freshwater cap caused a major reduction in the Atlantic Ocean conveyor belt, the ocean current system that moderates temperatures in southern Greenland and Europe (see Chapter 12).[20]

> Massive, abrupt cooling over much of the Northern Hemisphere outside of the tropics, with a return to near ice-age climate.

What caused this unusual combination of abrupt ice-sheet melting and instability that coincided with climatic cooling? A theory has been proposed that explains the major and abrupt extinction of Pleistocene megafauna, the disappearance of the Clovis culture, and destabilization of the ice sheet were triggered by an extraterrestrial (cosmic) impact at the YDB 12,800 years ago.

Evaluation of several Clovis sites, selected because they were well documented and dated, supports the hypothesis of a major extraterrestrial impact ~12,800 years ago. Evidence of such an event is found on four continents (North America, South America, Western Europe, Western Asia as well as in the Greenland ice sheet (Figure 14.21); Many sites are in North America including Santa Rosa Island in the Santa Barbara Channel, California.[25,26]

Analysis of over 300 dates from 23 locations has pinned down the event to be with 95 percent confidence at 12,835–12,735 Years ago (CalB.P.)[27] This means that the date of the YDB event is likely a worldwide time marker, a so-called datum horizon, that is synchronous and can be traced over at least four continents.

Existing evidence suggests that the YDB Episode was triggered by the impacts of many low-density

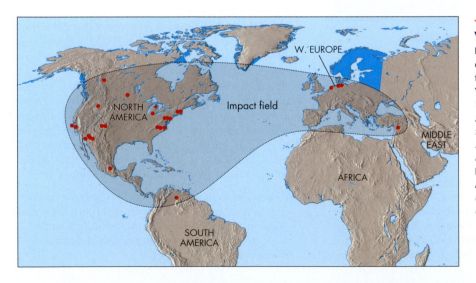

< FIGURE 14.21 Possible Younger Dryas Impact Field
Red dots show 27 sites where evidence has been collected for the ~12,800-year-old YDB impact. *(Modified after Wittke, J. H. and 27 others. 2013. Evidence for deposition of 10 million tonnes of impact spherules across four continents 12,800 years ago.* Proceedings of the Natural Academy of Sciences. *PNAS early edition. pnas.org/cgi/doi10.1073/pnas.)*

extraterrestrial objects (from a fragmented comet) that exploded widely in the atmosphere over the Northern Hemisphere including the Great Lakes Region. The Laurentide ice sheet covering much of Canada became partially destabilized by the proposed event, especially at its edges, and consequently released its meltwater from proglacial lakes to the oceans. Scientists hypothesize that many objects exploded over the continents and the ice sheet, setting off heat flashes and shock waves and generating intense winds (hundreds of kilometers per hour) over wide areas of Earth. The event also would have spawned fireballs leading to widespread biomass burning. This led to habit loss through destruction of forests and grasslands. Also, the soot and ash produced by the biomass burning was so abundant and dense that it blocked sunlight for days to weeks with loss of photosynthesis and food production. These factors in combination with the abrupt cooling led to the extinction of many of the large Pleistocene animals.[20]

Physical evidence for a cosmic impact at the Younger Dryas boundary includes the discovery of a variety of markers that resulted from cosmic impact, including high temperature meltglass and magnetic microspherules, carbon spherules, charcoal and soot, and other evidence of intense wildfires.[26] Also an abundance peak of nanodiamonds (including those thought to be formed only by cosmic impact) has been identified in the impact layer compared with their relative scarcity in sediments above and below. Crucially, analysis of ice from the Greenland ice sheet at precisely the beginning of the Younger Dryas identified a major platinum (Pt) abundance anomaly (peak) that was caused by the impacting comet. Following the discovery of the Pt anomaly in the Greenland ice sheet, several sites in the USA where the YDB had been reliably located by careful C-14 dating (see Appendix D) were revisited in search of the Pt anomaly. The Pt anomaly was identified in four well-dated sites in the western USA and Ohio. Also the anomaly was identified in another seven sites the along the U.S. eastern seaboard at a level considered to be the YD onset (and the YDB layer) based mostly on archaeological changes[28,29]

The discovery of the Pt anomaly in the Greenland ice sheet and subsequently in other terrestrial sites is the most convincing evidence yet of the YDB impact event. The previous careful work on the YDB sites in North America and Europe allowed researchers to test exactly the right levels in the ice sheet in search of cosmic impact evidence. The resulting fortuitous discovery of the Pt abundance anomaly in the ice sheet in turn prompted a search, and its subsequent discovery, in terrestrial sites. This research is a clear illustration of how scientific work progresses from one discovery leading to others.

As scientists continue to test YDB theory, more data will be forthcoming, and we will learn more about the catastrophic event that brought about the demise of the Pleistocene megafauna, the end of the Clovis culture, and the anomalous abrupt return to near-glacial conditions that lasted for about 1,000 years.[20,30] As the hypothesis continues to be thoroughly tested, it is likely that more direct physical evidence of the cosmic impact will be discovered including the possible discovery of large impact craters hidden on the ocean floor and/or under the ice sheets. Recently a young 30 km (19 mi.) diameter crater was discovered in northern Greenland. Although the date of this crater is not known, the impact likely occurred less than 100,000 years ago and could be as young as ~13,000 years old.

The discovery of the YDB event has important implications for recent human history. Evidence exists for human population declines in North America and Western Europe that coincided with significant cultural shifts in the hunter/gatherers of that time. In North America the broadly distributed successful Clovis culture suddenly disappeared, and, following a hiatus in human activity over broad areas, was replaced by a diversity of local cultures. In Western Europe, cultures known for their cave paintings declined; and in Syria, agriculture began as hunter-gatherer groups changed to hunter-cultivator societies. The harsh times of the cold and drier Younger Dryas evidently made food gathering more difficult and uncertain compared to the deliberate cultivation of crops.[26]

If the theory is confirmed, as the bulk of the evidence suggests, it perhaps may be considered as the most important theory advanced since that of plate tectonics revolution in the late 1960s. Should such a cataclysmic event occur in the future, civilization as we know it would almost certainly disappear (see Case Study 14.4: Professional Profile).

14.3 CHECK your understanding

1. Define *mass extinction*.

2. What are the hypotheses for the cause of mass extinctions?

3. When was the greatest mass extinction in Earth's history?

4. What is the cause of the K-Pg mass extinction, and what is the main evidence to support the event?

5. What is the main evidence supporting the hypothesized Younger Dryas impact event?

6. Assuming that the hypothesized Younger Dryas impact is correct, what would the consequences be if such an event were to happen in the future?

14.4 Professional Profile

Emeritus Professor James Kennett, University of California, Santa Barbara

Professor James Kennett (Figure 14.4A) was born and received his early education in Wellington, New Zealand. The wonderful geologic exposures of his native country, especially in the Nelson area where his grandparents farmed, fostered a boyhood interest in geology. Kennett received B.S. and Ph.D. degrees at Victoria University of Wellington where he began to meld his interests in marine geology and the history of life. Kennett and his wife Diana immigrated to the United States in 1966 expressly to participate in the great American science enterprise then blossoming.

A member of the U.S. National Academy of Sciences, Kennett is a world leader in marine geology. He has published over 250 papers and four books. In 1982 he published the now classic college textbook *Marine Geology*, which is a major synthesis within a plate tectonic framework of Earth's oceanographic and climatic evolution. Kennett is considered a pioneer in developing paleo-oceanography (integrative studies of past ocean history) as a new field. The thread tying his discoveries together is a fundamental understanding of the evolution of the global environment, in particular climate and ocean circulation and their effects on the development of life, especially marine life.

Jim Kennett says he has been blessed by being associated with many wonderful graduate and undergraduate students and by being closely linked with colleagues who have helped discover much about how our Earth has operated in its past. He has been a major advisor to 20 doctorate and numerous Master's students, many of whom have contributed enormously to our understanding of the Earth's environmental evolution.

Kennett's work is notable for creating and testing a number of innovative hypotheses, including the potential role of marine methane emissions in influencing abrupt climate change. His questions are often controversial, as is the case for most new scientific ideas that precipitate a paradigm shift. Kennett tests important hypotheses that change the way we perceive how Earth works, and he is a master of the scientific method that calls for careful collection and analysis of data.

◀ **FIGURE 14.4.A Professor James Kennett**
Kennett in the remarkable Takoro Canyon of Southern Taiwan, which has some of the highest rates of tectonic uplift in the world. While in Taiwan, Jim presented his first lecture at the National Taiwan University about the YDB Impact hypothesis. *(Photograph courtesy of James Kennett)*

Over the last 17 years, Kennett and his many colleagues (especially Allen West, Ted Bunch, and his son, Douglas Kennett) have been addressing the question (hypothesis) of whether or not a major cosmic impact (possibly consisting of many airbursts) 12,800 years ago caused a wide range of otherwise enigmatic environmental and biologic changes on Earth. This YDB cosmic impact hypothesis is posited to have triggered an abrupt and anomalous cooling at the onset of the Younger Dryas, the extinction of the majority of the large North American Ice Age animals, and abrupt human cultural and population changes in the Americas and Europe and parts of the Middle East. Such an event in the future would end civilization as we know it.

The hypothesis of a major cosmic impact occurring only a few thousand years ago and causing major environmental, biotic, and human consequences, is perhaps the most important geologic and anthropologic discovery in decades. It may well explain a wide range of enigmatic events that have puzzled scientists for years: abrupt climate cooling when Earth should have remained warm; abrupt extinction of most of the large and abundant mammals over North America, including mammoths, giant sloths, camels, and horses; abrupt changes in human culture (including the disappearance of the widely distributed Clovis culture over North America and the cave-painting Magdelanian culture of Western Europe), in association with clear reductions in human population over broad regions; and the rise of agriculture in Syria. Kennett continues to point out that, although a wide range of data now supports a major cosmic impact (e.g., peak abundances in nanodiamonds, high-temperature melt-glass and spherules, platinum) at many sites on land in North America, Western Europe, Syria, and the Greenland ice sheet (a giant depository of material settling out of the atmosphere), this hypothesis requires further testing and evaluation.

14.4 Linkages with Other Natural Hazards

The impact or airburst of an asteroid or comet is a direct cause for a number of other natural hazards, including tsunamis, wildfires, earthquakes, mass wasting, climate change, and, possibly, volcanic eruptions. Most impacting objects land in the world's oceans, and large objects will cause tsunamis. The asteroid that hit the Yucatán Peninsula close to 65 million years ago created large, complex tsunamis that spread across the Gulf of Mexico and are recorded in sedimentary deposits in Mexico, Texas, Alabama, and Cuba. Researchers have estimated that waves 200–300 m (~660–1000 ft.) high originated from the impact blast and subsequent landslides in the Gulf of Mexico.[19]

Wildfires of regional or global extent have resulted from three well-documented airbursts or impact events: the K-Pg boundary, the Younger Dryas boundary, and the Tunguska events. Superheated clouds of gas and debris apparently reach temperatures capable of drying out and then igniting living vegetation. Computer simulations suggest that wildfire patterns following a large impact are complex and involve rock blasted into space that then returns to Earth on the other side of the planet. If it is any consolation, the simulations suggest that these wildfires did not burn the entire land surface of Earth.[13]

Seismic waves from a large impact most likely activate numerous landslides, both on the land and under the water. For example, the asteroid that formed the Chicxulub impact on the Yucatán Peninsula appears to have produced a **M** 10 earthquake and caused the mass wasting of the continental slope of North America as far north as the Grand Banks of Newfoundland.[22]

Both the asteroid that created the Chicxulub Crater at the K-Pg boundary and the probable airburst at the end of the Pleistocene Epoch caused global changes in climate. Impacts on land can inject large quantities of dust into the atmosphere. This dust, combined with smoke from wildfires, would result in global cooling for a number of years after any major impact. The global cooling would then be followed by a prolonged period of global warming from the large amounts of carbon dioxide and other greenhouse gases produced by post-impact wildfires.[13]

Extraterrestrial impacts have also been hypothesized to produce large volcanic eruptions by causing melting and instability in Earth's mantle.[31] This melting could result in huge eruptions of lava, referred to as flood basalts, and create large igneous provinces on land. These provinces contain 100 times more lava than from any historic volcanic eruptions.[32] In the geologic past, these eruptions appear to have produced global changes to the atmosphere and oceans and contributed to major extinctions of life.

14.4 CHECK your understanding

1. Explain the various hazards linked to a large extraterrestrial impact in the future.

14.5 Minimizing the Impact Hazard

RISK RELATED TO IMPACTS

The total risk of a hazardous event is related both to the probability that it will occur and to the consequences, should it take place. Consequences include the concepts of resiliency, exposure, vulnerability, adaption, mitigation, hazard perception, and social media. Thus, risk can be thought of as the product of hazard, exposure, and vulnerability (see Figure 1.11). The consequences of a large airburst or direct impact from a large extraterrestrial object several kilometers in diameter would be catastrophic. Although seven out of ten such events would likely occur over or in the oceans, the enormous size of an asteroid or comet would mean that the effects would be felt worldwide. Certainly, there would be significant differences, depending on the site of impact or airburst, but the overall consequences would constitute a global catastrophe with high potential for mass extinction. On Earth, events of this magnitude probably have return periods of tens to hundreds of millions of years (Figure 14.22). Smaller objects, on the order of a few tens of meters, would cause a regional catastrophe if they produced an airburst or surface impact near a populated area. The size of the devastated area would be on the order of several thousand square kilometers. Even a small impact could result in millions of deaths if the event occurred over or in an urban region. Smaller but regionally significant events are sometimes called "Tunguska-type" events, after the 1908 airburst in Siberia.

Airbursts from asteroids with diameters of about 50 to 100 m (~165 to 330 ft.) are capable of causing catastrophic damages to a region about every 1000 years (Figure 14.22). Using the Tunguska event as typical of one that could occur on our planet every thousand years, an urban area is likely to be destroyed every few tens of thousands of years. The basis for this assessment comes from the distribution of 300 small asteroids known to have exploded in the atmosphere. Extrapolation of the data allows estimates to be made for larger, more damaging events.[7]

Predictions of the likelihood and type of future impacts contain enormous statistical uncertainty. Deaths caused by an impact during a typical century may range from zero to as high as several hundred thousand. Computer-run simulations suggest that during a given century, approximately 450 deaths per year may be attributed to an impact. A truly catastrophic event could kill millions of people; averaging this number over thousands of years produces a relatively high average annual death toll.

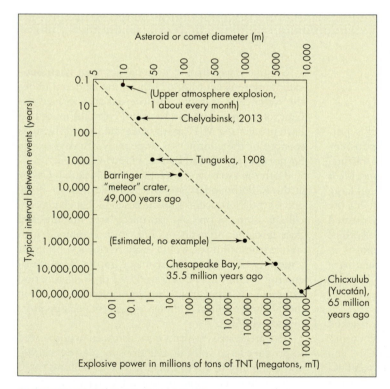

∧ FIGURE 14.22 Energy from Impact
Estimate of the relationship between the energy of Earth impact and time interval between events for various sizes of asteroids or comets. Also shown is the estimated diameter for the impacting comet or asteroid. *(Based on Weissman, P. R., McFadden, L., and Johnson, T. V., eds.*, Encyclopedia of the Solar System. *San Diego, CA: Academic Press, 1999; and Brown, P., Spalding, R. E., ReVelle, D. O., Tagliaferri, E., and Worden, S. P., "The Flux of Near-Earth Objects Colliding with the Earth,"* Nature *420: 294–96, 2002.)*

MINIMIZING THE IMPACT HAZARD

It is only recently that scientists and policy makers have appreciated the risk of an impact of an asteroid or comet. Are we helpless, or is there something that can be done to minimize the hazard? The first and foremost need is to identify nearby objects in our solar system that might threaten Earth. Identification and categorization of comets and asteroids that cross Earth's orbit are already in progress and could be scaled up to include objects in smaller size classes. The additional size classes would include objects with diameters less than 50 m (~160 ft.), those 50 meters to several hundred meters, and those with a diameter of several kilometers. A program known as Spacewatch, which has been operating since 1981, is attempting to inventory the region surrounding Earth, with expansion to the entire solar system. Another program, known as the Near-Earth Asteroid Tracking Project (NEAT), was started in 1996. The objective of this program, which is supported by the National Aeronautical and Space Administration (NASA), is to study the size distribution and dynamic processes associated with **near-Earth objects** (NEOs) and, specifically, to identify those objects with a diameter greater than 1 km (~3300 ft.).

NEOs either reside and orbit between Earth and the sun or have orbits that intersect with Earth's orbit. Although these objects include numerous meteoroids, asteroids, and comets, only the asteroids and comets that are tens of meters to a few kilometers in diameter are a significant hazard (Figure 14.23). If a NEO with a diameter of a few tens of meters were to burst in the atmosphere or impact Earth's surface, we would experience a Tunguska-type event of regional destruction. A NEO of several kilometers in diameter would cause a global catastrophe.

As mentioned earlier, most asteroids originate in the asteroid belt located between Mars and Jupiter (see Figure 14.5). If an asteroid remains in the belt, it does not become a NEO; however, if an asteroid is disturbed by a collision or by the gravitational pull of Jupiter, its orbit may become more elliptical, crossing Earth's orbit, or enter the space between the sun and Earth. The current estimate of Earth-crossing asteroids with diameters larger than 100 m (~330 ft.) is close to 85,000. Larger Earth-crossing asteroids are scarcer. An estimated 1,100 have a diameter greater than 1 km (~3300 ft.), and 30 are greater than 5 km (~3 mi.).[33]

Comets, the other type of NEO, are generally a few kilometers in diameter and have an expanding stream of gas and dust that produces a spherical cloud around the comet. A tail of gas and dust is "blown" out from the cloud by the force of solar radiation, sometimes called the solar wind; thus, the tail is always directed away from the sun. Many of the dust and gas particles burn up in the atmosphere when Earth's orbit passes through a comet's tail. This creates streams of meteors, referred to as a meteor shower. Halley's Comet is the most famous and perhaps the best studied of all comets because of a 1986 spacecraft mission to observe it. The expedition found that the comet is a fluffy, porous body with little strength. In fact, the entire nucleus of Halley's Comet is about 20 percent ice and, therefore, it must have 80 percent empty space, consisting of a network of cracks and voids between loosely cemented materials.[4] Halley's Comet visits the space above Earth approximately once every 76 years, giving every generation a chance to view it. Of the 10,000 NEOs identified as of July 2013, only 93 of the objects have a gaseous tail when they approach the sun and are classified as comets.[34] An estimated 15 percent of the other objects are thought to be extinct comet nuclei that no longer generate a tail when heated by solar radiation.[33]

NEOs, both asteroids and comets, apparently have a relatively short life span. Consequently, new NEOs must continuously be ejected from the asteroid belt to explain the large number thus far identified. Likewise, the orbits of some comets from the outer solar system are perturbed by planets or other objects and eventually become NEOs as well.

NEO observation programs utilize telescopes with digital imaging devices. Using these devices, astronomers take three to five images of a given area of the night sky at intervals of 10 to 60 minutes. Computer software is then used to compare the images to identify

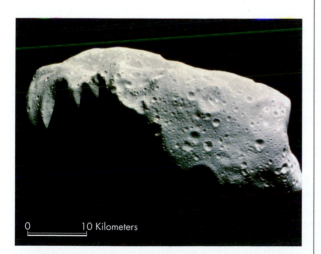

∧ FIGURE 14.23 Asteroid 243 Ida
Numerous impact craters on the surface of asteroid 243 Ida show that collisions between objects in the asteroid belt are common. Ida is approximately 52 km (~32 mi.) long and orbits the sun at a distance of 441 million km (~274 million mi.). It takes nearly five years for Ida to complete one orbit of the sun. Photograph taken by the NASA Galileo spacecraft in 1993. *(JPL/NASA)*

fast moving objects.[31–33,35] The programs and systems used to identify NEOs are expected to intensify in the future, and increasing numbers of objects will be cataloged. This is a first step toward evaluating the potential hazard from NEOs. However, evaluation will take a long time because many of the objects have orbits that may not bring them close to Earth for decades, and the average amount of time between potentially catastrophic impacts is at least thousands of years for the smallest objects. The good news is that most of the objects identified as being potentially hazardous to Earth will likely not collide with our planet until several thousand years after they have been discovered. Therefore, we will have an extended period of time to learn about a particular extraterrestrial object and attempt to develop appropriate technology to minimize the hazard.[4]

By one estimate, about 20 million extraterrestrial bodies in near-Earth orbits have the potential for a significant impact.[4] Only about 4 percent of these bodies are likely to penetrate Earth's atmosphere and excavate a crater. More than half are structurally weak and prone to explode at altitudes of about 30 km (~19 mi.) above Earth's surface. Although these explosions are spectacular, they are not a significant hazard at the surface of Earth. The remaining objects, which constitute about 40 percent of the total population of objects, are moderately strong, stony asteroids (see Table 14.1). These relatively slow moving asteroids can penetrate the atmosphere and produce a serious threat at the surface of Earth. Such bodies produce Tunguska-type events.

Identifying all these objects will be extremely difficult; with a diameter of only 25 m (~80 ft.), they are difficult to identify and track.[4] It is currently impossible for us to identify and catalog the estimated 10 million small asteroids or comets. We are much better prepared to identify and track objects of a few hundred meters to a kilometer or so in diameter. In 2007, NASA proposed to Congress that existing surveys be extended in order to identify 90 percent of potentially hazardous NEOs greater than 140 m (~460 ft.) in size by 2020.[33]

As survey efforts to identify smaller objects increase, more NEOs that are potentially hazardous to Earth will be detected. Communication of this hazard to the public and public officials must be both timely and authoritative. As with earthquake predictions, pending impacts must be evaluated by committees of scientists before predictions should be acted upon. The International Astronomical Union established a technical review process for predictions, which has had mixed results.[36] NASA's NEO Program Office maintains a website (http://neo.jpl.nasa.gov) that classifies potentially hazardous NEOs, based on the Torino Impact Hazard Scale (Table 14.3).

Once a large NEO is recognized to be on a collision path with our planet, options available to avoid or minimize the hazard from an airburst or crater-forming event

▼ TABLE 14.3 The Torino Impact Hazard Scale

No Hazard	0	Low collision hazard or object will burn up in atmosphere.
Normal	1	Object will pass near Earth with collision extremely unlikely.
	2	Somewhat close encounter; collision unlikely and does not merit public attention.
Merits Attention by Astronomers	3	Close encounter with localized destruction possible; merits public attention if collision less than a decade away.
	4	Close encounter with regional destruction possible; merits public attention if collision less than a decade away.
	5	Close encounter with serious, but uncertain, regional destruction threat; merits contingency planning if less than a decade away.
Threatening	6	Close encounter with serious, but uncertain, global catastrophe threat; merits contingency planning if less than three decades away.
	7	Very close encounter with unprecedented, but still uncertain, global catastrophe threat; merits international contingency planning if less than a century away.
	8	Collision will occur with object capable of localized destruction on land or tsunami if offshore; once every 50 to 100 years.
Certain Collisions	9	Collision will occur with object capable of regional devastation or major tsunami; once every 10,000 to 100,000 years.
	10	Collision will occur with object capable of global climatic catastrophe that threatens civilization; once every 100,000 years or more.

Source: Based on Morrison, D., Chapman, C. R., Steel, D., and Binzel, R. P. "Impacts and the Public: Communicating the Nature of the Impact Hazard," in *Mitigation of Hazardous Comets and Asteroids*, ed. M. J. S. Belton, T. H. Morgan, N. H. Samarasinha, and D. K. Yeomans, pp. 353–90. New York: Cambridge University Press, 2004.

are somewhat limited. The largest events are the real killers, and there would be no safe place on the planet. All living things within the blast area would be killed immediately; those farther away are likely to be killed in the ensuing months from the cold and the destruction of the food chain. Even if we could identify and intercept the object, blowing it apart into smaller pieces would likely cause more damage than would one impact from the larger body because each of the smaller pieces, some of which would become radioactive in the explosion, would rain down upon Earth. A more plausible approach would be to try to gently divert the object so that it misses Earth. Let's assume we identify a 400 m (~1300 ft.) diameter asteroid that we believe will strike Earth in approximately 100 years. In all likelihood, the body has been crossing Earth's orbit for millions of years without an impact, and if it were possible to slightly nudge it and change its orbit, it would miss, rather than strike, Earth. This scenario is not unlikely because the probability that we would identify the object at least 100 years before impact is about 99 percent. We have the technology to change the orbit of a threatening asteroid with small nuclear explosions that are close enough to the asteroid to nudge it, but far enough away to avoid breaking it up. NASA has determined that such a "standoff" explosion would be 10 to 100 times more effective than non-nuclear alternatives, such as pulsed lasers or focused solar radiation, in changing the orbit of a potentially hazardous object.[36] To do so would require a coordinated mission between the world's military and space agencies, and the cost of such an expedition would likely exceed $1 billion. However, this seems a small price to pay, considering the potential damages from a Tunguska-type event in an urban area.

Another option for smaller events might be evacuation. If, months in advance, we could precisely predict the point of impact, evacuation is theoretically possible. However, evacuating an area of several thousand square kilometers would be a tremendous and likely impossible undertaking.[4]

In summary, we continue to catalog extraterrestrial objects that intersect Earth's orbit. We are beginning to think about our options to minimize the hazard. Given the potential long-term warning before an object would actually strike the surface of Earth, we may be able to devise methods to intercept and minimize the hazard by nudging the object into a different orbit so that it misses Earth.

14.5 CHECK your understanding

1. How is the risk of extraterrestrial impact determined?

2. How might near-Earth objects (NEOs) be identified?

3. Explain how the risk of dying from an asteroid impact is greater than the risk of a car accident.

4. How might the hazard of an extraterrestrial impact be minimized?

APPLYING the 5 Fundamental Concepts

The Chelyabinsk Meteor, February 2013

1 **Science helps us predict hazards.**

2 **Knowing hazard risks can help people make decisions.**

3 **Linkages exist between natural hazards.**

4 **Humans can turn disastrous events into catastrophes.**

5 **Consequences of hazards can be minimized.**

The Chelyabinsk meteor that entered the atmosphere above Russia on February 15, 2013, was a small asteroid, about 18 m (~60 ft.) in size.

It entered the atmosphere at a high speed of about 19 km (~13 mi.) per second, exploding at an elevation of about 23 km (~15 mi.) (see chapter-opener photograph and Figure 14.1).[1] That object is the largest known asteroid to have entered Earth's atmosphere since the Tunguska event of 1908, which caused a much wider zone of damage in a remote area of Siberia. The Chelyabinsk meteor, fireball, and blast were video-recorded by numerous people on the ground and were also recorded by sensors monitored by the U.S. government.

Approaching Earth at a low angle (~17°), the Chelyabinsk meteor entered the upper atmosphere and began to slow. Compression of atmospheric gases in front of the advancing meteor generated massive amounts of heat, causing the meteor to erupt into a fireball as it streaked across the sky before finally exploding and producing a tremendous shock wave. Most of the damage to the more than 7,000 buildings affected by the explosion consisted of blown out windows (Figure 14.24), but the shockwave did cause the roof of a zinc factory to collapse. Numerous small meteorites from the blast fell to Earth; scientists collected some of these from an area near a 6 m (~18 ft.) diameter hole that was found penetrating a frozen lake (Figure 14.25a and b). Eight months later, a 1.5 m (~5 ft.) long meteorite fragment weighing about 600 kg (~1,300 lb.) was recovered from the bottom of the thawed lake. The recovered meteorites are chondritic in composition (see Table 14.1), similar to that of many other meteorites that have been collected around Earth. Meteorite density was about 3.6 g per cm^3 (3.6 times that of water), and the total weight of the Chelyabinsk meteor is estimated at about 11,000 tons.[37]

One of most important aspects of the event is not just the size of the fireball, which was huge, but also the point at which it exploded. The explosion occurred higher in the atmosphere than most airbursts we have observed in recent decades. With the explosive force equivalent to 20 to 30 World War II–type atomic bombs,[1] had the airburst been lower in the atmosphere, say,

less than 2500 m (~6500 ft.), catastrophic damage would have resulted. Should such an event occur lower in the atmosphere over a large urban area, millions of deaths could occur. This is a somewhat unnerving thought when you consider that the Chelyabinsk impact was completely unexpected.

Given this potential scenario, what can we do in the future to minimize the hazard from extraterrestrial impacts from objects on the order of a few tens to hundreds of meters or more in diameter? Unfortunately, the answer to that question at the present time is that we can do nothing. However, the effects of an impact could

∧ FIGURE 14.24 Damage from Chelyabinsk Airburst
Broken glass from blown-out windows produced most of the injuries following the Chelyabinsk airburst.
(Andrei Ladygin/Russian Look/MCT/Newscom)

(a)

(b)

∧ FIGURE 14.25 Chelyabinsk Impact
(a) Small fragments (note number in circle on paper) of the Chelyabinsk meteorite *(Krasilnikov/ZUMA Press/Newscom)*; (b) Holes in ice covering a lake where a meteorite fragment hit. *(epa European pressphoto agency by b.v./Alamy)*

be minimized if we were able to predict when and where a strike might occur and if there was enough time to evacuate.

As mentioned in the chapter opener, at the time of the Chelyabinsk impact, astrophysicists from around the world were watching asteroid 2012 DA14, which was to pass by Earth later that day. Since its discovery the previous year, scientists had been closely tracking 2012 DA14 because its orbit intersected Earth's. Nearly three times the size of the Chelyabinsk meteor, a direct impact from 2012 DA14 would be catastrophic. Fortunately, astrophysicists were able to quickly calculate its trajectory and determine that the asteroid would not impact Earth on this orbit, but its path would bring it within 30,000 km (~19,000 mi.) of Earth, close enough to pass between us and our global positioning satellites. Having dodged the proverbial bullet, scientists were focused on collecting as much data from the close fly by of 2012 DA14 as possible, when Chelyabinsk hit us from behind. Given the infrequency of large Earth-crossing extraterrestrial objects, it is difficult to believe that these two meteors passing Earth within 16 hours of each other are not somehow related. However, analysis of the Chelyabinsk meteor trajectory clearly shows that its orbit was unrelated to that of 2012 DA14 (Figure 14.26).

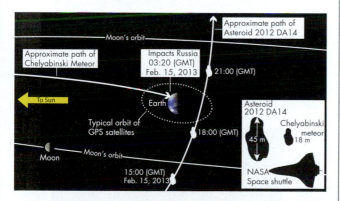

∧ FIGURE 14.26 Cosmic Coincidence
Just 16 hours after the Chelyabinski meteor exploded above Russia, Asteroid 2013 DA14 passed beneath the orbits of Earth's Global Positioning Satellite (GPS) network. Although closely timed, the distinctly different paths of the two asteroids indicates that the objects were not related to one another.

So how is it that we were able to track astroid 2012 DA14 for an entire year and completely miss the Chelyabinsk meteor? These smaller objects, although capable of catastrophic damage, are not easily identified. The U.S. Congress has mandated a program that calls for NASA to identify 90 percent of those objects that are 150 m and larger in diameter. However, funds appropriated to do so have been insufficient to date. The technology, however, is available, and as it improves, it could identify these larger objects and eventually allow for even smaller NEOs to be recognized. Once they are identified and if time were permitting, the effects of an impact could be minimized by evacuation of the impact area. It may even be possible to intercept the object and attempt to deflect its orbit such that it would not strike Earth.

APPLYING THE 5 FUNDAMENTAL CONCEPTS

1. Scientific advances allow us to monitor large asteroids crossing Earth's orbit. Discuss why the Chelyabinsk event came as a surprise.

2. Given what you know about our ability to identify and track asteroids that have potential to inflict catastrophic damage to Earth, discuss the risk we face from extraterrestrial impacts of varying sizes.

3. What other hazards could have been linked to an airburst or impact event on the scale of the Chelyabinsk meteor event?

4. The Chelyabinsk meteor is the only meteor known to have caused a large number of injuries. How is population and land use linked to human injuries, deaths, and economic losses from small extraterrestrial impacts?

5. Assuming that the next extraterrestrial impact can be predicted, how might loss of lives and property from these impacts be minimized?

CONCEPTS in review

14.1   Earth's Place in Space

LO:1   Differentiate between asteroids, meteoroids, and comets.

- The evolution of Earth most likely started about 14 billion years ago with the so-called explosion of the "Big Bang" that produced atomic particles that later formed galaxies, stars, and planets. Our sun formed about 5 billion years ago from a solar nebula, and Earth formed about 4.6 billion years ago.
- Asteroids, meteoroids, and comets are extraterrestrial objects that can intercept Earth's orbit. Small objects may burn up in the atmosphere and be visible as meteors at night. Depending on their size, velocity, and composition, large objects from a few cm to 1000 km in size may disintegrate in the atmosphere in an airburst or hit the surface of Earth as a meteorite.

KEY WORDS
asteroid (p. 560), **comet** (p. 562), **meteor** (p. 562), **meteoroid** (p. 562)

14.2   Airbursts and Impacts

LO:2   Describe the physical processes associated with airbursts and impact craters.

- A meteoroid entering the atmosphere at about 85 km (53 mi.) above Earth's surface at a velocity of 12–72 km per second (~7 to 45 mi per second) will become a meteor and emit light. It will then either explode in an airburst in the atmosphere at an altitude of 12–50 km (~7–31 mi.) or collide with Earth.
- Objects that impact Earth's surface produce both simple and complex craters. Many craters have yet to be identified because they are severely eroded, lie on the seafloor, or are buried by more recent deposits.

KEY WORDS
airburst (p. 564), **impact crater** (p. 566), **meteorite** (p. 564)

14.3 and 14.4   Mass Extinctions Linkages with Other Natural Hazards

LO:3   Suggest the possible causes of mass extinction and provide examples.

LO:4   Evaluate the evidence for the impact hypothesis that produced the mass extinction at the end of the Cretaceous Period.

- Large objects can cause local to global catastrophic damage and mass extinction of life. The best-documented impact occurred 65 million years ago at the end of the Cretaceous Period (K-Pg

KEY WORDS
mass extinction (p. 571)

boundary) and likely produced the mass extinction of many species, including the large dinosaurs.

- Recent studies suggest that an airburst may have been responsible for the extinction of many large Pleistocene mammals and the disappearance of the Clovis people at the start of the Younger Dryas period 12,800 years ago.
- Many linkages exist between impacts of comets and asteroids with Earth and other natural hazards. For example, a hit in the ocean can cause a tsunami. An impact on land would cause large-scale fires across the landscape. Linkages to atmospheric hazards would include cooling of the atmosphere, acid rain, and fallout of particulates that would be harmful to life. A truly large comet or asteroid could be what is sometimes termed a "planet cleansing" event that would cause mass extinction of much of the life on Earth.

14.5   Minimizing the Impact Hazard

LO:5   List the physical, chemical, and biological consequences of an impact from a large asteroid or comet.

LO:6   Analyze the risk of impact or airburst of extraterrestrial objects and suggest how that risk might be minimized.

- Risk assessment is difficult for events with long return periods (in this case, large objects striking Earth). Nevertheless, the potential consequences from an impact of a large comet or asteroid are so large that the product of the probability of occurring, which is very small, and the likelihood of major, worldwide exposure and vulnerability, which are regional to global, produces a relatively large risk, compared to other hazards that we normally think about. However, the probability of such an impact is so low in any one year that it is beyond our ability to comprehend the event ever happening.
- Sufficient studies of comets and asteroids and their interaction with planets enable us to make fairly straightforward predictions of the return time when an object of a particular size would impact Earth. For example, an event the size of Tunguska occurs on average every 1,000 years. On the other hand, the return period for an impact of an asteroid the size of the one that struck the Yucatán Peninsula 65 million years ago occurs approximately every 100 million years. Because the consequences of being struck by a large asteroid or comet are so great, there is an effort to identify near-Earth objects, and an impact hazard scale has been developed.
- Programs such as Spacewatch and NEAT (Near-Earth Asteroid Tracking) can, with high certainty, identify NEOs of diameters greater than a few hundred meters at least 100 years before possible impact. There are about 10 million smaller, potential Tunguska-type objects that could produce catastrophic damage to urban areas. Identifying all these objects will be extremely difficult. Thus, we are particularly vulnerable to impacts from these smaller objects.

0   10 Kilometers

KEY WORDS
near-Earth objects (p. 583)

- The human population of Earth today is about 7 billion, compared to about 1 billion in 1830. Thus, an event in 1830 that might have killed a million people would likely kill more than 10 million people today.
- The most encouraging note is that a large comet or asteroid would probably be seen a hundred years or more before it is likely to strike Earth. Therefore, we might have time to try to minimize the hazard through a variety of physical means. Smaller events that might catastrophically damage an urban area are more difficult to identify. If we assume that we might see such an object a year or so before impact, there might be time to evacuate if we can accurately predict where it will strike. However, evacuation of large urban centers with millions of people is difficult.

CRITICAL thinking questions

1. Summarize what has been learned from the 1908 impact in Russia.
2. Describe the likely results if a Tunguska-type event, were it to occur over or in central North America. If the event were predicted with 100 years' warning, what could be done to mitigate the effects, if changing the object's orbit were not possible? Outline a plan to minimize death and destruction.
3. How would the effects of an asteroid impact in water differ from those of an asteroid impact on land? Consider what would happen physically and chemically with water and how the impact craters might differ.
4. Compare the velocity of an asteroid or comet, seismic waves, and sound waves. Why do they differ?
5. How did the effects of the impact at the K-Pg boundary differ from the proposed airburst at the Younger Dryas boundary?

References

1. **Yeomans, D.,** and **Chodas, P.** 2013. Additional details on the large fireball event over Russia on Feb. 15, 2013. NASA Near Earth Object Program. http://neo.jpl.nasa.gov/news/fireball_130301.html.
2. **Cloud, P.** 1978. *Cosmos, Earth and man.* New Haven, CT: Yale University Press.
3. **Rubin, A. F.** 2002. *Disturbing the solar system.* Princeton, NJ: Princeton University Press.
4. **Lewis, J. S.** 1996. *Rain of iron and ice.* Redding, MA: Addison-Wesley.
5. **Davidson, J. P., Reed, W. E.,** and **Davis, P. M.** 1997. *Exploring Earth.* Upper Saddle River, NJ: Prentice Hall.
6. **Grieve, R. A. F.** 2006. *Impact structures in Canada.* St. Johns, Newfoundland and Labrador: Geological Association of Canada GEOtext 5.
7. **Brown, P., Spalding, R. E., ReVelle, D. O., Tagliaferri, E.,** and **Worden, S. P.** 2002. The flux of small near-Earth objects colliding with the Earth. *Nature* 420: 294–96.
8. **Lipschutz, M. E.,** and **Schultz, L.** 2007. Meteorites. In *Encyclopedia of the solar system,* 2nd ed., ed. L.-A. McFadden, P. R. Weissman, and T. V. Johnson, pp. 251–82. New York: Elsevier.
9. **Grieve, R. A. F., Cintala, M. J.,** and **Tagle, R.** 2007. Planetary impacts. In *Encyclopedia of the solar system,* 2nd ed., ed. L.-A. McFadden, P. R. Weissman, and T. V. Johnson, pp. 813–28. New York: Elsevier.
10. **Williams, S. J., Barnes, P.,** and **Prager, E. J.** 2000. U.S. Geological Survey coastal and marine geology research—Recent highlights and achievements. U.S. Geological Survey Circular 1199.
11. **Dott, Jr., R. H.,** and **Prothero, D. R.** 1994. *Evolution of the Earth,* 5th ed. New York: McGraw-Hill.
12. **Stanley, S. M.** 2005. *Earth system history.* New York: W. H. Freeman.
13. **Kring, D. A.,** and **Durda, D. D.** 2003. The day the world burned. *Scientific American* 289(6): 98–105.
14. **Alvarez, W.** 1997. *T. Rex and the crater of doom.* New York: Vintage Books, Random House.
15. **Alvarez, L. W., Alvarez, W., Asaro, F.,** and **Michel, H. V.** 1980. Extraterrestrial cause for Cretaceous-Tertiary extinction. *Science* 208: 1095–108.
16. **Pope, K. O., Ocampo, A. C.,** and **Duller, C. E.** 1991. Mexican site for the K/T impact crater? *Nature* 351: 105.
17. **Swisher, III, C. C., Grajales-Nishimura, J. N., Montanari, A., Margolis, S. V., Claeys, P., Alvarez,**

W., Ranne, P., Cedillo-Pardo, E., Maurrasse, F. J.-N. R., Curtis, G. H., Smit, J., and McWilliams, M. O. 1992. Coeval 40 Ar/39 Ar ages of 65.0 million years ago from Chicxulub crater melt rocks and Cretaceous-Tertiary boundary tektites. *Science* 257: 954–58.

18. Hildebrand, A. R., Penfield, G. T., Kring, D. A., Pilkington, N., Camargo, Z. A., Jacobsen, S. B., and Boynton, W. V. 1991. Chicxulub crater: A possible Cretaceous/Tertiary boundary impact crater on the Yucatan peninsula, Mexico. *Geology* 19: 867–71.

19. Matsui, T., Imamura, F., Tajika, E., Nakano, Y., and Fujisawa, Y. 2002. Generation and propagation of a tsunami from the Cretaceous-Tertiary impact event. In *Catastrophic events and mass extinctions*, ed. C. Koeberl, and K. G. MacLeod, pp. 69–77. Geological Society of America Special Paper 356.

20. Firestone, R. B., and 25 others. 2007. Evidence for an extraterrestrial impact 12,900 years ago that contributed to the megafauna extinctions and the Younger Dryas cooling. *Proceedings of the National Academy of Sciences (PNAS)* 104(41): 16016–21.

21. Lange, I. M. 2002. Ice Age mammals of North America: A guide to the big, the hairy, and the bizarre. Missoula, MT: Mountain Press.

22. Grayson, D. K. 1984. Nineteenth-century explanations of Pleistocene extinctions: A review and analysis. In *Quaternary extinctions: A prehistoric revolution*, ed. P. S. Martin, and R. G. Klein, pp. 5–39. Tucson: University of Arizona Press.

23. Haynes, Jr., C. V. 2008. Younger Dryas (black mats) and the Rancholabrean termination in North America. *Proceedings of the National Academy of Sciences (PNAS)* 105(18): 6520–25.

24. Waters, M. R., and Stafford, Jr., W. S. 2007. Redefining the age of Clovis: Implications for the peopling of the Americas. *Science* 315: 1122–26.

25. Kennett, D. J., and 10 others. 2008. Wildfire and abrupt ecosystem disruption on California's Northern Channel Islands at the Allerod–Younger Dryas boundary (13.0–12.9ka). *Quaternary Science Reviews* 27: 2528–43.

26. Wittke, J. H. and 27 others. 2013. Evidence for deposition of 10 million tons of impact spherules across four continents 12,800 years ago. *Proceedings of the National Academy of Sciences. PNAS* early edition. pnas.org/cgi/doi/10.1073/pnas.1301760110

27. Kennett, J. P., and 25 others. 2015. Bayesian chronological analysis consistent with synchronous age of 12,845–12,735 CalB.P. for Younger Dryas boundary on four continents. PNAS. E4344–E4353. www.pnas.org/cgi/doi/10.1073/pnas.1507145112.

28. Petaev, M. I. and 3 others. 2013. Large Pt anomaly in the Greenland ice core points to a cataclysm on the onset of the Younger Dryas. *Proceedings of the National Academy of Sciences. PNAS early edition*. pnas.org/cgi/doi101073/pnas1303924110.

29. Moore, C. R., and 12 others. 2017. Widespread platinum anomaly documented at the Younger Dryas onset in North American sedimentary sequences. www.nature.com/scientificreports/7:44032/DOI:10.1038/sre44031.

30. Kennett D. J., Kennett, J. P., West, A., Mercer, C., Que Hee, S. S., Bement, L., Bunch, T. E., Sellers, M., and Wolbach, W. S. 2009. Nanodiamonds in the Younger Dryas boundary sediment layer. *Science* 323: 94.

31. Jones, A. 2005. Meteorite impacts as triggers to large igneous provinces. *Elements* 1: 277–81.

32. Kerr, R. A. 2007. Humongous eruptions linked to dramatic environmental changes. *Science* 316: 527.

33. National Aeronautics and Space Administration. 2007. Near-Earth object survey and deflection analysis of alternatives: Report to Congress. Washington, DC: Author.

34. Near Earth Object Program, NASA. 2013. Near-Earth discovery statistics. http://neo.jpl.nasa.gov/stats/.

35. McFadden, L. A., and Binzel, R. P. 2007. Near-Earth objects. In *Encyclopedia of the solar system*, 2nd ed., ed. L. A. McFadden, P. R. Weissman, and T. V. Johnson, pp. 283–300. New York: Elsevier.

36. Morrison, D., Chapman, C. R., Steel, D., and Binzel, R. P. 2004. Impacts and the public: Communicating the nature of the impact hazard. In *Mitigation of hazardous comets and asteroids*, ed. M. J. S. Belton, T. H. Morgan, N. H. Smarasinha, and D. K. Yeomans, pp. 353–90. New York: Cambridge University Press.

37. Laboratory of Meteoritics. 2013. The Type of Meteorite "Chelyabinsk" Is Determined: Ordinary Chondrite Ll5 (S4, W0). www.meteorites.ru/menu/press-e/yuzhnouralsky2013-e.php.

APPENDIX A: Minerals

Characteristic Properties of Minerals

Minerals are naturally occurring substances with defined physical and chemical properties. Physical properties are commonly used for identification, and chemical composition is the basis for mineral classification. Mineral properties can also have an aesthetic or utilitarian value. For example, native copper, when discovered in the Great Lakes region by Native Americans, was first valued for its metallic properties of luster and malleability and was used for jewelry and tools. Many ancient cultures, in addition to our own, have valued gemstones such as rubies and sapphires for their beauty. Historically, one of the most valued minerals has been halite (NaCl), or common table salt, because all animals, including people, need it to live. Halite has been mined and obtained from coastal evaporation ponds for thousands of years. Some clay minerals have been highly valued because they can be molded into useful pots and decorative statues and painted with black, red, and orange paint made from still other minerals. Today, specific physical properties of a wide variety of minerals are used for everything from ceramics to electronics, metallurgy, agriculture, water treatment, and personal jewelry. An understanding of the physical properties will help us identify common minerals.

Identifying Minerals

Identification of mineral specimens that are large enough to fit in the palm of your hand, referred to as hand specimens, is a combination of pattern recognition and testing for particular properties or characteristics. These

▼ TABLE A.1 Mohs Hardness Scale

	Relative Hardness[1]	Mineral	Comment	Hardness of Common Materials
Softest	1	Talc	Softest mineral known; used to make powder	Graphite, "lead" in pencils (1–2)
	2	Gypsum	Used to make wall board	Fingernail (~2.2)
	3	Calcite	Main mineral in marble and limestone	Copper penny[2] (~3.2)
	4	Fluorite	Mined for fluorine, used in glass and enamel	
	5	Apatite	Enamel of your teeth	Pocketknife (~5.1), Glass (~5.5)
	6	Orthoclase	Common rock-forming mineral	Hard steel file or needle (~6.5)
	7	Quartz	When purple is gemstone amethyst, birthstone for February	
	8	Topaz	When transparent is gemstone, birthstone for November	
	9	Corundum	When red is ruby, gemstone, birthstone for July; when blue is sapphire, gemstone, birthstone for September	
Hardest	10	Diamond	Hardest known mineral; gemstones are extremely brilliant	

[1] Note: The scale is relative, and unit increases in hardness do not represent an equal increase in true hardness.

[2] U.S. penny minted before 1982 or Canadian penny minted before 1997.

properties include hardness, specific gravity, fracture, cleavage, crystal form, color, luster, as well as other properties that are diagnostic of a particular mineral or group of minerals.

HARDNESS

Most physical properties of minerals reflect characteristics of their internal atomic structure. For example, a mineral's *hardness*, that is, its resistance to scratching, is related to the size, spacing, and strength of bonding of the atoms within the mineral. Absolute hardness is not easy to measure, so in 1822 Austrian mineralogist Friedrich Mohs came up with an ingenious 1–10 scale of *relative hardness* (H) that is still widely used. On the Mohs scale the softest mineral is talc (H = 1) and the hardest mineral is diamond (H = 10) (Table A.1). The change in absolute hardness from one number to the next is not the same; any mineral or substance will scratch a mineral with a lower hardness on the Mohs scale. You can use common objects, such as a fingernail (H about 2.2), a copper penny (a U.S. penny minted before 1982, or a Canadian penny minted before 1997, H about 3.2), and a glass plate (H = 5.5) to approximate the relative hardness of minerals.

For example, you can determine if a mineral is harder or softer than 5.5 by taking a fresh, unweathered surface of the mineral and attempting to scratch a glass plate. You must be certain that the mineral is actually making a scratch and not just leaving a line of powdered mineral on the surface. Do not attempt to use a glass test plate if it becomes fractured or broken; be sure to follow proper procedures in disposing of broken or damaged glass. If the mineral is softer than glass, then you can determine if it is harder than a copper penny. By this process you can bracket the relative hardness (e.g., 73.2 and 65.5) of the unknown mineral sample. Hardness tests can sometimes be misleading, so never rely on hardness alone to identify a mineral.

SPECIFIC GRAVITY

Some minerals have unusually light or heavy specific gravity. The *specific gravity* of a mineral is the density of a mineral compared with the density of water. Water has a specific gravity of 1. The specific gravity of minerals varies from about 2.2 for halite to 19.3 for gold. Most minerals have specific gravities of about 2.5 to 4.5. In practice, we take a sample of a mineral in hand, heft it, and estimate if it has a light, medium, or heavy specific gravity.

CLEAVAGE

Many minerals have a distinctive way in which they break (Table A.2). Those that break along smooth, even planar surfaces are said to have cleavage. *Cleavage*

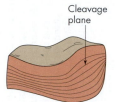
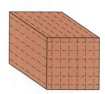

(a) One direction of cleavage. Mineral examples: muscovite and biotite micas

(b) Two directions of cleavage at right angles or nearly right angles. Mineral examples: feldspars, pyroxene

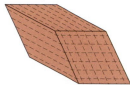
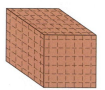

(c) Two directions of cleavage not at right angles. Mineral example: amphibole

(d) Three directions of cleavage at right angles. Mineral examples: halite, galena

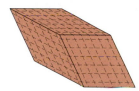

(e) Three directions of cleavage not at right angles. Mineral example: calcite

▲ FIGURE A.1 Common Types of Cleavage
Except for mica, where cleavage is indicated with solid lines, cleavage directions in this diagram are shown with dashed lines. (a) Mica has one cleavage direction and breaks into sheets. (b) Feldspar and pyroxene have two cleavage directions at nearly right angles. (c) Amphibole has two directions of cleavage that do not meet at right angles. (d) Halite and galena have three cleavage directions that meet at right angles. This relationship causes both minerals to break into cube-shaped fragments. (e) Calcite also has three cleavage directions, but they do not meet at right angles. This relationship causes calcite to break into rhombohedrons.

develops because the forces that bond atoms together in a crystalline structure are rarely equal in all directions. Most cleavage will appear as a series of small, reflective steps with each parallel step breaking along the same direction of weak bonding within the mineral. One technique for recognizing cleavage is to rotate the specimen in front of you with a light source, such as the Sun or a ceiling light, over your shoulders. Each time you encounter a direction of cleavage, the small, parallel cleavage steps will reflect light all at once.

▼ **TABLE A.2** **Properties and Environmental Significance of Selected Common Minerals**

Mineral Class	Mineral	Chemical Formula	Color	Hardness	Other Characteristics	Comment
Silicates	Plagioclase feldspar	$(Na, Ca)Al(Si, Al)$ Si_2O_8	Usually white or gray, but may be others	6	Two good cleavages at approximately 90°; and may have fine striations on one of the cleavage surfaces	One of the most common of the rock-forming minerals; is a group of feldspars ranging from sodium to calcium rich; important industrial minerals
	Alkali feldspar	$(Na, K)AlSi_3O_8$	Gray, white to pink or salmon color	6	Two good cleavages at 90°; may be translucent to opaque with glassy luster	One of the most common of the rock-forming minerals; used in porcelain and a wide variety of industrial processes
	Quartz	SiO_2	Varies from colorless to white, gray, pink, purple, and several others, depending on impurities	7	Often has good crystal shape with six sides; conchoidal fracture; coarse crystalline varieties have glassy luster, and microcrystalline varieties have dull to waxy luster	Very common rock-forming mineral; resistant to most chemical weathering; basic constituent of glasses and fluxes; commonly used as an abrasive; colored varieties, such as amethyst and agate, are semiprecious gemstones
	Pyroxene	$(Ca, Mg, Fe)_2Si_2O_6$	Usually greenish to black	5–6	Crystals are commonly short and stout; two cleavages at about 90°	Important group of rock-forming minerals; particularly common in igneous rocks; weathers rather quickly
	Amphibole	$(Na, Ca)_2(Mg, Al, Fe)_5Si_8O_{22}(OH)_2$	Generally black to green	5–6	Distinguished from pyroxene by the cleavage angle being 120° rather than 90°; generally has better cleavage with higher luster than pyroxene	Important group of rock-forming minerals particularly for igneous and metamorphic rocks; relatively nonresistant to weathering
	Olivine	$(Mg, Fe)_2SiO_4$	Generally green, but also may be yellowish	6.5–7	Conchoidal fracture; commonly occurs in aggregates of small glassy grains	Important rock-forming mineral, particularly for igneous and metamorphic rocks; relatively nonresistant to chemical weathering
	Clay	Various hydrous aluminum silicates with elements such as Ca, Na, Fe, Mg, and K	Generally white, but may vary due to impurities	1–2	Generally found as soft, earthy masses composed of very fine grains; may have an earthy odor when moist; often difficult to identify the particular clay mineral from hand specimen	Clay minerals are very important from an environmental viewpoint; many uses in society today; clay-rich soils often have many engineering geology problems

	Name	Composition	Color	Hardness	Cleavage/Breakage	Uses/Significance
	Chlorite	Hydrous Mg, Fe, Al silicate	Green to dark green	2–2.5	Commonly in layered or scaly masses with one direction of cleavage; thin sheets are flexible and nonelastic; glassy to pearly luster	Important group of rock-forming minerals in metamorphic rocks
	Talc	$Mg_3Si_4O_{10}(OH)_2$	Pale green to white or gray	1	In layered or compact masses with one direction of cleavage; thin sheets flexible and nonelastic; soapy feel with pearly to greasy luster	Used to make ceramics, paint, paper roofing, cosmetics, and ornamental stone
	Biotite (black mica)	$K(Mg, Fe)_3AlSi_3O_{10}(OH)_2$	Black to dark brown or dark green	2.5–3	Breaks apart in parallel sheets as a result of excellent cleavage in one direction	Important rock-forming mineral common in igneous and metamorphic rocks
	Muscovite (white mica)	$KAl_2(AlSi_3O_{10})(OH)_2$	White, light yellow, brown, pink, or green; colorless in cleavage sheets	2–3	Breaks apart into thin sheets due to excellent cleavage in one direction	Important rock-forming mineral, particularly for igneous and metamorphic rocks; used for several purposes, including industrial roofing materials, paint, and rubber
Carbonates	Calcite	$CaCO_3$	Colorless to white, but may have a variety of colors resulting from impurities	3	Effervesces strongly in dilute hydrochloric acid; often breaks apart to characteristic rhombohedral pieces as a result of two good cleavages at 78°; transparent varieties display double refraction, where a single dot on white paper looks like two dots if viewed through the mineral	Main constituent of the important sedimentary rock limestone and the metamorphic rock marble; associated with a variety of environmental problems with these rocks, including the development of sinkholes; chemically weathers rapidly; used in a variety of industrial processes including asphalts, fertilizers, insecticides, cement, and plastics
	Dolomite	$Ca, Mg(CO_3)_2$	Generally white, but may be a variety of other colors, including light brown and pink	3.5–4	When powdered, will slowly effervesce in dilute hydrochloric acid; two cleavages at 78°; may be transparent to translucent and have a vitreous to pearly luster	Common mineral in the sedimentary rock dolostone and dolomitic limestone
	Malachite	$Cu_2CO_3(OH)_2$	Bright to emerald green, may be dark green	3.5–4	Effervesces slightly in dilute hydrochloric acid and turns the acid solution green; often displays a swirling structure of light and darker green bands	Valued as a decorative stone; used in jewelry and as a copper ore

(Continued)

Mineral Class	Mineral	Chemical Formula	Color	Hardness	Other Characteristics	Comment
Oxides and Hydroxides	Hematite	Fe_2O_3	Usually various shades of reddish brown to red to dark gray	1–6.5	Streak is dark red	A common mineral found in small amounts in many igneous rocks, particularly basaltic rocks
	Magnetite	Fe_3O_4	Black	6	Magnetic	The most important ore of iron
	Goethite	$FeO(OH)$	Generally yellow to yellow-brown or black	1–5.5	Often present as earthy masses or in the form of a crust; streak is yellow-brown	Formerly referred to as limonite, which is now known to be relatively rare; often forms from the chemical weathering of iron minerals and is present as "rust"
Sulfides	Pyrite	FeS_2	Generally a pale brassy yellow tarnishing to brown	6–6.5	Often present as well-formed cubic crystals with striations on crystal faces	Has been used in the production of sulfuric acid, but mostly known for adding sulfur content to coal and contributing to the formation of acid-rich waters that result from the weathering of the mineral
	Chalcopyrite	$CuFeS_2$	Dark brassy to golden yellow, tarnishing to iridescent films that are reddish to blue-purple	3.5–4	Easily disintegrates; greenish black streak lacks cleavage	An important ore of copper; often associated with gold and silver ores
	Galena	PbS	Silver gray	2.5	Gray to black streak; high specific gravity; general metallic luster; cubic cleavage	Primary ore of lead
Sulfates	Gypsum	$CaSO_4 \cdot 2H_2O$	Generally colorless to white, but may have a variety of colors due to impurities	2	Often is transparent to opaque with one perfect cleavage; may form fibrous crystals, but often is an earthy mass	Several industrial uses in making of plaster of Paris for construction material, fertilizer, and flux for pottery
	Anhydrite	$CaSO_4$	Commonly white or gray, but may be colorless	3–3.5	Commonly observed as massive fine aggregates; may be translucent to transparent	Used for the production of sulfuric acid, as a filler material in paper, and occasionally as an ornamental stone

Group	Mineral	Formula	Color	Hardness	Distinguishing Characteristics	Uses
Halides	Fluorite	CaF_2	Variable colors, often purple, yellow, white, or green	4	Often observed as cubic crystals but may also be massive; four perfect cleavages	A number of industrial uses, including a flux in the metal industry, optical lenses, and in the production of hydrofluoric acid
	Halite	$NaCl$	Generally colorless to white, but may be a variety of colors due to impurities	2.5	Salty taste; often has a cubic form due to three perfect cleavages	Common table salt; makes up rock salt, which is used for ice removal and as a host for buried nuclear waste
Native Elements	Gold	Au	Gold yellow	2.5–3	Crystals are rare; extremely high specific gravity; very ductile and malleable	Primary use as a monetary standard; used in jewelry and in the manufacture of computer chips and other electronics
	Diamond	C	Generally colorless, but also occurs in various shades of yellow, brown, blue, pink, green, and orange	10	Commonly cut to display brilliant luster	Used in industrial abrasives and, of course, jewelry
	Graphite	C	Black to gray	1–2	Often occurs as foliated masses; black streak and marks paper as well; greasy feel	A variety of industrial uses including lubricants, dyes, and in pencils as the "lead"
	Native Sulfur	S	Usually yellow when pure, but may be brown to black from impurities	1.5–2.5	Characteristic sulfurous smell, similar to the smell of fireworks or gunpowder	Often a by-product of the oil industry; may be used to manufacture sulfuric acid

Modified after Davidson, J. P., Reed, W. E., and Davis, P. M. 1997. *Exploring Earth*. Upper Saddle River, NJ: Prentice Hall; and Birchfield, B. C., Foster, R. J., Keller, E. A., Melhom, W. N., Brookins, D. G., Mintt, L. W., and Thurman, H. V. 1982. *Physical geology*. Columbus, OH: Charles E. Merrill.

Two characteristics of cleavage are important in identifying minerals: the number of cleavage directions and the angle between the cleavage directions. Minerals may have one, two, three, four, or six cleavage directions (Figure A.1). When you count the number of cleavage directions, be sure not to count the same direction twice because cleavage surfaces commonly repeat themselves on opposite sides of a specimen. In determining the angle between two cleavages, it is generally sufficient to decide whether or not the angle is close to 90°. In some specimens there is the potential to confuse a cleavage surface with a crystal face that developed as the crystal grew. Unlike cleavage surfaces, crystal faces are generally not repetitive, nor do they form imperfect steps.

CRYSTAL FORM

Crystal faces are important, however, in recognizing *crystal form*, which is the geometric shape taken on by the mineral as it crystallized. Unfortunately, most mineral specimens do not show crystal form; however, when it is present, crystal form can be a useful diagnostic property for identifying some minerals. For example, the pointed and elongate hexagonal crystal form of quartz is diagnostic, as are the cube-shaped crystals of pyrite.

FRACTURE

Broken surfaces of a mineral that do not show cleavage are described as *fracture*. The type of fracture that is present is generally important only if the specimen has no cleavage. Although most fracture is uneven or irregular, there are two types of fracture that are distinctive: *conchoidal fracture* and *fibrous fracture*. In conchoidal fracture, the broken surface is curved and can be at least partially smooth. A broken glass bottle will exhibit conchoidal fracture. In many cases a conchoidal fracture surface will reflect light continually as a specimen is rotated.

COLOR

The *color* of a mineral can be misleading because a given mineral may have several different colors, or several different minerals may have the same color. For example, depending on the impurities present, normally clear quartz may be white, pink, purple, red, brown, yellow, green, or black.

A more reliable way of using color to identify some minerals is to powder part of a specimen on an unglazed porcelain plate. The resulting *streak* made by the powder is especially useful in identifying metallic minerals. For example, the hand specimen of the mineral hematite (Fe_2O_3) may be dull black or a shiny, dark metallic silver color, but its streak is always red. Streak tests can be done only for minerals that are softer than the porcelain "streak plate," which typically has a hardness of about 6.5.

LUSTER

Luster refers to the way light is reflected from a mineral. Most minerals can be described as having either a metallic or nonmetallic luster. Minerals with metallic luster appear silver, brass, or shiny, but not an enamel-like black, and are opaque, even on thin edges. Nonmetallic luster can be quite variable with descriptive terms for luster such as glassy, pearly, greasy, earthy, and resinous (like pine sap or dried turpentine). Light will also pass through thin edges of a nonmetallic mineral.

Some minerals have more than one luster. For example, some specimens of graphite have metallic luster and others are nonmetallic.

STEPS IN IDENTIFYING A MINERAL

1. Decide if the mineral has a metallic or a nonmetallic luster and refer to Table A.3. If the mineral is metallic, determine the color of its streak.

2. Determine the relative hardness of the specimen by first testing to see if it will scratch glass or alternatively, if it can be scratched by a knife (Table A.1). Note that if the mineral has a hardness close to that of glass (H about 5.5) or a pocket knife (H about 5.1), determining hardness may be difficult and the mineral may appear in Table A.3 in either the "harder than glass" or "softer than glass" sections.

3. Decide whether the mineral has cleavage. If cleavage is present, determine the number of directions and whether the angle between the directions is close to 90°. If cleavage is not present, determine the type of fracture.

4. Once you think you have identified the mineral, test for other physical or chemical properties (e.g., taste, smell, reaction with dilute hydrochloric acid) listed for that mineral in Tables A.2 and A.3 to confirm your identification.

The identification of minerals in the field or in the laboratory from small specimens, using simple tests and perhaps a hand lens to determine cleavage and other properties, is basically a pattern recognition exercise. After you have looked at a number of minerals over a period of time and have learned how particular minerals vary, you will become more proficient at mineral identification. When positive identification is necessary, mineralogists use a variety of sophisticated analytical equipment to determine the chemical composition and internal structure of a sample.

▼ TABLE A.3 Key to Assist Mineral Identification

To use this table; (1) decide whether the luster of the mineral to be identified is nonmetallic or metallic. If it is nonmetallic, decide whether it is a light-colored or a dark-colored mineral; (2) use a glass plate to determine if the mineral specimen is harder or softer than the glass plate; (3) look for evidence of cleavage; (4) compare your samples with the other properties shown here and use Table A.2 for final identification.

Luster	Hardness	Cleavage	Description	Mineral
Light-colored nonmetallic luster	Harder than glass	Shows cleavage	White, pink or salmon-colored; 2 cleavage planes at nearly right angles; hardness, 6.	Orthoclase (Alkali feldspar)
			White, gray, or green-gray; 2 cleavage planes at nearly right angles; hardness, 6; striations on one cleavage.	Plagioclase
		No cleavage	White, clear, or any color; glassy luster; transparent to translucent; hexagonal (six-sided) crystals; hardness, 7; conchoidal fracture.	Quartz
			Various shades of green and yellowish green; glassy luster; granular masses and crystals in rocks; Olivine hardness, 6.5–7 (apparent hardness may be much less in granular masses).	Olivine
	Softer than glass	Shows cleavage	Colorless to white; salty taste; 3 perfect cleavages forming cubic fragments; hardness, 2.5.	Halite
			White, yellow to colorless; 3 perfect cleavages forming rhombohedral fragments; hardness, 3; effervesces with dilute hydrochloric acid.	Calcite
			Pink, colorless, white, or brown; rhombohedral cleavage; hardness, 3.5–4; effervesces with dilute hydrochloric acid only if powdered.	Dolomite
			White to transparent; 1 perfect cleavage; hardness, 2.	Gypsum
			Pale green to white; feels soapy; 1 cleavage; hardness, 1.	Talc
			Colorless to light yellow, brown, or green; transparent in thin elastic sheets; 1 perfect cleavage; hardness, 2–3.	Muscovite
			Colorless to yellow to light blue, green, or purple; 4 cleavages not at 90°; hardness, 4.	Fluorite
		No cleavage	Pale green to white; feels soapy; hardness, 1.	Talc
			White to transparent; hardness, 2.	Gypsum
			Yellow; resinous luster; smells like fireworks; hardness, 1.5–2.5.	Sulfur
Dark-colored nonmetallic luster	Harder than glass	Shows cleavage	Black to dark green; 2 cleavage planes at nearly 90°; hardness, 5–6.	Pyroxene
			Black to dark green; 2 cleavage planes about 120°; hardness, 5–6.	Amphibole
		No cleavage	Various shades of green and yellow; glassy to dull or waxy luster; granular masses and crystals in rocks; hardness, 6.5–7 (apparent hardness may be much less).	Olivine
			White, clear, or any color; glassy luster; transparent to opaque; hexagonal (six-sided) crystals; hardness, 7; conchoidal fracture.	Quartz
			Steel gray to black; reddish brown streak; earthy appearance; hardness, 5.5–6.5.	Hematite
	Softer than glass	Shows cleavage	Black to brownish or greenish black; cleavage, 1 direction; hardness, 2.5–3.	Biotite
			Violet, yellow, green, pink, white, and colorless; 4 cleavages other than 90°; hardness, 4.	Fluorite
			Various shades of green; cleavage, 1 direction; in layered or scaly masses; hardness, 2–2.5.	Chlorite
		No cleavage	Dull to bright red; reddish brown streak; earthy appearance; hardness, 1–6.5.	Hematite
			Lead-pencil black; smudges fingers; hardness, 1–2; 1 cleavage that is apparent only in large crystals.	Graphite
			Yellow-brown to dark brown; may be almost black; streak yellow-brown; earthy; hardness (usually soft), 1–5.5.	Goethite
Metallic luster			Black; strongly magnetic; hardness, 6.	Magnetite
			Lead-pencil black; smudges Angers; hardness, 1–2; 1 cleavage that is apparent only in large crystals.	Graphite
			Light brass yellow; black streak; cubic crystals, commonly with striations; hardness, 6–6.5.	Pyrite
			Dark brass yellow; may be tarnished; greenish black streak; hardness, 3.5–4; massive.	Chalcopyrite
			Shiny gray; black streak; very heavy; cubic cleavage; hardness, 2.5.	Galena

Modified after Birchfield, B. C., Foster, R. J., Keller, E. A., Melhom, W. N., Brookins, D. G., Mintt, K. W., and Thurman, H. V. 1982. *Physical geology.* Columbus, OH: Charles E. Merrill; and Hamblin, W. K. and Howard, J. D. 2005. *Exercises in physical geology,* 12th ed. Upper Saddle River, NJ: Pearson Prentice Hall.

APPENDIX B: Rocks

In Chapter 1 we stated that a *rock* is an aggregate of one or more minerals. In traditional geologic investigations, this is the most commonly used definition. Another definition is used by geologists and geological engineers who are concerned with the properties of earth materials that affect engineering design and environmental problems associated with natural hazards. In environmental and engineering geology, the term *rock* is reserved for earth materials that cannot be removed without blasting using explosives, such as dynamite; earth materials that can be excavated with normal earth-moving equipment, for example a shovel or a bulldozer, are called *soil*. Thus, loosely compacted, poorly cemented sandstone may be considered a soil, whereas well-compacted clay may be called a rock.

This pragmatic definition of rock described above also applies to the use of the term *soil* in engineering and environmental geologic reports. To an engineer, soil is earth material that can be excavated without blasting, regardless of whether it is on land or under water. In traditional geologic investigations the term *soil* is restricted to surficial earth material that has developed on land by in-place weathering. Other loose earth material not formed by in-place weathering is generally considered *sediment*, distinguished from sedimentary rock, which requires a hammer to break it apart. These differences in definitions are more than academic; they can affect communication between engineers, scientists, architects, planners, and emergency responders.

Identifying Rocks

Rocks are composed of minerals, and in some cases other substances such as natural glass and pieces of organisms. Whereas a few rocks contain essentially only one mineral, most contain several. As with minerals, the identification of hand specimens of rocks is primarily through the recognition of patterns. However, there are some useful hints that will assist you. The first task is to decide if the rock is igneous, sedimentary, or metamorphic. Sometimes this decision is not as easy as you might think. It is particularly difficult to identify rocks that are fine grained. A specimen is considered fine grained if you cannot see individual mineral grains with your naked eye. Even if a specimen is not fine grained, it is advisable to use a magnifying glass or hand lens to examine a mineral or a rock.

Some general rules of thumb will assist you in getting started: (1) If the sample is composed of bits and pieces of other rocks, it is most likely sedimentary, especially if the pieces have rounded corners; (2) if the rock specimen has minerals that are all oriented in the same direction, such as parallel sheets of mica, then it is most likely metamorphic; (3) if individual mineral crystals are relatively coarse grained, meaning you can see them with the naked eye or easily with a hand lens, and the crystals are interlocking and composed of minerals such as quartz and feldspar, it is probably an igneous rock that cooled underground, such as granite; (4) if the rock is mostly fine grained but contains some larger crystals, it is probably an igneous rock that cooled on or just below Earth's surface; (5) if the rock is relatively soft and effervesces (fizzes) when dilute hydrochloric acid is applied, it is probably a carbonate sedimentary rock, such as limestone, or a metamorphic rock, such as marble; (6) if it is a really grungy, soft, highly weathered, or altered rock, as many in nature seem to be, you are going to have some problems identifying it!

Physical Properties of Rocks

Physical properties of rocks include color, specific gravity, relative hardness, porosity, permeability, texture, and strength.

COLOR

The color of a rock varies depending upon the minerals present and the amount of weathering that has occurred. Rocks encountered in their natural environment are various shades of light gray to brown to black.

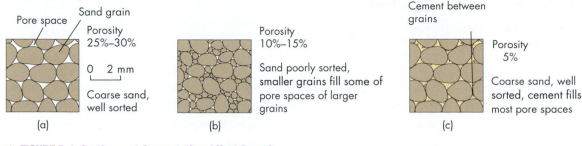

∧ FIGURE B.1 Sorting and Cementation Affect Porosity
(a) Porosity of a well-sorted coarse sand. (b) Porosity of a poorly sorted sand. (c) Porosity of a cemented, well-sorted coarse sand; examples include calcium carbonate ($CaCO_3$) and silica (SiO_2).

Chemical weathering can produce surface stains of black iron or manganese oxides, and brown to orange stains of iron oxides and hydroxides.

SPECIFIC GRAVITY AND RELATIVE HARDNESS

The same definition of specific gravity we gave for minerals applies to rocks and refers to the weight of the rock relative to the weight of water. Some rocks are lighter than water and will float. For example, pumice, a light-colored volcanic rock with many holes produced by gas bubbles, is lighter than water. Rocks become heavier as iron- and magnesium-bearing minerals become more abundant. There is no scale for relative rock hardness similar to the Mohs scale for minerals. Soft rocks can be broken with your fingers, and hard rocks require a sledgehammer to break apart.

POROSITY AND PERMEABILITY

Porosity is the percentage of a rock's volume that is empty space, that is void space between grains, or open space in fractures. *Permeability* is the capacity of a porous rock to transmit a fluid, such as oil or water. The properties of porosity and permeability are important in understanding many geologic hazards, such as landsliding and earthquakes.

TEXTURE

The *texture* of a rock refers to the size, shape, and arrangement of the crystals or grains within it. Crystal or grain size can be measured directly or described qualitatively as fine or coarse grained. A rock with crystals or grains that are visible with the naked eye or easily with a magnifying hand lens is considered coarse grained; one that is not is fine grained. The degree to which grains in a rock are similar in size is referred to as *sorting*. For example, a sandstone in which most sand grains are similar in size is described as well sorted (Figure B.1a, c), whereas one that contains a variety of grain sizes is poorly sorted (Figure B.1b). Crystal or grain shape can vary from spherical with rounded corners to irregular with sharp, angular corners. Likewise, the arrangement of crystals or grains can be variable. Crystals can be interlocking, such as in granite or marble; aligned, such as mica grains in schist; or small grains may occupy spaces between large grains, such as in poorly sorted sandstone (Figure B.1b). Other rock textures are characterized by vesicles, or cavities, formed by gas expansion during their formation. For example, vesicular rocks commonly occur near the surface of a lava flow where gas is escaping. The texture of a rock can also influence its porosity. Rocks that are poorly sorted or well cemented generally will have lower porosity than those that are well sorted and only partially cemented. For example, in sandstone, porosity decreases from about 30 percent to 5 percent as pore spaces are filled with smaller grains or cementing material such as calcite ($CaCO_3$) or quartz (SiO_2) (Figure B.1).

STRENGTH

The *strength* of a rock is commonly reported in terms of its compressive strength, which is the mechanical compression necessary to break or fracture the specimen. As measured in the laboratory, the strength of a rock is given as a force per unit area, such as Newtons per meter squared (N/m^2). Ranges of compressive strength for some of the common rock types are shown in Table B.1.

▼ TABLE B.1 Strength of Common Rock Types

	Rock Type	Range of Compressive Strength (10^6 N/m^2)	Comments
Igneous	Granite	100 to 280	Finer-grained granites with few fractures are the strongest; generally suitable for most engineering purposes.
	Basalt	50 to greater than 280	Zones of broken rock holes or fractures reduce the strength.
Metamorphic	Marble	100 to 125	Solution openings and fractures weaken the rock.
	Gneiss	160 to 190	Generally suitable for most engineering purposes.
	Quartzite	150 to 600	Very strong rock.
Sedimentary	Shale	Less than 2 to 215	May be a very weak rock for engineering purposes; careful evaluation is necessary.
	Limestone	50 to 60	May have clay layers, solution openings, or fractures that weaken the rock.
	Sandstone	40 to 110	Strength varies with nature and extent of fractures, composition of grains, degree and type of mineral cement.

Data primarily from Bolz, R. E., and Tuve, G. L. eds. 1973. *Handbook of tables for applied engineering science*. Cleveland, OH: CRC Press.

APPENDIX C: Maps and Related Topics

Topographic Maps

A *topographic map* depicts the shape and elevation of the land surface as well as the location of human-made features, such as roads, schools, and political boundaries. Elevations and some other natural features are shown by *contour lines*, which on topographic maps are lines of equal elevation above sea level. Individual contour lines on a topographic map are a fixed vertical distance apart, known as the *contour interval* (CI). Common contour intervals are 5, 10, 20, 40, 80, or 100 meters or feet. The contour interval selected for a particular map depends upon the topography being represented as well as the scale of the map. If there is relatively little difference in elevation, or relief, between the highest and lowest points on a map, then the contour interval selected may be relatively small; however, if there is higher relief, the map maker will choose a larger contour interval so that the reader can resolve individual contour lines on steep slopes. This brings up an important point: Where contour lines are close together, the inclination or slope of the land is relatively steep, compared to flatter areas where contour lines are farther apart.

The *scale* of a topographic map, or of any other map, may be communicated in several ways. First, a scale may be given as a ratio, such as 1 to 24,000 (1:24,000), which means that 1 inch on a map is equal to 24,000 inches (2000 ft.) on the ground. Second, a topographic map may have several graphic or bar scales that are subdivided in feet, kilometers (kilometres), and miles. These scales are generally found in the lower margin of the map and are useful for measuring distances. Finally, the scale of some maps is stated in terms of specific units of length on the map, such as 1 inch equals 200 feet. This means that 1 inch on the map is equivalent to 200 feet (2400 in.) on the ground. In this example we could also state that the scale is 1 to 2400. The most common scale for U.S. Geological Survey topographic maps is 1:24,000 and for Natural Resources Canada maps is 1:50,000, but scales of 1:125,000 or smaller are also used. Remember, 1 ÷ 24,000 is a larger number than 1 ÷ 125,000, so 1:125,000 is the smaller scale. Generally, the smaller the map scale, the more area is shown. In addition to contour lines, topographic maps also show a number of cultural features, such as

roads, houses, and other buildings. Features such as streams and rivers are often shown in blue. In fact, a whole series of symbols are commonly used on topographic maps. These symbols are shown in Figure C.1 for United States maps, and similar symbols are printed on the back of most Canadian maps.

READING TOPOGRAPHIC MAPS

Reading or interpreting topographic maps is as much an art as it is a science. After you have looked at many topographic maps that represent the variety of landforms and features found at the surface of Earth, you begin to recognize these forms by the shapes of the contours. This is a process that takes a fair amount of time and experience in looking at a variety of maps. However, there are some general rules for reading topographic maps:

> Valleys containing streams of any size have contours that form Vs, in which the apex of the V points in the upstream direction. This is sometimes known as the *rule of Vs*. Thus, if you are trying to draw the drainage pattern that shows all the streams, you should continue the stream in the upstream direction as long as the contours are still forming a V-pattern. The Vs will no longer be noticeable near the drainage divide, that is, the high point between two drainage basins.

> Where contour lines are spaced close together, the slope or inclination of the land surface is relatively steep, and where contour lines are spaced relatively far apart, the slope is relatively low and the land is flatter. As you travel up an incline or slope, you may notice that the contour lines are spaced relatively far apart and that at some point the angle of inclination or slope changes and the contour lines become closer together; this point is known as "break in slope." A break in slope is commonly observed at the foot of a mountain or where a valley side meets the surface of a floodplain.

> Contours near the upper parts of hills or mountains may show closure; that is, the contour line meets itself rather than extending to the edge of the map. These closed contours may be relatively oval or round for a conical peak, or longer and

Control data and monuments

Vertical control

Third order or better, with tablet	BM ×16.3
Third order or better, recoverable mark	× 120.0
Bench mark at found section corner	BM ×18.3
Spot elevation	× 5.3

Contours

Topographic

Intermediate	
Index	
Supplementary	
Depression	
Cut; fill	

Bathymetric

Intermediate	
Index	
Primary	
Index primary	
Supplementary	

Boundaries

National	
State or territorial	
County or equivalent	
Civil township or equivalent	
Incorporated city or equivalent	
Park, reservation, or monument	

Surface features

Levee	Levee
Sand or mud area, dunes, or shifting sand	(Sand)
Intricate surface area	(Strip mine)
Gravel beach or glacial moraine	(Gravel)
Tailings pond	(Tailings pond)

Mines and caves

Quarry or open pit mine	
Gravel, sand, clay, or borrow pit	
Mine dump	(Mine dump)
Tailings	(Tailings)

Vegetation

Woods	
Scrub	
Orchard	
Vineyard	
Mangrove	(Mangrove)

Glaciers and permanent snowfields

Contours and limits	
Form lines	

Marine shoreline

Topographic maps

Approximate mean high water	
Indefinite or unsurveyed	

Topographic-bathymetric maps

Mean high water	
Apparent (edge of vegetation)	

Coastal features

Foreshore flat	
Rock or coral reef	
Rock bare or awash	
Group of rocks bare or awash	
Exposed wreck	
Depth curve; sounding	
Breakwater, pier, jetty, or wharf	
Seawall	

Rivers, lakes, and canals

Intermittent stream	
Intermittent river	
Disappearing stream	
Perennial stream	
Perennial river	
Small falls; small rapids	
Large falls; large rapids	
Masonry dam	
Dam with lock	
Dam carrying road	
Perennial lake; Intermittent lake or pond	
Dry lake	(Dry lake)
Narrow wash	
Wide wash	Wide wash
Canal, flume, or aquaduct with lock	
Well or spring; spring or seep	

Submerged areas and bogs

Marsh or swamp	
Submerged marsh or swamp	
Wooded marsh or swamp	
Submerged wooded marsh or swamp	
Rice field	(Rice)
Land subject to inundation	Max pool 431

Buildings and related features

Building	
School; church	
Built-up area	
Racetrack	
Airport	
Landing strip	
Well (other than water); windmill	
Tanks	
Covered reservoir	
Gaging station	
Landmark object (feature as labeled)	
Campground; picnic area	
Cemetery: small; large	

Roads and related features

Roads on Provisional edition maps are not classified as primary, secondary, or light duty. They are all symbolized as light duty roads.

Primary highway		
Secondary highway		
Light duty road		
Unimproved road		
Trail		
Dual highway		
Dual highway with median strip		

Railroads and related features

Standard gauge single track; station	
Standard gauge multiple track	
Abandoned	

Transmission lines and pipelines

Power transmission line; pole; tower	
Telephone line	Telephone
Aboveground oil or gas pipeline	
Underground oil or gas pipeline	Pipeline

∧ **FIGURE C.1 Topographic Map Symbols**
This chart shows some of the common symbols that the U.S. Geological Survey uses on its topographic maps. Older maps may have slightly different symbols. Symbols on Canadian topographic maps are printed on the reverse side of the map. *(From U.S. Geological Survey)*

narrower for a ridge. Remember, the elevation of the top of a mountain or hill is higher than the last contour shown and may be estimated by taking half the contour interval and adding that value to the elevation of the highest contour line. That is, if the highest contour on a peak were 1000 m with a contour interval of 50 m, then the approximate elevation of the mountain top is 1025 m.

> Most topographic depressions are also shown with closed contours; the difference is that a depression contour has small *hachure* or tic marks on the contour lines that point toward the center of the depression.

> Sometimes the topography on a slope is unusually *hummocky*, that is, undulatory or uneven, when compared with the general topography of the area. Such hummocky topography may indicate mass wasting processes and landslide deposits.

In summary, after observing a variety of topographic maps and working with them for some time, you will begin to see the pattern of contours as an actual landscape consisting of hills, valleys, and other features.

LOCATING YOURSELF ON A MAP

The first time you take a topographic map outdoors you may have some difficulty locating where you are. Determining where you are is crucial to mapping floodplains, landslides, or other features. One way to locate yourself on a map is to identify two or three distinctive features that you can see—such as a mountain peak, road intersection, or a prominent bend in a road or river—that are also shown on the map. You can then use a compass to determine your location. This is done by taking a compass direction, that is, a bearing, to each of the prominent features and drawing these bearings on the map; your location is where they intersect.

Today we can also use orbiting *global positioning system (GPS)* satellites for locating ourselves and working with maps. Hand-held GPS satellite receivers are readily available at a very modest price and can generally locate your position on the ground with an accuracy of 5 to 10 m (16 to 33 ft.). GPS receivers work by receiving signals from at least four satellites and by calculating the distance from each satellite to your location. This is done with extremely accurate clocks that determine the time delay in receiving an expected radio signal from each satellite; the longer the delay, the further the satellite is from the receiver. With signals from at least four satellites, the computer in your GPS receiver can calculate a three-dimensional position on Earth's surface. Using correction signals transmitted from the Federal Aviation Administration *Wide Area Augmentation System* (WAAS), the accuracy of defining your position can be reduced to

less than 3 m (10 ft.) (Figure C.2). Another way to increase positional accuracy is to average multiple measurements taken at the same location over a period of 10 minutes or more.

Your GPS receiver can be linked directly to *geographic information system (GIS)* computer software, so when a position is known, it may be plotted directly on a map viewed on a computer display in the receiver, your vehicle, or a laptop computer. GPS and GIS technology have revolutionized the way we do mapping in the field. A note here concerning the expression *"in the field"*; this expression means "outside on the surface on Earth," not the "field of geology." For example, a geologist about to go out and study landslides triggered by heavy rains in the Appalachians would say to her colleagues, "I am going into the field."

AN EXAMPLE FROM A COASTAL LANDSCAPE

In this example we will look at topographic and cultural features in a coastal landscape (Figure C.3). If we were flying along the coast in a plane, we would have a sideways or oblique view of this landscape consisting of two hills with an intervening valley (Figure C.3a). The coast along the eastern hill to the right is a sea cliff, and in the center is a peninsula that was formed as a sand spit by longshore coastal processes. A hook on the end of the peninsula suggests that the direction of littoral sand transport in the surf zone and beach is from the east (right) to the west (left).

The topographic map for the area has a contour interval (CI) of 20 ft. (Figure C.3b). This map indicates that the elevation of the highest hill on the east side of the map is approximately 290 ft. above sea level; the estimated elevation is half way between 280 ft., the highest contour on the hill, and 300 ft., which would be the next contour if the hill were higher. Three streams drain the hill, two into the ocean and one into a river; notice that in all three streams the topographic contours create a V in the upstream direction toward the top of the hill.

Other information that may be "read" from the topographic map includes the following:

> The landform on the western portion of the map (left side) is a hill with elevation of about 275 ft., a gentle slope to the west, and a steep slope to the east (right) toward the ocean and the river valley. The eastern slope is particularly steep near the top of the mountain where the contours are very close together.

> A river in the center part of the map is flowing into a bay protected by a hooked sand bar. The relatively flat land next to the river is a narrow floodplain. Along the west side of the river is a light-duty road

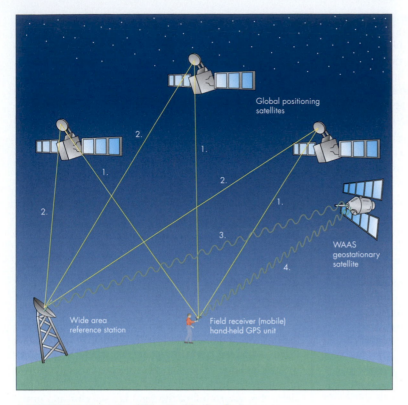

˄ FIGURE C.2 Using the Global Positioning System
U.S. Air Force NAVSTAR Global Positioning System (GPS) satellites orbit approximately 22,200 km (13,800 mi.) above Earth. Each satellite has its own distinct signal, shown with #1 yellow lines, which can be recognized by a hand-held GPS receiver. Using extremely accurate timekeeping, the receiver calculates the distance to a satellite by determining how long its signal took to arrive. A receiver must process signals from at least four satellites to calculate a three-dimensional position on Earth's surface. As long as the satellite signal is not altered, most GPS receivers will achieve a positional accuracy of within 5 to 10 m (16 to 33 ft.) of their true position on Earth's surface. Increased accuracy can be obtained with a correction signal obtained from a Wide Area Reference Station that is part of the Federal Aviation Administration's Wide Area Augmentation System (WAAS). The WAAS reference station receives signals from the same GPS satellites, shown as #2 yellow lines, at the same time as the mobile GPS receiver. The WAAS then compares the position calculated from the GPS satellites with its true position to establish a correction factor. The correction factor is then transmitted to a WAAS geostationary satellite, shown as the #3 wavy line, and then broadcast back to the mobile GPS receiver, shown as the #4 wavy line. A GPS receiver with WAAS correction typically has a positional accuracy of within 3 m (10 ft.) or less of the true position.

that parallels the river and then extends along the coastline. A second, unimproved road crosses the river and extends out to the head of the sand spit, providing access to a church and two other buildings. The floodplain is delineated on both sides of the river by the 20-ft. contour, above which to the 40-ft. contour is a break in slope at the edge of the valley.

> The eastern and southern slopes of the 275-foot hill on the west (left) side of the map have a number of small streams flowing mostly southward into the ocean. These streams appear to be relatively steep gullies that are eroding into the hillslope. In particular, the stream directly south of the 275-foot

elevation mark appears to be cutting headward into the steep hillside.

To continue a study of this area, we might construct an east–west *topographic profile* across the area along line A–A'. In constructing a profile we must decide how much to exaggerate the vertical scale of the view (Figure C.3c). The resulting *vertical exaggeration (VE)* is determined by comparing the same unit on both the vertical and the horizontal scale; for example, the number of feet represented by 1 inch on both scales. After constructing topographic profiles, your next task might be to obtain aerial photographs and geologic maps to draw more conclusions concerning the topography and geology.

(a)

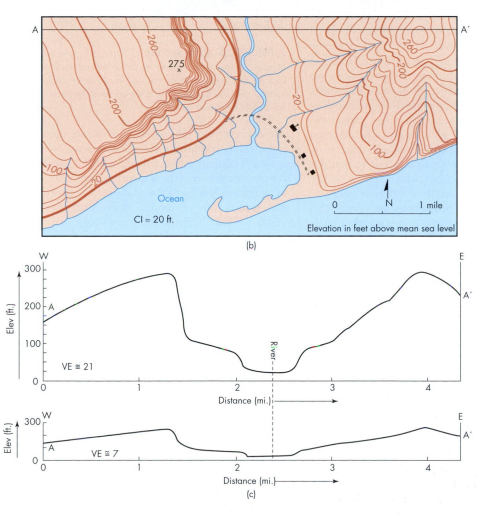

260

275
×

200

20

100

20

100

260

200

20

100

Ocean

CI = 20 ft.

0   N   1 mile

Elevation in feet above mean sea level

(b)

W   E

Elev (ft.)

300

A   A'

200

100

River

VE ≅ 21

0
0   1   2   3   4

Distance (mi.)

W   E

Elev (ft.)

300

A   A'

VE ≅ 7

0
0   1   2   3   4

Distance (mi.)

(c)

∧ FIGURE C.3 Topography of a Coastal Landscape

(a) An oblique view of the coastline mapped in part (b) of this figure. (b) Topographic map for the same area with a contour interval of 20 ft. (c) Topographic profiles along line A–A' of the topographic map shown with vertical exaggeration of approximately 21 times and 7 times. Vertical exaggeration is the ratio of the vertical to horizontal scale for the topographic profile. The ratio for the upper profile is approximately 21, so it has a vertical exaggeration of approximately 21 times. For the lower profile, the vertical exaggeration is about 7 times. In the real world, of course, there is no vertical exaggeration (the vertical and horizontal scales are the same). As an experiment, you might try to make a topographic profile along line A–A' with no vertical exaggeration. What do you conclude? *(From U.S. Geological Survey)*

GEOLOGIC MAPS

Geologists are interested in the types of earth materials found in a particular location and in their spatial distribution. Producing a geologic map is a very basic step in understanding the geology of an area. In constructing a *geologic map* a geologist must make interpretations that are consistent with the Earth history of the mapped area and the geologic processes that have been at work. Like any other scientific endeavor, each new observation becomes a test of the geologist's interpretations, sometimes referred to as a working hypothesis.

The first step in preparing a geologic map is to obtain a good base map (usually topographic) or aerial photograph on which the geologic information may be transferred. A geologist then goes into the field and makes observations of earth materials where they are exposed at or near the surface. From these exposures, called outcrops, the geologist groups distinctive rock or sediment types into separate units called *formations*. Each formation must be thick enough to be depicted on the scale that is being mapped. On the geologic map, each formation is separated from adjacent formations by boundaries known as *contacts*, which appear on the map as thin lines between swaths of different colors.

As the geologist maps formations, he or she also takes measurements of the three-dimensional orientation, or attitude, of the mapped units. These measurements, referred to as *strike and dip*, are generally made with a compass and a device for measuring vertical

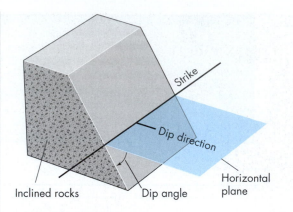

∧ FIGURE C.4 Strike and Dip
Idealized block diagram showing the strike and dip of inclined sedimentary rocks. Strike is the long direction of a T-shaped symbol, and the dip direction is shown with the short line perpendicular to strike.

angles known as a clinometer. Each strike and dip measurement is commonly recorded directly on the map with a T-shaped strike and dip symbol. Strike is the compass direction of the line formed by the intersection of layering in the earth material with a horizontal plane, and dip is the maximum angle that the layering makes with the horizontal (Figure C.4).

For example, a simple geologic map of an area of approximately 1350 km² (520 mi.²) might show three rock types—sandstone, conglomerate, and shale (Figure C.5a).

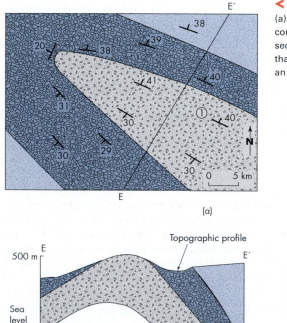

(a)

(b)

◄ FIGURE C.5 Geologic Map
(a) A very simple, idealized geologic map showing three formations, each consisting of a different type of sedimentary rock. (b) Geologic cross-section and topographic profile along line E–E′. The cross-section shows that the pattern of the formations on the geologic map is one made by an arch-like geologic structure known as an anticline.

⊢40

Strike and dip of sedimentary rock units (showing angle of dip); for example, at ① on the map. The strike is northwest and dip is 40° to the northeast.

▨ Sandstone

▨ Conglomerate

▨ Shale

∧ FIGURE C.6 Digital Elevation Model
This digital elevation model (DEM) of the Los Angeles area is constructed as an oblique view from the
southwest with a fair amount of vertical exaggeration. The flat, light blue area in the lower half of the image is
the Los Angeles Basin. The smaller basin above the Los Angeles Basin is the San Fernando Valley, site of the
epicenter of the 1994 Northridge earthquake. It is separated from the Los Angeles Basin by the Santa Monica
Mountains. To the left of the Los Angeles Basin are the Palos Verdes Hills on the Palos Verdes Peninsula, the
site of the Portuguese landslide discussed in Chapter 6. Finally, on the right side of the map is the San Andreas
Fault, a strike-slip fault that is one of the major geologic hazards in southern California. *(Courtesy of Robert
Crippin, NASA, Jet Propulsion Laboratory)*

On this map the arrangement of the strike and dip symbols suggests the presence of an arch-like underground
geologic structure known as an anticline. The nature of
this structure becomes apparent in a geologic cross-
section and topographic profile constructed along the line
E–E′ (Figure C.5b). Geologists often make a series of
cross-sections of a geologic map to better understand the
geology of the area. Geologic maps at a variety of scales
from 1:250,000 to 1:24,000 are generally available from a
number of sources, including the U.S. Geological Survey.

Digital Elevation Models

Topographic data for the United States and many other
parts of the world are now available as digital files. These
data sets contain arrays of elevation values at a specific
spacing, for example, the surface ground elevations on a
30 m grid for a 900 m^2 (9700 ft.^{2}) area. Computer programs are then used to synthesize and view the data; color
shading may be added to show the topography. The resulting graphic representation of Earth's surface is known
as a *digital elevation model* (DEM). DEMs can be examined
from a variety of angles using software such as Google

Earth, and you may view the topography obliquely from
any compass direction. The vertical dimension may also
be exaggerated so that minor topographic differences may
become more apparent. A DEM for the Los Angeles
Basin clearly illustrates that Los Angeles is nearly surrounded by mountains and hills, which have been uplifted
by recent tectonic activity (Figure C.6). DEMs are becoming important research tools in evaluating the topography of an area. They may be used to delineate and map
features such as fault scarps, landslides, karst topography,
floodplains, and volcanic and impact craters.

Summary

Several types of maps and images are useful in evaluating geology and Earth processes. Of particular importance are topographic maps and profiles, and geologic
maps and cross-sections. Digital elevation models may
be constructed from topographic data, and a variety of
other special purpose maps are also available. Examples
include maps of recent landslides, floodplain and coastal
flood hazard maps, and engineering geology maps that
show engineering properties of earth materials.

APPENDIX D: How Geologists Determine Geologic Time

In order to understand the history of Earth, we must determine the actual age of earth materials with the science of *geochronology*. In geochronology, scientists combine *relative and absolute dating* techniques to reconstruct Earth history. Relative dating establishes a chronological order of events using fossil evidence and geologic relationships, the use of which is governed by geologic principles and laws. The oldest of these laws, the law of superposition, was proposed by Nicholas Steno in 1668, and has been applied and tested by scientists for over 300 years. This law states that in any succession of sediment or sedimentary rock that has not been deformed, the oldest layer lies at the bottom, with successively younger layers above. Although a relative geochronology can be established based upon superposition and other laws and principles of relative dating, it is absolute dating that provides the information necessary to establish rates of geologic processes and the numeric ages of earth materials.

From a natural hazards perspective, it is important to establish rates of geologic processes and the timing of past geologic events such as volcanic eruptions, earthquakes, tsunamis, floods, and landslides. The chronology of these events is critical in estimating their return period or recurrence interval. Knowing how frequently an event takes place helps us better understand the hazard and predict when a similar event is likely to occur in the future.

Geologic time is much different, in a way, than our normal time framework. Although geologic time and "normal" time use the same units of measure—years—they differ vastly in duration and in the instruments we use to measure duration.[1] Normal time is counted in hours, days, seasons, or decades, and the instrument used to measure time is a clock. In contrast, geologic time—sometimes called "deep time"—is measured in tens of thousands to hundreds of millions to several billion years. To measure this vast amount of time, geologists take advantage of naturally occurring forms of uranium (U), potassium (K), carbon (C), and other chemical elements to date the earth materials in which they are found.

The natural forms of these chemical elements are called *isotopes* and are distinguished from each other by their weight. The weight of an isotope is the sum of the weight of smaller, subatomic particles called protons and neutrons that are found in the center of the atom, the nucleus. Each proton and neutron has a weight of 1, so the number of protons plus the number of neutrons determines the atomic weight of an isotope. All isotopes of a chemical element have the same number of protons, because it is the number of protons that defines a chemical element. Where isotopes vary is in their number of neutrons. For example, carbon has three common isotopes, carbon-12, carbon-13, and carbon-14. All three isotopes have 6 protons because they are carbon, but the number of neutrons varies from 6 to 7 to 8, respectively.

Isotopes also differ in their stability. Some isotopes, such as carbon-12 and carbon-13, have a stable number of neutrons and protons in their nucleus and do not split apart. Most isotopes, however, are unstable and will undergo spontaneous splitting, sometimes called decay, at some point in the future. Unstable isotopes are commonly called *radioactive isotopes* because radiation is released when they decay. Although it is impossible to predict when an individual unstable atom will decay, laboratory studies can determine the rate at which a large number of atoms of a radioactive isotope will decay. Each radioactive isotope has a constant rate of decay that is unaffected by physical forces.

Absolute dating is possible if a measurable quantity of a radioactive isotope is incorporated into a mineral or substance when it forms. The decay of the unstable radioactive isotope is irreversible and becomes a clock that is running at a constant rate. In absolute dating, the unstable radioactive isotope is known as the *parent* and the *stable isotope* that eventually forms from the decay of the parent is known as the *daughter product*. Radioactive isotopes, particularly those of very heavy elements, undergo a series of radioactive decay steps that finally ends when a stable, nonradioactive isotope is produced. For example, the isotope uranium-238 (U-238) undergoes 14 nuclear transformations in different steps to

▼ TABLE D.1 Half-Lives of Radioactive Isotopes Commonly Used in Absolute Dating

Parent Radioactive Isotope	Daughter Stable Isotope	Half-Life
Uranium-238	Lead-206	4.5 billion yr
Uranium-235	Lead-207	700 million yr
Potassium-40	Argon-40	1.3 billion yr
Carbon-14	Nitrogen-14	5730 yr

finally decay to stable lead-206 (Pb-206), which is not radioactive.

An important characteristic of a radioactive isotope such as U-238 is its *half-life*, which is the time required for one-half of a given amount of the isotope to decay to another form. Every radioactive isotope has a unique and characteristic half-life (Table D.1). As decay proceeds over time, the amount of parent radioactive isotope in a substance decreases and the amount of the stable daughter isotope proportionally increases (Figure D.1). Without knowing the actual quantity of radioactive isotope that was initially present, it is often possible to obtain a numeric date by comparing the relative proportions of parent and daughter isotope in a substance. In most cases, two important conditions must be met to obtain an accurate numeric date. First, no new atoms of the parent isotope have been added to the substance since it formed; this is true for many minerals in igneous rocks where the radioactive isotope becomes part of the atomic structure when a mineral crystallizes from magma. Second, all atoms of the daughter isotope produced by decay remain trapped in the substance; again this occurs for minerals in igneous rocks where atoms are strongly bonded in the crystal structure. Should any daughter product "leak" from the sample, it would give a numeric date that was too young. Because radioactive isotopes such as U-238, U-235, and K-40 have relatively long half-lives (Table D.1), their decay is useful in dating rocks on the order of millions to billions of years. For example, the ratio of Pb-207/Pb-206, the daughter products of U-235 and U-238, has been used to date the oldest piece of continental crust, a crystal of the mineral zircon from sandstone in Western Australia, at 4.4 billion years before the present. The actual methods for measuring amounts of parent and daughter isotopes and calculating numeric dates are complex and tedious, although the concept is easy to grasp. These methods have been successfully used to assign numeric dates to the geologic time table and to delineate important Earth history events such as mountain building, ice ages, and the appearance of life forms.

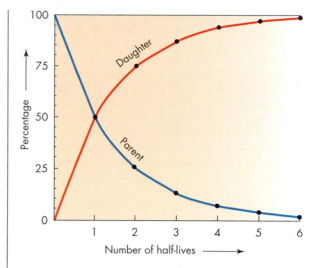

▲ FIGURE D.1 Relative Proportion of Parent and Daughter Isotopes Over Time
This graph shows the reduction of a parent radioactive isotope to a daughter stable isotope with each additional half-life.

Physical anthropologists and archeologists concerned with human history are interested in developing a geochronology for the last 8 million years of Earth history; that is the period from which human and ancestral human fossils are found. From a natural hazards perspective, we are often interested in establishing the geochronology of the past few hundred to few hundred thousand years of Earth's history, and for this we have several potential methods. For example, the radioactive isotope U-234 undergoes decay to thorium (Th)-230 at a known rate, and the ratio of these two isotopes is useful in numerically dating a variety of materials, such as coral, back to several hundred thousand years.

For sediment younger than about 40,000 years, C-14 is used extensively for numeric dating. The most common form of carbon is stable C-12, but C-14, which is radioactive, occurs in small quantities and undergoes radioactive decay to the stable daughter isotope nitrogen-14 (N-14). The half-life of the decay of C-14 to N-14 is 5730 years. The C-14 method works because C-14 is only incorporated into organic matter while an organism is alive; when the organism dies, the C-14 in its tissue undergoes radioactive decay without being replaced with new C-14. Common materials dated with C-14 include wood, bone, charcoal, and other types of buried organic material. C-14 has a relatively short half-life; as a result, after about 40,000 years, the amount of C-14 remaining is very small and difficult to measure. Consequently, use of this technique is limited.

The field of geochronology is expanding rapidly, and new techniques are constantly being developed. For example, it is now possible to directly date the amount of time that a landform has been exposed at Earth's surface. This is known as *exposure dating*. The basic idea is that certain isotopes, such as beryllium-10, aluminum-26, and chlorine-36, are produced when cosmic rays interact with Earth's atmosphere and accumulate in measurable quantities in surface materials such as soil, alluvium, and exposed rock surfaces. Thus, the amount of these isotopes that has accumulated is a measure of the minimum time of exposure to the surface environment.

Another innovative technique, known as *lichenometry*, uses the slow, but constant growth of *lichens* on rock surfaces to determine a minimum time that a rock is exposed. Lichens are plant-like growths of photosynthetic algae or bacteria with a fungus. These organisms tend to grow in circular patches with a rate of growth that is known for a particular species. Careful measurements of the size of lichen patches, coupled with known growth rates, provide minimum numeric ages for exposure of the rocks on which they are found. This method has been successfully used in California and New Zealand to date regional occurrences of rockfalls generated by large earthquakes, thus dating past seismic activity.[2,3] Lichenometry can provide dates to about 1,000 years before present.

Sediment may also be accurately dated through the use of *dendrochronology*, which is the analysis of annual growth rings of wood. Because of variations in annual precipitation, the width of tree rings differs from year to year. This produces a distinctive sequence of wide and thin rings similar to a bar code on merchandise at a store. Once a regional chronology is developed using growth rings in both living and dead trees, it then becomes possible to determine the age of a growth ring pattern from a piece of buried wood. This method has been used extensively by archeologists to date prehistoric sites of human habitation and also by climatologists as a method of reconstructing prehistoric patterns of precipitation. Growth rings record climate because during dry years, tree rings are narrow relative to wet years.

Very accurate numeric geochronology may also be obtained from counting varves. A *varve* is a layer of sediment representing one year of deposition, usually in a lake or an ocean. Careful counting of varves may extend the chronology back several thousand years.

Although the most accurate chronology is the historical record, its length varies from only a few hundred years in the United States and Canada to several thousand years in China. Given the brevity of human history compared with the great length of geologic history, we resort to numeric dating methods to establish geochronology. There are more than 20 methods useful in establishing geochronology that yield numeric dates. These methods are crucial to understanding rates of geologic processes and the recurrence interval of natural hazards.

REFERENCES

1. **Ausich, W. I.**, and **Lane, G. N.** 1999. *Life of the past*, 4th ed. Upper Saddle River, NJ: Prentice Hall.

2. **Bull, W. B.**, and **Brandon, M. T.** 1998. Lichen dating of earthquake-generated regional rockfall events. Southern Alps, New Zealand. *Geological Society of America Bulletin* 110: 608–84.

3. **Bull, W. B.** 1996. Dating San Andreas fault earthquakes with lichenometry. *Geology* 24: 111–14.

Glossary

Aa A cooler or thicker lava flow that solidifies with a blocky surface texture.

Abrupt climate change Large-scale change in the global climate system that takes place over a few decades or less.

Absolute dating Determination of the geologic age at which time a fossil organism, earth material, or geologic feature formed; the age is usually expressed in years; commonly established by the analysis of radioactive isotopes and their decay products.

Acceptible risk Exposure to danger that an individual or society is willing to take.

Acid rain Water droplets falling from clouds that have been made artificially acidic by pollutants. In volcanoes, this acidity commonly comes from sulfate aerosols in eruption clouds, or sulfur dioxide fumes or from hydrogen chloride gas produced when lava vaporizes seawater. Rainwater normally has a slight acidity from dissolved carbon dioxide.

Active fault A fault along which future movement is thought possible. Two criteria are mappable displacement along a fracture or fractures in the past 10,000 years or several mappable displacements in the past 35,000 years.

Adaptive management Structured management with the use of science that promotes sustainable coasts, and allows for changing conditions during management.

Advance With respect to a glacier, an interval of time during which the area occupied by glacial ice increased.

Aerosol With respect to climate, a suspension of microscopic liquid and solid particles, such as mineral dust and soot, in the atmosphere.

Aftershock A subsequent, less intense earthquake that occurs a few minutes to a year or so after a main earthquake, and in the same general area. Usually one of many quakes of lesser magnitude than the main shock.

Airburst With respect to collision with extraterrestrial objects, the explosion of a meteoroid or comet in the atmosphere, generally at an altitude of 12 to 50 km (~7.5 to 30 mi.).

Air mass thunderstorm An ordinary storm with lightning and thunder that develops from local convection within an unstable air mass. Commonly forms in the afternoon and is generally not severe.

Albedo A measure of the reflectivity of a material or surface to electromagnetic radiation, especially sunlight. Commonly expressed as the percentage of radiation reflected versus absorbed.

Alberta Clipper An extratropical cyclone that forms east of the Rocky Mountains in the Canadian province of Alberta and moves rapidly to the southeast across Canada and the northern United States. Commonly associated with snowstorms, especially blizzards.

Alluvial fan Fan-shaped deposit composed of coarse sediment that is dropped by a stream as it emerges from a mountain front onto flatter terrain; typically made of a variable proportion of stream and debris-flow deposits. The fan takes the shape of a segment of a cone.

Alluvium Unconsolidated sediment, such as sand, gravel, and silt, deposited by a stream.

Andesite A type of lava that has a silica content intermediate between basalt and rhyolite. Commonly, but not exclusively, associated with composite volcanoes where it forms by the partial melting of oceanic crust in a subduction zone below the volcano.

Angle of repose Maximum angle that loose material will sustain.

Anthropogenic Produced by humans or originating in human activity.

Anticipate In respect to natural hazards and disasters, advance planning for the effects of the natural process.

Aquifer A useable accumulation of subsurface water in earth material that is saturated with water.

Artificial bypass Human action that moves sand or other sediment around an obstruction, such as dredging and pumping sand from updrift to downdrift sides of a jetty or harbor.

Artificial control of natural processes An action or structure made by humans to influence phenomena that are not caused or produced by humans.

Ash fall Fine airborne material that settles onto the land surface. The material is unconsolidated volcanic debris (ash), less than 4 mm in diameter, that has been blown out of a volcano during an eruption; it may travel a considerable distance from the volcano.

Ash flow See *Pyroclastic flow.*

Asteroid Rock or metallic particle in space with a diameter between 10 m (~30 ft.) and 1000 km (~620 mi.).

Asteroid belt Region between orbits of Mars and Jupiter that contains most of the large rock or metallic particles in our solar system.

Asthenosphere Upper zone of Earth's mantle, located directly below the lithosphere; a hot, slowly flowing layer of relatively weak rock upon which tectonic plates move.

Atmosphere Layer of gases surrounding a planet, such as Earth.

Atmospheric pressure The force per unit area exerted by the weight of gas molecules that surround a planet; also called barometric pressure, because it is commonly measured with a barometer.

Atmospheric stability A condition of equilibrium in which a rising or sinking parcel of air tends to return to its original position. This equilibrium develops when a relatively small difference in air temperature exists between the surface and the air aloft.

Atoll A ring or horseshoe-shaped group of coral reefs, low islands, and shoals generally surrounding a large island, submerged island, bank, or underwater plateau.

Attenuation With respect to earthquakes, the loss of energy as seismic waves pass through earth materials. For example, this process reduces the energy of seismic waves from deep subduction-zone earthquakes before they reach Earth's surface.

Avalanche A type of landslide involving a large mass of snow, ice, or entrained rock debris that slides, flows, or falls rapidly down a mountainside.

Avulsion The process by which all or part of a river or distributary channel is abandoned in favor of a new channel.

Bankfull discharge Flow condition in which water completely fills a stream channel. Many stream channels are formed and maintained by bankfull discharge.

Barometric pressure See *Atmospheric pressure*.

Barrier island An elongate, offshore accumulation of sand that is above water level and is separated from the mainland by a bay or lagoon; stops large waves from reaching the shore of the mainland.

Basalt A common volcanic rock consisting of feldspar and other silicate minerals rich in iron and magnesium; has a relatively low silica content of about 50 percent; forms shield volcanoes.

Base level The theoretical lowest elevation to which a river may erode, generally at or about sea level.

Beach Associated with an ocean or lake, an accumulation of loose material, most commonly sand, gravel, or bits of shell, on a shoreline as the result of wave action. As used by river rafters, a wide sand accumulation along a low bank of the channel.

Beach budget Inventory of sediment influx and loss along a particular stretch of coastline.

Beach drift Continual movement of sand grains along the shore by the runup and backwash of waves.

Beach face The part of the shore that is normally exposed to swash action, that is, the runup and backwash of the waves; the sloping surface between the ocean or lake and the berm.

Beach nourishment Mechanical addition of sediment, generally sand, to a shore for recreational and aesthetic purposes, as well as to buffer coastal erosion.

Bedding plane A surface separating layers of sedimentary rocks. In mass wasting, this is commonly a weakness along which weathering occurs and landslides develop.

Berm The relatively flat or landward sloping part of a beach profile produced by wave-deposited sediment; the part of the beach where most people sunbathe.

Bermuda-Azores High A persistent high-pressure region in the North Atlantic during the summer and fall months that lies approximately between the Island of Bermuda, east of North Carolina, and Azores Islands, west of Portugal.

Blind fault A fracture or zone of rupture along which displacement does not extend to the Earth's surface.

Blizzard A severe winter storm producing large amounts of falling or blowing snow that create low visibilities for extended periods of time. In the United States, for a storm to be classified as a blizzard winds must exceed 56 km (~35 mi.) per hour with visibilities of less than 0.4 km (~0.25 mi.) for at least 3 hours in Canada, for a storm to be classified as a blizzard winds must exceed 40 km (~25 mi.) per hour with visibilities of less than 1 km (~0.6 mi.) for at least 4 hours.

Block With respect to a fragment ejected from a volcano, an angular particle that is longer than 64 mm (~2.5 in.).

Bluff As a landform, a high steep bank or cliff along a lakeshore or river valley commonly made of weakly consolidated earth material.

Bomb With respect to a volcanic fragment, a smooth-surfaced particle that is longer than 64 mm (~2.5 in.); commonly forms from a blob of magma ejected from a lava fountain; the magma cools and forms a streamlined shape as it falls to the surface.

Braided With respect to a river or stream, a channel pattern with numerous islands and sand and gravel bars that continually divide and subdivide the flow of water. This pattern is most evident during low and moderate stream flow.

Braided river See *Braided stream*.

Braided stream A flowing channel of water with numerous islands and sand and gravel bars that continually divide and subdivide the flow.

Breaker zone With respect to a shoreline, that part of the beach and nearshore environment where incoming waves peak up, become unstable, and collapse toward the shore.

Breakwater A structure, such as a wall, that may be attached to a beach or located offshore, designed to protect a beach or harbor from the force of waves.

Breccia A rock or zone within a rock composed of angular fragments; may have a sedimentary, volcanic, tectonic, or impact origin.

C soil horizon Lowest soil horizon, sometimes known as the *zone of partially altered parent material*.

Caldera A semi-circular region of low topography that develops at the top of a volcano or in the continental crust due to collapse of the surface due to removal of magma from the subsurface during a volcanic eruption.

Calving The separation of large blocks of ice from the terminus of a glacier that flows into a large body of water; calved blocks may form icebergs.

Carbon-14 Radioactive isotope of carbon with a half-life of approximately 5730 years; used in absolute dating of organic materials back to approximately 40,000 years.

Carbon dioxide (CO_2) A colorless, odorless gas that is denser than air and is produced by plant and fossil fuel combustion, animal and microbe respiration, and volcanic eruptions. In the atmosphere, it is the primary cause of global warming because, as a greenhouse gas, it heats the air by absorbing infrared radiation.

Carbon monoxide (CO) A colorless, odorless gas that forms in low-oxygen conditions with incomplete combustion or is released from volcanoes; in low concentrations it is highly toxic to humans and animals.

Carbon sequestration The removal and isolation of carbon compounds, particularly carbon dioxide, from chemical reactions in the carbon biogeochemical cycle; a term commonly used to describe the removal of atmospheric carbon dioxide and its isolation underground or in the ocean.

Carrying capacity Maximum number of a population of a species that may be maintained within a particular environment without reducing the ability of that environment to maintain that population in the future.

Catastrophe An event or situation causing sufficient damage to people, property, or society that recovery or rehabilitation is long and complex; natural processes most likely to produce a catastrophe include floods, hurricanes, tornadoes, tsunamis, volcanoes, and large fires.

Cave A natural subterranean void commonly consisting of a series of chambers that are large enough for a person to enter; formed most often in limestone or marble.

Cave system See *Cave*.

Cavern A large cave.

Cenote On the Yucatán Peninsula of Mexico, a steep-walled collapse sinkhole that extends below the water table. Some of these sinkholes are developed in fractured limestone on the perimeter of the Chicxulub impact crater.

Channel pattern The shape of a flowing stream as viewed from above ("bird's-eye view"); the three most common shapes or patterns are straight, meandering, and braided.

Channel restoration The process of returning a stream and adjacent areas to a more natural state.

Channelization Engineering technique to straighten, widen, deepen, or otherwise modify a natural stream channel.

Chute In mass wasting and landslides, a narrow path formed and maintained by the rapid downslope movement of avalanches, rockfalls, or debris flows.

Cinder cone A conical volcanic hill formed by an accumulation of coarse pyroclastic material erupted from the volcano; also known as a *scoria cone*.

Clay particles Very small particles less that 1/256 mm in diameter.

Clear-cutting Forestry practice of harvesting all trees from a tract of land.

Cleavage In a mineral, the smooth, easy fracture that takes place along parallel surfaces lying in one or more planes; these surfaces develop where bonds between atoms are weak or are especially long.

Climate Characteristic weather at a particular place or region during many seasons, years, or decades.

Climate forcing An imposed positive or negative change of Earth's energy balance measured in Watts per meter squared (W/m^2).

Coarse grained soils Type of soil where over half of the soil particles (by weight) is composed of particles greater than 0.074 mm in diameter.

Coastal Zone bioengineering Coastal defense that uses living or dead plants to assist in controlling coastal erosion.

Cold front An atmospheric boundary between two air masses of contrasting temperature along which the cooler air mass advances into the warmer air mass.

Collapse sinkhole Depression in sediment or bedrock caused by a subsurface cave-in.

Collapsible soils Some windblown dust deposits that are loosely bound with clay particles and water-soluble minerals that collapse when water percolates through

them, often resulting in land surface subsidence.

Colluvium Mixture of weathered rock, soil, and other, usually angular, material on a slope; produced by creep, landsliding, and other surface processes.

Color In a mineral, the name associated with the particular wavelengths of light that are reflected from the entire specimen or from its powder.

Combustion Second phase of a wildfire that follows every ignition. Includes flaming, where fine fuel and volatile gases are rapidly oxidized at high temperature, and glowing or smoldering, which take place later at lower temperature. Both flaming and glowing liberate energy as heat and light.

Comet Particle in orbit in space composed of a spongelike rocky core surrounded by ice and covered by carbon-rich dust with a tail of gas and dust that glows, as its orbit approaches the sun; ranges from a few meters to a few hundred kilometers in diameter.

Community Internet Intensity Map Online depiction of how people and structures have experienced an earthquake on the basis of email reports.

Composite volcano Steep-sided volcanic cone produced by alternating layers of pyroclastic debris and lava flows; also known as a *stratovolcano*.

Conchoidal fracture In a rock or mineral, a broken surface that is smoothly curved, like the inside of a clamshell.

Conduction With respect to physics, the transfer of heat through a substance by molecular interactions.

Constructive interference In an ocean or a lake, combining two or more waves on the water surface to form a resultant higher wave.

Contact Interface or boundary between two geologic mapping units that is generally shown as a thin line on a geologic map.

Continental collision boundary The interface between two tectonic plates of similar density that are moving toward each other. Deformation and mountain formation occur along the interface because neither plate is able to slide below the other.

Continental Drift Hypothesis proposed by Alfred Wegener, in the early twentieth century, that states the continental landmasses of the world "drift" about the surface of the globe and were once united to

form a single super-continent, which he called Pangea. The hypothesis was supported by geologic and paleontologic data.

Contour line On a topographic map, a series of connected points in which all points are the same elevation above sea level.

Contour interval (CI) On a topographic map, the vertical distance between each line of equal elevation; generally expressed in meters or feet.

Convection Transfer of heat by a mass of moving particles; for example, in boiling water, hot water rises to the surface and displaces cooler water that moves toward the bottom. Atmospheric heat may be transferred vertically. This is the primary process in which heat is transferred within the atmosphere and during wildfires.

Convection cell A self-contained circulation loop in a fluid in which hot, less dense fluid rises and cooler, denser fluid sinks; develops on a large scale in the atmosphere and within the Earth.

Convergence Coming together.

Convergent boundary Interface between two tectonic plates in which one plate generally descends below the other in a process known as *subduction*.

Core With respect to the interior of Earth, the central part of Earth below the mantle, divided into a solid inner core with a radius of approximately 1300 km and a molten outer core with a thickness of about 2000 km; the core is thought to be metallic and composed mostly of iron.

Coriolis effect An apparent deflection in the path of a moving object. This apparent deflection, to the right in the Northern Hemisphere and to the left in the Southern Hemisphere, results from Earth's rotation. The magnitude of this effect increases with increasing latitude from the equator to the poles.

Crater A type of volcanic vent that forms a semicircular depression near the summit of most volcanoes, particularly cinder cones and stratovolcanoes. Volcanic craters form in response to the forceful ejection of volcanic debris and gasses.

Creep Slow downslope movement of soil and other weakly consolidated earth materials; characterized by slow flowing, sliding, or slipping.

Crevasse On a glacier, a large, deep fracture in the surface of the glacial ice caused by stretching of ice at and near the glacier surface.

Crown fire With respect to wildfires, flaming combustion that carries through canopies of trees; commonly driven by high wind and aided by a steep slope.

Crust The thin outermost rock layer of Earth that is differentiated from the underlying mantle by having a lower density.

Crustal assimilation A process where crustal rocks are melted by and incorporated into a body of magma that is rising toward the surface. Addition of felsic crustal melt to the original magma body causes progressive increase of the silica content of the magma as it rises.

Crystal form In a mineral, the characteristic geometric shape assumed by that mineral because of its orderly, internal atomic structure.

Crystallization Process by which atoms precipitate from solution such that they organize themselves into the orderly internal atomic arrangement of a mineral. The mineral may or may not possess crystal faces.

Cumulonimbus Thunderstorm cloud that grows vertically until it reaches a stable layer of air; generally with an anvil-shaped top.

Cumulus Any cloud that develops vertically; generally has a flat base with a cauliflower-like top; one of the three basic cloud types.

Cumulus stage First phase of thunderstorm development characterized by the growth of domes and towers on the top of a cumulus cloud as it becomes a cumulonimbus cloud.

Cutbank Steep or nearly vertical slope on the outside of a bend in a stream channel; typically forms by stream erosion.

Cyclone An area of low atmospheric pressure characterized by rotating winds that, in the Northern Hemisphere, circulate in a counterclockwise direction.

Cyrosphere The portion of a planet consisting of ice; on Earth this includes permafrost in soil and below the seafloor, glaciers, ice sheets, ice caps and fields, ice shelves, and icebergs.

Daughter product A stable isotope that forms by radioactive decay of an unstable parent isotope.

Debris flow Rapid downslope movement of unconsolidated, water-saturated earth material that became unstable because of torrential rain, rapid melting of snow and ice, or sudden drainage of a pond or lake; sometimes restricted to flows of this type that contain mainly coarse material.

Decadal cycles Cycles of storm tracks and thus climate and weather in North America and Western Europe that may last several decades; also known as *Decadal oscillations*.

Decompression melting The process by which solid rock melts due to a decrease in pressure that is caused by upwelling of hot rocks within the interior of the planet.

Delta Low, nearly flat area of land formed near the mouth of a stream where it enters a lake or the ocean; commonly triangular or fan-shaped and crossed by branching distributary channels of the stream that created it.

Delta plain Generally flat land surface that develops where a stream enters an ocean or lake; commonly covered with marsh or swamp and cut by distributary channels.

Dendroclimatology Study of tree rings, sets of which usually form each year, to establish the absolute age of a tree or piece of wood from a tree.

Deposition In geology, the laying down or accumulation of earth material by processes that include the mechanical settling of sediment, the chemical precipitation of mineral matter, or the in-place accumulation of organic matter.

Derecho A windstorm with straight-line winds exceeding 90 km (~56 mi.) per hour along a line that is at least 400 km (~250 mi.) in length; damage produced by this windstorm is often equivalent to that caused by a tornado.

Desert An arid or semiarid region typified by mean annual precipitation less than 250 mm (~10 in.); native animals and vegetation adapted to conserve limited water and soil that may be rich in salts.

Desertification Conversion of land from a more biotically productive state to one that more closely resembles a desert.

Desiccation crack A rupture in fine-grained sediment, typically mud, produced by complete or nearly complete drying.

Differentiated With respect to meteoroids and asteroids, rock that has undergone igneous and sometimes metamorphic processes.

Digital elevation model (DEM) Three-dimensional computer depiction of a land surface as viewed from an oblique angle. Constructed with a computer program that samples land-surface elevation data on a specific spacing grid, such as every 30 m (~100 ft.).

Direct effect A change that follows an event without any intervening factors. In a natural accident, disaster, or catastrophe, such effects could include people killed, injured, or dislocated and property damaged or destroyed; also referred to as a *primary effect*.

Directivity With respect to an earthquake, the increased intensity of shaking in the direction toward which a fault ruptures; observed in some moderate-to-large earthquakes.

Disappearing stream A brook, creek, or river that flows from a surface channel into the ground and continues flowing underground.

Disaster One possible effect of a hazard on society. Usually a sudden event that causes great damage or loss of life during a limited time and in a limited geographic area.

Disaster preparedness With respect to a natural hazard, actions of individuals, families, communities, states, provinces, or entire nations to minimize losses from a hazardous event before it occurs.

Discharge Quantity of water flowing past a particular point on a stream, usually measured in cubic feet per second (cfs) or cubic meters per second (cms).

Dissipative stage Final phase of a thunderstorm in which the supply of moist air is blocked by downdrafts in the lower levels of the cloud. Without a source of moisture, precipitation decreases and the storm cloud begins to disappear.

Distant tsunami A series of gravity waves in the ocean originating from a source typically thousands of kilometers away from the shoreline that is inundated. These waves are produced by the displacement of the entire water column of the ocean by an underwater earthquake, landslide, volcanic eruption, or extraterrestrial impact.

Distributary channel A waterway that branches from another waterway in a downstream direction. Typically found on stream deltas and some alluvial fans.

Divergence Spreading apart.

Divergent boundary Interface between tectonic plates characterized by production of new lithosphere; found along mid-ocean ridges and continental rift zones. Equivalent reference points on each plate will move away from each other with time.

Doppler effect Change in wave frequency that takes place when either the emitter or the detector is moving toward or away from the other. This effect can be heard in the higher-pitched sound of an approaching ambulance and the lower-pitched sound of one that is moving away.

Doppler radar With respect to weather, an electronic system that emits electromagnetic waves to detect the velocity of precipitation that is falling either toward or away from the receiver. Used to detect rotation within a thunderstorm that often precedes formation of a tornado.

Downburst Localized area of very strong winds in a downdraft; commonly beneath a severe thunderstorm.

Downdrift Located in the direction toward which a current of water, such as a longshore current, is flowing.

Downstream flood Condition in which surface runoff from a relatively wide area has caused a stream to overflow its banks—more common in the lower part of a drainage basin where tributary streams have increased the discharge of the overflowing stream.

Drainage basin Area that contributes surface water to a particular stream network.

Driving force With respect to mass wasting, an influence that tends to make earth material move downslope.

Drought Extended period of unusually low precipitation that produces a temporary shortage of water for people, other animals, and plants.

Dryline An atmospheric boundary between two air masses of contrasting moisture content where the air mass with less moisture advances into the air mass with greater moisture. Common in the south-central United States in late spring and summer, it is a source of atmospheric instability for the development of severe weather.

Dry ravel After a fire, the gravity-driven process of moving a large volume of sand, gravel, and organic debris downslope. This debris was stored upslope of brush vegetation before a fire.

Dry slot A nearly cloud-free area behind a cold front in an extratropical cyclone or a wall cloud in a severe thunderstorm that is created by sinking dry air.

Duff A layer of compacted and partially decomposed leaves, twigs, and other organic matter in a forest soil that accumulates below leaf litter and above mineral soil; generally burns by glowing and smoldering combustion.

Dust storm A weather event in which visibility at eye level drops to less than 1 km (0.6 mi.) for hours or days because of fine airborne particles of silt and clay.

Dynamic equilibrium With respect to the behavior of streams and rivers, the overall balance between the work that a river does in transporting sediment and the amount of sediment that it carries.

E-line Long, narrow band inland from the shore that marks the distance to which coastal erosion is expected to have washed away dry land in a given number of years.

E soil horizon A light-colored horizon underlying the A horizon that is leached of iron-bearing compounds.

E-zone With respect to coastal erosion, a hazard area between sea level and an e-line that is expected to erode away within a particular time period.

Earth climate system Emphasizes the interactions among the atmospheric, cryospheric, and terrestrial, and marine biospheric processes that affect the climate of Earth.

Earth fissure A large, deep crack in loose sediment at the land surface; formed by the lowering of the water table and the consequent drying out of the earth material.

Earth's energy balance Condition in which incoming solar radiant energy generally equals outgoing radiant energy from our planet; this equality exists despite changes in the energy's form as it moves through the atmosphere, oceans, land, and living things before being radiated back into outer space.

Earth system science The study of Earth as a system.

Earthquake Sudden movement of a block of Earth's crust along a geologic fault; the movement releases accumulated strain in the rocks.

Earthquake cycle A hypothesis that explains the periodic nature of earthquakes on a given fault on the basis of a drop in elastic strain after an earthquake and reaccumulation of strain before the next event.

Easterly waves North–south oriented low-pressure troughs that migrate from east to west at equatorial latitudes causing local convergence and divergence of the trade winds giving rise to persistent thunderstorms, which can develop into tropical storms and hurricanes.

Eddy With respect to ocean currents, the circular movement of water that develops behind an obstruction, where two currents pass in opposite directions or along the edge of a permanent current; can have a central core of either warm or cold water. Can be large enough to be detectible with satellite sensors.

Eddy vortex A persistent, circular movement of warm water that develops on the edge of the Loop Current in the Gulf of Mexico that can affect the intensity of hurricanes that cross it.

EF scale A graduated range of values from EF0 to EF5 for describing tornado intensity based on the maximum three-second wind velocity inferred from damage to buildings, towers, poles, and trees. A modification of the F-scale developed by T. Theodore Fugita in 1971.

Effusive eruption A passive outpouring of magma onto Earth's surface generating a lava flow. Effusive eruptions are typical of low viscosity basaltic magmas, which readily allow volatiles to escape nonexplosively. Low-volatile Andesitic and rhyolitic magmas may erupt effusively.

Ejecta blanket Layer of debris surrounding an impact crater; typically forms an irregular terrain of low mounds and shallow depressions.

El Niño An event in the atmosphere and ocean during which trade winds weaken or even reverse, the eastern equatorial Pacific Ocean becomes anomalously warm, and the westward moving equatorial ocean current weakens or reverses; this change in ocean temperature can cause anomalous weather in adjacent landmasses.

Elastic rebound A process where elastic strain is replaced by permanent fault offset during an earthquake.

Elastic strain Deformation that returns to its original shape when stress is released.

Electrical resistivity Degree to which earth materials are unable to conduct an electrical current. Measurements of this property are used in the exploration for energy, mineral, and water resources.

Electromagnetic energy Property of a system that enables it to do work through oscillations in an electric and magnetic field.

Electromagnetic spectrum The collection of all possible wavelengths or frequencies of electromagnetic radiation.

Electromagnetic wave An oscillation in an electric and magnetic field that transfers radiant energy.

Energy A quantifiable physical property describing the state of an object or a system. The part of energy pertinent to this text and natural hazard is the force. See *Force*.

Enhanced Fujita scale See *EF scale*.

Environmental unity A principle of environmental studies that states that everything is connected to everything else.

Epicenter Point on Earth's surface directly above the focus of an earthquake; the focus is the area that first ruptures on a geologic fault in an earthquake.

Eustatic sea level The worldwide average vertical position of the ocean surface.

Evacuation With respect to a natural hazard, the movement of people away from the location of a probable dangerous or destructive natural phenomenon.

Expansive soil In-place, weathered earth materials that, upon wetting and drying, will alternately expand and contract, causing damage to foundations of buildings and other structures.

Exponential growth A type of compound growth in which a total amount or number increases at a certain percentage each year, and each year's rate of growth is added to the total from the previous year; characteristically stated in terms of a particular doubling time, that is, the time in years it will take the original number to double; commonly used in reference to population growth.

Exposure dating Determining the amount of time that sediment or rock has been at Earth's surface by measuring quantities of isotopes in the material that are produced by cosmic rays.

Extinction With reference to a wildfire, the final phase in which all combustion, including smoldering, ceases because of insufficient heat, fuel, or oxygen. With reference to organisms, the complete disappearance of a biological species.

Extratropical cyclone An area of low atmospheric pressure, characterized by rotating winds, which generally develops along a weather front between 30° and 70° latitude; commonly associated with heavy precipitation, strong winds, and other severe weather; can produce a storm surge and coastal erosion when crossing a shoreline.

Eye With respect to hurricanes and related intense tropical cyclones, the central circular area of light winds and partly cloudy skies where subsiding dry air encounters rising moist air.

Eyewall The portion of a hurricane immediately adjacent to a central circular area of nearly clear sky and light winds; usually the region having the highest winds and most intense rainfall.

F-scale See *EF scale*.

Falling With respect to mass wasting and landslides, earth materials such as rocks dropping through the air from steep slopes.

Fault A fracture or fracture system in earth materials along which rocks on opposite sides of the fracture have moved relative to one another.

Fault creep Slow, essentially continuous movement of blocks of Earth's crust principally along one side of a fracture; also called *tectonic creep*.

Fault gouge A clay-rich zone of pulverized rock produced by movement along a fracture or fracture system in Earth's crust; can act as an underground barrier or dam to the flow of groundwater.

Fault scarp Steep slope or small cliff that was formed during an earthquake by rupture and displacement of a block of Earth's crust.

Faulting The process of creating a rupture in Earth's crust by the movement of one crustal block in relation to another.

Felsic Refers to an igneous rock or melt composition with more than about 65 percent silica rich in, which is characteristic of granitic rock and rhyolitic magma.

Fetch Distance that wind blows over a body of water; a principal factor that determines the height of windblown waves; in a cyclone, a contributing factor to the height of the storm surge.

Fibrous fracture A distinctive pattern for the breakage of minerals in which long slivers of the mineral break away from the specimen.

Fine grained soils Type of soil where over half of the soil particles (by weight) is composed of particles less than 0.074 mm in diameter.

Fire management With respect to wildfires, the control of combustion to minimize loss of human life and damage to property. Accomplished through understanding the fire regime, public education, remote sensing of vegetation to determine fire potential, and prescribed burns.

Fire regime An ecosystem's potential for wildfire as inferred from fuel types, terrain, and past fire behavior.

Fissure Large fracture or crack in rock along which the opposing side have separated; may become a volcanic vent if lava or pyroclastic material is extruded.

Flash flood Overbank flow that results from a rapid increase in stream discharge; commonly occurs in the upstream part of a drainage basin and in small tributaries downstream.

Flashy discharge Stream flow characterized by a short lag time or response time between rainfall and peak stream discharge.

Flood basalt A laterally extensive, basaltic lava flow that generally is part of a thick sequence of similar flows. These flows commonly build up a plateau; also referred to as a *plateau basalt*.

Flood discharge The stream flow volume per unit of time (represented by the units cms, cubic meters per second, or cfs, cubic feet per second) at which damage to personal property is likely to begin.

Flooding From a hazards perspective, high water levels in a stream, lake, or ocean that may damage human facilities. As a natural process, overbank flow that may construct a floodplain adjacent to a stream channel or a higher-than normal water level along a coast that extends inland beyond the beach.

Floodplain Flat topography adjacent to a stream produced by overbank flow and by lateral migration of the channel and associated sand or gravel bars.

Floodplain regulation Governmental restriction of land use in an area likely to be inundated by a stream's overbank flow that could damage buildings and infrastructure.

Flood-proofing With respect to urban flooding refers to practices such as raising buildings above potential floodwater, constructing flood doors on buildings, placing flood walls around buildings or other structures, or installing improved drains to remove floodwater accumulation.

Flood stage Water level of a stream at which damage to personal property is likely to begin.

Flow With respect to mass wasting, downslope movement of earth materials that deform as a fluid, such as the fluid movement of grains of sand, rock, snow or ice, debris, and mud.

Flowage See *Flow*.

Flowstone General term for a mineral deposit, most commonly of calcium carbonate, formed by flowing water on the wall or floor of a cave.

Flux In a natural system, the rate at which a substance moves from one part of the system to another.

Focus In an earthquake, the point or location in Earth of the initial break or rupture, and from which energy is first released; during an earthquake, seismic energy radiates out from this point or location. Also known as the *hypocenter*.

Foliation plane With respect to mass wasting, a surface or zone of weakness in metamorphic rocks produced by the alignment of minerals during metamorphism.

Footwall Block of Earth's crust that is below a slanted geologic fault.

Fog A cloud that is in contact with the ground.

Force With respect to physics, an influence that tends to produce or to change the motion of a body, or one that produces stress within a body.

Forcing With respect to climate, a variable, such as solar radiation, volcanic aerosols, or greenhouse gases, that influences the direction of climate change.

Forecast With respect to a natural hazard, an announcement that states that a particular event, such as a flood or storm, is likely to occur during a particular time interval, often with some statement of the degree of its probability.

Foreshock Small to moderate earthquake that occurs before and in the same general area as the main earthquake.

Formation A body of rock or sediment composed of one or more distinctive types that can be mapped at the surface or traced in the subsurface—commonly given a geographic place name (e.g., Potsdam Sandstone) and formally defined in a scientific publication.

Fracture With respect to minerals, the breakage of a mineral in an irregular, curving or fibrous manner.

Free face A steep cliff of bare rock.

Frequency The number of waves passing a point of reference per second; commonly expressed in cycles per second or hertz (Hz); the inverse of the wave period.

Front With respect to weather, a boundary between two air masses that are distinguished by differences in temperature and, commonly, moisture content. See

Cold front, *Warm front*, *Stationary front*, and *Occluded front*.

Frostbite Injury to body tissues caused by exposure to extreme cold.

Frost heaving The upward and lateral movement of soil particles and the land surface as the result of ice formation in the soil.

Frost-susceptible soil Surficial earth material that is likely to move and expand when frozen as the result of the accumulation of ice between particles of the material; commonly unconsolidated material with a high content of silt.

Fugita scale See *EF-scale*.

Funnel cloud A narrow, rotating cloud extending downward from a thunderstorm. This cloud may become a tornado if it reaches the ground.

Gabion A metal wire basket, commonly made out of chain-link fencing that is filled with rocks. These baskets are stacked along stream banks to prevent erosion and are used as barriers to slow water flow in retention ponds.

Geochronology Science of dating earth materials, fossil organisms, and geologic features and events; and of determining the sequence of events in Earth history.

Geographic information system (GIS) Computer technology capable of storing, retrieving, transforming, and displaying spatial information about Earth and of making maps with these data.

Geologic cycle A group of interrelated sequences of Earth processes known as the hydrologic, rock, tectonic, and biogeochemical cycles.

Geologic map A two-dimensional representation of the age and composition of earth materials that are found beneath the soil within a given geographic area.

Geyser A particular type of hot spring that ejects hot water and steam above Earth's surface; perhaps the most famous is Old Faithful in Yellowstone National Park.

Glacial interval A portion of Earth history during which a cooler, wetter climate supported widespread continental glaciers.

Glacier A large mass of flowing ice that moves downslope under its own weight and persists from year to year; formed, at least in part, on land by the compaction and recrystallization of snow.

Global climate model (GCM) Linked computer programs that use environmental data in mathematical equations to predict global change, such as increases in mean temperature, changes in precipitation, or some other atmospheric variable.

Global positioning system (GPS) Interacting ground control, satellites, and radio receivers that obtain extremely accurate locations and elevations on Earth's surface. Especially useful for detecting land surface changes related to a volcanic eruption, earthquake, tectonic creep, the slow movement of a landslide, or land subsidence.

Global warming The increase in the mean annual temperature of the lower atmosphere and oceans in the past 150 years, primarily as a consequence of burning fossil fuels that emit greenhouse gases into the atmosphere.

Gradient Slope of the land over which a stream flows. Determined by calculating the vertical drop in elevation through some horizontal distance; commonly expressed as m per kilometer or ft. per mile.

Granular structure A soil structure consisting of individual grains or clumps of grains that form generally small semi-round soil peds.

Great earthquake A sudden rupture along a geologic fault in which the release of energy has a moment magnitude of 8 or greater.

Greenhouse effect Trapping of heat in the lower atmosphere by the absorption of infrared energy by water vapor, carbon dioxide, methane, nitrous oxides, halocarbons, and other gases.

Greenhouse gas A substance such as carbon dioxide, methane, nitrous oxides, or halocarbons (e.g., Freon) that absorb infrared radiation, and contributes to global warming of the lower atmosphere.

Groin A long, narrow structure of rock, concrete, wood, or other material generally constructed perpendicular to the shore to protect the coastline and trap sediment in the zone of littoral drift.

Groin field With respect to coastal processes, a group of long, narrow structures of rock, concrete, wood, or other material generally constructed perpendicular to the shore to protect the coastline and trap sediment in the zone of littoral drift.

Ground acceleration With respect to an earthquake, the amount that the speed or velocity of the shaking of the Earth's surface changes in a unit of time, such as a second. Ground acceleration is commonly given as a fraction or percentage of gravitational acceleration, which is 980 cm (~380 in.) per second. The greatest (or peak) ground acceleration during an earthquake is a principal control on the amount of damage to structures.

Ground fire With respect to wildfires, combustion that takes place below the surface; characterized by more smoldering and little flaming; fuels typically are roots and partially decayed organic matter, such as peat.

Ground penetrating radar An electronic system that uses 10 to 1000 MHz electromagnetic waves to image Earth's shallow subsurface to locate features such as natural cavities and fissures, permafrost, contaminated groundwater, and buried pipes and tanks.

Groundwater Subsurface accumulation of water that entirely fills all voids in earth materials.

Groundwater mining The pumping of subsurface water at a faster rate than it is being replenished.

Gust front Leading edge of strong and variable surface winds from a thunderstorm downdraft.

Habitat A particular set of environmental conditions in which adapted plants and animals thrive.

Hachure With respect to topographic contours, a small mark drawn perpendicular to a contour line to indicate the downward sloping of the land into a closed depression; also referred to as a *tic mark*.

Hail Large rounded or irregular piece of ice that has grown while moving up and down numerous times within the clouds of a severe thunderstorm.

Half-life The amount of time necessary for one-half of the atoms of a particular radioactive element to decay.

Halocarbon A chemical compound (e.g., Freon) containing a halogen element, such as chlorine, fluorine, or bromine, that is bonded to carbon; generally synthetic in origin and in the atmosphere contributes to global warming in troposphere and ozone depletion in stratosphere.

Hanging wall A block of Earth's crust that is above a slanted geologic fault.

Hardness With respect to minerals, the resistance of a mineral to scratching; see *Relative hardness*.

Hardpan A soil layer of hard material such as clay or silica oxide or calcium carbonate.

Hawaiian-type eruption An effusive to mildly explosive volcanic eruption that ranks between 0 and 1 on the volcanic explosivity index. This eruption type is characterized by fire fountains and lava flows and is typical of shield volcano eruptions. A central caldera, in contrast to an Icelandic-type eruption where lavas erupt from a fissure, typifies the volcanic vent.

Hazard See *Natural hazard*.

Headland A small peninsula on an irregularly shaped shoreline that is more resistant to erosion than the surrounding earth material.

Headwaters Tributaries of a river that feed into it near its uppermost reach.

Heat energy The property of a system that enables it to do work through random molecular motion. This property may be transferred by conduction, convection, or radiation.

Heat index A value on a temperature scale that describes the human body's perception of air temperature, taking into account relative humidity.

Heat waves A prolonged period of extremely high air temperature that is both longer and hotter than normal. The criteria for establishing what constitutes longer and hotter than normal varies geographically with climate and living conditions.

High pressure center An area of elevated atmospheric pressure where the wind blows clockwise in the Northern Hemisphere. Also called an *anticyclone* or a *high*.

Hot spot A hypothesized, nearly stationary heat source located below the lithosphere that feeds overlying volcanic processes near Earth's surface.

Hot spring A natural discharge of groundwater at a temperature higher than that of the human body.

Hot tower A rain cloud that reaches the top of the troposphere in a hurricane. The release of latent heat in its formation warms the air at the high altitude.

Hummocky An irregular land surface marked by numerous small mounds and depressions.

Hurricane Tropical cyclone characterized by sustained winds of 119 km (118 km in Canada or 74 mi. per hour); known as a typhoon in the western North Pacific and South China Sea, severe tropical cyclone in the southwest Pacific and southern Indian Oceans, and severe cyclonic storm in the northern Indian Ocean.

Hurricane warning An alert issued by weather forecasters for an area where hurricane conditions are expected in 24 hours or less. In Canada this alert is also issued for areas likely to experience high waves and coastal flooding, but not necessarily hurricane-strength winds.

Hurricane watch An alert issued by weather forecasters for an area that may experience hurricane conditions within 36 hours. In Canada this alert is not restricted to a 36-hour time frame.

Hydraulic chilling With respect to volcanoes, the use of water to cool and solidify flowing lava; a method used to divert lava flows threatening property.

Hydrogen sulfide (H$_2$S) A toxic, flammable gas that smells like rotten eggs and is emitted by volcanoes.

Hydrograph A graph of the discharge (flow) of a stream with time.

Hydrologic cycle Circulation of water from the oceans to the atmosphere and back to the oceans by way of precipitation, evaporation, runoff from streams, and groundwater flow.

Hydrology The study of surface and subsurface water.

Hydrophobic layer Lower layer of debris on forest floors or brushland that has accumulated water-repellant organic chemicals; especially common in drier climates after very hot fires.

Hypocenter See *Focus*.

Hypothermia A medical condition characterized by rapid loss of core body heat causing intense shivering, loss of muscle coordination, mental sluggishness, and confusion. This condition can develop in an improperly dressed person in cool, wet, windy weather or in one immersed in water that is 25°C (~77°F) or less; unless treated, the condition leads to unconsciousness and eventually death.

Ice storm A prolonged period of freezing rain during which thick layers of ice accumulate on all cold surfaces.

Iceberg Large block of floating glacial ice that has broken from the front of a glacier, dropping into the water by a process called *calving*.

Icelandic-type eruption An effusive to mildly explosive volcanic eruption that ranks between 0 and 1 on the volcanic explosivity index. This eruption type is characterized by fire curtains and lava flows typical of Icelandic shield volcano eruptions. An elongate fissure, in contrast to a Hawaiian-type eruption where lavas erupt from central caldera, typifies the volcanic vent.

Igneous rock An aggregate of minerals formed from the solidification or explosion of magma; extrusive if it crystallized on or very near the surface of Earth, and intrusive if it crystallized well beneath the surface.

Impact With respect to hazardous events, the effect or influence of an action.

Impact crater A circular-to-elliptical depression produced by collision of a large object, such as an asteroid, with a planet, moon, or other asteroid.

Impervious cover With respect to urban hydrology, surfaces covered with concrete, roofs, or other structures that impede the infiltration of water into the soil. In general, as urbanization proceeds, the percentage of the land that is under impervious cover increases.

Inactive fault A fracture in the Earth's crust along which rock has not moved in the past 2 million years.

Indirect effect A change that depends upon intervening factors. In a natural accident, disaster, or catastrophe, such effects could include emotional distress; the donation of money, goods, and services; or the payment of taxes to finance recovery; also called a *secondary effect*.

Injection well An artificial, generally cylindrical, hole excavated for pumping fluids, such as liquid hazardous waste, underground.

Inner core See *Core*.

Instrumental intensity With respect to earthquakes, the direct measurement of ground motion during an earthquake with a high-quality seismograph. Used to immediately produce a map showing perceived shaking and potential damage.

Insurance The guarantee of monetary compensation for a specified loss in return for payment of a premium.

Integrated coastal zone management Management that promotes ecological function of the coastal zone through sustainable solutions that minimizes conflicts from development in the coastal zone.

Intensity With respect to earthquakes, a measure of the effects of an earthquake at a specific place; measured as the effects on people and structures using the Modified Mercalli Scale or the direct measurement of ground motion by a sensitive seismograph; see *Instrumental intensity*. With respect to cyclones, the strength of the storm as indicated by sustained wind speeds and lowest atmospheric pressure.

Interglacial interval A portion of Earth history when continental glaciation was minimal or absent.

Intraplate earthquake Sudden movement along a fault caused by the abrupt release of accumulated strain in the interior of a lithospheric plate, far away from any plate boundary.

Ion An electrically charged atom or molecule formed by a loss or gain of electrons.

Iridium A platinum-group metal that is more abundant in meteorites than in earth materials. Its high concentration in certain clay deposits is attributed to their formation during or after impact of a large extraterrestrial body.

Isostasy The principle stating that thicker, more buoyant crust is topographically higher than crust that is thinner and denser. Also, with respect to mountains, the weight of rocks of the upper crust is compensated by buoyancy of the mass of deeper crystal rocks; that is, mountains have "roots" of lighter crustal rocks extending down into the denser mantle rocks, like icebergs in the ocean.

Isotope A variety of a chemical element distinguished by its number of neutrons; may be stable or radioactive.

Jet stream A concentrated flow of air near the top of the troposphere, characterized by strong winds.

Jetty A very long, narrow structure of rock, concrete, or other material generally constructed perpendicular to the shore; commonly constructed in pairs at the mouth of a river or inlet to a lagoon, estuary, or bay; designed to stabilize a channel, control deposition of sediment, and deflect large waves.

Joule The basic unit of energy in the metric system; equivalent to the work done by a force of one Newton when its point of application moves one meter in the direction that the force is acting.

Karst See *Karst topography*.

Karst plain A generally flat land surface containing numerous sinkholes.

Karst topography A landscape characterized by sinkholes, caverns, and diversion of surface water to subterranean routes that results from dissolution of subsurface rocks by groundwater.

Kinetic energy The property of a system that enables it to do work through motion.

K soil horizon A calcium carbonate-rich horizon in which the carbonate often forms laminar layers parallel to the surface; carbonate completely fills the pore spaces between soil particles.

K-T boundary The interface in the geologic record where earth materials that formed during the Cretaceous Period are in contact with earth materials that formed during the Tertiary Period; dated as approximately 65 million years old.

Kyoto Protocol International agreement to reduce emissions of greenhouse gases in an effort to manage global warming; now legally binding in at least 175 countries, including Canada, Russia, and the European Union. The United States has not honored its original agreement to implement this protocol.

Lag time The time interval between the greatest amount of precipitation and peak discharge of a stream. Urbanization generally decreases the lag time.

Lahar A type of volcanic mudflow or debris flow that is triggered by melting of surface snow and or ice, potentially in the form of large glaciers, due to heating by volcanic gasses or ash during an eruption.

Landslide Specifically, rapid downslope movement of rock or soil; also a general term for all types of downslope movement.

Landslide hazard map A two-dimensional representation of the land surface that identifies areas that are likely to experience landslides and the appropriate land uses in those areas.

Landslide risk map A two-dimensional representation of the land surface that identifies areas that are likely to experience a rapid mass movement of earth material, states the probability that one of these events will occur, and assesses potential losses.

Land-use planning The preparation of an overall master plan for future development of an area; the plan may recommend zoning restrictions and infrastructure both practical and appropriate for the community and its natural environment; based on mapping and classification of existing human activities and environmental conditions, including natural hazards.

Lapilli Gravel-size pyroclastic debris ranging in size from 2 to 64 mm (~0.08 to 2.5 in.).

Latent heat Energy created by random molecular motion that is either absorbed or released when a substance undergoes a phase change, such as during evaporation or condensation.

Latent heat of vaporization Energy absorbed by water molecules as they evaporate.

Lateral blast Type of volcanic eruption characterized by explosive activity that is more or less parallel to the surface of Earth. A lateral blast may take place when the side of a volcano fails.

Lava Molten material that flows onto the land surface from a volcano, or rock that solidifies from such molten material.

Lava dome Relatively small volcano type composed of highly viscous magma, usually rhyolitic in composition, that is often restricted to the vent region. Lava domes often found plugging the vent within the volcanic crater of a recently erupted stratovolcano.

Lava flow Eruption of magma at Earth's surface that generally moves as a liquid mass downslope from a volcanic vent.

Lava tube A natural enclosed conduit or tunnel through which magma moves down the slope of a volcano (sometimes many kilometers); eventually the magma may again emerge at the surface; after the volcanic eruption, a tube commonly remains as an open void and forms a type of cave.

Levee A mound or embankment parallel to a stream channel; it may consist of fine sediment deposited from overbank flow during a flood or be an earthen embankment constructed by humans to protect adjacent land from flooding.

Lichenometry An absolute dating technique using measurements of the diameter of lichens on a rock surface to determine how long the surface has been exposed to the atmosphere and sunlight.

Lightning A natural, high-voltage electrical discharge between a cloud and the ground, between clouds, or within clouds. The discharge takes a few tenths of a second and emits a flash of light that is followed by thunder.

Linkage With respect to natural hazards, a relationship between two phenomena.

Liquefaction Transformation of water-saturated granular material from the solid state to a liquid state.

Lithification The hardening of loose earth material into rock, commonly by cementation and compaction.

Lithosphere Outer layer of Earth, approximately 100 km (~60 mi.) thick, which comprises the tectonic plates that contain the ocean basins and continents; includes both Earth's crust and the solid, upper part of the mantle.

Littoral transport Movement of sediment in the nearshore environment as a result of return flow from waves that have washed up on the shore; sediment is moved by both the longshore current and by beach drift.

Local tsunami A series of gravity waves in the ocean originating from a source that is close to the shoreline that is inundated. These waves are produced by the displacement of the entire water column by either an underwater earthquake, landslide, volcanic eruption, or extraterrestrial impact.

Loess Deposit of windblown silt and clay.

Longitudinal profile With respect to streams, a graph that shows the decrease in elevation of a stream bed between its head and its mouth. This graph of elevation versus distance generally produces a line that is concave up.

Longshore bar and trough Elongated depression and adjacent underwater ridge of sand roughly parallel to shore; produced by wave action.

Longshore current Waterflow parallel to shore that develops in the surf zone as the result of waves that strike the land at an angle. Responsible for longshore drift.

Longshore drift The sediment transported in the nearshore environment parallel to the shore by a wave-generated current.

Longshore trough See *Longshore bar and trough*.

Loose-snow avalanche A rapid downslope flow of granular snow and ice.

Love wave A type of earthquake-generated surface oscillation in which molecular vibrations within the Earth are transverse to the direction the oscillation is moving.

Low pressure center A parcel of the atmosphere that has relatively low air pressure, often referred to as a *low*. see *Cyclone*.

Luster With respect to mineralogy, the way that light is reflected from a mineral; may be catagorized as metallic or nonmetallic.

Magma Naturally occurring molten rock material formed deep within the crust or in the mantle.

Magma chamber A large pool of molten rock that accumulates in the subsurface and is the primary source of lava, and volcanic gasses and debris that are extruded to the surface during an eruption.

Magma evolution Refers to the increase in silica content of a magma as it ascends toward the surface. The change results from crustal assimilation and chemical processes within the magma as it rises.

Magnetic reversal Involves the change of Earth's magnetic field between normal polarity and reverse polarity; also sometimes known as geomagnetic reversal.

Magnitude With respect to natural hazards, especially earthquakes, the amount of energy released.

Magnitude-frequency concept The idea that the intensity and extent of an event are inversely proportional to how often it occurs.

Mainshock The primary, highest magnitude earthquake in a series of earthquakes in a given area during a limited time interval.

Major earthquake A sudden rupture along a geologic fault in which the release of energy has a moment magnitude of 7 to 7.9.

Managed retreat With respect to coastal erosion, a strategy involving the landward movement of buildings and associated infrastructure away from a shoreline to remove the property from an area susceptible to flooding by seawater.

Mantle An internal layer of Earth approximately 3000 km (~1900 mi.) thick composed of rocks that are primarily iron and magnesium-rich silicates. The lower boundary of the mantle is with the core, and the upper boundary is with the crust. The boundary is known as the Mohorovičić discontinuity (also called the *Moho*).

Mass extinction Sudden loss of large numbers of plant and animal species relative to new species being added.

Mass wasting A comprehensive term for any type of downslope movement of earth materials.

Material amplification Increase in the amplitude of seismic shaking caused by some earth materials. Such increase is generally associated with soft sediments, such as silt and clay deposits.

Mature stage With respect to thunderstorm development, the second phase in which downdrafts move and precipitation falls from the base of a cumulonimbus cloud. With regard to tornado development, the phase in which a visible funnel extends from the thunderstorm to the ground.

Meander With respect to streams, a bend in the channel that migrates back and forth across the floodplain, depositing sediment on the inside of the bend on a point bar, and eroding the outside of the bend into a cutbank.

Meandering A stream channel pattern that is sinuous and is characterized by gentle bends that migrate back and forth across a floodplain.

Meandering river See *Meandering stream.*

Meandering stream A single, sinuous channel of flowing water that moves through gentle bends that migrate back and forth across a floodplain.

Megafloods Extremely large floods capable of causing a catastrophe that may be expected to reoccur every few hundred years.

Megathrust earthquake A term used to describe large earthquakes (Magnitude 8.0 and greater) that occur along low-angle faults at subduction zones.

Mesocyclone A region of rotating clouds, typically around 3 to 10 km (~2 to 6 mi.) in diameter on the flank of a supercell thunderstorm.

Mesoscale convective system (MCS) A large, circular, long-lived and interacting group of thunderstorms that are commonly identifiable on satellite images; generally contain severe thunderstorms and often produce tornadoes.

Mesosphere A zone in the upper atmosphere from about 50 to 80 km (~31 to 50 mi.) where the temperature decreases with altitude; directly above the stratosphere.

Metamorphic rock An aggregate of minerals that formed beneath Earth's surface from preexisting rock by the effects of heat, pressure, and chemically active fluids. In foliated metamorphic rocks, mineral grains are preferentially aligned parallel to one another or light and dark minerals are segregated into bands; non-foliated metamorphic rocks have neither characteristic.

Meteor Extraterrestrial particle from dust to several centimeters in size that is burned up from frictional heat upon entry into Earth's atmosphere; light emitted from the burning meteor forms a "shooting star."

Meteorite Extraterrestrial particle from dust to asteroid size that hits Earth's surface.

Meteoroid Extraterrestrial particle in space that is smaller than 10 m and larger than dust size; may form from breakup of asteroids.

Meteor shower Large numbers of meteoroids entering the atmosphere and producing light from frictional heating; commonly occurs when Earth's orbit passes through the tail of a comet.

Microburst A strong, localized downburst from a thunderstorm that affects an area of 4 km (~2.5 mi.) or less.

Microearthquake Sudden movement along a geologic fracture or fault in which the release of energy has a moment magnitude less than 3. Patterns of these very small quakes may be precursors to larger-magnitude earthquakes; also called a *very minor earthquake.*

Microzonation With respect to earthquakes, the detailed location and land-use planning of areas having various earthquake hazards.

Mid-ocean ridge A seafloor mountain range commonly found in the central part of oceans characterized by seafloor spreading. An example is the Mid-Atlantic Ridge.

Milankovitch cycles Natural variations in the intensity of solar radiation that reaches Earth's surface that recur at approximately 20,000-, 40,000-, and 100,000-year intervals.

Mineral A naturally occurring, inorganic crystalline substance with an ordered atomic structure and a specific chemical composition.

Mitigation The avoidance of, lessening, or compensation for anticipated harmful effects of an action, especially with respect to the natural environment.

Modified Mercalli Intensity scale A scale with 12 divisions on which the amount and severity of shaking and damage from an earthquake can be ranked.

Moho See *Mohorovičić discontinuity.*

Mohorovičić discontinuity Compositional boundary that separates the less dense crust from the underlying higher-density mantle.

Moisture content The amount of water in a soil as a percent (weight of water to weight of the solid soil particles as a percent). Also called the *water content.*

Moment magnitude scale A numerical assessment of the amount of energy released by an earthquake on the basis of its seismic moment, which is the product of the average amount of slip on the fault that produced the earthquake, the area that actually ruptured, and the shear modulus of the rocks that failed.

Mudflow A slurry of fine, unconsolidated earth materials and water that flows rapidly downslope or down a channel; also referred to as an *earth flow* or type of *debris flow.*

Multiple hypotheses Several proposed explanations for a phenomenon that are being considered concurrently; also referred to as *multiple working hypotheses.*

Natural hazard A potential danger that poses a threat to people or property that exists or is caused by nature; generally one that is not made or caused by humans.

Natural service function A benefit that arises from an event caused by nature that is also a hazard to people or the environment.

Near-Earth Asteroid Tracking Project (NEAT) NASA-supported program started in 1996 to study the size, distribution, and dynamic processes associated with near-Earth objects and to identify those objects having a diameter of about 1 km (~3300 ft.) or greater.

Near-Earth object (NEO) Asteroids that reside and orbit between Earth and the sun or have orbits that intersect Earth's orbit.

Newton Measure of force in the metric system in which 1 Newton is the force necessary to accelerate a 1 kg (~2.2 lb.) mass 1 m (~3.3 ft.) per second each second that it is in motion.

Nor'easter An extratropical cyclone that tracks along the East Coast of the contiguous United States and Canada and which has continuously blowing northeasterly winds just ahead of the storm; commonly characterized by hurricane-strength winds, heavy snows, intense precipitation, and large waves along the coast.

Normal fault A generally steep fracture or fracture system along which the hanging wall has moved down relative to the foot-wall in the direction that the fracture dips.

North Atlantic Oscillation A North Atlantic atmospheric pressure phenomenon that fluctuates in intensity affecting the path of westerly winds that flow from North America toward northern Europe and consequently the winter weather on these two continents.

Nowcasting With respect to weather, the near real-time prediction of the movement and behavior of weather systems, such as severe storms, that have formed or are in the process of forming.

Nuée ardente See *Pyroclastic flow*

O soil horizon Soil horizon that contains plant litter and other organic material; found above the *A* soil horizon.

Ocean acidification An increase in the acidity of seawater due primarily to dissolving additional atmospheric carbon dioxide. This condition results from higher levels of carbon dioxide in the atmosphere and is threatening the growth of corals and other marine organisms that make skeletal material from calcium carbonate.

Ocean conveyor belt A large-scale circulation pattern in the Atlantic, Indian, and southwestern Pacific Oceans that is driven by differences in water temperature and density; also referred to as a *thermohaline current*. Northward-flowing warm water in this current maintains the moderate climate of northern Europe; if this current were to slow or stop, ice-age conditions might return to portions of the Northern Hemisphere.

Occluded front An atmospheric boundary between two air masses of contrasting temperature along which the colder air mass advances into a cool air mass and elevates warmer air above the boundary.

Organic soil A surficial, in-place accumulation of partially decayed plant material, such as peat.

Organizational stage With regard to tornado development, the initial phase in which a funnel cloud forms from rotating winds within a severe thunderstorm.

Outer core See *Core*.

Outflow boundary A line separating thunderstorm-cooled air from the surrounding air. This line behaves like a cold front with both a wind shift and drop in temperature. In large thunderstorm complexes this boundary may persist for more than 24 hours and provide instability that fosters development of new storms.

Overbank flow Moving floodwaters that have exceeded the capacity of a stream channel and have spread outward beyond its sides.

Overwash Flooding of the beach, dunes, and in some cases an entire barrier island by a storm surge; commonly transports sand from the beach and dunes inland; see *Washover channel*.

Ozone (O3) A pungent, strongly reactive form of oxygen gas that is most abundant in the stratosphere, where it protects the Earth surface from high levels of ultraviolet radiation. In the lower atmosphere it is an air pollutant.

Ozone depletion Stratospheric loss of triatomic oxygen; related to release of halocarbons and other gases that destroy triatomic oxygen in the atmosphere.

P wave The fastest type of earthquake vibration within the body of Earth; also called a *primary or compressional wave*, this vibration oscillates in the direction that it travels.

Pahoehoe A type of lava with a smooth to ropy surface texture that forms when a flow solidifies; most common in basalt.

Paleomagnetism Refers to the study of magnetism of rocks and the intensity and direction of the magnetic field of Earth in the geologic past.

Paleoseismicity The occurrence of earthquakes in space and time in the geologic past.

Parcel With respect to weather, an imaginary balloonlike blob of air several hundred cubic meters in volume that acts independently of the surrounding air.

Parent isotope An unstable variety of a chemical element distinguished by its number of neutrons and its potential for radioactive decay. Measuring the proportions of parent and daughter isotopes in a substance is the basis for most absolute dating.

Particulate With respect to the atmosphere, a piece of dust, pollen, soot, ash, or other very fine material that is suspended in the air.

Peat Organic-rich sediment composed primarily of partially decayed plant material, such as moss, leaves, and sticks; typically forms in bogs and swamps and has a high water content.

Peds The natural soil fragment or clod, the soil naturally breaks into these fragments that children may call dirt clods.

Peléan-type eruption A severe volcanic eruption that ranks between 3 and 4 on the volcanic explosivity index. This eruption type is characterized by a moderate to large eruption column with plume heights between 3 and 15 km (~2–9 mi.) and are typical of lava dome eruptions triggered by dome collapse.

Permafrost Earth material in the ground with a temperature that is below freezing continuously for at least two years; commonly cemented with ice; underlies about 20 percent of the world's land area.

Permanent gas A substance in the atmosphere, such as nitrogen or oxygen, that cannot be compressed with pressure alone and is present in generally constant proportions.

Permeability A measure of the ability of an earth material to transmit fluids such as water or oil.

Phreatomagmatic See *Vulcanian-type eruption*.

Pineapple Express An atmospheric flow of tropical moisture and extratropical storms originating in the region of the Hawaiian Islands and directed toward the West Coast of the United States.

Plate tectonics An all-encompassing, global scientific theory that explains the characteristics and origin of most earthquakes, volcanoes, and other Earth features; based on the subdivision of the entire lithosphere into more than a dozen large, rigid blocks that move horizontally and interact to cause deformation, melting, and other changes.

Pleistocene Epoch Geologically recent interval of Earth history characterized by widespread continental glaciation and commonly referred to as the "Ice Age"; the subdivision of the Quaternary Period before the Holocene Epoch.

Plinian-type eruptions A cataclysmic volcanic eruption that ranks from 4 to 6 on the volcanic explosivity index. This eruption type is characterized by a towering eruption column with a plume height that may exceed 35 km (~22 mi.) and is typical of stratovolcano eruptions.

Plunging breaker A type of water wave in which the wave crest rises up and falls into its own trough as it strikes a relatively steep shoreline; tends to be associated with beach erosion.

Point bar Accumulation of sand and other sediment on the inside of a meander bend of a stream.

Polar desert An arid region near the North or South Pole where the annual precipitation averages less than 250 mm (~10 in.) per year.

Polar jet stream A concentrated flow of air near the top of the troposphere that, in the Northern Hemisphere, typically crosses southern Canada or the northern conterminous United States and is characterized by strong winds blowing from west to east.

Pool Common bed form produced by scour in meandering and straight stream channels with relatively low channel slope; characterized at low flow by slow-moving, deep water; generally, but not exclusively, found on the outside of meander bends.

Porosity The percentage of void or empty space in earth material such as soil or rock.

Potential energy The property of a system that enables it to do work by virtue of an object's position above some reference level.

Potentially active With respect to earthquakes, a term applied to a geologic fault that shows evidence of movement in the Pleistocene Epoch, but not in the past 10,000 years.

Power With respect to physics, the rate of doing work.

Precursor With respect to natural hazards, a physical, chemical, or biological phenomenon that occurs before a hazardous event such as an earthquake, volcanic eruption, or landslide.

Precursor event See *Precursor*.

Prediction With respect to a hazardous event, such as an earthquake, the advance determination of the date, time, and size of the event.

Preignition Initial phase of a wildfire during which fuel is brought to a temperature and water content that favors ignition; involves both preheating and pyrolysis.

Preparedness A state of readiness achieved by an individual or group of people, such as for a natural hazard or war; see *Disaster preparedness*.

Prescribed burn A fire purposely set and contained within a designated area to reduce the amount of fuel available for a wildfire; also called a *controlled burn*.

Primary effect See *Direct effect*.

Prismatic structure A soil structure consisting of clay-rich columns.

Process With respect to natural hazards, the physical, chemical, or biological ways by which events, such as volcanic eruptions, earthquakes, landslides, and floods, affect Earth's surface.

Proxy data With respect to climate, proxy data refers to data that is not strictly climate in nature, but can be used to infer, indirectly, changes in the climate record.

Punctuated uniformitarianism Concept that the geologic record of gradual mountain building, canyon erosion, and landscape construction was periodically interrupted by catastrophic events that caused rapid changes to topography or biota.

Pyroclastic activity Eruptive or explosive volcanic activity in which all types of volcanic debris, from ash to very large particles, are physically blown from a volcanic vent.

Pyroclastic debris Volcanic material that is explosively erupted from a volcano. The material ranges in size from fine ash to large blocks and may be composed of low-density magma, which cooled in the atmosphere during the eruption or of cold dense rock ripped from the interior of the volcano.

Pyroclastic deposit An accumulation of particles forcefully ejected from a volcanic vent, explosive in origin, containing volcanic ash particles and those of larger size such as blocks and bombs.

Pyroclastic flow Extremely hot and rapid flow of eruptive material down the flank of a volcano; consists of volcanic gases, ash, and other materials that move rapidly; commonly forms upon the collapse of an eruption column; may also be known as an *ash flow*, *fiery cloud*, or *nuée ardente*.

Pyroclastic rock Lithified igneous earth material composed of fragments of volcanic glass, mineral crystals, and fragments of other rocks that were ejected during a volcanic eruption.

Pyrolysis In a fire, a group of chemical processes by which heat divides or splits large fuel molecules into smaller ones. Products from these processes include volatile gases, mineral ash, tars, and carbonaceous char.

Quick clay Deposit of very fine sediment that when disturbed, such as by seismic shaking, may spontaneously liquefy and lose all shear strength.

R soil horizon Consolidated bedrock that underlies the soil.

Radiation With respect to physics, the transfer of energy as electromagnetic waves or as moving subatomic particles. A primary way that heat is transferred from the sun to Earth, the Earth to the atmosphere, and to a lesser extent from a wildfire to the atmosphere.

Radioactive isotope Unstable variety of a chemical element distinguished by its number of neutrons and its emission of ionizing radiation.

Rain band Precipitation-producing clouds that spiral around a hurricane.

Reactive With respect to natural hazards, a person's action or response to a hazard that occurs after an event, such as a natural accident, disaster, or catastrophe.

Recurrence interval The time between natural events, such as floods or earthquakes. Commonly given as the average recurrence interval, which is determined by averaging a series of intervals between events.

Red flag warning With respect to wildfires, a notice from the National Weather Service or other governmental agency that an extreme fire condition is occurring or is likely to occur within 24 hours.

Refraction Bending of a waveform. With respect to coastal processes, the bending of surface waves as they enter shallow water. This takes place where part of the wave "feels the bottom" and slows down while the remainder of the wave continues to move forward at a faster velocity.

Relative dating Placing geologic features, objects, earth materials, or events in chronological order in the geologic time scale without determining absolute ages.

Relative hardness With respect to minerals, the ability of one mineral or object to scratch another. Generally determined on the Mohs Hardness Scale in which diamond is assigned the highest value of 10 and talc the lowest value of 1. Other minerals, such as corundum, quartz, and calcite, have intermediate values of relative hardness.

Relative humidity A measure of the amount of water vapor in the air compared to the amount that would saturate the air at a given temperature and pressure; commonly expressed as a percentage.

Relative sea level The local or regional position of the sea at the shore that is influenced by uplift or subsidence of the land and changes in global eustatic sea level.

Relief With respect to topography, the difference in elevation between a higher point (as on a hilltop) and a lower one (as on a valley floor).

Residence time In natural systems, the duration that a substance spends at a specific place or in a specific compartment. A measure of how long a chemical element, compound, or other substance stays in a specific part of a system.

Resisting force With respect to mass wasting, an influence that tends to oppose downslope movement.

Resonance With respect to earthquakes, the matching of the frequency of shaking with the natural vibrational frequency of an object. With respect to storm surge and tsunamis, the superposition of waves reflecting from one side of a narrow bay or lake with waves of water moving in another direction.

Resurgent caldera Uplift by rising magma of the central collapsed block of a giant volcanic crater; the giant crater resulted from an earlier explosion or collapse of the volcano summit; see *Caldera*.

Retreat With respect to glaciers, an interval of time when the area occupied by glacial ice decreased.

Retrofitting With respect to earthquakes, the renovation and physical modification of existing structures to withstand ground shaking and other effects of an earthquake.

Return interval The recurrence interval of wildfires; see *Recurrence interval*.

Return stroke In cloud-to-ground lightning, the component that involves the downward flow of electrons. Electron flow starts near the ground and is then initiated at successively higher points in a channel of ionized air. This flow produces the first bright light in a lightning flash.

Reverse fault A generally steep fracture or fracture system in which the hanging wall has moved up relative to the footwall in a direction opposite to the direction of fault dip. A low-angle reverse fault is called a *thrust fault*.

Rhyolite A fine-grained, light-colored volcanic rock consisting of feldspar, ferromagnesian minerals, and quartz with a relatively high silica content of about 70 percent; associated with volcanic events that may be very explosive.

Richter scale A semi-qualitative measurement of the size of an earthquake based on the amount of ground shaking as determined by maximum S-wave amplitude measured on a seismogram. The scale is not absolute and is affected by local geologic conditions and distance from the earthquake epicenter.

Ridge With respect to weather, an elongated area of high atmospheric pressure.

Riffle A section of stream channel, which at low flow is shallow, fast-flowing water moving over a gravel streambed.

Rift In tectonics, a long, narrow trough that is bounded by normal faults; a surface rupture produced by extension of the lithosphere.

Ring of Fire A popular name given to the chain of volcanoes on islands and on the continents surrounding the Pacific Ocean; actual combustion is much less common than the red-hot glow of lava fountains and flows.

Rip current A seaward flow of water in a confined narrow area from a beach to beyond the breaker zone.

Riprap Layer or assemblage of large broken stones placed to protect an embankment or shoreline against erosion by running water or breaking waves.

Risk From a natural hazards viewpoint, risk may be considered as the product of the probability of an event times the consequences.

Risk analysis An evaluation that estimates the probability that an event will occur.

River A large, natural stream that carries a considerable volume of flowing surface water.

River system Consists of three zones from headwater to river mouth known respectively as the zone of sediment and water production, zone of transport of water and sediment, and zone of sediment deposition.

Rock In geology, a solidified natural aggregate of one or more minerals, natural glass, or solid organic remains. In engineering or design applications, a solidified natural material that requires blasting for excavation or removal.

Rock cycle Group of interrelated processes by which igneous, metamorphic, and sedimentary rocks can each be produced from the others.

Rogue wave A single crest of an oscillation of the surface of the ocean that is much higher than usual. Commonly caused by the constructive interference of smaller waves.

Rope stage With regard to tornado development, the final decaying phase in which the upward-spiraling air in the funnel contacts down-drafts from the thunderstorm, and the tornado begins to move erratically.

Rotational With respect to landslides, a type of sliding downslope movement that develops on a well-defined, upward-curving slip surface in homogeneous earth material.

Rule of Vs As applied to topographic maps, the principle that all contour lines

must form a shape similar to the letter V when they cross a stream valley and that the apex of the V points upstream.

Runoff Water moving over the surface of Earth as overland flow on slopes or as stream flow; that part of the hydrologic cycle represented by precipitation or snowmelt that results in overland flow to a stream.

Runup With respect to waves, the up-rush of the wave along the shore; also, the combined vertical and horizontal distance that a tsunami moves inland from the shoreline.

S wave One type of earthquake vibration that occurs within the body of Earth; also called a *secondary* or *shear wave*, vibrations from this type of oscillation are perpendicular to the direction that it travels.

Safety factor With respect to landsliding, the ratio of resisting to driving forces; a safety factor greater than 1 suggests that a slope is stable.

Sand storm High winds, generally in the desert, that carry fine (~0.05 to 2 mm) mineral particles to heights less than 2 m (~7 ft.) above the ground.

Saturated flow Type of flow in soil where all pores are filled with water.

Scale In maps, the relationship between the distance between features on the map and their actual distance apart on Earth's surface. Expressed either as a ratio, such as 1:24,000, or as a bar scale, a segmented line on the map.

Scarp Steep slope or cliff at the head of a landslide or along a geologic fault; produced by the downslope movement of a slump or slide or by the displacement of earth materials in an earthquake.

Scoria cone See *Cinder cone*.

Sea cliff Steep, commonly nearly vertical bluff adjoining a beach or other coastal environment; produced by a combination of wave erosion and erosion from the land, such as landsliding and runoff of surface water. Groundwater seepage may also contribute to its development.

Seafloor spreading The plate tectonics concept that new crust is continuously added to the edges of lithospheric plates at divergent plate boundaries as a result of upwelling of magma along mid-ocean ridges.

Sea wall A structure, commonly made of concrete, large stone blocks, wood or steel pilings, or cemented sand bags, built parallel to the coastline to protect buildings and infrastructure from wave and flood damage.

Secondary effect With respect to natural hazards, phenomena that are indirectly the result of a hazardous event, such as wildfires that are ignited by lava flows or water pollution that occurs after flooding has damaged a sewage treatment plant; applies to those phenomena that are caused by the primary or direct effects of a hazardous event.

Secondary succession With respect to ecology, the regular pattern of colonization and replacement of plant species in a biological community that was partially destroyed in a hazardous event, such as a wildfire, flood, or lengthy volcanic ash fall.

Sediment Transported particulate material created by weathering, biological activity, or chemical precipitation.

Sediment flushing With respect to aftermath of wildfire, the transport of sediment from slopes to stream channels and eventually downstream from even moderate storm events.

Sedimentary rock Hard, compacted, and cemented particulate material that was created by weathering, biological activity, or chemical precipitation.

Seismic gap Area along an active fault zone that is capable of producing a large earthquake but has not produced one recently.

Seismic source The point at which vibrations from an earthquake are generated.

Seismic wave A periodic disturbance of particles of Earth by an earthquake or other vibration that is propagated without the net movement of the particles.

Seismogram A written or digital graphic record made by a seismograph.

Seismograph Instrument that records earthquakes, also known as a seismometer.

Sensible heat Energy created by random molecular motion that may be physically sensed, such as by a thermometer.

Shake map A near-real-time, two-dimensional representation of the ground motion and potential damage from an earthquake.

Shear strength With respect to mass wasting, the internal resistance in earth material to shear stress; on slopes this strength resists failure by sliding or flowing.

Shield volcano A broad, convex volcanic mountain built up by successive lava flows; the largest type of volcanic mountain.

Shrinking stage With regard to tornado development, late phase in which the tornado thins and begins to tilt as its supply of warm moist air is reduced.

Shrink-swell clay Earth material composed of very fine mineral particles that decrease in volume (shrink) when they are dry and increase in volume (swell) when they are wet.

Silt particle Small particles 1/256 to 1/16 mm.

Sinkhole Surface depression formed by the dissolution of underlying sedimentary rock or collapse over a subterranean void such as a cave.

Slab avalanche Downslope movement of snow and ice that starts out as large cohesive blocks; generally more dangerous and damaging than loose-snow avalanches.

Sliding With respect to mass wasting, the deformation or downslope movement of a nearly intact block of earth material along a slip surface.

Slip rate Long-term rate of displacement along a fault; usually measured in millimeters or centimeters per year.

Slope The slant or incline of the land surface.

Slope stability map A two-dimensional representation of the land surface that shows its susceptibility to mass wasting; commonly based on a survey of past landslides that includes a measurement of the steepness of the slopes on which they occurred.

Slow earthquake An event in which rupture and movement along a geologic fault takes days to months to complete; also referred to as a *silent earthquake*.

Slump Type of landslide characterized by downward slip of a mass of earth material, generally along a curved slip surface.

Slump block A mass of landslide material that has moved downslope along a curved slip surface.

Slumping With respect to mass wasting, the downslope movement of rock, sediment, or soil along a curved slip surface.

Smectite A group of clay minerals with extremely small crystals that can absorb large amounts of water; common constituent of expansive soil.

Snow avalanche Rapid downslope movement of snow, ice, and rock.

Soil In geology, a generally loose accumulation of surface earth material that has been altered by chemical weathering in the place that it accumulated and can serve as a medium for plant growth. In engineering and design applications, unconsolidated surface material that can be excavated or removed without blasting.

Soil chronosequence A series of soils arranged from youngest to oldest.

Soil erosion Physical processes such as running water and blowing wind that removes soil particles.

Soil horizons Collection of distinct soil layers parallel to the surface (for example, A, B, and C) each produced by vertical and horizontal movements of the materials in a soil that creates the soil profile.

Soil slip A type of mass wasting event in which a thin layer of unconsolidated earth material slides downslope along a narrow path; the slip surface for this type of mass movement is commonly in weathered slope deposits that are below the soil, but above unweathered bedrock; in California shallow soil slips are commonly referred to as *mudslides*.

Soil taxonomy A classification of soils used by agriculture and soil science.

Solar nebula A flattened, pancake-like rotating disk of hydrogen and helium dust. Around 5 billion years ago a solar nebula condensed to form the sun by accretion of matter from its flattened disk.

Solutional sinkhole A depression that develops by the dissolution of underlying bedrock, such as limestone.

Sorting In sediment, size distribution of the particles; poorly sorted sediment has a large range in particle size, and in well-sorted sediment most particles are close to the same size.

Spacewatch A program started in 1981 to inventory the region surrounding Earth for asteroids and comets with the goal of identifying those whose orbits cross Earth's orbit.

Specific gravity The density of a mineral or rock as compared with the density of water.

Spilling breaker A type of water wave in which the wave crest tumbles down its own forward slope as it strikes a gently sloping shoreline; tends to be associated with deposition of sand on a beach.

Spit A claw-shaped, sandy to gravelly coastal landform at the end of a peninsula or island formed by accretion of beach ridges in the direction of longshore drift.

Spot fire A small fire that is commonly ignited by wind-blown burning embers ahead of the flaming front of a wildfire.

Spreading center Synonymous with mid-ocean ridges where new crust is continuously added to the edges of lithospheric plates.

Spring With respect to groundwater, the natural outflowing of subsurface water where the groundwater system intersects Earth's surface.

Squall line A line of thunderstorms often accompanied by high wind and heavy rain; commonly forms in advance of a cold front or dryline and can produce tornadoes.

Stable isotope A variety of a chemical element distinguished by its number of neutrons and one that does not naturally undergo radioactive decay. In absolute dating, the product of radioactive decay, the daughter isotope, is stable.

Stage With respect to flowing water, the height of the water in a stream channel. Typically used in reference to the water level at which flooding begins to occur. Flood stage is commonly the water level that is likely to cause damage to personal property.

Stalactite A cylindrical or conical mineral deposit, most commonly of calcium carbonate, that has grown downward from the roof of a cave or other overhanging feature.

Stalagmite A column or ridge of mineral deposits, most commonly calcium carbonate, on the floor of a cave; forms as water drips from stalactites on the cave roof.

Stationary front Transition zone between two different air masses, commonly with contrasting temperature, that moves very little.

Stepped leader In cloud-to-ground lightning, a channel of ionized air that approaches the ground in a series of nearly invisible bursts. This channel becomes the path for the luminous return stroke.

Storm surge Wind-driven ocean waves, usually accompanying a hurricane, nor'easter, or similar storm, that pile water up on a coastline.

Strain Change in shape or size of a material as a result of applied stress.

Stratosphere Zone in Earth's atmosphere above the troposphere where the air temperature is either constant or warms with increasing altitude; contains significant quantities of ozone, which protects life from high levels of ultraviolet radiation.

Stratovolcano See *Composite volcano.*

Streak The color of a mineral in powdered form; commonly produced by rubbing a mineral specimen on an unglazed ceramic plate.

Stream With respect to surface water, a body of water that flows in a channel; includes a brook, creek, and river.

Strength In soil, the ability to resist deformation; the resistance results from cohesive and frictional forces in the soil. In rock, the mechanically applied compression necessary to break or fracture a specimen.

Stress A type of physical force. In geological terms, stress is the force applied to rocks that causes them to deform. The stress results from relative movement of plates along tectonic boundaries can either be compressional, tensional, or shearing, corresponding to convergent, divergent, and transform plate boundaries, respectively.

Strike-slip fault A fracture or fracture system along which a block of Earth's crust moves horizontally relative to the block on the opposite side of the fracture.

Strike and dip Two parameters measured by geologists to describe the three-dimensional orientation of features, including layering in rocks, that intersect Earth's surface.

Strombolian-type eruption A mildly explosive volcanic eruption that ranks between 1 and 2 on the volcanic explosivity index. This eruption type is characterized by ejection of incandescent cinders that usually accumulate around the vent region, giving rise to the formation of a cinder cone volcano.

Strong earthquake Earthquake in which the release of energy has a moment magnitude of 6 to 6.9.

Subduction zone Convergence of tectonic plates where one plate dives beneath another and is consumed in the mantle.

Submarine landslide An underwater mass movement, often caused by an earthquake that can occur on a sloping seafloor, such as the wall of an underwater canyon, the continental slope, or the seafloor surrounding an island.

Submarine trench A relatively narrow, long (often several 1000s of kilometers), deep (often several kilometers) depression on the ocean floor that forms as a result of convergence of two tectonic plates with subduction of one.

Subsidence Sinking, settling, or other lowering of parts of the crust of Earth.

Subtropical jet stream A concentrated flow of air near the top of the troposphere that, in the Northern Hemisphere, typically crosses Mexico and Florida and is characterized by strong winds blowing from west to east.

Suction vortices Small, intense whirls within a larger tornado; these rapidly rotating winds may cause much of the intense damage done by tornadoes.

Sulfur dioxide (SO_2) A colorless gas with a pungent odor similar to a burnt match. This gas is emitted from volcanoes and burning of fossil fuels, and when dissolved in rainwater produces acid rain.

Supercell An unusually long-lived thunderstorm that has a rotating updraft that continually redevelops on the storm's flank.

Supercell storm See *Supercell.*

Surface fire Combustion in a wildfire that takes place primarily along the ground and consumes fuels such as grass, shrubs, dead and downed limbs, and leaf litter.

Surface wave One type of vibration produced by an earthquake in which the oscillation moves along the boundary between the air and land; generally strongest close to the epicenter where it causes much of the damage to buildings and other structures.

Surf zone The area between the breaking waves and the swash zone along the shore.

Swash zone Area along the shore where waves run up on the beach face and then back again into the water.

Swell With respect to coastal processes, the sorting out of storm-produced waves into wave sets having more or less uniform heights and lengths. This sorting allows groups of waves to move long distances from storms to coastal areas with relatively little loss of energy.

Talus Pieces of rock that have accumulated at the base of a cliff or steep slope; may form individual piles or a continuous slope composed of blocks of rock that have fallen from above.

Tectonic Pertaining to forces involved in large-scale crustal movement.

Tectonic creep Slow, more or less continuous, movement along a fault; also called *fault creep.*

Tectonic cycle A repetitive sequence of events and processes that create and destroy the Earth's crust and its ocean basins and mountain ranges.

Tectonic framework The geometry and spatial patterns of faults and seismic sources that define a specific region.

Tectonic plate A very large, fault-bounded block of Earth's crust and underlying upper mantle that moves as a coherent mass on top of the asthenosphere; also called a *lithospheric plate*. This type of block commonly forms at a mid-ocean ridge and is destroyed in a subduction zone.

Tectonic creep Slow, more or less continuous movement along a fault.

Tephra Any material ejected and physically blown out of a volcano; mostly ash.

Texture In rock, the size, shape, and arrangement of mineral grains and other particles. In soil, the size distribution of mineral and rock particles.

Thermal expansion With respect to the ocean, the increase in volume of seawater that occurs upon heating.

Thermal spring A place where heated groundwater flows from rock, sediment, or soil onto the land surface or into a body of surface water such as a river, lake, or ocean. Water temperature must be appreciably higher than the mean annual air temperature; includes both warm springs and hot springs.

Thermokarst An irregular terrain of sinkholes and mounds formed by the melting of permafrost.

Thermosphere With respect to weather, the outermost part of the atmosphere starting at about 80 km (~50 mi.) in altitude, characterized by an upward increase in temperature and very little gas; directly above the mesosphere.

Thrust fault A low-angle fracture or fractures in Earth's crust where older rocks are displaced over younger rocks.

Topographic map A two-dimensional representation of the relief of Earth's surface using contour lines of equal elevation.

Topographic profile A sideways view of the land surface with a curving line drawn to show how it changes in elevation from one point to another point.

Tornado A destructive, commonly funnel-shaped cloud of violently rotating winds that extends downward from a severe thunderstorm to reach Earth's surface.

Total load With respect to stream processes, the sum of the dissolved, suspended, and bed load that a stream or river carries.

Tower karst A landscape of steep-sided hills rising above a plain or above sinkholes. Commonly developed by extensive dissolution of limestone in a humid tropical climate.

Transform boundary Interface between two tectonic plates where one plate slides horizontally past another, such as along the San Andreas fault in California; synonymous with a *transform fault*.

Transform fault Type of fracture of Earth's crust associated with offsets of mid-ocean ridges where one plate is displaced horizontally; may form a plate boundary, such as the San Andreas fault in California.

Translational With respect to landslides, a type of sliding downslope movement that develops on a well-defined, planar slip surface, most commonly in layered or jointed rock or along a weak clay layer.

Triangulation The location of a point using distances from three points.

Tributary A small stream that flows into a larger stream.

Triple junction Areas where three tectonic plates and their boundaries join.

Troglobite An organism, such as a blind salamander, that lives its entire life underground in a cave.

Tropical cyclone General term for large thunderstorm complexes rotating around an area of low pressure that formed over warm water in the tropics or subtropics; includes low-intensity tropical depressions and tropical storms and high-intensity storms called hurricanes, typhoons, severe tropical cyclones, or severe cyclonic storms.

Tropical depression A circular atmospheric low-pressure center in the tropics whose winds are less than 63 km (~39 mi.) per hour; the threshold for a tropical storm.

Tropical disturbance An area of disorganized, but persistent thunderstorms and associated low-pressure trough that may be in the formative stages of a tropical depression.

Tropical storm An organized system of thunderstorms around a circular low-pressure center that derives its energy from warm ocean waters and has winds between 63 km (~39 mi.) and 119 km (~74 mi.) per hour; stronger than a tropical depression, but weaker than a hurricane.

Tropopause The boundary between the troposphere and the stratosphere.

Troposphere Lowermost layer of the atmosphere defined by a general decrease in temperature with increasing height above Earth's surface.

Trough With respect to weather, an elongated area of low atmospheric pressure.

Tsunami A series of gravity waves caused by the large-scale displacement of the entire water column of the ocean or rarely a large lake. Most commonly the result of a large earthquake deforming the seafloor, although may also be caused by a large underwater or seacliff landslide, collapse of part of a volcano into the ocean, a submarine volcanic explosion, or the ocean impact of an asteroid. These waves are not tidal waves and, in a few cases, are not seismic sea waves.

Tsunami ready A certification issued by the National Oceanic and Atmospheric Administration or other governmental agency that a community has taken appropriate preparedness measures for future tsunamis.

Tsunami runup map A two-dimensional representation of the level to which a tsunami has inundated or is projected to inundate a coastal zone.

Tsunami warning An emergency notice from a governmental agency that a tsunami has been detected and is heading toward the area receiving the alert.

Tsunami watch An emergency notice from a governmental agency that an event has taken place—such as an earthquake, landslide, or volcanic explosion—which could cause a tsunami to strike the alerted area.

Typhoon See *Hurricane*.

Ultra-Plinian-type eruption A super-colossal volcanic eruption that ranks between 6 and 7 on the volcanic explosivity index. These eruptions occur when stratovolcanoes or continental calderas collapse during eruption causing the remaining subsurface magma to be forcibly erupted.

Unified soil classification system Classification of soils, widely used in engineering practice, based on amount of coarse particles, fine particles, or organic material.

Uniformitarianism Concept that the present is the key to the past; that is, we can read the geologic record by studying present processes.

Unsaturated flow Type of flow in soil where all pores are not filled with water.

Updrift Located in the direction from which a water current, such as a longshore current, is flowing.

Uplift An increase in land elevation that is commonly the result of tectonic or volcanic processes.

Urban heat island effect A local climatic condition resulting from various design and land-use practices in a city, large town, or other extensively developed area. This condition can intensify heatwaves and cause temperatures in a metropolitan area to be up to 12°C (~22°F) warmer than the surrounding rural area.

Valley fever A potentially fatal respiratory illness produced by fungal spores found in desert soils. Mass wasting caused by earthquakes can produce dust containing these spores.

Variable gas With respect to the atmosphere, a substance such as carbon dioxide or water vapor that cannot be compressed with pressure alone and is present in changing proportions.

Varve A very thin sedimentary layer or couplet of layers deposited in a single year; in glacial lake deposits it consists of a coarse layer deposited when ice melts in the summer and a fine layer deposited when very fine suspended sediment finally settles out in the winter.

Vein In geology, the filling of a fracture with earth material, especially minerals, that differs from the surrounding rock.

Vertical exaggeration (VE) Vertical scale of a diagram or model that is larger than the horizontal scale; used in many geologic cross-sections to show relationships more clearly.

Vertical wind shear A change in direction or speed of the wind with changing altitude.

Very minor earthquake See *Microearthquake.*

Viscosity The ease of flow of a fluid; controlled by the magnitude of the internal friction within a fluid. Thicker fluids have greater viscosities.

Vog A volcanic fog or smog of sulfur dioxide and other volcanic gases combined with aerosols of sulfuric acid and dust. In Hawaii, vog is commonly an acrid blue haze produced by gases emitted from Kilauea volcano.

Volatiles Chemical compounds, such as water (H_2O) or carbon dioxide (CO_2), that evaporate easily and exist in a gaseous state at the Earth's surface.

Volcanic ash eruption An explosive volcanic event in which a tremendous quantity of fine-grained rock and volcanic glass is shattered and expelled in clouds of very fine particles into the atmosphere.

Volcanic dome Type of volcano characterized by very viscous magma with a high silica content; activity is generally explosive.

Volcanic Explosivity Index (VEI) A relative scale by which different volcanic eruptions can be compared based on quantitative and qualitative measurement of explosivity, including the plume height, volume of ejected material, and frequency of eruption.

Volcanic vent Circular or elongated opening through which lava and pyroclastic debris are erupted.

Vortex With regard to weather, a spinning column of wind.

Vulcanian-type eruption A mild to severe explosive eruption that ranks between 2 and 3 on the volcanic explosivity index. These eruptions are triggered by the interaction between groundwater and rising magma, which causes episodic violent explosions when the water flashes to steam, in a process known as a preatomagmatic eruption.

Wadati-Benioff Zone Inclined zone of earthquakes produced as a tectonic plate is subducted.

Wall cloud A localized, persistent, and often abrupt lowering of condensed water vapor from the rain-free base of a severe thunderstorm. These clouds may rotate and are commonly associated with tornado formation.

Warning With respect to natural hazards, the announcement of a possible hazardous event, such as a large earthquake or flood, that could occur in the near future.

Warm front An atmospheric boundary between two air masses of contrasting temperature along which the warmer air mass advances into the cooler air mass.

Washover channel A narrow conduit of water eroded through a beach or coastal dunes by a storm surge.

Watch With respect to weather, an alert issued by a forecast office that meteorological conditions are favorable for severe weather, such as a tornado, hurricane, severe thunderstorm, winter storm, or flash flood.

Water content The amount of water in a soil as a percent (weight of water to weight of the solid soil particles as a percent); also called the *moisture content.*

Water–energy–food nexus The concept that water, energy, and food are inextricably tied together such that changes in one will effect change on one or both of the other factors.

Watershed Land area that contributes water to a particular stream system; generally used to describe the catchment area for a smaller tributary stream; see *Drainage basin.*

Waterspout A weak, generally funnel-shaped, wind vortex that most commonly develops under fairweather conditions and has all or part of its track over the sea or a lake; some of these whirlwinds are tornadoes that form or track over the water.

Water vapor The gas phase of water.

Watt The basic unit of power; generally expressed as units of energy per unit of time.

Wave front A long, continuous crest of a single oscillation, such as one on the surface of a lake or ocean.

Wave height Vertical distance between the crest and preceding trough of an oscillation, such as one on the surface of a lake or ocean.

Wave normal An imaginary line perpendicular to the crest of an oscillation, such as one on the surface of a lake or ocean.

Wave period The time in seconds for successive crests of an oscillation, such as one on the surface of a lake or ocean, to pass a reference point; the inverse of the frequency of the oscillation.

Wavelength With respect to physics, the horizontal distance between two successive crests or troughs of an oscillation. Used to describe characteristics of seismic, electromagnetic, water, and other oscillations.

Weather Atmospheric conditions, such as air temperature, humidity, and wind speed, at any given time and place.

Weathering Changes that take place in rocks and minerals at or near Earth's surface in response to physical, chemical, and biological activity; the physical, chemical, and biological breakdown of rocks and minerals.

Wicked Problem A natural hazard management problem that is resistant to resolution due to strong social, political, or economic implications.

Wide Area Augmentation System (WAAS) A network of global positioning system (GPS) receivers at fixed locations that act as reference points for transmitting positional corrections by satellite to mobile GPS receivers. Operated by the Federal Aviation Administration, this system currently improves the accuracy of locations obtained from WAAS-enabled GPS receivers in the United States and parts of Canada.

Wildfire Uncontrolled combustion or burning of plants in a natural setting, such as a forest, grassland, brushland, or tundra.

Wildland–urban interface The area where natural undeveloped grassland, brushland, or forest is adjacent to land developed for homes, businesses, or recreation.

Wind chill The influence of wind on the perception and actual effect of cold air. Meteorological agencies in the United States and Canada have developed a wind chill index based on how the human body loses heat in the cold and wind.

Wind shear A change in the speed or direction of the wind along a given direction.

Work In physics, the exertion of a force to overcome resistance or to produce molecular change.

Younger Dryas A previously unexplained glacial interval from around 13,000 to 11,600 years ago that coincides with the extinction of many large mammals in the Northern Hemisphere and the disappearance of the Clovis culture. A recent hypothesis proposes that these events were caused by the airburst of a comet over the Canadian continental ice sheet.

Index